Panel Data Analysis Using EViews

Panel Data Analysis Using EViews

I Gusti Ngurah Agung
Graduate School of Management
Faculty of Economics and Business, University of Indonesia

WILEY

Library of Congress Cataloging-in-Publication Data

Agung, I Gusti Ngurah.
  Panel data analysis using eviews / I. Gusti Ngurah Agung.
    pages cm
  Includes bibliographical references and index.
  ISBN 978-1-118-71558-1 (hardback)
  1. Statistics. 2. EViews (Computer file) I. Title.
  HA29.A377 2014
  005.5'5–dc23

                                    2013024207

A catalogue record for this book is available from the British Library.

ISBN: 978-1-118-71558-1

Set in 10/12pt Times by Thomson Digital, Noida, India.

1 2014

Dedicated to my wife
Anak Agung Alit Mas,
as well as all my Generation

Contents

Preface

The main objectives of this book are to present (1) various general equation of panel data models, with some specific models; (2) various illustrative statistical results based on selected specific models with special notes and comments and (3) comparative studies between sets of special type models, such as heterogeneous regression, fixed-effects and random effects models, so that readers can be informed of a model's limitation(s) compared to the others in the set.

This book presents over 250 illustrative examples of panel data analysis using EViews, compared to the books of Baltagi (2009a,b) on *Econometric Analysis of Panel Data* and *A Companion to Econometric Analysis of Panel Data* which mainly present the mathematical concepts of the models with some data analysis. Referring to the fixed- and random effects models, Baltagi presented statistical results based on various additive models and none with the numerical time independent variable. However, Baltagi quotes a simple dynamic panel data model with heterogeneous coefficients on the lagged dependent variable and the time trend presented by Wansbeek and Knaap (1999, in Baltagi, 2009a, p. 168), and a random walk model with heterogeneous trend presented by Hardi (2000, in Baltagi, 2009a).

Similarly, this is the case for most of the panel data models presented in Gujarati (2003). Wooldridge (2002), and in more than 300 papers presented in five international journals, such as the *Journal of Finance* (JOF) from the years 2010 and 2011, *International Journal of Accounting* (IJA), *Journal of Accounting and Economics* (JAE), *British Accounting Review* (BAR), and *Advances in Accounting, incorporating Advances in International Accounting* (AA) from the years 2008, 2009 and 2010, which are additive models.

However, it is important to note that Wooldridge (2002) presented a random effect model with trend or the numerical time independent variable, Bansal (2005) presented the models with trend and Time-Related Effects (TRE), but based on time series data, and Agung (2009a) presented various models with trend and TRE. So I would say that various models, either additive or interaction models, with the numerical time independent variable or the time and time-period dummy variables, should be acceptable or valid and reliable panel data models.

I found that a very limited number of models with interaction independent variables or heterogeneous regressions models were presented. Only Giroud and Mueller (2011) presented several Year-Industry fixed effects interaction models (or Year-Industry FEMs with interaction independent variables). Referring to the dummy variables models, (Siswantoro and Agung, 2010) presented their findings that only 63 out of 268 papers in the four journals (*IJA*, *JAE*, *BAR* and *AA*), had dummy variables models, and only five of the models had interaction independent variables. In addition, Dharmapala, et al. (2011) presented interaction models or heterogeneous regressions using the Firm and Year dummies, and Park and Jang (2011) presented an interaction period-fixed-effects model with 34 parameters, besides the year dummies. In fact, the heterogeneous regressions model, which is an interaction model, was introduced by Johnson and Neyman in 1962 (cited in Huitema, 1980).

If a multiple regression panel data model does not have any dummy variable, then the regression model presents a single continuous model for whole individual-time observations. I would consider such a model to be inappropriate. On the other hand, a dummy variables model could also be the worst within its group with the same set of numerical and categorical independent variables, which are illustrated in this book.

Referring to various models indicated here, this book presents various models, either additive or interaction models, with the numerical time independent variable or time and time-period dummies variables. Note that the numerical time variable has been used to present classical growth models, namely the geometric and exponential growth models (Agung, 2009a, 2011b). Furthermore, the time t in fact represents an environmental variable, which is invariant over individuals or research objects.

The models presented in this book in fact are derived from my first two books in the data analysis using EViews (Agung, 2009a, 2011). For this reason, I recommend to readers to use the models in the first two books as the basic and main references to develop various alternative or more advanced models based on panel data, because this book only presents some of the models.

Furthermore, special statistical results using the object VAR and GLS are illustrated, which have not been presented in other books as well as papers in the international journals. A *manual stepwise selection method* is introduced, aside from the application of the STEPLS estimation method provide by EViews. Even though STEPLS regressions have been commonly applied, my book proposes and introduces how to apply the STEPLS, using a multistage stepwise selection method specifically for interaction models with numerical and categorical independent variables; such as continuous interaction models by groups of the research objects (firms or individuals) and time points or time periods.

Based on my own point of view, models based on panel data should be classified into three groups; namely (1) The group of models based on unstacked data, or the group of time-series models by states (firms or individuals); (2) The group of models based on stacked or pool data, especially for incomplete panel data; and (3) The group of models based on natural experimental or special structural balanced panel data. For this reason, this book contains 14 chapters, which are classified into three parts.

Part I presents the *Time Series Data Analyses by States*. In this part the panel data considered is unstacked data, where the units of the analysis are time observations. The sets or multi-dimensionals of exogenous, endogenous and environmental variables, respectively, for the state i can be presented using the symbols $X\_i_t = (X1\_i, \ldots, Xk\_i, \ldots)_t$, $Y\_i_t = (Y1\_i, \ldots, Yg\_i, \ldots)_t$, and $Z_t = (Z1, \ldots, Zj, \ldots)_t$, for $i = 1, \ldots, N$; and $t = 1, \ldots, T$.

Note that the scores of the environmental variables are constant for all states or individuals. All the time series models presented in Agung (2009a) are valid models for each state i. This part presents only the analyses *specifically* for the unstacked data with a small number of N. The models for a large N will be presented in Part II and Part III.

The four chapters contained in this part are as follows:

Chapter 1 presents multivariate data analyses based on a single time series by states, using various models, multivariate lagged-variable autoregressive growth models, namely MLVAR(p,q)_GM, seemingly causal models (SCMs) with trend or time-related effects, fixed-effects and random effects models, VAR and VEC models. In addition, this chapter also presents piece-wise models, various models having environmental independent variables, TGARCH(a,b,c) and instrumental variables models.

Chapter 2 presents multivariate data analyses based on bivariate time series by states, as the extension of all models presented in Chapter 1. In addition, this chapter also presents simultaneous causal models.

Chapter 3 presents multivariate data analyses based on multivariate time series by states, as the extension of all models presented in Chapter 2. In addition, this chapter also presents special VAR models with an environmental multivariate, which have not been found in other books and papers.

Chapter 4 presents the application of various SCMs, either additive or interaction models, based on a single time series Y_i_t, bivariate time series (X_i_t, Y_i_t), trivariate time series $(X1\_i_t, X2\_i_t, Y\_i_t)$ or $(X\_i_t,$

$Y1\_i_t, Y2\_i_t$), and the application of SCMs as the alternative VAR models with the environmental multivariate presented in Chapter 3.

Part II presents *Pool Panel Data Analyses*. In this part the panel data considered is stacked data where the units of the analysis are the individual-time or firm-time observations. So the sets or multi-dimensionals of exogenous, endogenous and environmental variables, respectively, for the firm i, at the time t can be presented using the symbols $X_{it} = (X1, \ldots, Xk, \ldots)_{it}$, $Y_{it} = (Y1, \ldots, Yg, \ldots)_{it}$, and $Z_t = (Z1, \ldots, Zj, \ldots)_t$, for $i = 1, \ldots, N_t$; and $t = 1, \ldots, T$. Note that the symbol N_t is used to indicate that the models presented in this part should be valid for incomplete or unbalanced pool panel data as well as balanced pool panel data. However, special models for the balanced panel data, as a natural experimental data, will be presented in Part III.

The statistical methods and models applied can directly be derived from the models based on cross-section data presented in Agung (2011), by using or inserting the time dummy independent variables. More complex models should be considered or defined based on pool panel data with a large N and large (or very large) T. With large time-point observations, then time T can be used as a numerical independent variable, so a defined model could be a continuous or discontinuous model of T, because at least two time periods have to be considered, as presented in Agung (2009a).

This part contains seven chapters as follows:

Chapter 5 presents the preliminary evaluation analysis, the applications of the object "*Descriptive Statistics and Tests*", for multi-dimensional problems by times, and the object "*N-way Tabulation*", multifactorial cell-proportion models, Kendall's tau, and multiple association between categorical variables.

Chapter 6 presents general choice models, specifically the binary choice models having categorical or numerical independent variables. Special findings and notes are presented based on alternative binary choice models having numerical independent variables.

Chapter 7 presents advanced general choice models as the extension of all models presented in Chapter 6. In addition, this chapter demonstrates data analysis based on multifactorial binary and ordered choice models with categorical or numerical independent variables, using the manual step-by-step process in modeling to compare to the STEPLS estimation method, new unexpected stepwise polynomial regressions, general choice models with trend and the time-related effects.

Chapter 8 presents the application of additive and interaction GLMs (univariate general linear models) by *Group* and *Time*, using the original measured variables or transformed variables, starting with simple models such as ANOVA and quantile models; bivariate correlation analysis and STEPLS regressions. In addition, this chapter also presents piece-wise autoregressive models by time-points with one, two and several numerical exogenous variables, the applications of the White and Newey–West options, polynomial effects models, general continuous linear models and ANCOVA models, including the worst ANCOVA model in a theoretical sense. Finally, this chapter also presents a discussion on the non-stationary problem.

Chapter 9 presents fixed-effects models and alternatives. The limitation or hidden assumption of each fixed-effects model, such as the individual fixed-effects model, time fixed-effects model, and the individual-time fixed-effects model, are discussed in detail. In addition, this chapter also presents extended fixed-effects models. Several fixed-effects models are selected from the *Journal of Finance* (2011) to present their specific characteristics to the readers. Note that the fixed effects are in fact ANCOVA models with specific hidden assumptions. For this reason, several alternative heterogeneous regression models are presented and recommended, such as the heterogeneous classical growth models by individuals or groups, piece-wise heterogeneous regressions and heterogeneous regressions with trend or time-related-effects by individuals or groups.

Chapter 10 presents special notes on selected problems, which are the impacts of misclassification of the research objects defined based on the whole or total firm-time observations, or by doing data analysis by treating the pool panel data as a random cross-section of data. For instance, a dummy of the return rate R_{it}, is defined as $DR = 1$ if R_{it} equals and less than zero, and $DR = 0$ if otherwise, would be misleading. Take note that this dummy variable DR_it does not represent two disjointed sets of observed firms or individuals over times, but it represents two disjointed sets of $R_{it}s$ scores, namely the sets: $1.\{R_{it} \leq 0\}$ and $2.\{R_{it} > 0\}$. So some or many firms should have both negative and positive scores of R_{it} over times, or based on all firm-time observations. In other words, the two sets of scores are not the firms' classifications. At the very extreme case, all firms may have both negative and positive observed scores, for a long time period of observations. Similarly, this applies to the other dummy variables and first difference variables.

On the other hand, the models with R_{it} as a numerical variable and using a ratio variable could also be misleading; these are demonstrated using scatter graphs with regression lines, the kernel fit or nearest neighbor fit curves. In fact, Agung (2009a) presented unexpected relationships between pairs of time-series variables in the form of their scatter graphs, because the pairs of time-series variables were presented as cross-section variables. In addition, some models from the international journals are selected to discuss their limitation or hidden assumptions, and present alternative or extended models.

Chapter 11 presents the application of various types of the Seemingly Causal Models (SCMs), wherein the most important and recommended models are the multivariate heterogeneous models by group and time or *time period* (TP). The illustrative multivariate linear-effect models, nonlinear-effect models, and bounded models by groups or times are presented with special notes and comments. We find that unexpected statistical results can be obtained because of outliers. For this reason, possible treatment of the outliers is presented.

Part III presents *Natural Experimental Data Analysis*. In this part the panel data considered is stacked by cross-section with $N \times T$ observations, where the units of the analysis are the individual-time or firm-time observations. So the sets or multidimensional of exogenous, endogenous and environmental variables, respectively, for the firm i, at the time t can be presented using the symbols $X_{it} = (X1, \ldots, Xk, \ldots)_{it}$, $Y_{it} = (Y1, \ldots, Yg, \ldots)_{it}$, and $Z_t = (Z1, \ldots, Zj, \ldots)_t$, for $i = 1, \ldots, N$; and $t = 1, \ldots, T$. In the natural experimental data analysis, environmental variables could be represented by the time or time period variable, say TP, such as before and after a critical time or event; before, during and after an economic crisis; and before, between and after two consecutive critical events. In addition, classification or cell-factor, namely *CF*, can be defined as the treatment factors, with the response variable Y_{it} and the covariates (cause or upstream variables, or predictors) are the lags $Y_{it}(-p)$ and $X_{it}(-q)$, at least for $p = q = 1$, if the data is annual, semi-annual or quarterly data set.

Take note that the cell-factor (*CF*) should be generated or defined as a reference or based group of the research objects (individuals or firms), which is invariant or constant over times; such as regions or states, the firm sectors, status of the business (public and nonpublic), and family-nonfamily business. The *CF* or an invariant *GROUP* variable also can be generated based on numerical variables, such as *SIZE, ASSET*, or *LOAN* of the firms at a certain time point, for instance at the time $T = 1$, using the median, quantile or percentile as the alternative cutting points.

$Yg_{it}(-1)$ should have different effects on a response variable, say Yg_{it}, between the groups generated by *CF* and *TP*, then the simplest recommended model considered involves the heterogeneous regression lines of Yg on $Xk(-1)$ by *CF* and *TP*. In practice, a set of heterogeneous regressions could be reduced to homogeneous regressions (ANCOVA or fixed-effects models), because we find that the covariate $Xk(-1)$ has insignificant different effects on Yg between the groups generated by *CF* and *TP* in a statistical sense. In a theoretical sense, however, the homogeneous regressions might not be appropriate and some could be the worst ANCOVA models. For these reasons, this part presents special notes and comments of the limitation of the fixed-effects models, and demonstrates the worst possible ANCOVA model in a theoretical sense.

This is similar for the simplest heterogeneous regressions of Yg on $Yg(-1)$, namely the LV(1) model of Yg_{it} by *CF* and *TP*. Furthermore, the models could easily be extended to LVAR(p,q) models with the exogenous variables X_{it}, and $X_{it}(-1)$, and the environmental variable Z_t.

All models based on a true experimental data, presented in Agung (2011, Chapter 8), should be valid based on balanced pool data, by using *CF* and *TP* as the environmental treatment factors.

This part contains three chapters as follows:

Chapter 12, at the first stage, presents how to develop special balanced pool data, and how to define and generate the reference or based group variable(s), namely the cell-factor (*CF*), which is invariant or constant over times or time periods. Then various types of heterogeneous regressions are presented, such as LV(p), AR(q), and LVAR(p,q), starting with $p = q = 1$, such as (1) The models by CF (Groups) and times, (2) The models by CF with Trend, and (3) The models by *CF* with time-related effects, with or without exogenous or environmental variables. More advanced models, such as bounded, polynomial and generalized LV(p) models, also are presented. For doing data analysis, in EViews 6 using the function *@Expand(CF,TP)*, is recommended, to write the equation specification of the model. For a model with a large number of various types of independent variables, applying the manual stepwise selection method is recommended, which is demonstrated using the general linear LV(1) models, binary choice LV(1) models and ordered choice LV(1) models. Finally, this chapter also presents quantile regressions, ANCOVA or fixed-effects models.

Chapter 13 presents the applications of various multivariate lagged variables autoregressive models, namely LVAR(p,q) SCMs (Seemingly Causal Models), where each of the multiple regressions in a model can have different sets of independent variables, as the extension of all models presented in previous chapter. In fact, all models presented in previous chapter are valid models for each endogenous variable Y_g or its transformed variable, namely $H_g(Y_g)$. Various illustrative SCMs based on the numerical variables *(Y1,Y2)*, *(Y1,Y2,Y3)*, *(X1,Y1,Y2)* and *(X1,X2,Y1,Y2)*, are presented in the form of selected illustrative path diagrams, and then the corresponding linear-effects models can easily be defined, either additive, two-way or three-way interaction models, with or without an environmental variable. Hence, analysis could be done based on the heterogeneous regressions by *CF* and *T* or *TP*, or the heterogeneous regressions by *CF* with trend and time-related effects.

For comparison, the application of the VAR and VEC models, and fixed-effects or MANCOVA models is presented. Finally, the Granger Causality Test based on the VAR model and additive SCMs is presented, with an extension for the interaction SCMs, called the *Generalized Granger Causality Tests*.

Chapter 14 presents the applications of *Generalized Least Squares* (GLS) estimation method, especially for the *Cross-Section Random Effects Model* (CSREM) and *Period Random Effects Model* (PEREM), based on a special structural balanced panel data. Since the models can easily be derived from the OLS regression models presented in previous chapters, this chapter presents only some illustrative examples, by using the same models which have been presented in previous chapters. In addition, the applications of two-way effects models, such as *two-way random effects model* (TWREM), *two-way fixed-effects model* (TWFEM), *two-way random-fixed-effects model* (TWRFEM), and *two-way fixed-random effects model* (TWFREM) are also presented.

I wish to express my gratitude to the Graduate School of Management, Faculty of Economics, University of Indonesia, and The Ary Suta Center, for providing a rich intellectual environment and facilities that were indispensable for writing this text. In the process of doing data analyses, I wish to thank Thomas Gareth, Senior Principal Economist, IHS EViews, as well as my colleagues, Ruslan Priyadi, PhD, Professor Nachrowi D. Nachrowi, PhD, Zaafri Ananto Husodo, PhD, and Bambang Hermanto, PhD, for their input on selected illustrative examples presented in this book. I also would like to thank Tridianto Subagio, a member of the computing staff at the Graduate School of Management, who has given great help whenever I had problems with software.

In the process of writing this book in English, I am indebted to my daughter, Ningsih Agung Chandra, and my son, Darma Putra, for their time in correcting my English. My daughter has a Bachelor of Science from the Department of Biostatistics, School of Public Health, the University of North Carolina at Chapel Hill, USA, and a Master's Degree in Communication Studies (MSi) from the London School of Public

Relations, Jakarta (LSPR). Now, she is a senior lecturer and thesis coordinator of the graduate program at LSPR. In addition, she is also the PR and Communication Manager of the Macau Government Tourist Office (MGTO) Representative in Indonesia, and her profile can be found through Google by typing her complete name – Martingsih Agung Chandra. My son has an MBA from De La Salle University, Philippines, and a BSc in Management from Adamson University, Phillipines. Now, he is the director of the Pure Technology Indonesia, Jakarta.

Finally, I would like to thank the reviewers, editors and staff at John Wiley & Sons, Ltd for their work in getting this book to publication.

About the Author

With regards to the request from John Wiley & Sons, Ltd: *"Please provide us with a brief biography including details that explain why you are the ideal person to write this book,"* I present my background, experiences and findings in doing statistical data analysis.

I have a PhD in Biostatistics (1981) and a Master's degree in Mathematical Statistics (1977) from the North Carolina University at Chapel Hill, NC, USA; a Master's degree in Mathematics from New Mexico State University, Las Cruces, NM, USA; a degree in Mathematical Education (1962) from Hasanuddin University, Makassar, Indonesia; and a certificate from "Kursus B-I/B-II Ilmu Pasti" (B-I/B-II Courses in Mathematics), Yogyakarta – a five-year non-degree program in advanced mathematics. So I would say that I have good background knowledge in theoretical as well as applied statistics. In my dissertation on biostatistics, I presented new findings, namely the Generalized Kendall's tau, the Generalized Pair Charts, and the Generalized Simon's Statistics, based on the data censored to the right.

Based on my knowledge in mathematics, mathematical functions in particular, I can evaluate the limitation, hidden assumptions or the unrealistic assumption(s), of all regression functions, such as the fixed-effects models, which are in fact ANCOVA models. For comparison, my book presents the best and the worst ANCOVA models.

Furthermore, based on my exercises and experiments in doing data analyses of various field of studies; such as finance, marketing, education and population studies since 1981 when I worked at the Population Research Center, Gadjah Mada University, 1985–1987; and while I have been at the University of Indonesia 1987 up until now, I have found unexpected or unpredictable statistical results based on alternative panel data models, compared to panel data models which are commonly applied.

Part One

Panel Data as a Multivariate Time Series by States

Abstract

Part I, containing the first four chapters, considers unstacked panel data where the units of the analysis are time observations. So the sets or multidimensional exogenous, endogenous and environmental variables, respectively, for the state i can be presented using the symbols; $X\_i_t = (X1\_i, \ldots, Xk\_i, \ldots)_t$, $Y\_i_t = (Y1\_i, \ldots, Yg\_i, \ldots)_t$, and $Z_t = (Z1, \ldots, Zj, \ldots)_t$, for $i = 1, \ldots, N$; and $t = 1, \ldots, T$. Note that the scores of the environmental variables are constant for all states or individuals. Using these symbols, panel data is considered as the data of multivariate time series by states (countries, regions, agencies, firms, industries, households or individuals).

It is noted that all of the time series models presented in Agung (2009a) can easily be applied to conduct the data analysis based on each state in the panel data; as well as the general multivariate models by states and time periods, presented in Section 3.7. This part presents just the specific analyses for unstacked data with a small number of N. The models for a large N will be presented in Part II and Part III.

Panel Data Analysis Using EViews, First Edition. I Gusti Ngurah Agung.
© 2014 John Wiley & Sons, Ltd. Published 2014 by John Wiley & Sons, Ltd.
Companion website: www.wiley.com/go/panel_data

1

Data Analysis Based on a Single Time Series by States

1.1 Introduction

Panel data can be viewed as a finite set of time-series data. As an illustration Table 1.1 presents part of the data in POOLG7.wf1, namely *unstacked data*, consisting of a single time series *GDP* from seven countries. Note that this table shows seven time series variables, namely GDP_CAN_t to GDP_US_t.

Based on each time series of GDP by states, various growth models can be considered as presented in Agung (2009a, Chapter 2), starting with classical growth models, namely geometric and exponential growth models, and their extensions. Therefore, based on the seven states, the multivariate growth models should be applied as presented in the following sections.

1.2 Multivariate Growth Models

1.2.1 Continuous Growth Models

In general, let Y_{it} be the observed value of the variable Y for the i-th individual (a country, state, region, agency, community, household or person) at time t, for $i = 1, \ldots, N$, and $t = 1, \ldots, T$. In panel data analysis, the symbol $Y\_i(t)$, Y_i_t, or $Y\_$"*Name*"$_t$ will be used to indicate the time series variable Y_{it}, such as the variable GDP_Can_t to GDP_US_t in POOL7.wf1. In this chapter, the panel data set will be considered as a finite set of time-series variables. For this reason, the simplest model considered is a multivariate classical growth model with the following general equation.

$$\log(Y\_i_t) = C(i1) + C(i2)^* t + \mu_{it} \tag{1.1}$$

where *C(i2)* indicates the exponential growth rate of Y_i, that is the growth rate of the variable Y for the i-th individual (country, state, region, community, household, firm or agency), *C(i1) is* the intercept parameter, and μ_{it} their residuals which, in general, should be autocorrelated (see to Agung 2009a, Chapter 2).

Panel Data Analysis Using EViews, First Edition. I Gusti Ngurah Agung.
© 2014 John Wiley & Sons, Ltd. Published 2014 by John Wiley & Sons, Ltd.
Companion website: www.wiley.com/go/panel_data

Table 1.1   A subset of the unstacked data in POOLG7.wf1

Year	GDP_CAN	GDP_FRA	GDP_GER	GDP_ITA	GDP_JPN	GDP_UK	GDP_US
1950	6209	4110	3415	2822	1475	5320	8680
1951	6385	4280	3673	3023	1649	5484	9132
1952	6752	4459	4013	3131	1787	5517	9213
1953	6837	4545	4278	3351	1884	5791	9450
1954	6495	4664	4577	3463	1972	5971	9177
1955	6907	4861	5135	3686	2108	6158	9756
1956	7349	5195	5388	3815	2249	6238	9756
1957	7213	5389	5610	3960	2394	6322	9724
1958	7061	5463	5787	4119	2505	6340	9476
1959	7180	5610	6181	4351	2714	6569	9913

Therefore, the basic growth model considered should be a multivariate autoregressive growth model, namely $MAR(q_1, \ldots, q_i, \ldots)\_GM = MAR(q)\_GM$, with the following general equation, where the error terms ε_{it} would be assumed or accepted to have an $i.i.d.N(0, \sigma_i^2)$, in a theoretical sense. Refer to the special notes presented in Section 2.14.3 (Agung, 2009a).

$$\log(Y\_i_t) = C(i1) + C(i2)^* t + \mu_{it}$$
$$\mu_{it} = \rho_{i1}\mu_{i,t-1} + \rho_{i,t-2}\mu_{i,t-2} + \ldots + \rho_{i,t-q_i}\mu_{i,t-q_i} + \varepsilon_{it} \tag{1.2}$$

However, for a multivariate GLM, the vector of the error terms $(\varepsilon_1, \varepsilon_2, \ldots, \varepsilon_N)$, in general, would have a residual correlation matrix, namely $CM(\varepsilon)$, or a residual covariance matrix, namely $\Sigma(\varepsilon)$, which is not a diagonal matrix, and should indicate that the endogenous variables $log(Y\_i)$ or Y_i, for the states $i = 1, 2, \ldots, N$, are correlated in a statistical sense, even though they may not be correlated in a theoretical sense. In other words, the quantitative correlations between all $log(Y\_i)$ are taken into account in the estimation process.

Example 1.1   Illustrative growth curves
As an illustration, Figure 1.1 presents the growth curves GDP_t of two pairs of neighboring countries, namely (a) *GDP_CAN* and *GDP_US*, and (b) *GDP_FRA* and *GDP_GER*, which clearly show differential characteristics. Corresponding to growth curves, we find that each pair of the five variables *GDP_CAN,*

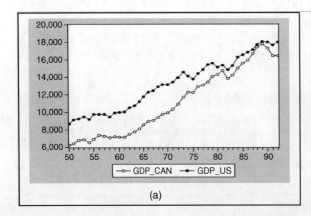
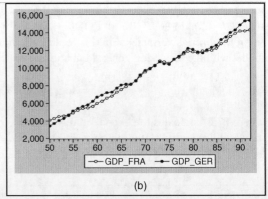

(a)   (b)

Figure 1.1   Growth curves of GDP_CAN, GDP_US, GDP_FRA and GDP_GE

GDP_US, *GDP_FRA*, *GDP_GER* and the time *t* variable are significantly positively correlated with a *p*-value $= 0.0000$. However, unexpected statistical results are obtained based on the model in (1.2), as presented in Example 1.3.

Growth curves are important descriptive statistics in any time series, as well as panel data analyses. Many findings and conclusions can be derived based on descriptive statistical summaries. See various continuous and discontinuous growth curves and time series models presented in Agung (2009a), and the descriptive statistical summaries presented in Agung (2004, 2009b, 2011). For additional illustrations, see the graphical presentations in Leary (2009), and Chambers and Dimson (2009).

Example 1.2   A multivariate classical growth model (MCGM)

Figure 1.2 presents the statistical results based on a MCGM of *GDP_Can*, *GDP_US*, *GDP_Fra*, and *GDP_Ger*. Its residuals graphs are obtained by selecting View/Residuals/Graphs, as presented in Figure 1.3. Based on these results, the following notes are presented.

1. Note that the four regressions in the model in fact represent a growth model by states, which has been presented as a multiple regression model or a single time series model using dummy variables of the states in Agung (2009a).
2. Using the standard *t*-test statistic in the output, it can be concluded that *GDP_Can*, *GDP_US*, *GDP_Fra* and *GDP_Ger*, have significant positive exponential growth rates of

$$\hat{C}(11) = 0.0273339, \hat{C}(21) = 0.018282, \hat{C}(31) = 0.030681, \text{and } \hat{C}(41) = 0.032058.$$

3. The null hypothesis H_0: $C(11) = C(21) = C(31) = C(41)$ is rejected based on the Chi-square statistic of $\chi_0^2 = 242.8469$ with $df = 3$ and a *p*-value $= 0.0000$. Therefore, it can be concluded that the growth rates of *GDP* of the four countries have significant differences. The other hypotheses on the growth rates differences can easily be tested using the Wald test.
4. However, note that the MCGM is an inappropriate time series model indicated by the very small Durbin–Watson statistics values of the four regressions, as well as their residuals graphs in Figure 1.3. For this reason, a modified GM will be presented in the following example. Refer also to Chapter 2 in Agung (2009a).

Estimation Method: Least Squares
Date: 08/22/09   Time: 18:49
Sample: 1950 1992
Included observations: 43
Total system (balanced) observations 172

	Coefficient	Std. Error	t-Statistic	Prob.
C(10)	8.683015	0.017182	505.3611	0.0000
C(11)	0.027339	0.000704	38.81228	0.0000
C(20)	9.080149	0.011628	780.8541	0.0000
C(21)	0.018282	0.000477	38.34938	0.0000
C(30)	8.414288	0.025252	333.2182	0.0000
C(31)	0.030681	0.001035	29.63753	0.0000
C(40)	8.403292	0.030639	274.2682	0.0000
C(41)	0.032058	0.001256	25.52230	0.0000

Determinant residual covariance	1.82E-11

Equation: LOG(GDP_CAN)=C(10)+C(11)*T
Observations: 43

R-squared	0.973504	Mean dependent var	9.257129
Adjusted R-squared	0.972858	S.D. dependent var	0.347921
S.E. of regression	0.057320	Sum squared resid	0.134708
Durbin-Watson stat	0.323229		

Equation: LOG(GDP_US)=C(20)+C(21)*T
Observations: 43

R-squared	0.972878	Mean dependent var	9.464070
Adjusted R-squared	0.972216	S.D. dependent var	0.232736
S.E. of regression	0.038794	Sum squared resid	0.061702
Durbin-Watson stat	0.452012		

Equation: LOG(GDP_FRA)=C(30)+C(31)*T
Observations: 43

R-squared	0.955405	Mean dependent var	9.058594
Adjusted R-squared	0.954317	S.D. dependent var	0.394138
S.E. of regression	0.084241	Sum squared resid	0.290960
Durbin-Watson stat	0.060788		

Equation: LOG(GDP_GER)=C(40)+C(41)*T
Observations: 43

R-squared	0.940785	Mean dependent var	9.076509
Adjusted R-squared	0.939340	S.D. dependent var	0.415012
S.E. of regression	0.102214	Sum squared resid	0.428355
Durbin-Watson stat	0.084715		

Figure 1.2   *Statistical results based on a multivariate classical growth model*

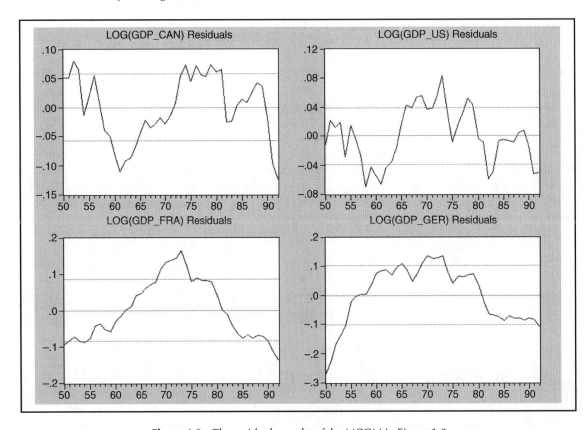

Figure 1.3   *The residuals graphs of the MCGM in Figure 1.2*

5. On the other hand, by observing the residual graphs in Figure 1.3, then it can be said that a polynomial growth model should be explored for each state, such as quadratic regressions of *log(GDP_FRA)* and *log(GDP_GER)* on the time *t*, and at least third degree polynomials of *GDP_CAN* and *GDP_US* on the time *t*. Do this as an exercise.

6. As an additional illustration, Table 1.2 presents the correlations between the time *t* with the dependent variables of each model. Note that each parameter *C(i2)* has exactly the same value of the *t*-statistic, as well as Prob(*t*-stat). Compared to the results in Figure 1.2, the following notes and conclusions are made.

    6.1   The testing hypothesis on each *C(i2)*, either a two- or one-sided hypothesis, can be done using the corresponding bivariate correlation. To generalize the results, the set of simple linear regressions can be presented using a correlation matrix of the set of variables considered.

Table 1.2   *Bivariate correlations of time* t *with each of the dependent variables of the multivariate model in Figure 1.2*

	LOG(GDP_CAN)	LOG(GDP_US)	LOG(GDP_FRA)	LOG(GDP_GER)
Time *t*	0.986 663	0.98 635	0.97 745	0.96 994
***t*-stat**	38.81 228	38.34 938	29.63 753	25.52 230
Prob.	0.00 000	0.00 000	0.00 000	0.00 000

6.2   On the other hand, by doing a series of *state-by-state analyses*, we obtain exactly the same set of four regressions as presented in Figure 1.2. For this reason the model presented in Figure 1.2 will be referred to as the *system of independent states*.

Example 1.3   A MAR(1)_GM unexpected result

Figure 1.4 presents the statistical results based on a MAR(**1**)_GM = MAR(1,1,1,1)_GM of the four time series *GDP_CAN, GDP_US, GDP_FRA*, and *GDP_GER*. Based on these results, the following findings and notes are presented.

1. The estimate of $C(31) = -0.212\,090$ with a *p*-value $= 0.9358$, which should indicate the (adjusted) growth rate of *GDP_FRA*, is an *unexpected result*, since $r(log(GDP\_FRA), t) = 0.97\,448$ with a *p*-value $= 0.0000$ and obtains a simple linear regression function of $LOG(GDP\_FRA) = 8.3841 + 0.0307^*t$ as presented in Figure 1.2, with an exponential growth rate of *GDP_FRA* as $r = 0.0307$.

2. This finding indicates the impact of using an AR(1) on the parameter estimates is in fact unpredictable. Nothing is wrong with the model, but the structure of the data set cannot provide acceptable estimates. Compared to the growth curve of *GDP_FRA* in Figure 1.1, the AR(1)_GM of *GDP_FRA* should be considered as an unacceptable or inappropriate time series model for representing the *GDP* of France. The results of the author's experimentation based on the variable *GDP_FRA*, are presented in the following examples.

3. On the other hand, we find the residual matrix correlation, says $M(\varepsilon)$, is not a diagonal matrix. For comparison, application to the WLS or SUR estimation methods is recommended. Do this as an exercise.

4. For a comparison study, Table 1.3 presents a summary of the statistical results using the series of state-by-state analyses based on the LS AR(1)_GMs. Note that this table shows the coefficients of the time *t* and the AR(1) terms are exactly the same as those in Figure 1.4, but they have different intercepts. Compare this to the other statistics.

System: SYS01
Estimation Method: Iterative Least Squares
Date: 07/07/09   Time: 15:05
Sample: 1951 1992
Included observations: 43
Total system (balanced) observations 168
Convergence achieved after 2 iterations

	Coefficient	Std. Error	t-Statistic	Prob.
C(10)	8.675757	0.117154	74.05426	0.0000
C(11)	0.025588	0.003869	6.613016	0.0000
C(12)	0.881066	0.094300	9.343263	0.0000
C(20)	9.080932	0.045963	197.5692	0.0000
C(21)	0.017432	0.001607	10.84940	0.0000
C(22)	0.788836	0.104235	7.567884	0.0000
C(30)	86.20318	1629.510	0.052901	0.9579
C(31)	-0.212090	2.627543	-0.080718	0.9358
C(32)	0.996671	0.035945	27.72798	0.0000
C(40)	8.768484	0.188648	46.48057	0.0000
C(41)	0.020326	0.004748	4.281062	0.0000
C(42)	0.891366	0.035255	25.28374	0.0000

Determinant residual covariance	2.00E-14

Equation: LOG(GDP_CAN)=C(10)+C(11)*T+[AR(1)=C(12)]
Observations: 42

R-squared	0.991522	Mean dependent var	9.269590
Adjusted R-squared	0.991087	S.D. dependent var	0.342288
S.E. of regression	0.032314	Sum squared resid	0.040724
Durbin-Watson stat	1.510506		

Equation: LOG(GDP_US)=C(20)+C(21)*T+[AR(1)=C(22)]
Observations: 42

R-squared	0.988209	Mean dependent var	9.473482
Adjusted R-squared	0.987604	S.D. dependent var	0.227124
S.E. of regression	0.025287	Sum squared resid	0.024938
Durbin-Watson stat	1.762500		

Equation: LOG(GDP_FRA)=C(30)+C(31)*T+[AR(1)=C(32)]
Observations: 42

R-squared	0.997723	Mean dependent var	9.076151
Adjusted R-squared	0.997607	S.D. dependent var	0.381517
S.E. of regression	0.018665	Sum squared resid	0.013587
Durbin-Watson stat	1.414278		

Equation: LOG(GDP_GER)=C(40)+C(41)*T+[AR(1)=C(42)]
Observations: 42

R-squared	0.996821	Mean dependent var	9.098903
Adjusted R-squared	0.996658	S.D. dependent var	0.392866
S.E. of regression	0.022710	Sum squared resid	0.020114
Durbin-Watson stat	1.596264		

Figure 1.4   *Statistical results based on a MAR(**1**)_GM of the GDP of four countries*

Table 1.3   *Summary of the statistical results based on the four LS AR(1)_GMs in Figure 1.4*

Variable	Dependent Variables							
	log(Gdp_Can)		log(Gdp_US)		log(Gdp_Fra)		log(Gdp_Ger)	
	Coef.	*t*-Stat.	Coef.	*t*-Stat.	Coef.	*t*-Stat.	Coef.	*t*-Stat.
C	8.701 345	76.608 12	9.098 363	204.4720	85.991 09	0.052 856	8.788 810	47.760 36
T	0.025 588	6.613 016	0.017 432	10.849 40	−0.212 090	−0.080 718	0.020 326	4.281 062
AR(1)	0.881 066	9.343 263	0.788 836	7.567 884	0.996 671	27.727 98	0.891 366	25.283 74
R-squared	0.991 522		0.988 209		0.997 723		0.996 821	
Adjusted R-squared	0.991 087		0.987 604		0.997 607		0.996 658	
S.E. of regression	0.032 314		0.025 287		0.018 665		0.022 710	
F-statistic	2280.633		1634.300		8545.536		6115.340	
Prob(F-statistic)	0.000 000		0.000 000		0.000 000		0.000 000	
Durbin–Watson stat	1.510 506		1.762 500		1.414 278		1.596 264	

1.2.2   Discontinuous Growth Models

Corresponding to the inappropriate estimate of $C(31) = -0.212\,090$ in Figure 1.4, experimentation should be done based on the data of the *GDP_FRA*. See the following examples.

Example 1.4   An experimentation based on *GDP_FRA*
By using trial-and-error methods, we finally obtain the statistical results in Figure 1.5, based on two sub-samples of sizes 29 and 30, respectively, for $T < 31$ and $T < 32$. Based on these results the following findings and notes are presented.

1. Based on the results in Figure 1.5(a), the null hypothesis H_0: $C(2) \leq 0$ is rejected, based on the *t*-statistic of $t_0 = 1.819\,125$ with a *p*-value $= 0.0804/2 = 0.0402 < 0.05$. Therefore, it can be

Dependent Variable: LOG(GDP_FRA)
Method: Least Squares
Date: 07/08/09 Time: 09:55
Sample: 1950 1992 IF T< 31
Included observations: 29
Convergence achieved after 4 iterations

	Coefficient	Std. Error	t-Statistic	Prob.
C	8.494952	0.705860	12.03490	0.0000
T	0.031080	0.017085	1.819125	0.0804
AR(1)	0.941059	0.107208	8.777880	0.0000

R-squared	0.997041	Mean dependent var	8.904769
Adjusted R-squared	0.996814	S.D. dependent var	0.335020
S.E. of regression	0.018911	Akaike info criterion	-5.000398
Sum squared resid	0.009299	Schwarz criterion	-4.858954
Log likelihood	75.50578	Hannan-Quinn criter.	-4.956100
F-statistic	4380.578	Durbin-Watson stat	1.490196
Prob(F-statistic)	0.000000		

| Inverted AR Roots | .94 | | |

(a)

Dependent Variable: LOG(GDP_FRA)
Method: Least Squares
Date: 07/08/09 Time: 09:56
Sample: 1950 1992 IF T< 32
Included observations: 30
Convergence achieved after 4 iterations

	Coefficient	Std. Error	t-Statistic	Prob.
C	9.809473	18.01907	0.544394	0.5906
T	0.076572	0.197390	0.387924	0.7011
AR(1)	1.019645	0.102863	9.912682	0.0000

R-squared	0.996862	Mean dependent var	8.920418
Adjusted R-squared	0.996629	S.D. dependent var	0.340169
S.E. of regression	0.019750	Akaike info criterion	-4.916720
Sum squared resid	0.010531	Schwarz criterion	-4.776600
Log likelihood	76.75080	Hannan-Quinn criter.	-4.871895
F-statistic	4288.174	Durbin-Watson stat	1.527986
Prob(F-statistic)	0.000000		

Inverted AR Roots	1.02	
Estimated AR process is nonstationary		

(b)

Figure 1.5   *Statistical results based on an AR(1)_GM of* GDP_FRA *using two sub-samples*

concluded that *GDP_FRA* has a significant positive growth rate of 3.11% within the time period $t = 1$ to $t = 30$.

2. On the other hand, Figure 1.5(b) shows a note "*Estimated AR process is nonstationary*", which indicates that the AR(1)_GM is an inappropriate time series model within the period $t = 1$ to $t = 31$. Finally, based on the whole data set, the Inverted AR Root $= 1.00$ is obtained, however, without the statement "*Estimated AR process is nonstationary*".

Example 1.5   A piece-wise growth model of *GDP_FRA*

As the complement of the AR(1)_GM of *GDP_FRA* for $t < 31$, Figure 1.6(a) presents another piece of AR(1)_GM of *GDP_FRA* for $t >= 31$, which should be considered an acceptable time series model, in a statistical sense. Note that this model shows that GDP_FRA has a significant positive growth rate of 2.18% based on the *t*-statistic of $t_0 = 7.425\,573$ with a *p*-value $= 0.0000$, for $t >= 31$, compared to the growth rate of 3.11% for $t < 31$. Therefore, based on these findings the growth model of *GDP_FRA* could be presented by a two-piece GM using dummy variables *Dt1* and *Dt2*, which should be generated for the two time periods.

Figure 1.6(b) presents the statistical results based on an acceptable two-piece AR(2)_GM of *log(GDP_FRA)*, in a statistical sense. Based on these results, the following pair of regression functions can be derived.

$$\log(GDP\_FRA) = 8.3274 + 0.0359^*t + [AR(1) = 1.2142, AR(2) = -0.3301], \quad \text{for } t < 31$$

$$\log(GDP\_FRA) = 9.0101 + 0.0127^*t + [AR(1) = 1.2142, AR(2) = -0.3301], \quad \text{for } t \geq 31$$

Based on these findings, the MAR**(1)**_GM presented in Figure 1.4 should be modified to a MAR (1,1,2,1)_GM, with the statistical results presented in Figure 1.7.

Dependent Variable: LOG(GDP_FRA)				
Method: Least Squares				
Date: 07/08/09 Time: 16:45				
Sample: 1950 1992 IF T> 30				
Included observations: 13				
Convergence achieved after 4 iterations				
	Coefficient	Std. Error	t-Statistic	Prob.
C	8.639295	0.118147	73.12313	0.0000
T	0.021810	0.002937	7.426573	0.0000
AR(1)	0.610004	0.158493	3.848772	0.0032
R-squared	0.976634	Mean dependent var		9.458464
Adjusted R-squared	0.971960	S.D. dependent var		0.078091
S.E. of regression	0.013076	Akaike info criterion		-5.636837
Sum squared resid	0.001710	Schwarz criterion		-5.506464
Log likelihood	39.63944	Hannan-Quinn criter.		-5.663634
F-statistic	208.9827	Durbin-Watson stat		1.417305
Prob(F-statistic)	0.000000			
Inverted AR Roots	.61			

(a)

Dependent Variable: LOG(GDP_FRA)				
Method: Least Squares				
Date: 07/08/09 Time: 17:56				
Sample (adjusted): 1952 1992				
Included observations: 41 after adjustments				
Convergence achieved after 12 iterations				
LOG(GDP_FRA)=(C(11)+C(12)*T)*DT1+(C(21)+C(22)*T)*DT2				
+[AR(1)=C(1), AR(2)=C(2)]				
Variable	Coefficient	Std. Error	t-Statistic	Prob.
C(11)	8.327340	0.098475	84.56308	0.0000
C(12)	0.035929	0.004369	8.223822	0.0000
C(21)	9.010104	0.211207	42.65998	0.0000
C(22)	0.012724	0.005679	2.240634	0.0315
C(1)	1.214156	0.161654	7.510837	0.0000
C(2)	-0.330056	0.173660	-1.900584	0.0656
R-squared	0.998026	Mean dependent var		9.093576
Adjusted R-squared	0.997744	S.D. dependent var		0.368947
S.E. of regression	0.017522	Akaike info criterion		-5.116229
Sum squared resid	0.010746	Schwarz criterion		-4.865462
Log likelihood	110.8827	Hannan-Quinn criter.		-5.024913
Durbin-Watson stat	2.000139			
Inverted AR Roots	.80	.41		

(b)

Figure 1.6   *Statistical results based on (a) an AR(1)_GM of GDP_FRA for t >= 31, and (b) a two-piece (discontinuous) AR(1)_GM of GDP_FRA*

```
System: SYS04
Estimation Method: Iterative Least Squares
Date: 07/08/09   Time: 18:20
Sample: 1951 1992
Included observations: 43
Total system (unbalanced) observations 167
Convergence achieved after 12 iterations
```

	Coefficient	Std. Error	t-Statistic	Prob.
C(10)	8.675757	0.117154	74.05426	0.0000
C(11)	0.025588	0.003869	6.613016	0.0000
C(12)	0.881066	0.094300	9.343263	0.0000
C(20)	9.080932	0.045963	197.5692	0.0000
C(21)	0.017432	0.001607	10.84940	0.0000
C(22)	0.788836	0.104235	7.567884	0.0000
C(30)	8.327353	0.098426	84.60564	0.0000
C(31)	0.035928	0.004367	8.227148	0.0000
C(32)	9.010076	0.211183	42.66470	0.0000
C(33)	0.012725	0.005678	2.240981	0.0265
C(34)	1.214152	0.161655	7.510770	0.0000
C(35)	-0.330039	0.173659	-1.900506	0.0593
C(40)	8.768484	0.188648	46.48057	0.0000
C(41)	0.020326	0.004748	4.281062	0.0000
C(42)	0.891366	0.035255	25.28374	0.0000

Determinant residual covariance	1.85E-14

```
Equation: LOG(GDP_CAN)=C(10)+C(11)*T+[AR(1)=C(12)]
Observations: 42
R-squared             0.991522   Mean dependent var    9.269590
Adjusted R-squared    0.991087   S.D. dependent var    0.342288
S.E. of regression    0.032314   Sum squared resid     0.040724
Durbin-Watson stat    1.510506

Equation: LOG(GDP_US)=C(20)+C(21)*T+[AR(1)=C(22)]
Observations: 42
R-squared             0.988209   Mean dependent var    9.473482
Adjusted R-squared    0.987604   S.D. dependent var    0.227124
S.E. of regression    0.025287   Sum squared resid     0.024938
Durbin-Watson stat    1.762500

Equation: LOG(GDP_FRA)=(C(30)+C(31)*T)*DT1+(C(32)+C(33)*T)*DT2
          +[AR(1)=C(34),AR(2)=C(35)]
Observations: 41
R-squared             0.998026   Mean dependent var    9.093576
Adjusted R-squared    0.997744   S.D. dependent var    0.368947
S.E. of regression    0.017522   Sum squared resid     0.010746
Durbin-Watson stat    2.000128

Equation: LOG(GDP_GER)=C(40)+C(41)*T+[AR(1)=C(42)]
Observations: 42
R-squared             0.996821   Mean dependent var    9.098903
Adjusted R-squared    0.996658   S.D. dependent var    0.392866
S.E. of regression    0.022710   Sum squared resid     0.020114
Durbin-Watson stat    1.596264
```

Figure 1.7   *Statistical results based on a MAR(1,1,2,1)_GM of GDP_Can, GDP_US, GDP_Fra and GDP_Ger*

1.3   Alternative Multivariate Growth Models

As an extension of all the continuous and discontinuous growth models presented in Agung (Agung, 2009a, Chapters 2 and 3), various multivariate growth models can easily be derived. However, only some selected models will be presented in the following sub-sections.

1.3.1   A Generalization of MAR(p)_GM

As an extension of the MAR(p)_GM in (1.2), the following growth model is presented.

$$\log(Y\_i) = F_i(t, C(i^*)) + \mu_{it}$$
$$\mu_{it} = \rho_{i1}\mu_{i,t-1} + \rho_{i,t-2}\mu_{i,t-2} + \ldots + \rho_{i,t-p_i}\mu_{i,t-p_i} \tag{1.3}$$

where $F_i(t, C(i^*))$ can be any functions of t, such as the polynomial and the natural logarithmic of t, either continuous or discontinuous functions, as well as nonlinear with a finite number of parameters, namely $C(i^*)$, for each $i = 1, \ldots, N$. Note that any continuous and discontinuous growth models in Agung (2009a) could be inserted for the function $F_i(t, C(i^*))$. For example, as follows:

1.3.1.1   A Polynomial Growth Model

The independent-states system of polynomial growth models has the following equation for $i = 1, \ldots, N$.

$$\log(Y\_i_t) = c(i0) + c(i1)^* t + \ldots + c(ik_i)^* t^{k_i} + \mu_{it} \tag{1.4a}$$

1.3.1.2   A Translog Linear Model

The independent-states system of the translog linear growth models has the following equation for $i = 1, \ldots, N$.

$$\log(Y\_i_t) = c(i0) + c(i1)^* \log(t) + \mu_{it} \tag{1.4b}$$

1.3.1.3   The Simplest Nonlinear Growth Model

The independent-states system of the simplest nonlinear growth model has the following equation for $i = 1, \ldots, N$.

$$\log(Y\_i_t) = c(i0) + c(i1)^* t^{c(i2)} + \mu_{it} \tag{1.4c}$$

1.3.1.4   A Two-Piece Growth Model

The independent-states system of two-piece growth models has the following equation for $i = 1, \ldots, N$.

$$\log(Y\_i_t) = (c(i10) + c(i11)^* t)^* Dt1 + (c(i20) + c(i21)^* t)^* Dt2 + \mu_{it} \tag{1.4d}$$

where $Dt1$ and $Dt2$ are dummy variables of two time periods considered, such as $t < t_0$ and $t \geq t_0$, which are defined to be valid for all states or individuals, or all $i = 1, \ldots, N$. To generalize, the dummy variables would be dependent on i, namely $Dt(i1)$ and $Dt(i2)$, the model can easily be extended to three or more time periods, and the linear function of t within each time period could be replaced by other functions of t. For further illustration, refer to various discontinuous growth models presented in Chapter 3, (Agung, 2009a), specifically the multivariate models by states and time periods in the general models (3.79) to (3.87).

1.3.2   Multivariate Lagged Variables Growth Models

Corresponding to the MAR(p)_GM in (1.2), a multivariate lagged variables growth model, namely MLV(q)_GM, may be considered an alternative growth model with the following general equation, where the error terms should also be assumed or accepted in a theoretical sense to have an *i.i.d.*$N(0, \sigma^2)$.

$$\log(Y\_i_t) = C(i0) + \sum_{j=1}^{q_i} C(ij)^* \log(Y\_i_{t-j}) + C(i, q_i + 1)^* t + \varepsilon_{it} \tag{1.5}$$

Note that the lag variable $log(Y\_i_{t-j})$ is not a cause factor of $log(Y\_i_t)$, but is an up-stream or a predictor variable. Also, the exogenous variables, namely X_i_t and X_i_{t-j}, used in most models are not really the true cause factors of the dependent variable of these models. See the models presented in Section 1.4.

All lagged variables and autoregressive models, in fact, are *dynamic models* (Gujarati, 2003, Gourierroux and Manfort, 1997, Hamilton, 1994, and Kmenta, 1986). Therefore, various models in (1.5) should be considered as *multivariate dynamic growth models* (MDGM), or multivariate *dynamic models with trend*, for $i = 1, \ldots, N$. Wooldridge (2002; 493) presents another type of dynamic model, called *dynamic unobserved effects models*.

Example 1.6   A MLV(1)_GM of GDP? in Figure 1.7

As an alternative multivariate growth model of GDP in Figure 1.7, Figure 1.8 presents the statistical results based on an MLV(1)_GM, where the regression of *GDP_Fra* is a two-piece LV(1)_GM. Based on the results in Figures 1.7 and 1.8, the following findings and notes are presented.

1.   The estimates of the parameter C(12) in both models have exactly the same values of 0.881 066, which indicates the first-order autocorrelation of *log(GDP_Can)*. Similarly for the parameters C(22) and C(42), respectively, there is first-order autocorrelation of *log(GDP_US)* and *log(GDP_Ger)*.

Figure 1.8 (left panel):

System: SYS05_LAG
Estimation Method: Least Squares
Date: 08/21/09   Time: 15:10
Sample: 1951 1992
Included observations: 42
Total system (balanced) observations 168

	Coefficient	Std. Error	t-Statistic	Prob.
C(10)	1.057433	0.815685	1.296374	0.1968
C(11)	0.003043	0.002651	1.147831	0.2528
C(12)	0.881066	0.094300	9.343263	0.0000
C(20)	1.934999	0.944323	2.049086	0.0422
C(21)	0.003681	0.001951	1.886825	0.0611
C(22)	0.788836	0.104235	7.567884	0.0000
C(310)	-0.116130	0.793076	-0.146430	0.8838
C(311)	-0.001504	0.003765	-0.399557	0.6900
C(312)	1.019645	0.095786	10.64503	0.0000
C(320)	4.084122	2.518689	1.621527	0.1070
C(321)	0.010557	0.005974	1.733656	0.0850
C(322)	0.529470	0.288767	1.833553	0.0687
C(40)	0.972882	0.295011	3.297777	0.0012
C(41)	0.002208	0.001180	1.871744	0.0632
C(42)	0.891366	0.035255	25.28374	0.0000

Determinant residual covariance	2.04E-14

Figure 1.8 (right panel):

Equation: LOG(GDP_CAN)=C(10)+C(11)*T+C(12)*LOG(GDP_CAN(-1))
Observations: 42

R-squared	0.991522	Mean dependent var	9.269590
Adjusted R-squared	0.991087	S.D. dependent var	0.342288
S.E. of regression	0.032314	Sum squared resid	0.040724
Durbin-Watson stat	1.510506		

Equation: LOG(GDP_US)=C(20)+C(21)*T+C(22)*LOG(GDP_US(-1))
Observations: 42

R-squared	0.988209	Mean dependent var	9.473482
Adjusted R-squared	0.987604	S.D. dependent var	0.227124
S.E. of regression	0.025287	Sum squared resid	0.024938
Durbin-Watson stat	1.762500		

Equation: LOG(GDP_FRA)=(C(310)+C(311)*T+C(312)*LOG(GDP_FRA(-1)))
    *DT1+(C(320)+C(321)*T+C(322)*LOG(GDP_FRA(-1)))*DT2
Observations: 42

R-squared	0.997960	Mean dependent var	9.076151
Adjusted R-squared	0.997676	S.D. dependent var	0.381517
S.E. of regression	0.018391	Sum squared resid	0.012176
Durbin-Watson stat	1.557845		

Equation: LOG(GDP_GER)=C(40)+C(41)*T+C(42)*LOG(GDP_GER(-1))
Observations: 42

R-squared	0.996821	Mean dependent var	9.098903
Adjusted R-squared	0.996658	S.D. dependent var	0.392866
S.E. of regression	0.022710	Sum squared resid	0.020114
Durbin-Watson stat	1.596264		

Figure 1.8 *Statistical results based on a MLV(1)_GM of* GDP_Can, GDP_US, GDP_Fra *and* GDP_Ger

2. Corresponding to the regression model of *log(GDP_Fra)*, Figures 1.7 and 1.8 present different types of two-piece growth models. Figure 1.7 presents a two-piece AR(2)_GM, where the autocorrelation of the error terms AR(1) and AR(2) should be valid for the whole time period. On the other hand, Figure 1.8 presents a two-piece LV(1)_GM, where $\hat{C}(312) = 1.019645$ is the AR(1) of *log(GDP_Fra)* for $t < 31$, and $\hat{C}(312) = 0.382358$ is its AR(1) for $t \geq 31$.

3. However, Figure 1.8 presents a negative adjusted growth rate of log(GDP_Fra), for $t < 31$, namely $\hat{C}(311) = -0.001504$ which is an inappropriate estimate. For this reason, the statistical results based on a MLV(1,1,2,1) are presented in Figure 1.9, where the two-piece regressions of *log(GDP_Fra)* is

Figure 1.9 (left panel):

Estimation Method: Least Squares
Date: 08/21/09   Time: 15:21
Sample: 1951 1992
Included observations: 42
Total system (unbalanced) observations 167

	Coefficient	Std. Error	t-Statistic	Prob.
C(10)	1.057433	0.815685	1.296374	0.1968
C(11)	0.003043	0.002651	1.147831	0.2529
C(12)	0.881066	0.094300	9.343263	0.0000
C(20)	1.934999	0.944323	2.049086	0.0422
C(21)	0.003681	0.001951	1.886825	0.0611
C(22)	0.788836	0.104235	7.567884	0.0000
C(310)	0.300243	0.872774	0.344011	0.7313
C(311)	0.000601	0.004205	0.142842	0.8866
C(312)	1.221207	0.192555	6.342122	0.0000
C(313)	-0.253122	0.209560	-1.207873	0.2290
C(320)	3.984467	2.545501	1.565297	0.1196
C(321)	0.009331	0.006145	1.518485	0.1310
C(322)	0.876620	0.492666	1.779340	0.0772
C(323)	-0.333157	0.381130	-0.874129	0.3834
C(40)	0.972882	0.295011	3.297777	0.0012
C(41)	0.002208	0.001180	1.871744	0.0632
C(42)	0.891366	0.035255	25.28374	0.0000

Determinant residual covariance	1.96E-14

Figure 1.9 (right panel):

Equation: LOG(GDP_CAN)=C(10)+C(11)*T+C(12)*LOG(GDP_CAN(-1))
Observations: 42

R-squared	0.991522	Mean dependent var	9.269590
Adjusted R-squared	0.991087	S.D. dependent var	0.342288
S.E. of regression	0.032314	Sum squared resid	0.040724
Durbin-Watson stat	1.510506		

Equation: LOG(GDP_US)=C(20)+C(21)*T+C(22)*LOG(GDP_US(-1))
Observations: 42

R-squared	0.988209	Mean dependent var	9.473482
Adjusted R-squared	0.987604	S.D. dependent var	0.227124
S.E. of regression	0.025287	Sum squared resid	0.024938
Durbin-Watson stat	1.762500		

Equation: LOG(GDP_FRA)=(C(310)+C(311)*T+C(312)*LOG(GDP_FRA(-1))
    +C(313)*LOG(GDP_FRA(-2)))*DT1+(C(320)+C(321)*T+C(322)
    *LOG(GDP_FRA(-1))+C(323)*LOG(GDP_FRA(-2)))*DT2
Observations: 41

R-squared	0.997910	Mean dependent var	9.093576
Adjusted R-squared	0.997467	S.D. dependent var	0.368947
S.E. of regression	0.018568	Sum squared resid	0.011378
Durbin-Watson stat	1.954962		

Equation: LOG(GDP_GER)=C(40)+C(41)*T+C(42)*LOG(GDP_GER(-1))
Observations: 42

R-squared	0.996821	Mean dependent var	9.098903
Adjusted R-squared	0.996658	S.D. dependent var	0.392866
S.E. of regression	0.022710	Sum squared resid	0.020114
Durbin-Watson stat	1.596264		

Figure 1.9 *Statistical results based on a MLV(1,1,2,1)_GM of* GDP_Can, GDP_US, GDP_Fra *and* GDP_Ger

LV(2)_GM, and the two regressions represent positive growth rates of *GDP_Fra*, namely $\hat{C}(311) = 0.000601$ and $\hat{C}(321) = 0.009331$, respectively, for $t < 31$ and $t \geq 31$, adjusted for *log(GDP_Fra(−1))* and *log(GDP_Fra(−2))*.

1.3.3 Multivariate Lagged-Variable Autoregressive Growth Models

As an extension of LVAR(1,1)_GM presented in Agung (2009a), data analysis based on a multivariate lagged variables autoregressive model, MLVAR(*p;q*)_GM, where $p = (p_i)$ and $q = (q_i)$, of the time series Y_i_t, for $i = 1, \ldots, N$, will have the following general equation.

$$\log(Y\_i_t) = C(i0) + \sum_{j=1}^{p_i} C(ij)^* \log(Y\_i_{t-j}) + C(i, p_i + 1)^* t + \mu_{it}$$

$$\mu_{it} = \sum_{k=1}^{q_i} \rho_{ik} \mu_{i,t-k} + \varepsilon_{it}$$

(1.6)

Note that for $q = 0$, the MLV(*p*)_GM will be obtained, and the MAR(*q*)_GM obtained for $p = 0$. Various special cases would be obtained, where $p_i = p$ and $q_i = q$ for all $i = 1, \ldots, N$.

1.3.4 Bounded MLVAR(*p;q*)_GM

As an extension of the general MLVAR(*p;q*)_GM in (1.6) as well as the bounded growth model presented in Agung (2009a), the bounded MLVAR(*p;q*)_GM, of the time series Y_i_t, $i = 1, 2, \ldots, N$, has the following general equation.

$$\log\left(\frac{Y\_i - Li}{Ui - Y\_i}\right) = C(i0) + \sum_{j=1}^{p_i} C(ij)^* \log(Y\_i_{t-j}) + C(i, p_i + 1)^* t + \mu_{it}$$

$$\mu_{it} = \sum_{k=1}^{q_i} \rho_{ik} \mu_{i,t-k} + \varepsilon_{it}$$

(1.7)

where *Li* and *Ui* are the lower and upper bounds of *Y_i*, which are theoretically selected fixed numbers.

1.3.5 Special Notes

Based on the findings presented previously, the following special notes are presented.

1. Unexpected parameter estimates can be obtained by using autoregressive or lagged variables growth models. In general, by inserting an additional independent variable to a model, we can never predict its impact on the parameter estimates. Refer to the special notes in Agung (2009a, Section 2.14.2). For this reason, one should use the trial-and-error method to develop several acceptable growth models, in both theoretical and statistical senses. Note that this statement also should be applicable for any statistical model.
2. Graphic representations between each of the independent variables and the corresponding dependent variable should be analyzed to evaluate their possible patterns of relationship. Specifically, whether a linear or non-linear model would acceptable. Refer to Chapter 1 in Agung (2009a).

3. Furthermore, residuals analysis should be done to identify the limitation of each model. Refer to Agung (2009a).
4. Corresponding to the relationship between Y_i_t and its lag Y_i_{t-s}, specific for $s = 1$, $s = 4$, and $s = 12$, respectively, if and only if the time series data are annually, quarterly, and monthly data sets, the following notes are presented.

    4.1   The observed values of Y_i_{t-s} and Y_i_t can be considered as the observations before and after a natural-experiment for the i-th individual, with a set of environmental variables could be the treatment or experimental factors, namely $Z_t = (Z1, Z2, \ldots, Zk)_t$. Refer to Section 1.8.

    4.2   The lag variable Y_i_{t-s} should be considered as a *covariate* in any time series models having Y_i_t as the dependent variable. So the "*classical growth model of Y_i_t with a covariate Y_i_{t-s}*" for the i-th individual may have the following alternative equations.

$$\log(Y\_i_t) = C(i0) + C(i1)^* t + C(i2)^* Y\_i_{t-s} + \mu_{it} \tag{1.8a}$$

$$\log(Y\_i_t) = C(i0) + C(i1)^* t + C(i2)^* \log(Y\_i_{t-s}) + \mu_{it} \tag{1.8b}$$

1.4   Various Models Based on Correlated States

It is known that stock prices of selected countries have a causal relationship. In this section, as an extension of the previously mentioned models, I consider the models based on correlated states. The definition is that two states are correlated if, and only if, their endogenous variables have a causal relationship. Note that if all variables are assumed or defined to be correlated, then all the time series models presented in Agung (2009a) can easily be applied.

With regards to the time series data by states or unstacked data considered, it is acceptable that growth of a problem indicator or variable of a state (country, region, firm or agency) should be theoretically influenced by the factors of the other state(s). For illustrative examples, at the first stage the *GDP* of two states, namely *GDP_US*, and *GDP_Can* in POOLG7.wf1, are defined to have a causal relationship. Here, two alternative causal relationships are considered, as presented in Figure 1.10, out of a lot of possible models.

Note that Figure 1.10(a) presents the path diagram where *GDP_US* is defined as the cause factor of *GDP_Can*. Based on this path diagram, the *simplest causal model* with trend would have the following system specification.

$$GDP\_US = C(10) + C(11)^* GDP\_US(-1) + C(12)^* t$$
$$GDP\_Can = C(20) + C(21)^* GDP\_Can(-1) + C(22)^* GDP\_US \tag{1.9}$$
$$+ C(23)^* GDP\_US(-1) + C(24)^* t$$

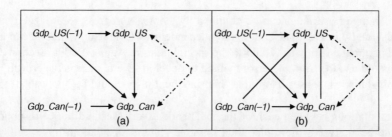

Figure 1.10   *Two alternative causal relationships between* GDP_US *and* GDP_Can

Note that each multiple regression in the model is an additive regression model of its independent variables. For instance, the first regression is an additive model of *GDP_US* on *GDP_US(−1)*, and the time *t*. In other words, this model represents the linear adjusted effects of *GDP_US(−1)*, and the time *t* on *GDP_US*. In fact, there are a lot more models that could be subjectively defined by the researchers. Refer to various time series models presented in Agung (2009a).

On the other hand, Figure 1.10(b) presents the path diagram where *GDP_US* and *GDP_Can* are defined to have a simultaneous causal linear effects or two-way causal effects. Based on this path diagram, the simplest causal model with trend would have the following system specification.

$$
\begin{aligned}
GDP\_US &= C(10) + C(11)^* GDP\_US(-1) + C(12)^* GDP\_Can \\
&\quad + C(13)^* GDP\_Can(-1) + C(14)^* t \\
GDP\_Can &= C(20) + C(21)^* GDP\_Can(-1) + C(22)^* GDP\_US \\
&\quad + C(23)^* GDP\_US(-1) + C(24)^* t
\end{aligned}
\tag{1.10}
$$

Note that the models (1.9) and (1.10) are not the VAR (Vector Autoregressive) models, since they do not have the same set of independent variables. For this reason, Agung (2009a) has introduced the MAR (Multivariate Autoregressive Model) and the SCM (Seemingly Causal Model) instead of the System Equation Model (SEM), because the term SEM is already used for the structural equation model. The following subsections present empirical examples of SCM and VAR Models.

To generalize, a problem indicator by states may be presented as *Y_s*, for the states $s = 1, \ldots, S$. Then, the relationship between the indicators *Y_s*, $s = 1, \ldots, S$, would be a matter of subjective or expert judgment by the researchers. It could be very difficult to define the path diagram of an SCM based on the *GDP* of the seven states as presented in POOLG7.wf1, even more so for the number of states greater than seven. For this reason, I recommend to all students planning to write theses or dissertations, select only two or three states for the data analysis, since they can apply various MLVAR(*p,q*) models and study the limitations of each model using residual analysis. Note that with a single variable *Y*, one would have to consider the variable *Y*, the time *t*-variable and the categorical state variable, as well as the lagged of *Y*, say $Y(-1), \ldots, Y(-p)$ for a selected integer *p*, as well as the indicators *AR(1), ..., AR(q)*.

1.4.1 Seemingly Causal Models with Trend

For illustrative purposes, Figure 1.10 presents two alternative theoretically defined SCMs between *GDP_US* and *GDP_Can*. Note that the arrows with dotted lines from the time *t* indicate that this is not a real causal factor. However, the following example presents data analysis based on the model (1.10) only.

Example 1.7   SCMs with trend
Figure 1.11(a) presents statistical results based on a bivariate first-order lagged-variable SCM, namely LV(**1**)_SCM, of *GDP_US* and *GDP_Can*, which show that the error terms of each regression have the first autocorrelation problem, indicated by the small value of its Durbin–Watson statistic. For this reason, Figure 1.11(b) presents statistical results based on its AR(**1**) model, namely LVAR(**1,1**)_SCM, which is acceptable, in both theoretical and statistical senses. Note that these models are not growth models. Based on this output, the following conclusions are derived.

1. The *p*-value $= 0.0000$ of the parameter *C(12)* in the first regression indicates that *GDP_Can* has a significant positive adjusted linear effect on *GDP_US*, and the *p*-value $= 0.0000$ of the parameter *C(22)* in the second regression indicates that *GDP_US* also has a significant positive adjusted linear effect on *GDP_Can*.

```
System: UNTITLED
Estimation Method: Least Squares
Date: 08/13/09  Time: 10:33
Sample: 1951 1992
Included observations: 42
Total system (balanced) observations 84
```

	Coefficient	Std. Error	t-Statistic	Prob.
C(10)	2340.532	678.8094	3.447996	0.0009
C(11)	0.703819	0.097711	7.203047	0.0000
C(12)	0.580407	0.100906	5.751936	0.0000
C(13)	-0.554104	0.099596	-5.563521	0.0000
C(14)	62.09836	20.53535	3.023974	0.0034
C(20)	-1888.823	869.9183	-2.171264	0.0331
C(21)	0.873552	0.070032	12.47358	0.0000
C(22)	0.813339	0.141403	5.751936	0.0000
C(23)	-0.500667	0.159265	-3.143598	0.0024
C(24)	-35.34991	26.51819	-1.333044	0.1866

Determinant residual covariance	1.78E+09

Equation: GDP_US = C(10)+C(11)*GDP_US(-1)+C(12)*GDP_CAN+C(13)
  *GDP_CAN(-1)+C(14)*T
Observations: 42

R-squared	0.994199	Mean dependent var	13335.29
Adjusted R-squared	0.993572	S.D. dependent var	2944.502
S.E. of regression	236.0718	Sum squared resid	2062006.
Durbin-Watson stat	1.432991		

Equation: GDP_CAN = C(20)+C(21)*GDP_CAN(-1)+C(22)*GDP_US+C(23)
  *GDP_US(-1)+C(24)*T
Observations: 42

R-squared	0.995001	Mean dependent var	11224.90
Adjusted R-squared	0.994460	S.D. dependent var	3754.577
S.E. of regression	279.4560	Sum squared resid	2889540.
Durbin-Watson stat	1.229405		

(a)

```
System: UNTITLED
Estimation Method: Iterative Least Squares
Date: 08/13/09  Time: 10:36
Sample: 1952 1992
Included observations: 42
Total system (balanced) observations 82
Convergence achieved after 8 iterations
```

	Coefficient	Std. Error	t-Statistic	Prob.
C(10)	3705.368	1399.393	2.647840	0.0100
C(11)	0.513887	0.213813	2.403435	0.0189
C(12)	0.598526	0.099965	5.987379	0.0000
C(13)	-0.556744	0.120949	-4.603130	0.0000
C(14)	105.4857	39.25613	2.687115	0.0090
C(15)	0.432322	0.254010	1.701991	0.0932
C(20)	-2473.391	1294.467	-1.910741	0.0601
C(21)	0.769470	0.193263	3.981463	0.0002
C(22)	0.845435	0.141269	5.984574	0.0000
C(23)	-0.395862	0.221602	-1.786366	0.0784
C(24)	-38.84161	43.44929	-0.893953	0.3744
C(25)	0.477109	0.264250	1.805520	0.0753

Determinant residual covariance	1.21E+09

Equation: GDP_US = C(10)+C(11)*GDP_US(-1)+C(12)*GDP_CAN+C(13)
  *GDP_CAN(-1)+C(14)*T+[AR(1)=C(15)]
Observations: 41

R-squared	0.994937	Mean dependent var	13437.80
Adjusted R-squared	0.994214	S.D. dependent var	2904.200
S.E. of regression	220.9119	Sum squared resid	1708072.
Durbin-Watson stat	1.921317		

Equation: GDP_CAN = C(20)+C(21)*GDP_CAN(-1)+C(22)*GDP_US+C(23)
  *GDP_US(-1)+C(24)*T+[AR(1)=C(25)]
Observations: 41

R-squared	0.995644	Mean dependent var	11342.95
Adjusted R-squared	0.995022	S.D. dependent var	3721.474
S.E. of regression	262.5740	Sum squared resid	2413079.
Durbin-Watson stat	1.876369		

(b)

Figure 1.11 *Statistical results based on the models with trends in (1.10), namely (a) a LV(**1**)_SCM, and (b) a LVAR(**1,1**)_SCM*

2. Therefore, based on the SCM in Figure 1.11(b), it can be concluded that the data supports the hypothesis that *GDP_US* and *GDP_Can* have simultaneous causal linear effects, adjusted for the other independent variables in the model.
3. In order to conduct the unadjusted simultaneous causal effects, the bivariate correlation analysis can easily be applied (Agung, 2006, 2009a, 2011). In this case, $H_0: \rho(GDP\_US, GDP\_Can) = 0$ is rejected based on the *t*-statistic of $t_0 = 36.43270$ with a *p*-value = 0.0000.
4. Various univariate and multivariate hypotheses could easily be tested using the Wald test.

Example 1.8 Translog linear SCMs with trend

The alternative models of those in Figure 1.11, Figure 1.12(a) and (b) present statistical results based on a translog linear LV(**1**)_SCM, and LVAR(**1,1**)_SCM. Based on these results the following notes are presented.

1. The translog linear LVAR(1,1)_SCM is an unacceptable model, in a statistical sense, based on the data set used, since the AR(1) of both regressions are insignificant with such a large *p*-values of 0.82 and 0.42, respectively.
2. In this case, the translog linear LV(**1**)_SCM would be a better model, supported by the fact that each independent variable has a significant adjusted effect on its corresponding dependent variable with sufficiently large DW statistics and their residual graphs, as shown in Figure 1.13. It would not be the best out of all possible models, which have not been explored.
3. Note that this translog-linear LV(**1**)_SCM can be viewed as a bivariate growth model, where *C(14)* indicates the adjusted exponential growth rate of *GDP_US*, and *C(24)* indicates the adjusted exponential growth rate of *GDP_Can*.

System: SYS02
Estimation Method: Least Squares
Date: 08/13/09 Time: 10:53
Sample: 1951 1992
Included observations: 42
Total system (balanced) observations 84

	Coefficient	Std. Error	t-Statistic	Prob.
C(10)	2.609122	0.648857	4.021103	0.0001
C(11)	0.735920	0.086255	8.531956	0.0000
C(12)	0.572822	0.088428	6.477804	0.0000
C(13)	-0.597241	0.089866	-6.645922	0.0000
C(14)	0.005485	0.001553	3.532318	0.0007
C(20)	-2.724012	0.882777	-3.085729	0.0029
C(21)	0.916529	0.077391	11.84283	0.0000
C(22)	0.927723	0.143216	6.477805	0.0000
C(23)	-0.546423	0.166390	-3.283987	0.0016
C(24)	-0.004779	0.002146	-2.227028	0.0290

Determinant residual covariance   4.96E-08

Equation: LOG(GDP_US) = C(10)+C(11)*LOG(GDP_US(-1))+C(12)
    *LOG(GDP_CAN)+C(13)*LOG(GDP_CAN(-1))+C(14)*T
Observations: 42

R-squared	0.994925	Mean dependent var	9.473482
Adjusted R-squared	0.994376	S.D. dependent var	0.227124
S.E. of regression	0.017033	Sum squared resid	0.010735
Durbin-Watson stat	1.814607		

Equation: LOG(GDP_CAN) = C(20)+C(21)*LOG(GDP_CAN(-1))+C(22)
    *LOG(GDP_US)+C(23)*LOG(GDP_US(-1))+C(24)*T
Observations: 42

R-squared	0.996381	Mean dependent var	9.269590
Adjusted R-squared	0.995990	S.D. dependent var	0.342288
S.E. of regression	0.021677	Sum squared resid	0.017385
Durbin-Watson stat	1.688250		

(a)

System: SYS03
Estimation Method: Iterative Least Squares
Date: 08/13/09 Time: 10:57
Sample: 1952 1992
Included observations: 42
Total system (balanced) observations 82
Convergence achieved after 6 iterations

	Coefficient	Std. Error	t-Statistic	Prob.
C(10)	2.670910	0.721082	3.704031	0.0004
C(11)	0.751711	0.101318	7.419300	0.0000
C(12)	0.561529	0.087604	6.409872	0.0000
C(13)	-0.609972	0.090577	-6.734285	0.0000
C(14)	0.005971	0.001652	3.613867	0.0006
C(15)	0.042296	0.185503	0.228008	0.8203
C(20)	-2.799240	1.021972	-2.739057	0.0078
C(21)	0.903637	0.108494	8.328919	0.0000
C(22)	0.951495	0.149107	6.381282	0.0000
C(23)	-0.549361	0.186524	-2.945253	0.0044
C(24)	-0.004892	0.002609	-1.875427	0.0649
C(25)	0.157224	0.195886	0.802631	0.4249

Determinant residual covariance   4.56E-08

Equation: LOG(GDP_US) = C(10)+C(11)*LOG(GDP_US(-1))+C(12)
    *LOG(GDP_CAN)+C(13)*LOG(GDP_CAN(-1))+C(14)*T+[AR(1)=C(15)]
Observations: 41

R-squared	0.995076	Mean dependent var	9.482115
Adjusted R-squared	0.994373	S.D. dependent var	0.222860
S.E. of regression	0.016718	Sum squared resid	0.009782
Durbin-Watson stat	1.844620		

Equation: LOG(GDP_CAN) = C(20)+C(21)*LOG(GDP_CAN(-1))+C(22)
    *LOG(GDP_US)+C(23)*LOG(GDP_US(-1))+C(24)*T+[AR(1)=C(25)]
Observations: 41

R-squared	0.996343	Mean dependent var	9.281978
Adjusted R-squared	0.995821	S.D. dependent var	0.336874
S.E. of regression	0.021778	Sum squared resid	0.016599
Durbin-Watson stat	1.773246		

(b)

Figure 1.12   *Statistical results based on (a) translog linear LV(**1**)_SCM, and (b) translog linear LVAR(**1,1**)_SCM; with trend*

4. To generalize, the variable *GDP* could easily be replaced by a variable *Y*. Then, as a modified model, the SCM will be written as pair of nonlinear models as follows:

$$Y\_US = Y\_US(-1)^{C(11)} Y\_Can^{C(12)} Y\_Can(-1)^{C(13)} Exp(C(10) + C(14)^* t)$$
$$Y\_Can = Y\_Can(-1)^{C(21)} Y\_US^{C(22)} Y\_US(-1)^{C(23)} Exp(C(20) + C(24)^* t)$$

(1.11)

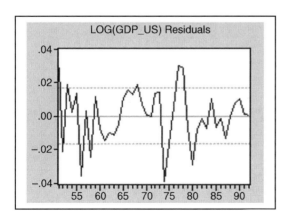
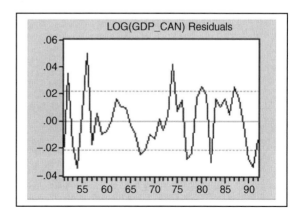

Figure 1.13   *Residual graphs of the LV(**1**)_SCM in Figure 1.11(a)*

5. Finally, for the seven states in POOLG7.wf1, a much more complex path diagram should be developed or defined to represent the theoretical causal model between the seven time series variables. Therefore, based on the path diagram, the system specification of a SCM would easily be written, either as translog linear or nonlinear models. However, by using many independent variables, the error messages of *"near singular matrix"* or *"overflow"*, as well as the unexpected estimates of parameters, may be obtained. Refer to special notes in Agung (2009a, Section 2.14).

1.4.2   The Application of the Object "VAR"

EViews provides the object *"VAR"* for conducting the data analysis based on a *vector autoregressive* (VAR) and *vector error correction* (VEC) models, which are special cases of the *multivariate autoregressive* (MAR) models and SCMs, (Agung, 2009a). See the following example.

Example 1.9   A VAR model
Figure 1.14 presents the statistical result based on a VAR model of *log(GDP_US)* and *log(GDP_Can)*, using the lag interval of endogenous "1 1", and exogenous variables "*C T*" with the default options. Based on this result the following notes are presented.

1. Note that this VAR model in fact is a special case of the MAR(1)_GM, where all regressions have exactly the same independent variables. Compared to the path diagram in Figure 1.10, the path diagram of this VAR model presented in Figure 1.15 shows the causal relationship between *log(GDP_US)* and *log(GDP_Can)* is not taken into account.
2. However, the quantitative coefficient of correlations of the independent variables *log(GDP_US(−1))*, *log(GDP_Can)* and the time *t* should be taken into account in the regression analysis, and it is well-known that they have an unpredictable impact on the estimate of the model parameters. Refer to Section 2.14.2 in Agung (2009a).

Vector Autoregression Estimates
Date: 08/13/09   Time: 12:31
Sample (adjusted): 1951 1992
Included observations: 42 after adjustments
Standard errors in ( ) & t-statistics in [ ]

	LOG(GDP_US)	LOG(GDP_...
LOG(GDP_US(-1))	0.902551	0.290894
	(0.11868)	(0.15103)
	[7.60498]	[1.92602]
LOG(GDP_CAN(-1))	-0.154154	0.773517
	(0.08402)	(0.10692)
	[-1.83475]	[7.23425]
C	2.238143	-0.647635
	(0.93168)	(1.18568)
	[2.40226]	[-0.54621]
T	0.005864	0.000661
	(0.00224)	(0.00285)
	[2.62140]	[0.23215]

(a)

R-squared	0.989169	0.992276
Adj. R-squared	0.988313	0.991666
Sum sq. resids	0.022909	0.037102
S.E. equation	0.024553	0.031247
F-statistic	1156.763	1627.292
Log likelihood	98.19675	88.07141
Akaike AIC	-4.485560	-4.003401
Schwarz SC	-4.320067	-3.837908
Mean dependent	9.473482	9.269590
S.D. dependent	0.227124	0.342288

Determinant resid covariance (dof adj.)		2.76E-07
Determinant resid covariance		2.26E-07
Log likelihood		202.1872
Akaike information criterion		-9.247008
Schwarz criterion		-8.916024

Estimation Proc:
==================================
LS 1 1 LOG(GDP_US) LOG(GDP_CAN) @ C T

VAR Model:
==================================
LOG(GDP_US) = C(1,1)*LOG(GDP_US(-1)) + C(1,2)*LOG(GDP_CAN(-1)) + C(1,3) + C(1,4)*T

LOG(GDP_CAN) = C(2,1)*LOG(GDP_US(-1)) + C(2,2)*LOG(GDP_CAN(-1)) + C(2,3) + C(2,4)*T

(b)

Figure 1.14   *Statistical results based on a VAR Model of* log(GDP_US) *and* log(GDP_Can)

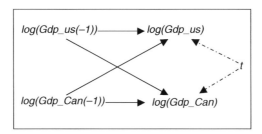

Figure 1.15   The path diagram of the VAR model in Figure 1.14

3. Furthermore, note that the symbol *C(i,j)* is used to present the model parameters, and test the hypotheses using the Wald test, that is the *Block Exogeinity Wald Test*.
4. For a model with many endogenous or exogenous variables, applying the object "*System*" is recommended instead of the object "*VAR*", because in general, the good fit multiple regressions in the model would have different sets of independent variables.
5. Refer to Chapter 6 in Agung, (2009a; p. 316), for more detailed notes on various VAR models, as well as their residual analysis, and a special causality test; the VAR Granger Causality/Block Exogeneity Wald Tests.
6. Furthermore, in order to match conditions in previous and recent years, Agung (2009a) proposes special VAR models using the lag interval of endogenous "4 4" for a quarterly data set, and "12 12" for a monthly data set.

Example 1.10   A vector error correction (VEC) model
Figure 1.16 presents the statistical result based on a VEC model by inserting the endogenous variables "*log(GDP_US) log(GDP_Can)*", the lag interval of endogenous "1 1", and exogenous variables "*T*" with the default options. Based on this result the following notes are presented.

1. By inserting the endogenous variables *log(GDP_US)* and *log(GDP_Can)*, the output directly presents two regressions with the first differences *Dlog(GDP_US)* and *Dlog(GDP_Can)* as their dependent variables.
2. Note that, in general, the first difference of *log(Y_s)*, in fact indicates the exponential growth rate of *Y_s*, which can be presented as follows:

$$\text{Dlog}(Y\_s) = \log(Y\_s_t) - \log(Y\_s_{t-1}) = R_t(Y\_s) \tag{1.12}$$

Then, the two independent variables Dlog(Y_US(−1)) and Dlog(Y_Can(−1)), can be presented as follows:

$$\text{Dlog}(Y\_s(-1)) = \log(Y\_s_{t-1}) - \log(Y\_s_{t-2}) = R_{t-1}(Y\_s) \tag{1.13}$$

For these reasons, the VEC model in fact presents a bivariate LV(1) model of $R_t(GDP\_US)$ and $R_t(GDP\_Can)$ with exogenous variables.
3. Beside the independent or exogenous variables *C* and *T*, both regressions in the VEC model have a special independent variable, called the *Cointegrating Equation*, namely:

$$CointEg1 = \log(GDP\_US(-1)) + 0.286109 \log(GDP\_Can) - 12.1466$$

4. For more detailed notes on various VEC models, as well as the characteristics of alternative cointegrating equations, refer to Section 6.3 in Agung (2009a).

Vector Error Correction Estimates Date: 08/13/09 Time: 14:48 Sample (adjusted): 1952 1992 Included observations: 41 after adjustments Standard errors in ( ) & t-statistics in [ ]		

Cointegrating Eq:	CointEq1	
LOG(GDP_US(-1))	1.000000	
LOG(GDP_CAN(-1))	0.286109 (0.26786) [1.06813]	
C	-12.11466	

Error Correction:	D(LOG(GDP...	D(LOG(GDP...
CointEq1	-0.229705 (0.10087) [-2.27722]	-0.041189 (0.13528) [-0.30447]
D(LOG(GDP_US(-1)))	0.118220 (0.20449) [0.57812]	0.203617 (0.27426) [0.74243]
D(LOG(GDP_CAN(-1)))	0.013610 (0.19289) [0.07056]	0.107740 (0.25869) [0.41648]
C	-0.117662 (0.06156) [-1.91131]	-0.002766 (0.08256) [-0.03350]
T	0.005992 (0.00270) [2.21965]	0.000896 (0.00362) [0.24751]

R-squared	0.163536	0.060650
Adj. R-squared	0.070596	-0.043723
Sum sq. resids	0.022341	0.040184
S.E. equation	0.024911	0.033410
F-statistic	1.759578	0.581090
Log likelihood	95.87921	83.84437
Akaike AIC	-4.433132	-3.846067
Schwarz SC	-4.224160	-3.637095
Mean dependent	0.016525	0.023027
S.D. dependent	0.025840	0.032703

Determinant resid covariance (dof adj.)	3.02E-07
Determinant resid covariance	2.33E-07
Log likelihood	196.7108
Akaike information criterion	-9.010285
Schwarz criterion	-8.508751

Estimation Proc:
==================================
EC(C,1) 1 1 LOG(GDP_US) LOG(GDP_CAN) @ T

VAR Model:
==================================
D(LOG(GDP_US)) = A(1,1)*(B(1,1)*LOG(GDP_US(-1)) + B(1,2)*LOG(GDP_CAN(-1)) + B(1,3)) + C(1,1)*D(LOG(GDP_US(-1))) + C(1,2)*D(LOG(GDP_CAN(-1))) + C(1,3) + C(1,4)*T

D(LOG(GDP_CAN)) = A(2,1)*(B(1,1)*LOG(GDP_US(-1)) + B(1,2)*LOG(GDP_CAN(-1)) + B(1,3)) + C(2,1)*D(LOG(GDP_US(-1))) + C(2,2)*D(LOG(GDP_CAN(-1))) + C(2,3) + C(2,4)*T

VAR Model - Substituted Coefficients:
==================================
D(LOG(GDP_US)) = - 0.229705011793*(LOG(GDP_US(-1)) + 0.286109023785*LOG(GDP_CAN(-1)) - 12.1146586662) + 0.118220324*D(LOG(GDP_US(-1))) + 0.0136103060211*D(LOG(GDP_CAN(-1))) - 0.117662329679 + 0.00599184978411*T

D(LOG(GDP_CAN)) = - 0.0411894448722*(LOG(GDP_US(-1)) + 0.286109023785*LOG(GDP_CAN(-1)) - 12.1146586662) + 0.203617013369*D(LOG(GDP_US(-1))) + 0.107740204007*D(LOG(GDP_CAN(-1))) - 0.00276581170121 + 0.000896101575108*T

Figure 1.16   *Statistical results based on a VEC Model of* log(GDP_US) *and* log(GDP_Can)

1.4.3   The Application of the Instrumental Variables Models

It is not an easy task to define a "good fit" instrumental variables model, since there is no general guide on how to select an acceptable set of instrumental variables corresponding to any defined statistical model. For this reason, Agung (2009a, p. 382) suggests everyone has *two-stages of problems* (TSOP), in demonstrating or developing an instrumental model. First, he/she should develop a model with at least one exogenous variable which is significantly correlated with the residual of the model. Second, he/she has to search to find the best possible set of instrumental variables. For various examples with special notes on instrumental variables models, refer to Chapter 7 in Agung (2009a).

Example 1.11   (A two-stage LSE method)
Figure 1.17 presents the statistical results based on an instrumental variable model with a trend of *log(GDP_US)* and *log(GDP_Can)*. Based on this result the following notes are presented.

1. Figure 1.17(a) presents the statistical results based on a bivariate AR(1,1)_SCM, where both regressions in the model are the simplest AR(1) linear regressions, with the same set of instrumental variables. These

System: SYS05
Estimation Method: Iterative Two-Stage Least Squares
Date: 08/13/09   Time: 15:55
Sample: 1951 1992
Included observations: 43
Total system (balanced) observations 84
Convergence achieved after 12 iterations

	Coefficient	Std. Error	t-Statistic	Prob.
C(10)	3.627138	0.616561	5.882853	0.0000
C(11)	0.632266	0.065137	9.706704	0.0000
C(12)	0.842680	0.090564	9.304761	0.0000
C(20)	-4.491572	1.349184	-3.329102	0.0013
C(21)	1.451370	0.140951	10.29701	0.0000
C(22)	0.851882	0.082311	10.34959	0.0000

Determinant residual covariance	8.01E-09

Equation: LOG(GDP_US) = C(10)+C(11)*LOG(GDP_CAN)+[AR(1)=C(12)]
Instruments: C T LOG(GDP_US(-1)) LOG(GDP_CAN(-1))
Observations: 42

R-squared	0.993172	Mean dependent var	9.473482
Adjusted R-squared	0.992822	S.D. dependent var	0.227124
S.E. of regression	0.019243	Sum squared resid	0.014441
Durbin-Watson stat	1.622016		

Equation: LOG(GDP_CAN) = C(20)+C(21)*LOG(GDP_US)+[AR(1)=C(22)]
Instruments: C T LOG(GDP_CAN(-1)) LOG(GDP_US(-1))
Observations: 42

R-squared	0.993355	Mean dependent var	9.269590
Adjusted R-squared	0.993014	S.D. dependent var	0.342288
S.E. of regression	0.028609	Sum squared resid	0.031921
Durbin-Watson stat	1.577089		

(a)

Estimation Method: Iterative Two-Stage Least Squares
Date: 08/13/09   Time: 16:04
Sample: 1952 1992
Included observations: 42
Total system (unbalanced) observations 81
Convergence achieved after 7 iterations

	Coefficient	Std. Error	t-Statistic	Prob.
C(10)	4.360923	0.645500	6.755880	0.0000
C(11)	-0.254102	0.109895	-2.312221	0.0236
C(12)	0.812034	0.086502	9.387500	0.0000
C(13)	1.245268	0.153931	8.089757	0.0000
C(14)	-0.417304	0.152645	-2.733820	0.0079
C(20)	-1.022586	0.510377	-2.003591	0.0489
C(21)	0.771526	0.093052	8.291338	0.0000
C(22)	0.333348	0.141338	2.358516	0.0211
C(23)	0.199423	0.186186	1.071095	0.2877

Determinant residual covariance	1.39E-07

Equation: LOG(GDP_US) = C(10)+C(11)*LOG(GDP_US(-1))+C(12)
    *LOG(GDP_CAN)+[AR(1)=C(13), AR(2)=C(14)]
Instruments: C T LOG(GDP_US(-1)) LOG(GDP_US(-2)) LOG(GDP_CAN(
    -1)) LOG(GDP_US(-3)) LOG(GDP_CAN(-2))
Observations: 40

R-squared	0.993566	Mean dependent var	9.490958
Adjusted R-squared	0.992831	S.D. dependent var	0.218292
S.E. of regression	0.018483	Sum squared resid	0.011957
Durbin-Watson stat	1.781571		

Equation: LOG(GDP_CAN) = C(20)+C(21)*LOG(GDP_CAN(-1))+C(22)
    *LOG(GDP_US)+[AR(1)=C(23)]
Instruments: C T LOG(GDP_CAN(-1)) LOG(GDP_CAN(-2)) LOG(GDP_US(
    -1))
Observations: 41

R-squared	0.994549	Mean dependent var	9.281978
Adjusted R-squared	0.994107	S.D. dependent var	0.336874
S.E. of regression	0.025861	Sum squared resid	0.024746
Durbin-Watson stat	1.874247		

(b)

Figure 1.17   *Statistical results based on bivariate models (a) AR(1,1)_SCM, and (b) LVAR(1,1;2,1)_SCM, with sets of instrumental variables*

results show that *log(GDP_Can)* and *log(GDP_US)* have significant simultaneous causal effects, since *log(GDP_Can)* has a significant positive effect on *log(GDP_US)* based on the *t*-statistic of $t_0 = 9.706\,704$ with a *p*-value $= 0.0000/2 = 0.0000$, and *log(GDP_US)* also has a significant positive effect on *log(GDP_Can)* based on the *t*-statistic of $t_0 = 10.29\,701$ with a *p*-value $= 0.0000/2 = 0.0000$. Note that in this case the *p*-values in the output should be divided by 2 for testing the one-sided hypotheses.

2. Figure 1.17(b) presents the statistical results based on a LVAR(1,1;2,1)_SCM, where the first regression is a LVAR(1,2) model with an exogenous variable *log(GDP_Can)* and the second regression is a LVAR(1,1) model with an exogenous variable *log(GDP_US)*. Compared to the model in Figure 1.14(a), the regressions in this model have different sets of instrumental variables

1.5   Seemingly Causal Models with Time-Related Effects

As the extension of the SCMs with trends in (1.9) and (1.10), the following system equations present SCMs with time-related effects.

1.5.1   SCM Based on the Path Diagram in Figure 1.10(a)

As an extension of the additive model in (1.9), a SCM with time-related effects based on the path diagram in Figure 1.10(a), will have the following system specification. Note that the two-way interaction t^*GDP_US

is inserted as an additional independent variable of the second regression to indicate the time-related effect of *DGP_US* on *GDP_Can*. In other words, the effect of *DGP_US* on *GDP_Can* depends on the time *t*.

$$
\begin{aligned}
GDP\_US &= C(10) + C(11)^* GDP\_US(-1) + C(12)^* t \\
GDP\_Can &= C(20) + C(21)^* GDP\_Can(-1) + C(22)^* GDP\_US \\
&\quad + C(23)^* GDP\_US(-1) + C(24)^* t + C(25)^* t^* GDP\_US
\end{aligned}
\tag{1.14}
$$

Note that this model is indicating that the effect of *GDP_US* on *GDP_Can* depends on the time *t*, indicated by the following partial derivative:

$$
\frac{\partial GDP\_Can}{\partial GDP\_US} = c(24) + c(25)^* t
$$

1.5.2   SCM Based on the Path Diagram in Figure 1.10(b)

As an extension of the interaction model in (1.14), a SCM with the time-related effects based on the path diagram in Figure 1.10(b), will have the following system specification. Note that the second regressions of the SCMs in (1.14) and (1.15) are identical models.

$$
\begin{aligned}
GDP\_US &= C(10) + C(11)^* GDP\_US(-1) + C(12)^* GDP\_Can \\
&\quad + C(13)^* GDP\_Can(-1) + C(14)^* t + C(15)^* t^* GDP\_Can \\
GDP\_Can &= C(20) + C(21)^* GDP\_Can(-1) + C(22)^* GDP\_US \\
&\quad + C(23)^* GDPGDP\_US(-1) + C(24)^* t + C(25)^* t^* GDP\_US
\end{aligned}
\tag{1.15}
$$

Example 1.12   A translog linear SCM with time-related effects

We find that the statistical results based on the model in (1.15) present several insignificant independent variables, including the two-way interactions *t*GDP_Can* and *t*GDP_US*. So, by using the trial-and-error method we can finally obtain a good fit translog linear SCM with time-related effects as presented in Figure 1.18,

System: UNTITLED
Estimation Method: Least Squares
Date: 08/13/09   Time: 17:12
Sample: 1951 1992
Included observations: 42
Total system (balanced) observations 84

	Coefficient	Std. Error	t-Statistic	Prob.
C(10)	2.520206	0.638031	3.949973	0.0002
C(11)	0.769750	0.083439	9.225269	0.0000
C(12)	0.568780	0.088966	6.393227	0.0000
C(13)	-0.617770	0.092585	-6.672465	0.0000
C(15)	0.000562	0.000163	3.455584	0.0009
C(20)	-2.759283	0.875206	-3.152724	0.0023
C(21)	0.928523	0.079587	11.66681	0.0000
C(22)	0.934923	0.143415	6.519011	0.0000
C(23)	-0.561499	0.166267	-3.377087	0.0012
C(25)	-0.000504	0.000220	-2.291717	0.0248

Determinant residual covariance	4.94E-08

Equation: LOG(GDP_US) = C(10)+C(11)*LOG(GDP_US(-1))+C(12)
   *LOG(GDP_CAN)+C(13)*LOG(GDP_CAN(-1)) +C(15)*T
   *LOG(GDP_CAN)
Observations: 42

R-squared	0.994869	Mean dependent var	9.473482
Adjusted R-squared	0.994314	S.D. dependent var	0.227124
S.E. of regression	0.017126	Sum squared resid	0.010852
Durbin-Watson stat	1.858029		

Equation: LOG(GDP_CAN) = C(20)+C(21)*LOG(GDP_CAN(-1))+C(22)
   *LOG(GDP_US)+C(23)*LOG(GDP_US(-1))+C(25)*T*LOG(GDP_US)
Observations: 42

R-squared	0.996406	Mean dependent var	9.269590
Adjusted R-squared	0.996017	S.D. dependent var	0.342288
S.E. of regression	0.021601	Sum squared resid	0.017265
Durbin-Watson stat	1.723052		

Figure 1.18   *Statistical results based on a reduced model of a modified model in (1.15)*

which is in fact a nonhierarchical reduced model of the modified model in (1.15). Based on these results the following notes are presented.

1. The *p*-value of the two-way interaction $T^*log(GDP\_Can)$ in the first regression indicates that the adjusted effect $log(GDP\_Can)$ on $log(GDP\_US)$ is significantly dependent on the time t, and the *p*-value of the two-way interaction $T^*log(GDP\_US)$ in the second regression also indicates that the adjusted effect $log(GDP\_US)$ on $log(GDP\_Can)$ is significantly dependent on the time t.
2. Therefore, based on this SCM, it can be concluded that the data support the hypothesis $log(GDP\_US)$ and $log(GDP\_Can)$ have simultaneous causal effects dependent on the time t.

1.6   The Application of the Object POOL

Many students, as well as less experienced analysts, have used the object *POOL* to present statistical results or outputs based on either fixed or random effects models, without considering or discussing the characteristics of the models, not to mention their limitations. For this reason, the following examples present illustrative statistical results with special notes.

The steps of the analysis using the object "*POOL*" are as follows:

1. By selecting *Object/New Objects/Pool ... OK*, the window in Figure 1.19(a) appears.
2. By inserting Cross-Section Identifiers, namely the series *_CAN _US _FRA _GER*, and then clicking "*Estimate*", then options in Figure 1.19(b) appear.

1.6.1   What is a Fixed-Effect Model?

The following example presents the statistical results based on simple multivariate growth models with special notes on the acceptability of the models.

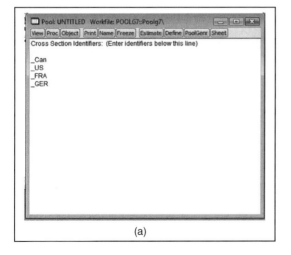

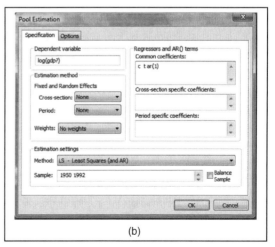

(a)   (b)

Figure 1.19   *The windows and options for using the object "POOL"*

Example 1.13   A fixed-effect MAR(1)_GM

For comparison with the statistical results in Figure 1.2, Figure 1.20(a) presents the statistical results by selecting cross-section: *Fixed*, inserting log(*GDP?*) as the dependent variable, and the series "*c t ar(1)*" as the *Regressors and AR() term Common Coefficients*.

Then by selecting *View/Representation*, the estimation equation in Figure 1.20(b) is obtained. Based on these results, the following conclusions and comments are presented.

1. Note that this fixed-effect AR(1) model has special characteristics, where the four regressions in the model have special intercept parameters, namely $C(4) + C(1)$, $C(5) + C(1)$, $C(6) + C(1)$, and $C(7) + C(1)$, respectively, where the parameters $C(4)$ to $C(7)$ are named as the cross-section fixed-effect parameters.
2. However, we find the equality of the parameters $C(4)$, $C(5)$, $C(6)$ and $C(7)$ cannot tested using the Wald test.
3. The growth rates of *GDP* are presented by a single parameter of $C(2)$ in the four countries, as well as a single autocorrelation of $C(3)$ for the four regressions, which should be inappropriate or unrealistic estimates, in a theoretical sense. For this reason, compared to the model in Figure 1.2 along with the model in Figure 1.20, this MAR(1)_GM should be considered unacceptable for representing growth rates of the *GDP* in the four countries.

Example 1.14   A fixed effect MLV(1)_GM

For a comparison with the statistical results in Figure 1.9, Figure 1.21(a) presents the statistical results by selecting Cross-section: *Fixed*, inserting log(*GDP?*) as the dependent variable, and the series "*c t log(GDP?*

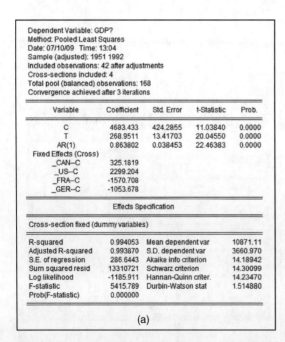

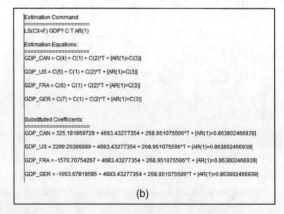

(a)   (b)

Figure 1.20 *Statistical results based on a fixed effect MAR(1)_GM of GDP_Can, GDP_US, GDP_Fra and GDP_Ger*

```
Dependent Variable: LOG(GDP?)
Method: Pooled Least Squares
Date: 07/10/09   Time: 14:11
Sample (adjusted): 1951 1992
Included observations: 42 after adjustments
Cross-sections included: 4
Total pool (balanced) observations: 168
```

Variable	Coefficient	Std. Error	t-Statistic	Prob.
C	0.684159	0.172521	3.965655	0.0001
T	0.001406	0.000575	2.445828	0.0155
LOG(GDP?(-1))	0.925098	0.020089	46.04897	0.0000
Fixed Effects (Cross)				
_CAN--C	-7.15E-05			
_US--C	0.009831			
_FRA--C	-0.008606			
_GER--C	-0.001153			

Effects Specification

Cross-section fixed (dummy variables)

R-squared	0.995548	Mean dependent var		9.229532
Adjusted R-squared	0.995410	S.D. dependent var		0.375014
S.E. of regression	0.025406	Akaike info criterion		-4.472593
Sum squared resid	0.104566	Schwarz criterion		-4.361023
Log likelihood	381.6978	Hannan-Quinn criter.		-4.427312
F-statistic	7244.806	Durbin-Watson stat		1.612986
Prob(F-statistic)	0.000000			

(a)

```
Estimation Command:
=====================
LS(CX=F) LOG(GDP?) C T LOG(GDP?(-1))

Estimation Equations:
=====================
LOG(GDP_CAN) = C(4) + C(1) + C(2)*T + C(3)*LOG(GDP_CAN(-1))

LOG(GDP_US) = C(5) + C(1) + C(2)*T + C(3)*LOG(GDP_US(-1))

LOG(GDP_FRA) = C(6) + C(1) + C(2)*T + C(3)*LOG(GDP_FRA(-1))

LOG(GDP_GER) = C(7) + C(1) + C(2)*T + C(3)*LOG(GDP_GER(-1))

Substituted Coefficients:
=====================
LOG(GDP_CAN) = -7.15294200332e-05 + 0.684159304375 + 0.00140608433156*T + 0.925097700701*LOG(GDP_CAN(-1))

LOG(GDP_US) = 0.009830871 3251 + 0.684159304375 + 0.00140608433156*T + 0.925097700701*LOG(GDP_US(-1))

LOG(GDP_FRA) = -0.00860598498651 + 0.684159304375 + 0.00140608433156*T + 0.925097700701*LOG(GDP_FRA(-1))

LOG(GDP_GER) = -0.00115345691855 + 0.684159304375 + 0.00140608433156*T + 0.925097700701*LOG(GDP_GER(-1))
```

(b)

Figure 1.21 *Statistical results based on a fixed effect MLV(1)_GM of* GDP_Can, GDP_US, GDP_Fra *and* GDP_Ger

(-1)" as the *Regressors and AR() term Common Coefficients*. Then by selecting *View/Representation*, the estimation equation in Figure 1.21(b) can be obtained. This model also has the same problem as the fixed-effect MAR(1)_GM, so it should be considered inappropriate for representing the growth rates of the *GDP* in the four countries.

1.6.2   What is a Random Effect Model?

The AR() terms cannot be used for a random effect model. For this reason, the following examples only present results based on the classical growth model and a random effect MLV(1)_GM.

Example 1.15   A random effect multivariate classical growth model: REMCGM
Figure 1.22 presents the statistical results based on a REMCGM of *GDPs* for the four countries, as well as the four regression functions having the same growth rates of *C(2)*. This model is also an inappropriate model in a theoretical sense, aside from the very small value of the weighted DW statistic of 0.078 665. For this reason, a random effect MLV(1)_GM is presented in the following example.

Example 1.16   A random effect MLV(1)_GM
For comparison with the statistical results based on the fixed effect MLV(1)_GM in Figure 1.21, Figure 1.23 presents the results based on a random effect MLV(1)_GM, which shows that the four regression functions have exactly the same coefficients of *C(1)*, *C(2)* and *C(3)*; and *C(4) = C(5) = C(6) = C(7) = 0*. In a theoretical sense, this is the worst model.

```
Dependent Variable: LOG(GDP?)
Method: Pooled EGLS (Cross-section random effects)
Date: 07/10/09   Time: 15:19
Sample: 1950 1992
Included observations: 43
Cross-sections included: 4
Total pool (balanced) observations: 172
Swamy and Arora estimator of component variances
```

Variable	Coefficient	Std. Error	t-Statistic	Prob.
C	8.618096	0.095592	90.15486	0.0000
T	0.027090	0.000616	43.97470	0.0000
Random Effects (Cross)				
_CAN—C	0.042773			
_US—C	0.248363			
_FRA—C	-0.154467			
_GER—C	-0.136669			

Effects Specification			
		S.D.	Rho
Cross-section random		0.188634	0.7797
Idiosyncratic random		0.100261	0.2203

Weighted Statistics			
R-squared	0.919193	Mean dependent var	0.744397
Adjusted R-squared	0.918718	S.D. dependent var	0.351667
S.E. of regression	0.100261	Sum squared resid	1.708871
F-statistic	1933.774	Durbin-Watson stat	0.078665
Prob(F-statistic)	0.000000		

Unweighted Statistics			
R-squared	0.755259	Mean dependent var	9.214075
Sum squared resid	6.299071	Durbin-Watson stat	0.021341

```
Estimation Command:
=====================
LS(CX=R) LOG(GDP?) C T

Estimation Equations:
=====================
LOG(GDP_CAN) = C(3) + C(1) + C(2)*T

LOG(GDP_US) = C(4) + C(1) + C(2)*T

LOG(GDP_FRA) = C(5) + C(1) + C(2)*T

LOG(GDP_GER) = C(6) + C(1) + C(2)*T


Substituted Coefficients:
=====================
LOG(GDP_CAN) = 0.0427728268614 + 8.61809606213 + 0.0270899687917*T

LOG(GDP_US) = 0.248362979351 + 8.61809606213 + 0.0270899687917*T

LOG(GDP_FRA) = -0.154467059659 + 8.61809606213 + 0.0270899687917*T

LOG(GDP_GER) = -0.136668746553 + 8.61809606213 + 0.0270899687917*T
```

Figure 1.22   *Statistical results based on a RECGM of GDP_Can, GDP_US, GDP_Fra and GDP_Ger*

```
Dependent Variable: LOG(GDP?)
Method: Pooled EGLS (Cross-section random effects)
Date: 07/10/09   Time: 14:58
Sample (adjusted): 1951 1992
Included observations: 42 after adjustments
Cross-sections included: 4
Total pool (balanced) observations: 168
Swamy and Arora estimator of component variances
```

Variable	Coefficient	Std. Error	t-Statistic	Prob.
C	0.452666	0.087556	5.169991	0.0000
T	0.000666	0.000323	2.059916	0.0410
LOG(GDP?(-1))	0.952062	0.010187	93.45688	0.0000
Random Effects (Cross)				
_CAN—C	0.000000			
_US—C	0.000000			
_FRA—C	0.000000			
_GER—C	0.000000			

Effects Specification			
		S.D.	Rho
Cross-section random		0.000000	0.0000
Idiosyncratic random		0.025406	1.0000

```
Estimation Command:
=====================
LS(CX=R) LOG(GDP?) C T LOG(GDP?(-1))

Estimation Equations:
=====================
LOG(GDP_CAN) = C(4) + C(1) + C(2)*T + C(3)*LOG(GDP_CAN(-1))

LOG(GDP_US) = C(5) + C(1) + C(2)*T + C(3)*LOG(GDP_US(-1))

LOG(GDP_FRA) = C(6) + C(1) + C(2)*T + C(3)*LOG(GDP_FRA(-1))

LOG(GDP_GER) = C(7) + C(1) + C(2)*T + C(3)*LOG(GDP_GER(-1))


Substituted Coefficients:
=====================
LOG(GDP_CAN) = 0 + 0.452666208449 + 0.000665617512811*T + 0.952061931017*LOG(GDP_CAN(-1))

LOG(GDP_US) = 0 + 0.452666208449 + 0.000665617512811*T + 0.952061931017*LOG(GDP_US(-1))

LOG(GDP_FRA) = 0 + 0.452666208449 + 0.000665617512811*T + 0.952061931017*LOG(GDP_FRA(-1))

LOG(GDP_GER) = 0 + 0.452666208449 + 0.000665617512811*T + 0.952061931017*LOG(GDP_GER(-1))
```

Weighted Statistics			
R-squared	0.995433	Mean dependent var	9.229532
Adjusted R-squared	0.995377	S.D. dependent var	0.375014
S.E. of regression	0.025497	Sum squared resid	0.107270
F-statistic	17980.39	Durbin-Watson stat	1.614198
Prob(F-statistic)	0.000000		

Unweighted Statistics			
R-squared	0.995433	Mean dependent var	9.229532
Sum squared resid	0.107270	Durbin-Watson stat	1.614198

Figure 1.23   *Statistical results based on a random effect MLV(1)_GM of GDP_Can, GDP_US, GDP_Fra and GDP_Ger*

1.6.3   Special Notes

Based on the statistical results of the multivariate fixed and random effects growth models using the object "POOL" presented previously, please note the following special points.

1. In general, a multivariate fixed and random effects growth model is the worst multivariate growth model by states, in a theoretical sense. This is even more so if it is known that a state should have a discontinuous or piece-wise growth curve, as this may be because of some external factors. Therefore, in general one should use models with heterogeneous slopes or *heterogeneous regressions* (Agung, 2006, 2011). For an additional illustration, Chandrasekaran and Tellis (in Malhotra, 2007, p. 45) present the findings of Golder and Tellis (2004) on piece-wise mean growth rates of a product's life cycle over six time periods; namely during the introduction, takeoff, growth, slowdown, early maturity and late maturity.
2. Referring to the ANCOVA models, in a statistical sense, models that have homogeneous slopes or *homogeneous regressions* (Agung, 2006, 2011) with various intercepts are acceptable. The main objectives of ANCOVA are to test the hypotheses on the *adjusted means differences* of the corresponding dependent variables, which in fact are the hypotheses on *the intercept differences* of the homogeneous regressions considered. However, analysis should be conducted using the object "*System*", instead of the object "*POOL*" – refer to point (2) in Example 1.13. See the following example.

Example 1.17   A MAR(1) ANCOVA growth model

For illustration, Figure 1.24 presents the statistical results based on a MAR(1) ANCOVA growth model, using the object "*System*". Based on these results the following conclusions and notes are presented.

1. Note that the growth rates of *GDP* of the four states are assumed to be equal to $\hat{C}(11) = 0.011266$ which are unacceptable in a theoretical sense. Similarly so for the first autoregressive indicator $\hat{C}(12) = 0.960457$.

System: SYS09
Estimation Method: Iterative Least Squares
Date: 08/19/09   Time: 09:53
Sample: 1951 1992
Included observations: 43
Total system (balanced) observations 168
Convergence achieved after 2 iterations

	Coefficient	Std. Error	t-Statistic	Prob.
C(10)	9.378796	0.344686	27.20967	0.0000
C(11)	0.011266	0.006555	1.718618	0.0876
C(12)	0.960457	0.009552	100.5540	0.0000
C(20)	9.452469	0.438635	21.54973	0.0000
C(30)	9.057210	0.362113	25.01213	0.0000
C(40)	9.573194	0.549645	17.41705	0.0000

Determinant residual covariance	5.73E-14

Equation: LOG(GDP_US)=C(10)+C(11)*T+[AR(1)=C(12)]
Observations: 42

R-squared	0.987376	Mean dependent var	9.473482
Adjusted R-squared	0.986729	S.D. dependent var	0.227124
S.E. of regression	0.026165	Sum squared resid	0.026699
Durbin-Watson stat	1.957912		

Equation: LOG(GDP_GER)=C(20)+C(11)*T+[AR(1)=C(12)]
Observations: 42

R-squared	0.996285	Mean dependent var	9.098903
Adjusted R-squared	0.996094	S.D. dependent var	0.392866
S.E. of regression	0.024552	Sum squared resid	0.023510
Durbin-Watson stat	1.457365		

Equation: LOG(GDP_UK)=C(30)+C(11)*T+[AR(1)=C(12)]
Observations: 42

R-squared	0.993413	Mean dependent var	9.064748
Adjusted R-squared	0.993075	S.D. dependent var	0.260995
S.E. of regression	0.021719	Sum squared resid	0.018397
Durbin-Watson stat	1.526476		

Equation: LOG(GDP_JPN)=C(40)+C(11)*T+[AR(1)=C(12)]
Observations: 42

R-squared	0.998088	Mean dependent var	8.728910
Adjusted R-squared	0.997990	S.D. dependent var	0.698799
S.E. of regression	0.031331	Sum squared resid	0.038284
Durbin-Watson stat	1.075140		

Wald Test:
System: SYS09

Test Statistic	Value	df	Probability
Chi-square	8.464132	3	0.0373

Null Hypothesis Summary:

Normalized Restriction (= 0)	Value	Std. Err.
C(10) - C(40)	-0.194398	0.275429
C(20) - C(40)	-0.120724	0.187951
C(30) - C(40)	-0.515984	0.256885

Restrictions are linear in coefficients.

Figure 1.24   *Statistical results based on an MAR(1) ANCOVA growth model*

```
System: SYS10
Estimation Method: Iterative Least Squares
Date: 08/19/09  Time: 11:06
Sample: 1951 1992
Included observations: 43
Total system (balanced) observations 168
Convergence achieved after 7 iterations

                 Coefficient    Std. Error    t-Statistic    Prob.

     C(10)         9.078944     0.038474      235.9787      0.0000
     C(11)         0.018204     0.001356       13.42234     0.0000
     C(12)         0.784511     0.103526        7.577923    0.0000
     C(20)         8.872253     0.077353      114.6988      0.0000
     C(22)         0.903524     0.020657       43.73899     0.0000
     C(30)         8.689235     0.051048      170.2162      0.0000
     C(32)         0.853583     0.092179        9.260062    0.0000
     C(40)         9.303788     0.251365       37.01301     0.0000
     C(42)         0.962359     0.007906      121.7262      0.0000

Determinant residual covariance        3.87E-14


Equation: LOG(GDP_US)=C(10)+C(11)*T+[AR(1)=C(12)]
Observations: 42
R-squared            0.988132  Mean dependent var     9.473482
Adjusted R-squared   0.987523  S.D. dependent var     0.227124
S.E. of regression   0.025370  Sum squared resid      0.025101
Durbin-Watson stat   1.743993


Equation: LOG(GDP_GER)=C(20)+C(11)*T+[AR(1)=C(22)]
Observations: 42
R-squared            0.996809  Mean dependent var     9.098903
Adjusted R-squared   0.996645  S.D. dependent var     0.392866
S.E. of regression   0.022756  Sum squared resid      0.020196
Durbin-Watson stat   1.607262
```

```
Equation: LOG(GDP_UK)=C(30)+C(11)*T+[AR(1)=C(32)]
Observations: 42
R-squared            0.993821  Mean dependent var     9.064748
Adjusted R-squared   0.993504  S.D. dependent var     0.260995
S.E. of regression   0.021035  Sum squared resid      0.017257
Durbin-Watson stat   1.471386


Equation: LOG(GDP_JPN)=C(40)+C(11)*T+[AR(1)=C(42)]
Observations: 42
R-squared            0.998109  Mean dependent var     8.728910
Adjusted R-squared   0.998012  S.D. dependent var     0.698799
S.E. of regression   0.031155  Sum squared resid      0.037854
Durbin-Watson stat   1.089900


Wald Test:
System: SYS10

Test Statistic       Value            df        Probability

Chi-square          130.9457          3          0.0000


Null Hypothesis Summary:

Normalized Restriction (= 0)            Value        Std. Err.

C(10) - C(40)                         -0.224815     0.241215
C(20) - C(40)                         -0.431517     0.242261
C(30) - C(40)                         -0.614530     0.240215

Restrictions are linear in coefficients.
```

Figure 1.25   *Statistical results based on an alternative MAR(1) ANCOVA growth model*

2. However, in a statistical sense, this model is an acceptable MANCOVA model of the variables *log(GDP_US)*, *log(GDP_Ger)*, *log(GDP_UK)*, and *log(GDP_JPN)*, where the time t is considered covariate, with the intercept parameters: $C(10)$, $C(20)$, $C(30)$, and $C(40)$.

3. Therefore, various hypotheses on the adjusted means differences of the *log(GDP?)* between any subsets of the four states can easily be tested using the Wald test. For example, H_0: $C(10) = C(20) = C(30) = C(40)$ is rejected based on the Chi-square statistic of $\chi_0^2 = 8.464132$, with $df = 3$ and a p-value $= 0.0373$, which indicates that the four states have significant adjusted means differences of the *log(GDP?)*.

4. As a comparison, Figure 1.25 presents the statistical results based on an alternative MAR(1)_GM, under the assumption the time t has the same slopes of $C(11)$, but various intercepts as well as AR(1). For this model, H_0: $C(10) = C(20) = C(30) = C(40)$ is rejected based on the Chi-square statistic of $\chi_0^2 = 130.9457$, with $df = 3$ and a p-value $= 0.0000$.

5. On the other hand, the null hypothesis H_0: $C(12) = C(22) = C(32) = C(42)$ should be considered in comparing this model with the model in Figure 1.24. The null hypothesis is rejected based on the Chi-square statistic of $\chi_0^2 = 10.43420$, with $df = 3$ and a p-value $= 0.0152$. Then, in a statistical sense, this model is a better fit compared to the model in Figure 1.24.

6. Note that the Durbin–Watson statistics of the regressions in Figures 1.21 and 1.24 indicate that other models should be explored, such as the higher order AR models. However, try it as an exercise.

Example 1.18   A MAR(1) heterogeneous growth model

Building on the model in Figure 1.25, as well as for further comparison, Figure 1.26 presents the statistical results based on a MAR(1) heterogeneous growth model. Based on these results, note the following:

1. The main objectives of this model are to test the hypotheses of the exponential growth rate differences between the GDPs of the four states, indicated by the parameters $C(11)$, $C(21)$, $C(31)$ and $C(41)$.

```
System: SYS11
Estimation Method: Iterative Least Squares
Date: 08/19/09   Time: 12:12
Sample: 1951 1992
Included observations: 43
Total system (balanced) observations 168
Convergence achieved after 2 iterations
```

	Coefficient	Std. Error	t-Statistic	Prob.
C(10)	9.098363	0.044497	204.4720	0.0000
C(11)	0.017432	0.001607	10.84940	0.0000
C(12)	0.788836	0.104235	7.567884	0.0000
C(20)	8.788810	0.184019	47.76036	0.0000
C(21)	0.020326	0.004748	4.281062	0.0000
C(22)	0.891366	0.035255	25.28374	0.0000
C(30)	8.623544	0.025492	338.2892	0.0000
C(31)	0.020563	0.000956	21.51224	0.0000
C(32)	0.709349	0.116626	6.082249	0.0000
C(40)	15.62487	27.31062	0.572117	0.5681
C(41)	-0.041003	0.188326	-0.217721	0.8279
C(42)	0.984128	0.030234	32.55033	0.0000

Determinant residual covariance		3.27E-14		

```
Equation: LOG(GDP_US)=C(10)+C(11)*T+[AR(1)=C(12)]
Observations: 42
```

R-squared	0.988209	Mean dependent var	9.473482
Adjusted R-squared	0.987604	S.D. dependent var	0.227124
S.E. of regression	0.025287	Sum squared resid	0.024938
Durbin-Watson stat	1.762500		

```
Equation: LOG(GDP_GER)=C(20)+C(21)*T+[AR(1)=C(22)]
Observations: 42
```

R-squared	0.996821	Mean dependent var	9.098903
Adjusted R-squared	0.996658	S.D. dependent var	0.392866
S.E. of regression	0.022710	Sum squared resid	0.020114
Durbin-Watson stat	1.596264		

```
Equation: LOG(GDP_UK)=C(30)+C(31)*T+[AR(1)=C(32)]
Observations: 42
```

R-squared	0.994162	Mean dependent var	9.064748
Adjusted R-squared	0.993863	S.D. dependent var	0.260995
S.E. of regression	0.020446	Sum squared resid	0.016304
Durbin-Watson stat	1.376546		

```
Equation: LOG(GDP_JPN)=C(40)+C(41)*T+[AR(1)=C(42)]
Observations: 42
```

R-squared	0.998137	Mean dependent var	8.728910
Adjusted R-squared	0.998041	S.D. dependent var	0.698799
S.E. of regression	0.030927	Sum squared resid	0.037302
Durbin-Watson stat	1.130107		

```
Wald Test:
System: SYS11
```

Test Statistic	Value	df	Probability
Chi-square	2.923708	3	0.4035

Null Hypothesis Summary:

Normalized Restriction (= 0)	Value	Std. Err.
C(11) - C(41)	0.058434	0.188333
C(21) - C(41)	0.061329	0.188386
C(31) - C(41)	0.061565	0.188329

Restrictions are linear in coefficients.

Figure 1.26   *Statistical results based on an alternative MAR(1) heterogeneous growth model*

2. The null hypothesis H_0: $C(11) = C(21) = C(31) = C(41)$ is accepted based on the Chi-square statistic of $\chi_0^2 = 2.923708$, with $df = 3$ and a p-value $= 0.4035$, which indicates that the four growth rates are insignificantly different in the corresponding populations. Based on this finding then, the model in Figure 1.25 can be considered to be a better fit, in a statistical sense, compared to the model in Figure 1.26. However, in a theoretical sense, would you be very confident in saying that the four growth rates of the GDPs are equal?

3. Compare the growth curve of *log(GDP_JPN)* in Figure 1.27 to the negative estimate of its growth rate, namely $\hat{C}(41) = -0.014003$ in Figure 1.26: this indicates that the model is inappropriate for representing the *GDP_JPN*. So a modified model should be explored. Do this as an exercise: refer to the case of the *GDP_FRA* presented in Examples 1.3 to 1.5.

1.7   Growth Models of Sample Statistics

In many studies, we should consider the time series of sample statistics, such as the mean, median and SD (standard deviation), of groups of individuals based on sample surveys as well as experiments. In general, the symbol *Y_gi(t)* will be used to indicate the time series of a single endogenous variable *Y* of the *i*-th individual within the *g*-th group, for $g = 1, \ldots, G$, and $i = 1, \ldots, I_g$. The panel data file, a set of time series *Y_gi* with the format shown in Table 1.4, where the first group ($g = 1$) contains five individuals and the second group ($g = 2$) contains eight individuals.

Based on this data set the time series of the mean, median and SD of the *Y*-variable can easily be generated, namely *MY_g, MedY_g*, and *SDY_g*, for $g = 1, \ldots, G$, either using EViews or Excel.

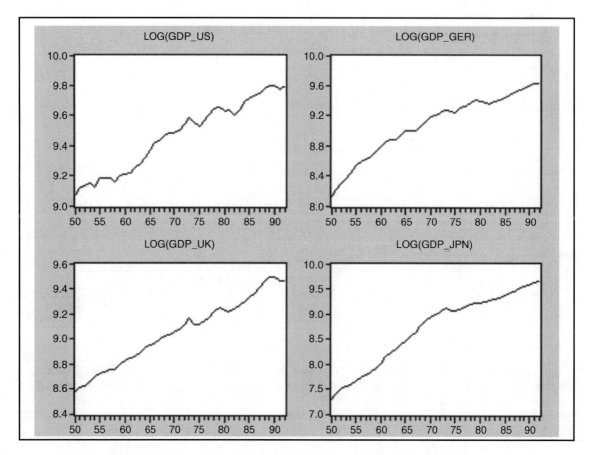

Figure 1.27   *The growth curves of the endogenous variables in Figure 1.26*

A latent variable, or a set of either independent or dependent factors or latent variables, can easily be generated for each group of individuals in order to reduce the dimension of the multivariate considered. For a detailed stepped analysis, refer to Chapter 10 in Agung (2011).

Therefore, all growth models previously presented, as well as their extensions, should be applicable for the time series of the sample statistics *MY_g, MedY_g* and *SDY_g*, as well as latent variables.

Table 1.4   *Illustrated format of a set of time series by two groups of individuals*

Time		g = 1			G = 2	
	Y_11	...	Y_15	Y_21	...	Y_28
1	*Y_11(1)*	...	*Y_15(1)*	*Y_21(1)*	...	*Y_28(1)*
.	...	...	...	...	...	...
t	*Y_11(t)*	...	*Y_51(t)*	*Y_1G(t)*	...	*Y_8G(t)*
.	...	...	...	...	...	...
T	*Y_11(T)*	...	*Y_51(T)*	*Y_1G(T)*		*Y_8G(T)*

Example 1.19   Generating sample statistics using the object "POOL"

For an illustration, the group of *GDP_FRA, GDP_GER* and *GDP_ITA* will be considered for data analysis. The steps of the analysis are as follows:

1. By selecting *Object/New Object/Pool . . . OK*, Figure 1.28(a) appears on the screen.
2. By entering "*_fra _ger _ita*" and selecting *View/Descriptive Statistics . . .* , Figure 1.28(b) is shown on the screen. Then by entering "*gdp?*" and selecting "*Time period specific*", the sample statistics in Figure 1.29 are obtained.
3. Each of the sample statistics can easily be copied to the file. For example, by using the copy-paste method of *Mean GDP?*, the POOLG7.wf1 will have an additional variable *Series01*. Then this variable can be renamed, for example as *M_GDP* or *Mean GDP?* Similarly, this can be done for each of the others.
4. The other copy-paste method can be done using Excel, by copying all sample statistics in Figure 1.29 as an Excel file, then opening the Excel file as an EViews work file.
5. As a result, data analysis based on various models of sample statistics of each group can be easily performed.

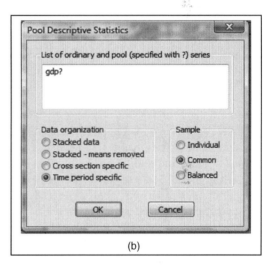

(a)	(b)

Figure 1.28   *The cross section identifiers and options of the pool descriptive statistics*

obs	Mean GDP?	Med GDP?	Sd GDP?	Min GDP?	Max GDP?
1950	3449.000	3415.000	644.6728	2822.000	4110.000
1951	3658.667	3673.000	628.6226	3023.000	4280.000
1952	3867.667	4013.000	675.8234	3131.000	4459.000
1953	4058.000	4278.000	626.6650	3351.000	4545.000
1954	4234.667	4577.000	669.6972	3463.000	4664.000
1955	4560.667	4861.000	769.7729	3686.000	5135.000
1956	4799.333	5195.000	857.9023	3815.000	5388.000
1957	4986.333	5389.000	895.6731	3960.000	5610.000
1958	5123.000	5463.000	884.4524	4119.000	5787.000
1959	5380.667	5610.000	936.3068	4351.000	6181.000
1960	5740.667	5948.000	1012.055	4641.000	6633.000

Figure 1.29   *Sample statistics of* GDP_FRA, GDP_GER *and* GDP_ITA

1.8   Special Notes on Time-State Observations

Corresponding to time-series models for independent states, as well as the models for dependent or correlated states illustrated previously, the following notes are made.

1. The time-state or time-cross-section observations should have much larger time-point observations compared to the cross-sectional observations.
2. As *a rule of thumb* the time points should be at least five times the total number of the variables in the system specification considered. Most researchers make the number of the units of analysis at least 10 times the number of variables in the model.
3. On the other hand, if panel data has a much larger cross-section observation compared to time-point observations, the following alternative data analysis is suggested.
    3.1   Conduct the analysis based on time series models of sample statistics by groups of states or individuals, such as the mean and standard deviation of the groups, or latent variables as presented in Section 1.7, in addition to the descriptive statistical summaries by groups.
    3.2   If the panel data has only a few time-point observations, then the panel data should be presented or considered as a set of cross-section data by times or a cross-section over times (which will be discussed in Part II). As a special case for a two-year observation, the panel data can be considered a *natural-experimental* data set.

1.9   Growth Models with an Environmental Variable

Suppose Y_i_t is an endogenous time series, say the productivity and return rates of the i-th industry or firm in a state/country, then in general there is an environmental or external time series with the same scores/values for all industries, namely Z_t, such as income per capita, inflation rate, exchange rate of US$, *GDP* and others at the state/country level.

Referring to the *GDPs* of three states in Europe in POOL7 G.wf1, namely $Y\_1_t = GDP\_Fra_t$, $Y\_2_t = DGP\_Ger_t$ and $Y\_3 = GDP\_\_Ita_t$, to generalize, they can be presented as Y_1_t, Y_2_t, and Y_3_t; or Y_a_t, Y_b_t, and Y_c_t; respectively. There should be at least one environmental factor, say Z_t, such as the US$ exchange rate or an external factor out of Europe, which could be defined or judged as a causal factor of the *GDPs* of the three states. Would you consider *GDP_US* or *GDP_Jpn*, or both for use as external factors?

Therefore, corresponding to the MLVAR($p;q$)_GM in (1.6), the following general MLVAR($p;q$)_GM with an environmental variable Z_t is made.

$$\log(Y\_i_t) = C(i0) + \sum_{j=1}^{p_i} C(ij)^* \log(Y\_i_{t-j}) + C(i, p_i + 1)^* t + C(i, p_i + 2)^* Z_t + \mu_{it}$$

$$\mu_{it} = \sum_{k=1}^{q_i} \rho_{ik} \mu_{i,t-k} + \varepsilon_{it}$$

(1.16)

However, in general, we know the effect of Z_t on Y_i_t is dependent on t. For this reason, applying the model with a time-related-effect (TRE) is recommended, as follows:

$$\log(Y\_i_t) = C(i0) + \sum_{j=1}^{p_i} C(ij)^* \log(Y\_i_{t-j}) + C(i, p_i + 1)^* t$$
$$+ C(i, p_i + 2)^* Z_t + C(i, p_i + 3)^* t^* Z_t + \mu_{it}$$

$$\mu_{it} = \sum_{k=1}^{q_i} \rho_{ik} \mu_{i,t-k} + \varepsilon_{it}$$

(1.17)

Note that various two- and three-way interaction models have been demonstrated in Agung (2009a) if a vector of environmental variable, namely $Z_t = (Z1_t, Z2_t, \ldots)$, should be considered. Furthermore, under the assumption that Y_i_t for some $i = 1, 2, \ldots, N$, are correlated, then a lot of advanced models, (such as various VAR, VEC and Instrumental Variables Models (IVMs), as well as various TGARCH(a,b,c) models), could easily be subjectively defined by a researcher. However, not everyone can always be very sure as to which is the true population model, since unexpected estimates of the model parameters could be obtained as the impact of the multicollinearity of all variables in the model, and these are highly dependent on the data set that happens to be selected or available to the researcher. Refer to special notes presented in Section 2.14 (Agung, 2009a). See the following selected models.

1.9.1 The Simplest Possible Model

The simplest model is a MLVAR(**1,1**) model with an environmental variable and TRE as follows:

$$\log(Y\_i_t) = C(i0) + C(i1)^*\log(Y\_i_{t-1}) + C(i2)^*t + C(i3)^*Z_t + C(i4)^*t^*Z_t + \mu_{it}$$
$$\mu_{it} = \rho_{i1}\mu_{i.t-1} + \varepsilon_{it} \tag{1.18}$$

For an illustration, a hypothetical data set is generated based on the data in POOL7 G.wf1, where $X\_1 = GDP\_US$, $Y\_1 = GDP\_Can$, $X\_2 = GDP\_Fra$, $Y\_2 = GDP\_UK$, $X\_3 = GDP\_Ger$, and $Y\_3 = GDP\_Ita$, and the environmental variable $Z1 = GDP\_Jpn$ is taken. See the following examples.

For a vector of the environmental variable, namely $Z = (Z1, \ldots, Zk)$, the model in (1.18) can be extended to a more general M LVAR(**1,1**), as follows:

$$\log(Y\_i_t) = C(i0) + C(i1)^*\log(Y\_i_{t-1}) + F_i(t, Z1, \ldots, Zk) + \mu_{it}^{\cdot\cdot}$$
$$\mu_{it} = \rho_{i1}\mu_{i.t-1} + \varepsilon_{it} \tag{1.19}$$

where $F_i(t, Z1, \ldots, Zk)$ is a function of the time, t and an external or environmental vector $Z = (Z1, \ldots, Zk)$ with a finite number of parameter for each $i = 1, 2, \ldots, N$. Therefore, there would be a lot of possible functions of two-way interaction factors, namely t^*Zk and Zi^*Zj, and a few selected three-way interactions.

On the other hand, specific to the quarterly and monthly data sets, Agung (2009a) proposes two alternative models using the lags Y_i_{t-4} and Y_i_{t-12} respectively, in order to match the conditions in the previous and recent years.

Example 1.20 An application of the system equation

Figure 1.30 presents the statistical results based on a MLVAR(1,1) model in (1.17). The main objective of this model is to test the hypothesis that the effect of the environmental variable $Z1$ on the trivariate (Y_1, Y_2, Y_3) depends on the time, t. Based on these results, see the following notes and comments.

1. Note that the interaction t^*Z1 has a significant effect on each of the variables Y_1, Y_2 and Y_3, with a p-value of 0.0003, 0.0370 and 0.0102, respectively. It can then be directly concluded that the effect of $Z1$ on the trivariate (Y_1, Y_2, Y_3) is significantly dependent on the time t.
2. On the other hand, if the effects of t^*Z1 on Y_i are insignificant for the i-th individual, testing the null hypothesis is suggested H_0: $C(13) = C(14) = C(23) = C(24) = C(33) = C(34) = 0$, using the Wald test. Refer to the following example.

```
System: SYS15
Estimation Method: Seemingly Unrelated Regression
Date: 11/01/09  Time: 17:36
Sample: 1952 1992
Included observations: 42
Total system (balanced) observations 123
Iterate coefficients after one-step weighting matrix
Convergence not achieved after: 1 weight matrix, 584 total coef iterations
```

	Coefficient	Std. Error	t-Statistic	Prob.
C(10)	-2.013163	0.577742	-3.484536	0.0007
C(11)	0.649242	0.079103	8.207511	0.0000
C(12)	0.095458	0.026042	3.665576	0.0004
C(13)	0.544602	0.109193	4.987528	0.0000
C(14)	-0.009671	0.002562	-3.774172	0.0003
C(15)	0.118118	0.146734	0.804976	0.4227
C(20)	0.897583	1.802112	0.498073	0.6195
C(21)	0.123505	0.136029	0.907932	0.3660
C(22)	0.133851	0.057065	2.345572	0.0209
C(23)	0.716732	0.145310	4.932445	0.0000
C(24)	-0.012416	0.005878	-2.112352	0.0370
C(25)	0.734109	0.076490	9.597462	0.0000
C(30)	-0.816335	1.059428	-0.770543	0.4427
C(31)	0.484293	0.176536	2.743304	0.0072
C(32)	0.150537	0.058116	2.590258	0.0110
C(33)	0.544049	0.157995	3.443462	0.0008
C(34)	-0.014455	0.005522	-2.617849	0.0102
C(35)	0.291409	0.176220	1.653669	0.1012
Determinant residual covariance	1.03E-11			

Equation: Y_1=C(10)+C(11)*Y_1(-1)+ C(12)*T+C(13)*Z1+C(14)*T*Z1
   +[AR(1)=C(15)]
Observations: 41

R-squared	0.998530	Mean dependent var	9.093576
Adjusted R-squared	0.998320	S.D. dependent var	0.368947
S.E. of regression	0.015122	Sum squared resid	0.008004
Durbin-Watson stat	1.712889		

Equation: Y_2=C(20)+C(21)*Y_2(-1)+ C(22)*T+C(23)*Z1+C(24)*T*Z1
   +[AR(1)=C(25)]
Observations: 41

R-squared	0.997409	Mean dependent var	9.120614
Adjusted R-squared	0.997039	S.D. dependent var	0.371363
S.E. of regression	0.020209	Sum squared resid	0.014294
Durbin-Watson stat	1.378654		

Equation: Y_3=C(30)+C(31)*Y_3(-1)+ C(32)*T+C(33)*Z1+C(34)*T*Z1
   +[AR(1)=C(35)]
Observations: 41

R-squared	0.998144	Mean dependent var	8.889700
Adjusted R-squared	0.997879	S.D. dependent var	0.430718
S.E. of regression	0.019839	Sum squared resid	0.013775
Durbin-Watson stat	1.718009		

Figure 1.30   *Statistical results based on a MLVAR(1,1) Model in (1.18)*

3. On the other hand, a reduced model should be made by deleting either one of t and $Z1$, or both, since t, $Z1$ and t^*Z1 in many cases are highly correlated, and their impacts on the parameter estimates are unpredictable. Refer to special notes in Section 2.14.3 (Agung, 2009a). In many cases, then it would be found t^*Z1 would have a significant effect in the reduced model.

4. Considering the previous results, note that the AR(1) term is insignificant with a p-value $= 0.4227$, in the first regression, and $Y\_2(-1)$ is insignificant in the second regression with a p-value $= 0.3660$. Since their p-values > 0.20; a reduced model should be explored. Do it as an exercise. For the intercept of the third regression, namely $C(30)$, it is not a problem.

1.9.2   The Application of VAR and VEC Models

As an extension of the model in (1.17), the application of the VAR and VEC Models are presented in the following examples. Refer to various VAR and VEC models and their limitations presented in Chapter 6 (Agung, 2009a).

Example 1.21   A VAR model using the object "*System*"

Corresponding to the model in Example 1.19, since a single environmental variable $Z1$ is defined to be a cause of factors Y_1, Y_2 and Y_3, then Y_1, Y_2 and Y_3 should be correlated in a theoretical sense. For this reason a VAR model could be applied. Referring to various VAR models presented in Chapter 6 (Agung, 2009a), then based on the variables $Y\_1, Y\_2, Y\_3, Z1$ and t, a lot of VAR models could easily be derived or defined. However, Agung (2009a; 380) states that the system function (estimation method) is the preferred method used to develop alternative multivariate time series models, since it is more flexible to use for developing a multivariate model where multiple regressions could have different sets of exogenous variables.

System: SYS20
Estimation Method: Seemingly Unrelated Regression
Date: 11/03/09   Time: 13:39
Sample: 1951 1992
Included observations: 42
Total system (balanced) observations 126
Linear estimation after one-step weighting matrix

	Coefficient	Std. Error	t-Statistic	Prob.
C(10)	-2.767715	0.510882	-5.417524	0.0000
C(11)	0.509448	0.081353	6.262231	0.0000
C(12)	0.123211	0.048940	2.517595	0.0133
C(13)	0.182611	0.101597	1.797398	0.0751
C(14)	0.023549	0.026241	0.897414	0.3716
C(15)	0.485480	0.088796	5.467380	0.0000
C(16)	-0.002907	0.002519	-1.153915	0.2512
C(20)	-1.364387	0.824546	-1.654713	0.1010
C(21)	-0.309881	0.131300	-2.360093	0.0201
C(22)	0.944625	0.078987	11.95920	0.0000
C(23)	0.086933	0.163974	0.530161	0.5971
C(24)	0.024415	0.042352	0.576476	0.5655
C(25)	0.415637	0.143313	2.900193	0.0045
C(26)	-0.002543	0.004066	-0.625490	0.5330
C(30)	-1.566476	0.726523	-2.156129	0.0334
C(31)	-0.132370	0.115691	-1.144169	0.2552
C(32)	0.189123	0.069597	2.717401	0.0077
C(33)	0.621420	0.144481	4.301053	0.0000
C(34)	0.058630	0.037317	1.571123	0.1192
C(35)	0.458434	0.126276	3.630408	0.0004
C(36)	-0.005686	0.003583	-1.586888	0.1155

| Determinant residual covariance | 1.04E-11 | | | |

Equation: Y_1=C(10)+ C(11)*Y_1(-1)+C(12)*Y_2(-1)+C(13)*Y_3(-1)+C(14)
   *T+C(15)*Z1+C(16)*T*Z1
Observations: 42

R-squared	0.998976	Mean dependent var	9.076151
Adjusted R-squared	0.998800	S.D. dependent var	0.381517
S.E. of regression	0.013215	Sum squared resid	0.006112
Durbin-Watson stat	2.045999		

Equation: Y_2=C(20)+C(21)*Y_1(-1)+C(22)*Y_2(-1)+C(23)*Y_3(-1)+C(24)
   *T+C(25)*Z1+C(26)*T*Z1
Observations: 42

R-squared	0.997484	Mean dependent var	9.098903
Adjusted R-squared	0.997053	S.D. dependent var	0.392866
S.E. of regression	0.021328	Sum squared resid	0.015921
Durbin-Watson stat	1.612953		

Equation: Y_3=C(30)+C(31)*Y_1(-1)+C(32)*Y_2(-1)+C(33)*Y_3(-1)+C(34)
   *T+C(35)*Z1+C(36)*T*Z1
Observations: 42

R-squared	0.998487	Mean dependent var	8.868850
Adjusted R-squared	0.998228	S.D. dependent var	0.446376
S.E. of regression	0.018792	Sum squared resid	0.012360
Durbin-Watson stat	1.697621		

Figure 1.31   *Statistical results based on a VAR model using the object system*

As an illustration and an extension of the model in Figure 1.30, Figure 1.31 presents the results of a VAR model using the system function, or the object "*System*". Based on these results, the following notes and conclusions are made.

1. The model represents a VAR model by entering "1 1" as the lag interval of endogenous. Refer to alternative lag intervals of alternative VAR models presented in Chapter 6 (Agung, 2009a), as well as the limitation of a VAR model compared to the system equation.
2. Since it is defined that the effect of *Z1* on *(Y_1,Y_2,Y_3)* depends on time, *t*, then the null hypothesis H_0: $C(15) = C(16) = C(25) = C(26) = C(35) = C(36) = 0$ should be tested at the first stage of testing the hypothesis. The null hypothesis can then be rejected based on the Chi-square test of $\chi_0^2 = 40.50483$ with $df = 6$ and a *p*-value $= 0.000$. Then we can conclude that the effect of *Z1* on *(Y_1,Y_2,Y_3)* is significantly dependent on the time *t*, adjusted or conditional for all other variables in the model.
3. Since some of the independent variables have large *p*-values, a reduced model should be explored. So, in general, three multiple regressions having different sets of independent variables are obtained. Therefore, the reduced model is not a VAR model anymore. Try this as an exercise.
4. In order to keep having a reduced VAR model, then one or two of the variables *t, Z1* or *t*Z1* should be deleted from the three regressions. However, note that each of the variables has significant positive or negative adjusted effects on *Y_3*, at a significance level of $\alpha = 0.10$. So, in a statistical sense, it is not wise to delete one of the variables from the third regression.
5. Based on each of the regressions, the following findings are derived.
   5.1 Based on the first regression, at a significance level of $\alpha = 0.10$, *t*Z1* has insignificant effect on *Y_1*, however, the null hypothesis H_0: $C(15) = C(16) = 0$ is rejected based on the Chi-square test

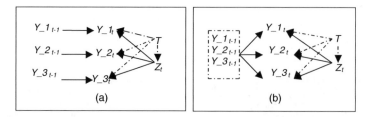

Figure 1.32   *Path diagrams of the models in (a) Figure 1.30, and (b) Figure 1.31*

of $\chi_0^2 = 537.53410$ with $df=2$ and a p-value $=0.000$. It can then be concluded that the effect of $Z1$ on Y_1 is significantly dependent on the time t, specifically the effect depends on a linear function of t, namely $[C(15)+C(16)^*t]$ adjusted for all variables in the model, not just the independent variable of the first regression.

5.2   On the other hand, even though each of t and t^*Z1 has an insignificant adjusted effect, it is found that the variables t, $Z1$ and t^*Z1 have significant joint adjusted effects on Y_1, since the null hypothesis H_0: $C(14)=C(15)=C(16)=0$ is rejected based on the Chi-square test of $\chi_0^2 = 42.67729 = 42.67\,729$ $df=3$ and a p-value $=0.000$. Based on this conclusion, if a reduced model should be obtained then, at most, two of the variables t, $Z1$ and t^*Z1 can be deleted.

5.3   Similar analysis can easily be done based on the other two regressions. Do it as an exercise.

6.   For a graphical illustration, Figure 1.32(a) and (b), respectively, presents the path diagrams of the models in Figures 1.30 and 1.31. Based on these diagrams, the following notes are made.

6.1   Note that Figure 1.32(a) shows that Y_i_{t-1} has a direct effect on each Y_i_t only, but Figure 1.32(b) shows that the trivariate $(Y\_1, Y\_2, Y\_3)_{t-1}$ has direct effect on Y_i_t.

6.2   The effect of Z_t on each endogenous variable Y_i_t, which is defined to be dependent on the time t, is represented as an arrow from t to Z_t, and then from Z_t to Y_i_t, and in the regression indicated by the term $(i3)+c(i4)^*t)^*Z$ in Figure 1.30, and in Figure 1.31 by $(C(i5)+c(i6)^*t)^*Z$.

6.3   The possible causal effects between Y_1, Y_2 and Y_3 are not identified, however, their quantitative correlations are taken into account in the estimation process. If they should have a type of causal effect, then a new model should be defined; either additive, two- or three-way interaction models. Refer to the models demonstrated in Agung (2009a), as well as the following chapter.

7.   As an extension of the model in Figure 1.31, we might consider Z_t as a function of t, then the following general model would also need to be considered.

$$y\_1=c(10)+c(11)^*y\_1(-1)+c(12)^*Y\_2(-1)+c(13)^*Y\_3(-1)+c(14)^*t+c(15)^*z1+c(16)^*t^*z1$$

$$y\_2=c(20)+c(21)^*Y\_1(-1)+c(22)^*Y\_2(-1)+c(23)^*Y\_3(-1)+c(24)^*t+c(25)^*z1+c(26)^*t^*z1$$

$$y\_3=c(30)+c(31)^*y\_1(-1)+c(32)^*Y\_2(-1)+c(33)^*Y\_3(-1)+c(34)^*t+c(35)^*z1+c(36)^*t^*z1$$

$$Z=c(40)+F(t)$$

$$(1.20)$$

where $F(t)$ is a function of the time, t with a finite number of parameters, without a constant parameter, such as $F(t)=C(41)^*log(t)$, and $F(t)=C(41)^*t+C(42)^*t^2+ \ldots +C(4k)^*t^k$.

Vector Error Correction Estimates
Date: 11/04/09 Time: 11:08
Sample (adjusted): 1952 1992
Included observations: 41 after adjustments
Standard errors in ( ) & t-statistics in [ ]

Cointegrating Eq:	CointEq1		
Y_1(-1)	1.000000		
Y_2(-1)	-0.188327 (0.10802) [-1.74350]		
Y_3(-1)	-0.697651 (0.17410) [-4.00729]		
C	-1.176237		
Error Correction:	**D(Y_1)**	**D(Y_2)**	**D(Y_3)**
CointEq1	-0.559820 (0.09342) [-5.99236]	-0.313246 (0.16058) [-1.95076]	-0.154371 (0.15937) [-0.96864]
D(Y_1(-1))	0.081726 (0.14843) [0.55061]	-0.135375 (0.25512) [-0.53063]	0.207249 (0.25320) [0.81852]
D(Y_2(-1))	0.121231 (0.11383) [1.06499]	0.116966 (0.19566) [0.59780]	0.163731 (0.19419) [0.84316]
D(Y_3(-1))	-0.276045 (0.11378) [-2.42613]	-0.307600 (0.19557) [-1.57286]	-0.164907 (0.19410) [-0.84962]
C	-3.352072 (0.64536) [-5.19409]	-2.258550 (1.10927) [-2.03607]	-1.128369 (1.10092) [-1.02493]
T	-0.004842 (0.01300) [-0.37242]	-0.063879 (0.02235) [-2.85840]	-0.006958 (0.02218) [-0.31372]
Z1	0.375213 (0.07155) [5.24404]	0.261613 (0.12298) [2.12724]	0.129550 (0.12206) [1.06139]
T*Z1	-0.000315 (0.00129) [-0.24487]	0.005822 (0.00221) [2.63512]	0.000392 (0.00219) [0.17875]
R-squared	0.691509	0.534601	0.315831
Adj. R-squared	0.626071	0.435880	0.170704
Sum sq. resids	0.005403	0.015962	0.015722
S.E. equation	0.012795	0.021993	0.021827
F-statistic	10.56746	5.415263	2.176242
Log likelihood	124.9793	102.7720	103.0817
Akaike AIC	-5.706306	-4.623024	-4.638131
Schwarz SC	-5.371951	-4.288668	-4.303776
Mean dependent	0.029314	0.034892	0.035789
S.D. dependent	0.020924	0.029282	0.023969
Determinant resid covariance (dof adj.)	2.36E-11		
Determinant resid covariance	1.23E-11		
Log likelihood	340.4857		
Akaike information criterion	-15.29199		
Schwarz criterion	-14.16354		

Figure 1.33 *Statistical results based on a VEC model*

Example 1.22 A VEC model

Figure 1.33 presents the statistical results based on a VEC model of the first differences between endogenous variables *DY_1, DY_2* and *DY_3*, exogenous variables *t, Z1* and t^*Z1, and “1 1” as the lag interval of endogenous variables. Refer to the characteristics of various VEC models and their limitations presented in Chapter 6 (Agung, 2009a).

1.9.3 Application of ARCH Model

Various TGARCH(a,b,c) time series models along with their limitations have been presented in Agung (2009a). For this reason, this section only presents the example of an ARCH(1) = TGARCH(1,0,0) model.

Example 1.23 A reduced ARCH(1) model

Figure 1.34 presents the statistical results based on a reduced ARCH(1) model, where its full mean model is presented in Figure 1.30. Based on these results, note the following:

1. Note that the regression of *Y_2* has only two independent variables, namely *Y_2(−1)* and t^*Z1, compared to the other two hierarchical regression models.
2. Based on the output, it can easily be derived that the effect of *Z1* on *(Y_1,Y_2,Y_3)* is significantly dependent on the time *t*. Otherwise, it can be tested using the Wald test.
3. The data supports that error terms have a multivariate Student’s *t*-distribution based on z-Statistic of $Z_0 = 0.108\,608$ with a *p*-value = 0.9135.

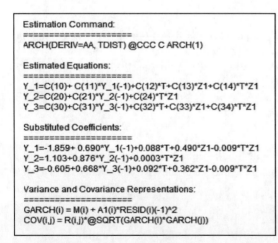

System: SYS27
Estimation Method: ARCH Maximum Likelihood (Marquardt)
Covariance specification: Constant Conditional Correlation
Date: 11/04/09   Time: 13:19
Sample: 1951 1992
Included observations: 42
Total system (balanced) observations 126
Disturbance assumption: Student's t distribution
Presample covariance: backcast (parameter =0.7)
Convergence achieved after 162 iterations

	Coefficient	Std. Error	z-Statistic	Prob.
C(10)	-1.858822	0.775986	-2.395431	0.0166
C(11)	0.690433	0.115008	6.003364	0.0000
C(12)	0.087609	0.040721	2.151449	0.0314
C(13)	0.490258	0.161171	3.041859	0.0024
C(14)	-0.008910	0.004033	-2.209131	0.0272
C(20)	1.103364	0.387836	2.844924	0.0044
C(21)	0.876077	0.046077	19.01343	0.0000
C(24)	0.000274	0.000157	1.746065	0.0808
C(30)	-0.605071	1.049288	-0.576649	0.5642
C(31)	0.667993	0.165157	4.044583	0.0001
C(32)	0.091821	0.063656	1.442456	0.1492
C(33)	0.361825	0.196797	1.838565	0.0660
C(34)	-0.008852	0.006192	-1.429544	0.1528

Variance Equation Coefficients				
C(35)	0.000268	0.000199	1.345405	0.1785
C(36)	-0.251133	0.520293	-0.482676	0.6293
C(37)	0.000375	0.000279	1.341809	0.1797
C(38)	0.231851	0.652811	0.355158	0.7225
C(39)	0.000392	0.000148	2.646134	0.0081
C(40)	-0.077905	0.445042	-0.175050	0.8610
C(41)	0.501524	0.259212	1.934804	0.0530
C(42)	0.511081	0.224333	2.278228	0.0227
C(43)	0.272723	0.253580	1.075494	0.2822

t-Distribution (Degree of Freedom)				
C(44)	52.61603	484.4564	0.108608	0.9135

Log likelihood	339.2191	Schwarz criterion	-14.10647
Avg. log likelihood	2.692215	Hannan-Quinn criter.	-14.70926
Akaike info criterion	-15.05805		

Equation: Y_1=C(10)+ C(11)*Y_1(-1)+C(12)*T+C(13)*Z1+C(14)*T*Z1

R-squared	0.998580	Mean dependent var	9.076151
Adjusted R-squared	0.998427	S.D. dependent var	0.381517
S.E. of regression	0.015133	Sum squared resid	0.008473
Durbin-Watson stat	1.495363		

Equation: Y_2=C(20)+C(21)*Y_2(-1)+C(24)*T*Z1

R-squared	0.996801	Mean dependent var	9.098903
Adjusted R-squared	0.996637	S.D. dependent var	0.392866
S.E. of regression	0.022782	Sum squared resid	0.020242
Durbin-Watson stat	1.560383		

Equation: Y_3=C(30)+C(31)*Y_3(-1)+C(32)*T+C(33)*Z1+C(34)*T*Z1

R-squared	0.998161	Mean dependent var	8.868850
Adjusted R-squared	0.997962	S.D. dependent var	0.446376
S.E. of regression	0.020149	Sum squared resid	0.015022
Durbin-Watson stat	1.516532		

Covariance specification: Constant Conditional Correlation
GARCH(i) = M(i) + A1(i)*RESID(i)(-1)^2
COV(i,j) = R(i,j)*@SQRT(GARCH(i)*GARCH(j))

Tranformed Variance Coefficients				
	Coefficient	Std. Error	z-Statistic	Prob.
M(1)	0.000268	0.000199	1.345405	0.1785
A1(1)	-0.251133	0.520293	-0.482676	0.6293
M(2)	0.000375	0.000279	1.341809	0.1797
A1(2)	0.231851	0.652811	0.355158	0.7225
M(3)	0.000392	0.000148	2.646134	0.0081
A1(3)	-0.077905	0.445042	-0.175050	0.8610
R(1,2)	0.501524	0.259212	1.934804	0.0530
R(1,3)	0.511081	0.224333	2.278228	0.0227
R(2,3)	0.272723	0.253580	1.075494	0.2822

Estimation Command:
=====================
ARCH(DERIV=AA, TDIST) @CCC C ARCH(1)

Estimated Equations:
=====================
Y_1=C(10)+ C(11)*Y_1(-1)+C(12)*T+C(13)*Z1+C(14)*T*Z1
Y_2=C(20)+C(21)*Y_2(-1)+C(24)*T*Z1
Y_3=C(30)+C(31)*Y_3(-1)+C(32)*T+C(33)*Z1+C(34)*T*Z1

Substituted Coefficients:
=====================
Y_1=-1.859+ 0.690*Y_1(-1)+0.088*T+0.490*Z1-0.009*T*Z1
Y_2=1.103+0.876*Y_2(-1)+0.0003*T*Z1
Y_3=-0.605+0.668*Y_3(-1)+0.092*T+0.362*Z1-0.009*T*Z1

Variance and Covariance Representations:
=====================
GARCH(i) = M(i) + A1(i)*RESID(i)(-1)^2
COV(i,j) = R(i,j)*@SQRT(GARCH(i)*GARCH(j))

Variance and Covariance Equations:
=====================
GARCH1 = C(35) + C(36)*RESID1(-1)^2
GARCH2 = C(37) + C(38)*RESID2(-1)^2
GARCH3 = C(39) + C(40)*RESID3(-1)^2

COV1_2 = C(41)*@SQRT(GARCH1*GARCH2)
COV1_3 = C(42)*@SQRT(GARCH1*GARCH3)
COV2_3 = C(43)*@SQRT(GARCH2*GARCH3)

Substituted Coefficients:
=====================
GARCH1 = 0.000267768110001 - 0.251132976834*RESID1(-1)^2
GARCH2 = 0.000374611760699 + 0.231851204332*RESID2(-1)^2
GARCH3 = 0.000392378253889 - 0.0779045947747*RESID3(-1)^2

COV1_2 = 0.501523819388*@SQRT(GARCH1*GARCH2)
COV1_3 = 0.511080942222*@SQRT(GARCH1*GARCH3)
COV2_3 = 0.272723253637*@SQRT(GARCH2*GARCH3)

Figure 1.34   *Statistical results based on an ARCH(1) model*

1.9.4   The Application of Instrumental Variables Models

Based on the variables Y_1, Y_2, Y_3, $Y\_1(-1)$, $Y\_2(-1)$, $Y\_3(-1)$, $Z1$ and the time t used in previous models, a lot of instrumental variables models can easily be subjectively defined. Corresponding to an instrumental variables model (IVM), Agung (2009a) states that there would be two stages of problems

(TSOP) in defining instrumental variables models, since the true population model can never be known and there is no general rule as to how to select the best possible set of instrumental variables.

Example 1.24   An application of the 2SLS estimation method

As an extension of the model in Figure 1.32, under the assumption that Y_1, Y_2 and Y_3 are correlated, and we define that the effect of $Z1$ on the trivariate (Y_1,Y_2,Y_3) depends on the time t, then by using *trial-and-error methods*, the statistical results presented in Figure 1.34 are obtained using the 2SLS. Based on these results, the following notes and conclusions are made.

1. Even though each of Y_2 and Y_3 has an insignificant adjusted effect on Y_1, the joint effects of Y_2 and Y_3 on Y_1 are significant, since the null hypothesis H_0: $C(11) = C(12) = 0$ is rejected based on a Chi-square test of $\chi_0^2 = 324.1802$ with $df = 2$ and a p-value $= 0.000$. The same conclusions are obtained based on the other two regressions. Therefore, we can conclude that the data supports the assumption that variables Y_1, Y_2 and Y_3 are correlated.
2. In a statistical sense, a reduced model should be explored, since one of the independent variables in each regression has a p-value > 0.20 (or 0.25). Do it as an exercise.

Example 1.25   An application of the 3SLS estimation method

As a modification of the MAR(1)_IVM in Figure 1.35, the following system specification is considered.

$$\begin{aligned}
y\_1 &= c(10) + c(11)^* y\_2 + c(12)^* Y\_3 + [ar(1) = c(13)] @\, c\, z1 @\, t\, t^* z1 \\
y\_2 &= c(20) + c(21)^* Y\_1 + c(22)^* Y\_3 + [ar(1) = c(23)] @\, c\, z1 @\, t\, t^* z1 \\
y\_3 &= c(30) + c(31)^* Y\_1 + c(32)^* Y\_2 + [ar(1) = c(33)] @\, c\, z1 @\, t\, t^* z1
\end{aligned} \quad (1.21)$$

However, an error message of *"Near Singular Matrix"* is obtained so trial-and-error methods should be applied to delete one or two of the variables from the model in (1.21). Finally, an unexpected good fit model is obtained, in a statistical sense, since each of the independent variables has significant adjusted effect with a p-value $= 0.000$, as presented in Figure 1.36.

System: SYS28
Estimation Method: Iterative Two-Stage Least Squares
Date: 11/04/09   Time: 15:51
Sample: 1951 1992
Included observations: 43
Total system (balanced) observations 126
Convergence achieved after 22 iterations

	Coefficient	Std. Error	t-Statistic	Prob.
C(10)	0.107289	1.086964	0.098706	0.9215
C(11)	0.626658	0.435041	1.440455	0.1525
C(12)	0.365074	0.339414	1.075602	0.2844
C(13)	0.824692	0.062984	13.09361	0.0000
C(20)	1.609625	0.672274	2.394299	0.0183
C(21)	0.201697	0.351989	0.573021	0.5678
C(22)	0.640546	0.303956	2.107366	0.0373
C(23)	0.776692	0.068356	11.36242	0.0000
C(30)	-2.091905	0.534275	-3.915405	0.0002
C(31)	0.205612	0.417156	0.492891	0.6230
C(32)	0.997769	0.435167	2.292841	0.0237
C(33)	0.753684	0.080540	9.357851	0.0000

| Determinant residual covariance | 7.79E-14 |

Equation: Y_1=C(10)+ C(11)*Y_2+C(12)*Y_3+[AR(1)=C(13)]
Instruments: C T Z1 T*Z1 Y_1(-1) Y_2(-1) Y_3(-1)
Observations: 42

R-squared	0.998316	Mean dependent var	9.076151
Adjusted R-squared	0.998183	S.D. dependent var	0.381517
S.E. of regression	0.016262	Sum squared resid	0.010049
Durbin-Watson stat	1.542501		

Equation: Y_2=C(20)+C(21)*Y_1 +C(22)*Y_3+[AR(1)=C(23)]
Instruments: C T Z1 T*Z1 Y_2(-1) Y_1(-1) Y_3(-1)
Observations: 42

R-squared	0.997385	Mean dependent var	9.098903
Adjusted R-squared	0.997179	S.D. dependent var	0.392866
S.E. of regression	0.020868	Sum squared resid	0.016548
Durbin-Watson stat	1.662941		

Equation: Y_3=C(30)+C(31)*Y_1+C(32)*Y_2+[AR(1)=C(33)]
Instruments: C T Z1 T*Z1 Y_3(-1) Y_1(-1) Y_2(-1)
Observations: 42

R-squared	0.997115	Mean dependent var	8.868850
Adjusted R-squared	0.996887	S.D. dependent var	0.446376
S.E. of regression	0.024905	Sum squared resid	0.023569
Durbin-Watson stat	1.713668		

Figure 1.35   *Statistical results based on a MAR(1) instrumental variables model, using the 2SLS estimation method*

	Coefficient	Std. Error	t-Statistic	Prob.
System: SYS29				
Estimation Method: Three-Stage Least Squares				
Date: 11/04/09 Time: 16:35				
Sample: 1951 1992				
Included observations: 43				
Total system (balanced) observations 126				
Iterate coefficients after one-step weighting matrix				
Convergence achieved after: 1 weight matrix, 467 total coef iterations				
C(10)	2.370430	0.412806	5.742231	0.0000
C(11)	-0.494364	0.142092	-3.479179	0.0007
C(12)	1.263649	0.123065	10.26818	0.0000
C(13)	0.752916	0.086334	8.721014	0.0000
C(20)	4.679244	0.765912	6.109373	0.0000
C(21)	-1.947988	0.347784	-5.601151	0.0000
C(22)	2.492620	0.320799	7.770038	0.0000
C(23)	0.750676	0.086492	8.679110	0.0000
C(30)	-2.230141	0.518800	-4.298655	0.0000
C(32)	1.217461	0.055994	21.74282	0.0000
C(33)	0.753724	0.053324	14.13487	0.0000
Determinant residual covariance	4.36E-15			

Equation: Y_1=C(10)+ C(11)*Y_2+C(12)*Y_3+[AR(1)=C(13)]
Instruments: C Z1 @ T T*Z1 Y_1(-1) Y_2(-1) Y_3(-1)
Observations: 42

R-squared	0.995587	Mean dependent var	9.076151
Adjusted R-squared	0.995239	S.D. dependent var	0.381517
S.E. of regression	0.026326	Sum squared resid	0.026336
Durbin-Watson stat	1.735812		

Equation: Y_2=C(20)+C(21)*Y_1 +C(22)*Y_3+[AR(1)=C(23)]
Instruments: C Z1 @ T T*Z1 Y_2(-1) Y_1(-1) Y_3(-1)
Observations: 42

R-squared	0.983786	Mean dependent var	9.098903
Adjusted R-squared	0.982506	S.D. dependent var	0.392866
S.E. of regression	0.051963	Sum squared resid	0.102606
Durbin-Watson stat	1.733474		

Equation: Y_3=C(30)+C(32)*Y_2+[AR(1)=C(33)]
Instruments: C Z1 @ T T*Z1 Y_3(-1) Y_2(-1)
Observations: 42

R-squared	0.996452	Mean dependent var	8.868850
Adjusted R-squared	0.996270	S.D. dependent var	0.446376
S.E. of regression	0.027261	Sum squared resid	0.028983
Durbin-Watson stat	1.678885		

Figure 1.36 *Statistical results based on a MAR(1) instrumental variables model, using the 3SLS estimation method*

Note that the regression of Y_3 only has a single independent variable Y_2. The reader could try deleting other variable(s) from the model in (1.20), including modifying the instrumental variables, but leaving the AR(1) terms as they are.

1.10   Models with an Environmental Multivariate

If an endogenous variable by states, namely Y_i_t, $i = 1,\ldots,N$, is known or defined to be effected by the same environmental multivariate, say $\mathbf{Z}_t = (Z1_t,\ldots,Zk_t,\ldots)$, then the set of Y_i_t, $i = 1,\ldots,N$, should be correlated, including the possibility of having causal relationships for some states. As an illustration, the following section presents selected models using two endogenous variables Y_1 and Y_2, which could easily be extended to three or more states.

1.10.1   Bivariate Correlation and Simple Linear Regressions

Data analysis based on the bivariate correlation of Y_i_t and Y_j_t, the simple linear regression of Y_i_t on Y_j_t, and the simple linear regression of Y_j_t on Y_i_t, would give exactly the same values of the t-statistic for testing the hypothesis that Y_j_t is a causal factor of Y_i_t, as well as Y_i_t and Y_j_t having simultaneous causal effects.

Based on these findings, it can be concluded that correlation analysis can be used to test the hypothesis stated earlier. On the other hand, it could be said that independent of a model, the independent variable may not be a causal factor of the corresponding dependent variable. Note that the causal relationship between any pair of variables should be derived based on a strong theoretical foundation: it is not based on the conclusion of testing a hypothesis. See the following example.

Example 1.26   Special findings
Figure 1.37(a), (b) and (c), respectively, present the statistical results based on the bivariate correlation of Y_1_t and Y_2_t, the simple linear regression (SLR) of Y_1_t on Y_2_t, and the SLR of Y_2_t on Y_1_t, which show exactly the same values of the t-statistic of $t_0 = 46.39045$.

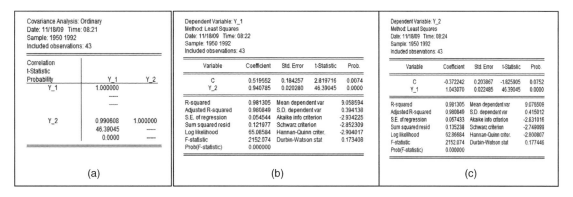

Figure 1.37 Statistical results based on (a) covariance analysis of Y_1$_t$ and Y_2$_t$, (b) the SLR of Y_1$_t$ on Y_2$_t$, and (c) the SLR of Y_1$_t$ on Y_2$_t$

Based on these results, the following notes and comments are presented.

1. Even though the regressions have very small DW-statistics, because their R-squares are very large, namely $R^2 = 0.981\,217$, then the SLR should be considered to be a good fit.
2. As an alternative analysis, Figure 1.38(a) presents the statistical results based on system equations of two SLRs using the LS estimation method, where both SLRs also show the same values of the t-statistic of $t_0 = 46.39\,045$. Thus the results of these system equations can be represented by the result of the covariance analysis in Figure 1.37(a). In other words, the simultaneous causal effects of Y_1 and Y_2 tested using the system equations in Figure 1.38(a) can be substituted by covariance analysis.

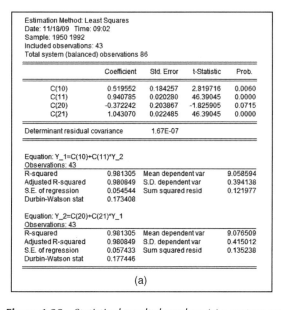

Figure 1.38 Statistical results based on (a) a system equation of SLRs of Y_1$_t$ and Y_2$_t$, and (b) a MAR(1) of Y_1$_t$ on Y_2

3. For a comparison, Figure 1.38(b) presents the statistical results using the SUR estimation method, which show different values of the t-statistic of $t_0 = 67.50334$ for both SLRs.
4. Compared to the regressions in Figure 1.37(b), the four regressions in Figure 1.38 also have very small DW-statistics, but very large R-squares.
5. Further analysis can easily be done based on MLV(p), MAR(q) or MLVAR(p,q) models using the variables Y_1 and Y_2, which will have much larger DW-statistics. Do it as an exercise.

1.10.2   Simple Models with an Environmental Multivariate

Since Y_1_t and Y_2_t are correlated, specifically linearly correlated, then simple models of Y_1_t and Y_2_t with an environmental multivariate Z_t and the time t as independent variables will have the following general equation.

$$
\begin{aligned}
Y\_1_t &= C(10) + C(11)^* Y\_2_{t-j} + F(t, Z_t, C(1^*)) + \mu_{1t} \\
Y\_2_t &= C(20) + C(21)^* Y\_1_{t-j} + F(t, Z_t, C(2^*)) + \mu_{1t}
\end{aligned}
\tag{1.22}
$$

for a subscript $j \geq 0$, where $F(t, Z_t, *)$ can be any functions $Z1_t, \ldots, Zk_t, \ldots$, and the time t, including some selected two- and three-ways of their interactions, with a finite number of parameters but no constant parameter. For instance, if the effect of $Z1$ on Y_i depends on $Z2$, then $Z1^*Z2$ should be used as an independent variable or a term of the function $F(t, Z_t, *)$. Note that there would be a lot of possible time-series models.

However, the following four groups of models will be considered, corresponding to selected forms of the function $F(t, Z_t, *)$, such as follows:

1. Additive models or two functions, namely $F(t, Z_t, *) = F_1(t, *) + F_2(Z_t, *)$.
   Refer to the models (1.4a) to (1.4d) for the alternative functions of $F_1(t, *)$, and $F_2(Z_t, *)$ can be additive or interaction functions of the components of Z_t.
2. Models with trend: $F(t, Z_t, C(i^*)) = C(i2)^* t + F_2(Z_t, C(ik))$, $i = 1,2$, and $k > 2$.
3. Models with Trend-Related Effects (TRE):

$$
F(t, Z_t, C(i^*)) = C(i2)^* t + F_2(Z_t, C(ik)) + t^* F_2(Z_t, C(i^*)), \quad \text{for } i = 1, 2, \text{ and } k > 2.
$$

4. Models without the time t: $F(t, Z_t, *) = F_2(Z_t, *)$. Refer to all seemingly causal models (SCMs) presented in Chapter 4 (Agung, 2009a).

Comparing these to the models in Figure 1.38, the models in (1.22), for $j = 0$ in fact show that Y_1_t and Y_2_t have simultaneous causal relationships.

To generalize, the following general model can be applied

$$
G_1(Y\_1_t) = C(10) + C(11)^* Y\_2_{t-j} + F(t, Z_t, C(1^*)) + \mu_{1t}
$$

$$
G_2(Y\_2_t) = C(20) + C(21)^* Y\_1_{t-j} + F(t, Z_t, C(2^*)) + \mu_{1t}
\tag{1.23}
$$

where $G_i(Y\_i_t)$ is a function of Y_i_t having no parameter, such as $G_i(Y\_i_t) = Y\_i_t$, $log(Y\_i_t)$ or $log[(Y\_i_t - Li)/(Ui - Y\_i_t)]$, where Li and Ui are the lower and upper bounds of Y_i_t, which should be subjectively selected by the researchers.

If there are correlated endogenous variables for three states, namely Y_1, Y_2 and Y_3, then the simple models of $G_i(Y\_i_t)$, $i = 1,2$ and 3 will have the general equation as follows:

$$
\begin{aligned}
G_1(Y\_1_t) &= C(10) + C(11)^* Y\_2_{t-j} + C(12)^* Y\_3_{t-j} + F(t, Z_t, C(1^*)) + \mu_{1t} \\
G_2(Y\_2_t) &= C(20) + C(21)^* Y\_1_{t-j} + C(22)^* Y\_3_{t-j} + F(t, Z_t, C(2^*)) + \mu_{1t} \\
G_3(Y\_3_t) &= C(30) + C(31)^* Y\_1_{t-j} + C(32)^* Y\_2_{t-j} + F(t, Z_t, C(3^*)) + \mu_{1t}
\end{aligned}
\tag{1.24}
$$

1.10.3   The VAR Models

1.10.3.1   Basic General VAR Models

For illustration, a VAR model of Y_1 and Y_2 with "1 p" as the lag intervals for the endogenous variables, and the time t and $\mathbf{Z}_t$ as exogenous variables will be considered. The model considered has the following general equation.

$$
\begin{aligned}
Y\_1_t &= C(110) + \sum_{j=1}^{p} C(11j)^* Y\_1_{t-j} + \sum_{j=1}^{p} C(12j)^* Y\_2_{t-j} + F(t, Z_t, C(13^*)) + \varepsilon_{1t} \\[2mm]
Y\_2_t &= C(210) + \sum_{j=1}^{p} C(21j)^* Y\_1_{t-j} + \sum_{j=1}^{p} C(22j)^* Y\_2_{t-j} + F(t, Z_t, C(23^*)) + \varepsilon_{1t}
\end{aligned}
\tag{1.25}
$$

1.10.3.2   Special VAR Interaction Models

With multivariate environmental variables, it is generally known that an effect of at least one of its components on the endogenous variables depends on the other component(s). Under these criteria, this section presents three alternative VAR interaction models of Y_1, Y_2 and Y_3 with the lag intervals for the endogenous: "1 1", and the environmental variables $Z1$ and $Z2$, such as follows:

1. A VAR interaction model with trend:

$$
\begin{aligned}
Y\_i = {}& C(i0) + C(i1)^* Y\_1(-1) + C(i2)^* Y\_2(-1) + C(i3)^* Y\_3(-1) \\
& + C(i4)^* t + C(i5)^* Z1 + C(i6)^* Z2 + C(i7)^* Z1^* Z2 \\
& for\ i = 1,2,3
\end{aligned}
\tag{1.26}
$$

2. A VAR interaction model with time-related effects:

$$
\begin{aligned}
Y\_i = {}& C(i0) + C(i1)^* Y\_1(-1) + C(i2)^* Y\_2(-1) + C(i3)^* Y\_3(-1) \\
& + C(i4)^* t + C(i5)^* Z1 + C(i6)^* Z2 + C(i7)^* Z1^* Z2 \\
& + C(i8)^* t^* Z1 + C(i9)^* t^* Z2 + C(i10)^* t^* Z1^* Z2 \\
& for\ i = 1,2,3
\end{aligned}
\tag{1.27}
$$

3. A VAR interaction model without the time t:

$$
\begin{aligned}
Y\_i = {}& C(i0) + C(i1)^* Y\_1(-1) + C(i2)^* Y\_2(-1) + C(i3)^* Y\_3(-1) \\
& + C(i4)^* Z1 + C(i5)^* Z2 + C(i6)^* Z1^* Z2 \\
& for\ i = 1,2,3
\end{aligned}
\tag{1.28}
$$

Corresponding to these VAR interaction models, note the following:

1. In practice, a reduced model obtained would be a good fit in a statistical sense, because the three variables $Z1$, $Z2$ and $Z1^*Z2$ are highly or significantly correlated in general. More so for the independent variables of the model in (2.27).
2. Since it is defined that the effects of $Z1$ $(Z2)$ on Y_i, $i = 1,2$ and 3 depend on $Z2$ $(Z1)$ in a theoretical sense, then the interaction $Z1^*Z2$ should be used in the model, as well as in the reduced model(s). So a reduced model should be obtained by deleting either $Z1$ or $Z2$, or both $Z1$ and $Z2$. Note that a model can be considered an acceptable or good fit, even though some of its independent variables have insignificant adjusted effects.
3. Note that three models here are hierarchical two- and three-way interaction models. However, corresponding to the earlier notes, an empirical acceptable model obtained would be non-hierarchical in general. See the following example.

Example 1.27 A reduced VAR interaction model
Figure 1.39 presents the statistical results based on two reduced models of the VAR interaction model in (1.26). Based on these results, the following conclusions and notes are made.

1. By using the full model in (1.26), each of the independent variables $Z1$, $Z2$ and $Z1*Z2$ has insignificant adjusted effects. By deleting either $Z1$ or $Z2$, the results in Figure 1.39(a) and (b) are obtained.

Vector Autoregression Estimates
Date: 11/17/09 Time: 13:53
Sample (adjusted): 1951 1992
Included observations: 42 after adjustments
Standard errors in ( ) & t-statistics in [ ]

	Y_1	Y_2	Y_3
Y_1(-1)	0.491574	-0.332933	-0.140681
	(0.07191)	(0.13066)	(0.11241)
	[6.83628]	[-2.54814]	[-1.25148]
Y_2(-1)	0.138289	0.957298	0.220475
	(0.04156)	(0.07551)	(0.06497)
	[3.32759]	[12.6773]	[3.39357]
Y_3(-1)	-0.085190	-0.214036	0.337287
	(0.10994)	(0.19976)	(0.17187)
	[-0.77488]	[-1.07144]	[1.96245]
C	2.285288	4.147653	4.394102
	(1.24927)	(2.26998)	(1.95300)
	[1.82929]	[1.82717]	[2.24992]
T	-0.002228	0.002815	0.004795
	(0.00155)	(0.00282)	(0.00242)
	[-1.43840]	[1.00008]	[1.97991]
Z1	-0.005993	-0.105799	-0.174082
	(0.11600)	(0.21078)	(0.18135)
	[-0.05166]	[-0.50194]	[-0.95994]
Z1*Z2	0.023509	0.025581	0.027959
	(0.00535)	(0.00972)	(0.00836)
	[4.39597]	[2.63254]	[3.34432]

(a)

Vector Autoregression Estimates
Date: 11/17/09 Time: 14:31
Sample (adjusted): 1951 1992
Included observations: 42 after adjustments
Standard errors in ( ) & t-statistics in [ ]

	Y_1	Y_2	Y_3
Y_1(-1)	0.493659	-0.332149	-0.128280
	(0.07233)	(0.13155)	(0.11174)
	[6.82556]	[-2.52493]	[-1.14802]
Y_2(-1)	0.136375	0.953908	0.205518
	(0.04279)	(0.07783)	(0.06611)
	[3.18684]	[12.2557]	[3.10854]
Y_3(-1)	-0.094441	-0.216255	0.283936
	(0.11487)	(0.20893)	(0.17747)
	[-0.82216]	[-1.03507]	[1.59991]
C	2.264351	3.244793	3.073037
	(0.40767)	(0.74149)	(0.62984)
	[5.55438]	[4.37606]	[4.87909]
T	-0.001988	0.003159	0.006566
	(0.00198)	(0.00360)	(0.00306)
	[-1.00489]	[0.87810]	[2.14866]
Z2	0.020480	0.096814	0.240974
	(0.11596)	(0.21091)	(0.17915)
	[0.17662]	[0.45904]	[1.34510]
Z1*Z2	0.021824	0.014612	0.004119
	(0.00872)	(0.01586)	(0.01347)
	[2.50330]	[0.92153]	[0.30583]

(b)

Figure 1.39 *Statistical results based on two reduced models of the model in (1.26)*

2. The model in Figure 1.39(a) is a better model, in a statistical sense, since the effect of the interaction $Z1^*Z2$ on each Y_i, $i = 1,2$ and *3* is significant, based on *t*-statistics greater than 2.6. Based on this model, the following conclusions are derived.

   2.1 The data supports the hypothesis stated that the effect of *Z1 (Z2)* on each Y_i depends on *Z2 (Z1)*.

   2.2 An disadvantage of this model is *Z1* has a negative adjusted effect on each Y_i, in fact *Z1* and Y_i are significantly positive correlated, which shows the unexpected impact of the multicollinearity between the independent variables, specifically between *Z1* and $Z1^*Z2$.

   2.3 Furthermore, since *Z1* has insignificant adjusted effect on each Y_i, based on such a small *t*-statistics, then *Z1* could be deleted. Try it as an exercise.

3. On the other hand, based on the results in Figure 1.39(b) we draw the following conclusions.

   3.1 Since the interaction $Z1^*Z2$ has a significant adjusted effect on Y_1, it cannot be deleted from the VAR model.

   3.2 Since *Z2* has an insignificant adjusted effect on each Y_i, based on such a small *t*-statistics, then *Z2* could be deleted. The reduced model obtained would be the same as the reduced model by deleting *Z1* from the model in Figure 1.39(a). We find the final reduced model can be considered the best fit, conditional for the data used.

Example 1.28   Additional analyses for a VAR model

As an illustration, the VAR model in Figure 1.39(a) will be referred to. EViews provides so many alternative options for doing additional analyses for a VAR model. By selecting *View/Residuals Tests*, the options in Figure 1.40(a) shown on the screen, and Figure 1.40(b) obtained by selecting *View/Lag Structure*. However, only several analyses will be demonstrated, such as follows:

1. *Residual Analysis*

   1.1 *Residual Autocorrelation Tests*

   Figure 1.41 presents the two statistics for testing the residual autocorrelation, which shows the null hypothesis, no residual autocorrelation up to lag 4, is accepted. As a result, the VAR model does not have the autocorrelation problem.

   1.2 *Basic Assumptions of Residuals*

   Figure 1.42 shows that the null hypothesis, residuals are multivariate normal, is accepted. So it can be concluded that the data supports a basic assumption of the residuals. The other assumption is the

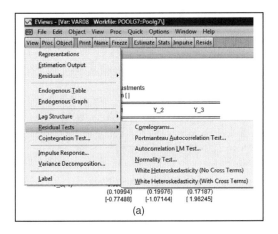
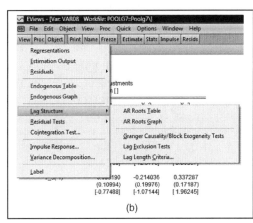

(a)   (b)

Figure 1.40 *Options for residual and lag structure, using EViews 6 or 7 Beta*

VAR Residual Portmanteau Tests for Autocorrelations
Null Hypothesis: no residual autocorrelations up to lag h
Date: 11/19/09   Time: 09:42
Sample: 1950 1992
Included observations: 42

Lags	Q-Stat	Prob.	Adj Q-Stat	Prob.	df
1	3.150951	NA*	3.227804	NA*	NA*
2	14.16143	0.5867	14.78880	0.5402	16
3	23.67465	0.5382	25.03382	0.4605	25
4	34.52254	0.4428	37.02358	0.3312	34

*The test is valid only for lags larger than the VAR lag order.
df is degrees of freedom for (approximate) chi-square distribution
*df and Prob. may not be valid for models with exogenous variables

VAR Residual Serial Correlation LM T...
Null Hypothesis: no serial correlation ...
Date: 11/19/09   Time: 09:37
Sample: 1950 1992
Included observations: 42

Lags	LM-Stat	Prob
1	5.867162	0.7531
2	12.27446	0.1983
3	8.502711	0.4844
4	10.25424	0.3303

Probs from chi-square with 9 df.

Figure 1.41   *The residual autocorrelation tests for the VAR model in Figure 1.39(a)*

VAR Residual Normality Tests
Orthogonalization: Cholesky (Lutkepohl)
Null Hypothesis: residuals are multivariate normal
Date: 11/19/09   Time: 09:52
Sample: 1950 1992
Included observations: 42

Component	Skewness	Chi-sq	df	Prob.
1	0.274744	0.528389	1	0.4673
2	-0.235376	0.387813	1	0.5335
3	-0.156196	0.170780	1	0.6794
Joint		1.086983	3	0.7802

Component	Kurtosis	Chi-sq	df	Prob.
1	2.234073	1.026626	1	0.3110
2	4.403388	3.446619	1	0.0634
3	2.912283	0.013465	1	0.9076
Joint		4.486710	3	0.2135

Component	Jarque-Bera	df	Prob.
1	1.555016	2	0.4595
2	3.834432	2	0.1470
3	0.184245	2	0.9120
Joint	5.573693	6	0.4726

Figure 1.42   *The residual normality tests for the VAR model in Figure 1.39(a)*

heterokedasticity of the residuals, which can easily be done by selecting the *Residual Tests/White hetero-kedasticity(*)*. Corresponding to the testing of the basic assumptions of the residuals, refer to the special notes and comments presented in Section 2.14.3 (Agung, 2009a).

2.   *The Lag Structure*

2.1   *The AR Roots*

By selecting *Lag Structure/AR Roots*, it is found that the three AR Roots are strictly less than one. Then we can conclude that the VAR satisfies the stability condition.

2.2   *Granger Causality Tests*

By selecting *Lag Structure/Granger Causality/Block Exogeneity Wald Tests*, the results in Figure 1.43 are obtained for making conclusions of the corresponding tests. For example, Y_1 and Y_2 have significant Granger causalities with the p-values of 0.0009 and 0.0108, respectively, but Y_1 and Y_3 have insignificant Granger causalities with p-values of 0.4384 and 0.2108, respectively.

However, the three variables Y_1, Y_2 and Y_3 have significant Granger causalities with p-values of 0.0029, 0.0015 and 0.0019, respectively.

2.3   *The VAR Lag Exclusion Wald Tests*

Based on the results in Figure 1.44 we can conclude that the first lags $Y\_1(-1)$, $Y\_2(-1)$ and $Y\_3(-1)$ have significant joint effects on each of Y_1, Y_2 and Y_3, as well as on the trivariate *(Y_1,Y_2,Y_3)*.

VAR Granger Causality/Block Exogeneity Wald Tests
Date: 11/18/09 Time: 15:28
Sample: 1950 1992
Included observations: 42

Dependent variable: Y_1

Excluded	Chi-sq	df	Prob.
Y_2	11.07285	1	0.0009
Y_3	0.600436	1	0.4384
All	11.66778	2	0.0029

Dependent variable: Y_2

Excluded	Chi-sq	df	Prob.
Y_1	6.493032	1	0.0108
Y_3	1.147988	1	0.2840
All	12.99838	2	0.0015

Dependent variable: Y_3

Excluded	Chi-sq	df	Prob.
Y_1	1.566190	1	0.2108
Y_2	11.51633	1	0.0007
All	12.56475	2	0.0019

Figure 1.43 *Statistical results for the VAR Granger causality tests*

VAR Lag Exclusion Wald Tests
Date: 11/19/09 Time: 15:54
Sample: 1950 1992
Included observations: 42

Chi-squared test statistics for lag exclusion:
Numbers in [ ] are p-values

	Y_1	Y_2	Y_3	Joint
Lag 1	92.81195	187.7680	28.13310	293.2393
	[0.000000]	[0.000000]	[3.41e-06]	[0.000000]
df	3	3	3	9

Figure 1.44 *The VAR lag exclusion Wald tests*

However, in general, the joint effects of the exogenous variables of a VAR cannot be tested using the VAR model. For this reason, Agung (2009a) recommends applying the object *System*, instead of the VAR model, since by using the object *System*, each regression in the model can have a different set of independent variables, and various hypotheses can easily be tested using Wald tests.

2.4 *The Lag Order Selection Criteria*

By selecting *Lag Structure/Lag Length Criteria* . . . , and then insert the lags to include $= 2$ . . . *OK*, the results in Figure 1.45 are obtained. These results show that 1 (one) is the lag order selected by the five criteria. Therefore, we can conclude that the VAR model is best based on these five criteria.

VAR Lag Order Selection Criteria
Endogenous variables: Y_1 Y_2 Y_3
Exogenous variables: C T Z1 Z1*Z2
Date: 11/19/09 Time: 11:01
Sample: 1950 1992
Included observations: 41

Lag	LogL	LR	FPE	AIC	SC	HQ
0	291.6335	NA	2.40e-10	-13.64066	-13.13912	-13.45803
1	355.7553	106.3485*	1.64e-11*	-16.32953*	-15.45185*	-16.00992*
2	359.8311	6.163353	2.13e-11	-16.08932	-14.83549	-15.63275

* indicates lag order selected by the criterion
LR: sequential modified LR test statistic (each test at 5% level)
FPE: Final prediction error
AIC: Akaike information criterion
SC: Schwarz information criterion
HQ: Hannan-Quinn information criterion

Figure 1.45 *The VAR lag order selection criteria using two lags*

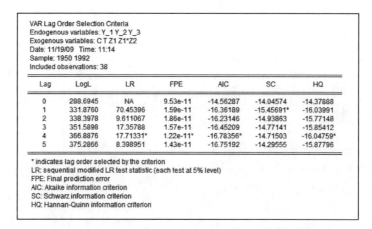

Figure 1.46   *The VAR lag order selection criteria using five lags*

However, we find that by inserting lags to include greater than 2, contradictory conclusions could be obtained. As an illustration, Figure 1.46 presents another result using five lags, which shows that 1 (one) is the lags' order selected by the SC criterion only. On the other hand, an error message is obtained, "*Near singular matrix*", using 10 lags. These contradictory findings lead to a great problem, since there are 9 (nine) alternatives results using 1–9 lags to consider. Based on the author's point of view, the simplest possible model should be the best selection. As an exercise, do the analysis based on a VAR model using the lags interval for the endogenous "4 4" and "1 4".

1.10.3.3 *Special Notes and Comments*

Corresponding to the environmental multivariate $Z_t = (Z1_t, \ldots, ZK_t)$, which has been defined or known to be the causal factor of the set of the endogenous variables Y_i_t, $i = 1, \ldots, N$ of the N-states (individuals), the following special notes and comments are made.

1. In theoretical sense, the variables Y_i_t, $i = 1, \ldots, N$ should be correlated variables. Therefore the whole set of $(N + K)$ variables, namely Y_i_t, $i = 1, \ldots, N$, and Zk, $k = 1, \ldots, K$, can be viewed as single time-series data containing $(N + K)$ variables.
2. For a small number of $(N + K)$, say 2–5, then all models presented in Agung (2009a) and Section 1.8, should be applicable. Following the step-by-step methods presented in Agung (2009a), everyone should have no difficulty in doing the data analysis.
3. On the other hand for a large N, reducing the dimension is recommended using the following alternative methods.
   3.1 To defined groups of states (individuals), using either the judgmental method or cluster analysis, then the groups' statistics, such as the means and SDs, can be considered as the derived time series for further time-series data analysis.
   3.2 To reduce the dimension using factor analysis. Then the time series latent variables models would be applied. Refer to Chapter 10 in Agung (2011).
4. Similarly, for a large K of the environmental multivariate. However, note that some of its components might not be correlated, in a theoretical sense.
5. Furthermore, the environmental variables can be dummy variables of the time periods, thereby piecewise time-series models should be applied.

1.11   Special Piece-Wise Models

As an illustration, the panel data used are the data of daily stock prices of 15 individuals (agencies or industries) consisting of eight banks and seven mining companies, used by one of the author's advisories, namely Mulia (2010), for her thesis. The symbols B_ and M_ respectively, are used to identify the stock prices (*SP*) of the banks and mining companies. Furthermore, we see there are two break time points, which represent the time points of the *auto-rejection regulations* or *price limitations* of the stock prices. The objectives of the analysis are to study the differences of the statistics, such as growth rates, variances (volatilities) and means of *SP*, between 15 days before and after each break point so that four time periods to be considered in the analyses. For a better graphical presentation of the statistical results, the break points are set at *Day* = 0 and *Day* = 40, so that the growth curves of each individual stock price (*SP*) are not very far apart. See the following examples.

1.11.1   The Application of Growth Models

For a preliminary information of the data set, and further data analysis, Figure 1.47(a) and (b) present the scatter graphs of *(Mean_Bank, Day)* and *(Mean_Mining,Day)* with their *Nearest Neighbor Fit Curves*. The individual time series, namely B_ and M_, can easily be presented. Try it as an exercise.

Example 1.29   A four-piece classical growth model
Figure 1.48 presents the statistical results based on a four-piece classical growth model of the mean stock prices of eight banks, namely *Mean_Bank*, using four dummy variables, namely *D1, D2, D3*, and *D4*. Based on these statistical results, the following notes and conclusions are made.

1.   The regression in Figure 1.48 represents four classical growth functions, as follows:

$$log(Mean\_Bank) = 7.089238 - 0.030904^*Day, \ for \ Period = 1$$
$$log(Mean\_Bank) = 7.463123 - 0.020809^*Day, \ for \ Period = 2$$
$$log(Mean\_Bank) = 7.401909 + 0.002723^*Day, \ for \ Period = 3$$
$$log(Mean\_Bank) = 7.724826 - 0.006886^*Day, \ for \ Period = 4$$

2.   By using the Wald test, we discover that the growth rate 15 days before the *Day* = 0 is smaller than after *Day* = 0, based on the *t*-statistic of $t_0 = -2.355\,063$ with $df = 54$ and a *p*-value = 0.0222/2 = 0.0111,

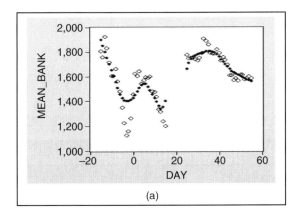
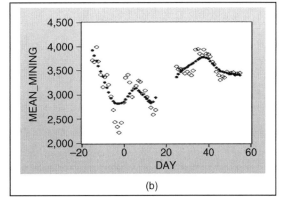

(a)   (b)

Figure 1.47   *Scatter graphs of (Mean_Bank, Day) and (Mean_Mining, Day) with their nearest neighbor fit curves*

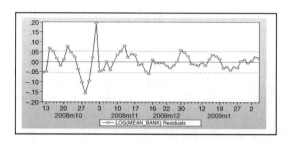

```
Dependent Variable: LOG(MEAN_BANK)
Method: Least Squares
Date: 12/06/09   Time: 15:31
Sample: 10/08/2008 2/04/2009
Included observations: 62
LOG(MEAN_BANK)=(C(10)+C(11)*DAY)*D1+(C(20)+C(21)*DAY)*D2
   +(C(30)+C(31)*DAY)*D3+(C(40)+C(41)*DAY)*D4
```

	Coefficient	Std. Error	t-Statistic	Prob.
C(10)	7.089238	0.025358	279.5657	0.0000
C(11)	-0.030904	0.002880	-10.72862	0.0000
C(20)	7.463123	0.028860	258.6009	0.0000
C(21)	-0.020809	0.003174	-6.555834	0.0000
C(30)	7.401909	0.094553	78.28350	0.0000
C(31)	0.002723	0.002880	0.945356	0.3487
C(40)	7.724826	0.152974	50.49751	0.0000
C(41)	-0.006886	0.003174	-2.169335	0.0345

R-squared	0.843170	Mean dependent var	7.376567
Adjusted R-squared	0.822840	S.D. dependent var	0.126189
S.E. of regression	0.053113	Akaike info criterion	-2.912860
Sum squared resid	0.152336	Schwarz criterion	-2.638391
Log likelihood	98.29865	Hannan-Quinn criter.	-2.805096
Durbin-Watson stat	1.161126		

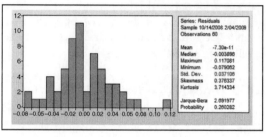

Figure 1.48   *Statistical results based on a four-piece growth model of* Mean_Bank

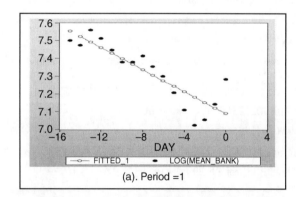

(a). Period =1

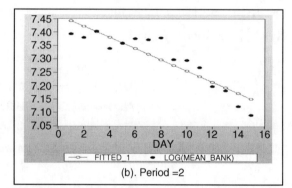

(b). Period =2

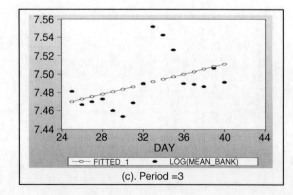

(c). Period =3

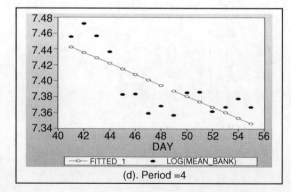

(d). Period =4

Figure 1.49   *Scatter graphs of the four-piece regression in Figure 1.48 and its fitted values, by time period*

and the growth rate 15 days before the $Day = 40$ is smaller than after $Day = 40$, based on the t-statistic of $t_0 = 2.241\,761$ with $df = 54$ and a p-value $= 0.0291/2 = 0.01\,455$. For a comparison, see the following example.

3. The residuals graph indicates that a nonlinear regression of $log(Mean\_Bank)$ on Day should be explored. On the other hand, Figure 1.49 clearly shows that a polynomial regression may be applied within each of the four time periods considered. See Example 1.30.

4. However, based on the $R^2 = 0.843\,170 > 80\%$, it could be concluded that the independent variables are good predictors for $log(Mean\_Bank)$.

5. Note that exactly the same analysis can easily be conducted based on the SP of each individual, as well as the $Mean\_Mining$.

Example 1.30   AR(2) four-piece growth model
By taking into account the autocorrelation of the classical growth model in Figure 1.49, Figure 1.50 presents the statistical results based on an AR(2) four-piece growth model, as a comparison.

Example 1.31   The nearest neighbor fit of *log(Mean_Bank)*
Figure 1.51 presents the scatter graph of *(log(Mean_Bank),Day)* with its *Nearest Neighbor Fit* by the time periods. The four graphs clearly show that nonlinear models should be applied within each time period. See the following example.

Example 1.32   A four-piece polynomial growth model
By using trial-and-error methods, statistical results are obtained based on a four-piece polynomial growth model presented in Figure 1.52. Based on these results, note the following:

1. Compared to the classical growth model in Figure 1.48 and the AR(2) growth model in Figure 1.49, this polynomial growth model has the largest value of $R^2 = 0.957\,224$. So, in a statistical sense, this model

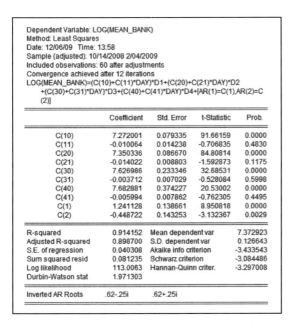
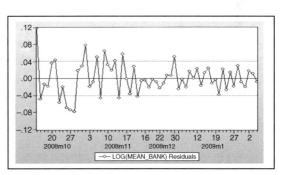
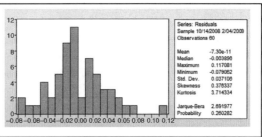

Figure 1.50   *Statistical results based on an AR(2) four-piece growth model of* Mean_Bank

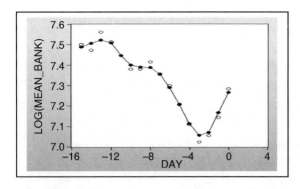
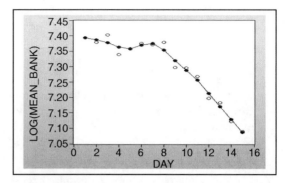
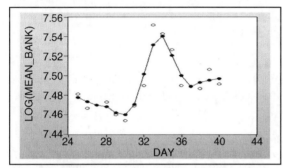
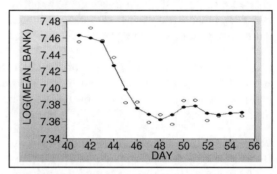

Figure 1.51   *Scatter graphs of (log(Mean_Bank),Day) with its nearest neighbor fit by time period*

```
Dependent Variable: LOG(MEAN_BANK)
Method: Least Squares
Date: 12/06/09   Time: 21:11
Sample: 10/08/2008 2/04/2009
Included observations: 62
LOG(MEAN_BANK)=(C(10)+C(11)*DAY+C(12)*DAY^2+C(13)*DAY^3+C(14)
    *DAY^4)*D1+(C(20)+C(21)*DAY+C(22)*DAY^2)*D2+(C(30)+C(31)*DAY
    +C(32)*DAY^2+C(33)*DAY^3+C(34)*DAY^4)*D3+(C(40)+C(42)*DAY^2
    +C(43)*DAY^3)*D4
```

	Coefficient	Std. Error	t-Statistic	Prob.
C(10)	7.273184	0.026987	269.5056	0.0000
C(11)	0.193225	0.026699	7.237188	0.0000
C(12)	0.055096	0.007600	7.249808	0.0000
C(13)	0.004689	0.000773	6.063391	0.0000
C(14)	0.000130	2.56E-05	5.081061	0.0000
C(20)	7.362563	0.026770	275.0267	0.0000
C(21)	0.014683	0.007699	1.907029	0.0628
C(22)	-0.002218	0.000468	-4.740573	0.0000
C(30)	46.35971	27.27787	1.699535	0.0960
C(31)	-4.792701	3.435079	-1.395223	0.1696
C(32)	0.218807	0.160933	1.359620	0.1806
C(33)	-0.004387	0.003325	-1.319543	0.1935
C(34)	3.26E-05	2.56E-05	1.276490	0.2082
C(40)	8.448153	0.361846	23.34736	0.0000
C(42)	-0.001244	0.000473	-2.628171	0.0116
C(43)	1.62E-05	6.53E-06	2.482523	0.0168

R-squared	0.957224	Mean dependent var	7.376567
Adjusted R-squared	0.943276	S.D. dependent var	0.126189
S.E. of regression	0.030054	Akaike info criterion	-3.953987
Sum squared resid	0.041550	Schwarz criterion	-3.405049
Log likelihood	138.5736	Hannan-Quinn criter.	-3.738460
Durbin-Watson stat	1.375620		

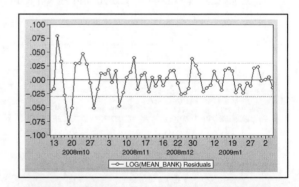
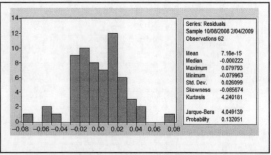

Figure 1.52   *Statistical results based on a four-piece polynomial growth model of* Mean_Bank

should be considered the best of the three growth models, even though it is a standard multiple regression. In other words, independent variables are the best predictors for *log(Mean_Bank)*.

2. The multiple regression in Figure 1.52 in fact represents the following four polynomial regressions within the four time periods, namely *Period* = 1, 2, 3 and 4, respectively.

$$\log(Mean\_Bank) = 7.273 + 0.193^*t + 0.055^*t^2 + 0.004689^*t^3 + 0.000130^*t^4$$
$$\log(Mean\_Bank) = 7.363 + 0.015^*t - 0.002218^*t^2$$
$$\log(Mean\_Bank) = 46.360 - 4.793^*t + 0.219^*t^2 - 0.004387^*t^3 + 3.3e - 05^*t^4$$
$$\log(Mean\_Bank) = 8.448 - 0.001^*t^2 + 1.6e - 05^*t^3$$

1.11.2   Equality Tests by Classifications

The option "*Equality Tests by Classifications*" provides the statistics for testing a hypothesis on the difference of the mean, median or variance of single variables between groups of individuals/objects generated by one or more classification or treatment factors.

Example 1.33   Test for equality of variances
As an illustration, Figure 1.53(a) and (b) presents the statistical results for testing the equality of variances of the variable *B_1* (*SP* for Bank-1), 15 days before and after the first and second break point, respectively, indicated by the *Period* = 1, 2, and *Period* = 3, 4. Based on these results, the following notes and conclusions are presented.

1. Based on the *F*-test, it can be concluded that the variances of *B_1* have significant differences between 15 days before and after each break point, namely at *Day* = 0 and *Day* = 40. Therefore, the *volatilities* of the *B_1s* stock prices before and after each break point have significant differences.
2. However, the Siegel–Tukey test should be questionable, since it has such a very small value compared to the others, specifically in Figure 1.53(b).

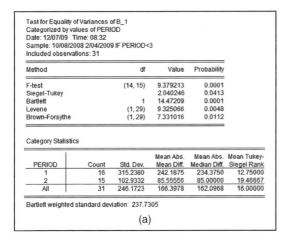

Figure 1.53   *Test for equality of the variances of* B_1, *15 days before and after two break points*

| Test for Equality of Means of B_1 |
| Categorized by values of PERIOD |
| Date: 12/07/09 Time: 08:39 |
| Sample: 10/08/2008 2/04/2009 IF PERIOD<3 |
| Included observations: 31 |

Method	df	Value	Probability
t-test	29	-1.780006	0.0856
Satterthwaite-Welch t-test*	18.35150	-1.828579	0.0838
Anova F-test	(1, 29)	3.168421	0.0856
Welch F-test*	(1, 18.3515)	3.343701	0.0838

*Test allows for unequal cell variances

Analysis of Variance

Source of Variation	df	Sum of Sq.	Mean Sq.
Between	1	179065.9	179065.9
Within	29	1638958.	56515.80
Total	30	1818024.	60600.81

Category Statistics

PERIOD	Count	Mean	Std. Dev.	Std. Err. of Mean
1	16	2631.250	315.2380	78.80950
2	15	2783.333	102.9332	26.57723
All	31	2704.839	246.1723	44.21385

(a)

| Test for Equality of Means of B_1 |
| Categorized by values of PERIOD |
| Date: 12/07/09 Time: 08:38 |
| Sample: 10/08/2008 2/04/2009 IF PERIOD>2 |
| Included observations: 31 |

Method	df	Value	Probability
t-test	29	7.525814	0.0000
Satterthwaite-Welch t-test*	21.00718	7.695656	0.0000
Anova F-test	(1, 29)	56.63788	0.0000
Welch F-test*	(1, 21.0072)	59.22312	0.0000

*Test allows for unequal cell variances

Analysis of Variance

Source of Variation	df	Sum of Sq.	Mean Sq.
Between	1	1086292.	1086292.
Within	29	556208.3	19179.60
Total	30	1642500.	54750.00

Category Statistics

PERIOD	Count	Mean	Std. Dev.	Std. Err. of Mean
3	16	3181.250	176.8945	44.22363
4	15	2806.667	78.75520	20.33450
All	31	3000.000	233.9872	42.02534

(b)

Figure 1.54   *Test for equality of the means of B_1, 15 days before and after two break points*

Example 1.34   Test for equality of means

In addition to the testing of variances presented in Figure 1.53(a) and (b), Figure 1.54(a) and (b) presents the statistical results for testing the equality of means of B_1, 15 days before and after the two break points, respectively. Corresponding to heterogeneity of the variances in a statistical sense, then the Welch F-test should be used to making the conclusion of the testing hypothesis on the means differences. In this case, however, the other tests also give exactly the same conclusion, at the significance level of either 5 or 10%.

On the other hand, the cell-mean model is not an appropriate time-series model generally – refer to Section 4.3.1 in Agung (2009a). So I cannot recommend conducting a test on the mean differences of a time series between long time periods: this is similarly so for testing equality of medians. We recommend the reader study and test their growth differences.

2

Data Analysis Based on Bivariate Time Series by States

2.1 Introduction

As an extension of the data in POOLG7wf1, in addition to the variable *GDP?*, many other variables could be observed or measured within seven states, such as the data of variables *M1?*, *PR?* and *SR?* similar to the data in DEMO.wf1, which have been used to demonstrate many time-series models. These include various continuous and discontinuous growth models, seemingly causal models, MLVAR(p,q) models, instrumental variables models, TGARCH(a,b,c), nonlinear and nonparametric models (Agung, 2009a).

Assuming that a new panel data of the seven states contains the four time-series variables *GDP, M1, PR*, and *RS*, then all models based on the DEMO.wf1 presented in Agung (2009a) should be applicable, and be good time-series models for each of the seven states. Therefore, there should be a set of various time-series models by the seven states.

To generalize, let's assume the *s*-th state has four variables *Y1_s*, *Y2_s*, *X1_s* and *X2_s* then many time-series models based on the four variables can be considered, which could easily be derived from all those presented in Agung (2009a), including models based on sub-sets of two, three and four variables as well as their lags, with or without time *t* as an independent variable. However, only selected models will be presented again in the following sections and chapters with their extensions.

Further generalization could be considered using a set of endogenous (impact, problem or down-stream) time-series variables, namely $Y\_i = (Y1\_i, \ldots, YG\_i)$ and exogenous (cause, source or up-stream) time-series variables or indicators, namely $X\_i = (X1\_i, \ldots, XK\_i)$ within the *i*-th individual (state, country, region, community, agency, household or person). However, this chapter presents only selected models based on the bivariate time series $(X\_i, Y\_i)_t = (X\_i_t, Y\_i_t)$, $i = 1, \ldots, N$; for a small number of states *N*. For illustration, hypothetical panel data is generated based on the data in POOL1.wf1, for $N = 4$, as follows: $X\_1 = ivmdep$, $X\_2 = ivmhoa$, $X\_3 = ivmmae$, $X\_4 = ivmmis$, $Y\_1 = ivmaut$, $Y\_2 = ivmbus$, $Y\_3 = ivmcon$ and $Y\_4 = ivmcst$.

Panel Data Analysis Using EViews, First Edition. I Gusti Ngurah Agung.
© 2014 John Wiley & Sons, Ltd. Published 2014 by John Wiley & Sons, Ltd.
Companion website: www.wiley.com/go/panel_data

2.2   Models Based on Independent States

Based on the hypothetical bivariate time series by states or individuals, namely (Y_i,X_i) as well as true data sets, many time series models can easily be developed. Some selected models will be presented in the following sections under the assumption that variables within a state are independent or uncorrelated with those in other states, which will be described as independent or uncorrelated states.

2.2.1   MAR(p) Growth Model with an Exogenous Variable

A MAR(p)_GM with an exogenous variable has the following general equation.

$$\log(Y\_i_t) = c(i0) + C(i1)^* t + C(i2)^* X\_i_{t-m} + \mu_{it}$$
$$\mu_{it} = \rho_{1i}\mu_{i(t-1)} + \ldots + \rho_{pi}\mu_{i(t-p)} + \varepsilon_{it} \tag{2.1}$$

for $m=0$ or a selected value of $m>0$, which could depends on the individuals or states, say $m=m(i)$. Similarly for the autoregressive: $p=p(i)$. Note that the appropriate values of $p(i)$, for $i=1,\ldots,N$, should be selected using trial-and-error methods, which depend highly on the data used.

Note that X_i_{t-m}, in general, is not a pure cause factor of the dependent variable, but it is an upstream variable, covariate or predictor variable. This model can be considered to be a MAR(p) model with trend with an extension, is a MAR(p) model with time-related effects, as follows:

$$\log(Y\_i_t) = c(i0) + C(i1)^* t + C(i2)^* X\_i_{t-m} + C(i3)^* t^* X\_i_{t-m} + \mu_{it}$$
$$\mu_{it} = \rho_{1i}\mu_{i(t-1)} + \ldots + \rho_{pi}\mu_{i(t-p)} + \varepsilon_{it} \tag{2.2}$$

2.2.2   A General MAR(p) Model with an Exogenous Variable

The MAR(p)_GM with an exogenous variable will have the following general equation.

$$G_i(Y\_i_t) = F_i(t, X\_i, X\_i_{t-m}) + \mu_{it}$$
$$\mu_{it} = \rho_{1i}\mu_{i(t-1)} + \ldots + \rho_{pi}\mu_{i(t-p)} + \varepsilon_{it} \tag{2.3}$$

where $G_i(Y\_i_t)$ is a constant function of Y_i_t, such as $G_i(Y\_i_t)=Y\_i$, $log(Y\_i)$ or $log[(Y\_i-Li)/(Ui-Y\_i)]$ where Li and Ui are fixed lower and upper bounds of Y_i_t; $F_i(t,X\_i,X\_i_{t-m})$ is a function of t, X_i and X_i_{t-m} for some selected $m \geq 1$, with a finite number of parameters for each state or individual i.

2.2.2.1   *General MAR(p) Additive Model with an Exogenous Variable*

Corresponding to the models in (2.1), the general MAR(p) additive model in the time t and an exogenous variable can be presented as follows:

$$G_i(Y\_i_t) = F_{1i}(t) + F_{2i}(X\_i, X\_i_{t-m}) + \mu_{it}$$
$$\mu_{it} = \rho_{1i}\mu_{i(t-1)} + \ldots + \rho_{pi}\mu_{i(t-p)} + \varepsilon_{it} \tag{2.4}$$

where $F_{1i}(t)$ and $F_{2i}(X\_i, X\_i_{t-m})$, respectively, are functions of t and (X_i, X_i_{t-m}) for some selected $m \geq 1$, having finite numbers of parameters. For instance, the simplest or basic function of $F_{2i}(X\_i, X\_i_{t-m}) = C(2i1)^*X\_i + C(2i2)^*X\_i(-1)$, where $m = 1$, and for the function of $F_{1i}(t)$ refer to the special functions or models in (1.4a) to (1.4d).

Example 2.1 An AR(1)_GM of independent states

As an illustration, Figure 2.1 presents the statistical results based on an AR(**1**)_GM = AR(1,1)_GM with an exogenous variable, or an AR(**1**) model with trend, under the assumption the two states are independent and use a hypothetical data set. In other words the endogenous variables Y_1 and Y_2 are not correlated in a theoretical sense. Based on these results, the following notes are presented.

1. The symbol AR(**1**) =AR(1,1) indicates that both models in Figure 2.1 are AR(1) models.
2. Note that both regressions in fact have exactly the same dependent and independent variables. The results should be different. However, there are theoretical and empirical concepts to be considered, such as follows:
    2.1 The results show that $X\_?$ have significantly positive adjusted effect on $Y\_?$ in both regressions. On the other hand, $X\_1(-1)$ has an insignificant negative adjusted effect on Y_1, and $X\_2(-1)$ has a significant negative adjusted effect on Y_2. In fact, $X\_1(-1)$ and Y_1, as well as $X\_2(-1)$ and Y_2, are significantly positively correlated.
    2.2 A bivariate correlation of two variables X and Y and the adjusted effect of X on Y based on any multiple regressions could lead to contradictory conclusions if X is highly or significantly correlated with at least one of the other independent variables. In general, the impact of multi-collinearity is unexpected.
3. In this case, the regression of Y_2 could be considered an acceptable regression, in a statistical sense, since each of the independent variables has significant adjusted effects. On the other hand, the regression of Y_1 should be reduced by deleting either one of X_1 and $X\_1(-1)$. Do it as an exercise using your own data set.

```
Estimation Method: Seemingly Unrelated Regression
Date: 09/08/09  Time: 10:04
Sample: 1968M03 1994M10
Included observations: 321
Total system (balanced) observations 640
Iterate coefficients after one-step weighting matrix
Convergence achieved after: 1 weight matrix, 10 total coef iterations
```

	Coefficient	Std. Error	t-Statistic	Prob.
C(10)	8.857636	1.040570	8.512290	0.0000
C(11)	-0.000458	0.002607	-0.175644	0.8606
C(12)	6.74E-06	2.77E-06	2.435094	0.0152
C(13)	-1.35E-06	2.77E-06	-0.486825	0.6266
C(14)	0.991595	0.005515	179.7990	0.0000
C(20)	8.376390	0.740724	11.30838	0.0000
C(21)	0.009309	0.004103	2.268827	0.0236
C(22)	3.44E-05	4.08E-06	8.418793	0.0000
C(23)	-8.74E-06	4.08E-06	-2.143099	0.0325
C(24)	1.006167	0.004549	221.2024	0.0000

Determinant residual covariance		4.02E-08

Equation: LOG(Y_1)=C(10)+C(11)*T+C(12)*X_1+C(13)*X_1(-1)
+[AR(1)=C(14)]
Observations: 320

R-squared	0.998286	Mean dependent var	8.411218
Adjusted R-squared	0.998264	S.D. dependent var	0.388015
S.E. of regression	0.016165	Sum squared resid	0.082311
Durbin-Watson stat	1.969874		

Equation: LOG(Y_2)=C(20)+C(21)*T+C(22)*X_2+C(23)*X_2(-1)
+[AR(1)=C(24)]
Observations: 320

R-squared	0.999509	Mean dependent var	9.413448
Adjusted R-squared	0.999502	S.D. dependent var	0.571793
S.E. of regression	0.012756	Sum squared resid	0.051256
Durbin-Watson stat	1.626705		

Figure 2.1 *Statistical results based on an AR(**1**)_GM of two independent Y_1 and Y_2*

	Coefficient	Std. Error	t-Statistic	Prob.
C(10)	8.754515	0.796046	10.99749	0.0000
C(11)	-0.000329	0.002141	-0.153446	0.8781
C(12)	6.61E-06	2.68E-06	2.463325	0.0140
C(13)	0.990597	0.005262	188.2418	0.0000
C(20)	8.853773	2.695825	3.284253	0.0011
C(21)	0.011293	0.009062	1.246123	0.2132
C(22)	3.64E-05	4.02E-06	9.058263	0.0000
C(23)	-1.19E-05	4.00E-06	-2.980967	0.0030
C(24)	1.200882	0.054815	21.90783	0.0000
C(25)	-0.197255	0.055280	-3.568281	0.0004

Estimation Method: Seemingly Unrelated Regression
Date: 09/08/09 Time: 12:47
Sample: 1968M02 1994M10
Included observations: 322
Total system (unbalanced) observations 640
Iterate coefficients after one-step weighting matrix
Convergence achieved after: 1 weight matrix, 12 total coef iterations

Determinant residual covariance   3.91E-08

Equation: LOG(Y_1)=C(10)+C(11)*T+C(12)*X_1+[AR(1)=C(13)]
Observations: 321

R-squared	0.998299	Mean dependent var	8.408489
Adjusted R-squared	0.998283	S.D. dependent var	0.390481
S.E. of regression	0.016181	Sum squared resid	0.082994
Durbin-Watson stat	1.968824		

Equation: LOG(Y_2)=C(20)+C(21)*T+C(22)*X_2+C(23)*X_2(-1)
  +[AR(1)=C(24),AR(2)=C(25)]
Observations: 319

R-squared	0.999523	Mean dependent var	9.416848
Adjusted R-squared	0.999515	S.D. dependent var	0.569442
S.E. of regression	0.012542	Sum squared resid	0.049235
Durbin-Watson stat	2.046425		

Figure 2.2   *Statistical results based on an AR(1,2)_GM of two independent states*

Example 2.2   An AR(1,2)_GM of independent states

As a modification of the model in Figure 2.1, Figure 2.2 presents the statistical results based on an AR(1,2) _GM with an exogenous variable, or an AR(1,2) model with trend, under the assumption the two states are independent. Based on these results, the following notes are presented.

1. The symbol AR(1,2) indicates that the first regression in the model is an AR(1) model, and the second is an AR(2) model.
2. Even though the time t has insignificant effect Y_1 with such a large p-value $= 0.8781$, it is still kept in the model in order to present an AR(1) model with trend, with the conclusion that t has an insignificant negative effect on Y_1 with a p-value $= 0.8781/2 = 0.43\,415$. If t is deleted, then a model without trend will be obtained.
3. Note that the first regression is an AR(1)_GM of *log(Y_1)* with an exogenous variable *X_1*, and the second is an AR(2)_GM of *log(Y_2)* with two exogenous variables *X_2*, and *X_2(−1)*.

2.2.2.2   *Various Additive Models with Trend*

A lot of additive models with trend can easily be proposed or defined based on a bivariate time series *(X_i, Y_i)_t*, such as bivariate LVAR(*p,q*) models for various *p* and *q*, and instrumental variables models. Refer to alternative models presented in Chapter 1, and find the following illustrative examples.

Example 2.3   LV(1)_GM of independent states

Compared to the AR(**1**) model in Figure 2.1, Figure 2.3 presents the statistical results based on an LV(**1**) model with trend, which show that *X_?* and *X_?(−1)* have significant adjusted effects on *Y_?* (in both regressions). For this reason, this model should be considered the better one in a statistical sense.

Example 2.4   LVAR(1,1;1,1)_GM of independent states

As an extension of the AR(**1**) model in Figure 2.1 and LV(**1**) model in Figure 2.3, Figure 2.4 presents the statistical results based on an LVAR(**1,1**) = LVAR(1,1;1,1) model. Based on these results, note the following:

1. The indicator AR(1) in the regression of *Y_1* is insignificant with such a large p-value, so it should be deleted from the model.

System: SYS19
Estimation Method: Seemingly Unrelated Regression
Date: 09/08/09   Time: 16:33
Sample: 1968M02 1994M10
Included observations: 321
Total system (balanced) observations 642
Linear estimation after one-step weighting matrix

	Coefficient	Std. Error	t-Statistic	Prob.
C(10)	0.116050	0.047987	2.418382	0.0159
C(11)	4.08E-05	3.70E-05	1.105149	0.2695
C(12)	8.08E-06	2.86E-06	2.824296	0.0049
C(13)	-8.25E-06	2.89E-06	-2.850189	0.0045
C(14)	0.986147	0.006190	159.3211	0.0000
C(20)	-0.008936	0.038793	-0.230351	0.8179
C(21)	-0.000216	5.03E-05	-4.286009	0.0000
C(22)	3.29E-05	3.87E-06	8.506620	0.0000
C(23)	-3.03E-05	3.94E-06	-7.683124	0.0000
C(24)	0.999473	0.005000	199.8914	0.0000
Determinant residual covariance		3.93E-08		

Equation: LOG(Y_1)=C(10)+C(11)*T+C(12)*X_1+C(13)*X_1(-1)+C(14)
  *LOG(Y_1(-1))
Observations: 321

R-squared	0.998310	Mean dependent var	8.408489
Adjusted R-squared	0.998289	S.D. dependent var	0.390481
S.E. of regression	0.016154	Sum squared resid	0.082456
Durbin-Watson stat	1.965341		

Equation: LOG(Y_2)=C(20)+C(21)*T+C(22)*X_2+C(23)*X_2(-1)+C(24)
  *LOG(Y_2(-1))
Observations: 321

R-squared	0.999523	Mean dependent var	9.410050
Adjusted R-squared	0.999517	S.D. dependent var	0.574135
S.E. of regression	0.012613	Sum squared resid	0.050269
Durbin-Watson stat	1.748408		

Figure 2.3 *Statistical results based on a LV(1)_GM of two independent states*

System: SYS20
Estimation Method: Seemingly Unrelated Regression
Date: 09/07/09   Time: 15:46
Sample: 1968M03 1994M10
Included observations: 321
Total system (balanced) observations 640
Iterate coefficients after one-step weighting matrix
Convergence achieved after: 1 weight matrix, 11 total coef iterations

	Coefficient	Std. Error	t-Statistic	Prob.
C(10)	0.111887	0.049831	2.245347	0.0251
C(11)	4.32E-05	3.81E-05	1.133174	0.2576
C(12)	8.18E-06	2.88E-06	2.841491	0.0046
C(13)	-8.37E-06	2.91E-06	-2.874499	0.0042
C(14)	0.986633	0.006425	153.5661	0.0000
C(15)	0.029476	0.055895	0.527335	0.5981
C(20)	-0.000815	0.044599	-0.018278	0.9854
C(21)	-0.000217	5.76E-05	-3.777169	0.0002
C(22)	3.55E-05	3.99E-06	8.898043	0.0000
C(23)	-3.28E-05	4.06E-06	-8.072890	0.0000
C(24)	0.998405	0.005748	173.6958	0.0000
C(25)	0.135131	0.055650	2.428226	0.0155
Determinant residual covariance		3.85E-08		

Equation: LOG(Y_1)=C(10)+C(11)*T+C(12)*X_1+C(13)*X_1(-1)+C(14)
  *LOG(Y_1(-1))+[AR(1)=C(15)]
Observations: 320

R-squared	0.998297	Mean dependent var	8.411218
Adjusted R-squared	0.998270	S.D. dependent var	0.388015
S.E. of regression	0.016141	Sum squared resid	0.081805
Durbin-Watson stat	2.029292		

Equation: LOG(Y_2)=C(20)+C(21)*T+C(22)*X_2+C(23)*X_2(-1)+C(24)
  *LOG(Y_2(-1))+[AR(1)=C(25)]
Observations: 320

R-squared	0.999528	Mean dependent var	9.413448
Adjusted R-squared	0.999521	S.D. dependent var	0.571793
S.E. of regression	0.012518	Sum squared resid	0.049204
Durbin-Watson stat	2.040438		

Figure 2.4 *Statistical results based on a LVAR(1,1;1,1)_GM of two independent states*

2. If it is deleted, then an LVAR(1,1;0,1)_GM would be obtained, with $X\_?$ and $X\_?(-1)$ have significant adjusted effects on $Y\_?$ (in both regressions). Do it as an exercise.

2.2.2.3 General MAR(p) Model with Time-Related Effects

As an extension of the model in (2.4), the general MAR(p) model with time-related effects (TRE) can be presented as follows:

$$G_i(Y\_i_t) = F_{1i}(t) + F_{2i}(X\_i, X\_i_{t-m}) + t^* F_{3i}(X\_i, X\_i_{t-m}) + \mu_{it}$$

$$\mu_{it} = \rho_{1i}\mu_{i(t-1)} + \ldots + \rho_{pi}\mu_{i(t-p)} + \varepsilon_{it}$$

(2.5)

```
System: SYS22
Estimation Method: Seemingly Unrelated Regression
Date: 09/08/09   Time: 17:33
Sample: 1968M02 1994M10
Included observations: 321
Total system (unbalanced) observations 641
Iterate coefficients after one-step weighting matrix
Convergence achieved after: 1 weight matrix, 13 total coef iterations
```

	Coefficient	Std. Error	t-Statistic	Prob.
C(10)	0.102083	0.056908	1.793818	0.0733
C(11)	2.99E-05	5.71E-05	0.522852	0.6013
C(12)	8.35E-06	2.93E-06	2.852447	0.0045
C(13)	-8.65E-06	3.10E-06	-2.793571	0.0054
C(14)	0.988055	0.007627	129.5390	0.0000
C(15)	5.14E-10	2.29E-09	0.224182	0.8227
C(20)	0.202400	0.084877	2.384608	0.0174
C(21)	5.61E-05	0.000112	0.500549	0.6169
C(22)	3.55E-05	3.96E-06	8.974525	0.0000
C(23)	-3.04E-05	4.12E-06	-7.380888	0.0000
C(24)	0.970792	0.011380	85.30799	0.0000
C(25)	-8.96E-09	3.18E-09	-2.818363	0.0050
C(26)	0.128701	0.056791	2.266213	0.0238
Determinant residual covariance		3.78E-08		

Equation: LOG(Y_1)=C(10)+C(11)*T+C(12)*X_1+C(13)*X_1(-1)+C(14)
 *LOG(Y_1(-1))+C(15)*T*X_1

Observations: 321			
R-squared	0.998311	Mean dependent var	8.408489
Adjusted R-squared	0.998285	S.D. dependent var	0.390481
S.E. of regression	0.016173	Sum squared resid	0.082388
Durbin-Watson stat	1.969563		

Equation: LOG(Y_2)=C(20)+C(21)*T+C(22)*X_2+C(23)*X_2(-1)+C(24)
 *LOG(Y_2(-1))+C(25)*T*X_2+[AR(1)=C(26)]

Observations: 320			
R-squared	0.999542	Mean dependent var	9.413448
Adjusted R-squared	0.999533	S.D. dependent var	0.571793
S.E. of regression	0.012357	Sum squared resid	0.047795
Durbin-Watson stat	2.041695		

Figure 2.5 *Statistical results based on a LVAR(1,1;0,1)_GM with TRE, of two independent states*

where $t^*F_{3i}(X\_i, X\_i_{t-m})$ indicates the time-related effects of any selected functions of $F_{3i}(X\_i, X\_i_{t-m})$ on $log(Y\_i_t)$. Note that the function $F_{3i}(X\_i, X\_i_{t-m})$ could be $C(3i1)^*X\_i + \ldots + C(3ik)^*X\_i^k$; $C(3i1)^*log(X\_i) + C(3i1)^*log^2(X\_i) + \ldots + C(3i1)^*log^k(X\_i)$, for $k \geq 1$ and others. Refer to similar models in (1.4a) to (1.4d).

Example 2.5   A LVAR(1,1;0,1) model with TRE

As an extension of the model in Figure 2.4, Figure 2.5 presents the statistical results based on a LVAR (1,1;0,1) Model with TRE. Based on these results, note the following:

1. It can be concluded that the two-way interaction t^*X_2 has a significant negative adjusted effect on $log(Y\_2)$, since the null hypothesis H_0: $C(25) \leq 0$ is rejected based on the *t*-statistic of $t_0 = -2.818636$ with a *p*-value $= 0.0050/2 = 0.0025$. Therefore, the data supports that the LVAR(1,1) model with TRE is a good fit for $log(Y\_2)$. So this model with TRE should be acceptable, in both theoretical and statistical senses.

2. On the other hand, the two-way interaction t^*X_1 has an insignificant adjusted effect on $log(Y\_1)$, with a large *p*-value. Therefore, it can be concluded that data does not support the LVAR(1,1) model with TRE for $log(Y\_1)$. In this case, a reduced model should be explored. The problem is, which independent variable(s) should be deleted from the model? Do it as an exercise. Note the impact of the multicollinearity between the independent variables.

2.3   Time-Series Models Based on Two Correlated States

A pair of states will be defined as correlated if and only if the endogenous variables within the two states are correlated at least, including the possibility of having causal or up- and down-stream relationships. Therefore, based on the variables of any two states, namely (X_i, Y_i) and (X_j, Y_j), a SCM, multiple association model or structural equation model (SEM) should be defined for the four time-series variables X_i, X_j, Y_i, and Y_j. Therefore, a lot of possible theoretical SCMs could be obtained depending on the true empirical or

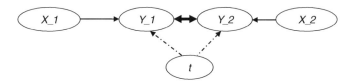

Figure 2.6 *The path diagram of a simultaneous SCM with trend of (X_1,Y_1), and (X_2,Y_2)*

observed variables. Some illustrative models will be presented in the following sections, under a restriction that only the endogenous variables are correlated. If the all endogenous and exogenous variables of the states considered are correlated, then the whole data set can be treated as a single time series data set. Hence, all models presented in Agung (2009a) will be applicable to the data set.

2.3.1   Analysis using the Object System

2.3.1.1   *The Simplest SCM with an Exogenous Variable*

Based on the variables *(X_1,Y_1)* and *(X_2,Y_2)*, under the assumption that *Y_1* and *Y_2* are correlated in a theoretical sense, the simplest SCM can be presented using the path diagram in Figure 2.6. This path diagram shows that *Y_1* and *Y_2* have simultaneous causal effects, indicated by thick double arrows. Note that the causal relationship between *Y_1* and *Y_2* could be a one-way causal relationship.

The simplest general model considered is as follows:

$$Y\_1_t = C(10) + C(11)^* t + C(12)^* Y\_2_{t-i_1} + C(13)^* X\_1_{t-j_1} + \mu 1_t$$
$$Y\_2_t = C(20) + C(21)^* t + C(22)^* Y\_1_{t-i_2} + C(23)^* X\_2_{t-j_2} + \mu 2_t$$

(2.6)

for subjectively selected set of integers i_1, j_1, i_2 and j_2. The simplest set is a set of zeros and/or ones. Two special cases are proposed in Agung (2009a) are $i_1=j_1=i_2=j_2=4$ for a quarterly time series, and $i_1=j_1=i_2=j_2=12$ for a monthly time series.

Note that this model is an additive model, where the first regression is an additive model in t, $Y\_2_{t-i_1}$, and $X\_1_{t-j_1}$, and the second regression is an additive in the time t, $Y\_1_{t-i_1}$, and $X\_2_{t-j_1}$. To generalize, the exogenous variable X_i_{t-k} could be replaced by any functions $F_i(X\_i_{t-k})$ having a finite number of parameters, using the endogenous variable $G_i(Y\_i)=Y\_i$, $log(Y\_i)$ or $log[(Y\_i\text{-}Li)/(Ui\text{-}Y\_i)$, where Li and Ui are the lower and upper bounds of Y_i. Therefore, there are a lot of alternative bivariate LVAR(p,q)_SCMs that could be subjectively defined or proposed by any researcher. Further extensions would use various instrumental variables models, as well as the models with time-related effects. However, researchers do not know which one is the true population or best fit model, since statistical results are highly dependent on the data set used – refer to special notes and comments presented in (Agung 2009a, Section 2.14).

Example 2.6   An AR(1,2)_SCM in (2.6)

For illustration, Figure 2.7 presents the statistical results based an AR(1,2)_SCM in (2.6), using the SUR estimation method. Based on these results, note the following:

1. This model assumed that *log(Y_1)* and *log(Y_2)* have simultaneous causal effects, indicated by their statuses as dependent as well as independent variables.
2. Note that, in the first regression, *log(Y_2)* has a positive significant adjusted effect on *log(Y_1)*, indicated by the parameter *C(11)*, and *log(Y_1)* also has a positive significant adjusted effect on *log(Y_2)* in the second regression. Therefore, it can be concluded that the data supports the hypothesis that *log(Y_1)* and *log(Y_2)* have simultaneous causal effects (*based on this model*).

Estimation Method: Seemingly Unrelated Regression
Date: 09/07/09 Time: 15:11
Sample: 1968M03 1994M10
Included observations: 321
Total system (unbalanced) observations 639
Iterate coefficients after one-step weighting matrix
Convergence achieved after: 1 weight matrix, 19 total coef iterations

	Coefficient	Std. Error	t-Statistic	Prob.
C(10)	4.008693	1.111908	3.605239	0.0003
C(11)	0.516824	0.063779	8.103354	0.0000
C(12)	-0.001330	0.002085	-0.637782	0.5238
C(13)	2.91E-06	2.71E-06	1.073554	0.2834
C(14)	-1.12E-06	2.58E-06	-0.431636	0.6662
C(15)	0.991699	0.007708	128.6667	0.0000
C(20)	65.01563	1213.840	0.053562	0.9573
C(21)	0.300320	0.042440	7.076283	0.0000
C(22)	-0.031712	0.372504	-0.085132	0.9322
C(23)	3.19E-05	4.11E-06	7.755354	0.0000
C(24)	-9.51E-06	3.89E-06	-2.442340	0.0149
C(25)	1.198033	0.052663	22.74889	0.0000
C(26)	-0.198544	0.053083	-3.740247	0.0002

Determinant residual covariance	2.56E-08

Equation: LOG(Y_1)=C(10)+C(11)*LOG(Y_2)+C(12)*T+C(13)*X_1+C(14)
  *X_1(-1)+[AR(1)=C(15)]

Observations: 320

R-squared	0.998328	Mean dependent var	8.411218
Adjusted R-squared	0.998301	S.D. dependent var	0.388015
S.E. of regression	0.015992	Sum squared resid	0.080307
Durbin-Watson stat	1.899769		

Equation: LOG(Y_2)=C(20)+C(21)*LOG(Y_1)+C(22)*T+C(23)*X_2+C(24)
  *X_2(-1)+[AR(1)=C(25),AR(2)=C(26)]

Observations: 319

R-squared	0.999513	Mean dependent var	9.416848
Adjusted R-squared	0.999504	S.D. dependent var	0.569442
S.E. of regression	0.012688	Sum squared resid	0.050229
Durbin-Watson stat	2.032616		

Figure 2.7   *Statistical results based on a AR(1,2)_SCM of the theoretical SCM in Figure 2.6*

3. Since three independent variables of the first regressions have insignificant adjusted effects, then a reduced model should be explored. Do it as an exercise.

2.3.1.2 *LVAR(p,q) SCM with an Exogenous Variable*

Under the assumption that Y_1 and Y_2 are correlated, then it could be derived that lags $Y\_1(-p)$ and $Y\_2(-p)$ for some values of p are the cause (source or up-stream) variables or factors for both Y_1 and Y_2. Figure 2.8 presents the simplest theoretical SCM, for $p = 1$, as a modification of the path diagram in Figure 2.7.

Based on this path diagram, the equation of the LVAR(*1,q*)_SCM will be as follows:

$$\begin{aligned}
Y\_1 = {}& C(10) + C(11)^* t + C(12)^* Y\_1(-1) + C(13)^* Y\_2(-1) \\
& + C(14)^* X\_1 + [AR(1) = C(15), \ldots, AR(q_1) = C(1, q_1 + 4)] \\
Y\_2 = {}& C(20) + C(21)^* t + C(22) Y\_1(-1) + C(23)^* Y\_2(-1) \\
& + C(24)^* X\_2 + [AR(1) = C(25), \ldots, AR(q_2) = C(2, q_2 + 4)]
\end{aligned}$$

(2.7)

Note that in this case, the exogenous (explanatory, source or up-stream) variable X_i is defined to be an independent variable of regression with dependent variable Y_i, for the *i*-th state or individual only. To generalize based on any data set, a LVAR(p,q)_model could be explored for each of the multiple regressions in the SCM using *trial-and-error methods* based on a particular data set.

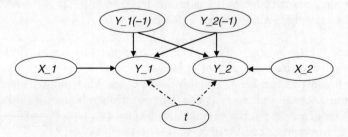

Figure 2.8   *The path diagram of a LV(1,1)_SCM with trend of (X_1,Y_1), and (X_2,Y_2)*

Furthermore, more advanced models could easily be defined using the endogenous variables $G_i(Y\_i) =$ $log(Y\_i)$ or $log[(Y\_i-Li)/(Ui-Y\_i)]$, where Li and Ui are the lower and upper bounds of Y_i, and various exogenous variables $F_i(X\_i)$ with a finite number of parameters.

Example 2.7 A LVAR(1,1)_SCM based on Figure 2.8

Figure 2.9 presents statistical results based a LVAR(**1,1**)_SCM using the SUR estimation method. Based on these results, note the following:

1. In a theoretical sense, there is nothing wrong with the model in Figure 2.9. However, using highly correlated independent variables, an unexpected estimate of the parameters will certainly be obtained. Therefore, in this case, various reduced models could be obtained based on the results in Figure 2.9. Refer to Section 2.14.2 in Agung (2009a).
2. Note that AR(1) has an insignificant adjusted effect with a large *p*-value in both regressions. In a statistical sense, AR(1) should be deleted from both regressions, since the lags of the endogenous have been used as the independent variables. Figure 2.10(a) presents the statistical results based on a LV(1,1)_SCM.
3. However, the AR(1) and *Y_1(−1)* are highly correlated in the first regression, as well as AR(1) and *Y_2(−1)* in the second regression, so another type of reduced model can be obtained by deleting the first lags *Y_1(−1)* and *Y_2(−1)* respectively from the first and second regressions instead of AR(1). The statistical results in Figure 2.10(b) are obtained based on an AR(1,2)_SCM, using trial-and-error methods.
4. Therefore, corresponding to the last three models in Figures 2.9 and Figure 2.10(a) and (b), which one would you consider to be the best model? Note that the model in Figure 2.10(b) has the smallest SE of regression. Based on this model, the following notes and conclusions are presented.
   4.1 Even though *Y_2(−1)* has an insignificant adjusted effect on *Y_1*, and *Y_1(−1)* has significant adjusted effect on *Y_2*, this model can be considered to show that *Y_1* and *Y_2* are correlated, which could be simultaneous causal relationships.
   4.2 The null hypothesis H_0: $C(13) = C(22) = 0$ is rejected based on the Chi-square statistic of $\chi_0^2 = 10.54370$ with $df = 2$ and a *p*-value $= 0.0051$. It can then be concluded that the data supports the hypothesis that *Y_1* and *Y_2* have a type of simultaneous causal relationship. Compare this model with the model in Figure 2.7.

Estimation Method: Seemingly Unrelated Regression
Date: 09/12/09 Time: 19:09
Sample: 1968M03 1994M10
Included observations: 321
Total system (balanced) observations 640
Iterate coefficients after one-step weighting matrix
Convergence achieved after: 1 weight matrix, 12 total coef iterations

	Coefficient	Std. Error	t-Statistic	Prob.
C(10)	41.46566	21.21395	1.954642	0.0511
C(11)	0.155837	0.225467	0.691174	0.4897
C(12)	0.996801	0.009589	103.9510	0.0000
C(13)	-0.002401	0.003973	-0.604268	0.5459
C(14)	-0.000248	0.000991	-0.249697	0.8029
C(15)	0.005555	0.054676	0.101598	0.9191
C(20)	-392.8686	104.5017	-3.759446	0.0002
C(21)	-3.051528	0.895086	-3.409202	0.0007
C(22)	0.033809	0.025304	1.336103	0.1820
C(23)	0.975678	0.009386	103.9556	0.0000
C(24)	0.056723	0.014112	4.019369	0.0001
C(25)	0.004585	0.054660	0.083874	0.9332

Determinant residual covariance	3.12E+08

Equation: Y_1 = C(10)+C(11)*T+C(12)*Y_1(-1)+C(13)*Y_2(-1)+C(14)*X_1
  +[AR(1)=C(15)]
Observations: 320

R-squared	0.997090	Mean dependent var	4806.125
Adjusted R-squared	0.997044	S.D. dependent var	1558.433
S.E. of regression	84.73477	Sum squared resid	2254514.
Durbin-Watson stat	2.069071		

Equation: Y_2 = C(20)+C(21)*T+C(22)*Y_1(-1)+C(23)*Y_2(-1)+C(24)*X_2
  +[AR(1)=C(25)]
Observations: 320

R-squared	0.998891	Mean dependent var	14121.02
Adjusted R-squared	0.998873	S.D. dependent var	6586.377
S.E. of regression	221.0992	Sum squared resid	15349844
Durbin-Watson stat	2.001177		

Figure 2.9 *Statistical results based on a LVAR(**1,1**)_SCM in Figure 2.8*

System: SYS25 Estimation Method: Seemingly Unrelated Regression Date: 09/12/09 Time: 19:07 Sample: 1968M02 1994M10 Included observations: 321 Total system (balanced) observations 642 Linear estimation after one-step weighting matrix

	Coefficient	Std. Error	t-Statistic	Prob.
C(10)	42.17134	20.83316	2.024241	0.0434
C(11)	0.148539	0.223172	0.665581	0.5059
C(12)	0.996703	0.009417	105.8382	0.0000
C(13)	-0.002363	0.003943	-0.599345	0.5492
C(14)	-0.000215	0.000982	-0.218635	0.8270
C(20)	-391.4929	103.7128	-3.774778	0.0002
C(21)	-3.033553	0.886955	-3.420190	0.0007
C(22)	0.034360	0.025091	1.369430	0.1714
C(23)	0.975745	0.009217	105.8676	0.0000
C(24)	0.056319	0.014000	4.022816	0.0001

Determinant residual covariance		3.11E+08

Equation: Y_1 = C(10)+C(11)*T+C(12)*Y_1(-1)+C(13)*Y_2(-1)+C(14)*X_1
Observations: 321

R-squared	0.997121	Mean dependent var	4796.988
Adjusted R-squared	0.997084	S.D. dependent var	1564.584
S.E. of regression	84.48175	Sum squared resid	2255345.
Durbin-Watson stat	2.057758		

Equation: Y_2 = C(20)+C(21)*T+C(22)*Y_1(-1)+C(23)*Y_2(-1)+C(24)*X_2
Observations: 321

R-squared	0.998898	Mean dependent var	14089.85
Adjusted R-squared	0.998884	S.D. dependent var	6599.741
S.E. of regression	220.4258	Sum squared resid	15353664
Durbin-Watson stat	1.991264		

(a)

System: SYS26 Estimation Method: Seemingly Unrelated Regression Date: 09/12/09 Time: 17:48 Sample: 1968M03 1994M10 Included observations: 321 Total system (unbalanced) observations 639 Iterate coefficients after one-step weighting matrix Convergence achieved after: 1 weight matrix, 20 total coef iterations

	Coefficient	Std. Error	t-Statistic	Prob.
C(10)	3824.137	1629.730	2.346486	0.0193
C(11)	5.032353	5.477925	0.918661	0.3586
C(13)	0.003608	0.020625	0.174927	0.8612
C(14)	0.036448	0.013615	2.677065	0.0076
C(15)	0.987317	0.007921	124.6446	0.0000
C(20)	7776.267	52950.80	0.146858	0.8833
C(21)	88.65012	135.4012	0.654722	0.5129
C(22)	-0.426327	0.131643	-3.238521	0.0013
C(24)	0.586803	0.065657	8.937460	0.0000
C(25)	1.174070	0.056546	20.76300	0.0000
C(26)	-0.170687	0.057401	-2.973587	0.0031

Determinant residual covariance		2.68E+08

Equation: Y_1 = C(10)+C(11)*T+C(13)*Y_2(-1)+C(14)*X_1+[AR(1)=C(15)]
Observations: 320

R-squared	0.997175	Mean dependent var	4806.125
Adjusted R-squared	0.997139	S.D. dependent var	1558.433
S.E. of regression	83.35231	Sum squared resid	2188496.
Durbin-Watson stat	2.035600		

Equation: Y_2 = C(20)+C(21)*T+C(22)*Y_1(-1)+C(24)*X_2+[AR(1)=C(25),A
  R(2)=C(26)]
Observations: 319

R-squared	0.999069	Mean dependent var	14152.30
Adjusted R-squared	0.999054	S.D. dependent var	6572.867
S.E. of regression	202.1966	Sum squared resid	12796525
Durbin-Watson stat	2.041411		

(b)

Figure 2.10   *Statistical results based on (a) a LV(1,1)_SCM, and (b) an AR(1,2)_SCM*

4.3   Otherwise, if $Y\_2(-1)$ should be deleted then another type of SCM would be obtained, where only the second regression shows a type of causal relationship between Y_1 and Y_2.

2.3.1.3   *Advanced LVAR_SCM with an Exogenous Variables*

As an illustration and as an extension of the model in (2.2), the path diagram in Figure 2.11 presents the following characteristics:

1. The recent time series Y_1 and Y_2 are defined as having simultaneous causal relationships.
2. The first lags $Y\_1(-1)$ and $Y\_2(-1)$ are defined to be the up-stream variables for the both recent variables Y_1 and Y_2.
3. The exogenous variables X_i and $X\_i(-1)$ are the explanatory (source, up-stream or cause) variables only for Y_i. Note that the broken line from $X\_i(-1)$ to Y_i indicates that $X\_i(-1)$ does not have direct effect on Y_i, if X_i is used as an independent variable.
4. The first lag $X\_i(-1)$ should be the up-stream variable of X_i. Note that in some cases only one of X_i and $X\_i(-1)$ may be used as an explanatory or cause factor. If a researcher has to choose one out of the two variables, then it is suggested to use $X\_i(-1)$, since it can be guaranteed that $X\_i(-1)$ is a cause or source factor of Y_i.

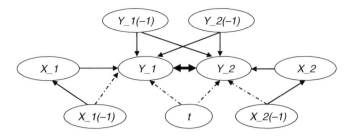

Figure 2.11   *The path diagram of the system specification in (2.8)*

Based on this path diagram, an LVAR(**1,1**)_SCM would have the following system equation.

$$Y\_1 = C(10) + C(11)^*t + C(12)^*Y\_1(-1) + C(13)^*Y\_2 + C(14)^*Y\_2(-1)$$
$$\quad + C(15)^*X\_1 + C(16)^*X\_1(-1) + [AR(1) = C(17)]$$
$$Y\_2 = C(20) + C(21)^*t + C(22)^*Y\_1 + C(23)^*Y\_1(-1) + C(24)^*Y\_2(-1)$$
$$\quad + C(25)^*X\_2 + C(26)^*X\_2(-1) + [AR(1) = C(27)]$$
$$X\_1 = C(30) + C(31)^*X\_1(-1) + [AR(1) = C(32)]$$
$$X\_2 = C(40) + C(41)^*X\_2(-1) + [AR(1) = C(42)]$$

(2.8)

Furthermore, note that many other SCMs can easily be defined based on a set of five variables: *t, X_1, X_2, Y_1* and *Y_2*, some of which will be presented in the following examples.

Example 2.8   An LVAR(1,1)_SCM in (2.8)

Figure 2.12 presents the statistical results based on the model in (2.8). Based on these results, the following notes and conclusions are derived.

Estimation Method: Seemingly Unrelated Regression
Date: 09/09/09   Time: 07:09
Sample: 1968M03 1994M10
Included observations: 321
Total system (balanced) observations 1280
Iterate coefficients after one-step weighting matrix
Convergence achieved after: 1 weight matrix, 14 total coef iterations

	Coefficient	Std. Error	t-Statistic	Prob.
C(10)	48.09809	20.50425	2.345762	0.0191
C(11)	0.150859	0.232456	0.648981	0.5165
C(12)	0.987143	0.009146	107.9345	0.0000
C(13)	0.160056	0.020437	7.831851	0.0000
C(14)	-0.161355	0.020166	-8.001306	0.0000
C(15)	0.014919	0.015161	0.984039	0.3253
C(16)	-0.014241	0.015106	-0.942694	0.3460
C(17)	-0.005723	0.054696	-0.104636	0.9167
C(21)	0.060744	0.026275	2.311877	0.0209
C(22)	0.992405	0.003411	290.9215	0.0000
C(23)	0.233387	0.053508	4.361716	0.0000
C(30)	-267.4423	102.5820	-2.607106	0.0092
C(31)	-1.625205	0.874189	-1.859099	0.0632
C(32)	1.014750	0.126603	8.015229	0.0000
C(33)	-0.962844	0.127811	-7.533321	0.0000
C(34)	0.984103	0.009564	102.9000	0.0000
C(35)	0.441694	0.062907	7.021432	0.0000
C(36)	-0.415804	0.064162	-6.477423	0.0000
C(37)	0.106753	0.053718	1.987292	0.0471
C(40)	-105025.1	330272.4	-0.317995	0.7505
C(41)	0.216235	0.119201	1.814029	0.0699
C(42)	-0.027622	0.030037	-0.919615	0.3580
C(43)	1.000573	0.001522	657.3921	0.0000

Determinant residual covariance	5.13E+17

Equation: Y_1 = C(10)+C(11)*T+C(12)*Y_1(-1)+C(13)*Y_2+C(14)*Y_2(-1)
+C(15)*X_1+C(16)*X_1(-1)+[AR(1)=C(17)]
Observations: 320

R-squared	0.997301	Mean dependent var	4806.125
Adjusted R-squared	0.997240	S.D. dependent var	1558.433
S.E. of regression	81.86856	Sum squared resid	2091168.
Durbin-Watson stat	1.952221		

Equation: X_1 = C(20)+C(21)*Y_1(-1)+C(22)*X_1(-1)+[AR(1)=C(23)]
Observations: 320

R-squared	0.999270	Mean dependent var	16880.57
Adjusted R-squared	0.999263	S.D. dependent var	11987.86
S.E. of regression	325.4529	Sum squared resid	33470599
Durbin-Watson stat	2.121665		

Equation: Y_2 = C(30)+C(31)*T+C(32)*Y_1(-1)+C(33)*Y_1(-1)+C(34)*Y_2(-1)
+C(35)*X_2+C(36)*X_2(-1)+[AR(1)=C(37)]
Observations: 320

R-squared	0.999080	Mean dependent var	14121.02
Adjusted R-squared	0.999060	S.D. dependent var	6586.377
S.E. of regression	201.9665	Sum squared resid	12726630
Durbin-Watson stat	1.998193		

Equation: X_2 = C(40)+C(41)*Y_1(-1)+C(42)*X_1(-1)+[AR(1)=C(43)]
Observations: 320

R-squared	0.999271	Mean dependent var	19793.87
Adjusted R-squared	0.999264	S.D. dependent var	6677.297
S.E. of regression	181.1238	Sum squared resid	10366647
Durbin-Watson stat	1.438998		

Figure 2.12   *Statistical results based on the LVAR(1,1)_SCM in (2.8)*

1. Each regression has a sufficient DW-statistic value indicating that there is no first autocorrelation problem. Therefore the model can be considered to be a good time-series model corresponding to the path diagram in Figure 2.11. However, it may have other limitations. Do the residual analysis as an exercise.
2. The parameter $C(17)$ has such a large p-value $= 0.7616$, that in a statistical sense the indicator AR(1) can be deleted from the regression. Try it as an exercise.

2.3.2   Two-SLS Instrumental Variables Models

Based on the nine variables in Figure 2.12, alternative instrumental variables models can be easily defined. Refer to Chapter 7 in Agung (2009a), for special notes on various instrumental variables models. Then refer to the following additional illustrations.

Example 2.9   LV(1) instrumental variables models

Figure 2.13 presents the statistical results based on two alternative 2SLS instrumental variables models using the nine variables in Figure 2.12. Corresponding to the simultaneous causal effects of Y_1 and Y_2, the second model should be considered better, since at a significance level of $\alpha = 0.05$, Y_2 has a significant negative adjusted effect on Y_1 in the first regression, with a p-value $= 0.0516/2 = 0.0258$, and in the second regression Y_1 has a significant positive adjusted effect on Y_2, with a p-value $= 0.0006/2 = 0.0003$.

1. The results in Figure 2.13(a) are obtained using the following system specification. Based on this model, note the following:

$$Y\_1 = c(10) + c(11)^* t + c(12)^* Y\_1(-1) @ c\ Y\_2\ Y\_2(-1) X\_1\ X\_1(-1)$$
$$Y\_2 = c(20) + c(21)^* t + c(22)^* Y\_2(-1) @ c\ Y\_1\ Y\_1(-1) X\_2\ X\_2(-1)$$

(2.9)

System: SYS25
Estimation Method: Two-Stage Least Squares
Date: 09/14/09 Time: 14:04
Sample: 1968M02 1994M10
Included observations: 321
Total system (balanced) observations 642

	Coefficient	Std. Error	t-Statistic	Prob.
C(10)	-76.72293	48.63708	-1.577458	0.1152
C(11)	-0.926096	0.361540	-2.561532	0.0107
C(12)	1.050263	0.022036	47.66144	0.0000
C(20)	333.3281	85.89623	3.880591	0.0001
C(21)	7.023509	2.149346	3.267743	0.0011
C(22)	0.899019	0.030433	29.54101	0.0000

Determinant residual covariance	6.01E+08

Equation: Y_1=C(10)+C(11)*T+C(12)*Y_1(-1)
Instruments: C Y_2 Y_2(-1) X_1 X_1(-1)
Observations: 321

R-squared	0.996698	Mean dependent var	4796.988
Adjusted R-squared	0.996677	S.D. dependent var	1564.584
S.E. of regression	90.19407	Sum squared resid	2586920.
Durbin-Watson stat	1.895642		

Equation: Y_2=C(20)+C(21)*T+C(22)*Y_2(-1)
Instruments: C Y_1 Y_1(-1) X_2 X_2(-1)
Observations: 321

R-squared	0.998271	Mean dependent var	14089.85
Adjusted R-squared	0.998260	S.D. dependent var	6599.741
S.E. of regression	275.2970	Sum squared resid	24100721
Durbin-Watson stat	1.193279		

(a)

System: SYS26
Estimation Method: Two-Stage Least Squares
Date: 09/14/09 Time: 14:06
Sample: 1968M02 1994M10
Included observations: 321
Total system (balanced) observations 642

	Coefficient	Std. Error	t-Statistic	Prob.
C(10)	-38.11944	43.06360	-0.885189	0.3764
C(11)	-0.011322	0.005805	-1.950370	0.0516
C(12)	1.044177	0.025624	40.75022	0.0000
C(20)	-69.52328	53.56226	-1.297990	0.1948
C(21)	0.098199	0.028560	3.438330	0.0006
C(22)	0.975223	0.006879	141.7777	0.0000

Determinant residual covariance	3.54E+08

Equation: Y_1=C(10)+C(11)*Y_2+C(12)*Y_1(-1)
Instruments: C T Y_2(-1) X_1 X_1(-1)
Observations: 321

R-squared	0.996834	Mean dependent var	4796.988
Adjusted R-squared	0.996814	S.D. dependent var	1564.584
S.E. of regression	88.31375	Sum squared resid	2480183.
Durbin-Watson stat	1.987628		

Equation: Y_2=C(20)+C(21)*Y_1+C(22)*Y_2(-1)
Instruments: C T Y_1(-1) X_2 X_2(-1)
Observations: 321

R-squared	0.998843	Mean dependent var	14089.85
Adjusted R-squared	0.998836	S.D. dependent var	6599.741
S.E. of regression	225.1910	Sum squared resid	16126097
Durbin-Watson stat	1.903862		

(b)

Figure 2.13   *Statistical results based on two LV(1,1) instrumental variables models, using 2SLS*

1.1 Both models are LV(1)_GMs with t and $Y\_i(-1)$ as the independent variables, and each has a significant adjusted effect on corresponding dependent variables.

1.2 With the DW $= 1.193\,279$ for the second regression, a modified model should be explored to obtain a model with a DW around 2.0. Do it as an exercise.

1.3 The simultaneous causal effects of Y_1 and Y_2 are presented using the instrumental variables.

2. For a comparison, the results in Figure 2.13(b) are obtained using the following system specification. Based on this model, note the following:

$$
\begin{aligned}
Y\_1 &= c(10) + c(11)^*Y\_2 + c(12)^*Y\_1(-1) \ @ \ c \ t \ Y\_2(-1)X\_1 \ X\_1(-1) \\
Y\_2 &= c(20) + c(21)^*Y\_1 + c(22)^*Y\_2(-1) \ @ \ c \ t \ Y\_1(-1)X\_2 \ X\_2(-1)
\end{aligned}
\tag{2.10}
$$

2.1 The two equations show that Y_1 and Y_2 have simultaneous causal effects, and they are not growth models.

2.2 At the significance level $\alpha = 0.05$, the null hypothesis H_0: $C(11) \geq 0$, is rejected based on the t-statistic: $t_0 = -1.590\,370$ with a p-value $= 0.0516/2 = 0.0258 < 0.05$. It can then be concluded that Y_2 has a significant negative adjusted effect on Y_1. On the other hand, corresponding to the parameter $C(21)$, it can be concluded that Y_1 has a significant positive adjusted effect on Y_2, based on the t-statistic: $t_0 = 3.438\,330$ with a p-value $= 0.0006/2 = 0.0003$. These contradictory conclusions are the impacts of the other independent variables in each model.

Example 2.10 AR(1) instrumental variables models

As an alternative model to the one in Figure 2.13(b), Figure 2.14(a) presents the statistical results using the following system specification, with the estimation setting "*Add lagged regressors to instruments for linear equations with AR terms*," which shows the convergence achieved after 20 iterations. As a comparison, Figure 2.14(b) presents the output, which shows that convergence was not achieved after one weighted matrix

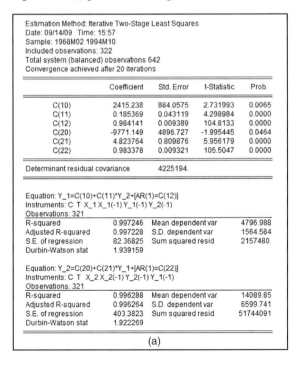

Figure 2.14 *Statistical results based on two AR(1,1) instrumental variables models, using 2SLS*

and 1000 total coefficient iterations. This output has demonstrated that unexpected statistics might be obtained using an instrumental variables model. See also the instrumental variables models presented in Chapter 14.

$$Y\_1 = c(10) + c(11)^* Y\_2 + [ar(1) = c(12)] @ c \, t \, X\_1 X\_1(-1)$$
$$Y\_2 = c(20) + c(21)^* Y\_1 + [ar(1) = C(22)] @ c \, t \, X\_2 X\_2(-1)$$

(2.11)

Example 2.11   LVAR(1,1) instrumental variables models

Figure 2.15(a) presents the statistical results based on a LVAR(1,1;1,1) instrumental variables model, using the following system specification, with the estimation setting "*Add lagged regressors to instruments for linear equations with AR terms*".

$$Y\_1 = c(10) + c(11)^* Y\_2 + c(12)^* Y\_1(-1) + [ar(1) = c(13)] @ c \, t \, X\_1 X\_1(-1)$$
$$Y\_2 = c(20) + c(21)^* Y\_1 + c(22)^* Y\_2(-1) + [ar(1) = C(23)] @ c \, t \, X\_2 X\_2(-1)$$

(2.12)

On the other hand, Figure 2.15(b) presents statistical results based on a reduced model, namely a LVAR (0,1;1,0) instrumental variables model using the following system specification. Note the first regression is an AR(1) model, and the second regression is a LV(1) model.

$$Y\_1 = c(10) + c(11)^* Y\_2 + [ar(1) = c(13)] @ c \, t \, X\_1 X\_1(-1)$$
$$Y\_2 = c(20) + c(21)^* Y\_1 + c(22)^* Y\_2(-1) @ c \, t \, X\_2 X\_2(-1)$$

(2.13)

(a)

System: SYS32
Estimation Method: Iterative Two-Stage Least Squares
Date: 09/15/09   Time: 11:20
Sample: 1968M03 1994M10
Included observations: 321
Total system (balanced) observations 640
Convergence achieved after 20 iterations

	Coefficient	Std. Error	t-Statistic	Prob.
C(10)	2304.403	985.3020	2.338778	0.0197
C(11)	0.184558	0.042539	4.338518	0.0000
C(12)	0.020412	0.056967	0.358318	0.7202
C(13)	0.984059	0.010131	97.13234	0.0000
C(20)	-39.64565	52.21966	-0.759209	0.4480
C(21)	0.075035	0.025031	2.997719	0.0028
C(22)	0.981017	0.005907	166.0699	0.0000
C(23)	0.038274	0.056455	0.677952	0.4981

Determinant residual covariance   3.20E+08

Equation: Y_1=C(10)+C(11)*Y_2+C(12)*Y_1(-1)+ [AR(1)=C(13)]
Instruments: C T X_1 X_1(-1) Y_1(-1) Y_2(-1) Y_1(-2)
Observations: 320

R-squared	0.997222	Mean dependent var	4806.125
Adjusted R-squared	0.997195	S.D. dependent var	1558.433
S.E. of regression	82.53143	Sum squared resid	2152414.
Durbin-Watson stat	1.976308		

Equation: Y_2=C(20)+C(21)*Y_1+C(22)*Y_2(-1)+[AR(1)=C(23)]
Instruments: C T X_2 X_2(-1) Y_2(-1) Y_1(-1) Y_2(-2)
Observations: 320

R-squared	0.998838	Mean dependent var	14121.02
Adjusted R-squared	0.998827	S.D. dependent var	6586.377
S.E. of regression	225.5662	Sum squared resid	16078118
Durbin-Watson stat	2.007981		

(b)

System: SYS33
Estimation Method: Iterative Two-Stage Least Squares
Date: 09/15/09   Time: 11:22
Sample: 1968M02 1994M10
Included observations: 322
Total system (balanced) observations 642
Convergence achieved after 13 iterations

	Coefficient	Std. Error	t-Statistic	Prob.
C(10)	2415.307	883.9578	2.732378	0.0065
C(11)	0.185366	0.043116	4.299234	0.0000
C(13)	0.984142	0.009390	104.8127	0.0000
C(20)	-1095.353	321.7815	-3.404028	0.0007
C(21)	0.739873	0.198632	3.724844	0.0002
C(22)	0.829009	0.045450	18.24013	0.0000

Determinant residual covariance   1.09E+09

Equation: Y_1=C(10)+C(11)*Y_2+ [AR(1)=C(13)]
Instruments: C T X_1 X_1(-1) Y_1(-1) Y_2(-1)
Observations: 321

R-squared	0.997246	Mean dependent var	4796.988
Adjusted R-squared	0.997228	S.D. dependent var	1564.584
S.E. of regression	82.36808	Sum squared resid	2157471.
Durbin-Watson stat	1.939163		

Equation: Y_2=C(20)+C(21)*Y_1+C(22)*Y_2(-1)
Instruments: C T X_2 X_2(-1)
Observations: 321

R-squared	0.996007	Mean dependent var	14089.85
Adjusted R-squared	0.995982	S.D. dependent var	6599.741
S.E. of regression	418.3310	Sum squared resid	55650253
Durbin-Watson stat	0.458511		

Figure 2.15 *Statistical results based on (a) a LVAR(1,1;1,1) instrumental variables model, and (b) a LVAR(0,1;1,0) instrumental variables model, using 2SLS*

Based on these results, note the following:

1. The model in Figure 2.15(a) is an uncommon model, since $Y\_1(-1)$ is used as independent as well as instrumental variables in the first regression. Likewise for $Y\_2(-1)$ in the second regression. Therefore, a modified model should be explored.
2. The reduced model in Figure 2.15(b) is obtained by deleting $Y\_1(-1)$ and $AR(1)$ from the first and second regressions, respectively.
3. Note that the second regression of the reduced model has a very small DW-statistic, so, in a statistical sense, a modified model should be explored. A $DW = 1.903\,862$ has been found by inserting an additional instrumental variable, namely $Y\_1(-1)$.

2.3.3 Three-SLS Instrumental Variables Models

There is no general rule or special guide for selecting the best possible set of instrumental variables for the 3SLS estimation method, even for the 2SLS instrumental variables models. So one may obtain good or acceptable statistical results based on a bad or unintentionally poor model, such as those in Figures 2.15(a) and Figure 2.15(b).

Example 2.12   3SLS instrumental variables models
As an illustration, Figure 2.16(a) presents the statistical results based on a standard instrumental variables model with a very small DW-statistic value. However, by using the following AR(1,1) model, the error

(a)

Estimation Method: Three-Stage Least Squares
Date: 09/15/09 Time: 13:13
Sample: 1968M02 1994M10
Included observations: 321
Total system (balanced) observations 642
Linear estimation after one-step weighting matrix

	Coefficient	Std. Error	t-Statistic	Prob.
C(10)	12331.65	987.3247	12.48997	0.0000
C(11)	321.6936	30.81305	10.44017	0.0000
C(12)	-4.233474	0.420810	-10.06031	0.0000
C(20)	1562.087	379.1307	4.120180	0.0000
C(21)	62.46779	2.657164	23.50920	0.0000
C(22)	0.501978	0.155993	3.217951	0.0014

Determinant residual covariance	7.26E+12

Equation: Y_1=C(10)+C(11)*T+C(12)*Y_2
Instruments: C Y_1(-1) Y_2(-1) X_1 @ C X_1(-1)
Observations: 321

R-squared	-18.521144	Mean dependent var	4796.988
Adjusted R-squared	-18.643918	S.D. dependent var	1564.584
S.E. of regression	6934.466	Sum squared resid	1.53E+10
Durbin-Watson stat	0.020968		

Equation: Y_2=C(20)+C(21)*T+C(22)*Y_1
Instruments: C Y_1(-1) Y_2(-1) X_2 @ C X_2(-1)
Observations: 321

R-squared	0.951604	Mean dependent var	14089.85
Adjusted R-squared	0.951300	S.D. dependent var	6599.741
S.E. of regression	1456.441	Sum squared resid	6.75E+08
Durbin-Watson stat	0.023118		

(b)

Estimation Method: Three-Stage Least Squares
Date: 09/15/09 Time: 14:19
Sample: 1968M02 1994M10
Included observations: 322
Total system (balanced) observations 642
Iterate coefficients after one-step weighting matrix
Convergence achieved after: 1 weight matrix, 3 total coef iterations

	Coefficient	Std. Error	t-Statistic	Prob.
C(10)	2899.994	2421.713	1.197497	0.2316
C(11)	-0.636914	7.090811	-0.089823	0.9285
C(12)	0.119475	0.139516	0.856353	0.3921
C(13)	0.035260	0.100715	0.350097	0.7264
C(14)	0.987790	0.011270	87.64593	0.0000
C(20)	-4460.508	5441.546	-0.819713	0.4127
C(21)	-54.66174	37.03895	-1.475791	0.1405
C(22)	0.678124	0.292211	2.320667	0.0206
C(23)	0.583591	2.069690	0.281970	0.7781
C(24)	0.981167	0.010727	91.46857	0.0000

Determinant residual covariance	3.21E+08

Equation: Y_1=C(10)+C(11)*T+C(12)*X1+C(13)*Y_2+[AR(1)=C(14)]
Instruments: C @ C Y_1(-1) T(-1) X1(-1) Y_2(-1)
Observations: 321

R-squared	0.997517	Mean dependent var	4796.988
Adjusted R-squared	0.997486	S.D. dependent var	1564.584
S.E. of regression	78.44698	Sum squared resid	1944641.
Durbin-Watson stat	2.047650		

Equation: Y_2=C(20)+C(21)*T+C(22)*X2+C(23)*Y_1+[AR(1)=C(24)]
Instruments: C @ C Y_2(-1) T(-1) X2(-1) Y_1(-1)
Observations: 321

R-squared	0.998769	Mean dependent var	14089.85
Adjusted R-squared	0.998753	S.D. dependent var	6599.741
S.E. of regression	233.0379	Sum squared resid	17160903
Durbin-Watson stat	1.927039		

Figure 2.16 *Statistical results based on (a) a standard instrumental variables model, and (b). An AR(1,1) instrumental variables model, using 3SLS*

message of "*Near Singular Matrix*" is obtained.

$$
\begin{aligned}
Y\_1 &= C(10) + C(11)^*T + C(12)^*Y\_2 + [AR(1) = C(14)] \\
&\quad @\,CY\_1(-1)\,Y\_2(-1)\,X\_1\,@\,CX\_1(-1) \\
Y\_2 &= C(20) + C(21)^*T + C(22)^*Y\_1 + [AR(1) = C(24)] \\
&\quad @\,CY\_1(-1)\,Y\_2(-1)\,X\_1\,@\,CX\_1(-1)
\end{aligned}
\tag{2.14}
$$

Therefore, by using trial-and-error methods, several estimable models could be obtained. As an illustration, unexpected statistical results are presented in Figure 2.16(b), using the following system specification, with the estimation setting "*Add lagged regressors to instruments for linear equations with AR terms*".

$$
\begin{aligned}
Y\_1 &= C(10) + C(11)^*T + C(12)^*X1 + C(13)^*Y\_2 + [AR(1) = C(14)]\,@\,C\,@\,C \\
Y\_2 &= C(20) + C(21)^*T + C(22)^*X2 + C(23)^*Y\_1 + [AR(1) = C(24)]\,@\,C\,@\,C
\end{aligned}
\tag{2.15}
$$

Based on the results in Figure 2.16(b) note the following:

1. Compare this to the system specification in (2.26), four instrumental variables are added based on the output, where one of them is the first lag: $T(-1)$, which has never been used in any time-series model, to my knowledge.
2. For this reason, this model should be considered inappropriate or unacceptable as a time-series model. See the following example.

Example 2.13   AR(1,1) 3SLS instrumental variables models

As an extension of the 2SLS instrumental variables model in (2.7), Figure 2.17 presents the statistical results using two alternative system specifications as follows:

$$
\begin{aligned}
Y\_1 &= c(10) + c(11)^*Y\_2 + [ar(1) = c(12)]\,@\,ct\,Y\_1(-1)\,Y\_2(-1)\,X\_1\,@\,cX\_1(-1) \\
Y\_2 &= c(20) + c(21)^*Y\_1 + [ar(1) = c(22)]\,@\,ct\,Y\_1(-1)\,Y\_2(-1)\,X\_2\,@\,CX\_2(-1)
\end{aligned}
\tag{2.16}
$$

$$
\begin{aligned}
Y\_1 &= c(10) + c(11)^*Y\_2 + [ar(1) = c(12)]\,@\,ct\,X\_1\,@\,cX\_1(-1) \\
Y\_2 &= c(20) + c(21)^*Y\_1 + [ar(1) = c(22)]\,@\,ct\,X\_2\,@\,cX\_2(-1)
\end{aligned}
\tag{2.17}
$$

Note that, even though Y_2 has an insignificant effect on Y_1 with such a large p-value, the null hypothesis H_0: $C(11) = C(21) = 0$ is rejected based on the Chi-square statistic of $\chi_0^2 = 17.38743$ with df $= 2$ and a p-value $= 0.0002$. Therefore, we can conclude that the data supports the hypothesis that Y_1 and Y_2 have simultaneous causal effects.

2.3.4   Analysis using the Object "VAR"

EViews provides an object "*VAR*" for conducting data analysis based on the VAR and VEC models. The VAR and VEC models are special cases of the MAR model, where all multiple regressions in the models have exactly the same set of lagged endogenous variables and the same set of exogenous variables. For this

	Coefficient	Std. Error	t-Statistic	Prob.

System: SYS29
Estimation Method: Three-Stage Least Squares
Date: 09/14/09   Time: 15:16
Sample: 1968M02 1994M10
Included observations: 321
Total system (balanced) observations 642
Linear estimation after one-step weighting matrix

	Coefficient	Std. Error	t-Statistic	Prob.
C(10)	33.02915	38.63857	0.854823	0.3930
C(11)	-0.060291	0.270934	-0.222530	0.8240
C(12)	0.997995	0.016848	59.23526	0.0000
C(20)	-130.4627	60.19223	-2.167435	0.0306
C(21)	-5.307504	1.310949	-4.048595	0.0001
C(22)	1.074382	0.018788	57.18432	0.0000

Determinant residual covariance   4.27E+08

Equation: Y_1=C(10)+C(11)*T+C(12)*Y_1(-1)
Instruments: C T Y_2 X_1 @ C Y_2(-1) X_1(-1)
Observations: 321

R-squared	0.997113	Mean dependent var	4796.988
Adjusted R-squared	0.997095	S.D. dependent var	1564.584
S.E. of regression	84.32644	Sum squared resid	2261282.
Durbin-Watson stat	2.058218		

Equation: Y_2=C(20)+C(21)*T+C(22)*Y_2(-1)
Instruments: C T Y_1 X_2 @ C Y_1(-1) X_2(-1)
Observations: 321

R-squared	0.998483	Mean dependent var	14089.85
Adjusted R-squared	0.998473	S.D. dependent var	6599.741
S.E. of regression	257.8674	Sum squared resid	21145607
Durbin-Watson stat	1.619996		

(a)

System: SYS35
Estimation Method: Three-Stage Least Squares
Date: 09/15/09   Time: 13:06
Sample: 1968M02 1994M10
Included observations: 322
Total system (balanced) observations 642
Iterate coefficients after one-step weighting matrix
Convergence achieved after: 1 weight matrix, 110 total coef iterations

	Coefficient	Std. Error	t-Statistic	Prob.
C(10)	6963.893	5131.864	1.356991	0.1753
C(11)	0.033597	0.138568	0.242461	0.8085
C(12)	0.995526	0.003216	309.5652	0.0000
C(20)	-11171.84	8262.461	-1.352120	0.1768
C(21)	5.060514	1.364370	3.709048	0.0002
C(22)	0.984598	0.009493	103.7207	0.0000

Determinant residual covariance   2.26E+08

Equation: Y_1=C(10)+C(11)*Y_2+[AR(1)=C(12)]
Instruments: C T X_1 @ C X_1(-1) Y_1(-1) Y_2(-1)
Observations: 321

R-squared	0.997243	Mean dependent var	4796.988
Adjusted R-squared	0.997226	S.D. dependent var	1564.584
S.E. of regression	82.40421	Sum squared resid	2159364.
Durbin-Watson stat	2.040785		

Equation: Y_2=C(20)+C(21)*Y_1+[AR(1)=C(22)]
Instruments: C T X_2 @ C X_2(-1) Y_2(-1) Y_1(-1)
Observations: 321

R-squared	0.995972	Mean dependent var	14089.85
Adjusted R-squared	0.995947	S.D. dependent var	6599.741
S.E. of regression	420.1734	Sum squared resid	56141534
Durbin-Watson stat	1.931551		

(b)

Figure 2.17 *Statistical results based on (a) the model in (2.12), and (b) the model in (2.13)*

reason, Agung (2009a) recommends using the object "*System*", instead of "*VAR*", since generally it is not appropriate that all multiple regressions of a set of endogenous variables by states or individuals have exactly the same set of independent (cause, source, up-stream or explanatory) variables. Another disadvantage in data analysis using the object VAR is that the Wald test cannot be applied. As an illustration, see the following examples using the object VAR based on the nine variables in Figure 2.12.

Example 2.14   A VAR(1,1) model

Corresponding to the variables in Figure 2.11, and comparing with the models presented in the previous examples, Figure 2.18 presents the statistical results based on a Basic VAR model of *(Y_1,Y_2)* with lag intervals for endogenous "1 1", and exogenous variables "*C t X_1 X_1(−1) X_2 X_2(−1)*". Based on these results, note the following.

1. Everyone can easily identify the differences between this VAR model and the SCMs presented in previous examples, as well as their path diagrams. Note that the path diagram of any VAR model does not present the simultaneous causal effects of *Y_1* and *Y_2* (refer to Figure 2.11, and all exogenous variables by states or individuals have direct effects on each of dependent variables.
2. However, in a statistical sense a VAR model should be an acceptable model with its own limitations. Similarly for the VEC models.
3. Exactly the same model can easily be applied using the object "*System*". Do it as an exercise.

Vector Autoregression Estimates
Date: 09/15/09   Time: 16:02
Sample (adjusted): 1968M02 1994M10
Included observations: 321 after adjustments
Standard errors in ( ) & t-statistics in [ ]

	Y_1	Y_2
Y_1(-1)	0.990596	0.033582
	(0.00948)	(0.02288)
	[104.451]	[1.46744]
Y_2(-1)	-0.006019	0.976526
	(0.00453)	(0.01093)
	[-1.32835]	[89.3204]
C	-4.091897	-223.7577
	(44.5803)	(107.572)
	[-0.09179]	[-2.08008]
T	-0.182553	-1.563567
	(0.40566)	(0.97884)
	[-0.45002]	[-1.59737]
X_1	0.037446	0.175073
	(0.01601)	(0.03864)
	[2.33854]	[4.53116]
X_1(-1)	-0.036389	-0.172175
	(0.01607)	(0.03877)
	[-2.26474]	[-4.44089]
X_2	0.115420	0.500372
	(0.02686)	(0.06481)
	[4.29695]	[7.72001]
X_2(-1)	-0.107863	-0.469598
	(0.02711)	(0.06541)
	[-3.97882]	[-7.17884]

Estimation Proc:
================================
LS 1 1 Y_1 Y_2 @ C T X_1 X_1(-1) X_2 X_2(-1)

VAR Model:
================================
Y_1 = C(1,1)*Y_1(-1) + C(1,2)*Y_2(-1) + C(1,3) + C(1,4)*T + C(1,5)*X_1 + C(1,6)*X_1(-1) + C(1,7)*X_2 + C(1,8)*X_2(-1)

Y_2 = C(2,1)*Y_1(-1) + C(2,2)*Y_2(-1) + C(2,3) + C(2,4)*T + C(2,5)*X_1 + C(2,6)*X_1(-1) + C(2,7)*X_2 + C(2,8)*X_2(-1)

VAR Model - Substituted Coefficients:
================================
Y_1 = 0.990596486328*Y_1(-1) - 0.00601854011252*Y_2(-1) - 4.0918971938 - 0.182553132236*T + 0.0374456151834*X_1 - 0.0363886251491*X_1(-1) + 0.115419904332*X_2 - 0.107862745861*X_2(-1)

Y_2 = 0.0335815268404*Y_1(-1) + 0.976526325225*Y_2(-1) - 223.757707252 - 1.56356736029*T + 0.175073173445*X_1 - 0.172175280098*X_1(-1) + 0.500371893905*X_2 - 0.469597559708*X_2(-1)

R-squared	0.997417	0.999155
Adj. R-squared	0.997359	0.999136
Sum sq. resids	2023282.	11780548
S.E. equation	80.40001	194.0040
F-statistic	17266.91	52858.80
Log likelihood	-1859.660	-2142.418
Akaike AIC	11.63651	13.39824
Schwarz SC	11.73050	13.49223
Mean dependent	4796.988	14089.85
S.D. dependent	1564.584	6599.741

Determinant resid covariance (dof adj.)	2.37E+08
Determinant resid covariance	2.25E+08
Log likelihood	-3997.966
Akaike information criterion	25.00913
Schwarz criterion	25.19712

Figure 2.18   *Statistical results based on a VAR(1,1) model*

2.4   Time-Series Models Based on Multiple Correlated States

As an extension of the time-series models based on two correlated states, this section presents the time-series models based on three correlated states, which could easily be extended to multiple correlated states. Find the following alternative models based on three bivariate time series *(X_1,Y_1)*, *(X_2,Y_2)* and *(X_3, Y_3)*, of three correlated states.

2.4.1   Extension of the Path Diagram in Figure 2.6

2.4.1.1   *A Simultaneous AR(1)_SCM with Trend*

As an extension of the simultaneous SCM presented in Figure 2.6, Figure 2.19 presents the path diagram of a simultaneous SCM with trend, based on *(X_1,Y_1)*, *(X_2,Y_2)* and *(X_3,Y_3)*.

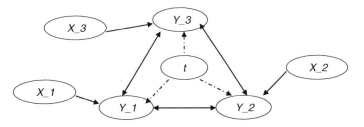

Figure 2.19 *The path diagram of a simultaneous SCM of (X_1,Y_1), (X_2,Y_2), and (X_3,Y_3)*

As an extension of the simplest model in (2.6), the simplest general model will be considered is as follows:

$$Y\_1_t = C(10) + C(11)t + C(12)Y\_2_{t-i_1} + C(13)Y\_3_{t-j_1} + C(14)X\_1_{t-k_1} + \mu1_t$$

$$Y\_2_t = C(20) + C(21)t + C(22)Y\_1_{t-i_2} + C(23)Y\_3_{t-j_2} + C(14)X\_2_{t-k_2} + \mu2_t \qquad (2.18)$$

$$Y\_3_t = C(30) + C(31)t + C(32)Y\_1_{t-i_3} + C(33)Y\_2_{t-j_3} + C(34)X\_3_{t-k_3} + \mu3_t$$

for a subjectively selected set of integers i_1, j_1, k_1, i_2, j_2, k_2, i_3, j_3 and k_3. Refer to the notes presented for the model in (2.6).

As a special case, for $i_1 = j_1 = i_2 = j_2 = i_3 = j_3 = 0$, the relationship between the endogenous variables Y_1, Y_2 and Y_3 is defined as *pair-wise correlated*. If Y_1, Y_2 and Y_3 are *completely correlated*, then each of the three regressions in (2.18) should have the interaction $Y\_2^*Y\_3$, $Y\_1^*Y\_3$ or $Y\_1^*Y\_2$, as an additional independent variable, respectively. However, in a theoretical sense, it is not an easy task to identify whether a set of three variables are pair-wise or completely correlated.

On the other hand, other types of correlations between Y_1, Y_2 and Y_3 should be considered, such as only one or two of the three pairs are defined to be correlated. In this case, the model can easily be derived from the model in (2.18) by deleting two or one of the corresponding endogenous independent variable(s).

2.4.1.2 *A Circular AR(1)_SCM with Trend*

As a modification of the path diagram in Figure 2.19, Figure 2.20 presents the path diagram of a special model, namely a *Circular SCM with trend*, based on *(X_1,Y_1)*, *(X_2,Y_2)* and *(X_3,Y_3)*. Note that a circular, seemingly causal relationship of a set of variables should be defined based on a strong theoretical base.

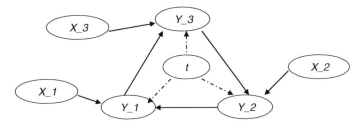

Figure 2.20 *The path diagram of a circular SCM of (X_1,Y_1), (X_2,Y_2) and (X_3,Y_3)*

```
Estimation Method: Seemingly Unrelated Regression
Date: 09/18/09   Time: 21:54
Sample: 1968M02 1994M10
Included observations: 322
Total system (balanced) observations 963
Iterate coefficients after one-step weighting matrix
Convergence achieved after: 1 weight matrix, 28 total coef iterations
```

	Coefficient	Std. Error	t-Statistic	Prob.
C(10)	2969.053	1215.426	2.442808	0.0148
C(11)	-2.010165	4.406429	-0.456189	0.6484
C(12)	0.159136	0.020733	7.675418	0.0000
C(13)	0.020272	0.013308	1.523240	0.1280
C(14)	0.985998	0.008024	122.8831	0.0000
C(20)	-4200.650	2490.372	-1.686756	0.0920
C(21)	-19.56644	10.47106	-1.868621	0.0620
C(22)	1.165492	0.059511	19.58439	0.0000
C(23)	0.111469	0.052062	2.141071	0.0325
C(24)	0.989461	0.008136	121.6172	0.0000
C(30)	1743.694	1265.108	1.378297	0.1684
C(31)	15.30655	5.664642	2.702122	0.0070
C(32)	0.568195	0.082048	6.925130	0.0000
C(33)	0.156739	0.011361	13.79673	0.0000
C(34)	0.985015	0.011678	84.34945	0.0000

| Determinant residual covariance | 1.43E+12 |

```
Equation: Y_1=C(10)+C(11)*T+C(12)*Y_2+C(13)*X_1+[AR(1)=C(14)]
Observations: 321
```

R-squared	0.997310	Mean dependent var	4796.988
Adjusted R-squared	0.997276	S.D. dependent var	1564.584
S.E. of regression	81.65719	Sum squared resid	2107056.
Durbin-Watson stat	1.944159		

```
Equation: Y_2=C(20)+C(21)*T+C(22)*Y_3+C(23)*X_2+[AR(1)=C(24)]
Observations: 321
```

R-squared	0.999400	Mean dependent var	14089.85
Adjusted R-squared	0.999392	S.D. dependent var	6599.741
S.E. of regression	162.6968	Sum squared resid	8364601.
Durbin-Watson stat	1.638043		

```
Equation: Y_3=C(30)+C(31)*T+C(32)*Y_1+C(33)*X_3+[AR(1)=C(34)]
Observations: 321
```

R-squared	0.999600	Mean dependent var	16076.63
Adjusted R-squared	0.999595	S.D. dependent var	6258.103
S.E. of regression	125.9524	Sum squared resid	5013028.
Durbin-Watson stat	1.806760		

Figure 2.21   *Statistical results based on an AR(1) Circular_SCM in (2.19)*

The equation of an AR(1)_Circular_SCM with trend and an exogenous variable is as follows (see Figure 2.21 for statistical results based on this):

$$
\begin{aligned}
Y\_1 &= C(10) + C(11)^*t + C(12)^*Y\_2 + C(13)^*X\_1 + [AR(1) = C(14)] \\
Y\_2 &= C(20) + C(21)^*t + C(22)^*Y\_3 + C(23)^*X\_2 + [AR(1) = C(24)] \\
Y\_3 &= C(30) + C(31)^*t + C(32)^*Y\_1 + C(33)^*X\_3 + [AR(1) = C(34)]
\end{aligned}
\tag{2.19}
$$

Example 2.15   An AR(1)_Circular_SCM in (2.19)

As an illustration, Figure 2.22 presents the statistical results based on the model in (2.19) using a hypothetical data set. Based on these results the following notes and conclusions are presented.

1. At the significance level of 0.05, Y_2 has a significant positive adjusted effect on Y_1 with a *p*-value = 0.0000/2 = 0.0000, based on the first regression; based on the second regression, Y_3 has a significant positive adjusted effect on Y_2 with a *p*-value = 0.0000/2 = 0.0000; and based on the third regression, Y_1 has a significant positive adjusted effect on Y_3 with a *p*-value = 0.0000/2 = 0.0000.
2. In this illustrative example, the conclusion will be the data supports the hypothesis that Y_1, Y_2, and Y_3 have a circular causal or up- and down-stream relationship.
3. This model could easily be extended to various alternative SCMs, such as LVAR(p,q)_Circular_SCM. Further extensions would use the endogenous variables $G(Y\_i) = log(Y\_i)$, and the exogenous variables are any functions $F_i(X\_i)$ or $H_i(t)$ with a finite number of parameters.

2.4.2   SCMs as VAR Models

A VAR model is a special type of SCM. However, use of the object "*System*" or the SCM is recommended, instead of the VAR Model. See the following illustrative examples.

```
System: SYS36
Estimation Method: Seemingly Unrelated Regression
Date: 09/16/09   Time: 12:14
Sample: 1968M03 1994M10
Included observations: 321
Total system (balanced) observations 960
Iterate coefficients after one-step weighting matrix
Convergence achieved after: 1 weight matrix, 11 total coef iterations
```

	Coefficient	Std. Error	t-Statistic	Prob.
C(10)	21.24507	23.28406	0.912429	0.3618
C(11)	-0.113852	0.265676	-0.428537	0.6684
C(12)	0.979821	0.013978	70.09866	0.0000
C(13)	-0.012821	0.007103	-1.805067	0.0714
C(14)	-0.016646	0.009764	-1.704900	0.0885
C(15)	0.001165	0.001235	0.942680	0.3461
C(16)	-0.009850	0.054451	-0.180898	0.8565
C(20)	-119.1014	94.11855	-1.265440	0.2060
C(21)	-1.349717	0.833624	-1.619096	0.1058
C(22)	-0.046303	0.037088	-1.248481	0.2122
C(23)	0.941734	0.014425	65.28608	0.0000
C(24)	0.097169	0.025585	3.797950	0.0002
C(25)	-0.006343	0.013555	-0.467939	0.6399
C(26)	0.033697	0.046559	0.723754	0.4694
C(30)	11.90912	44.66934	0.266606	0.7898
C(31)	0.695614	0.483681	1.438167	0.1507
C(32)	-0.001659	0.027027	-0.061386	0.9511
C(33)	-0.046837	0.010979	-4.265902	0.0000
C(34)	1.038231	0.017626	58.90240	0.0000
C(35)	-0.000185	0.002474	-0.074812	0.9404
C(36)	0.004515	0.046494	0.097109	0.9227
Determinant residual covariance	3.50E+12			

Equation: Y_1 = C(10)+C(11)*T+C(12)*Y_1(-1)+C(13)*Y_2(-1)+C(14)*Y_3*(
    -1)+C(15)*X_1+[AR(1)=C(16)]
Observations: 320

R-squared	0.997178	Mean dependent var	4806.125
Adjusted R-squared	0.997124	S.D. dependent var	1558.433
S.E. of regression	83.57407	Sum squared resid	2186188.
Durbin-Watson stat	2.020873		

Equation: Y_2 = C(20)+C(21)*T+C(22)*Y_1(-1)+C(23)*Y_2(-1)+C(24)*Y_3(
    -1)+C(25)*X_2 +[AR(1)=C(26)]
Observations: 320

R-squared	0.998870	Mean dependent var	14121.02
Adjusted R-squared	0.998848	S.D. dependent var	6586.377
S.E. of regression	223.5450	Sum squared resid	15641350
Durbin-Watson stat	2.083048		

Equation: Y_3 = C(30)+C(31)*T+C(32)*Y_1(-1)+C(33)*Y_2(-1)+C(34)*Y_3(
    -1)+C(35)*X_3+[AR(1)=C(36)]
Observations: 320

R-squared	0.999352	Mean dependent var	16109.83
Adjusted R-squared	0.999340	S.D. dependent var	6239.535
S.E. of regression	160.3099	Sum squared resid	8043869.
Durbin-Watson stat	2.024927		

Figure 2.22   *Statistical results based on a LVAR(**1,1**)_SCM in (2.20)*

Example 2.16   The simplest VAR model with trend

Based on the three bivariate time series, the simplest SCM considered is a trivariate LV(1)_SCM with trend as follows:

$$Y\_1 = C(10) + C(11)^*t + C(12)^*Y\_1(-1) + C(13)^*Y\_2(-1) + C(14)^*Y\_3^*(-1)$$
$$Y\_2 = C(20) + C(21)^*t + C(22)^*Y\_1(-1) + C(23)^*Y\_2(-1) + C(24)^*Y\_3(-1) \qquad (2.20)$$
$$Y\_3 = C(30) + C(31)^*t + C(32)^*Y\_1(-1) + C(33)^*Y\_2(-1) + C(34)^*Y\_3(-1)$$

Note that this model is in fact a VAR model of the time series Y_1, Y_2 and Y_3, with "1 1" as the lag intervals for endogenous variables, and "$C\ t$" as the exogenous variables, or a multivariate GLM (with the same set of independent variables).

Example 2.17   An extension of the SCM in (2.7)

As an extension of the model in (2.7), the model considered is a trivariate LVAR(**p,1**)_SCM with trend as follows:

$$
\begin{aligned}
Y\_1 &= C(10) + C(11)^*t + C(12)^*Y\_1(-1) + C(13)^*Y\_2(-1) + C(14)^*Y\_3^*(-1) \\
&\quad + C(15)^*X\_1 + [AR(1) = C(16),\ldots,AR(p) = C(1^*)] \\
Y\_2 &= C(20) + C(21)^*t + C(22)^*Y\_1(-1) + C(23)^*Y\_2(-1) + C(24)^*Y\_3(-1) \\
&\quad + C(25)^*X\_2 + [AR(1) = C(26),\ldots,AR(p) = C(2^*)] \\
Y\_3 &= C(30) + C(31)^*t + C(32)^*Y\_1(-1) + C(33)^*Y\_2(-1) + C(34)^*Y\_3(-1) \\
&\quad + C(35)^*X\_3 + [AR(1) = C(36),\ldots,AR(p) = C(3^*)]
\end{aligned}
\qquad (2.21)
$$

As an illustration, Figure 2.22 presents the statistical results based on a LVAR(**1,1**)_SCM with trend. Based on these results, note the following:

1. Note that the AR(1) term is insignificant with such a large *p*-values in the three multiple regressions. The main reason of this case is that each regression of *Y_i* has the first lags *Y_i(−1)* as an independent variable. *AR(1)* and *Y_i(−1)* are highly correlated, within each regression of *Y_i*. For this reason, in order to obtain a reduced model, either one of *AR(1)* and *Y_i(−1)* can be deleted from each regression.
2. For an illustration, Figure 2.23 presents the statistical results based on an AR(**1**)_SCM, which is a reduced model of the model in Figure 2.22 by deleting the *Y_i(−1)* from each regression of *Y_i*. Based on these results, note the following.

   2.1   Contradictory to the results in Figure 2.22, these results show that the AR(1) term is significant within each of the regressions.

   2.2   Three of the independent variables in the SCM have *p*-values > 0.20 (or 0.25), therefore, in a statistical sense, the suggestion is to explore a further reduced model.

   2.3   On the other hand, the negative signs of some of the independent variables also should be taken into consideration, since it has been found that each of the independent variables has a significant positive correlation with each *Y_i*. The negative signs are the unpredictable impact of multicollinearity. In fact, we find that each independent variable has a significant positive correlation with each of the other independent variables. Do you prefer to have statistical results showing positive estimates of the parameters?

   2.4   In order to increase the DW-statistic of the second and third regressions, the AR(1) and AR(2) terms could be used. Even though AR(2) terms are significant with *p*-values of 0.0002 and 0.0033, respectively, they have negative signs.

Estimation Method: Seemingly Unrelated Regression
Date: 09/17/09   Time: 09:59
Sample: 1968M03 1994M10
Included observations: 321
Total system (balanced) observations 960
Iterate coefficients after one-step weighting matrix
Convergence achieved after: 1 weight matrix, 32 total coef iterations

	Coefficient	Std. Error	t-Statistic	Prob.
C(10)	1956.014	993.0065	1.969790	0.0492
C(11)	-5.321536	3.976980	-1.338085	0.1812
C(13)	0.009570	0.019041	0.502599	0.6154
C(14)	-0.212329	0.027264	-7.787852	0.0000
C(15)	0.024180	0.012698	1.904190	0.0572
C(16)	0.983466	0.010265	95.80303	0.0000
C(20)	-130.6877	2802.811	-0.046627	0.9628
C(21)	58.76515	24.43463	2.404994	0.0164
C(22)	-0.387715	0.144357	-2.685811	0.0074
C(24)	0.052846	0.074380	0.710484	0.4776
C(25)	0.485832	0.063250	7.681169	0.0000
C(26)	1.009860	0.008370	120.6480	0.0000
C(30)	4074.708	1382.414	2.947531	0.0033
C(31)	29.68955	5.623247	5.279788	0.0000
C(32)	0.097871	0.088980	1.099918	0.2716
C(33)	-0.104126	0.033191	-3.137155	0.0018
C(35)	0.141974	0.011616	12.22277	0.0000
C(36)	0.983621	0.009198	106.9333	0.0000
Determinant residual covariance	3.50E+12			

Equation: Y_1 = C(10)+C(11)*T+C(13)*Y_2(-1)+C(14)*Y_3*(-1)+C(15)*X_1
+[AR(1)=C(16)]
Observations: 320

R-squared	0.997535	Mean dependent var	4806.125
Adjusted R-squared	0.997495	S.D. dependent var	1558.433
S.E. of regression	77.99381	Sum squared resid	1910073.
Durbin-Watson stat	2.062205		

Equation: Y_2 = C(20)+C(21)*T+C(22)*Y_1(-1)+C(24)*Y_3(-1)+C(25)*X_2
+[AR(1)=C(26)]
Observations: 320

R-squared	0.999045	Mean dependent var	14121.02
Adjusted R-squared	0.999029	S.D. dependent var	6586.377
S.E. of regression	205.1999	Sum squared resid	13221592
Durbin-Watson stat	1.776659		

Equation: Y_3 = C(30)+C(31)*T+C(32)*Y_1(-1)+C(33)*Y_2(-1)+C(35)*X_3
+[AR(1)=C(36)]
Observations: 320

R-squared	0.999582	Mean dependent var	16109.83
Adjusted R-squared	0.999576	S.D. dependent var	6239.535
S.E. of regression	128.5206	Sum squared resid	5186506.
Durbin-Watson stat	1.681992		

Figure 2.23   *Statistical results based on a reduced model of the model in Figure 2.22*

2.5   As an exercise, find other possible reduced models, including by deleting the AR(1) terms from all regressions in Figure 2.22. By deleting the AR(1) terms, the model in (2.20) could be one of the possible reduced models.

3.   These models could easily be extended to various alternative SCMs, such as LVAR(*p,q*)_SCM. Further extensions would use the endogenous variables $G(Y\_i) = log(Y\_i)$, or $log((Y\_i-Li)/(Ui-Y\_i))$ where *Li* and *Ui* are the fixed lower and upper bounds of *Y_i*. and the exogenous variables can be any functions $F_i(X\_i)$ or $H_i(t)$ with a finite number of parameters.

Example 2.18   An extension of the SCM in (2.8)

As an extension of the SCM in (2.8), under the assumption that *Y_1*, *Y_2* and *Y_3* have simultaneous causal effects, then the following SCM should be considered, with the statistical results in Figure 2.24. Based on

```
System: SYS39
Estimation Method: Seemingly Unrelated Regression
Date: 09/17/09   Time: 16:34
Sample: 1968M03 1994M10
Included observations: 321
Total system (balanced) observations 1920
Iterate coefficients after one-step weighting matrix
Convergence achieved after: 1 weight matrix, 12 total coef iterations
```

	Coefficient	Std. Error	t-Statistic	Prob.
C(10)	24.26686	23.91828	1.014574	0.3104
C(11)	-0.228772	0.291504	-0.784801	0.4327
C(12)	0.984931	0.013676	72.02094	0.0000
C(13)	-0.045146	0.028009	-1.611800	0.1072
C(14)	0.047862	0.027862	1.711351	0.0872
C(15)	0.264753	0.037363	7.086044	0.0000
C(16)	-0.262211	0.037883	-6.921545	0.0000
C(17)	0.015864	0.013995	1.133588	0.2571
C(18)	-0.015408	0.014073	-1.094837	0.2737
C(19)	-0.016789	0.052675	-0.318724	0.7500
C(20)	-49.14850	59.65122	-0.823931	0.4101
C(21)	-1.607505	0.527756	-3.045924	0.0024
C(22)	-0.114565	0.095583	-1.198602	0.2308
C(23)	0.067763	0.097136	0.697609	0.4855
C(24)	1.008114	0.009538	105.6922	0.0000
C(25)	1.102964	0.043837	25.16045	0.0000
C(26)	-1.070045	0.046771	-22.87847	0.0000
C(27)	0.048716	0.033355	1.460561	0.1443
C(28)	-0.054950	0.033290	-1.650637	0.0990
C(29)	0.070505	0.045601	1.546142	0.1222
C(30)	72.15766	30.89503	2.335575	0.0196
C(31)	1.514912	0.326762	4.636131	0.0000
C(32)	0.514272	0.069152	7.436813	0.0000
C(33)	-0.475974	0.070166	-6.783569	0.0000
C(34)	0.649429	0.026894	24.14797	0.0000
C(35)	-0.661591	0.027504	-24.05446	0.0000
C(36)	0.981095	0.012011	81.68144	0.0000
C(37)	-0.035498	0.009161	-3.874850	0.0001
C(38)	0.035496	0.009144	3.881866	0.0001
C(39)	0.044983	0.045467	0.989344	0.3226
C(41)	0.998918	0.002041	489.5260	0.0000
C(42)	0.262272	0.051480	5.094690	0.0000
C(50)	96.26715	46.54635	2.068200	0.0388
C(51)	0.998814	0.002230	447.8257	0.0000
C(52)	0.356770	0.047084	7.577294	0.0000
C(60)	342.8065	116.6258	2.939370	0.0033
C(61)	0.998188	0.001828	546.0369	0.0000
C(62)	0.272944	0.043398	6.289296	0.0000
Determinant residual covariance	4.21E+25			

Equation: Y_1 = C(10)+C(11)*T+C(12)*Y_1(-1)+C(13)*Y_2+C(14)*Y_2(-1)
+C(15)*Y_3+C(16)*Y_3(-1)+C(17)*X_1+C(18)*X_1(-1)+[AR(1)=C(19)]
Observations: 320

R-squared	0.997503	Mean dependent var	4806.125
Adjusted R-squared	0.997430	S.D. dependent var	1558.433
S.E. of regression	79.00317	Sum squared resid	1934865.
Durbin-Watson stat	2.054879		

Equation: Y_2 = C(20)+C(21)*T+C(22)*Y_1+C(23)*Y_1(-1)+C(24)*Y_2(-1)
+C(25)*Y_3+C(26)*Y_3(-1)+C(27)*X_2+C(28)*X_2(-1)+[AR(1)=C(29)]
Observations: 320

R-squared	0.999458	Mean dependent var	14121.02
Adjusted R-squared	0.999442	S.D. dependent var	6586.377
S.E. of regression	155.5938	Sum squared resid	7504923.
Durbin-Watson stat	1.986003		

Equation: Y_3 = C(30)+C(31)*T+C(32)*Y_1+C(33)*Y_1(-1)+C(34)*Y_2
+C(35)*Y_2(-1)+C(36)*Y_3(-1)+C(37)*X_3+C(38)*X_3(-1)
+[AR(1)=C(39)]
Observations: 320

R-squared	0.999671	Mean dependent var	16109.83
Adjusted R-squared	0.999662	S.D. dependent var	6239.535
S.E. of regression	114.7205	Sum squared resid	4079849.
Durbin-Watson stat	2.029304		

Equation: X_1 = C(40)+C(41)*X_1(-1)+[AR(1)=C(42)]
Observations: 320

R-squared	0.999257	Mean dependent var	16880.57
Adjusted R-squared	0.999252	S.D. dependent var	11987.86
S.E. of regression	327.7890	Sum squared resid	34060264
Durbin-Watson stat	2.163277		

Equation: X_2 = C(50)+C(51)*X_2(-1)+[AR(1)=C(52)]
Observations: 320

R-squared	0.999331	Mean dependent var	19793.87
Adjusted R-squared	0.999327	S.D. dependent var	6677.297
S.E. of regression	173.2333	Sum squared resid	9513095.
Durbin-Watson stat	2.252035		

Equation: X_3 = C(60)+C(61)*X_3(-1)+[AR(1)=C(62)]
Observations: 320

R-squared	0.999418	Mean dependent var	58541.32
Adjusted R-squared	0.999414	S.D. dependent var	25362.11
S.E. of regression	613.8812	Sum squared resid	1.19E+08
Durbin-Watson stat	2.118175		

Figure 2.24   *Statistical results based on the SCM in (2.22)*

these results, the following notes and conclusions are presented.

$$
\begin{aligned}
Y\_1 =\ & C(10) + C(11)^*t + C(12)^*Y\_1(-1) + C(13)^*Y\_2 + C(14)^*Y\_2(-1) + C(15)^*Y\_3 \\
& + C(16)^*Y\_3(-1) + C(17)^*X\_1 + C(18)^*X\_1(-1) + [AR(1) = C(19)] \\
Y\_2 =\ & C(20) + C(21)^*t + C(22)^*Y\_1 + C(23)^*Y\_1(-1) + C(24)^*Y\_2(-1) + C(25)^*Y\_3 \\
& + C(26)^*Y\_3(-1) + C(27)^*X\_2 + C(28)^*X\_2(-1) + [AR(1) = C(29)] \\
Y\_3 =\ & C(30) + C(31)^*t + C(32)^*Y\_1 + C(33)^*Y\_1(-1) + C(34)^*Y\_2 + C(35)^*Y\_2(-1) \\
& + C(36)^*Y\_3(-1) + C(37)^*X\_3 + C(38)^*X\_3(-1) + [AR(1) = C(29)] \\
X\_1 =\ & C(40) + C(41)^*X\_1(-1) + [AR(1) = C(42)] \\
X\_2 =\ & C(50) + C(51)^*X\_2(-1) + [AR(1) = C(52)] \\
X\_3 =\ & C(60) + C(61)^*X\_3(-1) + [AR(1) = C(62)]
\end{aligned}
\tag{2.22}
$$

1. Compare this to the results in Figure 2.19, only the AR(1) terms of the first and third regressions have large p-values > 0.20, and at a significance level of $\alpha = 0.10$, the AR(1) of the second regression is significantly positive with a p-value $= 0.1222/2 = 0.0611$. So a reduced model should be explored, however, the analysis will not be presented here.
2. Refer to the illustration and notes in Example 2.17

2.5 Time-Series Models with an Environmental Variable Z_t, Based on Independent States

As an extension of the models with an environmental variable presented in Section 1.8, this section presents the time series models of Y_i_t on the time t, an exogenous variable X_i_t, and an environmental variable Z_t. The time-series model would have the following general equation.

$$
G_i(Y\_i_t) = F_i(t, Y\_i_{t-j}, X\_i_{t-k}, Z_{t-l}) + \mu_{it}
\tag{2.23}
$$

where $G_i(Y\_i_t)$ is a function of Y_i_t without a parameter, and $F_i(t, Y\_i_{t-j}, X\_i_{t-k}, Z_{t-l})$ is a function of the lag variable Y_i_{t-j}, and the exogenous variables X_i_{t-k} and Z_{t-l}, for some integers (j,k,l) with a finite number of parameters. In general, $j \geq 1$, $k \geq 0$ and $l \geq 0$, which should be subjectively selected by researchers, so the scores of Y_i_{t-j}, X_i_{t-k} and Z_{t-l} are observed before the scores of the endogenous variable Y_i_t? Therefore, Y_i_{t-j}, X_i_{t-k} and Z_{t-l} can be declared - the up-stream variable of Y_i_t.

To generalize, the function $F_i(^*)$ can be a function of many lagged variables, such as $F_i(t, Y\_i_{t-1}, \ldots, X\_i_t, X\_i_{t-1}, \ldots, Z_t, Z\_i_{t-1}, \ldots)$, with a special case $\{F_{1i}(t) + F_{2i}(Y\_i_{t-1}, \ldots) + F_{3i}(X\_i_t, X\_i_{t-1}, \ldots) + F_{4i}(Z_t, Z\_i_{t-1}, \ldots)\}$. Find the simplest general model in the following section.

2.5.1 The Simplest Possible Model

The simplest possible model is an additive model with trend as follows:

$$
\begin{aligned}
Y\_1_t =\ & C(10) + C(11)^*t + \sum_{i=1}^{I_1} C(11i)^*Y\_1_{t-i} + \sum_{j=0}^{J_1} C(12j)^*X\_1_{t-j} + \sum_{k=0}^{K_1} C(13k)^*Z_{t-k} + \mu 1_t \\
Y\_2_t =\ & C(20) + C(21)^*t + \sum_{i=1}^{I_2} C(21i)^*Y\_2_{t-i} + \sum_{j=0}^{J_2} C(22j)^*X\_2_{t-j} + \sum_{k=0}^{K_2} C(11k)^*Z_{t-k} + \mu 2_t
\end{aligned}
\tag{2.24}
$$

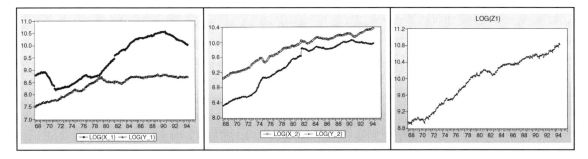

Figure 2.25   *The growth curves of the variables* log(X_i),log(Y_i), *for* i = 1,2; *and* log(Z1)

In many cases, $t - n = t - 1$, for all n, is used, so the values or scores of the independent variables are observed or measured before the dependent variables. Two special cases may be considered, where Y_i_{t-s}, for $s = 4$ and $s = 12$ are applied, if and only if the time-series data are quarterly and monthly data sets, respectively.

Example 2.19   Preliminary analysis based on (X_1,Y_1), (X_2,Y_2), and Z1

For the illustration, the bivariates *(X_1,Y_1)* and *(X_2,Y_2)* considered earlier will be used, and an additional environmental variable is generated as *Z1 = mmmae* based on the data in POOL1.wf1. The growth curves of the natural logarithmic of the five variables are presented in Figure 2.25, and their correlation matrix is presented in Figure 2.26. Based on this correlation matrix, the following notes and conclusions are made.

1. The environmental variable *log(Z1)* has a significant positive linear effect on each of the variables *log(X_1)*, *log(Y_1)*, *log(X_2)* and *log(Y_2)*.

```
Covariance Analysis: Ordinary
Date: 11/10/09   Time: 14:58
Sample (adjusted): 1968M01 1994M10
Included observations: 322 after adjustments
Balanced sample (listwise missing value deletion)
```

Correlation t-Statistic Probability	LOG(Z1)	LOG(X_1)	LOG(Y_1)	LOG(X_2)	LOG(Y_2)
LOG(Z1)	1.000000				

LOG(X_1)	0.877353	1.000000			
	32.70755	-----			
	0.0000	-----			
LOG(Y_1)	0.966175	0.798201	1.000000		
	67.01925	23.70328	-----		
	0.0000	0.0000	-----		
LOG(X_2)	0.994931	0.871461	0.968159	1.000000	
	176.9909	31.78442	69.18282	-----	
	0.0000	0.0000	0.0000	-----	
LOG(Y_2)	0.982882	0.893793	0.964772	0.986994	1.000000
	95.43353	35.65086	65.59940	109.8313	-----
	0.0000	0.0000	0.0000	0.0000	-----

Figure 2.26   *The matrix correlation of the variables* log(Z1), log(X_i) *and* log(Y_i)

2. As an illustration, note that the effect of $log(Z1)$ on $log(Y\_1)$ is significant based on the *t*-statistic of $t_0 = 67.01925$ with a *p*-value $= 0.0000$. To compare, exactly the same values of *t*-statistics are also obtained, based on the simple linear regression of $log(Y\_1)$ on $log(Z1)$, as well as the simple linear regression of $log(Z1)$ on $log(Y\_1)$. These findings indicate that the correlation analysis is clearly sufficient to test a hypothesis on the causal linear effect between any pairs of numerical variables.

3. Note that by using a MLVAR(p,q) model with trend and a set of exogenous variables unexpected estimates, or an error message can be obtained because of the unpredictable impact of the multicollinearity of all variables in the model. See the following examples.

Example 2.20   Application of additive models in (2.24)

Figure 2.27(a) presents the statistical results based on the simplest model in (2.24) with $t - n = t - 1$ for all *n*. Even though $X\_1(-1)$ has an insignificant adjusted effect on Y_1, this model can be considered a good fit, since its DW-statistic is sufficiently close to 2.0. Note that this model is in fact a MLV(**1**) model with trend and exogenous variables $X\_1(-1)$ and $Z1(-1)$. As a comparison, Figure 2.27(b) presents the statistical results based on a MLVAR(**1,1**) model, which shows the AR(1) terms in both regressions have large *p*-values. It is also shown that an MAR(**1**) is even worse, conditional for the data used: So, these findings cannot be generalized.

Figure 2.27   *Statistical results based on (a) a MLV(**1**) model, and (b) a MLVAR(**1,1**) model, with trend and exogenous variables* $log(X\_i(-1))$ *and* $log(Z1(-1))$

Figure 2.28 *Statistical results based on (a) a MAR(1) Model, and (b) a MLV(1) model, with trend and exogenous variables* $\log(Y\_i(-12))$, $\log(X\_i(-1)$ *and* $\log(Z1(-1))$

Example 2.21   Application of special models in (2.24)

Since the time-series data considered is a monthly data set, then try to use the lag $Y\_i(-12)$ as a covariate in the model (2.24). Figure 2.28(a) presents the statistical results based on a MAR(1) model with trend and exogenous variables $log(Y\_i(-12))$, $log(X\_i(-1)$ and $log(Z1(-1))$, which show seven out of 10 independent variables have p-values > 0.25. So this model is not a good fit in a statistical sense. Various reduced models have been developed by deleting one or two of the independent variables $log(Y\_i(-12))$, $log(X\_i(-1)$ and $log(Z1(-1))$ from each regression, but each of the remaining independent variables has an insignificant effect. Finally, by using the trial-and-error method, a good fit model can be obtained by replacing the AR(1) terms with $log(Y\_i(-1))$ with the statistical results presented in Figure 2.28(b), which is a MLV(**1**) with trend and exogenous variables $log(Y\_i(-12))$, $log(X\_i(-1)$ and $log(Z1(-1))$.

2.5.2   Interaction Models Based on Two Independent States

Based on the time series $(X\_1, Y\_1_t$ and $(X\_2, Y\_2)_t$ of two states or individuals, and an environmental factor Z_t, including their lags and the time t, a lot of possible time-series models can easily be derived or proposed. This depends on the causal or structural relationship defined based on the set of variables X_1_t, Y_1_t, X_2_t, Y_2_t, Z_t, and the time t, including the lags of the endogenous and exogenous variables. Everyone should select a set of variables and then define their relationships, supported by best possible judgment (knowledge

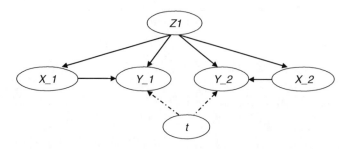

Figure 2.29   *Theoretical SCM of* $(X\_1, Y\_1)_t$, $(X\_2, Y\_2)_t$, *the time* t, *and* Z_t.

and experience). As an illustration, see the following theoretical causal relationships based on the set of main variables.

2.5.2.1   Interaction Models with Trend Based on Independent States

Figure 2.29 presents the theoretical causal relationships between the variables $(X\_1, Y\_1)_t$, $(X\_2, Y\_2)_t$, the time t, and an environmental factor Z_t. In this case, Z_t is defined as having a direct effect on the four variables of the two states, since it is an environmental or external factor of the states, but the time t is used only to present the trend or growth of the endogenous variables Y_1 and Y_2.

The simplest two-way interaction model considered is as follows:

$$
\begin{aligned}
Y\_1_t &= C(10) + C(11)^*t + C(12)^*X\_1_{t-i_1} + C(13)^*Z1_{t-j_1} \\
&\quad + C(14)^*X\_1_{t-i_1}{}^*Z1_{t-j_1} + \mu1_t \\
Y\_2_t &= C(20) + C(21)^*t + C(22)^*X\_2_{t-i_2} + C(23)^*Z1_{t-j_2} \\
&\quad + C(24)^*X\_2_{t-i_2}{}^*Z1_{t-j_2} + \mu2_t
\end{aligned}
\tag{2.25}
$$

for subjectively selected set of integers i_1, j_1, i_2 and j_2. The simplest set is a set of zeros and/or ones. Note that the two-way interaction factors should be used as the independent variables, since $Z1$ also has an indirect effect on Y_1 and Y_2 through X_1 and X_2, respectively. Compare this to the models previously.

If the effects of $Z1$ on each X_1 and X_2 should be considered, then the following model with trend is obtained.

$$
\begin{aligned}
Y\_1_t &= C(10) + C(11)^*t + C(12)^*X\_1_{t-i_1} + C(13)^*Z1_{t-j_1} \\
&\quad + C(14)^*X\_1_{t-i_1}{}^*Z1_{t-j_1} + \mu1_t \\
Y\_2_t &= C(20) + C(21)^*t + C(22)^*X\_2_{t-i_2} + C(23)^*Z1_{t-j_2} \\
&\quad + C(24)^*X\_2_{t-i_2}{}^*Z1_{t-j_2} + \mu2_t \\
X\_1_t &= C(30) + C(31)^*t + C(32)^*Z1_{t-j_3} + \mu3_t \\
X\_2_t &= C(40) + C(41)^*t + C(42)^*Z1_{t-j_4} + \mu3_t
\end{aligned}
\tag{2.26}
$$

2.6   Models Based on Correlated States

In this section, two cases of two correlated states will be considered. The first case is the two states, presented by $(X\_1, Y\_1)_t$ and $(X\_2, Y\_2)_t$, defined to be correlated if and only if the two endogenous variables

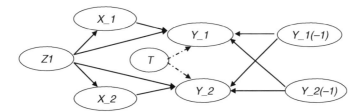

Figure 2.30 *The path diagram of a MLV(1,1)_SCM of (X_1,Y_1), and (X_2,Y_2), with trend and an environmental factor* Z_t

Y_1_t and Y_2_t are correlated, including the possibility of having a causal relationship, in a theoretical sense. Refer to the notes presented in Section 1.9.3.3 if the four variables $(X\_1,Y\_1)_t$ and $(X\_2,Y\_2)_t$ are completely correlated.

2.6.1 MLV(1) Interaction Model with Trend

As an extension of the path diagram in Figure 2.29, Figure 2.30 presents the path diagram of a MLV(**1**) model with trend based on correlated states, which is indicated by the direct effects of the first lags $Y\_1(-1)$ and $Y\_2(-1)$ to both Y_1 and Y_2.

Then, as the extension of the model in (2.25) based on this path diagram, the following model with trend is obtained.

$$
\begin{aligned}
Y\_1_t &= C(10) + C(11)^* t + C(12)^* Y\_1_{t-1} + C(13)^* Y\_2_{t-1} \\
&\quad + C(14)^* X\_1_{t-i_1} + C(15)^* Z1_{t-j_1} + C(16)^* X\_1_{t-i_1}{}^* Z1_{t-j_1} + \mu 1_t \\
Y\_2_t &= C(20) + C(21)^* t + C(22)^* Y\_1_{t-1} + C(23)^* Y\_2_{t-1} \\
&\quad + C(24)^* X\_2_{t-i_2} + C(25)^* Z1_{t-j_2} + C(26)^* X\_1_{t-i_2}{}^* Z1_{t-j_2} + \mu 2_t
\end{aligned}
\tag{2.27}
$$

2.6.2 Simultaneous SCMs with Trend

As an extension of the path diagram in Figure 2.30, Figure 2.31 presents the path diagram where Y_1 and Y_2 have simultaneous causal relationships. Then the equation of the model with trend can easily be derived

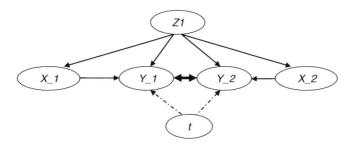

Figure 2.31 *The path diagram of a simultaneous SCM of (X_1,Y_1)_t, and (X_2,Y_2)_t, with trend and an environmental variable* Z_t

from the model in (2.25), and is obtained with the following equation.

$$
\begin{aligned}
Y\_1_t = {}& C(10) + C(11)^* t + C(12)^* Y\_2_{t-i_1} + C(13)^* X\_1_{t-j_1} + C(14)^* Z1_{t-k_1} \\
& + C(15)^* Y\_2_{t-i_1}{}^* X\_2_{t-j_1} + C(16)^* Y\_2_{t-i_1}{}^* Z1_{t-k_1} + C(17)^* X\_2_{t-j_1}{}^* Z1_{t-k_1} + \mu 1_t \\
Y\_2_t = {}& C(20) + C(21)^* t + C(22)^* Y\_1_{t-i_2} + C(23)^* X\_2_{t-j_2} + C(24)^* Z1_{t-k_2} \\
& + C(25)^* Y\_1_{t-i_2}{}^* X\_1_{t-j_2} + C(26)^* Y\_1_{t-i_2}{}^* Z1_{t-k_2} + C(27)^* X\_1_{t-j_2}{}^* Z1_{t-k_1} + \mu 2_t
\end{aligned}
\tag{2.28}
$$

where, in many cases, all i, j and k either equal to zero or one are used. See the following example.

Note that the first regression has three main factors, namely Y_2, X_1 and $Z1$, since each has direct effects on Y_1, and three two-way interaction factors, namely $Y\_2{*}X\_2$, to represent that X_2 has an indirect effect on Y_1 through Y_2, and the other two interactions indicate that $Z1$ has an indirect effect on Y_1 through X_1 and X_2. This is similar for the second regression.

If the effects of $Z1$ on each of X_1 and X_2 should be considered, then the following model with trend is obtained.

$$
\begin{aligned}
Y\_1_t = {}& C(10) + C(11)^* t + C(12)^* Y\_2_{t-i_1} + C(13)^* X\_1_{t-j_1} + C(14)^* Z1_{t-k_1} \\
& + C(15)^* Y\_2_{t-i_1}{}^* X\_2_{t-j_1} + C(16)^* Y\_2_{t-i_1}{}^* Z1_{t-k_1} + C(17)^* X\_2_{t-j_1}{}^* Z1_{t-k_1} + \mu 1_t \\
& + C(18)^* t^* Y\_2_{t-i_1} + C(19)^* t^* X\_1_{t-j_1} + C(110)^* t^* * Z1_{t-k_1} + \mu 1_t
\end{aligned}
\tag{2.29}
$$

$$
\begin{aligned}
Y\_2_t = {}& C(20) + C(21)^* t + C(22)^* Y\_1_{t-i_2} + C(23)^* X\_2_{t-j_2} + C(24)^* Z1_{t-k_2} \\
& + C(25)^* Y\_1_{t-i_2}{}^* X\_1_{t-j_2} + C(26)^* Y\_1_{t-i_2}{}^* Z1_{t-k_2} + C(27)^* X\_1_{t-j_2}{}^* Z1_{t-k_1} + \mu 2_t \\
& + C(28)^* t^* Y\_1_{t-i_2} + C(29)^* t^* X\_2_{t-j_2} + C(210)^* t^* Z1_{t-k_2} + \mu 2_t \\
X\_1_t = {}& C(30) + C(31)^* t + C(32)^* Z1_{t-k_3} + \mu 3_t \\
X\_2_t = {}& C(40) + C(41)^* t + C(42)^* Z1_{t-k_4} + \mu 4_t
\end{aligned}
\tag{2.30}
$$

Example 2.22   A simultaneous SCM

Figure 2.32 presents the statistical results based on a model in (2.28) using the AR(1) terms, since the model without AR(1) terms gives a small DW-statistic. Note that this model uses the first lagged independent variables. Based on these results, note the following:

1. Each of the independent variables of the first regression has a large p-value. In a statistical sense, the independent variables are highly correlated. Therefore, by using some of the independent variables, the independent variables of the reduced model have presented the effects of the omitted or deleted variables. However, the problem is, which variable(s) should be deleted? It cannot be generalized, since it is highly dependent on the data used. Try it as an exercise based on your own data set.
2. For a comparison, instead of deleting the independent variable(s), by deleting the AR(1) terms from both regressions, the results in Figure 2.33 are obtained. These results show the only independent variable, $log(Y\_2(-1))*log(X\_2(-1))$, has a large p-value $= 0.6081$, but each of the regressions has very small DW-statistic. If you have to choose one out of these two models, which one do you prefer?
3. On the other hand, Figure 2.34 presents the statistical results based on a LV(**1**) Simultaneous SCM, which shows two of the independent variables have large p-values. Despite that, this model should be considered the best fit, conditional for the data set. The limitations of the model should be studied using residual and other analyses. Refer to the residual analyses present in Agung (2009a).

System: SYS14_A
Estimation Method: Seemingly Unrelated Regression
Date: 11/12/09 Time: 10:23
Sample: 1968M03 1994M10
Included observations: 321
Total system (balanced) observations 640
Iterate coefficients after one-step weighting matrix
Convergence achieved after: 1 weight matrix, 27 total coef iterations

	Coefficient	Std. Error	t-Statistic	Prob.
C(10)	7.908403	5.693832	1.388942	0.1653
C(11)	-0.001767	0.006170	-0.286453	0.7746
C(12)	0.000826	0.529552	0.001559	0.9988
C(13)	0.064123	0.420761	0.152398	0.8789
C(14)	0.206090	0.434892	0.473889	0.6357
C(15)	0.011719	0.011722	0.999764	0.3178
C(16)	-0.014953	0.051572	-0.289949	0.7720
C(17)	-0.006985	0.042921	-0.162747	0.8708
C(18)	0.994194	0.006475	153.5480	0.0000
C(20)	12.37631	17.49233	0.707528	0.4795
C(21)	0.014890	0.025702	0.579312	0.5626
C(22)	0.011397	0.516359	0.022071	0.9824
C(23)	-0.176040	0.090073	-1.954424	0.0511
C(24)	-0.494534	0.442045	-1.118742	0.2637
C(25)	0.008337	0.004964	1.679556	0.0935
C(27)	0.064769	0.008418	7.694492	0.0000
C(28)	1.002575	0.005793	173.0530	0.0000

Determinant residual covariance	4.12E-08

Equation: LOG(Y_1)=C(10)+C(11)*T+C(12)*LOG(Y_2(-1))+C(13)*LOG(X_1(
     -1))+C(14)*LOG(Z1(-1))+C(15)*LOG(Y_2(-1))*LOG(X_2(-1))+C(16)
     *LOG(Y_2(-1))*LOG(Z1(-1))+C(17)*LOG(X_1(-1))*LOG(Z1(-1))
     +[AR(1)=C(18)]
Observations: 320

R-squared	0.998245	Mean dependent var	8.411218
Adjusted R-squared	0.998200	S.D. dependent var	0.388015
S.E. of regression	0.016462	Sum squared resid	0.084277
Durbin-Watson stat	2.033614		

Equation: LOG(Y_2)=C(20)+C(21)*T+C(22)*LOG(Y_1(-1))+C(23)*LOG(X_2(
     -1))+C(24)*LOG(Z1(-1))+C(25)*LOG(Y_1(-1))*LOG(X_1(-1))+C(16)
     *LOG(Y_1(-1))*LOG(Z1(-1))+C(27)*LOG(X_2)*LOG(Z1(-1))
     +[AR(1)=C(28)]
Observations: 320

R-squared	0.999503	Mean dependent var	9.413447
Adjusted R-squared	0.999491	S.D. dependent var	0.571793
S.E. of regression	0.012905	Sum squared resid	0.051797
Durbin-Watson stat	1.673528		

Figure 2.32   *Statistical results based on an AR(1) simultaneous SCM in (2.28)*

System: SYS14_B
Estimation Method: Seemingly Unrelated Regression
Date: 11/12/09 Time: 10:43
Sample: 1968M02 1994M10
Included observations: 321
Total system (balanced) observations 642
Linear estimation after one-step weighting matrix

	Coefficient	Std. Error	t-Statistic	Prob.
C(10)	-33.53372	2.722220	-12.31852	0.0000
C(11)	-0.001702	0.000330	-5.153069	0.0000
C(12)	5.370476	0.289016	18.58193	0.0000
C(13)	-1.834286	0.245536	-7.470528	0.0000
C(14)	4.383550	0.292918	14.96511	0.0000
C(15)	0.005438	0.010599	0.513029	0.6081
C(16)	-0.571908	0.029308	-19.51364	0.0000
C(17)	0.197262	0.024695	7.988079	0.0000
C(20)	-53.26475	2.800689	-19.01845	0.0000
C(21)	-0.000327	0.000232	-1.411893	0.1585
C(22)	5.174467	0.285582	18.11902	0.0000
C(23)	1.317695	0.302414	4.357250	0.0000
C(24)	6.038771	0.351269	17.19132	0.0000
C(25)	0.031875	0.001270	25.09145	0.0000
C(27)	-0.085447	0.031902	-2.678403	0.0076

Determinant residual covariance	8.26E-06

Equation: LOG(Y_1)=C(10)+C(11)*T+C(12)*LOG(Y_2(-1))+C(13)*LOG(X_1(
     -1))+C(14)*LOG(Z1(-1))+C(15)*LOG(Y_2(-1))*LOG(X_2(-1))+C(16)
     *LOG(Y_2(-1))*LOG(Z1(-1))+C(17)*LOG(X_1(-1))*LOG(Z1(-1))
Observations: 321

R-squared	0.980760	Mean dependent var	8.408489
Adjusted R-squared	0.980330	S.D. dependent var	0.390481
S.E. of regression	0.054765	Sum squared resid	0.938746
Durbin-Watson stat	0.285980		

Equation: LOG(Y_2)=C(20)+C(21)*T+C(22)*LOG(Y_1(-1))+C(23)*LOG(X_2(
     -1))+C(24)*LOG(Z1(-1))+C(25)*LOG(Y_1(-1))*LOG(X_1(-1))+C(16)
     *LOG(Y_1(-1))*LOG(Z1(-1))+C(27)*LOG(X_2)*LOG(Z1(-1))
Observations: 321

R-squared	0.990259	Mean dependent var	9.410050
Adjusted R-squared	0.990041	S.D. dependent var	0.574135
S.E. of regression	0.057294	Sum squared resid	1.027464
Durbin-Watson stat	0.173354		

Figure 2.33   *Statistical results based on a standard simultaneous SCM in (2.28)*

```
System: SYS14_C
Estimation Method: Seemingly Unrelated Regression
Date: 11/12/09  Time: 11:07
Sample: 1968M02 1994M10
Included observations: 321
Total system (balanced) observations 642
Linear estimation after one-step weighting matrix
```

	Coefficient	Std. Error	t-Statistic	Prob.
C(10)	-3.362283	0.822480	-4.087981	0.0000
C(11)	-8.04E-05	0.000100	-0.801368	0.4232
C(12)	0.419924	0.095769	4.384776	0.0000
C(13)	-0.099669	0.074421	-1.339267	0.1810
C(14)	0.427221	0.092094	4.638952	0.0000
C(15)	0.004869	0.003105	1.568029	0.1174
C(16)	-0.052970	0.009553	-5.544795	0.0000
C(17)	0.011670	0.007490	1.558101	0.1197
C(18)	0.945549	0.014433	65.51352	0.0000
C(20)	0.222809	0.856279	0.260207	0.7948
C(21)	-0.000241	5.20E-05	-4.626012	0.0000
C(22)	0.531810	0.090476	5.877939	0.0000
C(23)	-0.452473	0.079275	-5.707604	0.0000
C(24)	-0.061577	0.102799	-0.598999	0.5494
C(25)	0.000966	0.000449	2.150679	0.0319
C(27)	0.051995	0.007885	6.594094	0.0000
C(28)	0.965796	0.012007	80.43431	0.0000
Determinant residual covariance	3.78E-08			

```
Equation: LOG(Y_1)=C(10)+C(11)*T+C(12)*LOG(Y_2(-1))+C(13)*LOG(X_1(
    -1))+C(14)*LOG(Z1(-1))+C(15)*LOG(Y_2(-1))*LOG(X_2(-1))+C(16)
    *LOG(Y_2(-1))*LOG(Z1(-1))+C(17)*LOG(X_1(-1))*LOG(Z1(-1))+C(18)
    *LOG(Y_1(-1))
Observations: 321
```

R-squared	0.998368	Mean dependent var	8.408489
Adjusted R-squared	0.998326	S.D. dependent var	0.390481
S.E. of regression	0.015974	Sum squared resid	0.079612
Durbin-Watson stat	2.017004		

```
Equation: LOG(Y_2)=C(20)+C(21)*T+C(22)*LOG(Y_1(-1))+C(23)*LOG(X_2(
    -1))+C(24)*LOG(Z1(-1))+C(25)*LOG(Y_1(-1))*LOG(X_1(-1))+C(16)
    *LOG(Y_1(-1))*LOG(Z1(-1))+C(27)*LOG(X_2)*LOG(Z1(-1))+C(28)
    *LOG(Y_2(-1))
Observations: 321
```

R-squared	0.999524	Mean dependent var	9.410050
Adjusted R-squared	0.999511	S.D. dependent var	0.574135
S.E. of regression	0.012691	Sum squared resid	0.050252
Durbin-Watson stat	1.803415		

Figure 2.34   *Statistical results based on a LV(**1**) simultaneous SCM in (2.28)*

2.7 Piece-Wise Time-Series Models

Piece-wise or discontinuous time series models should be applied, as illustrated in Agung (2009a). For this reason, corresponding to the general model in (2.23), an *m*-piece time-series model would have the general equation as follows:

$$G_i(Y\_i_t) = \sum_{m=1}^{M} F_{im}(t, Y\_i_{t-j}, X\_i_{t-k}, Z_{t-l})^* Dt_m + \mu_{it} \tag{2.31}$$

where $Dt_m = Dtm$ is a dummy variable of the time *t*, defined as $Dtm = 1$ for the *m*-th time period, and $Dt_m = 0$ otherwise, $G_i(Y\_i_t)$ is a function of Y_i_t without a parameter, and $F_{im}(t, Y\_i_{t-j}, X\_i_{t-k}, Z_{t-l})$ is a function of the lags variables Y_i_{t-j}, X_i_{t-k} and Z_{t-l}, having a finite number of parameters, for all *i* and *m*. For an illustration, the simplest two-piece multivariate additive model with trend is as follows:

$$\begin{aligned}
\log(Y\_i_t) = {} & (c(i10) + c(i11)^* t + c(i12)^* Y\_i_{t-1} + c(i13)^* X\_i_t + c(i14)^* Z_t)^* Dt_1 \\
& + (c(i20) + c(i21)^* t + c(i22)^* Y\_i_{t-1} + c(i23)^* X\_i_t + c(i24)^* Z_t)^* Dt_2 + \mu_{it}
\end{aligned} \tag{2.32}$$

The other models can easily be derived from all models presented earlier, as well as presented in Agung (2009a). Furthermore, note that each time period can have different sets of variables, and various functions of $F_{im}(t, Y\_i_{t-j}, X\_i_{t-k}, Z_{t-l})$ as well as the endogenous variable $G_i(Y\_i_t)$ by time periods.

3

Data Analysis Based on Multivariate Time Series by States

3.1 Introduction

All time series models presented in Agung (2009a), as well as their extensions, are valid or applicable models for multivariate time series data of many states. Therefore, following the models previously presented, specifically their path diagrams, models based on multivariate time series by states could easily be derived or defined as a set of time-series models or SCMs by states, demonstrated in Chapter 2 under the assumption or precondition that the states are either uncorrelated or correlated.

To generalize, a set of endogenous (impact, problem or down-stream) time-series variables, namely $Y\_i = (Y1\_i, \ldots, YG\_i)$ and exogenous (cause, source or up-stream) time-series variables or indicators, namely $X\_i = (X1\_i, \ldots, XK\_i)$ within the i-th state (individual, country, region, community, agency, household or person) could be considered, for $i = 1, \ldots, N$, and $t = 1, \ldots, T$. Based on these variables and the time t the equation of a model for the i-th state, for independent states, can be presented in a general form as follows:

$$G_i(Y\_i_t) = F_i(t, X\_i_{t-s}, Y\_i_{t-r}) + \mu_{it} \tag{3.1}$$

where $G_i(Y\_i_t)$ is a function of Y_i_t without parameters, and $F_i(t, X\_i_{t-s}, Y\_i_{t-r})$ is a function of the variables t, $X\_i_t, X\_i_{t-1}, \ldots, X\_i_{t-s}$, and $Y\_i_{t-1}, \ldots, Y\_i_{t-r}$ with a finite number of parameters. As a result, there would be a lot of possible models defined, even based on only three to five variables besides time t, since models can be additive, or two- and three-way interaction models, instrumental variables models, VAR and VEC models, as well as the TGARCH(a,b,c) and nonlinear models.

In fact, all models presented in Agung (2009a) should be applicable for every state considered. Therefore, the data analysis can easily be done using any models in Agung (2009a) under the assumption that the states considered are independent states. The data analysis could easily be done using either a *state-by-state analysis* or a *system-specification-of-all-states analysis*.

Note that by using a system-specification-of-all-states analysis, then the model will have a large number of variables. Having a large number of exogenous and endogenous variables in a model would lead to the uncertainty of parameter estimates, because of the unpredictable impact of multicollinearity between all

Panel Data Analysis Using EViews, First Edition. I Gusti Ngurah Agung.
© 2014 John Wiley & Sons, Ltd. Published 2014 by John Wiley & Sons, Ltd.
Companion website: www.wiley.com/go/panel_data

variables in the model. Refer to special notes and comments presented in Agung (2009a, Section 2.14). Even though the states are independent, since the variables of all states are presented in single system equations, then the correlations or multicollinearity of all variables should be taken into account in the estimation process. So if the states are defined to be independent then state-by-state analysis should be applied.

On the other hand, if some of the states should be correlated or associated, then the states should be presented in system equations. Refer to the models for correlated states presented in Chapters 1 and 2. However, note that to define causal or structural relationships between such a large number of variables would be messy. For instance, referring to the path diagram in Figure 2.35, suppose there are five endogenous and five exogenous variables within each state, how would the path diagram look for the causal or structural relationship model based on two correlated states, let alone based on three or more correlated states?

For this reason, every analyst should use his or her best subjective judgment to select sub-sets of exogenous (source, up-stream or cause) variables by levels of their importance. For instance, the first sub-set is the most important cause or up-stream sub-set of variables, and the last sub-set would be the least important set. So data analysis should be done using multistage regression analyses by inserting the sub-sets of the variables one-by-one. Refer to the multistage stepwise regression analysis presented in Chapter 11 (Agung, 2011). Every researcher should present statistical results based on several regression models in all data analyses, beside descriptive statistical summaries.

Corresponding to this idea, the following sections will present illustrative examples based on simple models using only two exogenous time series, namely $(X1\_i,X2\_i) = X\_i$, and an endogenous time series, namely $Y1\_i$, for a small number of states (N). This will be extended to $X\_i = (X1\_i,X2\_i,X3\_i)$ and $Y\_i = (Y1\_i,Y2\_i)$, as the base models.

3.2   Models Based on (X_i,Y_i,Z_i) for Independent States

Referring to various time-series models presented in Agung (2009a) using three time series, namely $M1_t$, GDP_t and PR_t in DEMO.wf1, as an extension based on panel data we will consider M_i_t, GDP_i_t and PR_i_t, for $i = 1, \ldots, N$. To generalize, a set of three time series for the i-th state can be presented as $(X1\_i, X2\_i,Y1\_i)_t$, $(X1\_i,Y1\_i,Y2\_i)_t$ or $(X\_i,Y\_i,Z\_i)_t$.

Therefore, all models using the variables $M1_t$, GDP_t and PR_t, presented in Agung (2009a), can easily be transformed to the models for the i-th states using the time series $(X\_i,Y\_i,Z\_i)_t$. For this reason, only a few selected general models will be presented in the following sections, without the empirical examples.

3.2.1   MLVAR(p, q) Model with Trend Based on (X_i,Y_i,Z_i)

As an illustration, Figure 3.1 presents a path diagram of a LVAR(1,1) with trend for the i-th states, which is derived from the path diagram in Figure 2.28 (Agung, 2009a).

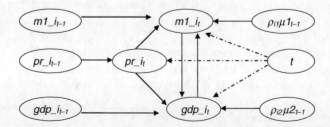

Figure 3.1   *A path diagram derived from Figure 2.28 in Agung (2009a)*

Note that the path diagram shows that $M1\_i$ and GDP_i have simultaneous causal effects.

3.2.1.1 MLVAR(1,1) Additive Model with Trend

Based on the path diagram in Figure 3.1, the simplest MLVAR(p,q) is a MLVAR(**1,1**) additive model with trend, with the following system specification.

$$
\begin{aligned}
m1\_i &= c(10) + c(11)^* t + c(12)^* m1\_i(-1) + c(13)^* gdp\_i \\
&\quad + c(14)^* pr\_i + [ar(1) = c(15)] \\
gdp\_i &= c(20) + c(21)^* t + c(22)^* m1\_i + c(23)^* gdp\_i(-1) \\
&\quad + c(24)^* pr\_1 + [ar(1) = c(25)] \\
pr\_i &= c(30) + c(31)^* t + c(32)^* pr\_i(-1) + [ar(1) = c(33)]
\end{aligned}
\tag{3.2}
$$

Note that by using the variables $log(m1\_i)$, $log(gdp\_i)$ and $log(pr\_i)$, the growth model, namely MLVAR (**1,1**)_GM, is obtained with c(11), c(21) and (31), respectively, being the adjusted growth rates of $m1\_i$, gdp_i and pr_i.

3.2.1.2 MLVAR(1,1) Interaction Model with Trend

Having two or more exogenous variables, the effect of an exogenous variable on the endogenous variable depends on the other exogenous variable(s). For this reason, based on the path diagram in Figure 3.1, applying the following two-way interaction system specification is recommended.

$$
\begin{aligned}
m1\_i &= c(10) + c(11)^* t + c(12)^* m1\_i(-1) + c(13)^* gdp\_i \\
&\quad + c(14)^* pr\_i + c(15)^* gdp\_i^* pr\_i + [ar(1) = c(16)] \\
gdp\_i &= c(20) + c(21)^* t + c(22)^* m1\_i + c(23)^* gdp\_i(-1) \\
&\quad + c(24)^* pr\_i + c(25)^* m1\_i^* pr\_i + [ar(1) = c(26)] \\
pr\_i &= c(30) + c(31)^* t + c(32)^* pr\_i(-1) + [ar(1) = c(33)]
\end{aligned}
\tag{3.3}
$$

Note that the first two regressions in this model are hierarchical two-way interaction regressions indicated by the independent variables gdp_i, pr_i and $gdp\_i^* pr\_i$ in the first regression, and $m1\_i$, pr_i and $m1\_i^* pr\_i$ in the second regression.

To generalize, for a set of independent states, a MLVAR(p,q) model with trend based on the time series $(X\_i, Y\_i, Z\_i)_t$, for $i = 1, \ldots, N$, will have the following general system specification.

$$
G_i(Y\_i_t) = c(i10) + c(i11)^* t + F_i(X\_i_{t-s}, Z\_i_{t-r}) + \sum_{j=1}^{p_i} c(g1j)^* Y\_i_{t-j}
$$
$$
+ \sum_{k=1}^{q_k} [AR(k) = c(g2k)]
\tag{3.4}
$$

where $G_i(Y\_i)$ is a function of Y_i without a parameter, and $F_i(X\_i_{t-s}, Z\_i_{t-r})$ is a function of X_i_{t-s} and Z_i_{t-r}, for some s and r with a finite number of parameters.

Note that by using the variables $G_i(Y\_i) = log(Y\_i)$, the growth model, namely MLVAR(p,q)_GM is obtained, with $c(i11)$ being the adjusted growth rate of Y_i.

3.2.1.3   MLVAR(1,1) Model with Time-Related Effects

As an extension of the model in (3.3), model (3.5) presents a MLVAR(1,1) model with time-related effects (TRE). Note that the first two regressions in this model are hierarchical three-way interaction regressions, indicated by the three-way interaction independent variables $t*gdp\_i* pr\_i$ with its two-way interactions and main factors in the first regression, and $t*ml\_i* pr\_i$ with its two-way interactions and main factors in the second regression.

$$
\begin{aligned}
ml\_i = \, & c(10) + c(11)^* t + c(12)^* ml\_i(-1) + c(13)^* gdp\_i + c(14)^* pr\_i \\
& + c(15)^* gdp\_i^* pr\_i + c(16)^* t^* gdp\_i + c(17)^* t^* pr\_i \\
& + c(18)^* t^* gdp\_i^* pr\_i + [ar(1) = c(19)] \\
gdp\_i = \, & c(20) + c(21)^* t + c(22)^* ml\_i + c(23)^* gdp\_i(-1) + c(24)^* pr\_i \\
& + c(15)^* ml\_i^* pr\_i + c(16)^* t^* ml\_i + c(17)^* t^* pr\_i \\
& + c(18)^* t^* ml\_i^* pr\_i + [ar(1) = c(29)] \\
pr\_i = \, & c(30) + c(31)^* t + c(32)^* pr\_i(-1) + [ar(1) = c(33)]
\end{aligned}
\tag{3.5}
$$

To generalize, for a set of independent states, a MLVAR(p,q) model with TRE based on the time series $(X\_i, Y\_i, Z\_i)_t$ for $i = 1, \ldots, N$, will have the following general system specification.

$$
\begin{aligned}
G_i(Y\_i_t) = \, & c(i10) + c(i11)^* t + t^* F_i(X\_i_{t-r}, Z\_i_{t-s}) + \sum_{j=1}^{p_i} c(g1j)^* Y\_i_{t-j} \\
& + \sum_{k=1}^{q_k} [AR(k) = c(g2k)]
\end{aligned}
\tag{3.6}
$$

By using the variables $G_i(Y\_i) = log(Y\_i)$, the growth model: MLVAR(p,q)_GM is obtained. Note that this model has independent variables $[(ci1) + F_i(X\_i_{t-r}, Z\_i_{t-s})]*t$ which indicate that the growth rate of Y_i depends on the function $[(ci1) + F_i(X\_i_{t-r}, Z\_i_{t-s})]$, by taking into account (adjusted for) other variables in the model for selected integers $r \geq 0$ and $s \geq 0$. Then, alternative good fit models may be obtained, which are highly dependent on the sampled data used.

A more general model has the equation specification as follows:

$$
\begin{aligned}
G_i(Y\_i_t) = \, & F_i(t, X\_i_{t-r}, Z\_i_{t-s}) + \sum_{j=1}^{p_i} c(g1j)^* Y\_i_{t-j} \\
& + \sum_{k=1}^{q_k} [AR(k) = c(g2k)]
\end{aligned}
\tag{3.7}
$$

3.3   Models Based on (X_i, Y_i, Z_i) for Correlated States

Note that the models based on two correlated states using the variables (X_1, Y_1), (X_2, Y_2) and an environmental variable $Z1$ presented in Section 2.6, can be viewed as the special cases of the models using variables $(X\_i, Y\_i, Z\_i)_t$ for $i = 1$ and 2, where $Z\_1_t = Z\_2_t = Z1_t$. For this reason the models based on two correlated states using the variables (X_i, Y_i, Z_i) could easily be derived from all models presented in

Section 2.6. However, the models presented are under the assumption only the endogenous of the states are correlated, which has been mentioned in the previous chapters. As an illustration, hypothetical panel data have been generated based on the data in POOL1.wf1, for $N=4$, as follows: $X\_1 = ivmdep$, $X\_2 = ivmhoa$, $X\_3 = ivmmae$, $X\_4 = ivmmis$, $Y\_1 = ivmaut$, $Y\_2 = ivmbus$, $Y\_3 = ivmcon$ and $Y\_4 = ivmcst$ and in addition it is generated $Z\_1 = mmaut$, $Z\_2 = mmbus$, $Z\_3 = mmcon$ and $Z\_4 = mmcst$.

Another hypothetical data set could easily be generated based on the seven states in POOLG7.wf1, such as for the four states in Europe *(GDP_Fra,GDP_Ger,GDP_ITA, GDP_UK)* $= (X1\_1, X2\_1, Y\_1, Z\_1)$ and for three states outside Europe *(GDP_Can, GDP_US,GDP_JPN)* $= (X\_2, Y\_2, Z\_2)$. The others could easily be generated based on other time-series data sets in the Example Files of EViews.

3.3.1 MLV(1) Interaction Model with Trend

As a modification of the path diagram in Figure 2.30, Figure 3.2 presents the path diagram of a MLV(**1**) model with trend based on two correlated states, which is indicated by the direct effects of the first lags Y_1 (-1) and $Y\_2(-1)$ on both Y_1 and Y_2.

$$
\begin{aligned}
Y\_1_t &= C(10) + C(11)^* t + C(12)^* Y\_1_{t-1} + C(13)^* Y\_2_{t-1} \\
&\quad + C(14)^* X\_1_t + C(15)^* Z\_1_t + C(16)^* X\_1_t{}^* Z\_1_t + \mu 1_t \\
Y\_2_t &= C(20) + C(21)^* t + C(22)^* Y\_1_{t-1} + C(23)^* Y\_2_{t-1} \\
&\quad + C(24)^* X\_2_t + C(25)^* Z\_2_t + C(26)^* X\_2_t{}^* Z\_2_t + \mu 2_t
\end{aligned}
\tag{3.8}
$$

If the effects of Z_i on X_i are taken into consideration, the following SCM model with trend is obtained.

$$
\begin{aligned}
Y\_1_t &= C(10) + C(11)^* t + C(12)^* Y\_1_{t-1} + C(13)^* Y\_2_{t-1} \\
&\quad + C(14)^* X\_1_t + C(15)^* Z\_1_t + C(16)^* X\_1_t{}^* Z\_1_t + \mu 1_t \\
Y\_2_t &= C(20) + C(21)^* t + C(22)^* Y\_1_{t-1} + C(23)^* Y\_2_{t-1} \\
&\quad + C(24)^* X\_2_t + C(25)^* Z\_2_t + C(26)^* X\_2_t{}^* Z\_2_t + \mu 2_t \\
X\_1_t &= C(30) + C(31)^* t + C(32)^* Z\_1_t + \mu 3_t \\
X\_2_t &= C(40) + C(41)^* t + C(42)^* Z\_2_t + \mu 4_t
\end{aligned}
\tag{3.9}
$$

Note the possibility of using the lags X_i_{t-r} and Z_i_{t-s} for some selected integers r > 0 and s > 0, as the independent variables of each model in the system Eqs. (3.8) and (3.9).

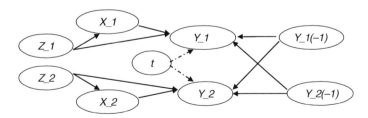

Figure 3.2 *The path diagram of a SCM based on (X_i,Y_i,Z_i), for two correlated states*

```
System: SYS15_A
Estimation Method: Seemingly Unrelated Regression
Date: 11/23/09   Time: 09:52
Sample: 1968M02 1994M10
Included observations: 321
Total system (balanced) observations 642
Linear estimation after one-step weighting matrix
```

	Coefficient	Std. Error	t-Statistic	Prob.
C(10)	-44.27498	26.06444	-1.698673	0.0899
C(11)	0.206221	0.398561	0.517414	0.6050
C(12)	0.976793	0.010025	97.43532	0.0000
C(13)	-0.011050	0.005331	-2.072888	0.0386
C(14)	0.015008	0.003626	4.138601	0.0000
C(15)	0.046486	0.008830	5.264384	0.0000
C(16)	-1.83E-06	3.97E-07	-4.618022	0.0000
C(20)	-443.6528	116.5814	-3.805518	0.0002
C(21)	-3.907772	1.020978	-3.827480	0.0001
C(22)	-0.035650	0.035616	-1.000962	0.3172
C(23)	0.933773	0.013482	69.26090	0.0000
C(24)	0.061011	0.015596	3.912086	0.0001
C(25)	0.125243	0.030953	4.046294	0.0001
C(26)	-1.65E-06	6.58E-07	-2.510298	0.0123

Determinant residual covariance	2.72E+08

Equation: Y_1=C(10)+C(11)*T+C(12)*Y_1(-1)+C(13)*Y_2(-1)+C(14)*X_1
+C(15)*Z_1+C(16)*X_1*Z_1
Observations: 321

R-squared	0.997283	Mean dependent var	4796.988
Adjusted R-squared	0.997231	S.D. dependent var	1564.584
S.E. of regression	82.32419	Sum squared resid	2128064.
Durbin-Watson stat	2.106516		

Equation: Y_2=C(20)+C(21)*T+C(22)*Y_1(-1)+C(23)*Y_2(-1)+C(24)*X_2
+C(25)*Z_2+C(26)*X_2*Z_2
Observations: 321

R-squared	0.998951	Mean dependent var	14089.85
Adjusted R-squared	0.998931	S.D. dependent var	6599.741
S.E. of regression	215.7785	Sum squared resid	14619951
Durbin-Watson stat	2.022317		

Figure 3.3   *Statistical results based on a reduced model in (3.8)*

Example 3.1   Application of the model in (3.8)

Figure 3.3 presents the results based on a model in (3.8), using the following system equation. Based on these results the following notes and conclusions are made.

$$y\_1 = c(10) + c(11)^* t + c(12)^* y\_1(-1) + c(13)^* y\_2(-1) + c(14)^* x\_1 + c(15)^* z\_1 + c(16)^* x\_1^* z\_1$$

$$y\_2 = c(20) + c(21)^* t + c(22)^* y\_1(-1) + c(23)^* y\_2(-1) + c(24)^* x\_2 + c(25)^* z\_2 + c(26)^* x\_2^* z\_2$$

1. Since each of X_i, Z_i and $X\_i^* Z\_i$ has a significant adjusted effect on Y_i, then it can be concluded the three independent variables have a significant joint effect on each Y_i.
2. Furthermore, it can be concluded that the data supports the hypothesis stated that the effect of X_i (Z_i) on Y_i depends on Z_i (X_i), adjusted for the other variables in the model. Compare this to the following example.
3. Note that the time t has an insignificant effect on each Y_i with a large p-value. In a statistical sense, the model might be reduced by deleting the time t. Do as an exercise.
4. In fact, the time t and each Y_i, $i = 1$ and 2 have a significant positive correlation of 0.932 352 and 0.972 359, respectively with p-values $= 0.0000$. So these findings show the unexpected impacts of multicollinearity between the independent variables. Refer to the special notes and comments on the multicollinearity problems in Section 2.14.2 (Agung, 2009a).
5. Corresponding to the findings in point (4), the time t could be used as an independent variable to indicate the trends of each Y_i is insignificant, adjusted for the other variables in the model. Compare this to the following example.

Example 3.2   Unexpected growth model

Figure 3.4 presents the results based on a growth model (translog linear model with trend), as a special case of the model in (3.6). Based on these results the following notes and conclusions are presented.

1. Compared to the model in Figure 3.3, this model gives so many variables that have insignificant adjusted effects with p-values > 0.20 (or 0.25). There is nothing wrong with the model but there are unexpected impacts of multicollinearity between the independent variables.

System: SYS15_B
Estimation Method: Seemingly Unrelated Regression
Date: 11/23/09   Time: 10:42
Sample: 1968M02 1994M10
Included observations: 321
Total system (balanced) observations 642
Linear estimation after one-step weighting matrix

	Coefficient	Std. Error	t-Statistic	Prob.
C(10)	-0.382069	0.401020	-0.952744	0.3411
C(11)	-9.61E-05	6.75E-05	-1.424196	0.1549
C(12)	0.973859	0.011728	83.03977	0.0000
C(13)	0.004862	0.012596	0.386005	0.6996
C(14)	0.037068	0.050918	0.728005	0.4669
C(15)	0.070018	0.052767	1.326930	0.1850
C(16)	-0.004613	0.005780	-0.798112	0.4251
C(20)	-0.276597	0.618248	-0.447389	0.6547
C(21)	-0.000218	5.31E-05	-4.110844	0.0000
C(22)	-0.010900	0.011400	-0.956205	0.3393
C(23)	0.934889	0.012208	76.58266	0.0000
C(24)	0.055718	0.069721	0.799155	0.4245
C(25)	0.035513	0.079225	0.448254	0.6541
C(26)	0.001619	0.007819	0.207085	0.8360
Determinant residual covariance	4.06E-08			

Equation: LOG(Y_1)=C(10)+C(11)*T+C(12)*LOG(Y_1(-1))+C(13)*LOG(Y_2(
   -1))+C(14)*LOG(X_1)+C(15)*LOG(Z_1)+C(16)*LOG(X_1)*LOG(Z_1)
Observations: 321

R-squared	0.998300	Mean dependent var	8.408489
Adjusted R-squared	0.998267	S.D. dependent var	0.390481
S.E. of regression	0.016254	Sum squared resid	0.082952
Durbin-Watson stat	2.123139		

Equation: LOG(Y_2)=C(20)+C(21)*T+C(22)*LOG(Y_1(-1))+C(23)*LOG(Y_2(
   -1))+C(24)*LOG(X_2)+C(25)*LOG(Z_2)+C(26)*LOG(X_2)*LOG(Z_2)
Observations: 321

R-squared	0.999479	Mean dependent var	9.410050
Adjusted R-squared	0.999469	S.D. dependent var	0.574135
S.E. of regression	0.013233	Sum squared resid	0.054987
Durbin-Watson stat	1.967455		

Figure 3.4   *Statistical results based on a growth model based on the path diagram in Figure 3.2*

2. At a significance level of $\alpha = 0.10$, the time t has a significant negative effect on $log(Y\_1)$, based on the t-statistics of $t_0 = 1.424\ 196$ with a p-value $= 0.1549/2 = 0.07\ 745$, and it has an insignificant negative effect on $log(Y\_2)$ with a p-value $= 0.3393/2 = 0.16\ 965$. In fact, the time t and $log(Y\_i)$, $i = 1$ and 2 have significant positive correlations of 0.910 547 and 0.954 609, respectively, with p-values $= 0.0000$.

3. Each of the independent variables $log(X\_1)$, $log(Z\_1)$ and $log(X\_1)*log(Z\_1)$ has insignificant adjusted effects on $log(Y\_1)$ but their joint effect is significant based on the Chi-square statistic of (with $df = 3$ and a p-value $= 0.0005$. For this reason, only one or two of the variables can be deleted in order to obtain an acceptable or a good fit model in a statistical sense. Since the hypothesis stated that the effect of X_1 (Z_1) on Y_1 depends on Z_1 (X_1), then only the main factor(s) should be deleted, either one of $log(X\_1)$ and $log(Z\_1)$ or both.

4. Similarly, $log(X\_2)$, $log(Z\_2)$ and $log(X\_2)*log(Z\_2)$ have a significant joint effect on $log(Y\_2)$, based on the Chi-square statistic of $\chi_0^2 = 37.70134$ with $df = 3$ and a p-value $= 0.0000$.

5. By using trial-and-error methods the results in Figure 3.5 are obtained, which show that $log(X\_i)*log(Z\_i)$ has a significant positive adjusted effect on $log(Y\_i)$ for each i. Based on these results, we can conclude the data supports the hypothesis stated that the effect of X_i (Z_i) on Y_i depends on Z_i (X_i), for each i.

3.3.2   MLV(1) Interaction Model with Time-Related Effects

As an extension of the path diagram in Figure 3.2, Figure 3.6 presents a path diagram which indicates that the effects of X_i and Z_i on Y_i depend on the time t. In this case, a model with time-related effects, namely the MLV(**1**)_Model with TRE, should be considered.

As an extension of the model in (3.8), a model with TRE would have the following equation, for $i = 1$ and 2. See the following example.

$$Y\_i_t = C(i0) + C(i1)^* t + C(i2)^* Y\_1_{t-1} + C(i3)^* Y\_2_{t-1}$$
$$+ C(i4)^* X\_i_t + C(15)^* Z\_i_t + C(16)^* X\_i_t{}^* Z\_i_t$$
$$+ C(i5)^* t^* X\_i_t + C(i6)^* t^* Z\_i_t + C(26)^* t^* X\_i_t{}^* Z\_i_t + \mu_{it} \tag{3.10}$$

System: SYS15_C
Estimation Method: Seemingly Unrelated Regression
Date: 11/23/09   Time: 11:51
Sample: 1968M02 1994M10
Included observations: 321
Total system (balanced) observations 642
Linear estimation after one-step weighting matrix

	Coefficient	Std. Error	t-Statistic	Prob.
C(10)	0.143552	0.054328	2.642333	0.0084
C(11)	-0.000151	5.36E-05	-2.821517	0.0049
C(12)	0.975896	0.011642	83.82429	0.0000
C(13)	0.012970	0.010808	1.200033	0.2306
C(14)	-0.029446	0.007753	-3.797873	0.0002
C(16)	0.002987	0.000756	3.950447	0.0001
C(20)	-0.037123	0.193750	-0.191600	0.8481
C(21)	-0.000229	4.14E-05	-5.535030	0.0000
C(22)	-0.007229	0.008972	-0.805713	0.4207
C(23)	0.937574	0.011900	78.78820	0.0000
C(24)	0.029860	0.027780	1.074885	0.2828
C(26)	0.004786	0.001630	2.936246	0.0034

Determinant residual covariance   4.08E-08

Equation: LOG(Y_1)=C(10)+C(11)*T+C(12)*LOG(Y_1(-1))+C(13)*LOG(Y_2(
    -1))+C(14)*LOG(X_1)+C(16)*LOG(X_1)*LOG(Z_1)
Observations: 321

R-squared	0.998295	Mean dependent var	8.408489
Adjusted R-squared	0.998268	S.D. dependent var	0.390481
S.E. of regression	0.016252	Sum squared resid	0.083204
Durbin-Watson stat	2.133725		

Equation: LOG(Y_2)=C(20)+C(21)*T+C(22)*LOG(Y_1(-1))+C(23)*LOG(Y_2(
    -1))+C(24)*LOG(X_2)+C(26)*LOG(X_2)*LOG(Z_2)
Observations: 321

R-squared	0.999478	Mean dependent var	9.410050
Adjusted R-squared	0.999470	S.D. dependent var	0.574135
S.E. of regression	0.013222	Sum squared resid	0.055069
Durbin-Watson stat	1.968586		

Figure 3.5   *Statistical results based on a reduced model in Figure 3.4*

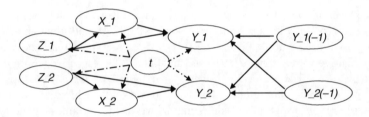

Figure 3.6   *The path diagram of a MLV(1) based on (X_i,Y_i,Z_i), for two correlated states*

Example 3.3   (MLV(1) with TRE)

Figure 3.7 presents the statistical results based on a MLV(1) with TRE (trend-related effects) in (3.10). Based on these results, the following notes and conclusions are made.

1. Since each of the variables t, t^*X_2, t^*Z_2 and $t^*X\_2^*Z\_2$ has a significant adjusted effect on Y_2, it can then be concluded that the variables have significant joint effects on Y_2. In other words, the joint effects of X_2, Z_2 and $X\_2^*Z\_2$ on Y_2 are significantly dependent on the time t.
2. On the other hand, each of the variables t, t^*X_1, t^*Z_1 and $t^*X\_1^*Z\_1$ has an insignificant adjusted effect on Y_1 with a large p-value, but their joint effects are significant based on the Chi-square statistic of $\chi_0^2 = 15.13211$ with $df=4$ and a p-value $=0.0044$. Therefore, it can be concluded that the joint effects of X_1, Z_1 and X_1^*Z1 on Y_1 are significantly dependent on time t.
3. However, in a statistical sense, a reduced model of the first regression should be explored. Several good fit reduced models might be obtained by using trial-and-error methods to delete alternative subsets of the main factors t, X_1 and Z_1 or their two-way interactions. In other words, the three-way interaction model should be kept to represent that the effects of X_i and Z_i on Y_i depend on the time t. One of the reduced models is presented in Figure 3.8, which is obtained by deleting the time t and Z_1 from the

System: SYS16_A
Estimation Method: Seemingly Unrelated Regression
Date: 11/23/09   Time: 12:36
Sample: 1968M02 1994M10
Included observations: 321
Total system (balanced) observations 642
Linear estimation after one-step weighting matrix

	Coefficient	Std. Error	t-Statistic	Prob.
C(10)	-12.07900	72.10217	-0.167526	0.8670
C(11)	-0.041813	0.868803	-0.048127	0.9616
C(12)	0.948630	0.013684	69.32254	0.0000
C(13)	-0.004184	0.006578	-0.636044	0.5250
C(14)	-0.001212	0.011153	-0.108659	0.9135
C(15)	0.080410	0.021451	3.748539	0.0002
C(16)	-9.39E-07	1.36E-06	-0.688664	0.4913
C(17)	6.35E-05	4.62E-05	1.372987	0.1703
C(18)	-0.000107	9.39E-05	-1.138481	0.2554
C(19)	-3.35E-09	5.78E-09	-0.579667	0.5623
C(20)	-244.8784	395.8168	-0.618666	0.5364
C(21)	0.942139	3.246312	0.290218	0.7717
C(22)	-0.079649	0.036626	-2.174663	0.0300
C(23)	0.899765	0.017596	51.13574	0.0000
C(24)	0.111499	0.050956	2.188133	0.0290
C(25)	-0.110012	0.067093	-1.639689	0.1016
C(26)	6.76E-06	3.52E-06	1.921141	0.0552
C(27)	-0.000368	0.000166	-2.216124	0.0270
C(28)	0.000807	0.000293	2.758300	0.0060
C(29)	-2.25E-08	8.45E-09	-2.663353	0.0079

Determinant residual covariance   2.50E+08

Equation: Y_1=C(10)+C(11)*T+C(12)*Y_1(-1)+C(13)*Y_2(-1)+C(14)*X_1
+C(15)*Z_1+C(16)*X_1*Z_1+C(17)*T*X_1+C(18)*T*Z_1+C(19)*T*X_1
*Z_1
Observations: 321

R-squared	0.997376	Mean dependent var	4796.987
Adjusted R-squared	0.997300	S.D. dependent var	1564.584
S.E. of regression	81.29839	Sum squared resid	2055532.
Durbin-Watson stat	2.201241		

Equation: Y_2=C(20)+C(21)*T+C(22)*Y_1(-1)+C(23)*Y_2(-1)+C(24)*X_2
+C(25)*Z_2+C(26)*X_2*Z_2+C(27)*T*X_2+C(28)*T*Z_2+C(29)*T*X_2
*Z_2
Observations: 321

R-squared	0.999011	Mean dependent var	14089.85
Adjusted R-squared	0.998983	S.D. dependent var	6599.741
S.E. of regression	210.4811	Sum squared resid	13778009
Durbin-Watson stat	2.034817		

Figure 3.7   *Statistical results based on a MLV(1) with TRE, for two correlated states*

System: SYS16_C
Estimation Method: Seemingly Unrelated Regression
Date: 11/23/09   Time: 14:11
Sample: 1968M02 1994M10
Included observations: 321
Total system (balanced) observations 642
Linear estimation after one-step weighting matrix

	Coefficient	Std. Error	t-Statistic	Prob.
C(10)	179.8312	41.20675	4.364119	0.0000
C(12)	0.967426	0.011739	82.41435	0.0000
C(13)	-0.008164	0.004396	-1.857065	0.0638
C(14)	-0.017625	0.006787	-2.596795	0.0096
C(16)	2.68E-06	7.84E-07	3.415779	0.0007
C(17)	0.000103	2.68E-05	3.832015	0.0001
C(18)	0.000138	3.51E-05	3.945152	0.0001
C(19)	-1.57E-08	3.25E-09	-4.821343	0.0000
C(20)	-157.6362	395.3295	-0.398746	0.6902
C(21)	1.577580	3.244070	0.486297	0.6269
C(22)	-0.082269	0.036397	-2.260289	0.0241
C(23)	0.896245	0.017525	51.14163	0.0000
C(24)	0.104545	0.050983	2.050579	0.0407
C(25)	-0.123284	0.066995	-1.840193	0.0662
C(26)	7.77E-06	3.51E-06	2.214305	0.0272
C(27)	-0.000370	0.000166	-2.228442	0.0262
C(28)	0.000806	0.000293	2.750454	0.0061
C(29)	-2.41E-08	8.43E-09	-2.857113	0.0044

Determinant residual covariance   2.62E+08

Equation: Y_1=C(10)+C(12)*Y_1(-1)+C(13)*Y_2(-1)+C(14)*X_1+C(16)*X_1
*Z_1+C(17)*T*X_1+C(18)*T*Z_1+C(19)*T*X_1*Z_1
Observations: 321

R-squared	0.997276	Mean dependent var	4796.987
Adjusted R-squared	0.997215	S.D. dependent var	1564.584
S.E. of regression	82.56507	Sum squared resid	2133718.
Durbin-Watson stat	2.091585		

Equation: Y_2=C(20)+C(21)*T+C(22)*Y_1(-1)+C(23)*Y_2(-1)+C(24)*X_2
+C(25)*Z_2+C(26)*X_2*Z_2+C(27)*T*X_2+C(28)*T*Z_2+C(29)*T*X_2
*Z_2
Observations: 321

R-squared	0.999012	Mean dependent var	14089.85
Adjusted R-squared	0.998983	S.D. dependent var	6599.741
S.E. of regression	210.4556	Sum squared resid	13774673
Durbin-Watson stat	2.026020		

Figure 3.8   *Statistical results based on a reduced model in Figure 3.7*

model. Do this as an exercise for the other reduced models. Note that an acceptable reduced model may have one or two independent variable(s) with insignificant adjusted effects.

4. On the other hand, stepwise estimation methods can be applied based on the first regression only. Refer to Chapter 11 in Agung (2011), for the alternative *"Equation Specification"* with *"List of search regressors"* in order to obtain a good fit model based on two- and three-way interaction models.

5. Note that this model can easily be extended to a more advanced model with TRE by inserting additional two-way interaction independent variables, namely $t^*Y\_i(-1)$ and $t^*Y\_2(-1)$.

3.4   Simultaneous SCMs with Trend

As an extension of the path diagram in Figure 2.6, Figure 3.9 presents the path diagram where Y_1 and Y_2 are defined as having simultaneous causal relationships, and the effect of X_i on Y_i depends on Z_i for each i. Then additive, two- and three-way interaction models could be derived as presented previously. However, for illustrative purposes, the following sections present several basic models with trend, and each of them can easily be extended to various models with time-related effects, MLVAR(p,q) models for $p \geq 0$ and $q \geq 0$, instrumental variables models and TGARCH(a,b,c) models (Agung, 2009a).

3.4.1   The Basic Simultaneous SCMs with Trend

3.4.1.1   The Simplest Basic Model with Trend

Based on the path diagram in Figure 3.9, the simplest basic model that could be considered is an additive multivariate model with four regressions as follows:

$$Y\_1_t = C(10) + C(11)^* t + C(12)^* X\_1_t + C(13)^* Z\_1_t + C(14)^* Y\_2_t + \mu 1_t$$
$$Y\_2_t = C(20) + C(21)^* t + C(22)^* X\_2_t + C(23)^* Z\_2_t + C(24)^* Y\_1_t + \mu 2_t$$
$$X\_1_t = C(30) + C(11)^* Z\_1_t + \mu 3_t$$
$$X\_2_t = C(30) + C(11)^* Z\_2_t + \mu 4_t$$

$$(3.11)$$

However, in general, the effect of $X\_i\ (Z\_i)$ on Y_i depends on $Z\_i\ (X\_i)$ for each i. For this reason, I recommend considering the interaction models, as follows.

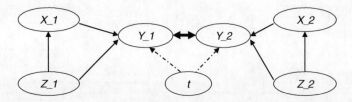

Figure 3.9   *The path diagram of a simultaneous SCM of (Y_1, Y_2) with trend on (X_1, Z_1) and (X_2, Z_2)*

3.4.1.2   The Basic Two-Way Interaction Models with Trend

Two types of two-way interaction basic models can be considered. The first basic model shows that the effect of $X\_i\ (Z\_i)$ on Y_i depends on $Z\_i\ (X\_i)$ for each i, with the equation as follows:

$$
\begin{aligned}
Y\_1_t ={}& C(10) + C(11)^*t + C(12)^*X\_1_t + C(13)^*Z\_1_t \\
&+ C(14)^*X\_1_t{}^*Z\_1_t + C(15)^*Y\_2_t + \mu 1_t \\
Y\_2_t ={}& C(20) + C(21)^*t + C(22)^*X\_2_t + C(23)^*Z\_2_t \\
&+ C(24)^*X\_2_t{}^*Z\_2_t + C(25)^*Y\_1_t + \mu 2_t
\end{aligned}
\tag{3.12}
$$

And the second basic model shows that the effect of $X\_i\ (Z\_i)$ on Y_i depends on $Z\_i\ (X\_i)$ for each i, the effect of Y_2 on Y_1 depends on (X_2, Z_2), and the effect of Y_1 on Y_2 depends on (X_1, Z_1), with the equation as follows:

$$
\begin{aligned}
Y\_1_t ={}& C(10) + C(11)^*t \\
&+ C(12)^*X\_1_t + C(13)^*Z\_1_t + C(14)^*X\_1_t{}^*Z\_1_t \\
&+ C(15)^*Y\_2_t + C(16)^*Y\_2_t{}^*X\_2_t + C(17)^*Y\_2_t{}^*Z\_2_t + \mu 1_t \\
Y\_2_t ={}& C(20) + C(21)^*t \\
&+ C(22)^*X\_2_t + C(23)^*Z\_2_t + C(24)^*X\_2_t{}^*Z\_2_t \\
&+ C(25)^*Y\_1_t + C(26)^*Y\_1_t{}^*X\_1_t + C(27)^*Y\_1_t{}^*Z\_1_t + \mu 2_t
\end{aligned}
\tag{3.13}
$$

Note that the two-way interactions $Y\_2^*X\_2$ and $Y\_2^*Z\_2$ in the first regression represent that both X_2 and Z_2 have indirect effects on Y_1 through Y_2. Similarly, the two-way interactions $Y\_1^*X\_1$ and $Y\_1^*Z\_1$ in the second regression show that both X_1 and Z_1 have indirect effects on Y_2 through Y_1.

3.4.1.3   The Basic Three-Way Interaction Model with Trend

As an extension of the model in (3.13), a three-way interaction basic model can be considered under the assumption that Y_i, X_i and Z_i are completely correlated for each i. Note that it is very difficult to identify whether or not a set of three variables is completely correlated in a theoretical sense. Therefore, it is a very subjective decision as to whether or not a three-way interaction model should be applied.

$$
\begin{aligned}
Y\_1_t ={}& C(10) + C(11)^*t \\
&+ C(12)^*X\_1_t + C(13)^*Z\_1_t + C(14)^*X\_1_t{}^*Z\_1_t \\
&+ C(15)^*Y\_2_t + C(16)^*Y\_2_t{}^*X\_2_t + C(17)^*Y\_2_t{}^*Z\_2_t \\
&+ C(18)^*Y\_2_t{}^*X\_2_t{}^*Z\_2_t + \mu 1_t \\
Y\_2_t ={}& C(20) + C(21)^*t \\
&+ C(22)^*X\_2_t + C(23)^*Z\_2_t + C(24)^*X\_2_t{}^*Z\_2_t \\
&+ C(25)^*Y\_1_t + C(26)^*Y\_1_t{}^*X\_1_t + C(27)^*Y\_1_t{}^*Z\_1_t \\
&+ C(28)^*Y\_1_t{}^*X\_1_t{}^*Z\_1_t + \mu 2_t
\end{aligned}
\tag{3.14}
$$

3.4.1.4   The Basic Simultaneous Causal Models with Time-Related-Effects

As an extension of the SCMs with trend in (3.11), the SCMs with time-related effects (TRE) can easily be derived. For instance, based on the path diagram in Figure 3.9 and the model in (3.12), the basic model with

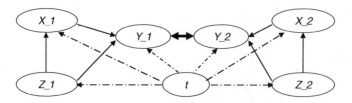

Figure 3.10 *The path diagram of a simultaneous SCM of (Y_1, Y_2) on (X_1, Z_1) and (X_2, Z_2), with trend*

TRE has the following equation

$$
\begin{aligned}
Y\_1_t = {}& C(10) + C(11)^* t + C(12)^* X\_1_t + C(13)^* Z\_1_t \\
& + C(14)^* X\_1_t {}^* Z\_1_t + C(15)^* Y\_2_t + C(16)^* t^* Y\_2_t + \mu 1_t \\
Y\_2_t = {}& C(20) + C(21)^* t + C(22)^* X\_2_t + C(23)^* Z\_2_t \\
& + C(24)^* X\_2_t {}^* Z\_2_t + C(25)^* Y\_1_t + C(26)^* t^* Y\_1_t + \mu 2_t
\end{aligned}
\tag{3.15}
$$

Note that the interaction t^*Y_2 in the first regression indicates that the effect of Y_2 on Y_1 depends on the time t, and the interaction t^*Y_1 in the second regression indicates that the effect of Y_1 on Y_2 depends on the time t. Write out alternative basic models with TRE based on the other models as an exercise.

Further extension of the basic models with TRE can be derived based on the path diagram in Figure 3.10. Note this figure shows that the time t has effects on all variables considered. As an illustration, based on the model in (3.15), the following model with TRE is obtained. Write out the other models as an exercise.

$$
\begin{aligned}
Y\_1_t = {}& C(10) + C(11)^* t + C(12)^* X\_1_t + C(13)^* Z\_1_t \\
& + C(14)^* X\_1_t {}^* Z\_1_t + C(15)^* t^* X\_1_t + C(16)^* t^* Z\_1_t \\
& + C(17)^* t^* X\_1_t {}^* Z\_1_t + C(18)^* Y\_2_t + C(19)^* t^* Y\_2_t + \mu 1_t \\
Y\_2_t = {}& C(20) + C(21)^* t + C(22)^* X\_2_t + C(23)^* Z\_2_t \\
& + C(24)^* X\_2_t {}^* Z\_2_t + C(25)^* t^* X\_2_t + C(26)^* t^* Z\_2_t \\
& + C(27)^* t^* X\_2_t {}^* Z\_2_t + C(28)^* Y\_1_t + C(29)^* t^* Y\_1_t + \mu 2_t
\end{aligned}
\tag{3.16}
$$

As a modified model the exogenous variables Z_1_t and Z_2_t can be replaced by an environmental variable Z_{t-s} for an integer $s \geq 0$, and then a system equation with an environmental variable can be obtained.

3.4.2 Alternative Time Series Models

Should the basic models in (3.11) to (3.16) have serial correlation problems, then the MLVAR($p;q$) models for selected for $p \geq 0$ and $q \geq 0$, should be applied. Alternative simple models would be MLV(1), MAR(1) and MLVAR(1,1). See the following example.

Example 3.4 An application of the simultaneous causal model in (3.14)
Figure 3.11 presents the results based on the MLV(1) model in (3.14), since by using the MLVAR(1;1), it is found that the AR(1) terms are insignificant in both regressions. Based on these results, the following notes and comments are made.

1. In order to test the hypothesis on the simultaneous causal effects of Y_1 and Y_2, which depend on exogenous variables, the following Wald tests should be done.

System: SYS17_A
Estimation Method: Seemingly Unrelated Regression
Date: 11/23/09   Time: 16:54
Sample: 1968M02 1994M10
Included observations: 321
Total system (balanced) observations 642
Linear estimation after one-step weighting matrix

	Coefficient	Std. Error	t-Statistic	Prob.
C(10)	-26.22428	29.29081	-0.895307	0.3710
C(11)	0.493571	0.433777	1.137843	0.2556
C(12)	0.979358	0.011655	84.02947	0.0000
C(13)	0.015548	0.003742	4.154598	0.0000
C(14)	0.037186	0.009883	3.762767	0.0002
C(15)	-1.77E-06	4.37E-07	-4.059494	0.0001
C(16)	-0.015555	0.008989	-1.730448	0.0840
C(17)	2.59E-07	3.77E-07	0.686339	0.4928
C(18)	-2.72E-07	4.24E-07	-0.640510	0.5221
C(20)	-772.3881	140.0536	-5.514945	0.0000
C(21)	-2.983498	1.023484	-2.915042	0.0037
C(22)	0.874968	0.017558	49.83303	0.0000
C(23)	0.123851	0.022074	5.610792	0.0000
C(24)	0.119548	0.035024	3.413342	0.0007
C(25)	-1.52E-06	1.01E-06	-1.498801	0.1344
C(26)	-0.033125	0.046541	-0.711724	0.4769
C(27)	2.48E-06	5.61E-07	4.418702	0.0000
C(28)	-1.34E-05	3.00E-06	-4.469588	0.0000

Determinant residual covariance   2.48E+08

Equation: Y_1=C(10)+C(11)*T+C(12)*Y_1(-1)+C(13)*X_1+C(14)*Z_1
+C(15)*X_1*Z_1+C(16)*Y_2+C(17)*Y_2*X_2+C(18)*Y_2*Z_2
Observations: 321

R-squared	0.997235	Mean dependent var	4796.987
Adjusted R-squared	0.997165	S.D. dependent var	1564.584
S.E. of regression	83.31194	Sum squared resid	2165555.
Durbin-Watson stat	2.076635		

Equation: Y_2=C(20)+C(21)*T+C(22)*Y_2(-1)+C(23)*X_2+C(24)*Z_2
+C(25)*X_2*Z_2+C(26)*Y_1+C(27)*Y_1*X_1+C(28)*Y_1*Z_1
Observations: 321

R-squared	0.999048	Mean dependent var	14089.85
Adjusted R-squared	0.999024	S.D. dependent var	6599.741
S.E. of regression	206.2047	Sum squared resid	13266357
Durbin-Watson stat	2.009420		

Figure 3.11   *Statistical results based on the MLV(1) of the model in (3.14)*

1.1 The null hypothesis: $C(16)=C(17)=C(18)=C(26)=C(27)=C(28)=0$ is rejected based on the Chi-square statistic of $\chi_0^2 = 36.27049$ with $df=6$ and a p-value $=0.0000$. It can then be concluded that Y_1 on Y_2 has significant simultaneous causal effects.

1.2 In the first regression, at a significance level of $\alpha=0.10$, the null hypothesis: $C(16)=C(17)= C(18)=0$ is rejected based on the Chi-square statistic of $\chi_0^2 = 6.385272$ with $df=3$ and a p-value $=0.0943$. It can be concluded that the effect of Y_2 on Y_1 is significantly dependent on (X_2,Z_2).

1.3 In the second regression, at a significance level of $\alpha=0.10$, the null hypothesis: $C(26)= C(27)=C(28)=0$ is rejected based on the Chi-square statistic of $\chi_0^2 = 32.57095$ with $df=3$ and a p-value $=0.0000$. Then, it can be concluded that the effect of Y_1 on Y_2 is significantly dependent on (X_1,Z_1).

2. Each of the variables X_1, Z_1 and $X\_1{}^*Z\_1$ in the first regression has a significant adjusted effect on Y_1, so it can directly be concluded the data supports the hypothesis stated: the effect of X_1 (Z_1) on Y_1 depends on Z_1 (X_1).

3. For the second regression, the null hypothesis: $C(23)=C(24)=C(25)=0$ is rejected based on the Chi-square statistic of $\chi_0^2 = 65.34636$ with $df=3$ and a p-value $=0.0000$. Then it can be concluded that X_2, Z_2 and $X\_2{}^*Z\_2$ have significant joint effects on Y_2.

4. Corresponding to each of the interactions $Y\_2{}^*X\_2$ and $Y\_2{}^*Z\_2$ in the first regression with such large p-values, a reduced model should be explored. Since, the null hypothesis: $C(17)=C(18)=0$ is accepted based on the Chi-square statistic of $\chi_0^2 = 0.517788$ with $df=2$ and a p-value $=0.7719$, then in a statistical sense, both interactions could be deleted to obtain a reduced model. In addition, $Y\_1{}^*Z\_1$ is deleted from the second regression to obtain a good fit reduced model presented in Figure 3.12

System: SYS17_B
Estimation Method: Seemingly Unrelated Regression
Date: 11/24/09　Time: 15:25
Sample: 1968M02 1994M10
Included observations: 321
Total system (balanced) observations 642
Linear estimation after one-step weighting matrix

	Coefficient	Std. Error	t-Statistic	Prob.
C(10)	-44.47758	26.24315	-1.694826	0.0906
C(11)	0.156217	0.398772	0.391746	0.6954
C(12)	0.977840	0.010107	96.74742	0.0000
C(13)	0.015587	0.003603	4.325632	0.0000
C(14)	0.047061	0.008809	5.342158	0.0000
C(15)	-1.85E-06	3.95E-07	-4.678703	0.0000
C(16)	-0.011611	0.005331	-2.177843	0.0298
C(20)	-713.2237	141.9860	-5.023198	0.0000
C(21)	-4.193502	1.007900	-4.160634	0.0000
C(22)	0.898307	0.017291	51.95370	0.0000
C(23)	0.113187	0.022277	5.080878	0.0000
C(24)	0.173655	0.033723	5.149530	0.0000
C(25)	-3.68E-06	8.98E-07	-4.101226	0.0000
C(26)	-0.120447	0.043238	-2.785690	0.0055
C(27)	1.79E-06	5.56E-07	3.216480	0.0014

Determinant residual covariance　　　2.63E+08

Equation: Y_1=C(10)+C(11)*T+C(12)*Y_1(-1)+C(13)*X_1+C(14)*Z_1
　　+C(15)*X_1*Z_1+C(16)*Y_2
Observations: 321

R-squared	0.997226	Mean dependent var	4796.987
Adjusted R-squared	0.997173	S.D. dependent var	1564.584
S.E. of regression	83.18301	Sum squared resid	2172696.
Durbin-Watson stat	2.108142		

Equation: Y_2=C(20)+C(21)*T+C(22)*Y_2(-1)+C(23)*X_2+C(24)*Z_2
　　+C(25)*X_2*Z_2+C(26)*Y_1+C(27)*Y_1*X_1
Observations: 321

R-squared	0.998978	Mean dependent var	14089.85
Adjusted R-squared	0.998955	S.D. dependent var	6599.741
S.E. of regression	213.2998	Sum squared resid	14240505
Durbin-Watson stat	2.011311		

Figure 3.12　*Statistical results based on a reduced MLV(1) model in Figure 3.11*

3.5　Models Based on $(X1\_i, X2\_i, X3\_i, Y1\_i, Y2\_i)$ for Independent States

3.5.1　Lagged Endogenous Variables: First Autoregressive Model with Exogenous Variables and Trend

For an illustration, the causal or structural relationships between the variables $(X1\_i, X2\_i, X\_3\_i)$ and $(Y1\_i, Y2\_i)$ are shown as the path diagram in Figure 3.13.

Based on this path diagram an LVAR(1)_SCM with trend for the i-th state can be presented as follows:

$$
\begin{aligned}
y1\_i_t = {} & c(1i0) + c(1i1)^* t + c(1i2)^* y1\_i_{t-1} + c(1i3)^* y2\_i_{t-1} \\
& + c(1i4)^* x1\_i_t + c(1i5)^* x2\_i_t + c(1i6)^* x3\_i_t + [ar(1) = c(1i7)] \\
y2\_i_t = {} & c(2i0) + c(2i1)^* t + c(2i2)^* y1\_i_{t-1} + c(2i3)^* y2\_i_{t-1} \\
& + c(2i4)^* x1\_i_1 + c(2i5)^* x2\_i_t + c(2i6)^* x3\_i_t + [ar(1) = c(2i7)]
\end{aligned}
\tag{3.17}
$$

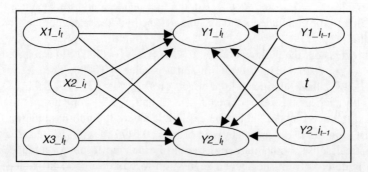

Figure 3.13　*The path diagram of a theoretical causal model*

Note that the characteristics of this model are as follows:

1. This multivariate model is an LVAR(**1**) additive linear model based on the model parameters $C(ij)$, it is also linear based on all independent variables.
2. The two regressions have dependent variables $Y1\_i_t$ and $Y2\_i_t$, with independent variables t, $Y1\_i_{t-1}$ and $Y2\_{t-1}$ represented by a set of six arrows on the right-hand side, and a set of three exogenous variables, namely $X1\_i_t$, $X2\_i_t$, and $X3\_i_t$ on the left-hand side.
3. Note the causal relationships between the independent variables are not defined in theoretical sense. However, their correlations should have unpredictable impacts on the parameter estimates.
4. The model in (3.17) can easily be extended to a hierarchical two-way interaction model by using at least one of the interactions $X1\_i^*X2\_i$, $X1\_i^*X3\_i$ and $X2\_i^*X3\_i$ as additional independent variables.
5. Finally, under the assumption that $X1\_i$, $X2\_i$ and $X3\_i$ are completely correlated in a theoretical sense, the hierarchical three-way interaction models would have variables $X1\_i$, X_2_i, $X3\_i$, $X1\_i^*X2\_i$, $X1\_i^*X3\_i$, $X\_2\_i^*X3\_i$ and $X1\_i^*X2\_i^*X3\_i$ as independent. However, in practice, a reduced model would be acceptable in a statistical sense, since the seven variables should be highly correlated.

3.5.2 A Mixed Lagged Variables First Autoregressive Model with Trend

By using the set of variables $X1\_i$, $X2\_i$, $X3\_i$, $Y1\_i$ and $Y2\_i$; and the time t-variable, a set of five AR(1) regressions would be found for the i-th state as follows:

$$
\begin{aligned}
y1\_i_t &= c(1i0) + c(1i1)^* t + c(1i2)^* y1\_i_{t-1} + c(1i3)^* y2\_i_{t-1} \\
&\quad + c(1i4)^* x1\_i_t + c(1i5)^* x2\_i_t + c(1i6)^* x3\_i_1 + [ar(1) = c(1i7)] \\
y2\_i_t &= c(2i0) + c(2i1)^* t + c(2i2)^* y1\_i_{t-1} + c(2i3)^* y2\_i_{t-1} \\
&\quad + c(2i4)^* x1\_i_1 + c(2i5)^* x2\_i_t + c(2i6)^* x3\_i_t + [ar(1) = c(2i7)] \\
x1\_i_t &= c(3i0) + c(3i1)^* x1\_i_{t-1} + c(3i2)^* x2\_i_{t-1} + c(3i3)^* x3\_i_{t-1} + [ar(1) = c(3i4)] \\
x2\_i_t &= c(4i0) + c(4i1)^* x1\_i_{t-1} + c(4i2)^* x2\_i_{t-1} + c(4i3)^* x3\_i_{t-1} + [ar(1) = c(4i4)] \\
x3\_i_t &= c(5i0) + c(5i1)^* x1\_i_{t-1} + c(5i2)^* x2\_i_{t-1} + c(5i3)^* x3\_i_{t-1} + [ar(1) = c(5i4)]
\end{aligned}
\tag{3.18}
$$

Note that the first two regressions are exactly the same as the model in (3.17). These regressions show that the effects of bivariate $\{Y1\_i_{1-1}, Y2\_i_{t-1}\}$ on both $Y1\_i_t$ and $Y2\_i_t$. These two regressions also show that the four exogenous variables: $X1\_i_t$, $X2\_i_t$, $X3\_i_t$ and the time t have effects on both $Y1\_i_t$ and $Y2\_i_t$; and the last three regressions show a trivariate model of the exogenous variables $X1\_i$, $X2\_i$ and $X3\_i$ without trend. This multivariate model is an AR(1) additive linear model based on the model parameters $C(ij)$, it is also linear based on all independent variables.

The association patterns between the variables in this model can be presented as a path diagram in Figure 3.14, which is exactly the same as the path diagram in Agung (2009a, Figure 2.108, p. 116). Note the different sets of arrows, as follows:

1. The first two regressions have dependent variables $Y1_t$ and $Y2_t$, with independent variables t, $Y1_{t-1}$ and $Y2_{t-1}$ represented by a set of six arrows in the right-hand box, and a solid/thick arrow from a set of three exogenous variables, namely $X1_t$, $X2_t$ and $X3_t$ in the left-hand box.
2. The last three regressions have dependent variables $X1_t$, $X2_t$ and $X3_t$, with their first lags, namely $X1_{t-1}$, $X2_{t-1}$ and $X3_{t-1}$ as independent variables, represented in the left-hand box.

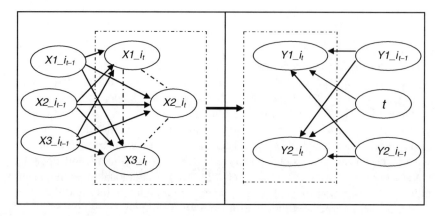

Figure 3.14   *The path diagram of the model in (3.18)*

3. These five regressions do not take into consideration the possible causal relationships between their independent variables. For example, the first two regressions do not consider the types of relationships between the six independent variables: $X1_t$, $X2_t$, $X3_t$, $Y1_{t-1}$, $Y2_{t-1}$ and t. However, their quantitative coefficient of correlations or multicollinearity should have an unexpected impact on the estimates of the model parameters.

4. The multivariate model or the system equations in (3.18) can easily be extended to two- or three-way interaction models of the exogenous variables $X1\_i$, $X2\_i$ and $X3\_i$, and then modified in order to have a lot of alternative time-series models by using the transformed variables, such as their natural logarithms, their higher lagged variables and their first differences.

3.5.3   Lagged Endogenous Variables: First Autoregressive Model with Exogenous Variables and Time-Related Effects

A further extension of the multivariate models with trend in (3.17) is a multivariate model with *time-related effects*. In this type of model, the two-way interactions between the time t and each of selected exogenous variables act as additional independent or exogenous variables of each regression in the system. Since the equations for this type of model could easily be derived from the previous illustrations, they will not be presented again in detail.

By using the set of variables $X1\_i$, $X2\_i$, $X3\_i$, $Y1\_i$ and $Y2\_i$; and the time t-variable, a set of two AR(1) regressions would be found for the i-th state as follows:

$$\begin{aligned}
y1\_i_t ={}& c(1i0) + c(1i1)^* t + c(1i2)^* y1\_i_{t-1} + c(1i3)^* y2\_i_{t-1} \\
&+ c(1i4)^* x1\_i_t + c(1i5)^* x2\_i_t + c(1i6)^* x3\_i_1 \\
&+ t^*(c(1i7)^* y1\_i_{t-1} + c(1i8)^* y2\_i_{t-1} + c(1i9)^* x1\_i_t \\
&+ c(1i10)^* x2\_i_t + c(1i11)^* x3\_i_1) + [ar(1) = c(1i12)] \\
y2\_i_t ={}& c(2i0) + c(2i1)^* t + c(2i2)^* y1\_i_{t-1} + c(2i3)^* y2\_i_{t-1} \\
&+ c(2i4)^* x1\_i_1 + c(2i5)^* x2\_{-t} + c(2i6)^* x3\_i_t \\
&+ t^*(c(2i7)^* y1\_i_{t-1} + c(2i8)^* y2\_i_{t-1} + c(2i9)^* x1\_i_1 \\
&+ c(2i10)^* x2\_i_t + c(2i11)^* x3\_i_t) + [ar(1) = c(2i12)]
\end{aligned} \tag{3.19}$$

3.5.4   Various Interaction Models

As an extension of those in (3.17)–(3.19), various interaction models can be derived, specifically based on the three exogenous variables $X1\_i$, $X2\_i$ and $X3\_i$, which in general can be presented as $F_i(x1\_i, x2\_i, x3\_i)$. Note that the effect of an exogenous variable on the corresponding endogenous variable, in general, depends on the other exogenous variable(s). For this reason, two alternative full interaction models are derived based on the ones mentioned previously. On the other hand, the model in (3.19) is in fact a two-way interaction model of the time t and each of the exogenous variables in the model.

3.5.4.1   Full Two-Way Interaction Model

The full two-way interaction model of the exogenous variables can be obtained by using the function in each of the models previously.

$$Fi(x1\_i, x2\_i, x3\_i) = \beta_{i1} x1\_i + \beta_{i2} x2\_i + \beta_{i3} x3\_i$$
$$+ \beta_{i4} x1\_i^* x2\_i + \beta_{i5} x1\_i^* x3\_i + \beta_{i6} x2\_i^* x3\_i \tag{3.20}$$

3.5.4.2   Full Three-Way Interaction Model

The full three-way interaction model can be obtained by using the function in the earlier models.

$$F_i(x1\_i, x2\_i, x3\_i) = \beta_{i1} x1\_i + \beta_{i2} x2\_i + \beta_{i3} x3\_i$$
$$+ \beta_{i4} x1\_i^* x2\_i + \beta_{i5} x1\_i^* x3\_i + \beta_{i6} x2\_i^* x3\_i + \beta_{i7}^* x1\_i^* x2\_i^* x3\_i \tag{3.21}$$

3.6   Models Based on (X_i, Y_i) for Correlated States

Based on multivariate time series for two correlated or dependent states, general multivariate linear models (MGLMs) can be presented as follows:

$$G_1(Y\_1) = F_1(t, X\_1, Y\_1_{t-p}, Y\_2_{t-p}) + \mu_{1t}$$
$$G_1(Y\_2) = F_2(t, X\_2, Y\_1_{t-p}, Y\_2_{t-p}) + \mu_{2t} \tag{3.22}$$

where $G_i(Y\_i)$ is a function of Y_i without a parameter, and $F_i(t, X\_i, Y\_1_{t-p}, Y\_2_{t-p})$ is a function of $(t, X\_i, Y\_1_{t-p}, Y\_2_{t-p})$ with a finite number of parameters for each i and some p. In many cases $p = 1$ is applied, however, in order to match the conditions in previous and recent years, and applying $p = 2, 4$ and 12, respectively is recommended for semester, quarterly and monthly time series. Special models considered are as follows:

3.6.1   Additive Models

A general additive model would have the three functions, namely $F_{i1}(t)$, $F_{i2}(X\_1)$, and $F_{i3}(Y\_1_{t-1}, Y\_2_{t-1})$. Then, corresponding to (3.20), the general additive model can be presented as follows:

$$G_1(Y\_1) = F_{11}(t) + F_{12}(X\_1) + F_{13}(Y\_1_{t-p}, Y\_2_{t-p}) + \mu_{1t}$$
$$G_2(Y\_2) = F_{21}(t) + F_{22}(X\_2) + F_{23}(Y\_1_{t-p}, Y\_2_{t-p}) + \mu_{2t} \tag{3.23}$$

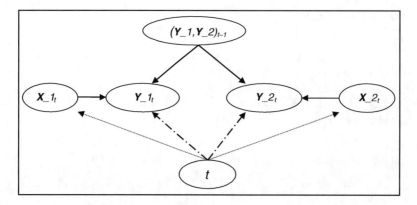

Figure 3.15   *The theoretical path diagram to represent the general model in (3.23)*

where $F_{ij}(^*)$ is a function of $(^*)$ having a finite number of parameters. Specific to the function $F_{it}(t)$, the models (1.4a–d) present four alternative functions, which could be used to modify or extend all models with trend presented in this chapter.

Selected models in (3.23) can be presented using the path diagram in Figure 3.15. Note that two types of arrows from the time t indicate three possible sets of models to consider, using either one or both.

For an illustration, the simplest additive model based on the time series $X\_i = (X1\_i, X2\_i, X3\_i,)$ and $Y\_i = (Y1\_i, Y2\_i)$, for $i = 1$ and 2, is a MLVAR(**1,1**) additive model of $Y1\_1$, $Y2\_1$, $Y1\_2$ and $Y2\_2$, with the exogenous variables: $Y1\_1(-1)$, $Y2\_1(-1)$, $Y1\_2(-1)$, $Y2\_2(-1)$, the time t, $X1\_i$, $X2\_i$ and $X3\_i$, with the following system specification. Note that this model is a modification of the additive model in (3.17) for two independent states, by inserting the four first lags $Y1\_1(-1)$, $Y2\_1(-1)$, $Y1\_2(-1)$ and $Y2\_2(-1)$ as independent variables.

$$y1\_i = c(1i0) + c(1i1)^* t$$
$$+ c(1i2)^* x1\_i_t + c(1i3)^* x2\_i_t + c(1i4)^* x3\_i_t$$
$$+ c(1i5)^* y1\_1(-1) + c(1i6)^* y2\_1(-1)$$
$$+ c(1i7)^* y2\_2(-1) + c(1i8)^* y2\_2(-1) + [ar(1) = c(1i9)]$$
$$y2\_i = c(2i0) + c(2i1)^* t \tag{3.24}$$
$$+ c(2i2)^* x1\_i_t + c(2i3)^* x2\_i_t + c(2i4)^* x3\_i_t$$
$$+ c(2i5)^* y1\_1(-1) + c(2i6)^* y2\_1(-1)$$
$$+ c(2i7)^* y2\_2(-1) + c(2i8)^* y2\_2(-1) + [ar(1) = c(2i9)]$$
$$for\ i = 1\&2.$$

3.6.2   Interaction Models

A two-way interaction model, that is a model with TRE, can easily be derived from the model in (3.19), by inserting the four first lags $Y1\_1(-1)$, $Y2\_1(-1)$, $Y1\_2(-1)$ and $Y2\_2(-1)$ as independent variables. The following MLV(1) with TRE is obtained.

$$
\begin{aligned}
y1\_i = {} & c(1i0) + c(1i1)^{*}t \\
& + c(1i2)^{*}y1\_1(-1) + c(1i3)^{*}y2\_1(-1) \\
& + c(1i4)^{*}y2\_1(-1) + c(1i5)^{*}y2\_2(-1) \\
& + c(1i6)^{*}x1\_i + c(1i7)^{*}x2\_i + c(1i8)^{*}x3\_i \\
& + t^{*}(c(1i9)^{*}y1\_1 + c(1i10)^{*}y2\_1(-1) \\
& + c(1i11)^{*}y1\_2(-1) + c(1i12)^{*}y2\_2(-1) \\
& + c(1i13)^{*}x1\_i + c(1i14)x2\_i + c(1i15)^{*}x3\_i) \\
y2\_i = {} & c(2i0) + c(2i1)^{*}t \\
& + c(2i2)^{*}y1\_1(-1) + c(2i3)^{*}y2\_1(-1) \\
& + c(2i4)^{*}y2\_1(-1) + c(2i5)^{*}y2\_2(-1) \\
& + c(2i6)^{*}x1\_i + c(2i7)^{*}x2\_i + c(2i8)^{*}x3\_i \\
& + t^{*}(c(2i9)^{*}y1\_1 + c(2i10)^{*}y2\_1(-1) \\
& + c(2i11)^{*}y1\_2(-1) + c(2i12)^{*}y2\_2(-1) \\
& + c(2i13)^{*}x1\_i + c(2i14)x2\_i + c(2i15)^{*}x3\_i) \\
& \text{for } i = 1, 2
\end{aligned}
\tag{3.25}
$$

The other interaction models could easily be derived from the general model in (3.22) and (3.23) using the functions in (3.20) and (3.21). Do it as an exercise.

3.6.3   Alternative Models for Two Correlated States

As a modification of the path diagram in Figure 3.15, Figure 3.16 presents another type of path diagram of the models for two correlated states, represented by double arrows between the endogenous variables Y_1 and Y_2. However, a special case might be considered, where Y_1 and Y_2 have a simultaneous causal relationship. Note that this path diagram also can be considered an extension of the path diagram in Figure 3.10 for a simultaneous SCM.

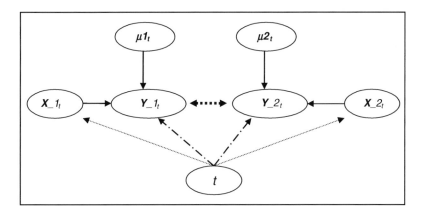

Figure 3.16   *The path diagram of a SCM based on (X _i, Y _i) for two correlated states*

For two correlated states, the basic models considered will have the general equation as follows:

$$G_1(Y\_1) = F_1(t, X\_1, Y\_2) + \mu_{1t}$$
$$G_2(Y\_2) = F_2(t, X\_2, Y\_1) + \mu_{2t}$$

(3.26)

However, a time-series model would be a MLVAR(p,q) model with the general equation as follows:

$$G_1(Y\_1_t) = F_1(t, X\_1, Y\_2) + \sum_{i=1}^{p_1} c(1^*)^* Y\_1_{t-i} + \sum_{j=1}^{q_1} \rho_{1j}\mu_{1t-j} + \varepsilon_{1t}$$

$$G_2(Y\_2_t) = F_2(t, X\_2, Y\_1) + \sum_{i=1}^{p_2} c(2^*)^* Y\_2_{t-i} + \sum_{j=1}^{q_2} \rho_{2j}\mu_{1t-j} + \varepsilon_{2t}$$

(3.27)

For an illustration, the simplest model based on the time series $X\_i = (X1\_i, X2\_i, X3\_i,)$ and $Y\_i = (Y1\_i, Y2\_i)$ for $i = 1$ and 2, is a MAR(1) model of $Y1\_1, Y2\_1, Y1\_2$ and $Y2\_2$, and exogenous variables: the time t, $X1\_i, X2\_i$ and $X3\_i$ with trend, which has the following general system specification.

$$y1\_1 = c(110) + c(111)^* y1\_2 + c(112)^* y2\_2 + F1(t, X\_1, c(11^*)) + [ar(1) = c(11^*)]$$
$$y2\_1 = c(210) + c(211)^* y1\_2 + c(212)^* y2\_2 + F1(t, X\_1, c(21^*)) + [ar(1) = c(21^*)]$$
$$y1\_2 = c(120) + c(121)^* y1\_1 + c(122)^* y2\_1 + F2(t, X\_2, c(12^*)) + [ar(1) = c(12^*)]$$
$$y2\_2 = c(220) + c(221)^* y1\_1 + c(222)^* y2\_1 + F2(t, X\_2, c(22^*)) + [ar(1) = c(22^*)]$$

(3.28)

where $F_i(t, X\_i, c(11^*))$ is a function of t and X_i; either additive or interaction functions. Refer to the models and functions in (3.17)–(3.25).

3.7   Discontinuous Time-Series Models

In discontinuous growth models, the time-series models in general should be applied because of effects of internal factors, as well as environmental factors, such as natural disasters, riots and bombs in countries, and world economic crises. Then, corresponding to the general model in (3.22) for two correlated states, the general discontinuous time series model by time periods can be presented as follows:

$$G_1(Y\_1_t) = \sum_{k=1}^{K} F_{1k}(t, X\_1, Y\_1_{t-s}, Y\_2_{t-s})^* Dt_k + \mu_{1t}$$

$$G_2(Y\_2_t) = \sum_{k=1}^{K} F_{2k}(t, X\_2, Y\_1_{t-s}, Y\_2_{t-s})^* Dt_k + \mu_{2t}$$

(3.29)

where Dt_k is the dummy variable of the k-th time period, for $k \geq 2$, which is intentionally selected based on environmental conditions, and the function $F_{ik}(^*)$ could be any function, such as those presented in (3.17)–(3.25). So there would be a lot of possible discontinuous models in a theoretical sense, and their estimates are highly dependent on the data used as well as the multicollinearity between the variables. Refer to special notes presented in Section 3.9, and all discontinuous time-series models presented in Agung (2009a).

On the other hand, note that this model in fact represents a set of K-sets of multiple regressions with heterogeneous or various slopes. For this reason, this model would be considered a *system of heterogeneous regressions*. On the other hand, the model in (3.31) is *a system of homogeneous regressions* in an ANCOVA model.

3.8   Additional Examples for Correlated States

Example 3.5   An extension of the model in Figure 2.4
As an extension or modification of the model for two independent states in Figure 2.4, Figure 3.17 presents the statistical results based on the MLVAR(1,1;0,1) translog linear model with trend for two correlated or dependent states.

Example 3.6   Application of the model in (3.24)
By using the MLVAR(**1,1**) model in (3.24) based on a hypothetical data generated based on the data in POOL1. wf1, the result that the AR(1) terms are insignificant with large p-values is obtained. Even using a true panel data set, the same problems would occur because each of the regressions has used the first lag of the dependent variable. Then the following MLV(**1**) additive model is applied, with the results presented in Figure 3.18.

$$
\begin{aligned}
y1\_1 = {}& c(110) + c(111)^*t + c(112)^*x1\_1 + c(113)^*x2\_1 + c(114)^*x3\_1 \\
& + c(115)^*y1\_1(-1) + c(116)^*y2\_1(-1) + c(117)^*y1\_2(-1) + c(118)^*y2\_2(-1) \\
y2\_1 = {}& c(210) + c(211)^*t + c(212)^*x1\_2 + c(213)^*x2\_2 + c(214)^*x3\_2 \\
& + c(215)^*y1\_1(-1) + c(216)^*y2\_1(-1) + c(217)^*y1\_2(-1) + c(218)^*y2\_2(-1) \\
y1\_2 = {}& c(120) + c(121)^*t + c(122)^*x1\_1 + c(123)^*x2\_1 + c(124)^*x3\_1 \\
& + c(125)^*y1\_1(-1) + c(126)^*y2\_1(-1) + c(127)^*y1\_2(-1) + c(128)^*y2\_2(-1) \\
y2\_2 = {}& c(220) + c(221)^*t + c(222)^*x1\_2 + c(223)^*x2\_2 + c(224)^*x3\_2 \\
& + c(225)^*y1\_1(-1) + c(226)^*y2\_1(-1) + c(227)^*y1\_2(-1) + c(228)^*y2\_2(-1)
\end{aligned}
\tag{3.30}
$$

System: SYS21
Estimation Method: Seemingly Unrelated Regression
Date: 09/08/09   Time: 17:24
Sample: 1968M02 1994M10
Included observations: 321
Total system (unbalanced) observations 641
Iterate coefficients after one-step weighting matrix
Convergence achieved after: 1 weight matrix, 13 total coef iterations

	Coefficient	Std. Error	t-Statistic	Prob.
C(10)	0.122280	0.052592	2.325076	0.0204
C(11)	5.15E-05	4.31E-05	1.195815	0.2322
C(12)	9.75E-06	3.05E-06	3.198733	0.0014
C(13)	-9.90E-06	3.07E-06	-3.229572	0.0013
C(14)	0.990679	0.009403	105.3547	0.0000
C(15)	-0.004923	0.009002	-0.546917	0.5846
C(20)	7.496187	1.256724	5.964865	0.0000
C(21)	0.008822	0.005151	1.712856	0.0872
C(22)	3.33E-05	4.27E-06	7.795475	0.0000
C(23)	-8.24E-06	4.10E-06	-2.009449	0.0449
C(24)	0.137447	0.045212	3.040045	0.0025
C(26)	1.005363	0.005015	200.4770	0.0000
Determinant residual covariance	4.00E-08			

Equation: LOG(Y_1)=C(10)+C(11)*T+C(12)*X_1+C(13)*X_1(-1)+C(14)
 *LOG(Y_1(-1))+C(15)*LOG(Y_2)
Observations: 321

R-squared	0.998315	Mean dependent var	8.408489
Adjusted R-squared	0.998289	S.D. dependent var	0.390481
S.E. of regression	0.016154	Sum squared resid	0.082202
Durbin-Watson stat	1.978419		

Equation: LOG(Y_2)=C(20)+C(21)*T+C(22)*X_2+C(23)*X_2(-1)+C(24)
 *LOG(Y_1)+C(15)*LOG(Y_2(-1))+[AR(1)=C(26)]
Observations: 320

R-squared	0.999520	Mean dependent var	9.413448
Adjusted R-squared	0.999511	S.D. dependent var	0.571793
S.E. of regression	0.012646	Sum squared resid	0.050058
Durbin-Watson stat	1.586337		

Figure 3.17   *Statistical results based on a MLVAR(1,1;0,1)_GM of two independent states*

	Coefficient	Std. Error	t-Statistic	Prob.
C(110)	-8.509602	46.39259	-0.183426	0.8545
C(111)	-0.332588	0.430336	-0.772855	0.4398
C(112)	-0.001418	0.002364	-0.599790	0.5488
C(113)	0.009912	0.007019	1.412111	0.1582
C(114)	-0.006228	0.001927	-3.231492	0.0013
C(115)	0.977647	0.015783	61.94252	0.0000
C(116)	0.013583	0.020217	0.671860	0.5018
C(117)	0.010953	0.014326	0.764586	0.4447
C(118)	0.000240	0.005494	0.043661	0.9652
C(210)	1.193761	77.30871	0.015441	0.9877
C(211)	-0.049827	1.079487	-0.046158	0.9632
C(212)	-0.021467	0.016151	-1.329147	0.1840
C(213)	0.020801	0.024259	0.857437	0.3914
C(214)	-0.007184	0.030023	-0.239268	0.8109
C(215)	-0.087828	0.041751	-2.103622	0.0356
C(216)	0.962076	0.020828	46.19179	0.0000
C(217)	0.108355	0.025355	4.273533	0.0000
C(218)	-0.020936	0.008308	-2.520150	0.0119

System: SYS18_A
Estimation Method: Seemingly Unrelated Regression
Date: 11/25/09   Time: 15:18
Sample: 1968M02 1994M10
Included observations: 321
Total system (balanced) observations 1284
Linear estimation after one-step weighting matrix

C(120)	-70.41029	78.92002	-0.892173	0.3725
C(121)	0.215536	0.736380	0.292697	0.7698
C(122)	-0.002414	0.003642	-0.662825	0.5076
C(123)	0.026353	0.010677	2.468200	0.0137
C(124)	-0.002172	0.003010	-0.721519	0.4707
C(125)	-0.004345	0.029110	-0.149252	0.8814
C(126)	-0.012695	0.031769	-0.399615	0.6895
C(127)	1.015696	0.023337	43.52232	0.0000
C(128)	-0.008486	0.008946	-0.948488	0.3431
C(220)	411.2895	200.9234	2.046996	0.0409
C(221)	5.351656	2.834753	1.887874	0.0593
C(222)	-0.063075	0.043750	-1.441718	0.1496
C(223)	0.142134	0.065051	2.184969	0.0291
C(224)	-0.143683	0.081158	-1.770425	0.0769
C(225)	-0.169628	0.108347	-1.565600	0.1177
C(226)	-0.100429	0.054397	-1.846230	0.0651
C(227)	0.216407	0.065971	3.280339	0.0011
C(228)	0.938387	0.021524	43.59787	0.0000

Determinant residual covariance   2.05E+17

Equation: Y1_1=C(110)+C(111)*T+C(112)*X1_1+C(113)*X2_1+C(114)
    *X3_1+C(115)*Y1_1(-1)+C(116)*Y2_1(-1)+C(117)*Y1_2(-1)+C(118)
    *Y2_2(-1)
Observations: 321

R-squared	0.997225	Mean dependent var	4796.987
Adjusted R-squared	0.997153	S.D. dependent var	1564.584
S.E. of regression	83.47549	Sum squared resid	2174065.
Durbin-Watson stat	2.165773		

Figure 3.18   *Statistical results based on a MLV(1)_GM of two independent states*

Based on these results, the following notes and conclusions are made.

1. The fourth regression of $Y2\_2$ is a good fit model, since each of the independent has a p-value < 0.20. Therefore, each of the independent variables has either a significant positive or negative effect on $Y2\_2$ at the significance level of 0.10.

2. On the other hand, the other regressions have several independent variables with large p-values. Therefore, in a statistical sense, reduced models should be explored. Since these regressions are additive models, the least important variable(s), in a theoretical sense, should be deleted step-by-step. Another alternative method is using the stepwise estimation method for each regression, if and only if the variables are equally important. Note that an unexpected reduced model might be obtained. Refer to the stepwise estimation methods demonstrated in Chapter 10 (Agung, 2011).

3. However, based on this model, the joint effects of the first lagged variables $Y1\_1(-1)$, $Y2\_1(-1)$, $Y1\_2(-1)$ and $Y2\_2(-1)$ on each variable or sub-sets of the four dependent variables, can be tested with the Wald test.

4. For illustration, $Y1\_1(-1)$, $Y2\_1(-1)$, $Y1\_2(-1)$ and $Y2\_2(-1)$ have significant joint effects on $Y1\_1$, since the null hypothesis: $C(115) = C(116) = C(117) = C(118) = 0$ is rejected based on the Chi-square statistics of $\chi_0^2 = 11524.61$ with $df = 4$ and a p-value $= 0.0000$.

Example 3.7   A case study by Indrawati (2002)

For an illustration of empirical results, the author's graduate student, Indrawati (2002) in her dissertation, considered a set of five time series such as *Stock Index*, *Money Supply*, *Consumer Price Index*, *Exchange Rate* and *Wall Street Index*, for four states, namely Indonesia, Thailand, South Korea and Taiwan, when presenting data analysis based on the VAR and VEC models. However, she did not present a model with the time t as an independent variable. Based her data set, a lot more time-series models by states could easily be defined, either using or not using time t as an independent variable.

For further data analyses, two time series in the data set can be considered or defined as *Y*-variables, namely *Y1_i* and *Y2_i*, and the others are the *X1_i*, *X2_i* and *X3_i*. Therefore, all models presented in this chapter as well as their extensions could easily be applied.

Example 3.8   A case study by Chalmers et al. (2006)

This study was based on the sample of 532 firm-years' observations, the Top 200 Australian Stock Exchange listed firms for each of the years 1999–2002, a sub-sample balanced panel data as a multivariate time series by firms was selected, namely (1) industry, (2) finance, (3) mining and (4) media, telecommunications and pharmaceutical classifications. The data contains 18 numerical variables, some of which can be defined as multivariate *Y*-variables. However, note that the data has only four years or time-point observations, so it could be extended to have a lot more time-point observations, such as monthly or quarterly. On the other hand, the panel data could be analyzed as cross-sectional data over time, which will be discussed in Part II of this book.

3.9   Special Notes and Comments

3.9.1   Extended Models

1. As an extension of all multivariate models presented previously, this can also be applied to various instrumental variables models as well as TGARCH(a,b,c) models.
2. Similar models in the forms of semilog, translog linear and quadratic models, as well as bounded growth models can be applied.
3. Furthermore, similar models can easily be derived using the first difference, namely dY_i_t, as well as $dlog(Y\_i_t) = log(Y\_i_t) - log(Y\_i_t) = R_t$, which is in fact the time series of exponential growth rates Y_i_t for $t = 1, \ldots, T$.
4. Finally, the models for three or more correlated states could easily be derived as the extension of previous models for two correlated states, by defining a path diagram that shows the causal relationships between the endogenous variables, so that various statistical models can be written.

3.9.2   Not-Recommended Models

1. Corresponding to all illustrative models that have been presented up this point, it can be concluded that the fixed-effect model is not recommended for application to panel data, since, in a theoretical sense, the slopes of the exogenous variable(s) should be different by states. The fixed-effect model with many exogenous variables would be the worst, even based on two or three states. Refer to Section 1.6.1.
2. This is more so the case for the random effect model – refer to Section 1.6.2.
3. Corresponding to the general discontinuous model (3.26), the following general model is also not recommended in a theoretical sense. However, note that this is an ANCOVA model with covariates $F_i(t, X\_i, Y\_1_{t-s}, Y\_2_{t-s}))$, which is acceptable in a statistical sense. Refer to special notes on the ANCOVA model presented in Agung (2011).

$$G_1(Y\_1_t) = \sum_{k=1}^{K} Dt_k + F_1(t, X\_1, Y\_1_{t-s}, Y\_2_{t-s}) + \mu_{1t}$$

$$G_2(Y\_2_t) = \sum_{k=1}^{K} Dt_k + F_2(t, X\_2, Y\_1_{t-s}, Y\_2_{t-s}) + \mu_{1t}$$

(3.31)

3.9.3   Problems with Data Analysis

Data analysis based on any model can easily be done using EViews. However, the problems are the statistical results show many independent variables with unexpected insignificant adjusted effects, even based on an interaction model with three independent variables, namely X_i, Z_i with $X\_i^*Z\_i$. Moreover, based on a model with a large number of variables, unexpected insignificant independent variables can be obtained such as those presented previously, because of multicollinearity between variables. If the error message "*Near singular matrix*" appears, this would be a great problem, since there is not a perfect way to modify the system equations. Refer to Section 2.14.2 in Agung (2009a).

Specific to interaction models, in general the main factors and their interactions are highly correlated. We have found that the results show at least one of the main factors and their interactions with a large *p*-value in many or most cases. Therefore, in a statistical sense, an acceptable reduced model should be explored using trial-and-error methods, or the stepwise estimation method (Agung, 2011).

Since an interaction model is used to represent the effect of an exogenous variable on the corresponding endogenous variable dependent on other exogenous variable(s), then any reduced model obtained should have interaction factor(s) as its independent variable(s). Refer to the reduced models presented in previous examples and the examples in Agung (2009a, 2011).

4

Applications of Seemingly Causal Models

4.1 Introduction

Many students and researchers present models without using time t as an independent variable in their panel data sets. It is well-known that the scatter graph of a bivariate time series (X_i_t, Y_i_t) for $t = 1, 2, \ldots, T$, of the i-state would be the same as the scatter graph of (X_i_j, Y_i_j) with $X\_i_j \leq X\_i_{j+1}$ for all $j = 1, 2, \ldots, T$. So models based on the time series (X_i_t, Y_i_t) without using the time t as a numerical independent variable would be models based on the cross-section data of (X_i_j, Y_i_j) with $X\_i_j \leq X\_i_{j+1}$ for all $j = 1, 2, \ldots, T$. Agung (2009a) has presented unexpected regressions obtained based on (X_i_j, Y_i_j), compared to the growth curves of X_i_t and Y_i_t for $t = 1, 2, \ldots, T$. Refer to those problems as well as various seeming causal models (SCMs) presented in Chapter 4 (Agung, 2009a).

Furthermore, all SCMs presented in Agung (2009a) can directly be considered SCMs for the i-th independent state by only using the symbol X_i and Y_i for exogenous and endogenous variables respectively. On the other hand all models with the time t as a numerical independent variable can easily be transformed into SCMs by using the following alternative methods.

4.1.1 Deleting Time t from Models

By deleting the time t, as well as its interactions with other exogenous variables from all models, SCMs can easily be obtained.

4.1.2 Replacing Time t with an Environmental Variable

By replacing the time t with an environmental variable, namely Z_t, then the SCMs with an environmental variable can easily be obtained based on all models with trend, or TREs. See the following illustrative examples. Try the others as an exercise.

For these reasons, this chapter will present only the SCMs for correlated states; specifically selected SCMs for two or three correlated states for illustration.

All SCMs presented in Chapter 4 (Agung, 2009a) can directly be used as SCMs for i-th independent states. For this reason, these will not be presented again in detail. Looking back to the problems of the

Panel Data Analysis Using EViews, First Edition. I Gusti Ngurah Agung.
© 2014 John Wiley & Sons, Ltd. Published 2014 by John Wiley & Sons, Ltd.
Companion website: www.wiley.com/go/panel_data

regressions of *log(M1)* on *RS* with and without dummy variables presented in Examples 4.6 and 4.7 (Agung, 2009a), the following examples provide additional illustrations.

4.2   SCMs Based on a Single Time Series Y_i_t

As seen in the additional examples in Section 4.3 (Agung, 2009a), the following examples present some of the problems in doing analysis based on a single time series Y_i_t for two correlated states using a SCM without time *t* as an independent variable.

Example 4.1   A continuous standard regression

By using *GDP_US* as a predictor of the *GDP_Can*, and *GDP_JPN* in POOLG7.wf1, Figure 4.1(a) and (b) respectively present their graphs with regression lines and nearest neighbor fit curves. Based on these graphs note the following:

1. The regression *GDP_Can* = *C(1)* + *C(2)**GDP_US, as well as any regression *Y_1* = *C(1)* + *C(2)**Y_2, where *Y_1* and *Y_2* are numerical variables, should be considered a continuous function, since its derivative *dY_1/dY_2* = *C(2)* always has a certain value.
2. The residuals graph of the simplest linear regression (SLR) of *GDP_Can* on *GDP_US* shows that the SLR is an inappropriate regression, by observing the DW-statistic and residuals graph of its SLR as presented in Figure 4.2.
3. In order to obtain model with a greater DW-statistic value, several alternative models, such as polynomial models, lagged variable autoregressive models, instrumental variables models or TGARCH(*a,b,c*) models (Agung, 2009a), should be explored in order to find the best possible model in a statistical sense.

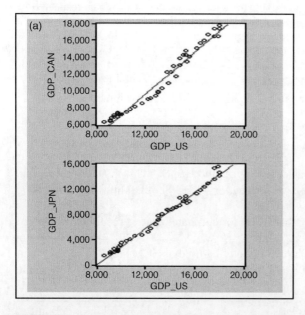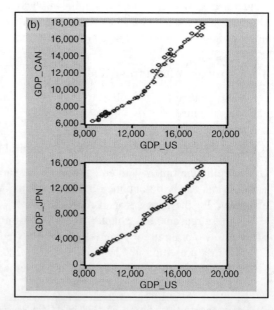

Figure 4.1   *Scatter graphs of GDP_CAN and GDP_JPN on GDP_US, with (a) their regression lines, and (b) their nearest neighbor fit curves*

Dependent Variable: GDP_CAN
Method: Least Squares
Date: 08/30/09   Time: 06:00
Sample: 1950 1992
Included observations: 43

	Coefficient	Std. Error	t-Statistic	Prob.
C	-5369.076	463.4508	-11.58500	0.0000
GDP_US	1.245732	0.034193	36.43270	0.0000

R-squared	0.970037	Mean dependent var	11108.26
Adjusted R-squared	0.969306	S.D. dependent var	3787.652
S.E. of regression	663.5863	Akaike info criterion	15.87859
Sum squared resid	18054219	Schwarz criterion	15.96051
Log likelihood	-339.3897	Hannan-Quinn criter.	15.90880
F-statistic	1327.342	Durbin-Watson stat	0.259844
Prob(F-statistic)	0.000000		

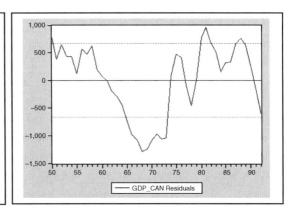

Figure 4.2   *Statistical results based on a SLR of* GDP_CAN *on* GDP_US

4. Furthermore, they could easily be extended using their transformed variables, such as $G(Y\_1)$ and $F(Y\_2)$, where $G(Y\_1) = log(Y\_1)$ or $log[(Y\_1 - L)/(U-Y\_1)]$, and $F(Y\_2)$ is any function of Y_2 with a finite number of parameters. Refer to all models in previous examples, by deleting time t, as well as all models in Aagung (2009a and 2011). However, additional selected models will be considered again in the following examples.

5. On the other hand, a special relationship between the bivariate correlation and SLRs could be demonstrated. Compared to the SLR of *GDP_CAN* on *GDP_US*, in Figure 4.3, specifically its t-statistic of $t_0 = 36.43\ 270$ with a p-value $= 0.0000$. Figure 4.3 presents the statistical results based on the SLR of *GDP_US* on *GDP_CAN*, and bivariate correlation between *GDP_US* and *GDP_CAN*, which show exactly the same values of the t-statistic. Based on these findings, we conclude the hypothesis stating that *GDP_US* and *GDP_CAN* have a causal relationship can be tested using correlation analysis. Refer also to Aagung (2006, 2009a and 2011a,b).

6. Finally, note that the simple linear regressions in Figures 4.2 and 4.3(a), as well as the correlation of *GDP_US* and *GDP_CAN* present exactly the same values for the t-statistic. For this reason, in general,

Dependent Variable: GDP_US
Method: Least Squares
Date: 08/30/09   Time: 06:21
Sample: 1950 1992
Included observations: 43

	Coefficient	Std. Error	t-Statistic	Prob.
C	4577.159	250.5387	18.26927	0.0000
GDP_CAN	0.778688	0.021373	36.43270	0.0000

R-squared	0.970037	Mean dependent var	13227.02
Adjusted R-squared	0.969306	S.D. dependent var	2994.604
S.E. of regression	524.6465	Akaike info criterion	15.40872
Sum squared resid	11285410	Schwarz criterion	15.49064
Log likelihood	-329.2875	Hannan-Quinn criter.	15.43893
F-statistic	1327.342	Durbin-Watson stat	0.263642
Prob(F-statistic)	0.000000		

(a)

Covariance Analysis: Ordinary
Date: 08/30/09   Time: 06:24
Sample: 1950 1992
Included observations: 43

Correlation t-Statistic Probability	GDP_US	GDP_CAN
GDP_US	1.000000	

GDP_CAN	0.984904	1.000000
	36.43270	-----
	0.0000	-----

(b)

Figure 4.3   *Statistical results based on (a) the SLR of* GDP_US *on* GDP_CAN, *and (b) bivariate correlation of* GDP_US *and* GDP_CAN

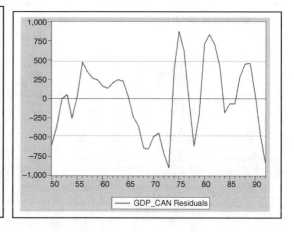

Dependent Variable: GDP_CAN				
Method: Least Squares				
Date: 08/27/09 Time: 16:07				
Sample: 1950 1992				
Included observations: 43				

	Coefficient	Std. Error	t-Statistic	Prob.
C	45809.61	10698.30	4.281952	0.0001
GDP_US	-10.14435	2.489739	-4.074462	0.0002
GDP_US^2	0.000815	0.000188	4.338662	0.0001
GDP_US^3	-1.88E-08	4.60E-09	-4.091727	0.0002

R-squared	0.985020	Mean dependent var	11108.26
Adjusted R-squared	0.983868	S.D. dependent var	3787.652
S.E. of regression	481.0732	Akaike info criterion	15.27832
Sum squared resid	9025827.	Schwarz criterion	15.44216
Log likelihood	-324.4840	Hannan-Quinn criter.	15.33874
F-statistic	854.8523	Durbin-Watson stat	0.648957
Prob(F-statistic)	0.000000		

Figure 4.4   *Statistical results based on a third-degree polynomial model of* GDP_Can *on* GDP_US

it is sufficient to conduct correlation analysis to test the hypothesis on the causal linear effect between any pairs of numerical variables X and Y.

Example 4.2   Polynomial standard regressions

Figure 4.4 presents the statistical results based on a third-degree polynomial of *GDP_CAN* on *GDP_US*, with its residuals graph. For comparison, Figure 4.5 presents the statistical results based on a fourth-degree polynomial of *GDP_CAN* on *GDP_US*, with its residuals graph. Which one would you consider to be better?

Example 4.3   WLS models of Figure 4.4

Figure 4.6 presents the statistical results based on the third-degree polynomial model of *(GDP_Can/E)* on *(GDP_US/E)*, where *E* is the error terms of the model in Figure 4.4. For a comparison, Figure 4.7 presents

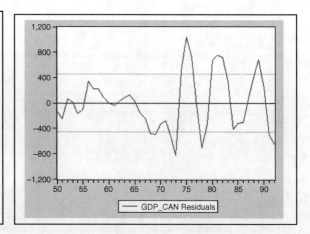

Dependent Variable: GDP_CAN				
Method: Least Squares				
Date: 08/30/09 Time: 08:46				
Sample: 1950 1992				
Included observations: 43				

	Coefficient	Std. Error	t-Statistic	Prob.
C	-67880.81	52564.14	-1.291390	0.2044
GDP_US	25.70452	16.43170	1.564325	0.1260
GDP_US^2	-0.003341	0.001893	-1.764723	0.0856
GDP_US^3	1.91E-07	9.53E-08	2.004921	0.0521
GDP_US^4	-3.91E-12	1.77E-12	-2.204841	0.0336

R-squared	0.986719	Mean dependent var	11108.26
Adjusted R-squared	0.985322	S.D. dependent var	3787.652
S.E. of regression	458.8922	Akaike info criterion	15.20445
Sum squared resid	8002120.	Schwarz criterion	15.40924
Log likelihood	-321.8957	Hannan-Quinn criter.	15.27997
F-statistic	705.8326	Durbin-Watson stat	0.860490
Prob(F-statistic)	0.000000		

Figure 4.5   *Statistical results based on a fourth-degree polynomial model of* GDP_Can *on* GDP_US

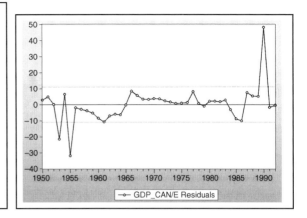

Figure 4.6   *Statistical results based on a third-degree polynomial model of* (GDG_Can/E) *on* (GDP_US/E)

the statistical results based on the third-degree polynomial model of *(GDP_Can/E^2)* on *(GDP_US/E^2)*. Refer to Agung (2011), and Neter and Wasserman (1974). Based on these results, note the following.

1. Even though these models are standard regression models, they have sufficiently large DW-statistics, which indicates their residuals do not have serial correlation problems. We also find the correlation of the residual of the model in Figure 4.6, Resid01, and Resid01($-$1), is insignificant based on the t-statistic of $t_0 = 0.624\,050$ with a p-value $= 0.5361$.
2. Compared to the model in Figure 4.4, the models in Figures 4.6 and 4.7 are better, since they have much greater DW-statistical values. However, which one is best? Since the model in Figure 4.6 has a greater DW-statistic value, then this one is preferred in a statistical sense.
3. However, their residual graphs show some outliers.
4. For comparison, do the analysis based on the fourth-degree polynomial model of *(GDG_Can/E)* on *(GDP_US/E)*, as well as *(GDG_Can/E^2)* on *(GDP_US/E^2)*. You will obtain unexpected estimates of the model parameters.

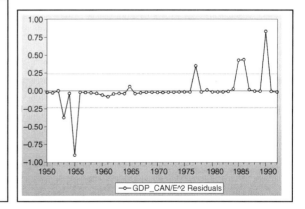

Figure 4.7   *Statistical results based on a third-degree polynomial model of* (GDG_Can/E^2) *on* (GDP_US/E^2)

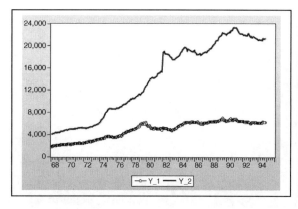 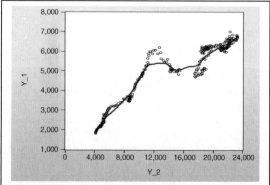

Figure 4.8   *Growth curves of* Y_1 *and* Y_2 *and their scatter graph with its kernel fit curve*

Example 4.4   (A Piece-wise regression based on (Y_1,Y_2))

Figure 4.8 presents the growth curves of Y_1 and Y_2, and their scatter graphs with a kernel fit curve. Note that the scatter graph of Y_1 on Y_2 clearly shows that a polynomial regression of Y_1 on Y_2 should be applied. In other words, the simplest linear regression of Y_1 on Y_2 should not be an acceptable model. However, if the linear effect of Y_2 on Y_1 should be considered then the *correlation analys* is as well as the simplest linear regression (SLR) of Y_1 on Y_2 or SLR of Y_2 on Y_1 can be used to test the hypothesis. Refer to the special findings presented in Example 1.25.

However, instead of a polynomial regression, three-piece regression lines will be applied using the following steps. As an exercise, try this for polynomial regression.

1. Identify the relative minimum and maximum of Y_1 with respect to Y_2, using Excel. In other words, the discontinuous or break points of Y_1 should be identified at the first stage. The relative maximum of Y_1 obtained for $Y\_2 = 12\,677$, and its relative minimum obtained for $Y\_2 = 17\,594$.
2. Then an ordinal variable (*OV*) of Y_2 with three categories should be generated using the following equation.

$$OV = 1 + 1^*(Y\_2 > 12\,677) + 1^*(Y\_2 > 17\,594) \tag{4.1}$$

3. Figure 4.9 presents the statistical results based on three-piece simple linear regressions, using the following equation specification. Based on these results the following notes are presented.

$$Y\_1@Expand(OV)Y\_2^*@Expand(OV) \tag{4.2}$$

3.1 Note that the model has the intercept parameters: *C(1)* to *C(3)*, and slope parameters: *C(4)* to *C(6)*. Therefore, the hypotheses considered would be the differences between slope parameters, which can easily be tested using the Wald test.

3.2 This model in fact represents an heterogeneous regressions with slopes of 0.463 867, −0.174 495, and 0.216 365, respectively, with equations as follows:

$$Y\_1 = -4.876756 + 0.463867^*Y\_2, \text{within the interval } OV = 1$$
$$Y\_1 = 7707.148057 - 0.174495^*Y\_2, \text{within the interval } OV = 2$$
$$Y\_1 = 1677.338451 + 0.2163165^*Y\_2, \text{within the interval } OV = 3$$

Dependent Variable: Y_1
Method: Least Squares
Date: 09/01/09   Time: 09:23
Sample (adjusted): 1968M01 1994M10
Included observations: 322 after adjustments

	Coefficient	Std. Error	t-Statistic	Prob.
OV=1	-4.876756	73.91035	-0.065982	0.9474
OV=2	7707.148	612.5788	12.58148	0.0000
OV=3	1677.338	292.6929	5.730711	0.0000
Y_2*(OV=1)	0.463867	0.009684	47.89950	0.0000
Y_2*(OV=2)	-0.174495	0.041300	-4.225105	0.0000
Y_2*(OV=3)	0.216316	0.014304	15.12326	0.0000
R-squared	0.966884	Mean dependent var		4787.699
Adjusted R-squared	0.966360	S.D. dependent var		1571.013
S.E. of regression	288.1417	Akaike info criterion		14.18324
Sum squared resid	26236103	Schwarz criterion		14.25357
Log likelihood	-2277.502	Hannan-Quinn criter.		14.21132
Durbin-Watson stat	0.129945			

Estimation Command:
=====================
LS Y_1 @EXPAND(OV) Y_2*@EXPAND(OV)

Estimation Equation:
=====================
Y_1 = C(1)*(OV=1) + C(2)*(OV=2) + C(3)*(OV=3) + C(4)*Y_2*(OV=1) + C(5)*Y_2*(OV=2) + C(6)*Y_2*(OV=3)

Substituted Coefficients:
=====================
Y_1 = -4.87675559559*(OV=1) + 7707.1480566*(OV=2) + 1677.33845098*(OV=3) + 0.463866885484*Y_2*(OV=1) - 0.174495112473*Y_2*(OV=2) + 0.216316484885*Y_2 *(OV=3)

Figure 4.9 *Statistical results based on the three-piece linear model in (4.2)*

In addition, Figure 4.10(a) presents the scatter graph Y_1 on Y_2, as well as three-piece regression lines.
3.3   For a further illustration and comparison, Figure 4.10(b) presents a similar graph based on the statistical results, using the following equation specification in EViews 6. For the previous versions of EViews, find the following example.

$$Y\_1\ C\ Y\_2^*@Expand(OV) \tag{4.3}$$

The results provide three-piece regressions as follows:

$$Y\_1 = 198.713 + 0.439^*Y\_2,\ \text{within the interval } OV = 1$$
$$Y\_1 = 198.713 + 0.330^*Y\_2,\ \text{within the interval } OV = 2$$
$$Y\_1 = 198.713 + 0.288^*Y\_2\ \text{within the interval } OV = 3$$

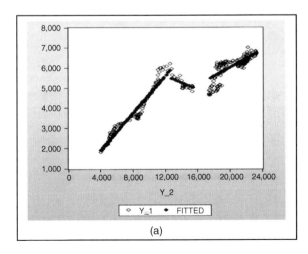

(a)

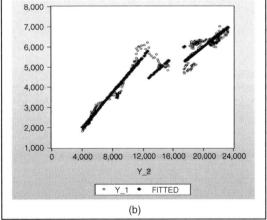

(b)

Figure 4.10 *Scatter graphs of Y_2 with Y_1 and fitted values of (a) the model in (4.2), and (b) the model in (4.3)*

(a)

Dependent Variable: Y_1
Method: Least Squares
Date: 09/01/09   Time: 14:28
Sample (adjusted): 1968M01 1994M10
Included observations: 322 after adjustments

	Coefficient	Std. Error	t-Statistic	Prob.
D1OV	-4.876756	73.91035	-0.065982	0.9474
D2OV	7707.148	612.5788	12.58148	0.0000
D3OV	1677.338	292.6929	5.730711	0.0000
D1OV*Y_2	0.463867	0.009684	47.89950	0.0000
D2OV*Y_2	-0.174495	0.041300	-4.225105	0.0000
D3OV*Y_2	0.216316	0.014304	15.12326	0.0000
R-squared	0.966884	Mean dependent var		4787.699
Adjusted R-squared	0.966360	S.D. dependent var		1571.013
S.E. of regression	288.1417	Akaike info criterion		14.18324
Sum squared resid	26236103	Schwarz criterion		14.25357
Log likelihood	-2277.502	Hannan-Quinn criter.		14.21132
Durbin-Watson stat	0.129945			

(b)

Dependent Variable: Y_1
Method: Least Squares
Date: 09/01/09   Time: 14:18
Sample (adjusted): 1968M01 1994M10
Included observations: 322 after adjustments
Y_1=(C(10)+C(11)*Y_2)*D1OV+(C(20)+C(21)*Y_2)*D2OV+(C(30)+C(31)
      *Y_2)*D3OV

Variable	Coefficient	Std. Error	t-Statistic	Prob.
C(10)	-4.876756	73.91035	-0.065982	0.9474
C(11)	0.463867	0.009684	47.89950	0.0000
C(20)	7707.148	612.5788	12.58148	0.0000
C(21)	-0.174495	0.041300	-4.225105	0.0000
C(30)	1677.338	292.6929	5.730711	0.0000
C(31)	0.216316	0.014304	15.12326	0.0000
R-squared	0.966884	Mean dependent var		4787.699
Adjusted R-squared	0.966360	S.D. dependent var		1571.013
S.E. of regression	288.1417	Akaike info criterion		14.18324
Sum squared resid	26236103	Schwarz criterion		14.25357

Figure 4.11   *Statistical results based on the three-piece regressions in (4.17), and (4.18)*

However, by looking at the scatter graph of Y_1 on Y_2, this model should be considered inappropriate or bad, since it presents three-piece regressions with the same intercepts of $\hat{C}(1) = 198.713$ and a positive slope of $\hat{C}(3) = 0.330$ within the interval $OV = 2$, which should be negative.

3.4   On the other hand, by using the equation specification "Y_1 @Expand(OV) Y_2", three simple linear regressions with the same slopes with various intercepts could be obtained. In other words, the set of regressions is homogeneous. In a statistical sense, this model is an ANCOVA model. Compare it with the regression line of $log(M1)$ on RS in Figure 4.32 (Agung, 2009a; p. 208).

Example 4.5   Alternative equation specifications of (4.4)

By using previous versions of EViews, namely EViews 4 or 5, the equation specification in (4.4) should be presented using dummy variables. Then, the equation specification can be as follows:

$$Y\_1 \ D1OV \ D2OV \ D3OV \ D1OV^*Y\_2 \ D2OV^*Y\_2 \ D3OV^*Y\_2 \tag{4.4}$$

or

$$Y\_1 = (C(10) + C(11)^*Y\_2)^*D1OV + (C(20) + C(21)^*Y\_2)^*D2OV \\ + (C(30) + C(31)^*Y\_2)^*D3OV \tag{4.5}$$

where $D1OV$, $D2OV$ and $D3OV$ are the three dummy variables of OV, with the statistical results presented in Figure 4.11(a) and (b).

4.3   SCMs Based on Bivariate Time Series (X_i_t, Y_i_t)

For two correlated states, a path diagram based on four time series: $X\_1_t, Y\_1_t X\_2_t$ and Y_2_t should be theoretically defined as the base for writing the equation of alternative SCMs; either additive, two- or three-way interaction models. In this book, two states are defined as correlated if and only if their endogenous variables are correlated, including their possible causal relationships.

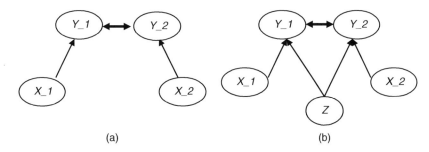

Figure 4.12 *The path diagram of a simultaneous SCM of (a) (X_1,Y_1) and (X_2,Y_2), and (b) (X_1,Y_1), (X_2, Y_2), and an environmental variable, Z*

For illustration, the path diagram in Figure 4.12(a) is obtained from the path diagram in Figure 2.6 by deleting the time t and Figure 4.12(b) by replacing the time t with an environmental variable Z_t. Therefore, the equations of alternative models based on four time series: X_1_t, Y_1_t, X_2_t and Y_2_t for two correlated states, with or without the environmental variable Z, could easily be written based on all models presented in Chapter 2. For this reason, the equations of the models will not be presented in this section.

However, as an extension, the environmental variable Z_t, which is an external factor of the states, should be considered to be the cause (source or up-stream) factor of the variables X_1 and X_2, in addition to the cause factors of Y_1 and Y_2. In this case, the general equation of a standard SCM can be presented as follows:

$$
\begin{aligned}
G_1(Y\_1_t) &= F_1(X\_1_t, Y\_2_t, Z_t) + \mu_{1t} \\
G_2(Y\_2_t) &= F_2(X\_2_t, Y\_1_t, Z_t) + \mu_{2t} \\
G_3(X\_1_t) &= F_3(Z_t) + \mu_{3t} \\
G_4(X\_2_t) &= F_4(Z_t) + \mu_{4t}
\end{aligned}
\tag{4.6}
$$

where $G_i(V)$ is a fixed function of V without a parameter, and $Fi(^*)$ is a function of $(^*)$ with a finite number of parameters.

4.3.1 Additive SCM

$$
\begin{aligned}
y\_1 &= c(10) + c(11)^* x\_1 + c(12)^* y\_2 + c(13)^* z + \mu 1 \\
y\_2 &= c(20) + c(11)^* x\_2 + c(12)^* y\_1 + c(23)^* z + \mu 2 \\
x\_1 &= c(30) + c(31)^* z + \mu 3 \\
x\_2 &= c(40) + c(41)^* z + \mu 4
\end{aligned}
\tag{4.7}
$$

4.3.2 A Two-Way Interaction Standard SCM

$$
\begin{aligned}
y\_1 &= c(10) + c(11)^* x\_1 + c(12)^* y\_2 + c(13)^* z \\
&\quad + c(14)^* x\_1^* z + c(15)^* y\_2^* z + \mu 1 \\
y\_2 &= c(20) + c(21)^* x\_2 + c(22)^* y\_1 + c(23)^* z \\
&\quad + c(24)^* x\_1^* z + c(25)^* y\_1^* z + \mu 2 \\
x\_1 &= c(30) + c(31)^* z + \mu 3 \\
x\_2 &= c(40) + c(41)^* z + \mu 4
\end{aligned}
\tag{4.8}
$$

4.3.3 A Three-Way Interaction Standard SCM

$$
\begin{aligned}
y\_1 =\ & c(10) + c(11)^*x\_1 + c(12)^*y\_2 + c(13)^*z \\
& + c(14)^*x\_1^*z + c(15)^*y\_2^*z + c(16)^*x\_1^*y\_2^*z + \mu1 \\
y\_2 =\ & c(20) + c(21)^*x\_2 + c(22)^*y\_1 + c(23)^*z \\
& + c(24)^*x\_2^*z + c(25)x\_2^*y\_1^*z + c(26)^*x\_2^*y\_1^*z + \mu2 \\
x\_1 =\ & c(30) + c(31)^*z + \mu3 \\
x\_2 =\ & c(40) + c(41)^*z + \mu4
\end{aligned}
\tag{4.9}
$$

4.4 SCMs Based on a Trivariate ($X1\_i,X2\_i,Y1\_i$)

All models with trend and TRE presented in the previous three chapters can easily be transformed to SCMs, either by deleting the time t or replacing it with an environmental variable. Note that corresponding to the two exogenous variables, the effect of an exogenous variable $X1\_i$ ($X2\_i$) on the endogenous variable $Y1\_i$ depends on the other exogenous variable $X2\_i$ ($X1\_i$). Based on this assumption, an SCM considered should have the interaction $X1\_i^*X2\_i$ as an independent variable. Refer to the models presented in Section 4.4 (Agung, 2009a).

4.4.1 Simple SCMs for Two Correlated States

Figure 4.13 shows the path diagram based on the variables $(X1\_i,X2\_i,Y\_i)$ for $i = 1$ and 2, and an environmental variable Z, with the condition that $Y1\_1$ and $Y1\_2$ have simultaneous causal effects, and assumes that the environmental variable Z only has effects on both $Y1\_1$ and $Y1\_2$. However, a more advanced model might be considered to show that the environmental variable can also have effects on selected or all exogenous variables.

4.4.1.1 A Two-Way Interaction Model

Based on the path diagram in Figure 4.13, consideration for the standard two-way interaction model is recommended, which can easily be extended to various time-series models presented in Agung (2009a). The system specification of the model will be as follows:

$$
\begin{aligned}
y1\_1 =\ & c(10) + c(11)^*x1\_1 + c(12)^*x2\_1 + c(13)^*x1\_1^*x2\_1 \\
& + c(14)^*y1\_2 + c(15)^*z + c(16)^*y1\_2^*z + \mu1 \\
y1\_2 =\ & c(20) + c(21)^*x1\_2 + c(22)^*x2\_2 + c(23)^*x1\_2^*x2\_2 \\
& + c(24)^*y1\_1 + c(25)^*z + c(26)^*y\_1^*z + \mu2 \\
x1\_1 =\ & c(30) + c(31)^*x2\_1 + \mu3 \\
x1\_2 =\ & c(40) + c(41)^*x2\_2 + \mu4
\end{aligned}
\tag{4.10}
$$

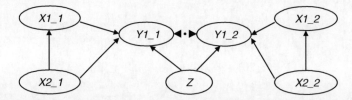

Figure 4.13 *The path diagram of a simultaneous SCM of (Y1_1,Y1_2) on (X1_1,X2_1) and (X1_2,X2_2) with an environmental variable Z*

4.4.1.2 A Three-Way Interaction SCM

Under the assumption that the variables *Y1_2*, *X1_2* and *X2_2* are completely correlated, then the effect of the three-way interaction *Y1_2*X1_2*X2_2* on *Y1_1* should be considered. Similarly, under the assumption that the variables *Y1_1*, *X1_1* and *X2_1* are completely correlated, then the three-way interaction *Y1_1*X1_1*X2_1* should be used as an independent of the regression of *Y1_2*. Then the standard three-way interaction SCM will have the system specification as follows:

$$
\begin{aligned}
y1\_1 = {} & c(10) + c(11)^{*}x1\_1 + c(12)^{*}x2\_1 + c(13)^{*}x1\_1^{*}x2\_1 \\
& + c(14)^{*}y1\_2 + c(15)^{*}z + c(16)^{*}y1\_2^{*}z \\
& + c(17)^{*}y1\_2^{*}x1\_2^{*}x2\_2 + \mu1 \\
y1\_2 = {} & c(20) + c(21)^{*}x1\_2 + c(22)^{*}x2\_2 + c(23)^{*}x1\_2^{*}x2\_2 \\
& + c(24)^{*}y1\_1 + c(25)^{*}z + c(26)^{*}y\_1^{*}z \\
& + c(17)^{*}y1\_1^{*}x1\_1^{*}x2\_1 + \mu2 \\
x1\_1 = {} & c(30) + c(31)^{*}x2\_1 + \mu3 \\
x1\_2 = {} & c(40) + c(41)^{*}x2\_2 + \mu4
\end{aligned}
\tag{4.11}
$$

4.4.2 Various Time-Series Models

The standard models in (4.10) and (4.11) should have autocorrelation problems, so modifying the models to a lot of possible time-series models is recommended, as presented in Agung (2009a), such as follows:

1. Various MLVAR(*p,q*)_SCM, for subjectively selected **p** ≥ 0, and **q** ≥ 0,
2. Various instrumental variables SCMs, using all alternative estimation methods provide by EViews.
3. Various TGARCH(*a,b,c*) models, as demonstrated in Chapter 8 (Agung, 2009a), and
4. Various models with dummy variables of time periods, corresponding to the four types of models mentioned earlier.

Example 4.6   A MLV(1) SCM based on Figure 4.13

Figure 4.14 presents the path diagram of MLV(1)_SCMs, based on the variables *(X1_i,X2_i,Y_i)* for *i* = 1 and 2, with an environmental variable *Z*.

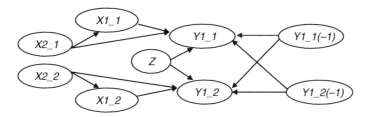

Figure 4.14   *The path diagram of a MLV(1,1)_SCM based on (X1_i,X2_i,Y1_i), for two correlated states with an environmental variable Z*

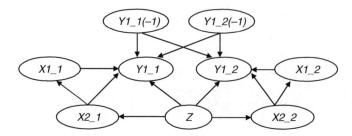

Figure 4.15   *The path diagram of an advanced MLV(1,1)_SCM of (X1_i,X2_i,Y1_i), for two correlated states, with an environmental variable Z*

The two-way interaction MLV(1) SCM has the equation as follows:

$$
\begin{aligned}
y1\_1 &= c(10) + c(11)^{*}x1\_1 + c(12)^{*}x2\_1 + c(13)^{*}x1\_1^{*}x2\_1 \\
&\quad + c(14)^{*}y1\_1(-1) + c(15)^{*}y1\_2(-1) + c(16)^{*}z + \mu 1 \\
y1\_2 &= c(20) + c(21)^{*}x1\_2 + c(22)^{*}x2\_2 + c(23)^{*}x1\_2^{*}x2\_2 \\
&\quad + c(24)^{*}y1\_1(-1) + c(25)^{*}y1\_2(-1) + c(26)^{*}z + \mu 2 \\
x1\_1 &= c(30) + c(31)^{*}x2\_1 + \mu 3 \\
x1\_2 &= c(40) + c(41)^{*}x2\_2 + \mu 4
\end{aligned}
\tag{4.12}
$$

Example 4.7   A more advanced LV(1) SCM

Figure 4.15 presents the path diagram of a more advanced LV(**1**)_SCM, based on the variables *(X1_i,X2_i, Y_i)* for *i* = 1 and 2, with an environmental variable *Z*.

$$
\begin{aligned}
y1\_1 &= c(10) + c(11)^{*}x1\_1 + c(12)^{*}x2\_1 + c(13)^{*}x1\_1^{*}x2\_1 \\
&\quad + c(14)^{*}y1\_1(-1) + c(15)^{*}y1\_2(-1) + c(16)^{*}z \\
&\quad + c(17)^{*}x2\_1^{*}z + \mu 1 \\
y1\_2 &= c(20) + c(21)^{*}x1\_2 + c(22)^{*}x2\_2 + c(23)^{*}x1\_2^{*}x2\_2 \\
&\quad + c(24)^{*}y1\_1(-1) + c(25)^{*}y1\_2(-1) + c(26)^{*}z \\
&\quad + c(27)^{*}x2\_2^{*}z + \mu 2 \\
x1\_1 &= c(30) + c(31)^{*}x2\_1 + c(32)^{*}x2\_1^{*}z + \mu 3 \\
x1\_2 &= c(40) + c(41)^{*}x2\_2 + c(42)^{*}x2\_2^{*}z + \mu 4
\end{aligned}
\tag{4.13}
$$

Example 4.8   A LVAR_SCM for three correlated states

Based on the data in POOLG7.wf1, under the assumption that *GDP_FRA*, *GDP_GER* and *GDP_ITA* are correlated, Figure 4.16 presents statistical results based on a simple LVAR(**1,1**)_SCM. The path diagram of this model is presented in Figure 4.17.

   Based on this example, the following notes and conclusions are made.

1.   In this case, it is theoretically defined that each of *GDP_FRA(−1)*, *GDP_GER(−1)* and *GDP_ITA(−1)* has a direct effect on *GDP_FRA*, *GDP_GER* and *GDP_ITA*. This model should be acceptable in a theoretical sense, under the assumption that *GDP_FRA_t*, *GDP_GER_t*, and *GDP_ITA_t* are correlated.

```
System: SYS13
Estimation Method: Iterative Least Squares
Date: 09/02/09  Time: 10:53
Sample: 1952 1992
Included observations: 42
Total system (balanced) observations 123
Convergence achieved after 53 iterations
```

	Coefficient	Std. Error	t-Statistic	Prob.
C(10)	221.2409	211.6618	1.045256	0.2982
C(11)	0.905371	0.132436	6.836303	0.0000
C(12)	0.148711	0.133620	1.112941	0.2682
C(13)	-0.066979	0.134156	-0.499265	0.6186
C(14)	0.295289	0.169399	1.743158	0.0842
C(20)	515.0191	285.3127	1.805104	0.0738
C(21)	0.058961	0.178053	0.331142	0.7412
C(22)	0.805648	0.219753	3.666151	0.0004
C(23)	0.139013	0.180192	0.771470	0.4421
C(24)	0.306650	0.208318	1.472033	0.1439
C(30)	-35.45110	172.0908	-0.206002	0.8372
C(31)	0.007913	0.113968	0.069430	0.9448
C(32)	0.172573	0.112794	1.529986	0.1289
C(33)	0.813933	0.117208	6.944372	0.0000
C(34)	0.148641	0.181629	0.818379	0.4149

Determinant residual covariance	1.16E+13

```
Equation: GDP_FRA=C(10)+C(11)*GDP_FRA(-1)+C(12)*GDP_GER(-1)
         +C(13)*GDP_ITA(-1)+[AR(1)=C(14)]
Observations: 41
```

R-squared	0.997001	Mean dependent var	9464.073
Adjusted R-squared	0.996668	S.D. dependent var	3138.903
S.E. of regression	181.1857	Sum squared resid	1181818.
Durbin-Watson stat	1.924589		

```
Equation: GDP_GER=C(20)+C(21)*GDP_FRA(-1)+C(22)*GDP_GER(-1)
         +C(23)*GDP_ITA(-1)+[AR(1)=C(24)]
Observations: 41
```

R-squared	0.995642	Mean dependent var	9727.756
Adjusted R-squared	0.995157	S.D. dependent var	3241.862
S.E. of regression	225.5971	Sum squared resid	1832185.
Durbin-Watson stat	1.828017		

```
Equation: GDP_ITA=C(30)+C(31)*GDP_FRA(-1)+C(32)*GDP_GER(-1)
         +C(33)*GDP_ITA(-1)+[AR(1)=C(34)]
Observations: 41
```

R-squared	0.997045	Mean dependent var	7891.610
Adjusted R-squared	0.996717	S.D. dependent var	3061.693
S.E. of regression	175.4282	Sum squared resid	1107902.
Durbin-Watson stat	1.937476		

Figure 4.16 *Statistical results based on a LVAR(1,1)_SCM*

2. In practice, at least one of *GDP_FRA(−1)*, *GDP_GER(−1)*, and *GDP_ITA(−1)* could have an insignificant adjusted effect on the corresponding dependent variables, since they are correlated. In fact, they are significantly pair-wise correlated. For instant, Figure 4.16 shows that *GDP_FRA(−1)* has insignificant adjusted effects on each of *GDP_GER* and *GDP_ITA* with large *p*-values of 0.7412 and 0.9448, respectively, even though *GDP_FRA(−1)* is significantly positively correlated with *GDP_GER* and *GDP_ITA*.

3. Therefore, in a statistical sense, a reduced model should be explored. However, the problem is, which one out of *GDP_FRA(−1)*, *GDP_GER(−1)*, and *GDP_ITA(−1)* should be deleted from each regression? It would be a matter of preference or personal judgment. Try it as an exercise.

4. For an illustration, Figure 4.18 presents the statistical results based on a special reduced model, where the lags *GDP_FRA(−1)*, *GDP_GER(−1)*, and *GDP_ITA(−1)*, respectively, are deleted from the first, second and third regressions, even though they have significant adjusted effects.

5. The main objective of this reduced model is to demonstrate that an acceptable reduced model, in a statistical sense, could be obtained by deleting an independent variable with significant adjusted effects. Refer to various reduced models presented in Agung (2009a and 2011).

6. Various models could be developed by taking into account the possible causal effects between variables *GDP_FRA_t*, *GDP_GER_t* and *GDP_ITA_t*. Do it as an exercise.

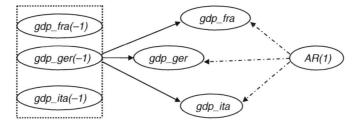

Figure 4.17 *The path diagram of the LVAR(1,1)_SCM in Figure 4.16*

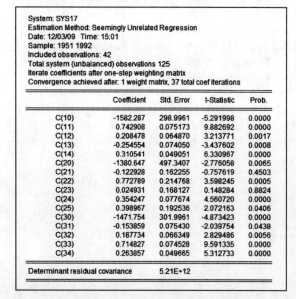

System: SYS14
Estimation Method: Iterative Least Squares
Date: 09/02/09   Time: 10:59
Sample: 1952 1992
Included observations: 42
Total system (balanced) observations 123
Convergence achieved after 13 iterations

	Coefficient	Std. Error	t-Statistic	Prob.
C(10)	139808.4	815437.6	0.171452	0.8642
C(12)	0.295242	0.144993	2.036251	0.0441
C(13)	-0.239771	0.193259	-1.240674	0.2173
C(14)	0.998347	0.010173	98.13695	0.0000
C(20)	1044.480	545.8836	1.913375	0.0583
C(21)	0.667709	0.225633	2.959268	0.0038
C(23)	0.340871	0.227921	1.495565	0.1376
C(24)	0.686862	0.113944	6.028071	0.0000
C(30)	-950.3028	505.5787	-1.879634	0.0628
C(31)	0.529898	0.187235	2.830129	0.0055
C(32)	0.417077	0.180530	2.310285	0.0227
C(34)	0.808972	0.099779	8.107630	0.0000

Determinant residual covariance   2.89E+13

(a)

Equation: GDP_FRA=C(10)+C(12)*GDP_GER(-1)+C(13)*GDP_ITA(-1)
    +[AR(1)=C(14)]
Observations: 41
R-squared	0.996920	Mean dependent var	9464.073
Adjusted R-squared	0.996671	S.D. dependent var	3138.903
S.E. of regression	181.1151	Sum squared resid	1213699.
Durbin-Watson stat	1.400861		

Equation: GDP_GER=C(20)+C(21)*GDP_FRA(-1)+C(23)*GDP_ITA(-1)
    +[AR(1)=C(24)]
Observations: 41
R-squared	0.994333	Mean dependent var	9727.756
Adjusted R-squared	0.993873	S.D. dependent var	3241.862
S.E. of regression	253.7542	Sum squared resid	2382473.
Durbin-Watson stat	1.829170		

Equation: GDP_ITA=C(30)+C(31)*GDP_FRA(-1)+C(32)*GDP_GER(-1)
    +[AR(1)=C(34)]
Observations: 41
R-squared	0.996661	Mean dependent var	7891.610
Adjusted R-squared	0.996390	S.D. dependent var	3061.693
S.E. of regression	183.9462	Sum squared resid	1251939.
Durbin-Watson stat	1.992041		

(b)

Figure 4.18   *Statistical results based on a reduced model of that in Figure 4.16*

7.   For an extension of the model, one or two environmental or external factors can be taken into consideration, such as the *GDP_US* and *GDP_Jpn*, and then a more complex path diagram should be developed. For illustration, see the following example.

Example 4.9   A SCM for correlated states with an external factor
Figure 4.19 presents the statistical results based on a model of *GDP_Fra*, *GDP_Ger* and *GDP_Ita*. The model presented is in fact a reduced model of the model MLVAR(**1,1**) based on the path diagram in

System: SYS17
Estimation Method: Seemingly Unrelated Regression
Date: 12/03/09   Time: 15:01
Sample: 1951 1992
Included observations: 42
Total system (unbalanced) observations 125
Iterate coefficients after one-step weighting matrix
Convergence achieved after: 1 weight matrix, 37 total coef iterations

	Coefficient	Std. Error	t-Statistic	Prob.
C(10)	-1582.287	298.9961	-5.291998	0.0000
C(11)	0.742908	0.075173	9.882692	0.0000
C(12)	0.208478	0.064870	3.213771	0.0017
C(13)	-0.254554	0.074050	-3.437602	0.0008
C(14)	0.310541	0.049051	6.330967	0.0000
C(20)	-1380.647	497.3407	-2.776058	0.0065
C(21)	-0.122928	0.162255	-0.757619	0.4503
C(22)	0.772789	0.214768	3.598245	0.0005
C(23)	0.024931	0.168127	0.148284	0.8824
C(24)	0.354247	0.077674	4.560720	0.0000
C(25)	0.398967	0.192536	2.072163	0.0406
C(30)	-1471.754	301.9961	-4.873423	0.0000
C(31)	-0.153859	0.075430	-2.039754	0.0438
C(32)	0.187734	0.066349	2.829486	0.0056
C(33)	0.714827	0.074528	9.591335	0.0000
C(34)	0.263857	0.049665	5.312733	0.0000

Determinant residual covariance   5.21E+12

Equation: GDP_FRA=C(10)+C(11)*GDP_FRA(-1)+C(12)*GDP_GER(-1)
    +C(13)*GDP_ITA(-1) +C(14)*GDP_US
Observations: 42
R-squared	0.998255	Mean dependent var	9340.643
Adjusted R-squared	0.998066	S.D. dependent var	3201.917
S.E. of regression	140.7970	Sum squared resid	733479.9
Durbin-Watson stat	1.648633		

Equation: GDP_GER=C(20)+C(21)*GDP_FRA(-1)+C(22)*GDP_GER(-1)
    +C(23)*GDP_ITA(-1)+C(24)*GDP_US+[AR(1)=C(25)]
Observations: 41
R-squared	0.996870	Mean dependent var	9727.756
Adjusted R-squared	0.996423	S.D. dependent var	3241.862
S.E. of regression	193.9023	Sum squared resid	1315934.
Durbin-Watson stat	1.858381		

Equation: GDP_ITA=C(30)+C(31)*GDP_FRA(-1)+C(32)*GDP_GER(-1)
    +C(33)*GDP_ITA(-1)+C(34)*GDP_US
Observations: 42
R-squared	0.998197	Mean dependent var	7775.690
Adjusted R-squared	0.998002	S.D. dependent var	3116.038
S.E. of regression	139.2973	Sum squared resid	717937.8
Durbin-Watson stat	2.145320		

Figure 4.19   *Statistical results based on a MLAR(1,1,1;0,1,0) of GDP_Fra, GDP_Ger and GDP_Ita, with an external factor GDP_US*

Figure 4.17 with an external factor: namely *GDP_US*. Based on these results, the following notes and conclusions are made.

1. The model is in fact a MLVAR(1,1,1;0,1,0), since the first and third regressions do not contain the AR(1) terms. Because they are insignificant in the full regressions, the first lags of the dependent variables have been used as the independent variables.
2. The second regression is a LVAR(1,1) with exogenous variables: *GDP_Fra(−1)*, *GDP_Ger(−1)*, *GDP_Ita(−1)* and *GDP_US*, where two of them are insignificant with large *p*-values. However, their joint effects on *GDP_Ger* are significant, based on the Chi-square statistic of $\chi_0^2 = 21.48844$ with $df = 3$ and a *p*-value $= 0.0001$. In a statistical sense, a reduced model should be explored. Do it as an exercise.

Example 4.10   The application of a VAR model

For comparison, Figure 4.20 presents the statistical results based on a VAR model of *GDP_FRA*, *GDP_GER* and *GDP_ITA*, using "1 1" as the lag intervals for endogenous with exogenous variables: *GDP_US, GDP_JPN* and *GDP_US*GDP_JPN*. Based on these results, the following notes and conclusions are made.

1. Compared to the results in Figure 4.19 using the object "*System*", individual regression in a VAR model cannot be reduced alone, since all regressions should have exactly the same set of independent variables. This is considered to be one of the limitations of data analysis using the object "*VAR*", since a system of regressions would have different sets of independent variables in practice.

```
Vector Autoregression Estimates
Date: 12/03/09  Time: 17:37
Sample (adjusted): 1951 1992
Included observations: 42 after adjustments
Standard errors in ( ) & t-statistics in [ ]
```

	GDP_FRA	GDP_GER	GDP_ITA
GDP_FRA(-1)	0.476795	-0.320457	-0.152371
	(0.08884)	(0.14695)	(0.11282)
	[5.36660]	[-2.18074]	[-1.35060]
GDP_GER(-1)	0.188395	1.024086	0.198438
	(0.05937)	(0.09820)	(0.07539)
	[3.17320]	[10.4287]	[2.63216]
GDP_ITA(-1)	-0.213285	-0.270204	0.632041
	(0.10042)	(0.16609)	(0.12751)
	[-2.12396]	[-1.62684]	[4.95668]
C	-843.0970	197.0901	-880.9308
	(429.923)	(711.089)	(545.923)
	[-1.96104]	[0.27717]	[-1.61365]
GDP_US	0.282211	0.184179	0.205555
	(0.05412)	(0.08952)	(0.06873)
	[5.21413]	[2.05738]	[2.99086]
GDP_JPN	0.453665	0.395202	0.096646
	(0.09475)	(0.15671)	(0.12031)
	[4.78813]	[2.52183]	[0.80329]
GDP_US*GDP_JPN	-1.33E-05	-5.62E-06	-3.11E-07
	(2.9E-06)	(4.8E-06)	(3.7E-06)
	[-4.54705]	[-1.16469]	[-0.08388]

R-squared	0.998982	0.997435	0.998268
Adj. R-squared	0.998808	0.996995	0.997971
Sum sq. resids	427723.1	1170120.	689675.8
S.E. equation	110.5471	182.8442	140.3746
F-statistic	5726.851	2268.305	3361.307
Log likelihood	-253.3952	-274.5293	-263.4279
Akaike AIC	12.39977	13.40616	12.87752
Schwarz SC	12.68938	13.69577	13.16713
Mean dependent	9340.643	9583.595	7775.690
S.D. dependent	3201.917	3335.595	3116.038

Determinant resid covariance (dof adj.)	5.18E+12
Determinant resid covariance	3.00E+12
Log likelihood	-782.1048
Akaike information criterion	38.24309
Schwarz criterion	39.11192

Figure 4.20 *Statistical results based on a VAR Model of* GDP_Fra, GDP_Ger *and* GDP_Ita, *with exogenous/ external variables* GDP_US, GDP_Jpn, *and* GDP_US*GDP_Jpn

2. As an illustration, note that *GDP_US*GDP_JPN* has a negative significant adjusted effect on *GDP_Fra* since $t_0 = -4.54\,704 < -2.0$, but it has insignificant adjusted effects on *GDP_Ger* and *GDP_Ita*. If this independent variable is deleted, then it should be deleted from all regressions. For this reason, applying the system equations is suggested, instead of the VAR model.

4.5   SCMs Based on a Trivariate $(X\_i_t, Y1\_i_t, Y2\_i_t)$

Under the assumption the two states considered are correlated, then this should be termed a theoretical seemingly causal model (SCM) or structural equation model (SEM) between the four endogenous variables *Y1_1*, *Y1_2*, *Y2_1* and *Y2_2*, in the form of a path diagram. Note that some pairs of the variables might have simultaneous causal relationships, as presented in Figures 4.12 and 4.13.

For illustration, see the path diagrams presented in Figure 4.21, which show a causal effect of $Y\_2 = (Y2\_1, Y2\_2)$ on $Y\_1 = (Y1\_1, Y1\_2)$, but more specific causal relationships between each of their components should be defined in order to have specific SCMs. There should be several possible path diagrams of the four endogenous variables, such as those presented in Chapter 8 (Agung, 2011). See the special path diagram presented in Figure 4.22.

Note that the environmental variable Z might have direct effects also on both *Y2_1* and *Y2_2*. Based on this path diagram, additive, two- and three-way interaction SCMs can easily be written, using the methods presented in Chapter 2, as well as Section 4.3. Do it as an exercise.

Further extensions of this path diagram would be where *Y1_1* and *Y1_2* have simultaneous causal effects, with multivariate exogenous or environmental variables.

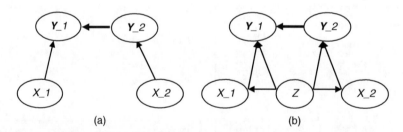

Figure 4.21   *The path diagrams of SCMs using the variables (a). (X_1, **Y**_1), and (X_2, **Y**_2), and (b). (X_1, **Y**_1), (X_2, **Y**_2), and an environmental variable Z*

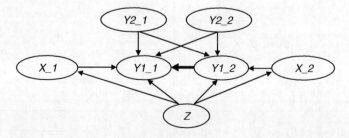

Figure 4.22   *A modified path diagram of the system specification*

4.6　SCMs Based on Multivariate Endogenous and Exogenous Variables

As an extension of Figure 4.21, Figure 4.23(a) presents a path diagram between the exogenous variables $X\_i = (X1\_i, \ldots, XK\_i)$ and endogenous variables $Y\_i = (Y1\_i, \ldots, YG\_i)$ for two correlated states, and Figure 4.23(b) presents a path diagram between the exogenous and endogenous variables, and an environmental variables $Z = (Z1, \ldots, ZM)$.

Note that if each multivariate X_i, Y_i, for $i = 1$ and 2, and Z, has two components, then one should define a path diagram based on 10 time series, with a total of 45 pairs of possible causal relationships. It will not be an easy task to select the best possible sub-set of the 45 pairs.

As an illustration, Figure 4.24 presents a special path diagram of the path diagram in Figure 4.23(a). For comparison, this path diagram is developed based on the path diagram in Figure 4.32 (Agung, 2009a). For this reason, either additive, two- or three-way SCMs, could easily be extended from the SCMs based on Figure 4.32 in Agung (2009a). As the extensions, the endogenous variables $Y1\_1$, $Y2\_1$, $Y2\_1$ and $Y2\_2$ could have alternative possible causal relationships in a theoretical sense. The following sections only present the standard SCMs.

4.6.1　Simple Additive SCM

The standard SCM has the system specification as follows:

$$
\begin{aligned}
y1\_1 &= c(110) + (111)^{*}y2\_1 + c(112)^{*}x1\_1 + c(113)^{*}y1\_2 \\
y2\_1 &= c(210) + c(211)^{*}x1\_1 + c(212)^{*}x2\_1 \\
x1\_1 &= c(310) + c(311)^{*}x2\_1 + c(312)^{*}x3\_1 \\
x3\_1 &= c(410) + c(411)^{*}x2\_1 \\
y1\_2 &= c(120) + (121)^{*}y2\_2 + c(122)^{*}x1\_2 \\
y2\_2 &= c(220) + c(221)^{*}x1\_2 + c(222)^{*}x2\_2 \\
x1\_2 &= c(320) + c(321)^{*}x2\_2 + c(322)^{*}x3\_2 \\
x3\_2 &= c(420) + c(421)^{*}x2\_2
\end{aligned}
$$

$$(4.14)$$

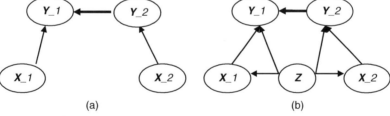

Figure 4.23　*The path diagram of SCMs using the variables (a). (**X**_ 1, **Y** _1), and (**X** _2, **Y** _2), and (b). (**X**_1, **Y**_1), (**X**_2, **Y**_2), and an environmental variable* **Z**

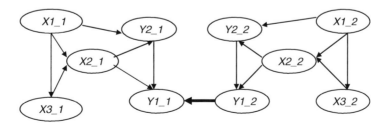

Figure 4.24　*A theoretical SCM based on (X1_i, X2_i, X3_i, Y1_i, Y2_i), for i = 1 and 2*

Note that the first four regressions are additive multivariate models for the first state, with an endogenous variable of the second state, namely $Y1\_2$, as an independent variable, and the last four regressions are the multivariate model for the second states.

4.6.2   A Two-Way Interaction SCM

The following system specifications only presents the regression set of endogenous variables: $Y1\_1$, $Y2\_1$, $Y1\_2$ and $Y2\_2$, where the first regression represents that $Y2\_2$ and $X2\_2$ have indirect effects on $Y1\_1$ through $Y1\_2$, indicated by the two interactions $Y1\_2*Y2\_2$ and $Y1\_2*X2\_2$ as additional independent variables.

$$
\begin{aligned}
\boldsymbol{y1\_1} =\ & c(110) + c(111)^*y2\_1 + c(112)^*x1\_1 + c(113)^*y2\_1^*x1\_1 + c(114)^*y2\_1^*x2\_1 \\
& + c(115)^*x1\_1^*x2\_1 + c(116)^*x1\_1^*x3\_1 + c(\boldsymbol{117})^*\boldsymbol{y1\_2} \\
& + c(\boldsymbol{118})^*\boldsymbol{y1\_2^*y2\_2} + c(\boldsymbol{119})^*\boldsymbol{y1\_2^*x2\_2} \\
y2\_1 =\ & c(210) + c(211)^*x1\_1 + c(212)^*x2\_1 + c(213)^*x1\_1^*x2\_1 \\
& + c(214)^*x1\_1^*x3\_1 \\
\boldsymbol{y1\_2} =\ & c(120) + c(121)^*y2\_2 + c(122)^*x1\_2 + c(123)^*y2\_2^*x1\_2 + c(124)^*y2\_2^*x2\_2 \\
& + c(125)^*x1\_2^*x2\_2 + c(126)^*x1\_2^*x3\_2 \\
y2\_2 =\ & c(220) + c(221)^*x1\_2 + c(22)^*x2\_2 + c(23)^*x1\_2^*x2\_2 \\
& + c(224)^*x1\_2^*x3\_2
\end{aligned}
\tag{4.15}
$$

4.6.3   A Three-Way Interaction SCM

Similar to the model in (4.11), the following system specification also presents the set of the regressions of endogenous variables only, namely $Y1\_1$, $Y2\_1$, $Y1\_2$ and $Y2\_2$. Note that the regression of $Y1\_1$ would have the interaction $Y1\_2*Y2\_2*X2\_2$ as an independent variable, under the assumption that the variables $Y1\_2$, $Y2\_2$ and $X2\_2$ are completely correlated. However, in a theoretical sense, it is not an easy task to identify whether or not a set of three numerical variables are completely correlated.

$$
\begin{aligned}
\boldsymbol{y1\_1} =\ & c(110) + c(111)^*y2\_1 + c(112)^*x1\_1 + c(113)^*y2\_1^*x1\_1 + c(114)^*y2\_1^*x2\_1 \\
& + c(115)^*x1\_1^*x2\_1 + c(116)^*x1\_1^*x3\_1 + c(117)^*y2\_1^*x1\_1^*x2\_1 \\
& + c(118)^*x1\_1^*x2\_1^*x3\_1 + c(\boldsymbol{119})^*\boldsymbol{y1\_2} + c(\boldsymbol{1110})^*\boldsymbol{y1\_2^*y2\_2} \\
& + c(\boldsymbol{1111})^*\boldsymbol{y1\_2^*x2\_2} + c(\boldsymbol{1112})^*\boldsymbol{y1\_2^*y2\_2^*x2\_2} \\
y2\_1 =\ & c(210) + c(211)^*x1\_1 + c(212)^*x2\_1 + c(213)^*x1\_1^*x2\_1 \\
& + c(214)^*x1\_1^*x3\_1 + c(215)^*x1\_1^*x2\_1^*x3\_1 \\
\boldsymbol{y1\_2} =\ & c(120) + c(121)^*y2\_2 + c(122)^*x1\_2 + c(123)^*y2\_2^*x1\_2 + c(124)^*y2\_2^*x2\_2 \\
& + c(125)^*x1\_2^*x2\_2 + c(126)^*x1\_2^*x3\_2 + c(127)y2\_2^*x1\_2^*x2\_2 \\
& + (128)^*x1\_2^*x2\_2^*x3\_2 \\
y2\_2 =\ & c(220) + c(221)^*x1\_2 + c(22)^*x2\_2 + c(23)^*x1\_2^*x2\_2 \\
& + c(224)^*x1\_2^*x3\_2 + c(225)^*x1\_2^*x2\_2^*x3\_2
\end{aligned}
\tag{4.16}
$$

4.6.4   Special Notes and Comments

Having a large number of variables in a model means that, in general, many of these variables would have insignificant adjusted effects on the corresponding dependent variable(s) because the independent variables are correlated in a statistical sense, even though some of them might not be correlated in a theoretical sense (such as sex and age). And their impacts on the estimates are unpredictable, an important cause factor can have a large or the largest *p*-value in a model but the least important cause factor has a significant adjusted effect. Refer to the special notes and comments in Section 2.14, and the process in obtaining a reduced model, presented in Agung (2009a).

4.6.5   Various Alternative SCMs

A lot of models, either with or without environmental variables, could be defined based on each path diagram presented in Sections 4.4–4.6 using lagged endogenous or exogenous variables, AR terms, as well as transformed variables, such as *log(Yg_i), log(Xk_i)*, and *log[(Yg_I – Lg_i)/(Ug_I – Yg_i)]*, where *Lg_i* and *Ug_i* are the lower and upper bounds of *Yg*.

More advanced models are instrumental variables models, TGARCH(*a,b,c*), and latent variables models, presented in Agung (2009a) with special notes and comments on the limitations of each.

On the other hand, the "*Near singular matrix*" or "*Over flow*" error message might be obtained, by using a large number of variables so then trial-and-error methods should be applied to solve the problem.

Example 4.11   The application of an AR(1) two-way interaction SCM
Figure 4.25 presents statistical results, based on the AR(1) model in (4.15). Based on these results, the following notes and conclusions are presented.

Estimation Method: Seemingly Unrelated Regression
Date: 12/03/09 Time: 15:36
Sample: 1968M02 1994M10
Included observations: 322
Total system (balanced) observations 1284
Iterate coefficients after one-step weighting matrix
Convergence achieved after: 1 weight matrix, 435 total coef iterations

	Coefficient	Std. Error	t-Statistic	Prob.
C(110)	685.9979	1323.675	0.518252	0.6044
C(111)	0.124722	0.067188	1.856323	0.0636
C(112)	-0.063643	0.100146	-0.635506	0.5252
C(113)	-3.51E-06	1.69E-06	-2.077716	0.0379
C(114)	3.92E-07	3.96E-07	0.991455	0.3217
C(115)	5.30E-06	2.07E-06	2.565657	0.0104
C(116)	-2.18E-06	3.73E-06	-0.583787	0.5595
C(117)	0.153058	0.039509	3.874028	0.0001
C(118)	0.983084	0.011275	87.19186	0.0000
C(120)	12699.34	2511.687	5.056101	0.0000
C(121)	0.123034	0.066393	1.853125	0.0641
C(122)	0.081856	0.075160	1.089094	0.2763
C(123)	-4.55E-06	2.51E-06	-1.811887	0.0702
C(125)	3.77E-06	4.70E-07	8.010518	0.0000
C(126)	0.992877	0.002625	378.2868	0.0000
C(210)	5030.105	9513.058	0.528758	0.5971
C(211)	0.038090	0.104601	0.364148	0.7158
C(212)	0.440040	0.036289	12.12585	0.0000
C(213)	1.51E-05	7.18E-06	2.103282	0.0356
C(214)	7.00E-07	1.74E-06	0.401582	0.6881
C(215)	-6.02E-06	2.55E-06	-2.359381	0.0185
C(216)	-5.19E-06	2.12E-06	-2.449495	0.0144
C(217)	0.998130	0.003820	261.2762	0.0000
C(220)	114782.9	105075.4	1.092387	0.2749
C(221)	-0.232462	0.241631	-0.962052	0.3362
C(222)	0.020444	0.236835	0.086323	0.9312
C(223)	4.08E-06	1.36E-05	0.298773	0.7652
C(225)	1.15E-05	9.71E-06	1.180860	0.2379
C(226)	0.998326	0.002176	458.8837	0.0000

Determinant residual covariance		2.42E+17

Equation: Y1_1=C(110)+C(111)*X2_1+C(112)*Y2_1+C(113)*X2_1*X1_1
  +C(114)*X2_1*X3_1+C(115)*X1_1*Y1_1+C(116)*X2_1*Y2_1+C(117)
  *Y1_2+C(118)*Y1_2*Y2_2+C(119)*Y1_2*X2_2+[AR(1)=C(1)]
Observations: 321

R-squared	0.997734	Mean dependent var	4796.987
Adjusted R-squared	0.997661	S.D. dependent var	1564.584
S.E. of regression	75.67654	Sum squared resid	1775351.
Durbin-Watson stat	2.099989		

Equation: Y1_2=C(120)+C(121)*X1_1+C(122)*X2_1+C(123)*X1_1*X2_1
  +C(125)*X2_1*X3_1+[AR(1)=C(126)]
Observations: 321

R-squared	0.999614	Mean dependent var	16076.63
Adjusted R-squared	0.999607	S.D. dependent var	6258.104
S.E. of regression	124.0024	Sum squared resid	4843631.
Durbin-Watson stat	1.876831		

Equation: Y2_1=C(210)+C(211)*X2_2+C(212)*Y2_2+C(213)*X2_2*X1_2
  +C(214)*X2_2*X3_2+C(215)*X1_2*Y2_2+C(216)*X2_2*Y2_2
  +[AR(1)=C(217)]
Observations: 321

R-squared	0.999762	Mean dependent var	14089.85
Adjusted R-squared	0.999757	S.D. dependent var	6599.741
S.E. of regression	102.9214	Sum squared resid	3315553.
Durbin-Watson stat	1.750326		

Equation: Y2_2=C(220)+C(221)*X1_2+C(222)*X2_2+C(223)*X1_2*X2_2
  +C(225)*X2_2*X3_2+[AR(1)=C(226)]
Observations: 321

R-squared	0.998669	Mean dependent var	36596.94
Adjusted R-squared	0.998648	S.D. dependent var	15966.73
S.E. of regression	587.1794	Sum squared resid	1.09E+08
Durbin-Watson stat	1.920218		

Figure 4.25   *Statistical results based on an AR(1) SCM in (4.15)*

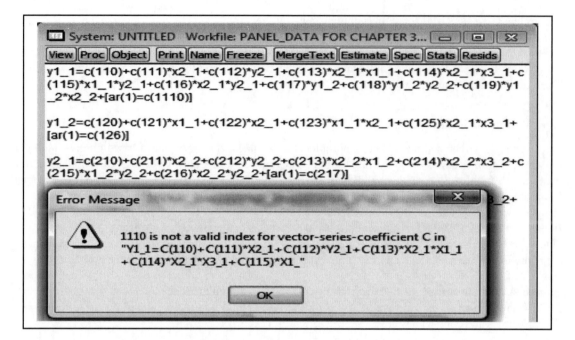

Figure 4.26   *An unexpected error message*

1. Note that the first regression uses $[AR(1) = C(1)]$, since using $[AR(1) = C(1110)]$ gets an unexpected error message, as shown in Figure 4.26. The results are also obtained by using other symbols $C(i)$ or $C(ij)$ for any integer i and j. The symbol $C(ijk)$, for any $i > 2$, also can be used since the parameter has not been used in the model.
2. In a statistical sense, an acceptable reduced model should be explored since several independent variables have insignificant adjusted effects with the *p*-values greater than 0.20 (or 0.25). Do it as an exercise.

Example 4.12   An TGARCH(1,1,0) model

By using the first regression in Figure 4.25 as an AR(1) mean model, a TGARCH(1,1,0) is applied with the statistical results presented in Figure 4.27(a); and Figure 4.27(b) presents the results based a modified model, namely a LV(1) mean model. For a more detail explanation on various TGARCH(*a,b,c*), refer to Agung (2009a). Based on these results the following notes are presented.

1. Figure 4.27(a) presents a note "*Estimated AR process is nonstationary*", so the estimates are not acceptable in a statistical sense. Note that in fact nothing is wrong with the AR(1) mean model but the data cannot provide an acceptable estimate. So, a modified model should be explored.
2. By using trial-and-error methods, an acceptable statistical result is obtained as presented in Figure 4.27(b) using a LV(1) mean model. In fact, by using $[AR(2) = C(2)]$, instead of $[AR(1) = C(1)]$, an acceptable result is also obtained, but the results in Figure 4.18(b) are better.
3. Since two of the independent variables have *p*-values greater than 0.25, then a reduced model should be obtained. By using trial-and error-methods, a reduced model with all independent variables having *p*-values less than 0.20 was obtained here by deleting the interaction *Y1_2*Y2_2*.
4. Note that a TGARCH(*a,b,c*) model should be using a single mean model. So for the system equations in Figure 4.25, the ARCH estimation process should be applied four times. Do it as an exercise.

```
Dependent Variable: Y1_1
Method: ML - ARCH (Marquardt) - Normal distribution
Date: 12/31/09  Time: 15:00
Sample (adjusted): 1968M02 1994M10
Included observations: 321 after adjustments
Convergence achieved after 46 iterations
Presample variance: backcast (parameter = 0.7)
Y1_1=C(110)+C(111)*X2_1+C(112)*Y2_1+C(113)*X1_1+C(114)
    *X2_1*X3_1+C(115)*X1_1*Y2_1+C(116)*X2_1*Y2_1+C(117)*Y1_2
    +C(118)*Y1_2*Y2_2+C(119)*Y1_2*X2_2+[AR(1)=C(1)]
GARCH = C(120) + C(121)*RESID(-1)^2 + C(122)*GARCH(-1)
```

Variable	Coefficient	Std. Error	z-Statistic	Prob.
C(110)	-1284.978	11410.82	-0.112611	0.9103
C(111)	0.097811	0.067725	1.444243	0.1487
C(112)	-0.039760	0.065383	-0.608111	0.5431
C(113)	-2.53E-06	2.14E-06	-1.181453	0.2374
C(114)	7.42E-07	5.74E-07	1.292022	0.1963
C(115)	5.40E-06	2.84E-06	1.899193	0.0575
C(116)	2.96E-09	2.23E-06	0.001327	0.9989
C(117)	0.274275	0.039017	7.001940	0.0000
C(118)	-3.36E-06	6.47E-07	-5.190883	0.0000
C(119)	-1.92E-06	1.14E-06	-1.674217	0.0941
C(1)	1.001460	0.011991	83.51814	0.0000

	Variance Equation			
C	32735.80	4598.544	7.118731	0.0000
RESID(-1)^2	0.177599	0.033002	5.381509	0.0000
GARCH(-1)	-1.001468	0.000581	-1722.777	0.0000

R-squared	0.997539	Mean dependent var	4796.988
Adjusted R-squared	0.997459	S.D. dependent var	1564.584
S.E. of regression	78.86580	Akaike info criterion	11.69046
Sum squared resid	1928143.	Schwarz criterion	11.85495
Log likelihood	-1862.319	Hannan-Quinn criter.	11.75614
F-statistic	9663.998	Durbin-Watson stat	2.130083
Prob(F-statistic)	0.000000		

Inverted AR Roots	1.00	
	Estimated AR process is nonstationary	

(a)

```
Dependent Variable: Y1_1
Method: ML - ARCH (Marquardt) - Normal distribution
Date: 12/31/09  Time: 15:20
Sample (adjusted): 1968M02 1994M10
Included observations: 321 after adjustments
Convergence achieved after 68 iterations
Presample variance: backcast (parameter = 0.7)
Y1_1=C(110)+C(111)*X2_1+C(112)*Y2_1+C(113)*X2_1*X1_1+C(114)
    *X2_1*X3_1+C(115)*X1_1*Y2_1+C(116)*X2_1*Y2_1+C(117)*Y1_2
    +C(118)*Y1_2*Y2_2+C(119)*Y1_2*X2_2+C(1)*Y1_1(-1)
GARCH = C(120) + C(121)*RESID(-1)^2 + C(122)*GARCH(-1)
```

Variable	Coefficient	Std. Error	z-Statistic	Prob.
C(110)	-147.7883	82.00193	-1.802254	0.0715
C(111)	0.031560	0.013677	2.307569	0.0210
C(112)	-0.041769	0.018799	-2.221944	0.0263
C(113)	-1.15E-06	3.53E-07	-3.261661	0.0011
C(114)	-2.46E-07	1.04E-07	-2.378487	0.0174
C(115)	1.44E-06	4.07E-07	3.537727	0.0004
C(116)	1.44E-06	8.82E-07	1.633138	0.1024
C(117)	0.032066	0.020904	1.533906	0.1251
C(118)	-1.58E-07	2.50E-07	-0.630074	0.5286
C(119)	-5.42E-07	5.71E-07	-0.949369	0.3424
C(1)	0.942412	0.017348	54.32363	0.0000

	Variance Equation			
C	3979.771	791.2111	5.029974	0.0000
RESID(-1)^2	0.410744	0.113453	3.620398	0.0003
GARCH(-1)	0.008649	0.109839	0.078739	0.9372

R-squared	0.997421	Mean dependent var	4796.988
Adjusted R-squared	0.997337	S.D. dependent var	1564.584
S.E. of regression	80.73423	Akaike info criterion	11.46357
Sum squared resid	2020585.	Schwarz criterion	11.62806
Log likelihood	-1825.904	Hannan-Quinn criter.	11.52925
F-statistic	9220.776	Durbin-Watson stat	2.132773
Prob(F-statistic)	0.000000		

(b)

Figure 4.27   *Statistical results based on a TGARCH(1,1,0) using (a) an AR(1) mean model, and (b) a LV(1) mean model*

Example 4.13   An instrumental variables model

Figure 4.28 presents the statistical results based on AR additive instrumental variables SCM, using the system specification as follows:

$$
\begin{aligned}
y1\_1 &= c(110) + c(112)^*x2\_1 + c(113)^*y1\_2 + c(114)^*y2\_1 \\
&\quad + [ar(1) = c(115), ar(2) = c(116)] @ c\,x1\_1\,x3\_1 \\
y1\_2 &= c(120) + c(121)^*x1\_1 + [ar(1) = c(122)] @ c\,x2\_1\,x3\_1 \\
y2\_1 &= c(210) + c(211)^*x2\_2 + [ar(1) = c(212)] @ c\,x1\_2\,x3\_2 \\
y2\_2 &= c(220) + c(221)^*x1\_2 + [ar(1) = c(222)] @ c\,x2\_2\,x3\_2
\end{aligned}
\tag{4.17}
$$

Based on these results, note the following.

1. Note that the results present additional instrumental variables corresponding to the selected option used. For a more detailed discussion, and special notes and comments on the application of the instrumental variables models, refer to Agung (2009a).

2. In fact, the system equations in Figure 4.25 would have been used as the basic model for instrumental variables model. However, the "*Singular error message*" was obtained, and it is not an easy task to modify the model. For this reason, the experimentation was done starting with a simple AR additive instrumental variables model. Finally, the results in Figure 4.29 were obtained.

System: SYS14_ISM
Estimation Method: Iterative Two-Stage Least Squares
Date: 12/31/09   Time: 14:47
Sample: 1968M02 1994M10
Included observations: 322
Total system (unbalanced) observations 1283
Convergence achieved after 10 iterations

	Coefficient	Std. Error	t-Statistic	Prob.
C(110)	3762.330	2447.612	1.537143	0.1245
C(112)	-0.019769	0.066420	-0.297640	0.7660
C(113)	0.248131	0.124836	1.987652	0.0471
C(114)	-0.124848	0.077128	-1.618720	0.1058
C(115)	0.939085	0.059058	15.90100	0.0000
C(116)	0.052321	0.059441	0.880209	0.3789
C(120)	65502.94	41543.79	1.576721	0.1151
C(121)	-0.212617	0.135169	-1.572976	0.1160
C(122)	0.998445	0.001314	760.0974	0.0000
C(210)	49205.12	33388.82	1.473701	0.1408
C(211)	-0.157199	0.177678	-0.884741	0.3765
C(212)	0.998124	0.001733	575.9126	0.0000
C(220)	144058.7	172204.5	0.836556	0.4030
C(221)	-0.001699	0.170515	-0.009962	0.9921
C(222)	0.998678	0.002083	479.3684	0.0000
Determinant residual covariance		3.82E+17		

Equation: Y1_1=C(110)+C(112)*X2_1+C(113)*Y1_2+C(114)*Y2_1
+[AR(1)=C(115), AR(2)=C(116)]
Instruments: C X1_1 X3_1 Y1_1(-1) X2_1(-1) Y1_2(-1) Y2_1(-1) Y1_1(-2)
X2_1(-2) Y1_2(-2) Y2_1(-2)
Observations: 320

R-squared	0.997205	Mean dependent var	4806.125
Adjusted R-squared	0.997160	S.D. dependent var	1558.433
S.E. of regression	83.05010	Sum squared resid	2165758.
Durbin-Watson stat	1.996559		

Equation: Y1_2=C(120)+C(121)*X1_1 +[AR(1)=C(122)]
Instruments: C X2_1 X3_1 Y1_2(-1) X1_1(-1)
Observations: 321

R-squared	0.999035	Mean dependent var	16076.63
Adjusted R-squared	0.999029	S.D. dependent var	6258.103
S.E. of regression	195.0442	Sum squared resid	12097437
Durbin-Watson stat	1.867183		

Equation: Y2_1=C(210)+C(211)*X2_2+[AR(1)=C(212)]
Instruments: C X1_2 X3_2 Y2_1(-1) X2_2(-1)
Observations: 321

R-squared	0.998780	Mean dependent var	14089.85
Adjusted R-squared	0.998772	S.D. dependent var	6599.741
S.E. of regression	231.2734	Sum squared resid	17008983
Durbin-Watson stat	1.836054		

Equation: Y2_2=C(220)+C(221)*X1_2 +[AR(1)=C(222)]
Instruments: C X2_2 X3_2 Y2_2(-1) X1_2(-1)
Observations: 321

R-squared	0.998638	Mean dependent var	36596.94
Adjusted R-squared	0.998629	S.D. dependent var	15966.73
S.E. of regression	591.1834	Sum squared resid	1.11E+08
Durbin-Watson stat	1.890292		

Figure 4.28   *Statistical results based on an AR(1) additive instrumental variables model*

System: SYS14
Estimation Method: Iterative Two-Stage Least Squares
Date: 12/31/09   Time: 16:42
Sample: 1968M02 1994M10
Included observations: 322
Total system (unbalanced) observations 1278
Convergence achieved after 97 iterations

	Coefficient	Std. Error	t-Statistic	Prob.
C(110)	-1269.330	612.9547	-2.070838	0.0386
C(112)	0.117033	0.073572	1.590737	0.1119
C(113)	0.343619	0.099846	3.441502	0.0006
C(114)	0.080228	0.092170	0.870435	0.3842
C(115)	1.18E-06	6.81E-07	1.739227	0.0822
C(116)	-3.27E-07	6.71E-07	-0.487832	0.6258
C(117)	-2.60E-06	1.13E-06	-2.295528	0.0219
C(118)	-5.19E-06	3.41E-06	-1.520548	0.1286
C(2)	1.085953	0.048159	22.54955	0.0000
C(5)	-0.172820	0.047717	-3.621792	0.0003
C(120)	-3727.533	3108.253	-1.199237	0.2307
C(121)	0.292929	0.265897	1.101665	0.2708
C(122)	0.978719	0.155463	6.295521	0.0000
C(123)	-1.13E-05	1.06E-05	-1.062507	0.2882
C(124)	0.980420	0.014734	66.54109	0.0000
C(210)	-109.8487	63.32586	-1.734657	0.0830
C(211)	0.956701	0.015658	61.09789	0.0000
C(212)	0.106978	0.023809	4.493185	0.0000
C(213)	-9.31E-07	2.16E-07	-4.304828	0.0000
C(214)	0.108098	0.058737	1.840365	0.0660
C(220)	-12126.77	15827.60	-0.766179	0.4437
C(221)	4.124703	2.845756	1.449423	0.1475
C(222)	4.061701	1.110998	3.655902	0.0003
C(223)	-0.000271	0.000167	-1.621230	0.1052
C(224)	0.973985	0.016386	59.44162	0.0000
Determinant residual covariance		6.94E+18		

Equation: Y1_1=C(110)+C(112)*X2_1+C(113)*Y1_2+C(114)*Y2_1+C(115)
*X2_1*X1_1+C(116)*X2_1*X3_1+C(117)*Y1_2*Y2_2+C(118)*Y1_2
*X2_2+[AR(2)=C(2), AR(5)=C(5)]
Instruments: C X1_1 X3_1 Y1_1(-2) X2_1(-2) Y1_2(-2) Y2_1(-2) X1_1(-2)
*X2_1(-2) X2_1(-2)*X3_1(-2) Y1_2(-2)*Y2_2(-2) X2_2(-2)*Y1_2(-2)
Y1_1(-5) X2_1(-5) Y1_2(-5) X1_1(-5)*X2_1(-5) X2_1(-5)*X3_1(
-5) Y1_2(-5)*Y2_2(-5) X2_2(-5)*Y1_2(-5)
Observations: 317

R-squared	0.995295	Mean dependent var	4833.574
Adjusted R-squared	0.995157	S.D. dependent var	1539.851
S.E. of regression	107.1603	Sum squared resid	3525380.
Durbin-Watson stat	1.205214		

Equation: Y1_2=C(120)+C(121)*X1_1+C(122)*X2_1 +C(123)*X1_1*X2_1
+[AR(1)=C(124)]
Instruments: C X3_1 Y1_2(-1) X1_1(-1) X2_1(-1) X1_1(-1)*X2_1(-1)
Observations: 321

R-squared	0.999413	Mean dependent var	16076.63
Adjusted R-squared	0.999405	S.D. dependent var	6258.103
S.E. of regression	152.6247	Sum squared resid	7361001.
Durbin-Watson stat	1.630280		

Equation: Y2_1=C(210)+C(211)*Y2_1(-1)+C(212)*X2_2+C(213)*Y2_2
*X2_2+[AR(2)=C(214)]
Instruments: X1_2 X3_2 C Y2_1(-2) Y2_1(-3) X2_2(-2) X2_2(-2)*Y2_2(-2)
Observations: 319

R-squared	0.998844	Mean dependent var	14152.30
Adjusted R-squared	0.998829	S.D. dependent var	6572.867
S.E. of regression	224.8966	Sum squared resid	15881639
Durbin-Watson stat	1.973920		

Equation: Y2_2=C(220)+C(221)*X1_2 +C(222)*X2_2+C(223)*X1_2*X2_2
+[AR(1)=C(224)]
Instruments: C X3_2 Y2_2(-1) X1_2(-1) X2_2(-1) X1_2(-1)*X2_2(-1)
Observations: 321

R-squared	0.996587	Mean dependent var	36596.94
Adjusted R-squared	0.996544	S.D. dependent var	15966.73
S.E. of regression	938.6148	Sum squared resid	2.78E+08
Durbin-Watson stat	2.304271		

Figure 4.29   *Statistical results based on an unexpected instrumental variables model*

3. Then the independent variables were inserted one-by-one to each regression. If the "*Near singular matrix*" error message is obtained, the variable will not be used. The statistical results have been obtained based on several instrumental models. Try this as an exercise using your own data set. One of these is presented in the following example.

Example 4.14   An unexpected instrumental variables model
By using the model in (4.17) as a base model, additional variables are selected from the variables in Figure 4.24, and then inserted one-by-one in order to have the mean models close as the regressions in Figure 4.25. Based on these results the following notes and comments are made.

1. Note that the first regression uses the AR(2) and AR(5) which are unexpected. This regression function is obtained using trial-and-error methods, which definitely cannot be generalized. However, estimates would be considered as acceptable results, in a statistical sense, since only 2 out 10 parameters are insignificant with p-values > 0.20.
2. In a statistical sense, a reduced model should be explored, but independent variables should be deleted, because of the unpredictable impact of multicollinearity of independent variables – refer to the special notes presented in Section 2.14.2 (Agung, 2009a), as well as the unpredictable impact of a new set of instrumental variables.
3. For comparison, each parameter of the third regression, namely the regression of $Y2\_1$, is significant. On the other hand, there is no good explanation as to *why* such a set of instrumental variables should be used. In fact, Agung (2009a) has recommended not applying them because there is no general rule as to how to select the best possible instrumental variables, and this demonstrates that models can be used without them.

4.7   Fixed- and Random Effects Models

Most students only consider using two alternative models for data analysis with a panel data set, namely fixed- and random effect models: then using a test statistic, such as the Hausman Test (Wooldridge, 2002, p. 288), for deciding which one would be the better model in a statistical sense. Refer to the pointers presented in Section 1.6.3 which show that the fixed- and random effect growth models are worse multivariate growth models by states in a theoretical sense. Therefore, the test should not be conducted by selecting one of the two worst models for the task.

In fact, EViews provides options for better models, namely those with *cross-section specific coefficients* or *period specific coefficients*, which will be presented in Section 4.9. However, for comparison between analysis using the objects "*POOL*" and "*System*", the following sections present again the fixed- and random effect models.

4.7.1   A Fixed-Effect Model and a MANCOVA Model

For a better discussion, first see the following two examples, which present statistical results based on a fixed-effect model and a MANCOVA model, respectively.

Example 4.15   An AR(2) fixed-effect interaction model
Figure 4.30 presents the results based on an AR(2) fixed-effect interaction model. The steps of the analysis are as follows (refer to Section 1.6):

1. With the work file on the screen, select *Object/New Objects . . . /Pool . . . OK*, then enter the *Cross Section Identifiers*: _1 _2 _3 _4.

Dependent Variable: Y?
Method: Pooled Least Squares
Date: 01/02/10   Time: 08:12
Sample (adjusted): 1968M03 1994M10
Included observations: 320 after adjustments
Cross-sections included: 4
Total pool (balanced) observations: 1280
Convergence achieved after 11 iterations

Variable	Coefficient	Std. Error	t-Statistic	Prob.
C	6236.762	2469.630	2.525383	0.0117
X?	0.318264	0.014919	21.33221	0.0000
Z?	-0.193388	0.028311	-6.830738	0.0000
X?*Z?	2.26E-06	3.18E-07	7.095881	0.0000
AR(1)	1.159602	0.028124	41.23224	0.0000
AR(2)	-0.163354	0.028096	-5.814090	0.0000
Fixed Effects (Cross)				
_1—C	-5387.027			
_2—C	11487.22			
_3—C	-12199.91			
_4—C	6099.726			

Effects Specification

Cross-section fixed (dummy variables)

R-squared	0.999748	Mean dependent var		17927.26
Adjusted R-squared	0.999746	S.D. dependent var		14829.11
S.E. of regression	236.1557	Akaike info criterion		13.77387
Sum squared resid	70883029	Schwarz criterion		13.81011
Log likelihood	-8806.274	Hannan-Quinn criter.		13.78747
F-statistic	630238.1	Durbin-Watson stat		2.046468
Prob(F-statistic)	0.000000			

Estimation Command:
=====================
LS(CX=F) Y? X? Z? X?*Z? AR(1) AR(2)

Estimation Equations:
=====================
Y_1 = C(7) + C(1) + C(2)*X_1 + C(3)*Z_1 + C(4)*X_1*Z_1 + [AR(1)=C(5),AR(2)=C(6)]

Y_2 = C(8) + C(1) + C(2)*X_2 + C(3)*Z_2 + C(4)*X_2*Z_2 + [AR(1)=C(5),AR(2)=C(6)]

Y_3 = C(9) + C(1) + C(2)*X_3 + C(3)*Z_3 + C(4)*X_3*Z_3 + [AR(1)=C(5),AR(2)=C(6)]

Y_4 = C(10) + C(1) + C(2)*X_4 + C(3)*Z_4 + C(4)*X_4*Z_4 + [AR(1)=C(5),AR(2)=C(6)]

Substituted Coefficients:
=====================
Y_1 = -5387.027015 + 6236.76242261 + 0.318263793206*X_1 - 0.193387867487*Z_1 + 2.2584140969e-06*X_1*Z_1 + [AR(1)=1.15960213348,AR(2)=-0.1633537201]

Y_2 = 11487.2153775 + 6236.76242261 + 0.318263793206*X_2 - 0.193387867487*Z_2 + 2.2584140969e-06*X_2*Z_2 + [AR(1)=1.15960213348,AR(2)=-0.1633537201]

Y_3 = -12199.914307 + 6236.76242261 + 0.318263793206*X_3 - 0.193387867487*Z_3 + 2.2584140969e-06*X_3*Z_3 + [AR(1)=1.15960213348,AR(2)=-0.1633537201]

Y_4 = 6099.72594446 + 6236.76242261 + 0.318263793206*X_4 - 0.193387867487*Z_4 + 2.2584140969e-06*X_4*Z_4 + [AR(1)=1.15960213348,AR(2)=-0.1633537201]

Figure 4.30   *Statistical results based on an AR(2) fixed effect interaction model*

2. Select the object "*Estimate*", then enter *Y?* as the dependent variable, and for the *Common Coefficients* enter the series: *X? Z? X?*Z? AR(1) AR(2)*.
3. Finally by selecting the option *Cross Section: Fixed . . . OK*, the results in Figure 4.30 are obtained.

Example 4.16   An AR(2) MANCOVA interaction model

Figure 4.31 presents the statistical results based on an AR(2) MANCOVA interaction model, using the system equations. Based on these results note the following:

1. Note that the symbols *C(71)*, *C(81)*, *C(91)* and *C(101)* represent the combination of the two intercept parameters of the fixed-effect model in Figure 4.30, since by using parameters *C(*) + C(1)*, the "*Near Nonsingular matrix*" error message is obtained. In fact, the design matrix is a nonsingular matrix.
2. The output presents an unexpected ordering, where *C(71)* is in the first line, and *C(81)*, *C(91)* and *C(101)* are in the last three lines.
3. For comparison with the fixed-effect model, see the following two regression functions of *Y_1*, which show that that the estimates of the slope parameters have the same values up to five decimal places: some of them even up to six decimal places. However, the estimates of the intercept parameters are 849.7 354 076 for the fixed-effect regression, compared to 849.76 696 611 for the MANCOVA.

    *   The first regression function in the fixed-effect model:

$$Y\_1 = -5387.027015 + 6236.76242261 + 0.31826^*X\_1 - 0.19339^*Z\_1$$
$$+ 2.25841e - 06^*X\_1^*Z\_1 + [AR(1) = 1.15960, AR(2) = -0.16335]$$

    *   The first regression function in the MANCOVA model:

$$Y\_1 = 849.76696611 + 0.31826^*X\_1 - 0.19339^*Z\_1 + 2.25841e - 06^*X\_1^*Z\_1$$
$$+ [AR(1) = 1.15960, AR(2) = -0.16335]$$

```
System: SYS16_ANCOVA
Estimation Method: Iterative Least Squares
Date: 01/02/10  Time: 08:40
Sample: 1968M03 1994M10
Included observations: 322
Total system (balanced) observations 1280
Convergence achieved after 9 iterations
```

	Coefficient	Std. Error	t-Statistic	Prob.
C(71)	849.7670	3584.832	0.237045	0.8127
C(2)	0.318264	0.014919	21.33222	0.0000
C(3)	-0.193388	0.028311	-6.830738	0.0000
C(4)	2.26E-06	3.18E-07	7.095879	0.0000
C(5)	1.159602	0.028124	41.23224	0.0000
C(6)	-0.163354	0.028096	-5.814094	0.0000
C(81)	17724.06	5627.278	3.149668	0.0017
C(91)	-5963.264	4342.310	-1.373293	0.1699
C(101)	12336.46	4695.421	2.627339	0.0087

| Determinant residual covariance | | 1.89E+18 | |

```
Equation: Y_1=C(71)+C(2)*X_1+C(3)*Z_1+C(4)*X_1*Z_1+[AR(1)=C(5),AR(
    2)=C(6)]
Observations: 320
```

R-squared	0.992629	Mean dependent var	4806.125
Adjusted R-squared	0.992512	S.D. dependent var	1558.433

```
Equation: Y_2=C(81)+C(2)*X_2+C(3)*Z_2+C(4)*X_2*Z_2+[AR(1)=C(5),AR(
    2)=C(6)]
Observations: 320
```

R-squared	0.998994	Mean dependent var	14121.02
Adjusted R-squared	0.998978	S.D. dependent var	6586.377
S.E. of regression	210.5631	Sum squared resid	13921764
Durbin-Watson stat	2.134122		

```
Equation: Y_3=C(91)+C(2)*X_3+C(3)*Z_3+C(4)*X_3*Z_3+[AR(1)=C(5),AR(
    2)=C(6)]
Observations: 320
```

R-squared	0.999294	Mean dependent var	16109.83
Adjusted R-squared	0.999282	S.D. dependent var	6239.535
S.E. of regression	167.1522	Sum squared resid	8773112.
Durbin-Watson stat	1.696004		

```
Equation: Y_4=C(101)+C(2)*X_4+C(3)*Z_4+C(4)*X_4*Z_4+[AR(1)=C(5),AR
    (2)=C(6)]
Observations: 320
```

R-squared	0.999476	Mean dependent var	36672.08
Adjusted R-squared	0.999467	S.D. dependent var	15934.79
S.E. of regression	367.8034	Sum squared resid	42477708
Durbin-Watson stat	2.083208		

Figure 4.31 *Statistical results based on an AR(2) MANCOVA interaction model*

4.7.2 A Random Effect Model and a Single Regression

For comparison, Example 4.16 presents the statistical results based on a random effect model of Y_i, $i = 1$, 2, 3 and 4, and Example 4.17 presents the statistical results based on a single regression for the four states. Even though the random effect model is considered the model for panel data, the following examples present illustrative statistical results of a random effect model, compared with the results using the object "*System*".

Example 4.17   A random effect model
Figure 4.32(a) presents the statistical results based on a random effect model of Y_i, $i = 1$, 2, 3 and 4, with the exogenous *X?, Z?* and *X?*Z?*, without an AR term, since it cannot be applied to the random effect model. However, Figure 4.32(b) presents the statistical results based on a random effect model of Y_i, $i = 1$, 2, 3 and 4, with the exogenous *Y?(−1), X?*, and *Z?*. The steps of analysis are the same as the fixed-effect model with the option *Cross-Section: Random*. Based on these results, the following findings and conclusions are presented.

1. By using the exogenous variables *X?, Z?* and *X?*Z?*, Figure 4.32(a) presents a very small DW-statistic. So, a lagged dependent variable should be used.
2. By using the exogenous variables *Y?(−1), X?, Z?* and *X?*Z?*, the error message in Figure 4.33(a) is obtained, which indicates that the number of cross-sections should be strictly greater than the number of the coefficients. So, one of the exogenous variables should be omitted, using trial-and-error methods, because there are only four cross-sections. However, the interaction *X?*Z?* should be kept in the model, since it is defined that the effect of *X? (Z?)* on *Y?* depends on *Z? (X?)* in a theoretical sense.
3. By using the exogenous *Y?(−1), X?, X?*Z?*, the error message in Figure 4.33(b) is obtained. However, the estimates obtained by using two sub-sets of exogenous variables, namely *{Y?(−1), Z?, X?*Z?}* have their results presented in Figure 4.32(b) and *{Y?(−1), X?, Z?}*.
4. Based on the results in Figure 4.32(b), the following conclusions are made.

Dependent Variable: Y?
Method: Pooled EGLS (Cross-section random effects)
Date: 01/02/10   Time: 11:30
Sample (adjusted): 1968M01 1994M10
Included observations: 322 after adjustments
Cross-sections included: 4
Total pool (balanced) observations: 1288
Swamy and Arora estimator of component variances

Variable	Coefficient	Std. Error	t-Statistic	Prob.
C	-515.3089	122.6302	-4.202139	0.0000
X?	0.128644	0.002740	46.95600	0.0000
Z?	0.917811	0.012669	72.44281	0.0000
X?*Z?	-1.42E-06	1.03E-07	-13.84732	0.0000
Random Effects (Cross)				
_1–C	0.000000			
_2–C	0.000000			
_3–C	0.000000			
_4–C	0.000000			

Effects Specification		
	S.D.	Rho
Cross-section random	0.000000	0.0000
Idiosyncratic random	1699.068	1.0000

Weighted Statistics			
R-squared	0.974212	Mean dependent var	17853.11
Adjusted R-squared	0.974152	S.D. dependent var	14816.06
S.E. of regression	2382.027	Sum squared resid	7.29E+09
F-statistic	16169.00	Durbin-Watson stat	0.033624
Prob(F-statistic)	0.000000		

Unweighted Statistics			
R-squared	0.974212	Mean dependent var	17853.11
Sum squared resid	7.29E+09	Durbin-Watson stat	0.033624

(a)

Dependent Variable: Y?
Method: Pooled EGLS (Cross-section random effects)
Date: 01/03/10   Time: 08:28
Sample (adjusted): 1968M02 1994M10
Included observations: 321 after adjustments
Cross-sections included: 4
Total pool (balanced) observations: 1284
Swamy and Arora estimator of component variances

Variable	Coefficient	Std. Error	t-Statistic	Prob.
C	34.47682	21.44920	1.607371	0.1082
Y?(-1)	0.993887	0.002787	356.5713	0.0000
Z?	0.009942	0.003617	2.748396	0.0061
X?*Z?	-1.04E-08	1.82E-08	-0.570459	0.5685
Random Effects (Cross)				
_1–C	0.000000			
_2–C	0.000000			
_3–C	0.000000			
_4–C	0.000000			

Effects Specification		
	S.D.	Rho
Cross-section random	0.000000	0.0000
Idiosyncratic random	325.8780	1.0000

Weighted Statistics			
R-squared	0.999503	Mean dependent var	17890.10
Adjusted R-squared	0.999502	S.D. dependent var	14822.58
S.E. of regression	330.6685	Sum squared resid	1.40E+08
F-statistic	858918.1	Durbin-Watson stat	1.877293
Prob(F-statistic)	0.000000		

Unweighted Statistics			
R-squared	0.999503	Mean dependent var	17890.10
Sum squared resid	1.40E+08	Durbin-Watson stat	1.877293

(b)

Figure 4.32   *Statistical results based on (a) a random effect model of Y_i, i = 1,2,3 and 4, and (b) a LV(1) RE model*

The joint effects of *Z?* and *X?*Z?* are significant, since the null hypothesis $C(3) = C(4) = 0$ is rejected, based on the *F*-statistic of $F_0 = 4.631\ 865$ with $df = (2.1280)$ and a *p*-value $= 0.0099$. So it can be concluded that the effect of *X?* on *Y?* is significantly dependent on *Z?*

- In a statistical sense, this model should be acceptable, even though *X?*Z?* is insignificant with such a large *p*-value.
- However, if a reduced model should be obtained, the variable *Z?* will be deleted, since it is defined that the effect of *X? (Z?)* on *Y?* depends on *Z? (X?)*, in a theoretical sense.

5. On the other, an additive model with the exogenous variables *{Y?(−1),X?,Z?}* could be applied

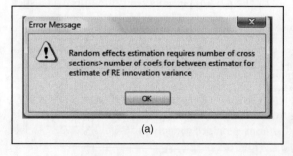

(a)

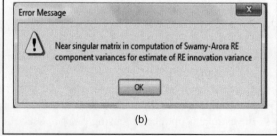

(b)

Figure 4.33   *Possible error messages obtained in doing analysis using the RE models*

System: SYS17
Estimation Method: Least Squares
Date: 01/03/10   Time: 10:58
Sample: 1968M02 1994M10
Included observations: 321
Total system (balanced) observations 1284

	Coefficient	Std. Error	t-Statistic	Prob.
C(1)	34.47682	21.76451	1.584084	0.1134
C(2)	0.993887	0.002828	351.4055	0.0000
C(3)	0.009942	0.003671	2.708579	0.0068
C(4)	-1.04E-08	1.85E-08	-0.562194	0.5741

| Determinant residual covariance | 5.10E+17 | | | |

Equation: Y_1=C(1)+C(2)*Y_1(-1)+C(3)*Z_1+C(4)*X_1*Z_1
Observations: 321

R-squared	0.995707	Mean dependent var	4796.988
Adjusted R-squared	0.995666	S.D. dependent var	1564.584
S.E. of regression	102.9963	Sum squared resid	3362809.
Durbin-Watson stat	1.411292		

Equation: Y_2=C(1)+C(2)*Y_2(-1)+C(3)*Z_2+C(4)*X_2*Z_2
Observations: 321

R-squared	0.998789	Mean dependent var	14089.85
Adjusted R-squared	0.998778	S.D. dependent var	6599.741
S.E. of regression	230.7227	Sum squared resid	16874842
Durbin-Watson stat	1.875142		

Equation: Y_3=C(1)+C(2)*Y_3(-1)+C(3)*Z_3+C(4)*X_3*Z_3
Observations: 321

R-squared	0.999285	Mean dependent var	16076.63
Adjusted R-squared	0.999278	S.D. dependent var	6258.103
S.E. of regression	168.1353	Sum squared resid	8961430.
Durbin-Watson stat	1.826227		

Equation: Y_4=C(1)+C(2)*Y_4(-1)+C(3)*Z_4+C(4)*X_4*Z_4
Observations: 321

R-squared	0.998642	Mean dependent var	36596.94
Adjusted R-squared	0.998629	S.D. dependent var	15966.73
S.E. of regression	591.0965	Sum squared resid	1.11E+08
Durbin-Watson stat	1.890359		

Figure 4.34   *Statistical results based on a LV(1) system equation (RE model in Figure 4.32b)*

Example 4.18   A system equation model

For a comparison to the model in Figure 4.32(b), Figure 4.34 presents the statistical results based on a system equation model, using the least squares estimation method. Compared to the results based on the random effect model in Figure 4.32(b), the following notes and conclusions are made.

1. The coefficients of the four regressions are exactly the same as the random effect model in Figure 4.32(b), but the other statistics are different, and the null hypothesis: $C(3) = C(4) = 0$ is accepted based on the Chi-square statistic of $\chi_0^2 = 1.545665$ with $df = 2$ and a p-value $= 0.4617$. Therefore, it can be concluded that $Z?$ and $X?^*Z?$ have insignificant joint effects on $Y?$, which is contradictory to the conclusion based on using the random effect model.

2. Note that Figure 4.34 presents a set of four regression functions with different statistics, such as R-squared and DW-statistics, compared to a single regression function for the random effect model. So the results using the object "*System*" provide more information on the four states by using four regressions instead of a single regression using the random effect model. If you have to choose one out of the two statistical results, which one do you prefer?

3. However, I would say that neither model is appropriate, since all firm-time observations are represented by a single regression function only.

4. On the other hand, it should be noted that by using the object "*System*", the parameters of the model do not strictly have to be less than the number of cross-sections, as was required for the random effect model. For illustration, see the statistical results in Figure 4.35, based on a MLV(2) SCM of Y_i, $i = 1,2,3$ and 4, with only six parameters. Based on these results the following notes and conclusions are made.

   4.1 In a statistical sense, this is a good time-series model, supported by the values of DW-statistics for each regression, and each of the independent variables has significant adjusted effects.

   4.2 However, in a theoretical sense, this model should be considered worse, since the four states are represented only by a single regression function. For this reason, applying the model with cross-section specific coefficients is recommended, as presented in the following section.

```
System: SYS18
Estimation Method: Least Squares
Date: 01/03/10   Time: 11:25
Sample: 1968M03 1994M10
Included observations: 320
Total system (balanced) observations 1280
```

	Coefficient	Std. Error	t-Statistic	Prob.
C(1)	-0.095100	24.05026	-0.003954	0.9968
C(2)	0.002395	0.000725	3.305394	0.0010
C(3)	0.017226	0.004290	4.015571	0.0001
C(4)	-4.06E-08	2.07E-08	-1.959908	0.0502
C(5)	1.047859	0.028039	37.37124	0.0000
C(6)	-0.062737	0.027819	-2.255201	0.0243

Determinant residual covariance	6.52E+17

```
Equation: Y_1=C(1)+C(2)*X_1+C(3)*Z_1+C(4)*X_1*Z_1+C(5)*Y_1(-1)+C(6)
        *Y_1(-2)
Observations: 320
```

R-squared	0.994157	Mean dependent var	4806.125
Adjusted R-squared	0.994064	S.D. dependent var	1558.433
S.E. of regression	120.0736	Sum squared resid	4527148.
Durbin-Watson stat	1.127988		

```
Equation: Y_2=C(1)+C(2)*X_2+C(3)*Z_2+C(4)*X_2*Z_2+C(5)*Y_2(-1)+C(6)
        *Y_2(-2)
Observations: 320
```

R-squared	0.998796	Mean dependent var	14121.02
Adjusted R-squared	0.998776	S.D. dependent var	6586.377
S.E. of regression	230.3985	Sum squared resid	16668210
Durbin-Watson stat	2.018433		

```
Equation: Y_3=C(1)+C(2)*X_3+C(3)*Z_3+C(4)*X_3*Z_3+C(5)*Y_3(-1)+C(6)
        *Y_3(-2)
Observations: 320
```

R-squared	0.999280	Mean dependent var	16109.83
Adjusted R-squared	0.999269	S.D. dependent var	6239.535
S.E. of regression	168.7559	Sum squared resid	8942265.
Durbin-Watson stat	1.935565		

```
Equation: Y_4=C(1)+C(2)*X_4+C(3)*Z_4+C(4)*X_4*Z_4+C(5)*Y_4(-1)+C(6)
        *Y_4(-2)
Observations: 320
```

R-squared	0.998666	Mean dependent var	36672.08
Adjusted R-squared	0.998645	S.D. dependent var	15934.79
S.E. of regression	586.6418	Sum squared resid	1.08E+08
Durbin-Watson stat	2.046974		

Figure 4.35   *Statistical results based on a MLV(2) system equation of* Y_i, $i = 1, 2, 3$ *and* 4

4.8   Models with Cross-Section Specific Coefficients

4.8.1   MAR(p) Model with Cross-Section Specific Coefficients

For discussion, see the statistical results in Figure 4.36, based on a pooled least squares multivariate autoregressive model, namely PLS_MAR(2) model, of Y_1 and Y_2 on the exogenous variables C, X_1, Z_1 and

```
Dependent Variable: Y?
Method: Pooled Least Squares
Date: 01/02/10   Time: 09:48
Sample (adjusted): 1968M03 1994M10
Included observations: 320 after adjustments
Cross-sections included: 2
Total pool (balanced) observations: 640
Convergence achieved after 16 iterations
```

Variable	Coefficient	Std. Error	t-Statistic	Prob.
_1--C	6009.172	2543.888	2.362200	0.0185
_2--C	1671.322	3671.680	0.455193	0.6491
_1--X_1	0.049440	0.030678	1.611559	0.1076
_2--X_2	0.763704	0.104915	7.279232	0.0000
_1--Z_1	-0.027596	0.050976	-0.541346	0.5885
_2--Z_2	0.144742	0.161581	0.895785	0.3707
_1--X_1*Z_1	-5.90E-07	1.73E-06	-0.340689	0.7335
_2--X_2*Z_2	-9.60E-06	6.42E-06	-1.495295	0.1353
_1--AR(1)	1.002403	0.109503	9.154088	0.0000
_1--AR(2)	-0.009326	0.109094	-0.085485	0.9319
_2--AR(1)	1.165958	0.043096	27.05492	0.0000
_2--AR(2)	-0.170437	0.043070	-3.957227	0.0001

R-squared	0.999470	Mean dependent var	9463.572
Adjusted R-squared	0.999461	S.D. dependent var	6677.908
S.E. of regression	155.0939	Akaike info criterion	12.94451
Sum squared resid	15105981	Schwarz criterion	13.02816
Log likelihood	-4130.243	Hannan-Quinn criter.	12.97698
F-statistic	107638.9	Durbin-Watson stat	2.045695
Prob(F-statistic)	0.000000		

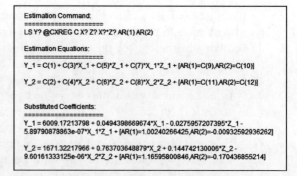

```
Estimation Command:
=====================
LS Y? @CXREG C X? Z? X?*Z? AR(1) AR(2)

Estimation Equations:
=====================
Y_1 = C(1) + C(3)*X_1 + C(5)*Z_1 + C(7)*X_1*Z_1 + [AR(1)=C(9),AR(2)=C(10)]

Y_2 = C(2) + C(4)*X_2 + C(6)*Z_2 + C(8)*X_2*Z_2 + [AR(1)=C(11),AR(2)=C(12)]

Substituted Coefficients:
=====================
Y_1 = 6009.17213798 + 0.0494398669674*X_1 - 0.0275957207395*Z_1 -
5.89790878863e-07*X_1*Z_1 + [AR(1)=1.00240266425,AR(2)=-0.00932592936262]

Y_2 = 1671.32217966 + 0.763703648879*X_2 + 0.144742130006*Z_2 -
9.60161333125e-06*X_2*Z_2 + [AR(1)=1.16595800846,AR(2)=-0.170436855214]
```

Figure 4.36   *Statistical results based on an AR(2) pool model of* Y_1 *and* Y_2

```
Estimation Method: Seemingly Unrelated Regression
Date: 01/06/10   Time: 10:46
Sample: 1968M02 1994M10
Included observations: 322
Total system (balanced) observations 642
Iterate coefficients after one-step weighting matrix
Convergence achieved after: 1 weight matrix, 8 total coef iterations
```

	Coefficient	Std. Error	t-Statistic	Prob.
C(1)	5859.167	1066.069	5.496050	0.0000
C(3)	0.045152	0.013493	3.346286	0.0009
C(7)	-1.39E-06	3.27E-07	-4.246919	0.0000
C(9)	0.992295	0.003574	277.6603	0.0000
C(2)	43724.96	28063.01	1.558100	0.1197
C(6)	-0.273880	0.061995	-4.417798	0.0000
C(8)	6.93E-06	2.06E-06	3.360032	0.0008
C(10)	0.998022	0.001868	534.3696	0.0000

Determinant residual covariance		2.98E+08

Equation: $Y\_1=C(1)+C(3)*X\_1+C(7)*X\_1*Z\_1+[AR(1)=C(9)]$
Observations: 321

R-squared	0.997364	Mean dependent var	4796.988
Adjusted R-squared	0.997339	S.D. dependent var	1564.584
S.E. of regression	80.70754	Sum squared resid	2064845.
Durbin-Watson stat	1.997819		

Equation: $Y\_2=C(2)+C(6)*Z\_1+C(8)*X\_1*Z\_1+[AR(1)=C(10)]$
Observations: 321

R-squared	0.998883	Mean dependent var	14089.85
Adjusted R-squared	0.998872	S.D. dependent var	6599.741
S.E. of regression	221.6197	Sum squared resid	15569550
Durbin-Watson stat	1.921796		

Figure 4.37 *Statistical results based on an AR(1) system equations of* Y_1 *and* Y_2

X_1*Z_1, with cross-section specific coefficients, using the object "POOL". Based on these results, the following notes and conclusions are made.

1. The steps of the analysis are as follows:
   1.1 With the work file on the screen, select *Object/New Objects . . . /Pool . . . OK*, then enter the *Cross Section Identifiers*: _1 _2
   1.2 Select the object "*Estimate*", then enter *Y?* as the dependent variable, and for the *Cross-Sections Specific Coefficients* enter the series: *C X? Z? X?*Z? AR(1) AR(2)*
   1.3 Finally by selecting the option *Cross Section: Non . . . OK*, the results in Figure 4.36 are obtained.
2. Note that the _1-AR(2) is insignificant with a large p-value $= 0.9319$, but the term _2-AR(2) is significant with a p-value $= 0.0001$. Therefore, there is a choice of either using an AR(1) model or the AR(2) model.
3. Similarly, this is the case for the pairs of exogenous variables. However, in a statistical sense, a reduced model should be explored by deleting one of *X?* and *Z?*, or both.
4. A good fit pooled LS model is obtained, by using an AR(1) model and omitting *Z?* with a DW-statistic of 1.743 968, and each of the independent variables has significant adjusted effects. Try this as an exercise using your own data set.
5. The PLS model with cross-section specific coefficients should be considered to be much better compared to the fixed-effect model, more so compared to the random effect model.
6. However, the PLS model, either full or reduced, should have the same set of independent variables as well as AR terms, for all states. This is a disadvantage of the PLS model compared to the system equations.
7. For comparison with the reduced model mentioned in point (3), Figure 4.37 presents statistical results based on an AR(1) model, but using the system equations. Note that these results demonstrate a good fit multivariate AR(1) model, where the regressions have different sets of independent variables.
8. Therefore, in general, two groups of multivariate models should be considered; namely PLS models with cross-section specific coefficients, and system equation models. If a MLVAR(p,q) model should have the same set of independent variables, then it is simpler to conduct the analysis using the PLS estimation method.

4.8.2 Advanced PLS Estimation Methods

Figure 4.38 presents alternative options and weights for doing a more advanced statistical analysis. In a statistical sense, all estimation methods are acceptable methods. However, it is not an easy task to select the

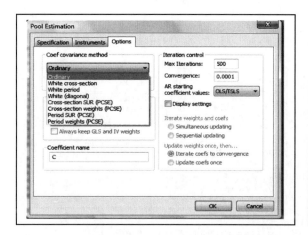 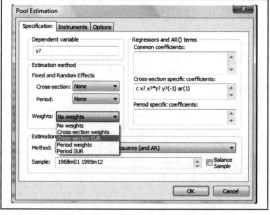

Figure 4.38   *Alternative options and weights for more advanced PLS estimation methods*

best possible estimation method, since estimates of parameters are highly dependent on the data used as well as the multicollinearity between the variables in the model. Blanchard (in Gujarati, 2003: 263) stated "Multicollinearity is God's will, not a problem with OLS (Ordinary Least Square) or statistical techniques in general". For this reason, trial-and-error methods should be applied and if there is no good knowledge or idea on what option to select, applying the default option(s) is suggested. See Example 4.20.

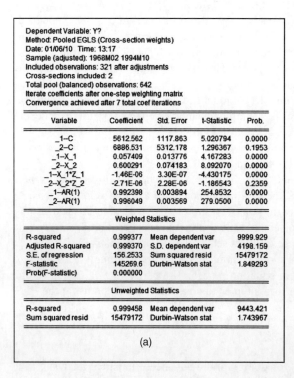 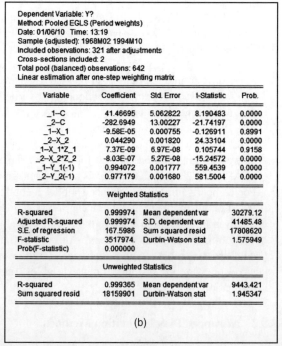

Figure 4.39   *Statistical results based on (a) a PWLS AR(1) model, and (b) a PWLS LV(1) model*

Example 4.19 Applications of the pooled WLS estimation method

Figure 4.39 presents the statistical results based on a reduced model of the model in Figure 4.36, and its modification using cross-section and period weights, respectively, with the ordinary coefficient covariance method. Based on these results, the following notes and comments are presented.

1. The results could be obtained based on a pooled weighted least squares PWLS LVAR(p,q) model, using the cross-section weights. For an illustration, a special case, namely the MAR(1) model, is presented in Figure 4.39(a).
2. However, for the period weights the AR term cannot be used. For this reason, Figure 4.39(b) presents the results based on a MLV(1) model.
3. Note that the results in Figure 4.39(a) and (b) lead to contradictory conclusions, those are the effects of _1-X_1, and the interactions _1-X_1*Z_1 and _1-X_2*X_2. Therefore, which one is a better model? By looking at the SE of regression and DW-statistic, then the MAR(1) model should be considered a better fit.
4. For extension, the other combinations between the options and weights could easily be applied. Try this as an exercise.

Example 4.20 Unexpected statistical results

Compared to the results in Figure 4.39, the illustrations in Figure 4.40(a) and (b) present two unexpected statistical results using selected options, which are unacceptable in a statistical sense. Based on these results the following notes and comments are presented.

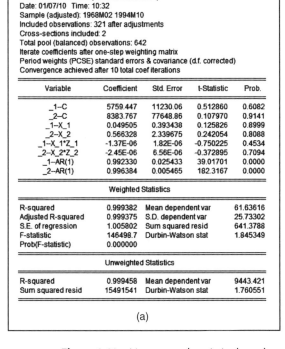

Dependent Variable: Y?
Method: Pooled EGLS (Cross-section SUR)
Date: 01/07/10 Time: 10:32
Sample (adjusted): 1968M02 1994M10
Included observations: 321 after adjustments
Cross-sections included: 2
Total pool (balanced) observations: 642
Iterate coefficients after one-step weighting matrix
Period weights (PCSE) standard errors & covariance (d.f. corrected)
Convergence achieved after 10 total coef iterations

Variable	Coefficient	Std. Error	t-Statistic	Prob.
_1--C	5759.447	11230.06	0.512860	0.6082
_2--C	8383.767	77648.86	0.107970	0.9141
_1--X_1	0.049505	0.393438	0.125826	0.8999
_2--X_2	0.566328	2.339675	0.242054	0.8088
_1--X_1*Z_1	-1.37E-06	1.82E-06	-0.750225	0.4534
_2--X_2*Z_2	-2.45E-06	6.56E-06	-0.372895	0.7094
_1--AR(1)	0.992330	0.025433	39.01701	0.0000
_2--AR(1)	0.996384	0.005465	182.3167	0.0000

Weighted Statistics

R-squared	0.999382	Mean dependent var	61.63616
Adjusted R-squared	0.999375	S.D. dependent var	25.73302
S.E. of regression	1.005802	Sum squared resid	641.3788
F-statistic	146498.7	Durbin-Watson stat	1.845349
Prob(F-statistic)	0.000000		

Unweighted Statistics

R-squared	0.999458	Mean dependent var	9443.421
Sum squared resid	15491541	Durbin-Watson stat	1.760551

(a)

Dependent Variable: Y?
Method: Pooled EGLS (Cross-section weights)
Date: 01/07/10 Time: 10:27
Sample (adjusted): 1968M02 1994M10
Included observations: 321 after adjustments
Cross-sections included: 2
Total pool (balanced) observations: 642
Iterate coefficients after one-step weighting matrix
White period standard errors & covariance (d.f. corrected)
Convergence achieved after 7 total coef iterations
WARNING: estimated coefficient covariance matrix is of reduced rank

Variable	Coefficient	Std. Error	t-Statistic	Prob.
_1--C	5612.562	NA	NA	NA
_2--C	6886.531	0.000201	34285882	0.0000
_1--X_1	0.057409	NA	NA	NA
_2--X_2	0.600291	6.48E-09	92657341	0.0000
_1--X_1*Z_1	-1.46E-06	3.81E-15	-3.83E+08	0.0000
_2--X_2*Z_2	-2.71E-06	4.80E-14	-56472515	0.0000
_1--AR(1)	0.992398	NA	NA	NA
_2--AR(1)	0.996049	1.87E-10	5.32E+09	0.0000

Weighted Statistics

R-squared	0.999377	Mean dependent var	9999.929
Adjusted R-squared	0.999370	S.D. dependent var	4198.159
S.E. of regression	156.2533	Sum squared resid	15479172
F-statistic	145269.6	Durbin-Watson stat	1.849293
Prob(F-statistic)	0.000000		

Unweighted Statistics

R-squared	0.999458	Mean dependent var	9443.421
Sum squared resid	15479172	Durbin-Watson stat	1.743967

(b)

Figure 4.40 *Unexpected statistical results using pooled EGLS method and selected options*

1. Contradictory to both estimates in Figure 4.39, Figure 4.40(a) shows that each of the independent varia-bles *X?* and *X?*Z?* have insignificant adjusted effects with a large value.
    1.1   Note that the estimates were obtained using the Pooled EGLS (Cross-section SUR) estimation method, with the option: *PCSE standard errors and covariance*.
    1.2   Using this data set, the model is not a good fit. However, it could be a good fit based on other data sets, which cannot be predicted.
2. On the other hand by using the Pooled EGLS (Cross-section SUR) estimation method, with the option: *White period standard errors and covariance*, the worst estimates in Figure 4.40(b) were obtained, which are unpredictable.
3. If one should use the weight and the coefficient covariance method, the results in Figures 4.39 and Figure 4.40 suggest that he/she should be using trial-and-error methods to select the best possible weight and coefficient covariance method. On the other hand, note that an error message might be obtained even from a good theoretical model.

4.8.3   Instrumental Variables Model

It has been mentioned that application of the instrumental variables model is not recommended, since it is not an easy task to select the best possible set of instrumental variables. However, for illustration purposes see the following example.

Example 4.21   Application of the instrumental variables models
Figure 4.41 presents two statistical results based on a MLVAR(1,1), using un-weighted and weighted two-stage least squares estimation methods, respectively. Based on these results the following notes and comments presented.

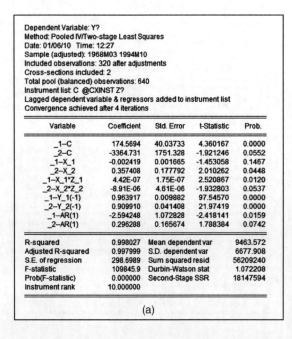

Figure 4.41   *Statistical results based on instrumental variables models, using the TSLS estimation method: (a) un-weighted, and (b) with the cross-section weights*

1. The estimates use the option: *"The lagged dependent and regressors added to the instrument list"*, otherwise the error message *"Order condition violated – Insufficient Instruments"* is obtained.
2. Both results show good fit models. However, in a statistical sense, the models should be modified in order to select those with greater DW-statistical values. Try this as an exercise.
3. A lot of instrumental variables models with acceptable estimates in a statistical sense can easily be derived using the options provided by EViews. However, no one ever knows which model is the true population model. Refer to the special notes and comments on the true population model in Section 2.14, as well as the illustrative statistical results based on various instrumental variables models, presented in Agung (2009a).

4.8.4 Special Notes and Comments

Any statistical model always has some limitations, besides its basic assumptions, even though the estimates of the model parameters are acceptable in a statistical sense. A sample may never be representative of the corresponding population (Agung, 2011, and 2004), and no one ever knows the true population model (Agung, 2009a, Section 2.14).

Corresponding to the variables *X?, Y?* and *Z?* of the models in Figures 4.39–4.41, Figure 4.42 presents the scatter graphs of *(X_1,Y_1)* and *(X_2,Y_2)* with their regression lines, and Figure 4.43 presents the

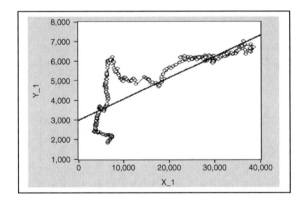 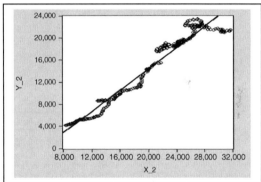

Figure 4.42 *Scatter graphs with regression lines of* (X_1,Y_1) *and* (X_2,Y_2)

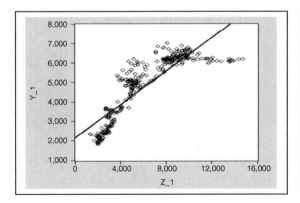 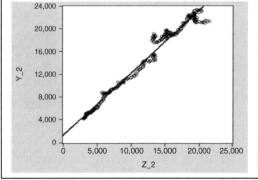

Figure 4.43 *Scatter graphs with regression lines of* (Y_1,Z_1) *and* (Y_2,Z_2)

scatter graphs of *(Y_1,Z_1)* and *(Y_2,Z_2)* with their regression lines. As such, the following notes and comments are presented.

1. In general the patterns of relationships between each exogenous variable, say X_i, with its corresponding endogenous variable Y_i will not be the same for all states or groups of individuals. These graphs have demonstrated that a homogenous set of regressions by states should make the worst statistical model. See the example that follows.
2. Note that the scatter graphs also show that the observed scores of Y_2 are clearly divided into two sub-sets, with a cutting point of 18 200. Then, two regression models of Y_2 could be presented, one regression for each sub-set, but it is a model with dummy variables, since for some scores of X_2, the scores of Y_2 are in different sub-sets.
3. Since the pooled least squares regression should have the same set of independent variables, the object "POOL" is not appropriate method for analysis if the regressions in system equations have different sets of independent variables. Therefore, applying the object "*System*" is recommended, such as in the various models presented previously.
4. Other limitations of a model can be identified using the DW-statistic and residual analysis. Various studies can be done based on the residuals as the variables. Refer to various residual analyses demonstrated in Agung (2011, and 2009a). For additional illustration corresponding to the models in Figure 4.40, the following example presents one of the methods in treating outliers.

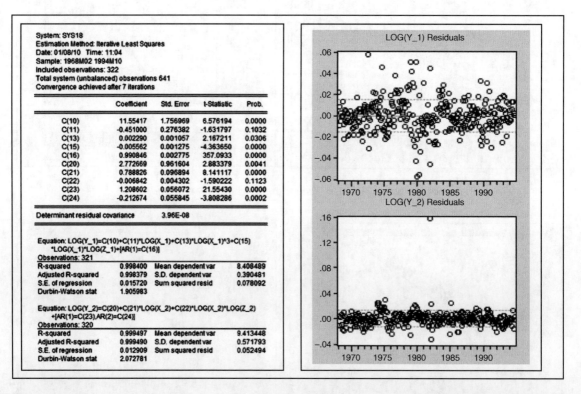

Figure 4.44 *Statistical results based on a system of translog polynomial and interaction regressions, and its residual graphs*

Example 4.22 A system of polynomial and interaction regressions

The objectives of this example are as follows:

1. To demonstrate one of the methods in treating an outlier(s).
2. To present a modified model, namely a system of translog regressions linear in their parameters, of the model in Figures 4.39–4.41, which also has the same outliers, since they use the same sets of endogenous and exogenous variables.

For discussion, Figure 4.44 presents statistical results based on a system of polynomial and interaction regressions of *log(Y?)* on *log(X?)* and *log(Z?)*, and their residuals graphs. Based on these results, the following notes and conclusions are made.

1. The results show that the model is a good fit in a statistical sense, since each of the independent variables has a significant adjusted effect, and the DW-statistics indicate that there are no autocorrelation problems. However, we find the residual graph of *log(Y_2)* clearly shows a far outlier. Note that the outlier(s) also can be found using box-plots.
2. Then, another data analysis was done based on a sub-sample without the outlier, with the statistical results presented in Figure 4.45. The steps of the analysis are as follows:
   - Having the output in Figure 4.44 on the screen, the residuals are generated by *Proc/Make Residuals . . . /Ordinary . . . OK*. In this case, we obtain the *RESID02* and *RESID03*.

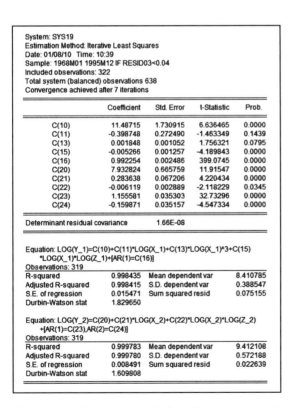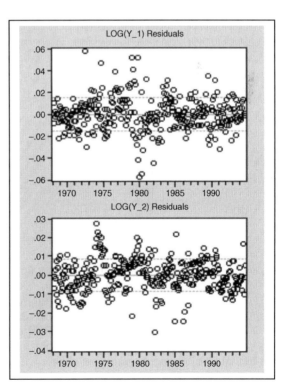

Figure 4.45 *Statistical results based on a system of translog polynomial and interaction regressions, with their residuals graphs, using a sub-sample* RESID03 < 0.04

- Select a sub-sample, using the "*If condition*": *RESID03* < 0.04.
- Then by using the same model, the results in Figure 4.45 are obtained.

3. Based on the results in Figure 4.45, the following conclusions and notes are made.

   3.1   Note that the regression of *log(Y_1)* is in fact a reduced model of the third-degree polynomial regression in *log(X_1)*. If *log(X_1)^2* is used as an independent variable then we find each of the independent variables is insignificant.

   3.2   *Log(X_2)* has a significant adjusted effect on *log(Y_2)* with a *p*-value = *0.0345*, but it is insignificant based on the results in Figure 4.44, using the outlier.

   3.3   In a statistical sense, the output indicates that the model is a good fit and its regression has a slightly greater R-squared value, compared to the output using the outlier.

   3.4   Furthermore, note that the output presents the note "Total system (*balanced*) observations 638", but Figure 4.44 presents the note "Total system (*unbalanced*) observations 641". Which one would you prefer?

   3.5   Corresponding to the DW-statistic of 1.609 808 for the regression of *log(Y_2)*, one might want to modify the model by using additional AR term(s) or lagged dependent variables in order to obtain a model with a greater DW-statistic. Similarly so for the regression of *log(Y_1)*. Do it as an exercise.

4.9   Cases in Industry

Industries' status can change over time; such as status of various return rates (positive/negatives, very low to very high), hedging, dividend (giving/not giving), before and after going public, being classified as public (LQ45/not LQ45) and changing environmental factor(s). In this case, binary choice models should be considered if a zero-one variable is taken as the problem indicator, and models with categorical independent variables, called the *dummy variables models* should also be considered. See the following sub-sections.

4.9.1   Dummy Variables Models

In this case, the multivariate time series by industries can be presented in general as (X_i_t, Y_i_t, CF_i_t), where *X_i* and *Y_i* are multidimensional exogenous and endogenous variables, and *CF_i* indicates a single *cell-factor* proposed in Agung (2011, 2009a, 2006), for the *i*-th industry. Note that that the factor *CF* could be generated based on one or more variables and these should be represented using dummy variables in time-series models. Therefore, the dummy variables models should be considered.

   All dummy variables SCMs using the cell-factor, *CF*, presented in Agung (2009a, Chapter 4) can easily be applied to the multivariate time series data for each industry. For this reason, for all readers, using all models presented in Agung (2009a) as the main reference is recommended, and they will not be discussed any more in this section. Therefore, the following sections only present some basic models, namely SCMs, specific to correlated industries in the sense that only the endogenous variables of the industries have causal relationships. Then they can easily be extended to more advanced models, such as LVAR(p,q) SCM, instrumental variables SCM and TGARCH(*a,b,c*).

4.9.1.1   *Simple Dummy Variables Model*

For a simple illustration, by replacing the environmental variable *Z* in Figure 4.12(b) by a cell-factor *CF*, the equations of dummy variables SCMs, such as additive, two- and three-way interaction SCMs, based on a trivariate time series *(X_i,Y_i,CF_i)_t* can easily be written based on the models in (4.9), (4.10) and (4.11), for two correlated industries. For instance, based on the model (4.10) the following SCM might be considered, where $Dc1$ is a dummy variable representing a dichotomous cell-factor, to replace the variable Z in (4.10) which can be generated based on the scores of Y_2, or the other numerical variable(s). However, note that this model is not using the numerical time-t as an independent variable, which could give unexpected time series models – refer to models illustrated in Chapter 4 (Agung, 2009a).

$$
\begin{aligned}
y1\_1 &= c(10) + c(11)^* x1\_1 + c(12)^* x2\_1 + c(13)^* x1\_1^* x2\_1 \\
&\quad + c(14)^* y1\_2 + c(15)^* Dc1 + c(16)^* y1\_2^* Dc1 + \mu 1 \\
y1\_2 &= c(20) + c(21)^* x1\_2 + c(22)^* x2\_2 + c(23)^* x1\_2^* x2\_2 \\
&\quad + c(24)^* y1\_1 + c(25)^* Dc1 + c(26)^* y\_1^* Dc1 + \mu 2 \\
x1\_1 &= c(30) + c(31)^* x2\_1 + c(32)^* Dc1 + c(33)^* x1\_1^* Dc1 + \mu 3 \\
x1\_2 &= c(40) + c(41)^* x2\_2 + c(42)^* Dc1 + c(43)^* x1\_2^* Dc1 + \mu 4
\end{aligned}
\tag{4.18}
$$

4.9.1.2 General Dummy Variables Model

To generalize the SCM in (4.18), the following general simultaneous SCM of two correlated industries are considered, specific to the endogenous variables.

$$
\begin{aligned}
G_1(y1\_1) &= F_1(x1\_1, x2\_1, y1\_2, CF) + \mu 1 \\
G_2(y1\_2) &= F_2(x1\_2, x2\_2, y1\_1, CF) + \mu 2
\end{aligned}
\tag{4.19}
$$

where $G_i(v)$ is a function of (v) having no parameter, $F_i(w)$ is a function of the multivariate w with a finite number of parameters, for the i-th industry over time, and CF is an environmental or cell-factor for all industries.

Further general simultaneous SCMs of a single endogenous variable Y_i_t, and exogenous multivariate $X\_i = (x1\_i, \ldots, xK\_i)_t$, and environmental or cell-factor CF for a small number of I-correlated industries, would be as follows:

$$
\begin{aligned}
G_i(y\_i) &= F_i(X\_1, y\_1, \ldots, y\_g, \ldots, CF) + \mu_i \\
&\text{for} \quad i = 1, \ldots, I; \quad \text{and} \quad g \neq i.
\end{aligned}
\tag{4.20}
$$

with a special case for $I = 3$, as follows:

$$
\begin{aligned}
G_1(y\_1) &= F_1(X\_1, y\_2, y\_3, CF) + \mu_1 \\
G_2(y\_2) &= F_2(X\_2, y\_1, y\_3, CF) + \mu_2 \\
G_3(y\_3) &= F_3(X\_3, y\_1, y\_2, CF) + \mu_3
\end{aligned}
\tag{4.21}
$$

Note that the exogenous Xk_i for each $k = 1, \ldots, K$, can be the main variable, lagged of a dependent variable, or two- and three-way interactions for the i-th industry, but the environmental factor CF is the same for all industries.

Part Two
Pool Panel Data Analysis

Abstract

This part, containing Chapters 5–11, mainly presents data analysis based on pool panel data, in the form of stacked panel data with a large individual observation N and small time-point observations T. So the statistical methods and models applied can directly be derived from the models based on cross-section data presented in Agung (2011), by using or inserting time dummy variables as additional independent variables, in addition to the possibility of using lagged endogenous or exogenous variables. A more complex model should be considered or defined based on pool panel data with a large N and large or very large T. Having large time-point observations, the time t can be viewed as a numerical variable, and then a defined model could be a continuous or discontinuous model of t because at least two time periods have to be considered, as presented in Agung (2009a).

Panel Data Analysis Using EViews, First Edition. I Gusti Ngurah Agung.
© 2014 John Wiley & Sons, Ltd. Published 2014 by John Wiley & Sons, Ltd.
Companion website: www.wiley.com/go/panel_data

5

Evaluation Analysis

5.1 Introduction

All studies based on pool panel data are in fact evaluation studies; however, most of them present only inferential statistical analysis to test hypotheses on causal relationships between a set of selected variables. In fact, causal relationships between a set of variables are often known even before data collection, supported by a good strong theoretical basis. Therefore, inferential statistical analysis would be done only to reconfirm whether or not the data supports the proposed hypothesis on causal relationships between the variables considered. Furthermore, note that testing hypotheses cannot provide information on a research object or a group of objects with problems or on certain levels of the problem considered.

This chapter will present descriptive statistical analysis as a means to conduct evaluation or policy analysis, with basic objectives as follows:

1. To identify the research object(s) or groups of objects with problem(s) or different levels of problems over time based on defined problem indicator(s).
2. To study possible cause or risk factors of problem(s), the controllable cause factors in particular.

In order to achieve the objectives, presenting descriptive statistical summariess (DSS) based on pool panel data is recommended, specifically pool panel data with a large number of individual observations, but few time-point observations.

In general pool panel data would have endogenous (impact, response or down-stream) multivariate $Y_{it} = (Y1, \ldots, Yg, \ldots)_{it}$, and exogenous (cause, source or up-stream) multivariate $X_{it} = (X1, \ldots, Xk, \ldots)_{it}$. In this chapter, data considered is panel data with a large number of individual observations but few time-point observations. In this case the dummy variables of time should be used as independent variables of any defined models, alongside the dummy variables of other categorical causes of the problem indicators, as well as the classification or groups of the observed individuals or objects. Note that, in general, a large number of individuals can easily be classified into groups, with different or distinct characteristics and behaviors based on one or more background variables of those research objects.

Panel Data Analysis Using EViews, First Edition. I Gusti Ngurah Agung.
© 2014 John Wiley & Sons, Ltd. Published 2014 by John Wiley & Sons, Ltd.
Companion website: www.wiley.com/go/panel_data

The simplest problem indicator is a zero-one problem indicator, which can be subjectively defined based on any problem variables. Note that either endogenous or exogenous variables could be considered problem variables, and they can be transformed to zero-one problem indicators.

DSSs of any zero-one problem indicators are the simplest methods of statistical analysis, but are also good for strategic decision-making (Agung, 2011, 2009a, 2008). However, since the output of data analysis can also give test statistics, then DSSs can be presented together with the relevant testing statistics.

All methods for cross-section data (presented in Agung, 2011), specifically the DSS in Chapter 2, can easily be applied by inserting or using an additional categorical time variable. For this reason, the following sections will present only some illustrative DSSs using the time T, which are very similar to the DSSs presented in Agung (2011).

Furthermore, to conduct a more detailed evaluation analysis, the research objects can also be classified into groups with different levels of problems, which can easily be generated using either parametric or nonparametric statistics.

5.2   Preliminary Evaluation Analysis

Most financial data of the firms have outliers, so before doing an advanced statistical analysis, conducting a preliminary evaluation on the scores of all research variables is suggested. For an illustrative example, Figure 5.1 presents a descriptive statistical summary of four variables in Rina_subset.wf1, with their box-plots, at the time $T = 1$. Rina is one of my students doing her dissertation. Based on these results the following notes and comments are presented.

1. The box-plots of the four variables clearly show that each variable has near and far outliers. These outliers, in general, would give problems in doing parametric statistical analysis, including data analysis based on modeling, more so in making generalizations based on sampled statistics since the mean of each variable can be misleading. Therefore, what should be done with the outliers?
    1.1   In order to form a general conclusion, it is common to delete the outliers so that then panel data will not be balanced.
    1.2   An alternative acceptable method is to replace the score using the average of the scores at the neighboring time points.
2. If an outlier is a true value, then the outlier should be a good indicator for doing further evaluation analysis, namely to find its cause factors, especially the controllable cause factors. For instance, why do

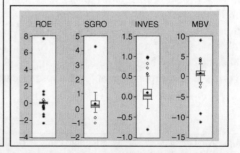

	ROE	SGRO	INVES	MBV
Mean	0.107931	0.303599	0.100528	0.891597
Median	0.078163	0.231095	0.034823	0.749158
Maximum	7.746926	4.305100	0.995888	9.090909
Minimum	-2.307558	-0.998844	-0.804476	-11.11111
Std. Dev.	0.760688	0.491703	0.256261	1.996290
Skewness	7.547437	3.954462	1.092931	-1.475686
Kurtosis	80.60036	35.49423	6.068317	19.94634
Jarque-Bera	33852.39	6058.141	76.87638	1602.733
Probability	0.000000	0.000000	0.000000	0.000000
Sum	14.03109	39.46790	13.06859	115.9076
Sum Sq. Dev.	74.64537	31.18862	8.471397	514.0872
Observations	130	130	130	130

Figure 5.1   *Descriptive statistical summary of ROE, SGRO, INVES and MBV*

ROE, INVES and *MBV* have very low scores or negative far outliers? On the other hand, why is the expenditure or expense of a firm (branch of a company) so high?

3. For a more detailed descriptive statistical analysis, at each time point, the suggestion is to use all illustrative examples presented in Chapter 2 (Agung, 2011).

4. Corresponding with the many outliers in Rina_subset.wf1, the following sections present only nonparametric statistical analysis, so outliers do not have any effects.

5.3 The Application of the Object "Descriptive Statistics and Tests"

5.3.1 Analysis Using the Option "Stats by Classification ..."

For illustration, based on the Rina_subset.wf1, three zero-one or dummy variables are generated, namely $DY_{it} = 1$, if $Y_{it} < 0$ and $DY_{it} = 0$ or otherwise, $DROE_{it}$, $DSGRO_{it}$ and $DINVES$, where ROE_{it}, $SGRO_{it}$ and $INVES$ are the returns on equity, sales growth and investment for the i-th individual at the time point t.

With a problem indicator on the screen, namely Y, then various descriptive statistical summaries of the means, variances or medians of Y by a single or multi-categorical factors, can easily be developed by selecting *View/Descriptive Statistics and Tests*, then selecting one of the six available options. In this case, however, two options will be considered in order to develop alternative descriptive statistical summaries of zero-one problem indicators by the time T.

5.3.1.1 One-Dimensional Problem over Time

For each zero-one problem indicator, the percentages or proportions of the research objects over time can easily be identified, as presented in the following example.

Example 5.1 One-way table of the proportions by time

Figure 5.2 presents one-way tables of the means of $DRoe_{it}$ and $DSgro_{it}$, or the proportions of $P(DRoe_{it} = 1)$ and $P(SGro_{it} = 1)$, by the categorical time T.

Based on these results the following findings and notes are presented.

1. The main objective of this type of analysis is to study the growth of the problems, indicated by the proportion $P(Y = 1)$ over times for all zero-one problem indicators.

Descriptive Statistics for DROE
Categorized by values of T
Date: 07/15/10   Time: 10:27
Sample: 1 390
Included observations: 390

T	Mean	Std. Dev.	Obs.
1	0.153846	0.362197	130
2	0.207692	0.407225	130
3	0.238462	0.427791	130
All	0.200000	0.400514	390

Descriptive Statistics for DSGRO
Categorized by values of T
Date: 07/15/10   Time: 10:32
Sample: 1 390
Included observations: 390

T	Mean	Std. Dev.	Obs.
1	0.146154	0.354627	130
2	0.392308	0.490153	130
3	0.253846	0.436894	130
All	0.264103	0.441420	390

Figure 5.2 *One-way tables of proportions* $P(DRoe_{it} = 1)$ *and* $P(DGro_{it} = 1)$ *by the time* T

Table 5.1   *A DSS based on the outputs in Figure 5.1*

	T = 1	T = 2	T = 3	OR(T = 2/T = 1)	OR(T = 3/T = 2)
Droe	0.153 846	0.207 692	0.238 462	1.441 747	1.194 543
DSgro	0.146 154	0.392 308	0.253 846	3.771 486	0.526 985

2. Based on the outputs in Figure 5.2, for the final research report, presenting a DSS as in Table 5.1 is suggested. The DSS also shows additional statistics, namely the Odds Ratios (ORs) between consecutive time points, which can easily be developed using Excel. For an illustration, the OR(T = 2/T = 1) = [(0.207 692/(1 − 0.207 692))/(0.153 846/(1 − 0.153 846))] = 1.441 747.

3. Then based on the proportions $P(Y=1)$ for each zero-one indicator Y, the growth of the problems of the research objects could easily be identified. For instance, based on the indicator *DROE*, the 130 firms have increasing problems.

4. Similarly, some findings can be presented based on OR statistics. For example, OR(T = 2/T = 1) = 1.441 717 indicates that an individual at $T=2$ has a *predicted risk* to have a negative *ROE (DROE = 1)*, so 1.441 717 times an individual at $T=1$. Refer to special notes on the reliability and validity of predicted risks or *predicted probabilities* presented in Chapters 2 and 7 (Agung, 2011). Note that a predicted probability is an abstract number and it is not valid for individuals in the sample.

5. *SGRO* (sales-growth) should be one of the cause factors (source or upstream variables) for *ROE* in a theoretical sense. For this reason, based on this DSS, the following notes and comments are made.

    5.1   At the time $T=1$, 14.62% out of 130 firms have negative *SGRO*, which increases to 39.23% at the time $T=2$ and then decreases to 25.38% at $T=3$.

    5.2   Note the percentages of *(SGRO < 0)* decrease from $T=2$ to $T=3$, but the percentages of *(ROE < 0)* increase from $T=2$ to $T=3$. Therefore, it could be concluded that there should be other cause factors for *ROE*, in addition to *SGRO*.

    5.3   In general, there are measured and unmeasured variables, either internal or external factors of the firms, which can be considered or defined as the cause factors for sales growth and they have joint effects on *ROE*. The following example presents the association between *DRoe* and *DSgro* by the time *T*.

    5.4   On the other hand, it should be noted that a negative $SGRO_{it}$ does not directly mean that the *i*-th firm is in a bad condition at the time *t*, since the $SALE_{it}$ can be very large and gives a large profit to the firm, but $SALE_{it} < SALE_{i,t-1}$. However, if a firm has negative sales growth for a series of time points, then the firm has a big problem.

5.3.1.2   Multi-Dimensional Problems by Time

With two or more zero-one problem indicators, descriptive statistical summaries would represent multi-dimensional problems by time and groups of the research objects. See the following examples.

Example 5.2   Three-way table of proportions
As an illustration, Figure 5.3 presents statistical results obtained by entering "*DSgro T DInves*" as the series/ group to classify. Based on this figure, the following notes and comments are presented.

1. This figure presents two conditional tables of *P(DRoe = 1)* by *DSGRO* and the time *T* for *DINVES = 0* and *DINVES = 1*, respectively.

```
Descriptive Statistics for DROE
Categorized by values of DSGRO and T and DINVES
Date: 08/01/10  Time: 09:34
Sample: 1 390
Included observations: 390
```

Table 1: Conditional table for DINVES=0:							Table 2: Conditional table for DINVES=1:					

Table 1: Conditional table for DINVES=0:

Mean Std. Dev. Obs.		1	2	T 3	All
	0	0.100000 0.047619		0.196429	0.119048
		0.302166 0.215540		0.400892	0.324813
		70	42	56	168
DSGRO	1	0.222222 0.407407		0.238095	0.315789
		0.440959 0.500712		0.436436	0.468961
		9	27	21	57
	All	0.113924 0.188406		0.207792	0.168889
		0.319749 0.393901		0.408388	0.375489
		79	69	77	225

Table 2: Conditional table for DINVES=1:

Mean Std. Dev. Obs.		1	2	T 3	All
	0	0.146341 0.189189		0.317073	0.218487
		0.357839 0.397061		0.471117	0.414967
		41	37	41	119
DSGRO	1	0.500000 0.291667		0.166667	0.304348
		0.527046 0.464306		0.389249	0.465215
		10	24	12	46
	All	0.215686 0.229508		0.283019	0.242424
		0.415390 0.424006		0.454776	0.429854
		51	61	53	165

Figure 5.3 *The proportions* P(Droe = 1) *by* DSGRO *and* T*, conditional for* DINVES

2. For a better presentation, Table 5.2 presents the DSS of a three-way table of the proportions *P(DRoe = 1)* by *DInves*, *DSgro* and *T*. Similar to the ORs presented in Table 5.1, this table also shows conditional ORs between selected pairs of groups, which can easily be computed using Excel.

3. The basic objectives of this table are to study or evaluate descriptively the growth of problems over time for the four groups of firms generated by *DSGRO* and *DINVES*, as well as their differences. For instance, *DINVES = DSGRO = DROE = 1* indicates a group with the worst problems since they have negative scores on the three variables *INVES*, *SGRO* and *ROA* over time.

4. Based on the output in Figure 5.3, we can compute the number of firms with the worst conditions at each time point. We obtain 5 (=10*0.5000), 7 (=24*0.291 667), and 2 (12*0.166 667) firms, respectively, with the worst condition at the time *T* = 1, 2 and 3.

5. In order to identify the firms with the worst conditions, the following steps should be taken:
   5.1 With the file on the screen, select a sub-sample by entering the "if" condition "*T = 1 and DRoe = 1 and DSgro = 1 and DInves = 1*" ... *OK*.
   5.2 Present the variables *FIRM_NAME* on the screen, then by selecting *View/One-way tabulation* ... *OK*, the following output is obtained. However, for confidentiality, the names of the firms are replaced using the symbols *A*, *B*, *C*, *D* and *E* (see Figure 5.4).

6. By using the same steps, the firms at the time *T* = 2 and *T* = 3 could easily be obtained. However, note that by using the sub-sample "*DRoe = 1* and *DSgro = 1* and *DInves = 1*" a list of the *FIRM_NAME* at the three time points can be obtained directly.

Table 5.2 *The proportions* P(DRoe = 1) *by* DInves, DSgro *and the time* T*, with conditional ORs between levels of each independent factor*

		T = 1	T = 2	T = 3	OR(T = 2/T = 1)	OR(T = 3/T = 2)
DInves = 0	DSgro = 0	0.100	0.048	0.196	0.450	4.889
	DSgro = 1	0.222	0.407	0.238	2.406	0.455
	OR(0/1)	0.389	0.073	0.782		
DInves = 1	DSgro = 0	0.146	0.189	0.317	1.361	1.990
	DSgro = 1	0.500	0.292	0.167	0.412	0.486
	OR(0/1)	0.171	0.567	2.321		
OR(Dinves(0/1)|DSgro = 0)		0.648	0.214	0.526		
OR(Dinves(0/1)|DSgro = 1)		0.286	1.670	1.562		

Tabulation of FIRM_NAME
Date: 07/23/10   Time: 09:15
Sample: 1 390 IF T=1 AND DROE=1 AND DSGRO=1 AND
     DINVES=1
Included observations: 5
Number of categories: 5

Value	Count	Percent	Cumulative Count	Cumulative Percent
A	1	20.00	1	20.00
B	1	20.00	2	40.00
C	1	20.00	3	60.00
D	1	20.00	4	80.00
E	1	20.00	5	100.00
Total	5	100.00	5	100.00

Figure 5.4   *List of firms with the worst condition at the time* T = 1

5.3.2   Analysis Using the Option "Equality Tests by Classification..."

The three-way table of proportions *P(Droe = 1)* in Figure 5.3 also can be developed using the object *"Equality Tests by Classification . . . "*. In addition, by using this option, we also can test the hypotheses on the mean, variance or median differences of an endogenous variable between the groups generated by several categorical variables considered, as presented in Figure 5.5.

   Based on this result, the following findings and notes are presented.

1. With the variable *DROE* on the screen, select *View/Descriptive Statistics . . .* , *Equality Tests by Classification . . .* , then by entering the variables *"DInves DSgro T"* as the series/group of classification, the output in Figure 5.5 is obtained.

Test for Equality of Means of DROE
Categorized by values of DSGRO and DINVES and T
Date: 08/02/10   Time: 15:02
Sample: 1 390
Included observations: 390

Method	df	Value	Probability
Anova F-test	(11, 378)	2.780886	0.0017
Welch F-test*	(11, 87.9536)	2.751028	0.0042

*Test allows for unequal cell variances

Analysis of Variance

Source of Variation	df	Sum of Sq.	Mean Sq.
Between	11	4.671679	0.424698
Within	378	57.72832	0.152720
Total	389	62.40000	0.160411

Figure 5.5   *Statistical results of the test for equality of means of* DRoe *by* T, DInves *and* DSgro

2. Note that this figure presents two statistical tests, namely the ANOVA F-test and Welch F-test for equality of the means of *ROE*, which can be presented as follows:

$$H_0 : \pi_{ijt} = \pi, \quad \forall i, j, t \quad vs \quad H_1 : Otherwise \tag{5.1}$$

where π_{ijt} is the proportion parameter of the subpopulation, indicated by $DINVES = i$, and $DSGRO = j$, at the time t. At the significance level of $\alpha = 0.10$, the null hypothesis is rejected based on both tests' statistics.

5.3.3   Analysis Using the Option "N-Way Tabulation..."

Agung (2011) presents various N-way tabulations of sets of categorical variables as either complete or incomplete frequency tables. However, for each set of variables, the tables presented are a set of two-dimensional frequency tables of the first two variables entered, conditional for the other variables, based on cross-sectional data. However, based on pool panel data, this section presents the distribution of multi-dimensional problems by or over time for research objects.

Example 5.3   Distribution of a two-dimensional problem by time
For illustration, Figure 5.6 presents the distribution of a two-dimensional problem, namely *CF_RS*, by the time *T*, where *CF_RS* is a cell-factor generated based on two zero-one indicators *DROE* and *DSGRO*, using

```
Tabulation of T and CF_RS
Date: 08/05/10  Time: 11:02
Sample: 1 390
Included observations: 390
Tabulation Summary
```

Variable	Categories
T	3
CF_RS	4
Product of Categories	12

Measures of Association	Value
Phi Coefficient	0.281834
Cramer's V	0.199287
Contingency Coefficient	0.271266

Test Statistics	df	Value	Prob
Pearson X2	6	30.97785	0.0000
Likelihood Ratio G2	6	30.90531	0.0000

Count % Row		CF_RS 11	12	21	22	Total
	1	7 5.38	13 10.00	12 9.23	98 75.38	130 100.00
T	2	18 13.85	9 6.92	33 25.38	70 53.85	130 100.00
	3	7 5.38	24 18.46	26 20.00	73 56.15	130 100.00
	Total	32 8.21	46 11.79	71 18.21	241 61.79	390 100.00

Figure 5.6   *Distribution of a two-dimensional indicator problem* CF_RS *by the time* T

the following equation.

$$CF\_RS = 11^*(DROE = 1\,AND\,DSGRO = 1) + 12^*(DROE = 1\,AND\,DSGRO = 0)$$
$$+ 21^*(DROE = 0\,AND\,DSGRO = 1) + 22^*(DROE = 0\,AND\,DSGRO = 0) \qquad (5.2)$$

Based on equation (5.2) and the output in Figure 5.6, the following notes and comments are presented.

1. Compared to the output in Figure 5.4, this figure presents a better picture on the number of firms with problems, starting from the worst two-dimensional problem indicated by $CF\_RS = 11$, those are the firms having negative scores for both *ROE* and *SGRO*.
2. The output presents three measures of associations between the two categorical variables CF_RS and T. However, the contingency coefficient is commonly presented. Refer to Agung (2011) and Conover (1980) for additional nonparametric measures of associations between pairs of variables.
3. The output also presents two statistics for testing the hypothesis as follows:

$$H_0 : CF\_RS\ \text{has the same distribution over times}$$

$$H_1 : \text{Otherwise}$$

The null hypothesis is rejected based on *Likelihood Ratio* $G_0^2 = 30.90531$ with $df = 6$ and a *p*-value $= 0.0000$. It can then be concluded that the probability distributions of CF_RS have a significant difference over times. In other words, it can be said that the time T has a significant effects on CF_RS.

5.4   Analysis Based on Ordinal Problem Indicators

As an extension of the zero-one problem indicator, in many cases, evaluators prefer to identify groups of research objects by levels of a problem. Referring to the zero-one indicator *DROE* which is defined based on a numerical variable *ROE* as one of the market performance indicators of the firms or banks, various ordinal variables can easily be generated to classify the firms into groups with different levels of the problem, with respect to *ROE*. In other words, starting from the group with the worst levels all the way up to the group with the best levels of *ROE*.

For illustration, the author's students have generated ordinal problem indicators in their dissertations to conduct evaluation studies, such as Ariastiadi (2010) who considers six ordinal variables to evaluate the performance of 119 banks in Indonesia. One of the variables is five levels of the measured variable *ROE*, which is generated as

$$(1)\,ROE < 0.00;\ (2)\,0 \le ROE < 0.50;\ (3)\,0.50 \le ROE < 1.25;$$
$$(4)\,1.25 \le ROE < 1.50;\ \text{and}\ (5)\,ROE \ge 1.50. \qquad (5.3)$$

He then uses DSSs to study the causal relationships between categorical variables.

On the other hand, Hamzal (2006) defines or generates ordinal indicators based on four unobservable latent variables (LV), namely the banks' performance, strategic flexibility, strategic consistency and perceived environment, where each of the indicators has five levels, such as

$$(1)\,LV < -1.50;\ (2)\,-1.50 \le LV < -0.50;\ (3)\,-0.50 \le LV < +0.50;$$
$$(4)\,0.50 \le LV < 1.50;\ \text{and}\ (5)\,LV \ge 1.5. \qquad (5.4)$$

Example 5.4   A special classification by levels of a problem
In the previous examples, a firm is defined as having a problem if and only if $ROE < 0$. Now, the firms can be classified by level of *ROE*, which can be subjectively selected or defined, based on its observed values.

```
Tabulation of T and O5ROE
Date: 07/22/10  Time: 11:52
Sample: 1 390
Included observations: 390
Tabulation Summary
```

Variable	Categories
T	3
O5ROE	5
Product of Categories	15

Measures of Association	Value
Phi Coefficient	0.163000
Cramer's V	0.115259
Contingency Coefficient	0.160877

Test Statistics	df	Value	Prob
Pearson X2	8	10.36194	0.2405
Likelihood Ratio G2	8	10.81767	0.2122

WARNING: Expected value is less than 5 in 40.00% of cells (6 of 15).

Count		O5ROE					
% Row		1	2	3	4	5	Total
	1	3	1	16	107	3	130
		2.31	0.77	12.31	82.31	2.31	100.00
	2	2	2	23	93	10	130
T		1.54	1.54	17.69	71.54	7.69	100.00
	3	3	5	23	91	8	130
		2.31	3.85	17.69	70.00	6.15	100.00
	Total	8	8	62	291	21	390
		2.05	2.05	15.90	74.62	5.38	100.00

Figure 5.7   *The distribution of an ordinal variable* O5ROE *by the time* T

Figure 5.7 presents the distribution of an ordinal variable *O5ROE* by the time *T*, where levels of *O5ROE* are generated as (1) $ROE < -1.0$; (2) $-1.0 \leq ROE < -0.5$; (3) $-0.5 \leq ROE < 0$; (4) $0 \leq ROE < +0.5$; and (5) $ROE \geq +0.5$. The main objective in generating these levels is to classify the firms with negative *ROEs* into three groups with three levels of the problem indicated by *ROE*.

Based on these results, the following notes and comments are presented.

1. The ordinal variable *O5ROE* is generated using the equation as follows:

$$O5ROE = 1 + 1^*(Roe >= -1.0) + 1^*(Roe >= -0.5) + 1^*(Roe >= 0.0) + 1^*(roe >= +0.5) \quad (5.5)$$

2. Then with the variables *T* and *O5ROE* on the screen, select *View/N_Way Tabulation . . . , Row % . . . OK*, the results in Figure 5.7 are obtained.
3. Note that the output shows that eight firms have the worst level of *ROE*, which can easily be identified using the steps presented in previous example. Then the cause factors should be identified.
4. A more detail classification could be made, especially for the 291 firms in *O5ROE* = 4. Do it as an exercise.

5. In addition to the two-way tabulation, specifically the distribution of the variable *O5ROE* by the time *T*, the output presents three measures of association, namely Phi Coefficient, Cramer V and Contingency Coefficient, and two test statistics: Pearson χ^2, and Likelihood Ratio G^2.

Example 5.5   Classifications using percentiles

By using the percentiles, namely the function *@Quantile(X,q)*, where *q* is a proportion or $q \in (0, 1.0)$, various levels of a problem can be generated based on a numerical problem *X*, such as follows:

1. A zero-one indicator or a dummy variable for the lowest 5% scores of *X*:

$$D1X = 1^*(X < @Quantile(X, 0.05)) \tag{5.6}$$

Then the group *D1X = 1* will contain 5% of the total number of the observations. Note that further analysis can be done based on the sub-sample "*D1X* = 1" only. Similarly, this can be done for the following zero-one indicator.

2. A zero-one indicator or a dummy variable for the highest 10% scores of *X*:

$$D2X = 1^*(X >= @Quantile(X, 0.10)) \tag{5.7}$$

3. Ordinal variables with various numbers of levels are generated based on a numerical variable *X*, such as follows for the illustration.

    3.1   An ordinal variable with three levels:

$$O3X = 1 + 1^*(X >= @Quantile(X, q1)) + 1^*(X >= @Quantile(X, q2)) \tag{5.8}$$

for $q1 \in [0.20, 0.30]$ & $q2 = 1 - q1$ is subjectively selected. Only the sub-sample "*O3X = 1 or O3X = 3*" might be considered for a more advanced analysis, to study the differences between the two-tailed sub-groups of individuals.

    3.2   An ordinal variable with four levels:

$$O4X = 1 + 1^*(X >= @Quantile(X, 0.25)) + 1^*(X >= @Quantile(X, 0.50))$$
$$+ 1^*(X >= @Quantile(X, 0.75)) \tag{5.9}$$

    3.3   An ordinal variable with five levels:

$$O5X = 1 + 1^*(X >= @Quantile(X, 0.20)) + 1^*(X >= @Quantile(X, 0.40))$$
$$+ 1^*(X >= @Quantile(X, 0.60)) + 1^*(X >= @Quantile(X, 0.80)) \tag{5.10}$$

    3.4   An ordinal variable with six levels:

$$O6X = 1 + 1^*(X >= @Quantile(X, 0.10)) + 1^*(X >= @Quantile(X, 0.30))$$
$$+ 1^*(X >= @Quantile(X, 0.50)) + 1^*(X >= @Quantile(X, 0.70)) \tag{5.11}$$
$$+ 1^*(X >= @Quantile(X, 0.90))$$

    3.5   Note the classifications presented here use the closed-open intervals, which are commonly applied. As the extension, various ordinal variables with alternative closed-open intervals or having more levels can easily be generated based on any numerical problem variables, if it is needed. In fact even, the rank of the numerical *X* could be used to differentiate all observed individuals.

5.5 Multiple Association between Categorical Variables

5.5.1 Applications of N-Way Tabulation

The multiple associations between any sets of categorical variables, including zero-one indicators, can easily be studied using the N-way tabulation procedure. However, this does not take the ordinal scale of the variables into account. See the following examples.

Example 5.6   Multiple association between *DRoe*, *DSgro*, *DInves* and *T*
With the four variables on the screen, by selecting *View/N-way Tabulation . . . , Row% . . . OK*, the statistical results are obtained. However, for illustration only the first four out of 12 tables are presented in Figure 5.8. Based on the output in this figure, the following findings and notes are presented.

1. It can be concluded that the four categorical variables have significant associations based on the LR statistics of $G_0^2 = 53.71209$ with $df = 18$ and a p-value $= 0.0000$. In other words, it can be said that the variables *T, DSgroe* and *DInves* have a significant joint effect on *DRoe*.
2. Table 1 in Figure 5.8 presents the association between *DRoe* and *DSgro* conditional for *DInves* $= 0$ and $T = 1$. At a significance level of $\alpha = 0.10$, they have an insignificant association, based on the LR

Figure 5.8 *Statistical results based on a four-way tabulation of* DRoe, DSgro, DInves *and* T

statistics of $G_0^2 = 0.987057$ with $df = 1$ and a p-value $= 0.3205$. Note that this table also presents three measures of associations between DRoe and *DSgro*, conditional for *DInves* $= 0$ and $T = 1$, namely the Phi Coefficient, Cramer's V and the contingency coefficient.

3. Table 2 in Figure 5.8 presents the association between *DRoe* and *DSgro* conditional for *DInves* $= 1$ and $T = 1$. At a significance level of $\alpha = 0.10$, they have a significant association, based on the LR statistics of $G_0^2 = 5.181757$ with $df = 1$ and a p-value $= 0.0228$.

4. Finally, Table 3 in Figure 5.8 presents the association between *DRoe* and *DSgro* conditional for $T = 1$. At a significance level of $\alpha = 0.10$, they have an significant association, based on the LR statistics of $G_0^2 = 6.442411$ with $df = 1$ and a p-value $= 0.0111$.

Example 5.7   Multiple associations between *DRoe*, *DSgro* and *DInves*, conditional for each time point *T*

Note that the complete statistical results in Example 5.6 cannot provide the association between the three variables *DROE*, *DSGRO* and *DINVES*, conditional for each time point. In order to study the association between these three variables, a cell factor, namely *CF_SI* with four levels, should be generated based on the variables *DSGRO* and *DINVES* and this can be generated using the following equation.

$$CF\_SI = 11^*(DSgro = 1 \text{ and } DInves = 1) + 12^*(DSgro = 1 \text{ and } DInves = 0)$$
$$+ 21^*(DSgro = 0 \text{ and } DInves = 1) + 22^*(DSgro = 0 \text{ and } DInves = 0) \tag{5.12}$$

Then Figure 5.9 presents part of the results of the three-way tabulation of variables *DROE*, *CF_SI* and *T*, namely two-way tabulations of *DROE* and *CF_SI*, conditional for $T = 1$ and $T = 2$ respectively, and an unconditional two-way tabulation of *DROE* and *CF_SI*. Based on these results, the following findings and notes presented.

Figure 5.9   *Statistical results based on a three-way tabulation of* DRoe, CF_SI *and* T

1. Compared to results in Figure 5.8, these results show that the four variables *DRoe*, *DSgro*, *DInves* and the time *T* also have a significant association, but the indicators *DSgro* and *DInves* are represented as a cell-factor, say *CF_SI*.
2. Based on Table 1, conditional for $T = 1$, it can be concluded that *DRoe*, *DSgro* and *DInves* have a significant multiple association. In other words, it can be said that *CF_SI* has a significant effect on *DRoe*, conditional for $T = 1$.
3. Similarly, based on Table 2, it can be concluded that *CF_SI* has a significant effect on *DRoe*, conditional for $T = 2$.
4. Finally, based on the unconditional table, it can be concluded that *CF_SI* has a significant effect on *DRoe*.

Example 5.8   A special association between two ordinal variables
For an illustration, Figure 5.10 presents the distribution of *O5ROE* in (5.5) by *O3SGRO* which is generated using the following equation, conditional for $T = 1$.

$$O3SGRO = 1 + 1^*(SGRO >= 0) + 1^*(SGRO >= @Quantile(SGRO, 0.5)) \qquad (5.13)$$

```
Tabulation of O3SGRO and O5ROE and T
Date: 08/06/10  Time: 13:08
Sample: 1 390 IF T=1
Included observations: 130
Tabulation Summary
```

Variable	Categories
O3SGRO	3
O5ROE	5
T	1
Product of Categories	15

Test Statistics	df	Value	Prob
Pearson X2	8	27.37524	0.0006
Likelihood Ratio G2	8	20.31905	0.0092

WARNING: Expected value is less than 5 in 73.33% of cells (11 of 15).

Table 1: Conditional table for T=1:

Count % Row		O5ROE					
		1	2	3	4	5	Total
	1	3 15.79	0 0.00	4 21.05	11 57.89	1 5.26	19 100.00
O3SGRO	2	0 0.00	1 4.00	4 16.00	19 76.00	1 4.00	25 100.00
	3	0 0.00	0 0.00	8 9.30	77 89.53	1 1.16	86 100.00
	Total	3 2.31	1 0.77	16 12.31	107 82.31	3 2.31	130 100.00

Measures of Association	Value
Phi Coefficient	0.458889
Cramer's V	0.324483
Contingency Coefficient	0.417072

Table Statistics	df	Value	Prob
Pearson X2	8	27.37524	0.0006
Likelihood Ratio G2	8	20.31905	0.0092

WARNING: Expected value is less than 5 in 73.33% of cells (11 of 15).

Figure 5.10   *The distribution of O5ROE by O3SGRO conditional for T = 1*

```
Covariance Analysis: Kendall's tau
Date: 08/05/10  Time: 16:49
Sample (adjusted): 1 130
Included observations: 130 after adjustments
```

tau-b Score (S) Probability	O5ROE	O3SGRO
O5ROE	1.000000 2588 ------	
O3SGRO	0.202109 671 0.0150	1.000000 4259 ------

(a)

```
Covariance Analysis: Kendall's tau
Date: 08/05/10  Time: 17:07
Sample (adjusted): 1 130
Included observations: 130 after adjustments
```

tau-b Score (S) Probability	ROE	SGRO
ROE	1.000000 8385 ------	
SGRO	0.202624 1699 0.0006	1.000000 8385 ------

(b)

Figure 5.11   *Kendall's tau indexes of the bivariates* (O5ROE,O3SGRO) *and* (ROE,SGRO)

Based on this output, the following notes and comments are made.

1. Note that the cell *(O5ROE = 1,SGRO = 3)* can be a structural empty cell, since it is impossible for a firm have the lowest level of negative $ROE < -1.0$ but a large positive $SGRO \geq @Quantile(SGRO,0.5)$ $= 0.231\,095$. Then the table is incomplete, so the test statistics do not have any value and cannot be applied.

2. On the other hand, even under the assumption that the table is complete, this one has four empty cells and 11 out of 15 cells have expected values less than 5. So it can be argued as to whether the measures of association and the test statistics are valid for making conclusions, based on the two-dimensional table of *O5ROE* and *O3SGRO*.

3. For these reasons, the following nonparametric measure should be applied.

5.5.2   Application of Kendall's Tau

Kendall (1938) developed a concordance-discordance measure of association between pairs of variables having at least ordinal scales, called the *Kendall's tau*. It has since been extended to Agung–Kendall's tau for the right censored observations (Agung, 1981). Based on pool panel data, then Kendall's tau can easily be applied to any pairs of ordinal problem indicators, as well as numerical problem indicators, conditional for the time variable *T*.

Example 5.9   An application of Kendall's tau

As the extension of the analysis presented in Figure 5.10, Figure 5.11(a) presents the Kendall's tau between two ordinal variables *O5ROE* and *O3SGRO* conditional for $T = 1$. For comparison, Figure 5.11(b) presents the Kendall's tau between the numerical variables *ROE* and *SGRO* conditional for $T = 1$.

6

General Choice Models

6.1 Introduction

Binary choice models based on pool panel data, specifically pool panel data with a few time-point cross-sectional observations, can easily be derived from all binary choice models based on cross-section data (Agung, 2011), by using a categorical time variable as an additional independent variable of binary choice models. For this reason, this chapter will present only some selected binary and ordered choice models, called *general choice models*.

General choice models here should be classified into two groups. The first group has *categorical variables*, including time, and the second has *numerical* as well as categorical variables.

6.2 Multi-Factorial Binary Choice Models

Binary choice models with a set of categorical independent variables are called *multi-factorial*. The following section starts with one-way binary choice models.

6.2.1 One-Way Binary Choice Model

Based on pool panel data, the simplest binary choice model considered will be a model of a zero-one problem indicator with a single categorical independent variable, either the time variable or a cause or classification factor as an independent variable. See the following example.

Example 6.1 Binary logit model of *DROE* on time *T*
As an extension of the proportion $P(Droe = 1)$ by time T presented in Figure 5.2, Figure 6.1 presents the statistical results using the following equation specification.

$$Y \quad @Expand(T) \tag{6.1}$$

Panel Data Analysis Using EViews, First Edition. I Gusti Ngurah Agung.
© 2014 John Wiley & Sons, Ltd. Published 2014 by John Wiley & Sons, Ltd.
Companion website: www.wiley.com/go/panel_data

```
Dependent Variable: DROE
Method: ML - Binary Logit (Quadratic hill climbing)
Date: 08/12/10   Time: 18:08
Sample: 1 390
Included observations: 390
Convergence achieved after 4 iterations
Covariance matrix computed using second derivatives
```

Variable	Coefficient	Std. Error	z-Statistic	Prob.
T=1	-1.704748	0.243086	-7.012937	0.0000
T=2	-1.338892	0.216208	-6.192620	0.0000
T=3	-1.161133	0.205813	-5.641682	0.0000

Mean dependent var	0.200000	S.D. dependent var	0.400514
S.E. of regression	0.400010	Akaike info criterion	1.008389
Sum squared resid	61.92308	Schwarz criterion	1.038898
Log likelihood	-193.6358	Hannan-Quinn criter.	1.020483
Avg. log likelihood	-0.496502		

Obs with Dep=0	312	Total obs	390
Obs with Dep=1	78		

```
Estimation Command:
=========================
BINARY(D=L) DROE @EXPAND(T)

Estimation Equation:
=========================
I_DROE = C(1)*(T=1) + C(2)*(T=2) + C(3)*(T=3)

Forecasting Equation:
=========================
DROE = 1-@CLOGISTIC(-(C(1)*(T=1) + C(2)*(T=2) + C(3)*(T=3)))

Substituted Coefficients:
=========================
DROE = 1-@CLOGISTIC(-(-1.70474809223*(T=1) - 1.33889212223*(T=2) - 1.16113264565*(T=3)))
```

Figure 6.1   *Statistical results using the equation specification in (6.1)*

Based on these results, the following notes and comments are presented.

1. The "*Estimate Equation*" of *I_DROE* in fact represents the following logit function, where $p = P(Droe = 1)$.

$$
\log\left(\frac{p}{1-p}\right) = C(1)^*(T=1) + C(2)^*(T=2) + C(3)^*(T=3) \tag{6.2}
$$
$$
= -1.704748^*(T=1) - 1.338892^*(T=2) - 1.161133^*(T=3)
$$

2. With the output on the screen, by using the "*Forecasting Equation*" the *predicted probabilities* can easily be generated, namely *DROEF*, using the following equation which can easily be copied from the output.

$$
DROEF = 1 - @CLOGISTIC(-(C(1)^*(T=1) + C(2)^*(T=2) + C(3)^*(T=3))) \tag{6.3}
$$

3. Figure 6.2 presents the one-way tabulation of *DROEF*, which is exactly the same as the mean of *DROE* or *P(Droe = 1)* by the time *T*, presented in Figure 5.2

```
Tabulation of DROEF
Date: 08/12/10   Time: 19:18
Sample: 1 390
Included observations: 390
Number of categories: 3
```

Value	Count	Percent	Cumulative Count	Cumulative Percent
0.153846	130	33.33	130	33.33
0.207692	130	33.33	260	66.67
0.238462	130	33.33	390	100.00
Total	390	100.00	390	100.00

Figure 6.2   *One-way tabulation of* DROEF *or* P(Droe = 1) *by the time* T

4. So it can be proven that the predicted probabilities are exactly the same as the sampled proportion of $P(DROE = 1)$, as presented in Figure 5.2. Furthermore, we can conclude that the estimate of the parameters C(1), C(2) and C(3) in (6.2) can be manually computed using the values of *DROEF* in Figure 6.2, that is, $\log(p_1/(1 - p_1)) = \log(0.153\ 846/(1 - 0.153\ 846)) = \hat{C}(1)$.

5. Since the predicted probabilities of a logit model are exactly the same as the sampled proportions by the time *T*, then it can be said that the logit should not be applied if the main objective is to compute predicted probabilities. For this reason, the main objective for conducting analysis using a logit model should be to test the hypotheses on growth of logit differences, indicated by $\log(p/(1 - p))$, over times. In other words, those hypotheses on the risk or chance differences in research objects as a single group between time points, which should be done using the Wald test. See the following alternative analysis.

Example 6.2   An alternative equation specification

As a modification of the binary logit model (BLM) presented in Figure 6.1, Figure 6.3 presents the statistical results based on the same BLM using the equation specification as follows:

$$Y \quad C \quad @Expand(T, @Droplast) \tag{6.4}$$

and Table 6.1 presents the parameters of the model or $\log(p_t/(1 - p_t))$, by times.

Based on the results in Figure 6.3 and the Table 6.1, the following findings are obtained.

1. Compared to the results in Figure 6.1, Figure 6.3 presents the test statistics, which can easily be used to test the hypotheses on the growth risk differences over times.

Table 6.1   *The parameters of the BLM in (6.4) by the time* T

	$T = 1$	$T = 2$	$T = 3$	$T(1 - 3)$	$T(2 - 3)$
$\log(p_t/(1 - p_t))$	$C(1) + C(2)$	$C(1) + C(3)$	$C(1)$	$C(2)$	$C(3)$

Figure 6.3   *Statistical results using the equation specification in (6.4)*

```
Wald Test:
Equation: Untitled

Test Statistic          Value          df      Probability

F-statistic           1.264690      (1, 387)     0.2615
Chi-square            1.264690          1        0.2608


Null Hypothesis Summary:

Normalized Restriction (= 0)         Value       Std. Err.

C(2) - C(3)                        -0.365856     0.325325

Restrictions are linear in coefficients.
```

Z-statistic:
$Z_0 = -0.365856/0.325325 = -1.12459$

t-statistic:
$t_0 = SQRT(F(1, 387))$
$= SQRT(1.264690) = 1.124584$
$Prob(t-Stat) = Prob(F-stat) = 0.2615$

Figure 6.4   *The Wald test for the null hypothesis* H_0: $C(2) = C(3)$, *and the derived values of the Z-statistic and* t-*statistic*

2. The null hypothesis H_0: $C(2) = C(3) = 0$ is accepted based on the Likelihood Ratio (LR) statistic of $G_0^2 = 3.0422390$ with $df = 2$ and a p-value $= 0.218\,462$. Therefore, we can conclude that the research objects (firms) as a group or subgroup in the corresponding population have insignificant risk differences for having a negative *ROE*.

3. Corresponding to the parameter $C(2)$, which indicates the risk difference of a firm in the population between the time $T = 1$ and $T = 3$, or the risk growth from $T = 1$ to $T = 3$, there are three alternative hypotheses can be considered, such as follows:

   a. A two-sided hypothesis: H_0: $C(2) = 0$, vs. H_1: $C(2) <> 0$

      At a significance level of $\alpha = 0.10$, the null hypothesis is rejected based on the z-statistic of $Z_0 = -1.706\,734$ with a p-value $= 0.0879$. If the hypothesis stated that $C(2) <> 0$, then the conclusion should be, "*the data supports the hypothesis*". In empirical studies, however, the hypothesis considered is either right- or left-sided.

   b. A right-sided hypothesis: H_0: $C(2) \leq 0$, vs. H_1: $C(2) > 0$

      Since, the z-statistic has a negative value, namely $Z_0 = -1.706\,734$, then it can be concluded directly that the null hypothesis is accepted. If the hypothesis stated that $C(2) > 0$, then the conclusion should be, "*the data does not support the hypothesis*".

   c. A left-sided hypothesis: H_0: $C(2) \geq 0$, vs. H_1: $C(2) < 0$

      At a significance level of $\alpha = 0.10$, the null hypothesis is rejected based on the z-statistic of $Z_0 = -1.706\,734$ with a p-value $= 0.0879/2 = 0.04\,395$. If the hypothesis stated that $C(2) < 0$, then the conclusion should be, "*the data supports the hypothesis*".

4. Another hypothesis on the risk differences between time $T = 1$ and $T = 2$, with the null hypothesis H_0: $C(2) = C(3)$, should be tested using the Wald test, with statistical results presented in Figure 6.4. Note that this figure also presents the derived values of the Z-statistic and t-statistic, which can be used to test one-sided hypotheses as presented previously.

5. Note that the value of the t-statistic of the Wald test can be obtained directly if the Wald test is done using EViews 7.

6.2.2   Two-Way Binary Choice Model

With large cross-sectional observations, in general, groups of research objects with different characteristics and behaviors can easily be identified or observed. For this reason, as an extension of the one-way binary choice models presented in previous section this section presents two-way binary choice models. To generalize, let Y be the zero-one problem indicator, the time T the *categorical variable*, and CF the categorical *cause-factor*, *classification-factor* or *cell-factor*, then the following alternative equation specification can be applied.

6.2.2.1   The Pure Interaction Model

The simplest equation specification applied is as follows:

$$Y @Expand(CF, T) \tag{6.5}$$

For a dichotomous CF and three time-points, the binary logit model has the general equation:

$$\log\left(\frac{p}{1-p}\right) = C(1)^*Dc1^*Dt1 + C(2)^*DCc1^*Dt2 + C(3)^*Dc1^*Dt3$$
$$+ C(4)^*Dc2^*Dt1 + C(5)^*Dc2^*Dt2 + C(6)^*Dc2^*Dt3 + \mu_{it} \tag{6.6}$$

where $Dci = (CF = i)$ and $Dtj = (T = j)$, respectively, is a dummy variable or a zero-one indicator of the i-th level of the cell-factor CF, and the j-th time point of T.

Based on this model the following notes and comments are presented.

1. The model in (6.5) is a *pure-interaction-model* without an intercept, since all independent variables are interaction factors.
2. The parameters $C(1)$ to $C(6)$ are the logit parameters within the six cells generated by CF and T, namely $log(p_{it}/(1 - p_{it}))$ for $i = 1$ and 2; and $t = 1$, 2 and 3. Compare them to the equation specification in (6.1), and the binary logit model in (6.2).
3. Based on the model in (6.5) all hypotheses on the differences between the six logits $log(p_{it}/(1 - p_{it}))$ can be easily tested using the Wald test. Refer to the Example 6.2, and do it as an exercise.
4. The following equation specification represents the model with an intercept. Note that $@Droplast$, could be replaced by $@Dropfirst$ or $@Drop(i,t)$:

$$Y\ \ C\ \ @Expand(CF, T, @Droplast) \tag{6.7}$$

6.2.2.2   Binary Logit Growth Model by Groups of the Research Objects

The equation specification applied is as follows:

$$Y\ \ C\ \ @Expand(CF, @Droplast)\ @Expand(CF)^* @Expand(T, @Droplast) \tag{6.8}$$

This equation specification represents the growth of the logit: $log(p_{it}/(1 - p_{it}))$, $t = 1,2, \ldots, T$; for each $CF = i$. As an extension of Table 6.1, Table 6.2 presents an illustrative example for CF having four levels at

Table 6.2   *A 4×3 table of the logit* $\log(p_{it}/(1 - p_{it}))$ *by* CF *and* T

CF	$T=1$	$T=2$	$T=3$	$T(1-3)$	$T(2-3)$
1	$C(1)+C(2)+C(5)$	$C(1)+C(2)+C(6)$	$C(1)+C(2)$	$C(5)$	$C(6)$
2	$C(1)+C(3)+C(7)$	$C(1)+C(3)+C(8)$	$C(1)+C(3)$	$C(7)$	$C(8)$
3	$C(1)+C(4)+C(9)$	$C(1)+C(4)+C(10)$	$C(1)+C(4)$	$C(9)$	$C(10)$
4	$C(1)+C(11)$	$C(1)+C(12)$	$C(1)$	$C(11)$	$C(12)$
CF(1 − 4)		$C(2)$			
CF(2 − 4)		$C(3)$			
CF(3 − 4)		$C(4)$			

three time-points. Note that the each parameter $C(5)$ to $C(12)$ represents a difference of the logit: $log(p_{it}/(1 - p_{it}))$ between two time points, which can easily be tested using the output using the Z-statistic: refer to Example 6.2. And note that the parameters $C(2)$ to $C(4)$ represent the logit differences between the levels of *CF*, conditional for $T = 3$.

6.2.2.3   Alternative Equation Specifications

For the same model obtained by using the equation specifications in (6.5), (6.7) and (6.8), there are two other alternative equation specifications, such as follows:

1.   The model with the main factor *T* and interaction factors:

$$Y \quad C \quad @Expand(T,@Droplast) \quad @Expand(CF,@Droplast)^*@Expand(T) \qquad (6.9)$$

2.   The model with both main factors and their interactions:

$$Y \quad C \quad @Expand(T,@Droplast) \quad @Expand(CF,@Droplast)$$
$$@Expand(CF,@Droplast)^*@Expand(T,@Droplast) \qquad (6.10)$$

   As an exercise, do the analysis using the binary choice model of *Y* on *CF* with four levels and three time points, and then develop a 4×3 table of the logit $log(p_{it}/(1 - p_{it}))$ by *CF* and *T*. Refer to Agung (2011), if needed.

6.2.2.4   Special Notes and Comments

Corresponding to all specification equations presented previously, the following special notes and comments are presented.

1.   The model in (6.10) is called a hierarchical model or full-factorial model, indicated by the two main factors and their interactions as the independent variables, but the others are nonhierarchical models.
2.   All models have the same number of parameters as the number of cells generated by the cause or classification factor *CF* and the time variable *T*. Refer to the model represented in Table 6.2, which has 12 parameters.
3.   The predicted probabilities obtained would be exactly the same as the sampled proportions $P(Y = 1)$. Refer to the predicted probabilities presented in Figure 6.2.
4.   If a model has fewer parameters than the number of cells generated by *CF* and *T*, the model should be considered a reduced model, but in most cases, it could be poor. For instance the additive model with the following equation specification is not recommended:

$$Y \quad C \quad @Expand(T,@Droplast) \quad @Expand(CF,@Droplast) \qquad (6.11)$$

   For illustration, Table 6.3 presents the parameters of a not-recommended or a poor model of *Y* on a dichotomous *CF* at three time points, for either a binary choice or a general linear model. Note that the

Table 6.3   *A 2×3 table of the logit* $log(p_{it}/(1 - p_{it}))$ *by* CF *and* T

CF	T = 1	T = 2	T = 3	T(1 − 3)	T(2 − 3)
1	C(1) + C(2) + C(4)	C(1) + C(3) + C(4)	C(1) + C(4)	C(2)	C(3)
2	C(1) + C(2)	C(1) + C(3)	C(1)	C(2)	C(3)
CF(1 − 2)	C(4)	C(4)	C(4)	0	0

position of each parameters $C(2)$ and $C(3)$ as the differences between two time points, conditional for $CF = 1$ and $CF = 2$, and the parameter $C(4)$ indicates the differences between $CF = 1$ and $CF = 2$, for all time points, which is too special a case to be found in empirical studies. In other words, it is impossible. Note that the table has six cells, but it only has four parameters, namely $C(1)$–$C(4)$.

6.2.3 Multi-Factorial Binary Choice Model

In order to make it simple, a multi-factorial binary choice model (BCM) will be presented as a two-way BCM using the time variable T, and a cell factor generated based on two or more categorical cause or classification factors. However, in practice, the cell factor does not have to be generated as a new variable. For instance, let A and B be two categorical cause or classification factors, in this case applying the following two equation specifications is recommended.

6.2.3.1 The Pure Interaction Model

The pure interaction model is obtained using the following simplest equation specification:

$$Y\ @Expand(A, B, T) \tag{6.12}$$

For an illustration, Table 6.4 presents the $2 \times 2 \times 3$ table of the logit $log(p_{ijt}/(1 - p_{ijt}))$ by A, B and T. Based on this table, the following notes and comments are presented.

1. The parameters $C(1)$ to $C(12)$ represent the logit: $log(p_{ijt}/(1 - p_{ijt}))$, for $i = 1, 2; j = 1, 2;$ and $T = 1, 2, 3$, in the corresponding sub-populations. And we find that the predicted probabilities are exactly the same as the sampled proportions $P(Y = 1)$. Refer to Agung (2011).
2. The conditional odds ratios (ORs) between two levels of a factor, conditional for the other two factors can easily be computed. For instance,

$$OR(T = 1/T = 2|A = 1, B = 1) = Exp(C(1) - C(2))$$

3. Various hypotheses on the logit differences should be tested using the Wald test. However, for the testing hypotheses, it might be simpler to apply the equation specification presented in Examples 6.4 or 6.5.

Example 6.3 Application of a pure interaction model
As an illustration, Figure 6.5 presents the statistical results based on a binary logit model of $Y = DROE$ on the dichotomous variables $A = DInves$, $B = DSgro$ and T, using the equation specification in (6.12). Based on this output, the following findings are obtained.

1. Note that the estimates of the parameters $C(1)$ to $C(12)$ of this binary logit regression, are the $log[p_{ijt}/(1 - p_{ijt})]$, where $p_{ijt} = P_{ijt}(DRoe = 1)$ are the sampled proportions of $DRoe = 1$ in each cell-(i,j,t). So they could be manually calculated using sampled proportions.

Table 6.4 *A $2 \times 2 \times 3$ table of the logit $log(p_{ijt}/(1 - p_{ijt}))$ by A, B and T*

A	B	T = 1	T = 2	T = 3
1	1	C(1)	C(2)	C(3)
1	2	C(4)	C(5)	C(6)
2	1	C(7)	C(8)	C(9)
2	2	C(10)	C(11)	C(12)

```
Dependent Variable: DROE
Method: ML - Binary Logit (Quadratic hill climbing)
Date: 07/20/10   Time: 11:06
Sample: 1 390
Included observations: 390
Convergence achieved after 5 iterations
Covariance matrix computed using second derivatives
```

Variable	Coefficient	Std. Error	z-Statistic	Prob.
DINVES=0,DSGRO=0,T=1	-2.197225	0.398410	-5.514990	0.0000
DINVES=0,DSGRO=0,T=2	-2.995732	0.724569	-4.134503	0.0000
DINVES=0,DSGRO=0,T=3	-1.408767	0.336350	-4.188397	0.0000
DINVES=0,DSGRO=1,T=1	-1.252763	0.801784	-1.562470	0.1182
DINVES=0,DSGRO=1,T=2	-0.374693	0.391675	-0.956644	0.3387
DINVES=0,DSGRO=1,T=3	-1.163151	0.512348	-2.270238	0.0232
DINVES=1,DSGRO=0,T=1	-1.763589	0.441858	-3.991306	0.0001
DINVES=1,DSGRO=0,T=2	-1.455287	0.419750	-3.467029	0.0005
DINVES=1,DSGRO=0,T=3	-0.767255	0.335615	-2.286118	0.0222
DINVES=1,DSGRO=1,T=1	5.87E-15	0.632456	9.29E-15	1.0000
DINVES=1,DSGRO=1,T=2	-0.887303	0.449089	-1.975786	0.0482
DINVES=1,DSGRO=1,T=3	-1.609438	0.774597	-2.077775	0.0377

Mean dependent var	0.200000	S.D. dependent var		0.400514
S.E. of regression	0.390795	Akaike info criterion		0.987353
Sum squared resid	57.72832	Schwarz criterion		1.109388
Log likelihood	-180.5338	Hannan-Quinn criter.		1.035728
Avg. log likelihood	-0.462907			

Obs with Dep=0	312	Total obs	390
Obs with Dep=1	78		

```
Estimation Command:
=========================
BINARY(D=L) DROE @EXPAND(DINVES,DSGRO,T)

Estimation Equation:
=========================
I_DROE = C(1)*(DINVES=0 AND DSGRO=0 AND T=1) + C(2)*(DINVES=0 AND
DSGRO=0 AND T=2) + C(3)*(DINVES=0 AND DSGRO=0 AND T=3) + C(4)*(DINVES=0
AND DSGRO=1 AND T=1) + C(5)*(DINVES=0 AND DSGRO=1 AND T=2) + C(6)*
(DINVES=0 AND DSGRO=1 AND T=3) + C(7)*(DINVES=1 AND DSGRO=0 AND T=1) +
C(8)*(DINVES=1 AND DSGRO=0 AND T=2) + C(9)*(DINVES=1 AND DSGRO=0 AND T=
3) + C(10)*(DINVES=1 AND DSGRO=1 AND T=1) + C(11)*(DINVES=1 AND DSGRO=1
AND T=2) + C(12)*(DINVES=1 AND DSGRO=1 AND T=3)

Forecasting Equation:
=========================
DROE = 1-@CLOGISTIC(-(C(1)*(DINVES=0 AND DSGRO=0 AND T=1) + C(2)*
(DINVES=0 AND DSGRO=0 AND T=2) + C(3)*(DINVES=0 AND DSGRO=0 AND T=3) +
C(4)*(DINVES=0 AND DSGRO=1 AND T=1) + C(5)*(DINVES=0 AND DSGRO=1 AND T=
2) + C(6)*(DINVES=0 AND DSGRO=1 AND T=3) + C(7)*(DINVES=1 AND DSGRO=0
AND T=1) + C(8)*(DINVES=1 AND DSGRO=0 AND T=2) + C(9)*(DINVES=1 AND
DSGRO=0 AND T=3) + C(10)*(DINVES=1 AND DSGRO=1 AND T=1) + C(11)*(DINVES=
1 AND DSGRO=1 AND T=2) + C(12)*(DINVES=1 AND DSGRO=1 AND T=3)))

Substituted Coefficients:
=========================
DROE = 1-@CLOGISTIC(-(-2.19722456393*(DINVES=0 AND DSGRO=0 AND T=1) -
2.99573226205*(DINVES=0 AND DSGRO=0 AND T=2) - 1.40876721693*(DINVES=0
AND DSGRO=0 AND T=3) - 1.25276296849*(DINVES=0 AND DSGRO=1 AND T=1) -
0.374693449441*(DINVES=0 AND DSGRO=1 AND T=2) - 1.1631508098*(DINVES=0
AND DSGRO=1 AND T=3) - 1.76358859133*(DINVES=1 AND DSGRO=0 AND T=1) -
1.45528723254*(DINVES=1 AND DSGRO=0 AND T=2) - 0.767255152714*(DINVES=1
AND DSGRO=0 AND T=3) + 5.87353698598e-15*(DINVES=1 AND DSGRO=1 AND T=
1) - 0.887303195001*(DINVES=1 AND DSGRO=1 AND T=2) - 1.60943791216*
(DINVES=1 AND DSGRO=1 AND T=3)))
```

Figure 6.5   *Statistical results based on a binary logit model of* DROE *on* DInves, DSgro *and* T

2. The advantage in using the binary logit model is testing the hypotheses on the predicted risks or probabilities between pairs of the cells using the Wald test. Similar to the DSS in Table 5.5, Table 6.5 presents the DSS of the $log[p_{ijt}/(1-p_{ijt})]$, and the F-statistics for testing the hypotheses on the differences of $log[p_{ijt}/(1-p_{ijt})]$, conditional for $DINVES=i$ and $DSGRO=j$. Note that the output of the Wald test also presents Chi-square statistics.

3. Based on the model parameters in Table 6.5, various hypotheses can easily be defined and tested using the Wald test, such as follows:

    3.1   Conditional for $DInves=DSgro=0$, at a significance level of $\alpha=0.10$, the null hypothesis $H_0: C(1)=C(2)=C(3)$ is rejected, based on the F-statistic: $F_0=2.484$ with $df=(2,378)$ and a p-value$=0.085$. Therefore, it can be concluded that the logit: $log(p_{ijt}/(1-p_{ijt}))$ has a significant growth, conditional for $DInves=DSgro=0$. Note that this conclusion is contradictory to the conclusion based on the cell-proportion model in Figure 5.6.

    3.2   Other hypotheses on the logit differences between levels of *DInves* conditional for $DSgro=j$ and $T=t$, can also be defined and tested using the Wald test. Similarly so for the hypotheses on the logit differences between levels of *DSgro* conditional for $DInves=i$ and $T=t$. Do it as an exercise.

Table 6.5   *DSS of the* $log[p_{ijt}/(1-p_{ijt})]$ *in Figure 6.5, and the* F-*statistics*

		T = 1	T = 2	T = 3	F-Stat	df	Prob
DInves = 0	DSgro = 0	−2.197	−2.996	−1.409	2.484	(2,378)	**0.085**
	DSgro = 1	−1.253	−0.375	−1.163	0.980	(2,378)	0.376
DInves = 1	DSgro = 0	−1.764	−1.455	−0.767	1.823	(2,378)	0.163
	DSgro = 1	5.87E − 15	−0.887	−1.609	1.362	(2,378)	0.257

Table 6.6　A $2 \times 2 \times 3$ table of the logit $\log(p_{ijt}/(1 - p_{ijt}))$ *by* CF *and* T

A	B	T = 1	T = 2	T = 3	T(1 − 3)	T(2 − 3)
1	*1*	*C(1)+ C(2)+ C(5)*	*C(1)+ C(2)+ C(6)*	*C(1)+ C(2)*	**C(5)**	**C(6)**
1	*2*	*C(1)+ C(3)+ C(7)*	*C(1)+ C(3)+ C(8)*	*C(1)+ C(3)*	**C(7)**	**C(8)**
2	*1*	*C(1)+ C(4)+ C(9)*	*C(1)+ C(4)+ C(10)*	*C(1)+ C(4)*	**C(9)**	**C(10)**
2	*2*	*C(1)+ C(11)*	*C(1)+ C(12)*	*C(1)*	**C(11)**	**C(12)**

4. The binary logit model presented is a pure three-way binary logit regression, out of many binary logit regressions, which can be written manually based on the $2 \times 2 \times 3$ table of proportions $P(Y = 1)$, including the *subjective-logit-regressions*, as presented in Agung (2011).
5. Various conditional ORs could easily be computed, and two-sided or one-sided hypotheses using the *F*-statistic in the output could be transformed to *Z*- and *t*-statistics, as presented in Figure 6.4.

6.2.3.2　*Binary Choice Growth Model*

As an extension of the growth model in (6.8), the following equation specification should be applied

$$Y \; C \; @Expand(A, B, @Droplast) \; @Expand(A, B)^* @Expand(T, @Droplast) \qquad (6.13)$$

For an illustration, Table 6.6 presents the $2 \times 2 \times 3$ table of the logit: $log(p_{ijt}/(1 - p_{ijt}))$ by *A*, *B* and *T*. Based on this table the following notes and comments are presented. Compared to Table 6.2, this table presents exactly the same parameters, so this table can be viewed as a two-dimensional table of the logit: $log(p_{ijt}/(1 - p_{ijt}))$

1. The parameters *C(5)–C(12)* represent the logit differences between two time-points, conditional for $A = i$ and $B = j$, and their corresponding ORs can easily be computed, such as OR(T = 2/T = 3|A = 1, B = 1) = Exp(C(6)). Furthermore, for each OR, two-sided and one-sided hypotheses can be tested using the *Z*-statistic in the output. Refer to Example 6.2.
2. Similarly for the conditional ORs:

$$OR(A = 1/A = 2|B = 2, T = 3) = Exp(C(3)), \text{and}$$
$$OR(B = 1/B = 2|A = 2, T = 3) = Exp(C(4))$$

3. Other ORs should be computed using two or more parameters, such as

$$OR(T = 1/T = 2|A = 1, B = 1) = Exp(C(5) - C(6)), \text{and}$$
$$OR(A = 1/A = 2|B = 1, T = 1) = Exp((C(2) + C(5)) - (C(4) + C(9)))$$

However, for each OR, two-sided and one-sided hypotheses can be tested using the Wald test – see Example 6.2.

Example 6.4　Application of a growth model in (6.13)

As a modification of the model presented in Figure 6.5, Figure 6.6 presents the statistical results of a binary logit growth model, using the equation specification in (6.13), for *Y = DROE*, *A = DInves*, *B = DSgro* and *T*. Based on this output, the following findings and notes are presented.

```
Dependent Variable: DROE
Method: ML - Binary Logit (Quadratic hill climbing)
Date: 08/16/10   Time: 18:34
Sample: 1 390
Included observations: 390
Convergence achieved after 5 iterations
Covariance matrix computed using second derivatives
```

Variable	Coefficient	Std. Error	z-Statistic	Prob.
C	-1.609438	0.774597	-2.077775	0.0377
DINVES=0,DSGRO=0	0.200671	0.844471	0.237629	0.8122
DINVES=0,DSGRO=1	0.446287	0.928709	0.480546	0.6308
DINVES=1,DSGRO=0	0.842183	0.844179	0.997636	0.3185
(DINVES=0 AND DSGRO=0)*(T...	-0.788457	0.521403	-1.512183	0.1305
(DINVES=0 AND DSGRO=0)*(T...	-1.586965	0.798831	-1.986609	0.0470
(DINVES=0 AND DSGRO=1)*(T...	-0.089612	0.951503	-0.094180	0.9250
(DINVES=0 AND DSGRO=1)*(T...	0.788457	0.644910	1.222585	0.2215
(DINVES=1 AND DSGRO=0)*(T...	-0.996333	0.554865	-1.795631	0.0726
(DINVES=1 AND DSGRO=0)*(T...	-0.688032	0.537427	-1.280233	0.2005
(DINVES=1 AND DSGRO=1)*(T...	1.609438	1.000000	1.609438	0.1075
(DINVES=1 AND DSGRO=1)*(T...	0.722135	0.895366	0.806524	0.4199

McFadden R-squared	0.074930	Mean dependent var		0.200000
S.D. dependent var	0.400514	S.E. of regression		0.390795
Akaike info criterion	0.987353	Sum squared resid		57.72832
Schwarz criterion	1.109388	Log likelihood		-180.5338
Hannan-Quinn criter.	1.035728	Restr. log likelihood		-195.1569
LR statistic	29.24626	Avg. log likelihood		-0.462907
Prob(LR statistic)	0.002079			

Obs with Dep=0	312	Total obs		390
Obs with Dep=1	78			

Figure 6.6   *Statistical results based on the equation specification in (6.13)*

1. The model parameters are exactly the same as the parameters presented in Table 6.6.
2. Referring to the $2 \times 2 \times 3$ table of the logits in Table 6.6, the output in Figure 6.6 presents the Z-statistic for testing one-sided hypotheses on ORs, as well as two-sided hypotheses, for each parameter $C(2)$–$C(12)$.
3. The null hypothesis H_0: $C(k) = 0$ for all $k = 2$ to 12 is rejected based on the LR statistic of $G_0^2 = 29.24626$ with a p-value $= 0.002\,079$. Then we can conclude that *DInves, DSgro* and the time T have significant joint effects on *DROE*.
4. Various other hypotheses can easily be tested using the Wald test, which is the same as the model presented in Figure 6.5, but the hypotheses should use different parameters. Do it as an exercise.

Example 6.5   An alternative equation specification
The following equation is applied with a specific objective, to present the logit $log(p_{ijt}/(1 - p_{ijt}))$ differences between two levels of the factor A, conditional for $B = j$ and $T = t$.

$$Y \ @Expand(B, T) \ @Expand(B, T)^* @Expand(A, @Droplast) \tag{6.14}$$

For an illustration, Table 6.7 presents a $2 \times 2 \times 3$ table of the $log(p_{ijt}/(1 - p_{ijt}))$ by the factors A, B and T. Note that each of the parameters $C(7)$ to $C(12)$ indicates the logit difference between the level $A = 1$ and $A = 2$. The corresponding two- or one-sided hypotheses can easily be tested using the Z-statistic in the statistical results based on the model.

Note that the model in (6.14) is a model without the intercept, as a modification of the equation specification in (6.15) that represents one of the many alternative models with intercept parameters and parameters

Table 6.7 A 2 × 2 × 3 table of the logit $\log(p_{ijt}/(1 - p_{ijt}))$ by CF and T, in (6.14)

A	B = 1			B = 2		
	T = 1	T = 2	T = 3	T = 1	T = 2	T = 3
1	C(1) + C(7)	C(2) + C(8)	C(3) + C(9)	C(4) + C(10)	C(5) + C(11)	C(6) + C(12)
2	C(1)	C(2)	C(3)	C(4)	C(5)	C(6)
Diff	C(7)	C(8)	C(9)	C(10)	C(11)	C(12)

Table 6.8 A 2 × 2 × 3 table of the logit $\log(p_{ijt}/(1 - p_{ijt}))$ by CF and T, in (6.15)

A	B = 1			B = 2		
	T = 1	T = 2	T = 3	T = 1	T = 2	T = 3
1	C(1) + C(2) + C(7)	C(1) + C(3) + C(8)	C(1) + C(4) + C(9)	C(1) + C(5) + C(10)	C(1) + C(6) + C(11)	C(1) + C(12)
2	C(1) + C(2)	C(1) + C(3)	C(1) + C(4)	C(1) + C(5)	C(1) + C(6)	C(1)
Diff	C(7)	C(8)	C(9)	C(10)	C(11)	C(12)

presented in Table 6.8, where $C(1)$ is the intercept.

$$Y \quad C \quad @Expand(B, T, @Droplast) \quad @Expand(B, T)^* @Expand(A, @Droplast) \tag{6.15}$$

6.2.3.3 Special Notes and Comments

Doing analysis using a binary choice model with many categorical independent variables in many cases gets an error message because the corresponding frequency table has an empty cell(s). Refer to various illustrative data analysis using frequency tables with empty cells, including incomplete tables as well as their special notes and comments, in Agung (2011). Statistical results based on binary choice models are also presented using a very special frequency table with empty cells.

6.3 Binary Logit Model of Y_{it} on a Numerical Variable X_{it}

6.3.1 A Comparative Study

Based on a binary choice model of a zero-one problem indicator Y_{it} on a numerical variable X_{it}, let's look at the following three alternative simple binary logit models as a comparative study. The main objective of this study is to get a basic understanding or view on the limitations of statistical models.

6.3.1.1 The Simplest Binary Logit Model

The simplest binary logit model of a zero-one problem indicator Y_{it} on a numerical variable X_{it} would be as follows;

$$\log\left(\frac{p}{1 - p}\right) = c(1) + c(2)^* X_{it} + \mu_{it} \tag{6.16}$$

Note that this model would indicate that all individuals or research objects, at all the time points, are presented as a single binary logit regression. For this reason, this model should be considered to be the worst based on panel data, because it is too simplified. This could never be accepted as a finding in all studies based on panel data with a large number of individuals, since in general, the individuals should consist of at least two groups possessing different characters and behaviors.

In addition to this limitation, refer to special notes and comments on the predicted probabilities presented in Agung (2011, Chapter 7), as well as various alternative binary logit models with a single numerical exogenous variable, such as logarithmic, hyperbolic, the power or the polynomial effects of X on $log(p/(1-p))$, as well as nonlinear binary choice models. For these reasons, I recommend to the reader review those models before reading the following sections, since the following sections present only some selected models.

6.3.1.2   *The Simplest Binary Logit Model by Time*

The simplest binary logit model of a zero-one problem indicator Y_{it} on a numerical variable X_{it} by a dichotomous time variable would be as follows:

$$\log\left(\frac{p}{1-p}\right) = (c(11) + c(12)^*X_{it})^*Dt1 + (c(21) + c(22)^*X_{it})^*Dt2 + \mu_{it} \tag{6.17}$$

where *Dt1* and *Dt2* are the two dummy variables of the time considered.

Compared to the model in (6.16), this model should be considered as better, since the model presents the differential linear effect of the numerical variable X over times, but is for all individuals in the sample.

6.3.1.3   *The Simplest Binary Logit Model by Groups of Individuals*

The simplest binary logit model of a zero-one problem indicator Y_{it} on a numerical variable X_{it} by a dichotomous cause, classification or cell-factor would be as follows:

$$\log\left(\frac{p}{1-p}\right) = (c(11) + c(12)^*X_{it})^*Dc1 + (c(21) + c(22)^*X_{it})^*Dc2 + \mu_{it} \tag{6.18}$$

where *Dc1* and *Dc2* are the two dummy variables of a cause, classification or cell-factor *CF*.

Note that this model can easily be extended to the cell factor with multi-levels, generated by one or more cause or classification factors, and estimates can be obtained as long as there is a sufficiently large number of observations within each cell, at least 10 as *a rule of thumb*, corresponding to the regression with a single numerical independent variable X within each cell.

Therefore, this extended model should be better compared to the model in (6.16), since it can present differential effects of X on the zero-one problem indicator Y between any number of groups of the research objects.

6.3.1.4   *The Simplest Binary Logit Model by* **CF** *and* **T**

The simplest binary logit model of a zero-one problem indicator Y_{it} on a numerical variable X_{it} by dichotomous variables *CF* and the time *T*, would be as follows:

$$\log\left(\frac{p}{1-p}\right) = (c(11) + c(12)^*X_{it})^*D11 + (c(21) + c(22)^*X_{it})^*D12$$
$$+(c(31) + c(32)^*X_{it})^*D21 + (c(41) + c(42)^*X_{it})^*D22 + \mu_{it} \tag{6.19}$$

where *Dit* is defined as $Dit = 1$ if the observed individuals in the cell *(CF = i,T = t)*, and $Dit = 0$ if otherwise.

Note that this model can easily be extended to cell factors with multi-levels. For this reason, this model should be considered the best, since it presents the differential linear effect of the numerical variable X by groups of individuals and times. However, corresponding to the linear effect of X on $log(p/(1 - p))$, refer to the statements presented earlier for the simplest binary logit model in (6.16).

Furthermore, note that the model in (6.19) shows a three-way interaction model without an intercept parameter, since the dummy variable *Dit* in fact represents the product of two dummy variables, namely *(CF = i)*(T = t)*, and then $X^*Dit = X^*(CF = i)^*T = t)$ is a three-way interaction variable or factor. Based on this finding, it can be concluded that a general binary choice model based on panel data would be at least a three-way interaction model, which will be presented in the following sections, using the dummy variables *Dit*, for $i = 1, 2$; and $t = 1, 2$ for illustration.

6.3.2   General Binary Logit Model with a Single Numerical Variable X

The binary logit model in (6.19) can be extended to many binary logit models, which can be estimated using the following general equation specification.

$$Y = F_1(X) \times D11 + F_2(X) \times D12 + F_3(X) \times D21) + F_4(X) \times D22 \tag{6.20}$$

where $F_k(X)$ can be any function of a numerical variable X having a finite number of parameters, for each index k. Therefore the function $F_k(X)$ can be presented as $F_k(X,c(k^*))$. Refer all possible functions presented in Chapter 7 (Agung, 2011). Some of the alternative functions are as follows:

$$\text{The simplest model}: \quad F_k(X) = c(k1) + c(k2)^*X \tag{6.20a}$$

$$\text{Logarithmic model}: \quad F_k(X) = c(k1) + c(k2)^*log(X) \tag{6.20b}$$

$$\text{Hyperbolic model}: \quad F_k(X) = c(k1) + c(k2)^*(1/X) \tag{6.20c}$$

$$\text{Exponential model}: \quad Fk(X) = c(k1) + c(k2)^*X\alpha^{(k)}, \alpha(k) \neq 0 \tag{6.20d}$$
$$= c(k1) + c(2)^*exp(\alpha(k)^*log(X))$$

$$\text{Polynomial model}: \quad F_k(X) = c(k1) + c(k2)^*X + c(k3)^*X^2 + \ldots + c(kn)^*X^n \tag{6.20e}$$

$$\text{Nonlinear model}: \quad F_k(X) = c(k1) + c(k2)^*X^{c(k3)} \tag{6.20f}$$

Note that the four functions $F_1(X)$, $F_2(X)$, $F_3(X)$, and $F_4(X)$ can be different types of functions. For instance, they can be four out of the six functions presented previously such as $F_1(X) = c(11) + c(12)^*X$; $F_2(X) = c(21) + c(22)^*log(X)$; $F_3(X) = c(31) + c(32)^*(1/X)$, and $F_4(X) = c(41) + c(42)^*X + c(43)^*X^2 + c(44)^*X^3$.

So, there are a lot of possible choices but we may never know the true population model. For this reason, when defining a model a researcher should use his/her best judgment, supported by knowledge and experience in various related fields. On the other hand, the estimates obtained are highly dependent on the data. Refer to unexpected estimates or statistical results based on various models, which have been presented as illustrative examples in Agung (2011, 2009a).

Example 6.6   A set of the simplest BLM by groups and times

Figure 6.7 presents the statistical results based on a set of the simplest binary logit model (BLM) using the following equation specification.

$$Y \ @Expand(A, T) \ MBV^*@Expand(A, T) \qquad\qquad (6.21)$$

Based on these results the following notes and findings are presented.

1. Applying the equation specification (ES) is suggested in (6.21), instead of the ES in (6.20), if the BLMs within all groups have the same numerical independent variable(s). See the following illustrative example.
2. The BLM in Figure 6.7 in fact represent a set of six *heterogeneous regression lines* (HRL) of $log(p/(1 - p))$ on *MBV* (market to book value) by the cause or classification factor A and the time T. Therefore, the specific objectives of this *HRL* are as follows:
   2.1   To test the hypotheses on the differences of the linear effects of *MBV* on $log(p/(1 - p))$ between the groups of individuals and time T, which can easily be tested using the Wald test. Do it as an exercise.
   2.2   To compute the scores of forecasting, namely *DINVESF*, using *MBV* as a numerical predictor by the factor A and the time T. For an illustration, Figure 6.8 presents the graphs of *DINVESF* conditional for $(A = 1, T = 1)$ and $(A = 2, T = 2)$ respectively. The steps of the analysis are as follows:
      - With the statistical results on the screen, select *Proc/Forecast . . . OK*, then a graph of *DINVESF* for the whole sample with its statistics shown on the screen.
      - By selecting *File/Save*, the variable *DINVESF* will be inserted in the file.
      - Select the object "*Sample*", a sub-sample of "$A = 1$ and $T = 1$" should be selected.

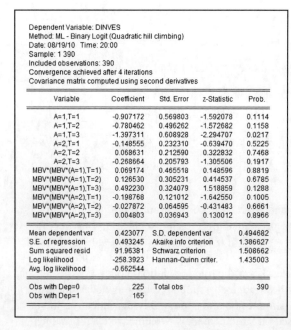

Figure 6.7   *Statistical results based on a BLM using the equation specification in (6.21)*

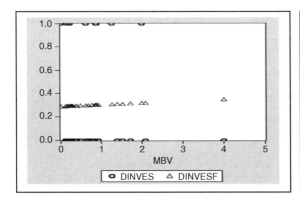 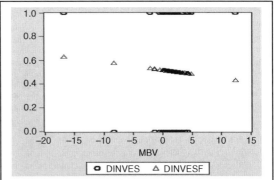

Figure 6.8 *The scatter graph of* DINVES *and* DINVESF *on* MBV *conditional for* (A = 1,T = 1) *and* (A = 2,T = 2)

- Then the scatter graphs of *DINVES* and *DINVESF on MBV*, conditional for *(A = 1,T = 1)* can easily be obtained, as shown in Figure 6.8.
- With the first scatter graph on the screen, the second graph can be obtained by selecting the object "*Sample*", then insert the if condition; "*A = 2 and T = 2*". Similarly, the other four scatter graphs could easily be obtained.

3. Based on the graphs in Figure 6.8, the following notes and comments made.
   3.1 The scores of *DINVESF*, are acceptable in a statistical or probabilistic sense, but in fact they are unobservable or abstract scores of the individuals in the population, with known scores of *MBV*. In fact, the scores of *DINVESF* are not valid for the individuals in the sample, but for the unobserved individuals in the population having the same scores of *MBV* with unknown score of *DINVES*.
   3.2 Note that each individual in the sample has a fixed or known score, that is either *DINVES* = 1 or *DINVES* = 0.
4. For writing and testing the hypotheses on the linear effect differences of *BMV* on *DINVES*, or *log(p/(1 − p))*, Table 6.9 presents the slope parameters of the model by the factors *A* and *T*, namely *C(7)* to *C(12)*. Note that this table in fact represents six simple regressions of *log(p/(1 − p))* on *MBV* with different intercepts, *C(1)* to *C(6)*, and different slopes, *C(7)* to *C(12)*

Example 6.7   Unfamiliar or unexpected statistical results
In order to present a generalized model, Figure 6.9 presents the statistical results based on a BLM of *Y* = *DINVES* on *X1* = *MBV* by the factor *A* and the time *T*. The functions $F_k(X1)$, *k* = 1 to 6 are intentionally selected just to demonstrate there are a lot of possible sets of predicted probabilities, which are unfamiliar for most readers. For this reason, Figure 6.10 presents two scatter graphs of *Y* and its predicted probabilities,

Table 6.9 *Parameters of the BLM in Figure 6.7*

	A = 1			A = 2		
	T = 1	T = 2	T = 3	T = 1	T = 2	T = 3
Intecept	C(1)	C(2))	C(3)	C(4)	C(5)	C(6)
Slope: BMV	C(7)	C(8)	C(9)	C(10)	C(11)	C(12)

Dependent Variable: Y
Method: ML - Binary Logit (Quadratic hill climbing)
Date: 08/20/10   Time: 17:41
Sample: 1 390
Included observations: 390
Convergence achieved after 10 iterations
Covariance matrix computed using second derivatives

	Coefficient	Std. Error	z-Statistic	Prob.
C(10)	-0.216876	1.349603	-0.160696	0.8723
C(11)	-4.508101	6.488637	-0.694769	0.4872
C(12)	5.282973	7.572362	0.697665	0.4854
C(13)	-1.487042	2.393673	-0.621239	0.5344
C(20)	1.178082	1.222908	0.963345	0.3354
C(21)	-3.241312	2.120395	-1.528636	0.1264
C(30)	-66.40238	43.42917	-1.528981	0.1263
C(31)	18.27886	12.08197	1.512904	0.1303
C(40)	-0.312744	0.208030	-1.503363	0.1327
C(41)	-0.004940	0.004739	-1.042390	0.2972
C(50)	105.0989	79.48680	1.322219	0.1861
C(51)	-46.07638	34.56305	-1.333111	0.1825
C(52)	1.302749	0.969798	1.343320	0.1792
C(60)	-0.686979	2.111201	-0.325398	0.7449
C(61)	0.070860	0.352032	0.201288	0.8405

Mean dependent var	0.423077	S.D. dependent var	0.494682
S.E. of regression	0.490664	Akaike info criterion	1.378490
Sum squared resid	90.28184	Schwarz criterion	1.531034
Log likelihood	-253.8056	Hannan-Quinn criter.	1.438959
Avg. log likelihood	-0.650784		

Obs with Dep=0	225	Total obs	390
Obs with Dep=1	165		

Estimation Command:
=========================
BINARY(D=L) Y=(C(10)+C(11)*X1+C(12)*X1^2+C(13)*X1^3)
*D11 + (C(20)+C(21)*1/(X1+1))*D12+(C(30)+C(31)*LOG(X1+
35))*D13+(C(40)+C(41)*X1^3)*D21+(C(50)+C(51)*LOG(X1+
35)+C(52)*LOG(X1+35)^3)*D22+(C(60)+C(61)*(X1+35)^0.5)
*D23

Estimation Equation:
=========================
I_Y = (C(10)+C(11)*X1+C(12)*X1^2+C(13)*X1^3)*D11 + (C
(20)+C(21)*1/(X1+1))*D12+(C(30)+C(31)*LOG(X1+35))*D13+
(C(40)+C(41)*X1^3)*D21+(C(50)+C(51)*LOG(X1+35)+C(52)
*LOG(X1+35)^3)*D22+(C(60)+C(61)*(X1+35)^0.5)*D23

Substituted Coefficients:
=========================
Y=(-0.216876054969-4.50810143568*X1+5.28297260959
*X1^2-1.48704174847*X1^3)*D11 + (1.17808205626-
3.24131157875*1/(X1+1))*D12+(-66.4023771752+
18.2788641139*LOG(X1+35)))*D13+(-0.312743932763-
0.00494035076779*X1^3)*D21+(105.098917471-
46.0763813994*LOG(X1+35)+1.30274877815*LOG(X1+35)^
3)*D22+(-0.686979470598+0.0708598088974*(X1+35)^0.5)
*D23

Figure 6.9 *Statistical results of the MBV of* Y = DINVES *on* X1 = MBV *by the factors* A *and* T

namely *PP_Y* on *X1*, which are quite different from the scatter graphs in Figure 6.8. Based on statistical results and scatter graphs, the following notes and findings are presented.

1. If (1/X1) is used as an independent variable, instead of *1/(X1 + 1)*, the error message "*Division by zero*" is obtained. In fact, it can be applied *1/(X1 + a)* for any $a \neq 0$.

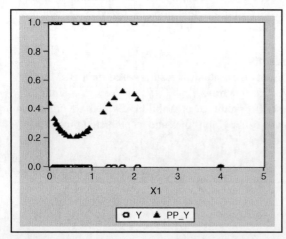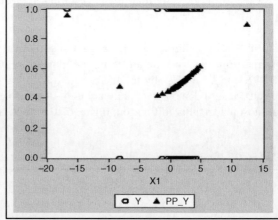

Figure 6.10 *The scatter graphs of* Y = DINVES *and* PP_Y *on* X1 = MBV *conditional for* (A = 1, T = 1) *and* (A = 2, T = 2)

2. If *log(X1)* is used, instead of *log(X1 + 35)*, the error message "*Log of non positive number*" is obtained. And if *X1^0.5* is used, instead of *(X1 + 35)^0.5*, the error message "*Attempt to raise a negative number to a non integer power*" is obtained. In fact, it can be used *(X1 − n)*, for any negative integer $n < Min(X1) = -33.3$.
3. With the statistical results on the screen, the predicted probability variable *PP_Y* is generated using the "*Forecasting Equation*" shown in Figure 6.9, that is:

$$PP\_Y = 1 - @CLOGISTIC(-((C(10) + C(11)^*X1 + C(12)^*X1^2 + C(13)^*X1^3)^*D11$$
$$+ (C(20) + C(21)^*1/(X1 + 1))^*D12 + (C(30) + C(31)^*LOG(X1 + 35))^*D13$$
$$+ (C(40) + C(41)^*X1^3)^*D21 + (C(50) + C(51)^*LOG(X1 + 35)$$
$$+ C(52)^*LOG(X1 + 35)^3)^*D22 + (C(60) + C(61)^*(X1 + 35)^0.5)^*D23))$$

4. Based on the sub-samples *(A = 1,T = 1)* and *(A = 2,T = 2)* respectively, the scatter graphs of *Y = DINVES* and *PROB_Y* on *X1 = MBV* in Figure 6.10 can easily be obtained. Refer to the previous example.
5. Referring to various binary choice models having numerical variables presented in Agung (2011) and these last two examples, then it is almost impossible to declare a certain binary logit models (BLM), which would give a reliable and valid set of predicted probabilities, since many other BLMs can easy be proposed by other researchers. On the other hand, a statistical user or a researcher might say that your BLM is bad or the worst BLM, because your model is (very) different from what he/she has been using so far.
6. Furthermore, note that the scatter graphs of the zero-one indicator *Y* and *PP_Y* on the exogenous numerical variable *X1*, conditional for *(A = i, T = t)*, in fact cannot be used to draw a conclusion as to whether or not *X1* is a good predictor, since *PP_Y* is valid for the unobserved individuals with the same scores of *X1* as the individuals in the sample, but unknown scores of *Y*. Similarly, for all BLMs having two or more numerical independent variables, which will be presented in the following section
7. For the reasons mentioned previously, it can be said that data analysis based on BLMs is important only for scientific exercises, since their predicted probabilities would not be empirically important.
8. For further illustration, Figure 6.11 presents the scatter graphs of *Y* and the forecast *YF* on *X1* conditional for *(A = 1,T = 1)* and *(A = 2,T = 2)*. Note that *YF* and PP_Y clearly show several far outliers. For

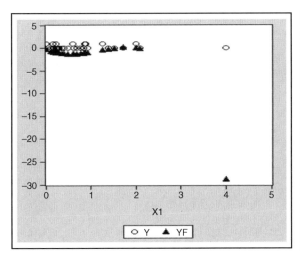
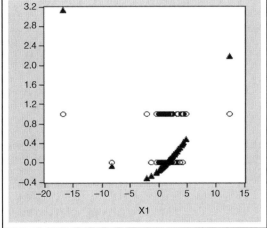

Figure 6.11 *The scatter graphs of* Y = DINVES *and its forecast* YF *on* X1 = MBV *conditional for* (A = 1,T = 1) *and* (A = 2,T = 2)

this reason, in a statistical sense, additional analysis should be done based on a sub-sample without outliers. However, the data analysis is not presented, here. Do it as an exercise.

6.3.3   Special Notes and Comments

1. The binary logit models presented previously are in fact also valid for cross-section over time data.
2. However, to represent the specific characteristics of a panel data, it applying the lag(s) of the exogenous variable $X1_{it}$, such as $X1_{i,t-1}$ for the simplest model is recommended.
3. The data analysis based on the binary logit models can easily be done for the binary probit and extreme value models. Do it as an exercise, and refer to the comparative study between the binary logit, probit and extreme value models, as well as assumptions of their error terms presented in Agung (2011).
4. These notes should be valid also for all binary choice models with two or more numerical exogenous variables, presented in the following examples.

6.4   Binary Logit Model of a Zero-One Indicator Y_{it} on $(X1_{it}, X2_{it})$

The binary logit model of a zero-one problem indicator Y on two numerical endogenous variables $X1$ and $X2$, by the dichotomous factors CF and T, can be estimated using the following general equation specification.

$$Y = F_1(X1, X2) \times D11 + F_2(X1, X2) \times D12$$
$$+ F_3(X1, X2) \times D21 + F_4(X1, X2) \times D22 \tag{6.22}$$

where $F_k(X1, X2)$ can be any function of the numerical variables $X1$ and $X2$ with a finite number of parameters, for each index k. So a lot of binary logit models could be subjectively defined or proposed by a researcher.

6.4.1   The Simplest Possible Function

Note that in a theoretical sense the *linear-effect* of a numerical exogenous (cause, source or upstream) variable on an endogenous variable depends on the other exogenous variable. For this reason, then the interaction $X1^*X2$ should be used as an independent variable, to indicate that the *linear effect of X1 (or X2) on Y depends on X2 (or X1)*. So, the simplest general function $F_k(X1, X2)$ has the following form.

$$F_k(X1, X2) = c(k0) + c(k1)^*X1 + c(k2)^*X2 + c(k3)^*X1^*X2 \tag{6.23}$$

In practice, however, a reduced function might be an acceptable model, in a statistical sense, which is highly dependent on the data that happens to be selected by or available to researchers. The possible reduced functions of the function (6.23) are as follows:

$$F_k(X1, X2) = c(k0) + c(k1)^*X1 + c(k3)^*X1^*X2 \tag{6.23a}$$

$$F_k(X1, X2) = c(k0) + c(k2)^*X2 + c(k3)^*X1^*X2 \tag{6.23b}$$

$$F_k(X1, X2) = c(k0) + c(k3)^*X1^*X2 \tag{6.23c}$$

$$F_k(X1, X2) = c(k0) + c(k1)^*X1 + c(k2)^*X2 \tag{6.23d}$$

Example 6.8   Application of BLMs in (6.23)

Table 6.10 presents a summary of the statistical results based on a full BLM of Y on $X1$, $X2$ and $X1^*X2$, by two dichotomous factors A and T, and three of its reduced models, using the model in (6.23).

Based on this summary, the following notes and comments are made.

1. Based on the full model, the following findings and notes can be presented,
    1.1   The four BLMs within the four cells or groups have shown that the parameter estimates or the regressions obtained are highly dependent on the data sets. Note that each of the independent variables of the model in $(A = 1, T = 1)$ has a significant adjusted effect on Y or $log(p/(1 - p))$. On the other hand, each of the independent variables of the model in $(A = 2, T = 2)$ has an insignificant adjusted effect on Y or $log(p/(1 - p))$.
    1.2   Without doing any testing, it can be said descriptively that interaction $X1^*X2$ has different effects on Y between the four cells considered. If the test hypothesis should be done, then it can easily be done using the Wald test. We should find that they have significant differences. Do it as an exercise.
    1.3   In the first two cells, namely cells $(A = 1, T = 1)$ and $(A == 1, T = 2)$, $X1^*X2$ has a significant effect on Y. So it can be concluded that the effect of $X1$ on Y is significantly dependent on $X2$. But in the other cells, $X1^*X2$ has an insignificant effect on Y.
    1.4   Note that there is nothing wrong with the model, even if $X1^*X2$ has an insignificant effect on Y within all cells considered.
    1.5   Special findings based on the regression within the cell $(A = 1, T = 1)$. The logit regression function can be written as follows:

$$log(p/(1 - p)) = 3.423 - 2.789X1 + (-31.923 + 25.327X1)X2$$

This function shows that the linear effect of $X2$ on $log(p/(1 - p))$ depends on $(-31.923 + 25.237X1)$, which increases with increasing positive values of $X1$. In other words,

Table 6.10   *A summary of the statistical results based on a full model in (6.22), and three out of its possible reduced models*

		Full Model		Red. Model 1		Red. Model 2		Red. Model 3	
Par	Variable	Coef.	Prob.	Coef.	Prob.	Coef.	Prob.	Coef.	Prob.
C(10)	Intercept	3.423	0.111	3.423	0.111	3.423	0.111	3.423	0.111
C(11)	X1	−2.789	0.073	−2.789	0.073	−2.789	0.073	−2.789	0.073
C(12)	X2	−31.923	0.039	−31.923	0.039	−31.923	0.039	−31.923	0.039
C(13)	X1*X2	25.327	0.049	25.327	0.049	25.327	0.049	25.327	0.049
C(20)	Intercept	−1.382	0.160	−1.670	0.047	−1.784	0.016	−1.848	0.010
C(21)	X1	−0.250	0.613			−0.113	0.768		
C(22)	X2	−1.926	0.570	−1.026	0.710				
C(23)	X1*X2	10.268	0.036	9.156	0.035	9.442	0.039	9.009	0.037
C(30)	Intercept	−0.154	0.520	−0.151	0.525	−0.333	0.102	−0.149	0.523
C(31)	X1	−0.197	0.115	−0.198	0.113			−0.199	0.101
C(32)	X2	0.026	0.913						
C(33)	X1*X2	−0.342	0.475	−0.336	0.481	−0.361	0.413		
C(40)	Intercept	0.098	0.658	0.078	0.712	0.081	0.707	0.061	0.764
C(41)	X1	−0.020	0.766			−0.020	0.771		
C(42)	X2	−0.436	0.363	−0.429	0.368				
C(43)	X1*X2	−0.496	0.308	−0.510	0.289	−0.514	0.275	−0.526	0.258

the effect of $X2$ on $log(p/(1-p))$ increases with increasing positive values of $X1$, and $X2$ has no effect for $X1 = 31.923/25.237$.

1.6   Note that for all discrete values of $X1$, the graph of the function in a two-dimensional coordinates with axes $log(p/(1-p))$ and $X2$, is a set of heterogeneous regression lines, with slopes $(-31.923 + 25.237X1)$, and intercepts $(3.423 - 2.789X1)$. In three-dimensional coordinates with axes $log(p/(1-p))$, $X1$ and $X2$, the curve would be an hyperboloid curve.

1.7   Other type of findings based on the regression within the cell $(A=2, T=1)$. The function can be written as follows:

$$log(p/(1-p)) = -0.154 - 0.197X1 + (0.026 - 0.342X1)X2$$

This function shows that the linear effect of $X2$ on $log(p/(1-p))$ depends on $(0.026 - 0.342X1)$ which decreases with increasing positive values of $X1$. In other words, the effect of $X2$ on $log(p/(1-p))$ decreases with increasing positive values of $X1$, up to the value of $X1 = 0.026/0.342 = 0.076\ 023$, where $X2$ has no effect. Afterward, the linear effect of $X2$ will be negative.

2.   The full model has seven out of 12 insignificant slope parameters, so a reduced model(s) should be explored. Since it is defined that the effect of $X2$ on Y is dependent on $X1$, then the interaction factor $X1^*X2$ should be used as the independent variable of the reduced models within the four groups $(A=i, T=t)$. Corresponding to the three reduced model presented in Table 6.10, the following notes and comments are presented.

2.1   For the full and reduced models within the cell $A=1$ and $T=2$.

Based on the full model, $X1^*X2$ has a significant effect on Y with a p-value $= 0.036$, then the full model is acceptable to show that the effect of $X2$ on Y is significantly dependent on $X1$. This is also the case based on the three reduced models. Which one would you prefer?

2.2   For the full and reduced models within the cell $A=2$ and $T=1$.

In this case, since $X1^*X2$ has an insignificant effect on Y based on the full model as well as the first two reduced models with large p-values > 0.4. However, the third reduced model is an additive model (without $X1^*X2$) with a single independent variable $X1$, which, at a significance level of $\alpha = 0.10$, has a significant negative effect on $log(p/(1-p))$ with a p-value $= 0.101/2 = 0.0505 < \alpha$.

2.3   For the full and reduced models within the cell $A=2$ and $T=2$.

In this case, $X1^*X2$ has insignificant effects on $log(p/(1-p))$ based on the full model as well as the three reduced models, however, the p-values < 0.30 for the reduced models. Note that using a significance level of $\alpha = 0.15$ (Lapin, 1973), it can be concluded that $X1^*X2$ has a significant negative effect on $log(p/(1-p))$ based on the three reduced models, with p-values less than 0.15, namely $0.289/2 = 0.1445$; $0.275/2 = 0.1375$; and $0.258/2 = 0.129$, respectively. So, which reduced model would you prefer?

Example 6.9   Alternative equation specification of a BLM

As an alternative equation specification of the full model presented in Table 6.10, the following equation could be applied.

$$Y\ @Expand(A, T)\ X1^*@expand(A, T)\ X2^*@Expand(A, T)$$
$$X1^*X2^*@Expand(A, T)$$

(6.24)

However, I recommend to apply this equation specification if and only if no reduced model(s) should be explored. For writing and testing the hypotheses using the Wald test, the parameters of a specific model are presented in Table 6.11.

Table 6.11 *Parameters of a specific model in the equation specification (6.24)*

	A = 1			A = 2		
	T = 1	T = 2	T = 3	T = 1	T = 2	T = 3
Intercept	C(1)	C(2)	C(3)	C(4)	C(5)	C(6)
X1	C(7)	C(8)	C(9)	C(10)	C(11)	C(12)
X2	C(13)	C(14)	C(15)	C(16)	C(17)	C(18)
X1*X2	C(19)	C(20)	C(21)	C(22)	C(23)	C(24)

6.4.2   Functions for Binary Logit Translog Models

6.4.2.1   *Binary Logit Translog Linear Model*

For the binary logit translog linear models, the functions $F_k(X1,X2)$, will be as follows:

$$F_k(X1, X2) = c(k0) + c(k1)^* log(X1) + c(k2)^* log(X2) \tag{6.25}$$

6.4.2.2   *Binary Logit Translog Quadratic Model*

For the binary logit translog quadratic models, the functions $F_k(X1,X2)$, will be as follows:

$$\begin{aligned} F_k(X1, X2) = c(k0) + c(k1)^* log(X1) + c(k2)^* log(X2) + c(k3)^* log(X1)^\wedge 2 \\ + c(k4)^* log(X1)^* log(X2) + c(k5)^* log(X)^\wedge 2 \end{aligned} \tag{6.26}$$

In practice, unpredictable statistical results, as well as unexpected reduced model(s) would be obtained, because of the impact of multicollinearity between the independent variables, which are highly dependent on the data used, as demonstrated in the following example. One of the simple reduced models uses the functions in (6.25).

Example 6.10   An application of a binary logit quadratic translog model
For an illustration, Table 6.12 presents a summary of the statistical results based on a reduced binary logit model using the function in (6.26), Based on these results, the following notes and comments are presented.

1. The model should be considered acceptable in a statistical sense, because each of the independent variables has either positive or negative adjusted effects at significance levels of 0.05, 0.10 or 0.15.

Table 6.12 *The statistical results summary of a reduced logit model using the function (6.26)*

	Group D11 = 1		Group D12 = 1		Group D21 = 1		Group D22 = 1	
Variable	Coef.	Prob.	Coef.	Prob.	Are Coef.	Prob.	Coef.	Prob.
Intercept	244 891.8	0.046	−7558.7	0.156	24 202.6	0.285	14 886.6	0.259
log(X1 + 20)	−80 125.3	0.047	4526.7	0.171	−3.8	0.103	−15.8	0.234
log(X2 + 40)	−66 336.8	0.046	161.9	0.227	−13 050.9	0.286	−8027.9	0.260
log(X1 + 20)^2			−735.8	0.174			3.1	0.200
log(X1 + 20)*log(X2 + 40)	21 704.5	0.047						
log(X2 + 40)^2					1760.1	0.288	1083.7	0.260

Dependent Variable: Y
Method: ML - Binary Logit (Quadratic hill climbing)
Date: 09/11/10   Time: 10:36
Sample: 1 390 IF T<3 AND D11=1
Included observations: 30
Convergence achieved after 7 iterations
Covariance matrix computed using second derivatives

Variable	Coefficient	Std. Error	z-Statistic	Prob.
C	11.63457	9.730875	1.195635	0.2318
LOG(X1)	15.74565	10.22613	1.539746	0.1236
LOG(X2)	11.63796	8.746444	1.330594	0.1833
LOG(X1)^2	1.529695	1.732723	0.882827	0.3773
LOG(X1)*LOG(X2)	7.207781	4.702046	1.532903	0.1253
LOG(X2)^2	2.643619	1.906863	1.386371	0.1656

McFadden R-squared	0.377219	Mean dependent var	0.300000
S.D. dependent var	0.466092	S.E. of regression	0.405995
Akaike info criterion	1.160869	Sum squared resid	3.955975
Schwarz criterion	1.441108	Log likelihood	-11.41303
Hannan-Quinn criter.	1.250520	Restr. log likelihood	-18.32593
LR statistic	13.82579	Avg. log likelihood	-0.380434
Prob(LR statistic)	0.016755		

Obs with Dep=0	21	Total obs	30
Obs with Dep=1	9		

Dependent Variable: Y
Method: ML - Binary Logit (Quadratic hill climbing)
Date: 09/11/10   Time: 10:41
Sample: 1 390 IF T<3 AND D11=1
Included observations: 30
Convergence achieved after 6 iterations
Covariance matrix computed using second derivatives

Variable	Coefficient	Std. Error	z-Statistic	Prob.
C	9.524785	8.332864	1.143039	0.2530
LOG(X1)	11.10992	6.883289	1.614043	0.1065
LOG(X2)	9.761047	7.582948	1.287236	0.1980
LOG(X1)*LOG(X2)	5.609891	3.617682	1.550687	0.1210
LOG(X2)^2	2.274446	1.672553	1.359865	0.1739

McFadden R-squared	0.354397	Mean dependent var	0.300000
S.D. dependent var	0.466092	S.E. of regression	0.406713
Akaike info criterion	1.122085	Sum squared resid	4.135384
Schwarz criterion	1.355618	Log likelihood	-11.83128
Hannan-Quinn criter.	1.196794	Restr. log likelihood	-18.32593
LR statistic	12.98930	Avg. log likelihood	-0.394376
Prob(LR statistic)	0.011328		

Obs with Dep=0	21	Total obs	30
Obs with Dep=1	9		

Figure 6.12   *Statistical results based on the binary logit translog quadratic model, and its reduced model, using the sub-sample* D11 = *1*

2. This table presents four different types of binary logit reduced model with unpredictable statistical results, which are highly dependent on the sub-data sets. The joint effects of the independent variables of each model can easily be tested using the Wald test.

3. The variable $log(X1 + 20)$ and $log(X2 + 40)$ should be used, because $X1$ and $X2$ have negative observed values of -16.7 and -38.4, respectively, based on the whole sample. However, if the data analysis is

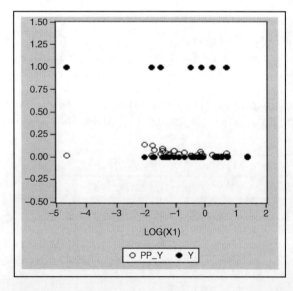
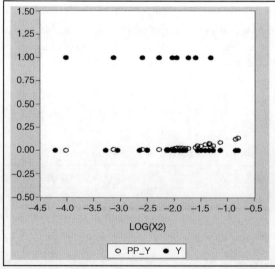

Figure 6.13   *Scatter graphs of the predicted probabilities* PP_Y *of the reduced model in Figure 6.14, and* Y, *on* log(X1) *and* log(X2), *respectively*

done using each of the sub-samples, namely the sub-samples $D11=1$, $D12=1$, $D21=1$ and $D22=1$, other findings should be obtained. For illustration, Figure 6.12 presents the statistical results using the variables $log(X1)$ and $log(X2)$, based on the sub-sample $D11=1$.

4. Corresponding to scatter graphs in Figure 6.10, for a more advanced illustration, Figure 6.13 presents the scatter graphs of the predicted probabilities of the reduced model, PP_Y, and Y on $log(X1)$ and $log(X2)$. In fact, the graph of the reduced model should be presented in a three-dimensional space. Based on these graphs, again the reliability and validity of the predicted probabilities can be argued (refer to Agung, 2011), even though each of the independent variables has a significant positive adjusted effect, at the significance level of 0.10.

6.4.2.3 *Other Types of General Functions*

The first type of general function is as follows:

$$F_k(X1, X2) = F_{k1}(X1) + F_{k2}(X1)^* X2 \tag{6.27a}$$

where $F_{k1}(X1)$ and $F_{k2}(X1)$ are the functions of a single variable $X1$ having a finite number of parameters.

Note that by using these functions, it is assumed that the linear effect of $X2$ on $log(p/(1-p))$ depends on $X1$ in the form of a function $F_{k2}(X1)$. Refer to the six specific functions presented in Section 6.3.1.2, out of a lot of possible functions for both $F_{k1}(X1)$ and $F_{k2}(X1)$.

Instead of the linear effect of $X2$, other types of its effects might be considered. For instance, the logarithmic, hyperbolic and exponential effects, by using the following functions, respectively.

$$F_k(X1, X2) = F_{k1}(X1) + F_{k2}(X1)^* log(X2) \tag{6.27b}$$

$$F_k(X1, X2) = F_{k1}(X1) + F_{k2}(X1)^* (1/X2) \tag{6.27c}$$

$$F_k(X1, X2) = F_{k1}(X1) + F_{k2}(X1)^* X2^{\alpha(k)}, for \ a \ certain \ number \ of \ \alpha(k) \neq 0 \tag{6.27d}$$

$$F_k(X1, X2) = F_{k1}(X1) + F_{k2}(X1)^* X2^{C(k^*))}, where \ C(k^*) \ is \ a \ parameter \tag{6.27e}$$

6.5 Binary Choice Model of a Zero-One Indicator Y_{it} on $(X1_{it}, X2_{it}, X3_{it})$

The binary choice model of a zero-one problem indicator Y on three numerical endogenous variables $X1$, $X2$ and $X3$, by the dichotomous factors CF and T, can be estimated using the following general equation specification

$$\begin{aligned} Y = {} & F_1(X1, X2, X3) \times D11 + F_2(X1, X2, X3) \times D12 \\ & + F_3(X1, X2, X3) \times D21 + F_4(X1, X2, X3) \times D22 \end{aligned} \tag{6.28}$$

where $F_k(X1,X2,X3)$ can be any function of the numerical variables $X1$, $X2$ and $X3$ having a finite number of parameters, for each index k. So there would be a lot of binary logit models that could be subjectively defined or proposed by a researcher, with several specific general functions as follows:

6.5.1   The Linear Effect of $X3$ on Y Depends on $X1$ and $X2$

If it is defined that the linear effect of $X3$ on Y depends of $X1$ and $X2$, the function $F_k(X1,X2,X3)$ in (6.28) can be presented as follows:

$$F_k(X1,X2,X3) = F_{k1}(X1,X2) + F_{k2}(X1,X2) \times X3 \qquad (6.29)$$

where $F_{k1}(X1,X2)$ and $F_{k2}(X1,X2)$ can be any functions of $(X1,X2)$ with a finite number of parameters. Refer to various alternative functions of $F(X1,X2)$, as well as $F(X1)$, presented in previous sections.

Example 6.11   An application of the function in (6.28)

For an illustration, Figure 6.14 presents the statistical results of a binary logit model using the following equation specification. Based on these results, the following notes and comments are presented.

$$
\begin{aligned}
Y = &((C(10) + C(11)^*X1 + C(12)^*X2 + C(13)^*X1^*X2) \\
&+ (C(14) + C(15)^*X1 + C(16)^*X2 + C(17)^*X1^*X2)^*X3)^*D11 \\
&+ ((C(20) + C(21)^*X1 + C(22)^*X2 + C(23)^*X1^*X2) \\
&+ (C(24) + C(25)^*X1 + C(26)^*X2 + C(27)^*X1^*X2)^*X3)^*D12 \\
&+ ((C(30) + C(31)^*X1 + C(32)^*X2 + C(33)^*X1^*X2) \\
&+ (C(34) + C(35)^*X1 + C(36)^*X2 + C(37)^*X1^*X2)^*X3)^*D21 \\
&+ ((C(40) + C(41)^*X1 + C(42)^*X2 + C(43)^*X1^*X2) \\
&+ (C(44) + C(45)^*X1 + C(46)^*X2 + C(47)^*X1^*X2)^*X3)^*D22
\end{aligned}
\qquad (6.30)
$$

Dependent Variable: Y
Method: Least Squares
Date: 09/09/10   Time: 17:56
Sample: 1 390 IF T<3
Included observations: 260
Y=((C(10)+C(11)*X1+C(12)*X2+C(13)*X1*X2)+(C(14)+C(15)*X1+C(16)*X2
+C(17)*X1*X2)*X3)*D11+((C(20)+C(21)*X1+C(22)*X2+C(23)*X1*X2)
+(C(24)+C(25)*X1+C(26)*X2+C(27)*X1*X2)*X3)*D12+((C(30)+C(31)
*X1+C(32)*X2+C(33)*X1*X2)+(C(34)+C(35)*X1+C(36)*X2+C(37)*X1
*X2)*X3)*D21+((C(40)+C(41)*X1+C(42)*X2+C(43)*X1*X2)+(C(44)
+C(45)*X1+C(46)*X2+C(47)*X1*X2)*X3)*D22

	Coefficient	Std. Error	t-Statistic	Prob.
C(10)	1.128820	0.653882	1.726337	0.0856
C(11)	-0.148872	0.281336	-0.529161	0.5972
C(12)	-2.840854	3.156724	-0.899937	0.3691
C(13)	-1.671839	3.608195	-0.463345	0.6436
C(14)	-0.315172	0.541391	-0.582152	0.5610
C(15)	-0.136682	0.275244	-0.496585	0.6200
C(16)	0.457701	2.299188	0.199071	0.8424
C(17)	2.566999	2.130600	1.204825	0.2295
C(20)	0.751273	0.500729	1.500360	0.1349
C(21)	-0.960445	0.484595	-1.981955	0.0487
C(22)	-4.218221	1.976177	-2.134536	0.0339
C(23)	7.101660	2.510476	2.828810	0.0051
C(24)	-0.173240	0.519491	-0.333481	0.7391
C(25)	0.790889	0.435175	1.817405	0.0705
C(26)	2.506440	1.475228	1.699019	0.0907
C(27)	-4.830421	1.924397	-2.510096	0.0128

C(30)	0.467902	0.072226	6.478294	0.0000
C(31)	-0.044814	0.025230	-1.776199	0.0770
C(32)	-0.157105	0.129658	-1.211691	0.2269
C(33)	-0.094388	0.168923	-0.558766	0.5769
C(34)	-0.046406	0.040153	-1.155744	0.2490
C(35)	0.013154	0.022781	0.577422	0.5642
C(36)	0.199348	0.135652	1.469555	0.1431
C(37)	0.073278	0.186961	0.391945	0.6955
C(40)	0.400980	0.068660	5.840037	0.0000
C(41)	-0.018181	0.018083	-1.005421	0.3158
C(42)	-0.011183	0.082443	-0.135644	0.8922
C(43)	-0.425119	0.145897	-2.913830	0.0039
C(44)	0.062990	0.036917	1.706233	0.0893
C(45)	0.022062	0.012195	1.809120	0.0717
C(46)	-0.004554	0.085818	-0.053065	0.9577
C(47)	0.313910	0.107712	2.914332	0.0039

R-squared	0.206924	Mean dependent var	0.430769
Adjusted R-squared	0.099094	S.D. dependent var	0.496139
S.E. of regression	0.470916	Akaike info criterion	1.446542
Sum squared resid	50.56163	Schwarz criterion	1.884780
Log likelihood	-156.0505	Hannan-Quinn criter.	1.622720
Durbin-Watson stat	2.156086		

Figure 6.14   *Statistical results based on a binary logit model using the equation specification in (6.30)*

1. Note that these results demonstrate that the regression with the same set of independent variables obtained is highly dependent on the data set used. Even though many independent variables have large *p*-values within each cell, if the differential effects of the same set of independent variables should be studied, then the Wald test has to be applied based on the full model in Figure 6.14.

2. With seven numerical independent variables within each of the four cells, it is common to have many independent variables with insignificant adjusted effects, since the main factors *X1*, *X2* and *X3*, and their interactions are highly correlated. So, in a statistical sense, a reduced model should be explored, using the trial-and-error method. However, it will not be done here.

3. In order to obtain a reduced model, it is easier (and recommended) to do the analysis based on each of the four sub-samples, namely the sub-samples $D11 = 1$, $D12 = 1$, $D21 = 1$ and $D22 = 1$, so that the *p*-value of each independent variable of the reduced model is strictly less than 0.20 or 0.25 (Hosmer and Lemeshow, 2000). Note that if an independent has a *p*-value < 0.20, then at the significance level of $\alpha = 0.10$, the independent variable has either a positive or negative adjusted effect on the dependent variable. For illustration, Figure 6.15 presents the statistical results based on a reduced model using the sub-sample $D11 = 1$. Note that corresponding to the parameter *C(11)*, at the significance level of $\alpha = 0.15$ (Lapin, 1973), it can be concluded that the independent variable *X1* has a positive adjusted effect on the dependent variable, with a *p*-value $= 0.2627/2 = 0.13\,135 < 0.15$.

4. Note the reduced model obtained by deleting the main factor *X3*, even has the smallest *p*-value compared to the interaction factors *X1*X3*, *X2*X3* and *X1*X2*X3*, since it is defined that the linear effect of *X3* on *Y* depends on *X1* and *X2*. This process has demonstrated that theoretical concept or personal judgment should be used to explore a reduced model or to delete an independent variable from a full model, it is not always based on the larger or largest *p*-value.

```
Dependent Variable: Y
Method: ML - Binary Logit (Quadratic hill climbing)
Date: 09/10/10   Time: 09:12
Sample: 1 390 IF T<3 AND D11=1
Included observations: 30
Convergence achieved after 7 iterations
Covariance matrix computed using second derivatives
```

	Coefficient	Std. Error	z-Statistic	Prob.
C(10)	7.614257	4.053038	1.878654	0.0603
C(11)	2.592260	2.314610	1.119956	0.2627
C(13)	-106.7565	61.29547	-1.741670	0.0816
C(15)	-9.217079	4.840045	-1.904338	0.0569
C(16)	-65.32421	35.97035	-1.816057	0.0694
C(17)	153.0837	75.58166	2.025408	0.0428

McFadden R-squared	0.535172	Mean dependent var	0.300000
S.D. dependent var	0.466092	S.E. of regression	0.349444
Akaike info criterion	0.967894	Sum squared resid	2.930668
Schwarz criterion	1.248133	Log likelihood	-8.518403
Hannan-Quinn criter.	1.057545	Restr. log likelihood	-18.32593
LR statistic	19.61505	Avg. log likelihood	-0.283947
Prob(LR statistic)	0.001476		

Obs with Dep=0	21	Total obs	30
Obs with Dep=1	9		

Figure 6.15   *Statistical results based on a reduced model using the sub-sample D11 = 1*

5. Furthermore, corresponding to the variables *X1*, *X2* and *X1*X2*, *X2* with the smallest *p*-value also is deleted from the model. On the other hand, one may consider finding other possible reduced model(s), such as by deleting *X1*, instead of *X2*, or by deleting both *X1* and *X2*. Do this as an exercise.
6. Finally, a summary of the statistical results based on all alternative models should be presented in a research report, such as that in Table 6.10.

6.5.2   Alternative Effects of *X3* on *Y* Depends on *X1* and *X2*

Similar to the functions in (6.27b to (6.27d)), alternative effects of *X3* on *Y* also can be considered. Then the function $F_k(X1,X2,X3)$ in (6.28) may have the following general forms.

$$F_k(X1,X2,X3) = F_{k1}(X1,X2) + F_{k2}(X1,X2)^* log(X3) \tag{6.31a}$$

$$F_k(X1,X2,X3) = F_{k1}(X1,X2) + F_{k2}(X1,X2)^*(1/X3) \tag{6.31b}$$

$$F_k(X1,X2,X3) = F_{k1}(X1,X2) + F_{k2}(X1,X2)^* X3^{\alpha(k)}, \alpha(k) \neq 0 \tag{6.31c}$$

$$F_k(X1,X2,X3) = F_{k1}(X1,X2) + F_{k2}(X1,X2)^* X3^{C(k^*))}, C(k^*) \; is \; a \; parameter \tag{6.31d}$$

6.6   Binary Choice Model of a Zero-One Indicator Y_{it} on $(X1_{it}, \ldots, Xh_{it}, \ldots)$

As the extension of binary choice models presented using the general equation specification in (6.28), the following general equation specification can be used to estimate the binary choice models of a zero-one indicator *Y* on a numerical multivariate *X* = (*X1*, . . . , *Xh*, . . .) by the dichotomous factors *CF* and the time *T*.

$$Y = F_1(X)^* D11 + F_2(X)^* D12 + F_3(X)^* D21 + F_4(X)^* D22 \tag{6.32}$$

where $F_k(X)$, for each *k* = 1, 2, 3 and 4; can be any function of the multivariate *X* = (*X1*, . . . , *Xh*, . . .) with a finite number of parameters.

Note that in fact the function $F_k(X_{it}) = F_k(X1_{it}, \ldots, Xh_{it}, \ldots)$ can be a function of the variable Xh_{it} and all its transformed or derived variables, for a certain Xh_{it}, some or all Xh_{it}s. Furthermore, referring to all possible functions $F_k(X)$, $F_k(X1,X2)$ and $F_k(X1,X2,X3)$, presented in previous sections, there would be a lot of possible explicit functions for $F_k(X) = F_k(X1, \ldots, Xh, \ldots)$.

6.7   Special Notes and Comments

Corresponding to a lot of possible functions of $F_k(X_{it})$ and $F_k(X_{it}, X_{i,t-1}, \ldots, X_{i,t-p})$, then a huge number of possible binary choice models could be defined or proposed. Therefore, it is not an easy task to select or define the best possible function, since we never know the true population model (see Section 2.14 in Agung, 2009a). For this reason, the following basic points should be taken into consideration.

By having a set of exogenous (cause, source or up-stream) variables, in general, there should be a variable, say Xh_1, which can easily be defined in a theoretical sense, and has its effect on the zero-one indicator *Y* which should depend on at least one of the other variables, say Xh_2. Therefore, the function $F_k(X)$, *k* = 1,2, . . . , should have at least a two-way interaction term, namely $Xh_1^* Xh_2$ in a mathematical sense.

In addition, if there are three variables, namely Xh_1, Xh_2 and Xh_3 which are defined to be completely correlated then the function $F_k(X)$ should have the term $Xh_1^*Xh_2^*Xh_3$ or the three two-way interactions: $Xh_1^*Xh_2$, $Xh_1^*Xh_3$ and $Xh_2^*Xh_3$ (see Agung, 2011, 2009a).

In order to have guidance for writing a statistical model, constructing a path diagram to present the multiple causal association, structural or seemingly causal association between any set of the research variables is recommended. Based on such path diagram statistical models, either additive or interaction, these models can be easily written. Refer to various alternative path diagrams presented in previous chapters, as well as in Agung (2011, 2009a), as the basis for writing alternative statistical models, including binary choice models.

However, one should always remember that unexpected statistical results and error messages might be obtained because of the unknown or unpredictable impact of multicollinearity between independent variables, highly dependent on the data used. It has been found that an important cause factor, in a theoretical sense, can have an insignificant adjusted effect on the corresponding dependent variable but the less or least important variable has a significant adjusted effect.

On the other hand, an error message could be obtained, even though the model is an acceptable model in a theoretical sense. So the trial-and-error method should be applied in order to obtain an acceptable model in both theoretical and statistical senses.

7

Advanced General Choice Models

7.1 Introduction

Based on my observations, I would say that most of the models presented in various journals aim to study the adjusted linear effects of each exogenous (independent) variable on the corresponding endogenous variable(s). For illustrations, refer to the models in (7.61) to (7.64), where all independent variables are linear or have a power of one. In fact, the effects of a numerical exogenous variable on a zero-one indicator should be nonlinear, or a piece-wise relationship, since the observed individuals should consist of at least two groups with different or distinct characteristics. Refer to Agung (2011, Chapter 7).

The nonparametric graph of a zero-one problem indicator on a numerical exogenous (cause, source or upstream) variable can be used as a guide to develop alternative polynomial or nonlinear models. For this reason, the following sections present the nonparametric and advanced parametric BCMs as the extension of the models presented in previous chapter, and in Agung (2011). For illustration, the panel data used is the Data_Sesi.wf1, as presented in Figure 7.1 with 2284 firm-years from 2002–2009. Triasesiarti (Sesi) is one of my advisees for her dissertation. Note that this data contains three dummy variables and 18 numerical variables, including the time t. In this case, the dummy variable: *D_Dividend*, can be used as an independent variable for a lot of binary choice models, by selecting sets of a certain number of independent variables out of the 20 other variables. Do you want to use all variables in a binary choice model? And do you think such a model would be the best or true population model?

Many students – perhaps almost all – think that by using a large number of variables in a model it can then explain or achieve all their study objectives. In fact, using a large number of independent variables will obtain uncertain parameter estimates because of the unpredictable impact of correlations or multicollinearity between independent variables – see Agung (2009a: Section 2.14) and various illustrative examples presented in Agung (2011, 2009a).

For comparison, many papers in various journals also present models with a large number of variables, such as Francis and Martin (2010) who present a model with 23 independent variables or 24 parameters, Hameed, et al. (2010) present a model with 23 parameters, Li, et al. (2010) present a model with 18 parameters, and Suriawinata (2004) presents a model with 17 parameters.

With regard to the models with such a large number of variables, it is not an easy task to present the best possible theoretical system equation model or seemingly causal model in the form of a path

Panel Data Analysis Using EViews, First Edition. I Gusti Ngurah Agung.
© 2014 John Wiley & Sons, Ltd. Published 2014 by John Wiley & Sons, Ltd.
Companion website: www.wiley.com/go/panel_data

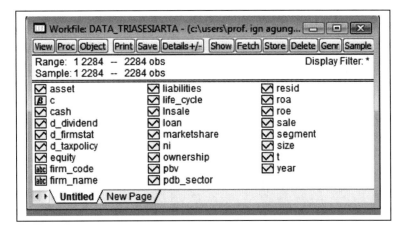

Figure 7.1 *The data Triasesiarti.wf1*

diagram, which is a very important guide diagram for writing interaction models (Agung, 2011, 2009a, 2006, 2004). Discussing interaction models, Siswantoro and Agung (2010) present a paper on "The Importance of the Effect of Exogenous Interaction Factors on Endogenous Variables in Accounting Models", and Wooldridge (2002; 132 and 170) states that "it is prudent to allow different intercepts across strata, and even different slopes in some cases". On page 170, he states "For example, in the equation (7.67) we can interact *female*, with the time dummy variables to see whether the productivity of *female* have changed over time, or we can interact *educ_{it}* and *computer_{it}* to allow the return to computer usage to depend on level of education".

So I am very confident that interaction models should be applied for almost all models that have at least two independent variables, more so for those with categorical and numerical independent variables. On the other hand, additive models should be considered inappropriate and are not recommended, some additive models can be the worst in a theoretical sense – see those in (6.11) and Agung (Agung 2009a; 183–184 and 190–191).

The following sub-sections present some special selected general choice models with time dummy variables.

7.2 Categorical Data Analyses

Nominal and ordinal categorical data analyses are considered to be nonparametric – see Agresti (1984, 1990), Bishop, et al. (1976), and Agung (2011, 2004). See the following nonparametric data analysis.

7.2.1 Multi-Factorial Binary Choice Models

Many variables in the data set could be considered as the cause (up-stream or source) factors of *D_Dividend*, either categorical or numerical variables. One of those variables is the numerical variable *SALE* which should be considered to be the most important cause (source or up-stream), since *SALE* should lead the sustainability of firms, and should be the up-stream or cause factor of various firms' indicators, such as net-income, cash-flow, ROA, ROE, ROI and others. However, for illustration, the numerical variable *SALE* will be transformed to an ordinal variable with two or more levels. The simplest one is an ordinal variable with two levels, namely *O2Sale*, which will be generated using the median as the cutoff point, as follows:

$$O2\ Sale = 1 + 1^*(Sale >= @quantile(Sale, 0.50)) \tag{7.1}$$

Example 7.1   The effect of O2Sale on D_Dividend by time
Figure 7.2 presents the statistical results based on a binary logit model (BLM) using the following equation specifications.

$$D\_Dividend\ @Expand(O2Sale, T) \tag{7.2}$$

$$D\_Dividend\ C\ @Expand(T, @Droplast)$$
$$@Expand(T)^*@Expand(O2Sale, @Droplast) \tag{7.3}$$

Based on these results the following findings and notes are presented.

1. Note that both equation specifications can be used also to conduct the data analysis based on the binary probit model (BPM), binary extreme value model (BEM) and the ordinal choice model.
2. Both outputs in Figure 7.2(a) and (b) present exactly the same BLMs. However, the model in (7.2) has disadvantages for testing hypotheses compared to the model in (7.3), since all hypotheses on the

Dependent Variable: D_DIVIDEND
Method: ML - Binary Logit (Quadratic hill climbing)
Date: 01/01/11   Time: 22:15
Sample: 1 2284
Included observations: 2284
Convergence achieved after 5 iterations
Covariance matrix computed using second derivatives

Variable	Coefficient	Std. Error	z-Statistic	Prob.
O2SALE=1,T=1	-1.347680	0.195389	-6.897401	0.0000
O2SALE=1,T=2	-1.690663	0.213433	-7.921276	0.0000
O2SALE=1,T=3	-1.656321	0.218251	-7.589060	0.0000
O2SALE=1,T=4	-1.819158	0.254096	-7.159347	0.0000
O2SALE=1,T=5	-2.251292	0.303488	-7.418047	0.0000
O2SALE=1,T=6	-2.255332	0.291526	-7.736290	0.0000
O2SALE=1,T=7	-2.114533	0.282930	-7.473702	0.0000
O2SALE=1,T=8	-1.888752	0.252901	-7.468344	0.0000
O2SALE=2,T=1	-0.577315	0.220893	-2.613548	0.0090
O2SALE=2,T=2	-0.340326	0.215076	-1.582349	0.1136
O2SALE=2,T=3	0.018349	0.191573	0.095781	0.9237
O2SALE=2,T=4	-0.358398	0.171142	-2.094154	0.0362
O2SALE=2,T=5	-0.350202	0.165809	-2.112082	0.0347
O2SALE=2,T=6	-0.869038	0.165213	-5.260111	0.0000
O2SALE=2,T=7	-0.955511	0.158666	-6.022165	0.0000
O2SALE=2,T=8	-0.584513	0.151336	-3.862353	0.0001

Mean dependent var	0.253065	S.D. dependent var	0.434863
S.E. of regression	0.417473	Akaike info criterion	1.059365
Sum squared resid	395.2757	Schwarz criterion	1.099531
Log likelihood	-1193.794	Hannan-Quinn criter.	1.074014
Avg. log likelihood	-0.522677		

Obs with Dep=0	1706	Total obs	2284
Obs with Dep=1	578		

(a)

Dependent Variable: D_DIVIDEND
Method: ML - Binary Logit (Quadratic hill climbing)
Date: 01/01/11   Time: 22:13
Sample: 1 2284
Included observations: 2284
Convergence achieved after 5 iterations
Covariance matrix computed using second derivatives

Variable	Coefficient	Std. Error	z-Statistic	Prob.
C	-0.584513	0.151336	-3.862353	0.0001
T=1	0.007198	0.267762	0.026882	0.9786
T=2	0.244188	0.262984	0.928527	0.3531
T=3	0.602862	0.244137	2.469360	0.0135
T=4	0.226116	0.228456	0.989756	0.3223
T=5	0.234311	0.224489	1.043753	0.2966
T=6	-0.284525	0.224049	-1.269922	0.2041
T=7	-0.370998	0.219266	-1.692003	0.0906
(T=1)*(O2SALE=1)	-0.770364	0.294908	-2.612216	0.0090
(T=2)*(O2SALE=1)	-1.350338	0.303004	-4.456498	0.0000
(T=3)*(O2SALE=1)	-1.674671	0.290403	-5.766712	0.0000
(T=4)*(O2SALE=1)	-1.460761	0.306356	-4.768177	0.0000
(T=5)*(O2SALE=1)	-1.901089	0.345829	-5.497189	0.0000
(T=6)*(O2SALE=1)	-1.386294	0.335086	-4.137125	0.0000
(T=7)*(O2SALE=1)	-1.159021	0.324383	-3.573007	0.0004
(T=8)*(O2SALE=1)	-1.304238	0.294723	-4.425306	0.0000

McFadden R-squared	0.076015	Mean dependent var	0.253065
S.D. dependent var	0.434863	S.E. of regression	0.417473
Akaike info criterion	1.059365	Sum squared resid	395.2757
Schwarz criterion	1.099531	Log likelihood	-1193.794
Hannan-Quinn criter.	1.074014	Restr. log likelihood	-1292.007
LR statistic	196.4244	Avg. log likelihood	-0.522677
Prob(LR statistic)	0.000000		

Obs with Dep=0	1706	Total obs	2284
Obs with Dep=1	578		

(b)

Figure 7.2   *Statistical results based on the BLM in (7.2) and (7.3)*

Table 7.1   *The parameters of the BLM in (7.2)*

O2Sale	$T=1$	$T=2$	$T=3$	$T=4$	$T=5$	$T=6$	$T=7$	$T=8$
1	C(1)	C(2)	C(3)	C(4)	C(5)	C(6)	C(7)	C(8)
2	C(9)	C(10)	C(11)	C(12)	C(13)	C(14)	C(15)	C(16)

Table 7.2   *The parameters of the BLM in (7.3)*

O2Sale	$T=1$	$T=2$	$T=3$	$T=4$	$T=5$	$T=6$	$T=7$	$T=8$
1	C(1)+C(2)+ C(9)	C(1)+C(3)+ C(10)	C(1)+C(4)+ C(11)	C(1)+C(5)+ C (12)	C(1)+C(6)+ C(13)	C(1)+C(7)+ C (14)	C(1)+C(8)+ C (15)	C(1)+ C (16)
2	C(1)+C(2)	C(1)+C(3)	C(1)+C(4)	C(1)+C(5)	C(1)+C(6)	C(1)+C(7)	C(1)+C(8)	C(1)
Diff.	C(9)	C(10)	C(11)	C(12)	C(13)	C(14)	C(15)	C(16)

differences between cells should be assessed with the Wald test using the parameters *C(1)–C(16)* as presented in Table 7.1.

3. On the other hand, based on the model in (7.3), 15 pair-wise comparisons represented by *C(2)–C(16)* (see Table 7.2) can be tested directly using the Z-statistic presented in the output. In addition, the output also presents the LR statistic of $G_0^2 = 196.4244$ with a *p*-value $= 0.000\,000$ which can be used to test the logit differences between the 16 cells, with the null hypothesis H_0: $C(k)=0$, for all $k=2$ to 16. So it can be concluded that *O2Sale* and the time *T* have significant joint effects on *D_Dividend*.

4. Note that the main objective of the model (7.3) is to test the hypotheses of *O2Sale* on *D_Dividend* by time *T*. Since the output shows that H_0: $C(k)=0$ for each $k=9$ to 16 is rejected based on the Z-statistic, it can then be concluded that *O2Sale* has significant effect on *D_Dividend*, conditional for each $T=1$ to 8.

5. The predicted probability, namely *PP_D_Dividend* or *PP(D_Dividend = 1)* generated using the *forecast equation*, is exactly equal to the sampled mean of *D_Dividend* by *O2Sale* and *T*, but their ordering in the outputs is different, as presented in Figure 7.3 (see Agung, 2011).

6. As extensions, the variable *SALE* could be transformed to ordinal variables with three or more levels, using nonparametric methods such as follows:

$$O3Sale = 1 + 1^*(Sale >= @Quantile(Sale, 0.3)) + 1^*(Sale >= @Quantile(Sale, 0.7))$$
$$O4Sale = 1 + 1^*(Sale >= @Quantile(Sale, 0.25))$$
$$+ 1^*(Sale >= @Quantile(Sale, 0.50)) + 1^*(Sale >= @Quantile(Sale, 0.75))$$
$$O5Sale = 1 + 1^*(Sale >= @Quantile(Sale, 0.2)) + 1^*(Sale >= @Quantile(Sale, 0.4))$$
$$+ 1^*(Sale >= @Quantile(Sale, 0.6)) + 1^*(Sale >= @Quantile(Sale, 0.8))$$

7. If there are a sufficiently large numbers of individual observations, then *O10Sale* can be considered for analysis. Then the analysis can easily be done using the equation specification in (7.3) by using the corresponding ordinal variable, under the condition that the three-way tabulation does not have an empty cell. We find that by using the *O5Sale*, the statistical results based on the BLM are obtained.

Example 7.2   BLM based on a four-dimensional table

For illustration, a BLM model of *D_Dividend* on *O2Sale*, *D_Firmstat* and *T* is considered to study the effect of *SALE* on *D_Dividend*, which depends on firm status and time. However, the error message obtained as

Tabulation of PP_DIVIDEND
Date: 01/04/11   Time: 12:43
Sample: 1 2284
Included observations: 2284
Number of categories: 16

Value	Count	Percent	Cumulative Count	Cumulative Percent
0.094891	137	6.00	137	6.00
0.095238	126	5.52	263	11.51
0.107692	130	5.69	393	17.21
0.131387	137	6.00	530	23.20
0.139535	129	5.65	659	28.85
0.155689	167	7.31	826	36.16
0.160256	156	6.83	982	42.99
0.206250	160	7.01	1142	50.00
0.277778	198	8.67	1340	58.67
0.295455	176	7.71	1516	66.37
0.357895	190	8.32	1706	74.69
0.359551	89	3.90	1795	78.59
0.411348	141	6.17	1936	84.76
0.413333	150	6.57	2086	91.33
0.415730	89	3.90	2175	95.23
0.504587	109	4.77	2284	100.00
Total	2284	100.00	2284	100.00

Test for Equality of Means of D_DIVIDEND
Categorized by values of T and O2SALE
Date: 01/04/11   Time: 12:49
Sample: 1 2284
Included observations: 2284

Method	df	Value	Probability
Anova F-test	(15, 2268)	13.94386	0.0000
Welch F-test*	(15, 807.479)	13.98525	0.0000

*Test allows for unequal cell variances

Category Statistics

O2SALE	T	Count	Mean	Std. Dev.	Std. Err. of Mean
1	1	160	0.206250	0.405882	0.032088
1	2	167	0.155689	0.363650	0.028140
1	3	156	0.160256	0.368025	0.029466
1	4	129	0.139535	0.347855	0.030627
1	5	126	0.095238	0.294715	0.026255
1	6	137	0.094891	0.294139	0.025130
1	7	130	0.107692	0.311191	0.027293
1	8	137	0.131387	0.339063	0.028968
2	1	89	0.359551	0.482588	0.051154
2	2	89	0.415730	0.495640	0.052538
2	3	109	0.504587	0.502288	0.048110
2	4	141	0.411348	0.493832	0.041588
2	5	150	0.413333	0.494081	0.040342
2	6	176	0.295455	0.457549	0.034489
2	7	198	0.277778	0.449039	0.031912
2	8	190	0.357895	0.480647	0.034870
	All	2284	0.253065	0.434863	0.009099

Figure 7.3   *Comparison between the predicted probability based on the BLM in (7.2), and the sampled mean of* D_Dividend *by* O2Sale *and* T

Figure 7.4   *Error message obtained using the equation specification (7.4)*

presented in Figure 7.4 indicates the four-dimensional table has at least one empty cell.

$$D\_Dividend \; C \; @Expand(O2Sale, D\_Firmstat, T, @Droplast) \qquad (7.4)$$

Example 7.3   Other BLM based on a four-dimensional table

As an extension of the model in (7.3), Figure 7.5 presents statistical results to study the effect of *O2Sale* on *D_Dividend* which depends on *O2Loan* and the time *T*, and Table 7.3 presents its parameters. Note that *O2Loan* or *Loan* is selected as a cause (source or up-stream) variable, since the effect of *O2Sale* on

Table 7.3 The parameters of the BLM in (7.5)

	O2Loan $=0$							
O2Sale	T $=1$	T $=2$	T $=3$	T $=4$	T $=5$	T $=6$	T $=7$	T $=8$
1	C(1)+C(2)+ C(17)	C(1)+C(3)+ C(18)	C(1)+C(4)+ C(19)	C(1)+C(5)+ C(20)	C(1)+C(6)+ C(21)	C(1)+C(7)+ C(22)	C(1)+C(8)+ C(23)	C(1)+C(9)+ C(24)
2	C(1)+C(2)	C(1)+C(3)	C(1)+C(4)	C(1)+C(5)	C(1)+C(6)	C(1)+C(7)	C(1)+C(8)	C(1)+C(9)
Diff.	**C(17)**	**C(18)**	**C(19)**	**C(20)**	**C(21)**	**C(22)**	**C(23)**	**C(24)**

	O2Loan $=1$							
O2Sale	T $=1$	T $=2$	T $=3$	T $=4$	T $=5$	T $=6$	T $=7$	T $=8$
1	C(1)+C(10)+ C(25)	C(1)+C(11)+ C(26)	C(1)+C(12)+ C(27)	C(1)+C(13)+ C(28)	C(1)+C(14)+ C(29)	C(1)+C(15)+ C(30)	C(1)+C(16)+ C(31)	C(1)+ C(32)
2	C(1)+C(10)	C(1)+C(11)	C(1)+C(12)	C(1)+C(13)	C(1)+C(14)	C(1)+(C15)	C(1)+C(16)	C(1)
Diff.	**C(25)**	**C(26)**	**C(27)**	**C(28)**	**C(29)**	**C(30)**	**C(31)**	**C(32)**

D_Dividend should depend on the loan. The equation specification applied is

$$D\_Dividend \ c \ @expand(O2loan.t, @droplast)$$
$$@expand(O2loan, t)^* @expand(o2sale, @droplast)$$

(7.5)

Based on the results In Figure 7.4 and Table 7.3, the following findings and notes are presented.

1. The table of the model parameters are developed as follows:
    1.1 Note the parameters of the model are *C(1)–C(32)*, starting from the symbols "*C*" to (*O2LOAN* $=1$ and $T=8)^*$(*O2SALE* $=2$) within 32 cells of a $2 \times 2 \times 8$ table, as presented in Table 7.3.
    1.2 Then enter *C(1)* in all cells.
    1.3 The parameter *C(2)* should be added in the column (*O2LOA* $=0,T=1$) for *O2SALE* $=1$ and *O2SALE* $=2$. Similarly, the parameters *C(3)* to *C(9)*, respectively, should be added in the columns (*O2LOA* $=0,T=t$) for $t=2$ to 8.
    1.4 The parameter *C(10)* should be added to the column (*O2LOA* $=1,T=1$) for *O2SALE* $=1$ and *O2SALE* $=2$. Similarly, the parameters *C(11)–C(16)*, respectively, should be added in the columns (*O2LOA* $=1,T=t$) for $t=2$ to 7.
    1.5 The parameters *C(17)–C(24)*, respectively, should be added in the cell ((O2LOA $=0,T=t$, O2SALE $=1$), for $t=1$ to 8.
    1.6 Finally, the parameters *C(25)* to *C(32)*, respectively, should be added in the cell ((O2LOA $=1$, T $=t$,O2SALE $=1$), for $t=1$ to 8.
2. The main objective of this model is to study the effects of *O2Sale* on *D_Dividend*, conditional for *O2Loan* $=i$, for $i=0,1$ and $T=t$, for $t=1$ to 8; which are 16 pair-wise comparisons, represented by each of the parameters *C(17)–C(32)*, and this can be tested directly using the Z-statistic in the output. Note that two of them have insignificant effects at the significance level $\alpha=0.10$, conditional for (*O2Loan* $=1$, $T=1$) and (*O2Loan* $=1$, $T=8$), respectively, with *p*-values of 0.1479 and 0.1329.

Dependent Variable: D_DIVIDEND
Method: ML - Binary Logit (Quadratic hill climbing)
Date: 01/02/11   Time: 18:31
Sample: 1 2284
Included observations: 2284
Convergence achieved after 6 iterations
Covariance matrix computed using second derivatives

Variable	Coefficient	Std. Error	z-Statistic	Prob.
C	-0.980829	0.204124	-4.805062	0.0000
O2LOAN=0,T=1	0.324050	0.387469	0.836324	0.4030
O2LOAN=0,T=2	1.152680	0.395978	2.910972	0.0036
O2LOAN=0,T=3	1.255266	0.366467	3.425316	0.0006
O2LOAN=0,T=4	0.980829	0.340207	2.883037	0.0039
O2LOAN=0,T=5	0.591364	0.348631	1.696246	0.0898
O2LOAN=0,T=6	0.287682	0.341565	0.842247	0.3996
O2LOAN=0,T=7	0.096627	0.340643	0.283660	0.7767
O2LOAN=0,T=8	1.009817	0.315674	3.198926	0.0014
O2LOAN=1,T=1	0.470004	0.361325	1.300779	0.1933
O2LOAN=1,T=2	0.287682	0.353553	0.813688	0.4158
O2LOAN=1,T=3	0.826679	0.321825	2.568720	0.0102
O2LOAN=1,T=4	0.389465	0.302955	1.285552	0.1986
O2LOAN=1,T=5	0.651350	0.289138	2.252731	0.0243
O2LOAN=1,T=6	0.015748	0.291241	0.054073	0.9569
O2LOAN=1,T=7	-0.010363	0.282407	-0.036695	0.9707

(O2LOAN=0 AND T=1)*(O2SALE=1)	-0.787334	0.409048	-1.924795	0.0543
(O2LOAN=0 AND T=2)*(O2SALE=1)	-1.642026	0.420269	-3.907086	0.0001
(O2LOAN=0 AND T=3)*(O2SALE=1)	-1.590114	0.387661	-4.101812	0.0000
(O2LOAN=0 AND T=4)*(O2SALE=1)	-1.419084	0.396068	-3.582932	0.0003
(O2LOAN=0 AND T=5)*(O2SALE=1)	-1.387027	0.431475	-3.214617	0.0013
(O2LOAN=0 AND T=6)*(O2SALE=1)	-1.691676	0.480640	-3.519630	0.0004
(O2LOAN=0 AND T=7)*(O2SALE=1)	-1.089879	0.433889	-2.511884	0.0120
(O2LOAN=0 AND T=8)*(O2SALE=1)	-2.135828	0.427326	-4.998119	0.0000
(O2LOAN=1 AND T=1)*(O2SALE=1)	-0.641854	0.445576	-1.440504	0.1497
(O2LOAN=1 AND T=2)*(O2SALE=1)	-1.504077	0.518188	-2.902572	0.0037
(O2LOAN=1 AND T=3)*(O2SALE=1)	-2.639057	0.644706	-4.093425	0.0000
(O2LOAN=1 AND T=4)*(O2SALE=1)	-2.201844	0.635497	-3.464757	0.0005
(O2LOAN=1 AND T=5)*(O2SALE=1)	-3.562341	1.030700	-3.456235	0.0005
(O2LOAN=1 AND T=6)*(O2SALE=1)	-1.114361	0.480265	-2.320302	0.0203
(O2LOAN=1 AND T=7)*(O2SALE=1)	-1.406703	0.557507	-2.523202	0.0116
(O2LOAN=1 AND T=8)*(O2SALE=1)	-0.628609	0.418330	-1.502662	0.1329

McFadden R-squared	0.092536	Mean dependent var	0.253065
S.D. dependent var	0.434863	S.E. of regression	0.415257
Akaike info criterion	1.054685	Sum squared resid	388.3313
Schwarz criterion	1.135017	Log likelihood	-1172.450
Hannan-Quinn criter.	1.083983	Restr. log likelihood	-1292.007
LR statistic	239.1132	Avg. log likelihood	-0.513332
Prob(LR statistic)	0.000000		

Obs with Dep=0	1706	Total obs	2284
Obs with Dep=1	578		

Figure 7.5   *Statistical results based on the BLM in (7.5)*

3. Based on the LR statistic of $G_0^2 = 239.1132$ with a p-value $= 0.000\,000$, we can present the following alternative conclusions. Which one would you consider the best?
   3.1   The effect of *O2Sale* on *D_Dividend* is significantly dependent on *O2Loan* and the time *T*.
   3.2   The joint effects of *O2Sale* and *O2Loan* on *D_Dividend* are significantly dependent on the time *T*.
   3.3   The three categorical variables *O2Sale*, *O2Loan* and the time *T* have significant joint effects on *D_Dividend*.
   3.4   The changing proportions $P(D\_Dividend = 1)$ over time are dependent on *O2Sale*, and *O2Loan* (see Figure 7.5).
4. In fact, the theoretical causal relationship between the variables *D_Dividend*, *O2Sale*, *O2Loan* and *T* can be presented in the form of a path diagram as in Figure 7.6. This path diagram has the following characteristics:
   - *O2Sale* and *O2Loan* have direct effects on *D_Dividend*, in a theoretical sense,
   - Even though the time *T* is not a cause factor, since *O2Sale*, and *O2Loan* should change over time, then the arrows with broken lines are used. Note that several or many other factors can be identified as having joint effects on SALE as well as LOAN, in a theoretical sense.
   - The time *T* has an indirect effect on *D_Dividend*, through *O2Sale* and *O2Loan*,
   - *O2Sale* and *O2Loan* are assumed to not have a causal relationship. On the other hand, they might have a causal relationship, such as LOAN has a positive indirect effect on SALE.

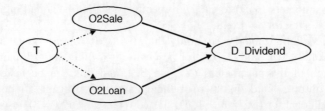

Figure 7.6   *Theoretical causal effects of O2Sale, O2Loan, and T on D_Dividend*

5. To generalize, this type of BCM is used to study the effects of any two categorical cause factor variables, and the time T, on a zero-one problem indicator. Then it can easily be extended to three or more categorical cause factors. However, we should note the possibility that the corresponding N-way tabulation has an empty cell. For a comparison, see the application of the N-way tabulation.

6. The effect of bivariate (*O2Sale,O2Loan*) on *D_Dividend*, conditional for each time $T=1$ to 8; can be tested using the Wald test. For an example, conditional for $T=1$, the null hypothesis is H_0: $C(2)+C(17)=C(2)=C(10)+C(25)=C(10)$, which indicates that the four cells within $T=1$ contains the same values of the logit. At a significance level of $\alpha=0.10$, we find that the null hypothesis is rejected based on the F-statistic of $F_0=2.446\,784$, with $df=(3,2252)$ and a p-value $=0.0621$, or the Chi-square statistic of $\chi_0^2=7.340353$ with $df=3$ and a p-value $=0.0618$.

Example 7.4   The simplest equation specification for BCM in (7.5)
If you feel that the model in Figure 7.5 is too complex or difficult, then you may applied the simplest equation specification as follows:

$$D\_Dividen @Expand(O2Sale, O2Loan, T) \tag{7.6}$$

with its parameters presented in Table 7.4. However, all hypotheses should be tested using the Wald test. Do it as an exercise.

Referring to the null hypothesis presented in point (3) in the previous example, based on this model the null hypothesis will be H_0: $C(1)=C(9)=C(17)=C(25)$. The other alternative hypotheses on the logit differences should be tested as follows:

1. Note that $C(k)=log[p_k/(1-p_k)]$, where p_k is the proportion of (*D_Dividend* $=1$) within the corresponding cell.
2. The growths of the logits over times by *O2Sale* and *O2Loan* can easily be seen in this table. They can then be graphically presented.
3. All hypotheses on the logit differences (risks or chances differences) between cells can easily be tested using the Wald test, such as follows:
   3.1 The different joint effects of *O2Sale* and *O2Loan* on *D_Dividend*, between $T=6$ and $T=7$, can be tested with H_0: $C(6)=C(7)$, $C(14)=C(15)$, $C(22)=C(23)$, $C(30)=C(31)$, to study the difference of their joint effects before and after a new tax policy.
   3.2 Corresponding to the new tax policy specific for *O2Sale* $=1$, and *O2Loan* $=0$, the logit differences between $T=6$, $T=7$, and $T=8$, can be tested using H_0: $C(6)=C(7)=C(8)$.
   3.3 The effect of *O2Sale* on *D_Dividend*, conditional for *O2Loan* $=0$, over times or at all the time points with H_0: $C(k)=C(k+16)$, for all $k=1$ to 8, which should be inserted for the Wald test as $C(1)=C(17)$, $C(2)=C(18)$, $C(3)=C(19)$, $C(4)=C(20)$, $C(5)=C(21)$, $C(6)=C(22)$, $C(7)=C(23)$, $C(8)=C(24)$

Table 7.4   *The parameters of the BCM in (7.6)*

O2Sale	O2Loan	$T=1$	$T=2$	$T=3$	$T=4$	$T=5$	$T=6$	$T=7$	$T=8$
1	0	C(1)	C(2)	C(3)	C(4)	C(5)	C(6)	C(7)	C(8)
1	1	C(9)	C(10)	C(11)	C(12)	C(13)	C(14)	C(15)	C(16)
2	0	C(17)	C(18)	C(19)	C(20)	C(21)	C(22)	C(23)	C(24)
2	1	C(25)	C(26)	C(27)	C(28)	C(29)	C(30)	C(31)	C(32)

7.2.2   Ordered Choice Models

It is recognized that the equation specification (ES) with the intercept "*C*" used for the binary choice models (BCMs) can be directly applied to ordered choice models (OCMs). See the following example.

Example 7.5   An ordered choice model

By using the equation specification in (7.5) the statistical results based on an ordered logit model (Quadratic hill climbing) are obtained, as presented in Figure 7.7. Based on these results the following findings and notes are presented.

1. The equation specification (ES) should be presented using the intercept "*C*". By using the ES in (7.6), a "*WARNING*" will be shown in the output. In fact, by using the ES in (7.5) but without the intercept "*C*", the same statistical results would also be obtained. Do it as an exercise.
2. It is interesting to note that the output presents exactly the same statistical values as presented in Figure 7.3, besides the intercept "*C*" and some others. However, there is a statistical value of 0.092 536 using two different names: *Pseudo R-Squared* and *McFadden R-squared*.
3. This result is obtained by using the option "*Quadratic hill climbing*", alternative statistical results can be obtained using the other options. However, we never know which one is best for a certain data set, in a theoretical sense.
4. As the extension of this model, a lot of multi-factorial ordered choice models could be presented based on any pool panel data, including the Data_Sesi.wf1, by defining or generating various ordinal dependent variables with three or more levels. For example, based on Net-income, Cash, ROA and ROE, ordinal variables with three or more levels can easily be generated then ordered choice models can be applied.

Example 7.6   An ordered choice model of O4NI

For an additional illustration, Figure 7.8 presents statistical results based on an ordered extreme value (OEV) model, using the following equation specification.

$$O4NI \; c \; @expand(O2loan, t, @droplast)$$
$$@expand(O2loan, t)^* @expand(o2sale, @droplast) \tag{7.7}$$

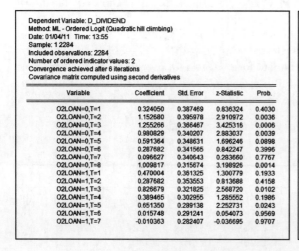

Figure 7.7   *Statistical results based on the OCM in (7.5)*

Dependent Variable: O4NI
Method: ML - Ordered Extreme Value (Quadratic hill climbing)
Date: 01/07/11   Time: 13:50
Sample: 1 2284
Included observations: 2284
Number of ordered indicator values: 4
Convergence achieved after 6 iterations
Covariance matrix computed using second derivatives

Variable	Coefficient	Std. Error	z-Statistic	Prob.
O2LOAN=0,T=1	0.272193	0.261039	1.042726	0.2971
O2LOAN=0,T=2	-0.061083	0.251963	-0.242428	0.8084
O2LOAN=0,T=3	0.000764	0.239219	0.003192	0.9975
O2LOAN=0,T=4	0.197647	0.232093	0.851587	0.3944
O2LOAN=0,T=5	0.068102	0.226322	0.300907	0.7635
O2LOAN=0,T=6	0.546973	0.242482	2.255728	0.0241
O2LOAN=0,T=7	0.685767	0.250669	2.735747	0.0062
O2LOAN=0,T=8	0.689774	0.238635	2.890504	0.0038
O2LOAN=1,T=1	0.327319	0.246814	1.326177	0.1848
O2LOAN=1,T=2	-0.029447	0.218612	-0.134701	0.8928
O2LOAN=1,T=3	-0.423048	0.196843	-2.149168	0.0316
O2LOAN=1,T=4	-0.251737	0.182399	-1.380141	0.1675
O2LOAN=1,T=5	-0.030894	0.182163	-0.169596	0.8653
O2LOAN=1,T=6	-0.027947	0.173841	-0.160762	0.8723
O2LOAN=1,T=7	-0.446971	0.161904	-2.760710	0.0058

(O2LOAN=0 AND T=1)*(O2SALE=1)	-1.258202	0.253527	-4.962787	0.0000
(O2LOAN=0 AND T=2)*(O2SALE=1)	-1.132711	0.243894	-4.644274	0.0000
(O2LOAN=0 AND T=3)*(O2SALE=1)	-1.490782	0.231527	-6.438900	0.0000
(O2LOAN=0 AND T=4)*(O2SALE=1)	-1.719545	0.232829	-7.385453	0.0000
(O2LOAN=0 AND T=5)*(O2SALE=1)	-1.445925	0.226894	-6.372704	0.0000
(O2LOAN=0 AND T=6)*(O2SALE=1)	-1.960515	0.240607	-8.148193	0.0000
(O2LOAN=0 AND T=7)*(O2SALE=1)	-2.174516	0.249498	-8.715573	0.0000
(O2LOAN=0 AND T=8)*(O2SALE=1)	-1.935125	0.236993	-8.165339	0.0000
(O2LOAN=1 AND T=1)*(O2SALE=1)	-1.226694	0.263752	-4.650935	0.0000
(O2LOAN=1 AND T=2)*(O2SALE=1)	-1.389463	0.228172	-6.089549	0.0000
(O2LOAN=1 AND T=3)*(O2SALE=1)	-1.373924	0.215267	-6.382432	0.0000
(O2LOAN=1 AND T=4)*(O2SALE=1)	-1.099627	0.202284	-5.436064	0.0000
(O2LOAN=1 AND T=5)*(O2SALE=1)	-1.609074	0.203518	-7.906300	0.0000
(O2LOAN=1 AND T=6)*(O2SALE=1)	-1.598534	0.191866	-8.331526	0.0000
(O2LOAN=1 AND T=7)*(O2SALE=1)	-1.247279	0.188073	-6.631881	0.0000
(O2LOAN=1 AND T=8)*(O2SALE=1)	-1.363245	0.191186	-7.130476	0.0000

Limit Points

LIMIT_2:C(32)	-2.128402	0.130780	-16.27464	0.0000
LIMIT_3:C(33)	-1.143429	0.126412	-9.045229	0.0000
LIMIT_4:C(34)	-0.205087	0.123001	-1.667365	0.0954

Pseudo R-squared	0.126044	Akaike info criterion	2.452851
Schwarz criterion	2.538203	Log likelihood	-2767.155
Hannan-Quinn criter.	2.483980	Restr. log likelihood	-3166.242
LR statistic	798.1732	Avg. log likelihood	-1.211539
Prob(LR statistic)	0.000000		

Figure 7.8   *Statistical results based on the OEVM in (7.7)*

where *O4NI* is generated as follows:

$$O4NI = 1 + 1^*(NI >= @Quantile(NI, 0.25)$$
$$+ 1^*(NI >= @Quantile(NI, 0.50) + 1^*(NI >= Quantile(NI, 0.75))$$

Based on this output, the following notes and comments are presented.

1. Similar to the model in (7.5), the main objectives of this model are to study the effects of *O2Sale* on *O4NI*, conditional for *O2Loan = i*, for *i* = 0,1 and *T = t*, for *t* = 1 to 8; which are the 16 pair-wise comparisons, represented by each of the parameters *C(16)–C(31)*, and can be tested directly using the Z-statistic in the output. Note that corresponding to the parameters, each of the three-way interaction independent variables has a significant adjusted effect on O4NI, based on the Z-statistic with a *p*-values = 0.0000.

2. Based on the LR statistic of $G_0^2 = 798.1732$ with a *p*-value = 0.000 000, we can conclude that *O2Sale*, *O2Loan*, and *T* have significant joint effects on the ordinal variable O4NI.

7.2.3   Not-Recommended Models

Referring to binary choice models (BCMs) based on four categorical variables *D_Dividend, O2Sale, O2Loan* and *T*, in order to generalize, let's consider the zero-one problem indicator *Y*, two dichotomous cause or classification factors *A* and *B*, and the time *T* = 1 to 8, then the proportions table of *P(Y = 1)* by *A*, *B* and *T* would have 32 cells. Suppose we present alternative models based on a data set with only the four variables *Y, A, B* and the time *T*, then the models presented in previous examples can easily be applied. On the other hand, this section presents illustrative examples of not-recommended or inappropriate models, because many papers discuss cell-proportion models with parameters less than the number of the cells generated by the categorical independent variables. However, the models are presented using their equation specifications, such as follows:

1. Model with 11 parameters, within 32 cells, generated by A, B and T:

$$Y \ C \ @Expand(A, B, @Droplast) \ @Expand(T, @Droplast) \tag{7.8}$$

2. Models with 10 parameters, within 32 cells, generated by A, B and T:

$$Y \ C \ @Expand(A, @Droplast) \ @Expand(B, @Droplast) \\ @Expand(T, @Droplast) \tag{7.9}$$

Note that each of these models has specific hidden assumptions, which are not realistic or acceptable in a theoretical sense, in addition to the basic assumptions of its error terms. See the following example.

Example 7.7   Characteristics of the model in (7.9)

Figure 7.9 presents the statistical results based on a binary probit model (BPM) in (7.9) with its table of model parameters. Based in this table the following notes and comments are presented.

1. Note that the output shows only three out of nine zero-one indicators have p-values > 0.20, so people may think that the model is an acceptable and good fit in a statistical sense, because he/she may never observe the table of its parameters.
2. Note that the logit differences of two cells, namely $[(A=i,T=t,B=1)-(A=i,T=t,B=2)]=$ $C(3)$ for all $i=1,2$ and $T=1$ to 8. This is similar for the logit differences of the cells $[(A=1,$

Dependent Variable: Y
Method: ML - Binary Probit (Quadratic hill climbing)
Date: 01/06/11 Time: 13:15
Sample: 1 2284
Included observations: 2284
Convergence achieved after 4 iterations
Covariance matrix computed using second derivatives

Variable	Coefficient	Std. Error	z-Statistic	Prob.
C	-0.438227	0.082871	-5.288051	0.0000
A=1	-0.864967	0.064364	-13.43868	0.0000
B=1	0.241016	0.061876	3.895169	0.0001
T=1	0.171630	0.117860	1.456221	0.1453
T=2	0.128062	0.118605	1.079735	0.2803
T=3	0.244427	0.114993	2.125583	0.0335
T=4	0.098785	0.114410	0.863433	0.3879
T=5	0.037226	0.114392	0.325427	0.7449
T=6	-0.175690	0.113285	-1.550869	0.1209
T=7	-0.193496	0.111457	-1.736065	0.0826

McFadden R-squared	0.078587	Mean dependent var	0.253065
S.D. dependent var	0.434863	S.E. of regression	0.416116
Akaike info criterion	1.051201	Sum squared resid	393.7483
Schwarz criterion	1.076305	Log likelihood	-1190.472
Hannan-Quinn criter.	1.060357	Restr. log likelihood	-1292.007
LR statistic	203.0699	Avg. log likelihood	-0.521222
Prob(LR statistic)	0.000000		

Obs with Dep=0	1706	Total obs	2284
Obs with Dep=1	578		

T	A=1 B=1	A=1 B=2	B (1-2)
1	C(1)+C(2)+C(3)+C(4)	C(1)+C(2)+C(4)	C(3)
2	C(1)+C(2)+C(3)+C(5)	C(1)+C(2)+C(5)	C(3)
3	C(1)+C(2)+C(3)+C(6)	C(1)+C(2)+C(6)	C(3)
4	C(1)+C(2)+C(3)+C(7)	C(1)+C(2)+C(7)	C(3)
5	C(1)+C(2)+C(3)+C(8)	C(1)+C(2)+C(8)	C(3)
6	C(1)+C(2)+C(3)+C(9)	C(1)+C(2)+C(9)	C(3)
7	C(1)+C(2)+C(3)+C(10)	C(1)+C(2)+C(10)	C(3)
8	C(1)+C(2)+C(3)	C(1)+C(2)	C(3)

T	A=2 B=1	A=2 B=2	B (1-2)
1	C(1)+C(3)+C(4)	C(1)+C(4)	C(3)
2	C(1)+C(3)+C(5)	C(1)+C(5)	C(3)
3	C(1)+C(3)+C(6)	C(1)+C(6)	C(3)
4	C(1)+C(3)+C(7)	C(1)+C(7)	C(3)
5	C(1)+C(3)+C(8)	C(1)+C(8)	C(3)
6	C(1)+C(3)+C(9)	C(1)+C(9)	C(3)
7	C(1)+C(3)+C(10)	C(1)+C(10)	C(3)
8	C(1)+C(3)	C(1)	C(3)

Figure 7.9   *Statistical results based on the BPM in (7.9), and its table of parameters*

$T=t,B=j)-(A=2,T=t,B=j)]=C(2)$ for all $j=1,2$ and $T=1$ to 8. This model uses the assumption that the joint effects of A and B on $D\_Dividend$ are constant over times. For these reasons, I would say that this condition presents an unrealistic situation, because we would never observe this condition in reality or based on a sample. Then I would declare this type of additive models is a not recommended, or even the worst statistical model.

3. For comparison, refer to the models in (7.61), presented in Li, et al. (2010), and the model in (7.62), presented in Suriawinata (2004), which have more than four additive dummy independent variables. So these models have hidden assumptions, but they are not stated by the researchers. Note that each defined model should have specific objectives with its own limitations, since all statistical models should have some limitations.

7.2.4 Application of N-Way Tabulation

Referring to the four categorical variables presented in Examples 7.4 and 7.5, and the path diagram in Figure 7.6, their associations also can be tested using the object "*N-Way Tabulation*". See the following examples.

Example 7.8 A multiple association of four categorical variables

Figure 7.10 presents a part of the statistical results based on the *N*-way tabulation of *D_Dividend*, *O2Sale*, *O2Loan* and *T*. Based on the complete output, the following findings and notes are presented.

```
Tabulation of D_DIVIDEND and O2SALE and O2LOAN and T
Date: 01/05/11   Time: 11:51
Sample: 1 2284
Included observations: 2284
Tabulation Summary

Variable                  Categories
D_DIVIDEND                    2
O2SALE                        2
O2LOAN                        2
T                             8
Product of Categories        64

Test Statistics        df      Value      Prob
Pearson X2             53     473.0200   0.0000
Likelihood Ratio G2    53     495.5015   0.0000


Table 1: Conditional table for O2LOAN=0, T=1:

                           O2SALE
Count                 1        2      Total
            0        89       27      116
D_DIVIDEND  1        21       14       35
            Total   110       41      151

Measures of Association       Value
Phi Coefficient             0.158679
Cramer's V                  0.158679
Contingency Coefficient     0.156718

Table Statistics       df      Value      Prob
Pearson X2             1      3.802016   0.0512
Likelihood Ratio G2    1      3.608644   0.0575
```

```
Table 26: Conditional table for O2LOAN=1:

                           O2SALE
Count                 1        2      Total
            0       376      483      859
D_DIVIDEND  1        44      239      283
            Total   420      722     1142

Measures of Association       Value
Phi Coefficient             0.252707
Cramer's V                  0.252707
Contingency Coefficient     0.245005

Table Statistics       df      Value      Prob
Pearson X2             1     72.92899    0.0000
Likelihood Ratio G2    1     80.29829    0.0000


Table 27: Unconditional table:

                           O2SALE
Count                 1        2      Total
            0       983      723     1706
D_DIVIDEND  1       159      419      578
            Total  1142     1142     2284

Measures of Association       Value
Phi Coefficient             0.261830
Cramer's V                  0.261830
Contingency Coefficient     0.253292

Table Statistics       df      Value      Prob
Pearson X2             1     156.5799    0.0000
Likelihood Ratio G2    1     161.0378    0.0000
```

Figure 7.10 *A part of the four-way tabulation based on* D_Dividend, O2Sale, O2Loan *and* T

Table 7.5   *Statistical summary based on the results in Figure 7.9*

Table#	Ind. Var	Conditional for		Contigency	LR stat.	df	p-value
0	A, B and T	–	–	–	495.5015	53	0.0000
1	A	B = 0	T = 1	0.156 718	3.60 844	1	0.0575
...	...	...	...	...	...	...	...
26	A	B = 0	–	0.245 005	80.29 829	1	0.0000
27	A	–	–	0.253 292	161.0378	1	0.000

1. For writing a paper, I suggest presenting a statistical summary as in Table 7.5 for the whole output, where $A = O2Sale$ and $B = O2Loan$.
2. The four categorical variables have significant associations based on the LR statistic of $G_0^2 = 495.5015$, with $df = 53$ and a p-value= 0.0000, compared to $G_0^2 = 239.1132$ with a p-value $= 0.000\,000$ based on BLM in Figure 7.4, and OCM in Figure 7.7. Then it can be said the data supports the hypothesis on the multiple association between the four variables. Furthermore, we can also conclude that the data supports the causal pattern presented in Figure 7.10.
3. Based on the three tables, namely Table 1, Table 26 and Table 27 in Figure 7.9; the following conclusions can be made, at a significance level $\alpha = 0.10$ or lower.
   3.1 Based on "Table 1 Conditional table for $O2Loan = 0$ and T = 1" it can be concluded that $O2Sale$ has a significant effect on $D\_Dividend$, conditional for $O2Loan = 0$ and T = 1, based on the LR statistic of $G_0^2 = 3.608644$, with $df = 1$ and a p-value $= 0.0575$.
   3.2 Based on "Table 26 Conditional table for $O2Loan = 1$" it can be concluded that $O2Sale$ has a significant effect on $D\_Dividend$, conditional for $O2Loan = 1$, based on the LR statistic of $G_0^2 = 80.29829$, with $df = 1$ and a p-value $= 0.0000$.
   3.3 Based on "Table 27 Unconditional table" it can be concluded that $O2Sale$ has a significant effect on $D\_Dividend$ based on the LR statistic of $G_0^2 = 161.0378$, with $df = 1$ and a p-value $= 0.0000$.

Example 7.9   An alternative multiple association of four categorical variables

An alternative analysis of the four categorical variables presented in Example 7.3. The main objective of this analysis is to study the joint effects of both ordinal variables $O2Sale$ and $O2Loan$ on $D\_Dividend$, conditional for $T = t$, for $t = 1$ to 8.

To achieve the objective, a cell factor CF should be generated based on $O2Sale$ and $O2Loan$, using the equation as follows:

$$CF = 1^*(O2Sale = 1 \text{ and } O2Loan = 0) + 2^*(O2Sale = 1 \text{ and } O2Loan = 1)$$
$$1^*(O2Sale = 2 \text{ and } O2Loan = 0) + 1^*(O2Sale = 2 \text{ and } O2Loan = 1)$$

Figure 7.11 presents a part of the statistical results based on three categorical variables $D\_Dividend$, CF and T, where CF is a categorical variable with four levels, generated by dichotomous $O2Sale$ and $O2Loan$. Based on this output, the following findings and notes are presented.

1. Compared to the path diagram in Figure 7.6, the causal relationships between $D\_Dividend$, $CF(O2Sale, O2Loan)$ and T, can be presented as in Figure 7.12.

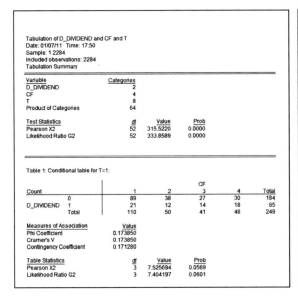
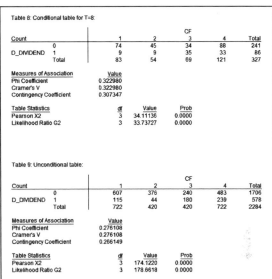

Figure 7.11   *A part of the three-way tabulation based on* D_Dividend, CF *and* T

Figure 7.12   *Theoretical causal effects of* O2Sale, O2Loan *and* T *on* D_Dividend

2. Compared to the LR statistic of $G_0^2 = 495.5015$, with $df = 53$ and a p-value $= 0.0000$ in Figure 7.9, this output present the LR statistic of $G_0^2 = 333.8589$, with $df = 52$ and a p-value $= 0.0000$ for testing the association between *D_Dividend, CF(O2Sale, O2Loan)* and *T*.

3. Note that this three-way tabulation and the four-way tabulation in Figure 7.10 in fact have the same number of cells; that is $8 \times 2 \times 2 \times 2 = 64$ cells, but both outputs present different LR statistics values and degrees of freedoms Why? For a further comparison, as an exercise do a two-way tabulation of the factor *CF(T,O2Sale,O2Loan)* with 32 cells and *D_Dividend*. The output should give the LR statistic with a $df = (32 - 1)(2 - 1) = 31$.

4. To generalize, the cell factor CF can be generated based on three or more categorical or numerical variables. However, one should note the possibility that the corresponding frequency table has empty cells, or, the table is incomplete. Refer to Agung (2011).

7.2.5   Applications of Ordinal-Ordinal Association

Note that in the ordered choice models, an ordinal independent variable is considered a nominal variable, more so in the *N*-way tabulations. This section presents applications of the Kendall's tau index to study the association between two variables with at least ordinal scales, called the *concordance-discordance measure of association* (CDMA), as well as other bivariate measures of associations for the comparison. Refer to Conover (1980) and Agung (2011, 2004).

Figure 7.13   *Kendall's tau of* O5Sale *and* O4NI, *conditional for* T = 6, 7 *and* 8

Figure 7.14   *Kendall's tau, Spearman's rank order and ordinary correlation of* SALE *and* NI, *conditional for* T = 8

Example 7.10   The effect of *O5Sale* on *O4NI* over time *T* = 6, 7 and 8

Figure 7.13 presents the Kendall's tau-b of two ordinal variables, *O5Sale* and *ONI*, conditional for each *T* = 6, 7 and 8, to study the effect of *O5Sale* on *O4NI* (*NI* = *net income*) at the three time points. Based on these outputs the following findings and notes are presented.

1. *O5SALE* has a positive significant effect on *O4NI*, conditional for each time point, with a *p*-value = 0.0000.
2. In a descriptive statistical sense, between the three time points, *O5SALE* has different effects on *O4NI*.

Example 7.11   The effect of *SALE* on *NI*, conditional for *T* = 8

As a modified analysis on the effect of *O5SALE* on *O4NI*, for illustration, Figure 7.14 presents the Kendall's tau-b, the Spearman's rank-order, and Ordinary correlation between two numerical variables: *SALE* and *NI*, conditional for *T* = 8.

Example 7.12   Special notes on the simplest OLS regressions of *NI* on *SALE*, and *SALE* on *NI*

To complete the discussion on the ordinary correlation analysis of *SALE* and *NI*, presented in Figure 7.14, Figure 7.15 presents the simplest OLS regressions of *NI* on *SALE*, and *SALE* on *NI*. Based on these outputs, it is very important to note that both regressions present exactly the same observed values of the *t*-statistic with ordinary correlation. Based on these findings, it can be concluded that a causal linear effect of all pairs in a set numerical variables, including the zero-one variables, can be tested using ordinary correlation analysis. See the statistical results demonstrated in Figure 7.16.

```
Dependent Variable: NI
Method: Least Squares
Date: 01/20/11   Time: 10:06
Sample: 1 2284 IF T=8
Included observations: 327
```

Variable	Coefficient	Std. Error	t-Statistic	Prob.
C	-87.40145	32.48477	-2.690536	0.0075
SALE	0.118549	0.003698	32.05483	0.0000

R-squared	0.759706	Mean dependent var	277.6095
Adjusted R-squared	0.758967	S.D. dependent var	1120.588
S.E. of regression	550.1540	Akaike info criterion	15.46437
Sum squared resid	98367565	Schwarz criterion	15.48755
Log likelihood	-2526.425	Hannan-Quinn criter.	15.47362
F-statistic	1027.512	Durbin-Watson stat	1.279448
Prob(F-statistic)	0.000000		

(a)

```
Dependent Variable: SALE
Method: Least Squares
Date: 01/20/11   Time: 10:07
Sample: 1 2284 IF T=8
Included observations: 327
```

Variable	Coefficient	Std. Error	t-Statistic	Prob.
C	1299.963	230.4666	5.640568	0.0000
NI	6.408376	0.199919	32.05483	0.0000

R-squared	0.759706	Mean dependent var	3078.989
Adjusted R-squared	0.758967	S.D. dependent var	8238.937
S.E. of regression	4044.915	Akaike info criterion	19.45441
Sum squared resid	5.32E+09	Schwarz criterion	19.47759
Log likelihood	-3178.795	Hannan-Quinn criter.	19.46366
F-statistic	1027.512	Durbin-Watson stat	2.477466
Prob(F-statistic)	0.000000		

(b)

Figure 7.15 *The simplest OLS regressions of: (a) NI on SALE, and (b) SALE on NI, for T = 8*

```
Covariance Analysis: Ordinary
Date: 01/21/11   Time: 11:31
Sample (adjusted): 4 2284
Included observations: 327 after adjustments
Pairwise samples (pairwise missing deletion)
```

Correlation t-Statistic Probability	Y	X1	X2
Y	1.000000		

X1	0.323405	1.000000	
	6.123339	-----	
	0.0000	-----	
X2	-0.088811	0.287578	1.000000
	-1.557171	5.218005	-----
	0.1205	0.0000	-----

```
Dependent Variable: Y
Method: Least Squares
Date: 01/21/11   Time: 11:14
Sample: 1 2284 IF T=8
Included observations: 323
```

Variable	Coefficient	Std. Error	t-Statistic	Prob.
C	-0.169497	0.074386	-2.278618	0.0233
X1	0.067760	0.011066	6.123339	0.0000

R-squared	0.104591	Mean dependent var	0.263158
Adjusted R-squared	0.101801	S.D. dependent var	0.441031
S.E. of regression	0.417979	Akaike info criterion	1.099404
Sum squared resid	56.08090	Schwarz criterion	1.122795
Log likelihood	-175.5537	Hannan-Quinn criter.	1.108741
F-statistic	37.49528	Durbin-Watson stat	2.311962
Prob(F-statistic)	0.000000		

```
Dependent Variable: X2
Method: Least Squares
Date: 01/21/11   Time: 11:17
Sample: 1 2284 IF T=8
Included observations: 307
```

Variable	Coefficient	Std. Error	t-Statistic	Prob.
C	0.066474	0.083506	0.796044	0.4266
Y	-0.245681	0.157774	-1.557171	0.1205

R-squared	0.007887	Mean dependent var	-0.002349
Adjusted R-squared	0.004635	S.D. dependent var	1.244286
S.E. of regression	1.241400	Akaike info criterion	3.276849
Sum squared resid	470.0272	Schwarz criterion	3.301128
Log likelihood	-500.9963	Hannan-Quinn criter.	3.286558
F-statistic	2.424781	Durbin-Watson stat	0.971633
Prob(F-statistic)	0.120467		

Figure 7.16 *Correlation analysis of a zero-one variable Y, and numerical variables X1 and X2, and two of their simplest OLS regressions, conditional for T = 8*

7.3   Multi-Factorial Choice Models with a Numerical Independent Variable

In theoretical sense, *SALE* or $X1 = lnSALE$, for *SALE* > 0 should have positive effect on the dummy variable $Y = D\_Dividend$. In other words, $p = P(Y = 1)$ will increase with increasing values of *lnSALE*, and $p = P(Y = 1)$ will decrease with increasing values of *LOAN* or $X2 = lnLOAN$, for *LOAN* > 0. So the simplest BCM of $Y = D\_Dividend$ on $X1 = lnSALE$ can be represented using the following equation specification.

$$D\_Dividend \ C \ lnSALE \tag{7.10}$$

and the simplest BCM of *Y* on $X2 = lnLOAN$ can be represented using the following equation specification.

$$D\_Dividend \ C \ lnLOAN \tag{7.11}$$

Note that both MCMs in (7.10) and (7.11) can be presented using the general equation specification (EQ) as follows:

$$Y \ C \ X \tag{7.12}$$

This ES can be applied for any dummy variable or zero-one indicator *Y*, with any numerical variables *X*, even though *X* and *Y* are not correlated at all in a theoretical sense. However, based on pooled panel data the model should be considered one of the worst, since the $N \times T$ observations or scores of *X* and *Y* are presented using a single BCM only, and their relationship should change over time, as well as by groups of individuals, especially for a large number of observed individuals or objects.

For these reasons, simple BCMs are suggested such as the heterogeneous BCM, with the following equation specifications:

$$Y \ C \ @Expand(CF, T, @Droplast) \ X^* @Expand(CF, T) \tag{7.13a}$$

or

$$Y \ C \ X @Expand(CF, T, @Droplast) \ X^* @Expand(CF, T, @Droplast) \tag{7.13b}$$

where *CF* represents a cell or classification factor, generated based on one or more cause (exogenous, source or up-stream) variables. Note that each of the models should be acceptable, in both theoretical and statistical senses for studying the *differential linear effects* of *X* on *log[p/(1−p)]*, where $p = P(Y = 1)$, between the cells generated by *CF* and the categorical time variable *T*. Note that the graph of this model represents a set of $(\#CF) \times T$ regression lines of *log[p/(1−p)]* on *X* with different slopes, where #*CF* indicates the number of levels of the cell factor CF. So this will be called the simplest *heterogeneous BCM*. In fact, we never know the true relationship between *X* and *Y*, it can be polynomial or nonlinear. Refer to the nonparametric graphs presented in Figures 7.17 and Figure 7.18, and the graphs presented in Agung (2011, Chapter 7). By using *Y* = *D_Dividend*, and each of *X1* = *lnSALE* and *X2* = *lnLOAN*, both statistical results can easily be obtained. Do it as an exercise.

To generalize, the heterogeneous BCMs with a single numerical independent variable would have the following general equation specifications:

$$Y \ C \ @Expand(CF, T, @Droplast) \ F(X)^* @Expand(CF, T) \tag{7.14a}$$

or

$$Y \ C \ F(X) \ @Expand(CF, T, @Droplast) \ F(X)^* @Expand(CF, T, @Droplast) \tag{7.14b}$$

where *F(X)* can be any functions of *X* with no parameter.

On the other hand, we find that homogeneous BCMs are applied in many studies with the following general equation specifications:

$$Y \ C \ F(X) \ @Expand(CF, T, @Droplast) \tag{7.15}$$

where *F(X)* is any function of *X* without a parameter. Note that these BCMs have the unrealistic assumption that *F(X)* has the same effects on *logit(Y)* or *log[p/(1−p)]* within all cells generated by *CF* and the time *T*. Note that this model can be considered a binary choice ANCOVA model, which in fact is a CF-Time (Group-Time) fixed-effects model.

For illustration, *lnsale* = *log(SALE)*, for *SALE* > 0, will be defined as a numerical cause factor of *D_Dividend*. To generalize, *D_Dividend* will be presented as *Y*, and the numerical cause (source or up-stream) variables will be presented as *X1* = *lnsale*, then the general equation specification in (7.14) and (7.15) can easily be applied. Referring to the function in (6.20a to 6.20f)) and other possible functions, then a huge number of multi-factorial binary or ordered choice models can be subjectively defined. Try it for some exercise.

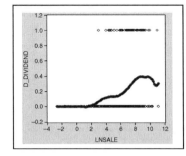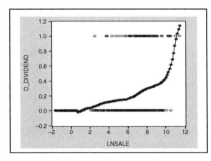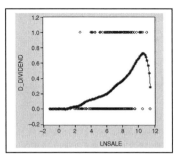

Figure 7.17 *Scatter graphs with kernel fit of* (LNSALE,D_Dividend), *for* T = 6, 7 *and* 8

Special parametric binary choice models with a numerical independent variable have been presented in Agung (2011, Chapter 7) with special notes and comments. For this reason, this section will present illustrative examples of nonparametric relationships between a numerical cause factor and a dummy variable, over time with some additional examples.

7.3.1 Nonparametric Estimation Methods

Nonparametric graph(s) of a zero-one problem indicator on a numerical exogenous variable can be used as a guide to develop alternative parametric BCMs, as well as to present the differential effects of the numerical exogenous variables on the zero-one indicator between the time points or groups of individuals. See the following examples.

Example 7.13 Scatter graphs with kernel fit curves

For illustration, Figure 7.17 presents the scatter graphs of *(LNSALE,D_Dividend)* with its kernel fit, for $T = 6, 7$ and 8, respectively.

Based on these graphs the following conclusions and notes are presented.

1. Even without testing hypothesis, based on the kernel fit graphs we can conclude that *lnSALE* do have different effects on *D_Dividend* before the new tax policy ($T = 6$) and after the new tax policy ($T = 7$ and $T = 8$).
2. At the first year ($T = 7$) after the new tax policy the proportions $p = P(D\_dividend = 1)$ increases with increasing *lnSale*, but with some estimates is greater than one. So it can be concluded that the kernel fit is inappropriate for $T = 7$. For comparison, its nearest neighbor fit shows a contradictory graph, which presents some negative estimates.
3. At the second year after the new tax policy ($T = 8$), the proportions suddenly decrease after a certain value of *lnSale*, as shown in the graph.

7.3.2 Parametric Estimation Methods

Compared to the polynomial BCMs presented in the following examples, I have found the only third-degree polynomial BCM presented by Korteweg (2010), with the equation as follows:

$$\ln\left(\frac{p_{i,t+1}}{1 - p_{i,t+1}}\right) = \beta_0 + \beta_1{}^*GAIN_{it} + \beta_2{}^*GAIN_{it}^2 + \beta_3{}^*GAIN_{it}^3 + \eta_{it} \qquad (7.16)$$

where $p_{i,t+1}$ is the probability of the firm i issuing debt or buying back equity worth more than 20% of its book assets in year $t+1$, and $GAIN_{it}$ is the change in B/V^L from rebalancing leverage from its current level to the optimum in year t.

Note that the limitation of this model is it only presents a single binary choice continuous function for all firm-time observations of 1213 undelivered firm-years in the SIC2. The following examples present alternative models.

Example 7.14   BCM by times with a numerical independent variable

Referring to the curves in Figure 7.17, alternative polynomial BCMs should be applied. By using trial-and-error methods, an acceptable (parametric) BCM is obtained by using the following equation specification, for $T = 6, 7$ and 8.

$$D\_DIVIDEND @EXPAND(T) \; LNSALE \wedge 3^* @EXPAND(T)$$
$$.LNSALE\wedge 4^* @EXPAND(T) \tag{7.17a}$$

The statistical results are presented in Figure 7.19, with its reduced model as follows:

$$D\_DIVIDEND @EXPAND(T) \; LNSALE \wedge 3^* @EXPAND(T)$$
$$LNSALE \wedge 4^* @EXPAND(T, @Drop(7)) \tag{7.17b}$$

Based on the outputs in Figure 7.19, the following findings and notes are presented.

1. Since both *either* $lnSALE^{3*}(T=7)$ *or* $lnSALE^3(T=7)$ has large *p*-values, and the null hypothesis $C(5) = C(8) = 0$ is rejected, then a reduced model can be obtained by deleting either one of the variables. It is found that both reduced models are good fits. But by deleting $lnSALE^4(T=7)$, the reduced model has a smaller SSR (sum-squared residuals).
2. By deleting *either* $lnSALE^{3*}(T=7)$ *or* $lnSALE^3(T=7)$, *the variable* $lnSALE^{4*}(T=7)$ *is obtained: in the reduced model it has a Z-statistic of 4.3 383 539, compared to a larger Z-statistic of 4.508 042 for the variable either* $lnSALE^{3*}(T=7)$ *or* $lnSALE^3(T=7)$.
3. Compared to the nonparametric curves in Figure 7.17, Figure 7.18 presents the scatter graphs of *(D_Dividend, P_Reduced)* on *lnSALE*, where *P_Reduced* is the predicted probability based on the

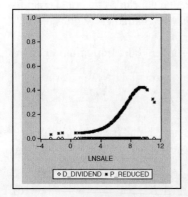 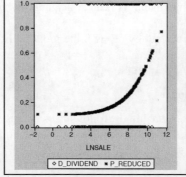 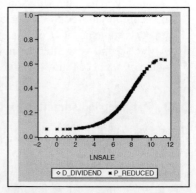

Figure 7.18   *Scatter graphs of* (LNSALE,D_Dividend, P_Reduced), *for* T = 6, 7 and 8

Dependent Variable: D_DIVIDEND
Method: ML - Binary Logit (Quadratic hill climbing)
Date: 01/15/11 Time: 19:01
Sample: 1 2284 IF T >5
Included observations: 959
Convergence achieved after 6 iterations
Covariance matrix computed using second derivatives

Variable	Coefficient	Std. Error	z-Statistic	Prob.
T=6	-3.056079	0.488608	-6.254662	0.0000
T=7	-2.365521	0.434973	-5.438316	0.0000
T=8	-2.678261	0.441396	-6.067709	0.0000
LNSALE^3*(T=6)	0.012521	0.004410	2.839030	0.0045
LNSALE^3*(T=7)	0.004373	0.003841	1.138572	0.2549
LNSALE^3*(T=8)	0.009349	0.003815	2.450515	0.0143
LNSALE^4*(T=6)	-0.000980	0.000420	-2.329820	0.0198
LNSALE^4*(T=7)	-0.000202	0.000361	-0.558903	0.5762
LNSALE^4*(T=8)	-0.000628	0.000358	-1.753370	0.0795

Mean dependent var	0.228363	S.D. dependent var		0.419997
S.E. of regression	0.401486	Akaike info criterion		1.000679
Sum squared resid	153.1313	Schwarz criterion		1.046344
Log likelihood	-470.8254	Hannan-Quinn criter.		1.018070
Avg. log likelihood	-0.490955			

Obs with Dep=0	740	Total obs	959
Obs with Dep=1	219		

Dependent Variable: D_DIVIDEND
Method: ML - Binary Logit (Quadratic hill climbing)
Date: 01/15/11 Time: 19:10
Sample: 1 2284 IF T>5
Included observations: 959
Convergence achieved after 6 iterations
Covariance matrix computed using second derivatives

Variable	Coefficient	Std. Error	z-Statistic	Prob.
T=6	-3.056079	0.488608	-6.254662	0.0000
T=7	-2.172832	0.251647	-8.634440	0.0000
T=8	-2.678261	0.441396	-6.067709	0.0000
LNSALE^3*(T=6)	0.012521	0.004410	2.839030	0.0045
LNSALE^3*(T=7)	0.002253	0.000500	4.508042	0.0000
LNSALE^3*(T=8)	0.009349	0.003815	2.450515	0.0143
LNSALE^4*(T=6)	-0.000980	0.000420	-2.329820	0.0198
LNSALE^4*(T=8)	-0.000628	0.000358	-1.753370	0.0795

Mean dependent var	0.228363	S.D. dependent var		0.419997
S.E. of regression	0.401314	Akaike info criterion		0.998918
Sum squared resid	153.1610	Schwarz criterion		1.039509
Log likelihood	-470.9812	Hannan-Quinn criter.		1.014377
Avg. log likelihood	-0.491117			

Obs with Dep=0	740	Total obs	959
Obs with Dep=1	219		

Figure 7.19   *Statistical results based on the BLM in (7.16) and its reduced model*

reduced model. Based on the nonparametric and parametric curves, the following notes and comments are presented.

3.1   For $T = 6$, both curves show that the proportions or predicted probabilities $p = P(Y=1)$ decreases for very large scores of *lnSALE*. So it is not the same as expected. Why? Note that the observed objects are a set of distinct firms, so they can have different problems or policies, even though they have greater values of *SALEs*.

3.2   For $T = 7$, both curves show that predicted probabilities $p = P(Y=1)$ increase with increasing scores of *lnSALE*.

3.3   For $T = 8$, the two curves look very different for large scores of *lnSALE*. In a theoretical sense, the curve in Figure 7.18 shows better that $p = P(Y=1)$ increases with increasing scores of *lnSALE*

Example 7.15   Alternative equation specifications with a numerical variable
By observing the graphs in Figure 7.18, it can be concluded that *lnSale* and *D_Dividend*, specifically *log[p/(1 − p)]*, should have different patterns of relationships for $T = 6$, 7 and 8. To achieve the objectives, the dummy time variables should be applied, namely *Dt6*, *Dt7* and *Dt8*, which can be generated using the equation $Dti = 1*(T = i)$ for each $i = 6$, 7 and 8, then the equation specification will have the general form as follows:

$$Y \quad F_1(X1)^*Dt6 \quad F_2(X1)^*Dt7 \quad F_3(X1)^*Dt8 \tag{7.18}$$

In this case $Y = D\_Dividen$, and $F_k(X1)$, $k = 1$, 2 and 3 are any functions of $X1 = lnSale$ with finite numbers of parameters. Many possible BCMs could give acceptable statistical results, which are highly dependent on the datasets. For an illustration the model in (7.17) can be represented as follows:

$$Y \quad Dt6\,X1^{\wedge}3^*Dt6\,X1^{\wedge}4^*Dt6 \quad Dt7\,X1^{\wedge}3^*Dt7 \quad Dt8\,X1^{\wedge}3^*Dt8\,X1^{\wedge}4^*Dt8 \tag{7.19a}$$

$$Y = (C(1) + C(2)^*X1^3 + C(3)^*X1^4)^*Dt6$$
$$+ (C(4) + C(5)^*X1^3)^*Dt7 + (C(6) + C(7)^*X1^3 + C(8)^*X1^4)^*Dt8 \qquad (7.19b)$$

$$Y \ C \ X1^3{}^*Dt6 \ X1^4{}^*Dt6 \ Dt7 \ X1^3{}^*Dt7 \ Dt8 \ X1^3{}^*Dt8 \ X1^4{}^*Dt8 \qquad (7.19c)$$

Note the last ES is representing a model with an intercept indicated by C, which can be applied for the BCM as well as the OCM (ordered choice model). See the following example.

7.3.3   Polynomial-Effect BCM with Three Numerical Exogenous Variables

As the extension of the polynomial-effect BCM with two numerical variables presented in Section 6.3.2, polynomial-effect BCM with three numerical exogenous variables could be developed using similar methods. See the following examples.

Example 7.16   Scatter and nonparametric graphs of Y on X3 by the time T
Figure 7.20 presents the scatter graphs of Y on X3 with their kernel fit curves, for $T = 6$, 7 and 8, to demonstrate that X3 has a polynomial-effect on $logit(Y) = log[p/(1 - p)]$. Based on these graphs, it can be concluded that a polynomial-effect *BCM* of $logit(Y)$ on X3 should be an appropriate model at each time point. Do it as an exercise.

For an additional illustration, Figure 7.21 presents the statistical results based on a fourth-degree polynomial BLM with the scatter graph of its predicted probability, namely *P_Y4th*, and Y on X3. Based on these outputs, the following notes and comments are presented.

1. In a mathematical sense, the binary logit regression is a continuous function of $log[p/(1 - p)]$ on X3, even though the graph of the predicted probability *P_Y4th* on X3 is not a smooth curve.
2. If the fifth-degree polynomial model is applied, we find that $X3^4$ and $X3^5$ have the *p*-values of 0.1680 and 0.4805, respectively.
   2.1   Take note that an insignificant result does directly mean a bad result. One should be using his/her best knowledge and experience to judge whether or not a model is acceptable. Talking about the important role of judgment in data analysis, even though it can be very subjective, Tukey (Tukey, 1962, quoted by Gifi, 1990, p. 22) presents the following statement.

In data analysis we must look to very heavy emphasis on judgment. At least three different sorts or sources of judgments are likely to be involved in almost every instance:

(a1) judgment based upon the experience of the particular field of subject matter from which the data come,

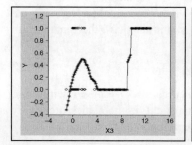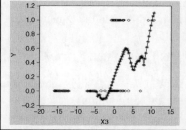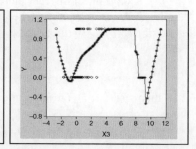

Figure 7.20   *Scatter graphs of* Y *on* X3, *for* T = 6, 7 *and 8, with their kernel fit curves*

Variable	Coefficient	Std. Error	z-Statistic	Prob.
C	-1.348227	0.144733	-9.315273	0.0000
X3	1.183947	0.350435	3.378506	0.0007
X3^2	0.326544	0.135579	2.408509	0.0160
X3^3	-0.119248	0.038537	-3.094358	0.0020
X3^4	0.007419	0.002692	2.755675	0.0059

Dependent Variable: Y
Method: ML - Binary Logit (Quadratic hill climbing)
Date: 02/03/11 Time: 12:20
Sample: 1 2284 IF T =8
Included observations: 327
Convergence achieved after 5 iterations
Covariance matrix computed using second derivatives

McFadden R-squared	0.096729	Mean dependent var	0.262997
S.D. dependent var	0.440936	S.E. of regression	0.416541
Akaike info criterion	1.071453	Sum squared resid	55.86893
Schwarz criterion	1.129404	Log likelihood	-170.1826
Hannan-Quinn criter.	1.094577	Restr. log likelihood	-188.4070
LR statistic	36.44882	Avg. log likelihood	-0.520436
Prob(LR statistic)	0.000000		

Obs with Dep=0	241	Total obs	327
Obs with Dep=1	86		

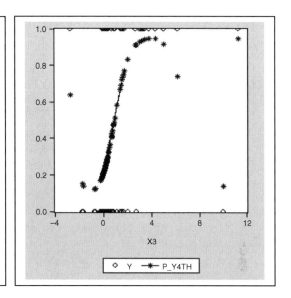

Figure 7.21   *Statistical results based on a fourth degree polynomial BLM of* Y *on* X3, *for* T = 8

(a2) judgment based upon a broad experience with how particular techniques of data analysis have worked out in variety of fields of applications,

(a3) judgment based on abstract results about the properties of particular techniques, whether obtained by mathematical proof or empirical sampling.

1. Bezzecri (1973, in Gifi, 1991, p.25) presents notes on the relationship between the data set and the model as follows:

The model must follow the data, and not the other way around. This is another error in the application of mathematics to the human sciences: the abundance of models, which are built a priori and then confronted with the data by what one calls a "test". Often the "test" is used to justify a model in which the parameters to be fitted is larger than the number of data points. And often is used, on the contrary, to reject as invalid the most judicial remarks of the experimenter. But what we need is a rigorous method to extract structure, starting from the data.

Furthermore, Hampel (1973, in Gifi, 1991, p. 27) presents notes as follows:

Classical (parametric) statistics derives results under the assumption that these models are strictly true. However, apart from some simple discrete models perhaps, such models are never exactly true. After all, a statistical model has to be simple (where "simple", of course has a relative meaning, depending on state and standards of the subject matter).

7.3.4   A Study of Li, et al. (2010)

Li, et al. (2010) present a panel data analysis based on a binary logit additive model of a zero-one endogenous variable, ADVERSE, on 16 exogenous variables using the following equation specification. However, a new ordering of the independent variables is made, specifically for dummy independent variables.

Based on this model, the following notes and comments are made.

1. The model is an *additive model*, since all independent variables are the main factors, including the eight dummy variables generated based on eight distinct dichotomous variables. In a statistical sense, this model is acceptable as the simplest linear regression, called the *hyperplane*, in 17-dimensional coordinates with the axes *log(p/(1−p))* and 16 independent variables.

2. However, one should note the limitations or hidden assumptions of this additive model, such as follows:

   2.1 The adjusted effects of a dichotomous independent variable, say *CFOACCT04*, on *ADVERSE*, conditional for each of all groups generated by the other seven dichotomous independent variables are the same, which is represented by a single parameter *C(2)*. This is one of the hidden assumption of this model. Note the eight dichotomous variables in the model can generate $2^8 = 256$ groups of individuals. So the model could represent a set of 256 homogeneous (parallel) multiple regressions with numerical independent variables that have a specific pattern of intercepts, which are not realistic or common. Refer to the model in (6.40).

   2.2 Take note that if there are two or more exogenous variables, it is common to have the effect of an exogenous variable on the endogenous variable as highly dependent on at least one of the other exogenous variables. So a proposed or defined model should have a two- or three-way interaction-independent variable at least. See the following studies using interaction models. Similar panel data analyses have been done by the author's students for their theses, using binary choice models with categorical independent variables, where all models presented have two- or three-way interaction-independent variables. Two of the studies examine the determinants of migration in Indonesia between the years 2000 and 2007 (Santoso, 2010), and a study on the determinants of divorce for women between 2000 and 2007 (Sinang, 2010).

   2.3 With regard to the model having a large number of categorical variables, Gifi (Gifi 1991; 45) presents the following notes.

In this sense discrete MVA (multivariate analysis) is useful if we want a fairly exhaustive analysis of the relationship between three or four variables, either because there are only three or four variables or because there are reasons to find three or four variables extremely important. Discrete MVA cannot be used for the simultaneous analysis of a large number of categorical variables.

7.3.5   A Study of Hameed, et al. (2010)

All general linear models (GLMs) with numerical endogenous (down-stream, impact or dependent) variables or problem indicators can easily be transformed into binary choice models with exactly the same set of exogenous (source, up-stream, cause or classification) factors or variables. For an illustration and a discussion, a two-way interaction GLM with a large number of independent variables presented in Hameed, et al. (2010) is quoted, such as follows.

$$
\begin{aligned}
\Delta ASPR_{p,t} = {}& \alpha_i + \sum_{k=1}^{4}\beta_{p,k}R_{m,t-k} + \sum_{k=1}^{4}\beta_{DOWN,p,k}R_{m,t-k}D_{DOWN,m,t-k} + \sum_{k=1}^{4}\gamma_{p,k}R_{p,t-k} \\
& + \sum_{k=1}^{4}\gamma_{DOWN,p,k}R_{p,t-k}D_{DOWN,p,t-k} + c_{1p}\Delta STD_{m,t} + c_{2p}\Delta STD_{p,t} \\
& + c_{3p}\Delta STD_{m,t-1} + c_{4p}\Delta STD_{p,t-1} + c_{5p}\Delta TURN_{p,t-1} \\
& + c_{6p}\Delta ROIB_{p,t-1} + \sum_{k=1}^{4}\phi_{p,k}\Delta ASPR_{p,t-k} + \varepsilon_{p,t}
\end{aligned}
\tag{7.20}
$$

Based on this model the following notes and comments are made.

1. This model has eight two-way interaction independent variables, namely $R_{m,t-k}{}^*D_{DOWN,m,t-k}$ and $R_{p,t-k}{}^*D_{DOWN,p,t-k}$ for $k = 1, 2, 3$ and 4, in which each is a two-way interaction of a numerical variable and a dummy variable.

2. However, the dummy variables $D_{DOWN,m,t-k}$ and $D_{DOWN,p,t-k}$ are not used as the main independent variables. Corresponding to these dummy variables, this model should have hidden assumptions on the intercept parameters, namely α_i, of the regressions generated by the dummy variables. So the intercept parameters α_i are free from the interaction factors $R_{m,t-k}{}^*D_{DOWN,m,t-k}$ and $R_{p,t-k}{}^*D_{DOWN,p,t-k}$. In other words, the regressions generated by the dummy variables $D_{DOWN,m,t-k}$ and $D_{DOWN,p,t-k}$ have the same or constant intercept. Compare this to the interaction model presented in the following section.

7.3.6 A Study of Francis and Martin (2010)

Francis and Martin (2010) present several interaction models; one of the models has three-way interaction independent variables, as follows:

$$
\begin{aligned}
X_{i,t-1} = &\ \beta_1 + \beta_2 D_{i,t-1} + \beta_3 R_{i,t-1} + \beta_4 D_{i,t-1}{}^*R_{i,t-1} + \beta_5 AcqCAR_{i,t} + \beta_6 AcqCAR_{i,t}{}^*D_{i,t-1} \\
&+ \beta_7 AcqCAR_{i,t}{}^*R_{i,t-1} + \beta_8 AcqCAR_{i,t}{}^*D_{i,t-1}{}^*R_{i,t-1} + \beta_9 Leverage_{i,t-1} + \beta_{10} Leverage_{i,t-1}{}^* \\
&D_{i,t-1} + \beta_{11} Leverage_{i,t-1}{}^*R_{i,t-1} + \beta_{12} Leverage_{i,t-1}{}^*D_{i,t-1}{}^*R_{i,t-1} + \beta_{13} Log(assets)_{i,t-1} \\
&+ \beta_{14} Log(assets)_{i,t-1}{}^*D_{i,t-1} + \beta_{15} Log(assets)_{i,t-1}{}^*D_{i,t-1}{}^* \\
&R_{i,t-1} + \beta_{16} MB_{i,t-1} + \beta_{18} MB_{i,t-1}{}^*R_{i,t-1} + \beta_{17} MB_{i,t-1}{}^*D_{i,t-1} + \beta_{19} MB_{i,t-1}{}^*R_{i,t-1}{}^*D_{i,t-1} \\
&+ \beta_{20} LIT_{i,t-1} + \beta_{21} LIT_{i,t-1}{}^*D_{i,t-1} + \beta_{22} LIT_{i,t-1}{}^*R_{i,t-1} + \beta_{23} LIT_{i,t-1}{}^*D_{i,t-1}{}^*R_{i,t-1} + e_{i,t}
\end{aligned}
\tag{7.21}
$$

where *D* is the only dummy variable defined based on *(R)* if it shows positive or negative return. Note that this model has two three-way interactions factors: $MB_{i,t-1}{}^*R_{ib,t-1}{}^*D_{i,t-1}$ and $LIT_{i,t-1}{}^*D_{i,t-1}{}^*R_{i,t-1}$, and can easily be transformed into binary choice models with exactly the same independent variables, by using various zero-one dependent variables generated based on the *X*-variable. Similarly, this is the case for various ordered choice models.

Based on this model the following notes and comments are presented.

1. This model is a special model, since the predicted variable uses a subscript *(t − 1)*, or a first lagged variable, namely $X_{i,t-1}$, and one of the predictor variable is using a subscript-*t*, namely $AcqCAR_{i,t}$.

2. Take note that the dummy variable $D_{i,t-1}$, for $i = 1, \ldots, N$; and $t = 1, \ldots, T$; in fact divides the $N(T − 1)$ individual scores of $R_{i,t-1}$ into two groups of scores. So a certain firm can a member of both groups, since the *i*-th firm can have both positive and negative values of *R* over times, say at time points $(t_1 − 1)$ and $(t_2 − 1)$, namely $R_{i,t_1-1} > 0$ and $R_{i,t_2-1} < 0$. Therefore, the objective of data analysis is not to compare two distinct groups of research objects or firms.

8

Univariate General Linear Models

8.1 Introduction

It is easy to see that the equation specifications for all binary choice models presented in previous Chapters 6 and 7 can be used to conduct pool panel data analysis using the univariate general linear models (UGLM or GLM for short). For this reason, this chapter will present only selected general equation specifications of GLMs, such as ANOVA, the Heterogeneous Regression and ANCOVA models. The multivariate general linear model (MGLM), system equation model (SEM) and seemingly causal model (SCM) will be presented later.

8.2 ANOVA and Quantile Models

Referring to the ANOVA model with alternative designs $[A^*B]$, $[A + A^*B]$, $[B + A^*B]$, and $[A + B + A^*B]$ for a numerical endogenous variable Y on two categorical factors A and B, as presented in Agung (2011, 2009a, 2006), then based on pool panel data, the time T should be considered an additional categorical classification factor. Then ANOVA and quantile models can be presented using the following equation specifications.

As an extension of the design $[A^*B]$, the equation specification (ES) of ANOVA and quantile models by the time points has the following equation, which represents the design $[A^*B^*T]$.

$$Y \text{ } @Expand(A, B, T) \tag{8.1}$$

For illustration, the parameters of this model for dichotomous factors A and B, with four time points of observations $(T = 1,\dots,4)$ are presented in Table 8.1. Based on this table various statistical hypotheses on the means differences of Y between cells can easily be written and tested using the Wald test. Some of those hypotheses are as follows:

1. To test the effect of T on Y, conditional for $(A = 1, B = 1)$, the null hypothesis is H_0: $C(1) = C(2) = C(3) = C(4) = 0$. However, if the effect of T on Y, conditional for $(A = i, B = j)$ should be tested for all i

Table 8.1 The mean parameters of the cell-mean model in (8.1)

		T = 1	T = 2	T = 3	T = 4
A = 1	B = 1	C(1)	C(2)	C(3)	C(4)
A = 1	B = 2	C(5)	C(6)	C(7)	C(8)
A = 2	B = 1	C(9)	C(10)	C(11)	C(12)
A = 2	B = 2	C(13)	C(14)	C(15)	C(16)

and j, then applying the following ES is recommended, which represents the design $[A^*B + A^*B^*T]$, because each of them can be tested directly using the *t*-statistic presented in the output.

$$Y \ C \ @Expand(A, B, @Droplast) \ @Expand(A, B)^* @Expand(T, @Droplast) \tag{8.2}$$

2. To test the effect of *B* on *Y*, conditional for *(A = 1, T = 1)*, the null hypothesis is H_0: $C(1) = C(5) = 0$. However, if the effect of *B* on *Y*, conditional for *(A = i, T = t)* should be tested for all *i and t*, then applying the following ES is recommended, which represents the design $[A^*T + A^*T^*B]$.

$$Y \ C \ @Expand(A, T, @Droplast) \ @Expand(A, T)^* @Expand(B, @Droplast) \tag{8.3}$$

3. To test the effect of *A* on *Y*, conditional for *(B = 1, T = 1)*, the null hypothesis is H_0: $C(1) = C(9) = 0$. However, if the effect of *A* on *Y*, conditional for *(B = j, T = t)* should be tested for all *j and t*, then applying the following ES is recommended, which represents the design $[B^*T + B^*T^*A]$.

$$Y \ C \ @Expand(B, T, @Droplast) \ @Expand(B, T)^* @Expand(A, @Droplast) \tag{8.4}$$

Note that these four ESs in fact represent exactly the same model. Construct each table of their parameters as an exercise, in order to identify the characteristics of the parameters of each model. For the other alternative equation specifications refer to various multi-factorial cell-mean models presented in Agung (2011, 2006).

Example 8.1   Two-way cell-mean (ANOVA) and quantile regressions
For illustration Table 8.2 presents the statistical results based on a two-way cell-mean and quantile regression using the following equation specification, for Y = ROA, and X1 = lnSALE, with T > 4, to present the characteristic differences between the group of firms generated by the dummy variable *D_Div* = *D_Dividend*, at two time points before and after the new tax policy. The parameters of the models are presented in Table 8.3. Based on this table, various hypotheses can easily be defined and tested using the *Prob (t-stat)*, the *F*-statistic and *Quasi-LR* statistic in Table 8.3, or the Wald test.

$$Y \ C \ @Expand(T, @droplast)$$
$$@Expand(T)^* @Expand(D\_Div, @Dropfirst) \tag{8.5}$$

Example 8.2   Three-way cell-mean (ANOVA) and quantile regressions
As an illustration Table 8.4 presents the statistical results based on a three-way cell-mean and quantile regression using the following equation specification, for $Y = \log(SALE)$ and $Y = \log(ASSET)$, with $T > 4$, to present the characteristic differences between the group of firms generated by the dichotomous A and B, at two time points before and after the new tax policy:

$$Y \ C \ @Expand(A, B, T, @Droplast) \tag{8.6}$$

Table 8.2 *Summary of the statistical results based on cell-mean and median regressions in (8.5)*

Dep. Var	ROA		X1	
Variable	Coef.	Prob.	Coef.	Prob.
C	0.068	0.769	5.979	0.000
T = 5	−0.070	0.838	−0.315	0.098
T = 6	−0.082	0.800	−0.122	0.499
T = 7	−0.551	0.087	0.193	0.282
(T = 5)*(D_Div = 1)	0.084	0.864	1.646	0.000
(T = 6)*(D_Div = 1)	0.097	0.846	1.387	0.000
(T = 7)*(D_Div = 1)	0.564	0.246	1.243	0.000
(T = 8)*(D_Div = 1)	0.028	0.951	1.544	0.000
R-squared	0.004		0.094	
Adjusted R-squared	−0.002		0.089	
S.E. of regression	3.591		1.989	
Sum squared resid	15 942.7		4855.0	
Log likelihood	−3351.7		−2597.7	
F-statistic	0.639		18.210	
Prob(F-statistic)	0.724		0.000	
C	0.0203	0.0000	6.154 858	0
T = 5	−0.0088	0.0528	−0.1862	0.3718
T = 6	−0.0023	0.5943	−0.0681	0.7544
T = 7	−0.0100	0.0304	0.1423	0.5223
(T = 5)*(Y = 1)	0.0526	0.0000	1.4313	0.0000
(T = 6)*(Y = 1)	0.0388	0.0000	1.3731	0.0000
(T = 7)*(Y = 1)	0.0502	0.0000	1.3152	0.0000
(T = 8)*(Y = 1)	0.0584	0.0000	1.3495	0.0000
Pseudo R-squared	0.0190		0.0585	
Adjusted R-squared	0.0134		0.0532	
S.E. of regression	3.5987		1.9993	
Quantile dependent var	0.0249		6.4362	
Sparsity	0.1031		4.3697	
Quasi LR statistic	172.36		106.42	
Prob(Quasi-LR start)	0.000		0.000	

Table 8.3 *The parameters of the regressions in Table 8.2*

D_Div	T = 5	T = 6	T = 7	T = 8	T(5-8)	T(6-8)	T(7-8)
1	C(1) + C(2) + C(5)	C(1) + C(3) + C(6)	C(1) + C(4) + C(7)	C(1) + C(8)	C(2) + C(5) − C(8)	C(2) + C(6) − C(8)	C(2) + C(7) − C(8)
0	C(1) + C(2)	C(1) + C(3)	C(1) + C(4)	*C(1)*	*C(2)*	*C(3)*	*C(4)*
D_Div(1-0)	*C(5)*	*C(6)*	*C(7)*	*C(8)*	C(5)-C(8)	C(6)-C(8)	C(7)-C(8)

Table 8.4 *Summary of the statistical results based on cell-mean and median regressions in (8.6)*

Dep. Variable	log(SALE), SALE > 0		log(SIZE), SIZE > 0	
Variable	Coef.	Prob.	Coef.	Prob.
C	7.627	0.000	2.082	0.000
A = 1, B = 1, T = 5	−3.212	0.000	−0.451	0.000
A = 1, B = 1, T = 6	−3.519	0.000	−0.480	0.000
A = 1, B = 1, T = 7	−3.254	0.000	−0.515	0.000
A = 1, B = 1, T = 8	−3.322	0.000	−0.496	0.000
A = 1, B = 2, T = 5	−2.325	0.000	−0.400	0.000
A = 1, B = 2, T = 6	−2.279	0.000	−0.441	0.000
A = 1, B = 2, T = 7	−2.191	0.000	−0.450	0.000
A = 1, B = 2, T = 8	−2.411	0.000	−0.442	0.000
A = 2, B = 1, T = 5	−0.549	0.036	−0.051	0.104
A = 2, B = 1, T = 6	−0.426	0.079	−0.044	0.125
A = 2, B = 1, T = 7	−0.438	0.061	−0.029	0.294
A = 2, B = 1, T = 8	−0.388	0.087	−0.024	0.382
A = 2, B = 2, T = 5	−0.081	0.710	−0.017	0.515
A = 2, B = 2, T = 6	−0.055	0.794	−0.003	0.909
A = 2, B = 2, T = 7	0.140	0.489	0.000	0.990
R-squared	0.452		0.594	
Adjusted *R*-squared	0.445		0.589	
S.E. of regression	1.553		0.186	
Sum squared resid	2938.6		42.3	
Log likelihood	−2287.7		336.5	
F-statistic	66.9		119.5	
Prob(*F*-statistic)	0.000		0.000	
C	7.485	0.000	2.072	0.000
A = 1, B = 1, T = 5	−2.785	0.000	−0.407	0.000
A = 1, B = 1, T = 6	−3.020	0.000	−0.416	0.000
A = 1, B = 1, T = 7	−2.758	0.000	−0.423	0.000
A = 1, B = 1, T = 8	−2.997	0.000	−0.439	0.000
A = 1, B = 2, T = 5	−1.925	0.000	−0.350	0.000
A = 1, B = 2, T = 6	−1.895	0.000	−0.362	0.000
A = 1, B = 2, T = 7	−1.739	0.000	−0.380	0.000
A = 1, B = 2, T = 8	−2.013	0.000	−0.378	0.000
A = 2, B = 1, T = 5	−0.625	0.016	−0.050	0.092
A = 2, B = 1, T = 6	−0.114	0.666	−0.043	0.197
A = 2, B = 1, T = 7	−0.268	0.273	−0.032	0.291
A = 2, B = 1, T = 8	−0.125	0.606	−0.033	0.246
A = 2, B = 2, T = 5	0.067	0.785	−0.023	0.463
A = 2, B = 2, T = 6	0.024	0.912	−0.004	0.903
A = 2, B = 2, T = 7	0.241	0.279	−0.001	0.966
Pseudo *R*-squared	0.268		0.411	
Adjusted *R*-squared	0.259		0.404	
S.E. of regression	1.567		0.190	
Quantile dependent var	6.436		1.903	
Sparsity	3.531		0.383	
Quasi-LR statistic	602.62		1166.74	
Prob(Quasi-LR stat)	0.000		0.000	

Table 8.5   *The parameters of the regressions in Eq. (8.17)*

A	B	T = 5	T = 6	T = 7	T = 8
1	1	C(1) + C(2)	C(1) + C(3)	C(1) + C(4)	C(1) + C(5)
1	2	C(1) + C(6)	C(1) + C(7)	C(1) + C(8)	C(1) + C(9)
2	1	C(1) + C(10)	C(1) + C(11)	C(1) + C(12)	C(1) + C(13)
2	2	C(1) + C(14)	C(1) + C(15)	C(1) + C(16)	C(1)

In addition, the following notes are presented.

1. The parameters of the models are presented in Table 8.5. Compare this with Table 8.1 for the model in (8.1). Based on this table, various hypotheses can easily be defined and tested using the *Prob(t-stat)*, the *F*-statistic and *Quasi-LR* statistic in Table 8.4, or the Wald test.
2. The dichotomous ordinal factors A and B, respectively, are generated using the following equations, based on the variables ASSET and LOAN using their medians as the cutting points.

$$A = 1 + 1^*(\text{ASSET} >= @\text{Quantile}(\text{ASSET}, 0.50)) \tag{8.7}$$

$$B = 1 + 1^*(\text{LOAN} >= @\text{Quantile}(\text{LOAN}, 0.50)) \tag{8.8}$$

3. If the two groups of objects should be generated based on a numerical variable for comparison, then selecting or defining the groups using the lowest 10–40% and the highest 10–40% scores of the numerical variable is recommended.

Example 8.3   Unexpected error message

Unexpectedly, the error message *"Near singular matrix"* is obtained, using the following equation specifications, for T > 4.

$$\log(SALE)\ C\ @Expand(A, @Droplast)\ @Expand(A)^*@Expand(B, T, @Droplast) \tag{8.9}$$

$$\log(SALE)\ C\ @Expand(B, @Droplast)\ @Expand(B)^*@Expand(A, T, @Droplast) \tag{8.10}$$

$$\log(SALE)\ C\ @Expand(A, B, @Droplast)\ @Expand(A, B)^*@Expand(T, @Droplast) \tag{8.11}$$

However, by using the following ES (8.12), the statistical results in Figure 8.1 are obtained.

$$\log(SALE)\ C\ @Expand(T, @Droplast)\ @Expand(T)^*@Expand(A, B, @Droplast) \tag{8.12}$$

In a theoretical sense, the ESs in (8.9) to (8.12), as well as the ES in (8.6) in fact represent the same model, that is the means of log(SALE) by the factors A, B and the time T > 4. Note that the values of R-squared, Adjusted R-Squared, up to the F-statistic, and Prob(F-Stat) in this figure are the same as their values in Table 8.4.

On the other hand, for the eight time points, the statistical results in Figure 8.2 were obtained by using the ES (8.9). The results are also shown that were obtained by using the ESs (8.10) and (8.11). These illustrative results have shown again that statistical results and multicollinearity problems are highly dependent on data used, with unexpected estimates of the parameters.

Variable	Coefficient	Std. Error	t-Statistic	Prob.
C	7.626964	0.142931	53.36102	0.0000
T=5	-0.080816	0.217289	-0.371929	0.7100
T=6	-0.054511	0.208827	-0.261032	0.7941
T=7	0.139588	0.201710	0.692021	0.4891
(T=5)*((T=5)*(A=1),B=1)	-3.131307	0.240190	-13.03680	0.0000
(T=5)*((T=5)*(A=1),B=2)	-2.243691	0.261435	-8.582206	0.0000
(T=5)*((T=5)*(A=2),B=1)	-0.467907	0.273859	-1.708573	0.0878
(T=6)*((T=6)*(A=1),B=1)	-3.464696	0.230896	-15.00546	0.0000
(T=6)*((T=6)*(A=1),B=2)	-2.224216	0.245493	-9.060212	0.0000
(T=6)*((T=6)*(A=2),B=1)	-0.371264	0.247879	-1.497763	0.1345
(T=7)*((T=7)*(A=1),B=1)	-3.393707	0.230826	-14.70246	0.0000
(T=7)*((T=7)*(A=1),B=2)	-2.330214	0.244493	-9.530798	0.0000
(T=7)*((T=7)*(A=2),B=1)	-0.577614	0.232832	-2.480818	0.0132
(T=8)*((T=8)*(A=1),B=1)	-3.321939	0.233200	-14.24500	0.0000
(T=8)*((T=8)*(A=1),B=2)	-2.411019	0.251946	-9.569568	0.0000
(T=8)*((T=8)*(A=2),B=1)	-0.387631	0.226573	-1.710841	0.0874

Dependent Variable: LOG(SALE)
Method: Least Squares
Date: 02/21/11   Time: 12:57
Sample: 1 2284 IF T>4 AND SALE > 0
Included observations: 1235

R-squared	0.451690	Mean dependent var	6.275546
Adjusted R-squared	0.444943	S.D. dependent var	2.084010
S.E. of regression	1.552632	Akaike info criterion	3.730651
Sum squared resid	2938.603	Schwarz criterion	3.796968
Log likelihood	-2287.677	Hannan-Quinn criter.	3.755597
F-statistic	66.94629	Durbin-Watson stat	0.499602
Prob(F-statistic)	0.000000		

Estimation Command:
=========================
LS LOG(SALE) C @EXPAND(T,@DROPLAST) @EXPAND(T)*@EXPAND (A,B,@DROPLAST)

Estimation Equation:
=========================
LOG(SALE) = C(1) + C(2)*(T=5) + C(3)*(T=6) + C(4)*(T=7) + C(5)*(T=5)*(A=1 AND B=1) + C(6)*(T=5)*(A=1 AND B=2) + C(7)*(T=5)*(A=2 AND B=1) + C(8)* (T=6)*(A=1 AND B=1) + C(9)*(T=6)*(A=1 AND B=2) + C(10)*(T=6)*(A=2 AND B=1) + C(11)*(T=7)*(A=1 AND B=1) + C(12)*(T=7)*(A=1 AND B=2) + C(13)* (T=7)*(A=2 AND B=1) + C(14)*(T=8)*(A=1 AND B=1) + C(15)*(T=8)*(A=1 AND B=2) + C(16)*(T=8)*(A=2 AND B=1)

Substituted Coefficients:
=========================
LOG(SALE) = 7.6269640099 - 0.0808161281989*(T=5) - 0.0545105768084* (T=6) + 0.139587756833*(T=7) - 3.13130746266*(T=5)*(A=1 AND B=1) - 2.24369145445*(T=5)*(A=1 AND B=2) - 0.467907399668*(T=5)*(A=2 AND B=1) - 3.46469615793*(T=6)*(A=1 AND B=1) - 2.22421605444*(T=6)*(A=1 AND B=2) - 0.371264193428*(T=6)*(A=2 AND B=1) - 3.39370717222*(T=7)* (A=1 AND B=1) - 2.33021363045*(T=7)*(A=1 AND B=2) - 0.577614007442* (T=7)*(A=2 AND B=1) - 3.32193904603*(T=8)*(A=1 AND B=1) - 2.41101854185*(T=8)*(A=1 AND B=2) - 0.387630605221*(T=8)*(A=2 AND B=1)

Figure 8.1   *Statistical results based on the ANOVA model in (8.23)*

Dependent Variable: LOG(SALE)
Method: Least Squares
Date: 02/21/11   Time: 13:39
Sample: 1 2284 IF SALE>0
Included observations: 2270

Variable	Coefficient	Std. Error	t-Statistic	Prob.
C	7.626964	0.140704	54.20587	0.0000
A=1	-2.411019	0.248020	-9.721082	0.0000
(A=1)*((A=1)*(B=1),T=1)	-0.741538	0.254649	-2.912004	0.0036
(A=1)*((A=1)*(B=1),T=2)	-0.851531	0.256515	-3.319620	0.0009
(A=1)*((A=1)*(B=1),T=3)	-0.890560	0.257501	-3.458471	0.0006
(A=1)*((A=1)*(B=1),T=4)	-0.599597	0.269934	-2.221273	0.0264
(A=1)*((A=1)*(B=1),T=5)	-0.801105	0.267706	-2.992485	0.0028
(A=1)*((A=1)*(B=1),T=6)	-1.108188	0.266303	-4.161374	0.0000
(A=1)*((A=1)*(B=1),T=7)	-0.843101	0.271510	-3.105230	0.0019
(A=1)*((A=1)*(B=1),T=8)	-0.910921	0.273165	-3.334690	0.0009
(A=1)*((A=1)*(B=2),T=1)	0.408020	0.297386	1.372022	0.1702
(A=1)*((A=1)*(B=2),T=2)	0.075592	0.288847	0.261701	0.7936
(A=1)*((A=1)*(B=2),T=3)	0.199951	0.286346	0.698284	0.4851
(A=1)*((A=1)*(B=2),T=4)	0.172839	0.290157	0.595675	0.5515
(A=1)*((A=1)*(B=2),T=5)	0.086511	0.286346	0.302121	0.7626
(A=1)*((A=1)*(B=2),T=6)	0.132292	0.278669	0.474728	0.6350
(A=1)*((A=1)*(B=2),T=7)	0.220393	0.282866	0.779142	0.4360

Variable	Coefficient	Std. Error	t-Statistic	Prob.
(A=2)*((A=2)*(B=1),T=1)	-1.431914	0.261661	-5.472403	0.0000
(A=2)*((A=2)*(B=1),T=2)	-1.301506	0.269983	-4.820694	0.0000
(A=2)*((A=2)*(B=1),T=3)	-1.074272	0.252735	-4.250585	0.0000
(A=2)*((A=2)*(B=1),T=4)	-0.524210	0.251115	-2.087530	0.0370
(A=2)*((A=2)*(B=1),T=5)	-0.548724	0.257914	-2.127544	0.0335
(A=2)*((A=2)*(B=1),T=6)	-0.425775	0.238492	-1.785277	0.0744
(A=2)*((A=2)*(B=1),T=7)	-0.438026	0.229566	-1.908065	0.0565
(A=2)*((A=2)*(B=1),T=8)	-0.387631	0.223042	-1.737929	0.0824
(A=2)*((A=2)*(B=2),T=1)	-1.038187	0.261661	-3.967683	0.0001
(A=2)*((A=2)*(B=2),T=2)	-0.925037	0.245102	-3.774084	0.0002
(A=2)*((A=2)*(B=2),T=3)	-0.608202	0.243706	-2.495640	0.0126
(A=2)*((A=2)*(B=2),T=4)	-0.275776	0.218193	-1.263907	0.2064
(A=2)*((A=2)*(B=2),T=5)	-0.080816	0.213902	-0.377818	0.7056
(A=2)*((A=2)*(B=2),T=6)	-0.054511	0.205573	-0.265165	0.7909
(A=2)*((A=2)*(B=2),T=7)	0.139588	0.198567	0.702977	0.4821

R-squared	0.423633	Mean dependent var	5.997284
Adjusted R-squared	0.415649	S.D. dependent var	1.999445
S.E. of regression	1.528433	Akaike info criterion	3.700360
Sum squared resid	5228.207	Schwarz criterion	3.781100
Log likelihood	-4167.908	Hannan-Quinn criter.	3.729816
F-statistic	53.06257	Durbin-Watson stat	0.533778
Prob(F-statistic)	0.000000		

Figure 8.2   *Statistical results based on the ANOVA model in (8.20), for the eight time points*

8.3   Continuous Linear-Effect Models

Based on pool panel data, three types of GLMs may be considered; namely continuous, piece-wise and discontinuous GLMs. This section will present continuous univariate GLMs, for each time point or each cell generated by groups of the research objects and the time variable.

8.3.1   Bivariate Correlation Analysis

I have found that most students and less experienced researchers do not pay much attention to, or do not even know that correlation analysis can be used to replace data analysis based on simple linear regression. Some even say that simple linear regression should be used to prove the effect of an independent variable on the corresponding dependent variable, and this cannot be done using the bivariate correlation. In fact, the causal linear effects between any pairs of variables are defined based on a good and strong theoretical concept. Refer to Example 7.12, which demonstrates that bivariate correlation analysis can certainly be used to replace the data analysis based on simple linear regressions.

For this reason, conducting the correlation analysis based on the set of variables, which will be used in the proposed models, is recommended. This recommendation is made for the following reasons.

1. The results can be used to test the direct linear effect of a cause (up-stream, source, independent, or exogenous) variable/factor on the impact (down-stream, dependent, problem, or endogenous) indicator/ variable.
2. To study or identify two or more cause factors which are very highly correlated, which would otherwise cause uncertainty of their parameter estimates, or multicollinearity problems. In such a case, the researcher should subjectively select and use the most important cause factor, in a theoretical sense.
3. By using selected two- and three-way interaction factors in analysis, the hypotheses can be tested on the linear effect of a cause factor on the down-stream variable(s) and how much this depends on the other variables. See the following example.

Example 8.4   Application of a correlation matrix

Figure 8.3 presents the output of the correlation analysis or correlation matrix based on the set of variables $Y1 = \text{Liability}/1000$, $Y2 = \text{ROA}$, $X1 = \text{lnSALE}$, $X2 = \text{lnLOAN}$ and $X3 = \text{Ni}/1000$ to study their seemingly causal effects or their up- and down-stream relationships, with their causal relationships defined as presented in Figure 8.4, which has been considered in Agung (2011, 2009a). Note that the symbols X1, X2, X3, Y1 and Y2 are used to indicate that the models presented should be applicable for specific set of variables in all studies.

This example presents an alternative analysis; that is, using a correlation matrix. Based on this output the following findings and notes are presented.

1. At the $\alpha < 0.10$ level of significance, each of the variables: $X1 = \text{lnSALE}$, $X2 = \text{lnLOAN}$, $X3 = \text{NI}/1000$, $X1^*X2$, $X1^*X3$, $X2^*X3$, and $X1^*X2^*X3$ has a positive linear effect on Y1, but it has either a positive or negative significant linear effect on $Y2 = \text{Liability}$. So, in a statistical sense, all variables can be used as independent variables of the univariate GLMs of each Y1 and Y2, as well as multivariate GLM of (Y1,Y2).

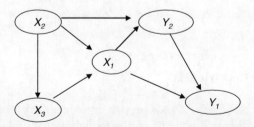

Figure 8.3   *A theoretical seemingly causal model at each time point*

Correlation t-Statistic Probability	Y1	Y2	X1	X2	X3	X1*X2	X1*X3	X2*X3	X1*X2*X3
Y1	1.000000 ----- -----								
Y2	0.048373 0.841614 0.4007	1.000000 ----- -----							
X1	0.470429 9.264308 0.0000	0.362601 6.761494 0.0000	1.000000 ----- -----						
X2	0.180301 3.185497 0.0016	-0.230819 -4.122527 0.0000	0.287578 5.218005 0.0000	1.000000 ----- -----					
X3	0.821414 13.78335 0.0000	0.359405 6.693009 0.0000	0.411439 7.844804 0.0000	-0.062538 -1.088923 0.2771	1.000000 ----- -----				
X1*X2	0.231560 4.136520 0.0000	-0.286792 -5.202461 0.0000	0.210922 3.749784 0.0002	0.927708 43.18671 0.0000	-0.109793 -1.919612 0.0558	1.000000 ----- -----			
X1*X3	0.642840 14.58404 0.0000	0.320297 5.875728 0.0000	0.398052 7.540546 0.0000	-0.047279 -0.822548 0.4114	0.996109 196.4160 0.0000	-0.087849 -1.532585 0.1264	1.000000 ----- -----		
X2*X3	0.215139 3.828370 0.0002	-0.096985 -1.693398 0.0914	-0.101670 -1.776047 0.0767	0.100810 1.760854 0.0793	0.056970 0.991650 0.3222	0.195539 3.464996 0.0006	0.059780 1.040730 0.2988	1.000000 ----- -----	
X1*X2*X3	0.259455 4.688727 0.0000	-0.117552 -2.057092 0.0405	-0.091294 -1.593166 0.1122	0.115481 2.020357 0.0442	0.061370 1.068508 0.2861	0.212641 3.781789 0.0002	0.068047 1.185283 0.2368	0.993961 157.4156 0.0000	1.000000 ----- -----

Figure 8.4 *An output matrix of a bivariate correlation analysis, conditional for T = 8*

2. Based on the previous point, we can conclude that the linear effect of X1 on each of Y1 and Y2 depends on X2, X3, and X2*X3. However, since X1 is significantly correlated with all variables X2, X3, X1*X2, X1*X3, X2*X3 and X1*X2*X3, then any model with X1 as an independent variable does not have to use all other variables as independent variables. Refer to the model in Figure 7.36, which does not have the main factors X2 and X3 as independent variables, and the following example.

3. In addition, take note that X1 has either positive or negative significant effects on X2, X3 and X2*X3, it can then be concluded that the linear effect of X2 (or X3) on X1 depends on X3 (or X2), as presented in the path diagram. On the other hand, X2 has an insignificant linear effect on X3, with a p-value $= 0.2771$.

4. Furthermore, it is also defined that the effect of Y2 on Y1 depends on X1, X2 and X3, as shown in the path diagram. In this case, the correlation matrix as presented in Figure 8.5 should be considered. Based on this correlation matrix, the following findings and notes are presented.

   4.1 At the $\alpha = 0.05$ level of significance, each of the interaction Y2*X1, Y2*X1*X3, Y2*X2*X3, and Y2*X1*X2*X3 has a positive significant effect on Y1, since its t-statistic has Prob < 0.10. So it can be concluded that the effect of Y2 on Y1 significantly depends on X1, X2 and X3, specifically X1, X1*X3, X2*X3 and X1*X2*X3.

   4.2 Since Y2 is significantly correlated with each of Y2*X1, Y2*X2, Y2*X1*X2, Y2*X1*X3, Y2*X2*X3, and Y2*X1*X2*X3, then the GLM of Y1 on Y2, Y2*X1, Y2*X2, Y2*X1*X2, Y2*X1*X3, Y2*X2*X3 and Y2*X1*X2*X3, does not have all of the variables as independent variables. See the following example.

5. Finally, take note that based on the five variables, there are $C_5^2 = 10$ pairs of possible up- and downstream relationships, where each could have either unidirectional or simultaneous causal effects. So many path diagrams would be possible by drawing an arrow or double arrows subjectively between all pairs of variables. Figure 8.4 presents only one out of many of the possible path diagrams – it is defined as having only five unidirectional up- and down-stream relationships. As an exercise, select your own five variables, and then construct the two best possible path diagrams.

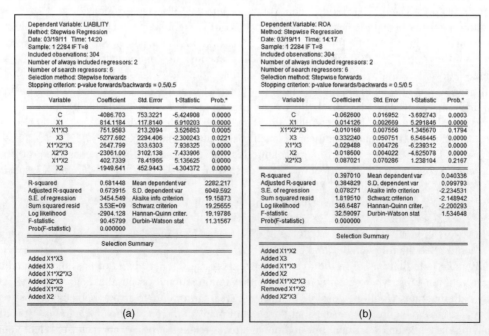

Correlation t-Statistic Probability	Y1	Y2	Y2*X1	Y2*X2	Y2*X1*X2	Y2*X1*X3	Y2*X2*X3	Y2*X1*X2...
Y1	1.000000							

Y2	0.048373	1.000000						
	0.841614	-----						
	0.4007	-----						
Y2*X1	0.109331	0.940158	1.000000					
	1.911437	47.94946	-----					
	0.0569	0.0000	-----					
Y2*X2	0.046109	0.241416	0.138541	1.000000				
	0.802145	4.323230	2.431034	-----				
	0.4231	0.0000	0.0156	-----				
Y2*X1*X2	0.050304	0.138177	0.037829	0.948479	1.000000			
	0.875298	2.424512	0.657868	52.02248	-----			
	0.3821	0.0159	0.5111	0.0000	-----			
Y2*X1*X3	0.479815	0.379380	0.562829	-0.097386	-0.172381	1.000000		
	9.503755	7.125635	11.83310	-1.700474	-3.041180	-----		
	0.0000	0.0000	0.0000	0.0901	0.0026	-----		
Y2*X2*X3	0.153198	-0.368240	-0.465130	0.217793	0.381862	-0.327923	1.000000	
	2.694100	-6.882987	-9.130955	3.877921	7.180171	-6.032245	-----	
	0.0075	0.0000	0.0000	0.0001	0.0000	0.0000	-----	
Y2*X1*X2*X3	0.158606	-0.342202	-0.445133	0.233345	0.395761	-0.336622	0.995884	1.000000
	2.791621	-6.328940	-8.638641	4.170236	7.489056	-6.212420	190.9329	-----
	0.0056	0.0000	0.0000	0.0000	0.0000	0.0000	0.0000	-----

Figure 8.5   *Correlation matrix of a set of variables, conditional for T = 8*

8.3.2   STEPLS Regressions

Example 8.5   Illustrative STEPLS regressions

As the continuation of the analysis presented in previous example, Figure 8.6 presents the statistical results based on two STEPLS regressions using "Yg C X1", for each g = 1, and 2; as the equation specification, and

Dependent Variable: LIABILITY
Method: Stepwise Regression
Date: 03/19/11   Time: 14:20
Sample: 1 2284 IF T=8
Included observations: 304
Number of always included regressors: 2
Number of search regressors: 6
Selection method: Stepwise forwards
Stopping criterion: p-value forwards/backwards = 0.5/0.5

Variable	Coefficient	Std. Error	t-Statistic	Prob.*
C	-4086.703	753.3221	-5.424908	0.0000
X1	814.1184	117.8140	6.910203	0.0000
X1*X3	751.9583	213.2094	3.526853	0.0005
X3	-5277.692	2294.406	-2.300243	0.0221
X1*X2*X3	2647.799	333.6303	7.936325	0.0000
X2*X3	-23061.00	3102.138	-7.433906	0.0000
X1*X2	402.7339	78.41965	5.135625	0.0000
X2	-1949.641	452.9443	-4.304372	0.0000

R-squared	0.681448	Mean dependent var		2282.217
Adjusted R-squared	0.673915	S.D. dependent var		6049.592
S.E. of regression	3454.549	Akaike info criterion		19.15873
Sum squared resid	3.53E+09	Schwarz criterion		19.25655
Log likelihood	-2904.128	Hannan-Quinn criter.		19.19786
F-statistic	90.45799	Durbin-Watson stat		11.31567
Prob(F-statistic)	0.000000			

Selection Summary

Added X1*X3
Added X3
Added X1*X2*X3
Added X2*X3
Added X1*X2
Added X2

(a)

Dependent Variable: ROA
Method: Stepwise Regression
Date: 03/19/11   Time: 14:17
Sample: 1 2284 IF T=8
Included observations: 304
Number of always included regressors: 2
Number of search regressors: 6
Selection method: Stepwise forwards
Stopping criterion: p-value forwards/backwards = 0.5/0.5

Variable	Coefficient	Std. Error	t-Statistic	Prob.*
C	-0.062600	0.016952	-3.692743	0.0003
X1	0.014126	0.002669	5.291846	0.0000
X1*X2*X3	-0.010168	0.007556	-1.345670	0.1794
X3	0.332240	0.050751	6.546445	0.0000
X1*X3	-0.029488	0.004726	-6.239312	0.0000
X2	-0.018600	0.004022	-4.625078	0.0000
X2*X3	0.087021	0.070286	1.238104	0.2167

R-squared	0.397010	Mean dependent var		0.040336
Adjusted R-squared	0.384829	S.D. dependent var		0.099793
S.E. of regression	0.078271	Akaike info criterion		-2.234531
Sum squared resid	1.819510	Schwarz criterion		-2.148942
Log likelihood	346.6487	Hannan-Quinn criter.		-2.200293
F-statistic	32.59097	Durbin-Watson stat		1.534648
Prob(F-statistic)	0.000000			

Selection Summary

Added X1*X2
Added X3
Added X1*X3
Added X2
Added X1*X2*X3
Removed X1*X2
Added X2*X3

(b)

Figure 8.6   *Statistical results based on two STEPLS regressions of Y1 = Liability and Y2 = ROA, on X1, X2, X3, X1*X2, X1*X3, X2*X3 and X1*X2*X3, conditional for T = 8*

"X2 X3 X1*X2 X1*X3 X2*X3 X1*X2*X3" as the list of search regressors. Based on these results the following notes and comments are made.

1. Take note that Figure 8.6(a) presents a regression of LIABILITY having a very large Durbin–Watson statistic of DW = 11.31 567. It is an unexpected value, since the data is a cross-section at T = 8. Compare this to the regression of ROA having a DW = 1.534 648, presented in Figure 8.6(b). See also the following regressions which have very small or large DW-statistic values, even when based on cross-section data. Should autoregressive models be applied? Refer to the autoregressive models presented next.

2. The regression with the dependent variable Y1 = Liability has the three main factors X1, X2 and X3, with all their two- and three-way interactions: it is a hierarchical model. On the other hand, the regression with the dependent variable Y2 = ROA is a nonhierarchical model, since it does not have X1*X2 in the model.

3. Both models are acceptable in showing that the linear effect of X1 on each of Y1 and Y2, depends on X2, X3 and X2*X3.

4. In addition to the models in Figure 8.6 based on the path diagram in Figure 8.5, Figure 8.7 presents the statistical results based on a model to show the linear effect of Y2 on Y1, which depends on X1, X2 and X3.

5. However, this is unexpected, the DW-statistics of the two regressions in Figure 8.7 are very large, greater than 13.0. In fact, at the time T = 8, the data is cross-section data, where the error terms of the models should not have autocorrelation problems. What happened with the DW-statistic or the regression cannot be explained yet.

6. Finally, Figure 8.8 presents statistical results based on two GLMs to show the causal or up- and downstream relationships between the exogenous variables X1, X2 and X3 as presented in Figure 8.3. The first GLM to present the linear effect of X3 on X1 depends on X2, with a large DW = 3.182 942, but the second presents the linear effect of X2 on X3, with a very small DW = 0.391 498.

Figure 8.7 *Statistical results based on a STEPLS regression to presents the linear effect of Y2 on Y1, depends on X1, X2 and X3; and its reduced model*

```
Dependent Variable: X1
Method: Least Squares
Date: 03/19/11  Time: 21:40
Sample: 1 2284 IF T=8
Included observations: 304
```

Variable	Coefficient	Std. Error	t-Statistic	Prob.
C	6.210537	0.104735	59.29742	0.0000
X2	0.551044	0.081336	6.774878	0.0000
X3	0.790700	0.087324	9.054790	0.0000
X2*X3	-0.514821	0.157098	-3.277059	0.0012

R-squared	0.293134	Mean dependent var	6.487666
Adjusted R-squared	0.286065	S.D. dependent var	2.072544
S.E. of regression	1.751189	Akaike info criterion	3.971538
Sum squared resid	919.9993	Schwarz criterion	4.020446
Log likelihood	-599.6738	Hannan-Quinn criter.	3.991102
F-statistic	41.46947	Durbin-Watson stat	3.182943
Prob(F-statistic)	0.000000		

(a)

```
Dependent Variable: X3
Method: Least Squares
Date: 03/19/11  Time: 21:41
Sample: 1 2284 IF T=8
Included observations: 307
```

Variable	Coefficient	Std. Error	t-Statistic	Prob.
C	0.287011	0.065694	4.368896	0.0000
X2	-0.056554	0.052883	-1.069429	0.2857

R-squared	0.003736	Mean dependent var	0.287144
Adjusted R-squared	0.000469	S.D. dependent var	1.151323
S.E. of regression	1.151053	Akaike info criterion	3.125724
Sum squared resid	404.1013	Schwarz criterion	3.150003
Log likelihood	-477.7987	Hannan-Quinn criter.	3.135433
F-statistic	1.143677	Durbin-Watson stat	0.391498
Prob(F-statistic)	0.285722		

(b)

Figure 8.8   *Statistical results based on two models to present (a) the linear effect of X3 on X1 depends on X2, and (b) the linear effect of X2 on X3, conditional for T = 8*

Example 8.6   Another illustrative STEPLS regression

Using the same forward stepwise selection method presented in previous example, Figure 8.9 presents the statistical results based on two STEPLS regressions, using the equation specification "Yg C X1" with Y3 = PBV and Y4 = PDB, and "X2, X3 X1*X2 X1*X3 X2*X3 X1*X2*X3" as the list of search regressors. Based these outputs, the following findings and notes are presented.

1. Based on the first output, the following regression function can be presented.

$$Y3 = [8.313 + 1.921X2 + 10.646X3 - 1.016X2 \times X3] - [0.999 + 0.878X3]X1$$

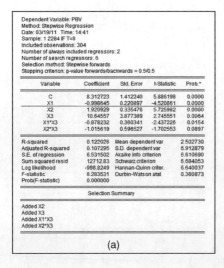

Figure 8.9   *Statistical results based on STEPLS regressions of Y3 = PBV and Y4 = PDB, on X1, X2, X3, X1*X2, X1*X3, X2*X3 and X1*X2*X3, conditional for T = 8, which are two-way interaction models*

(a)

Dependent Variable: PBV
Method: Stepwise Regression
Date: 03/20/11 Time: 07:49
Sample: 1 2284 IF T=8
Included observations: 304
Number of always included regressors: 4
Number of search regressors: 3
Selection method: Stepwise forwards
Stopping criterion: p-value forwards/backwards = 0.5/0.5

Variable	Coefficient	Std. Error	t-Statistic	Prob.*
C	8.393611	1.422388	5.901068	0.0000
X1*X2	-0.090638	0.148444	-0.610582	0.5419
X1*X3	-0.799775	0.370194	-2.160425	0.0315
X1*X2*X3	-0.095831	0.067038	-1.429511	0.1539
X1	-0.993171	0.220601	-4.502114	0.0000
X2	2.402213	0.857390	2.801774	0.0054
X3	9.777590	3.990794	2.450036	0.0149

R-squared	0.122881	Mean dependent var	2.502730
Adjusted R-squared	0.105161	S.D. dependent var	6.912879
S.E. of regression	6.539301	Akaike info criterion	6.616295
Sum squared resid	12700.45	Schwarz criterion	6.701884
Log likelihood	-998.8768	Hannan-Quinn criter.	6.650532
F-statistic	6.934753	Durbin-Watson stat	0.350409
Prob(F-statistic)	0.000001		

(b)

Dependent Variable: PBV
Method: Least Squares
Date: 03/20/11 Time: 07:53
Sample: 1 2284 IF T=8
Included observations: 304

Variable	Coefficient	Std. Error	t-Statistic	Prob.
C	8.292136	1.411159	5.876119	0.0000
X1*X3	-0.850569	0.360346	-2.360423	0.0189
X1*X2*X3	-0.107602	0.064139	-1.677644	0.0945
X1	-0.992378	0.220365	-4.503341	0.0000
X2	1.920574	0.335623	5.722421	0.0000
X3	10.34986	3.875090	2.670870	0.0080

R-squared	0.121780	Mean dependent var	2.502730
Adjusted R-squared	0.107045	S.D. dependent var	6.912879
S.E. of regression	6.532416	Akaike info criterion	6.610970
Sum squared resid	12716.39	Schwarz criterion	6.684333
Log likelihood	-998.8675	Hannan-Quinn criter.	6.640317
F-statistic	8.264536	Durbin-Watson stat	0.379764
Prob(F-statistic)	0.000000		

Figure 8.10 *Statistical results based on another STEPLS regression of Y3 = PBV, and its reduced model, conditional for T = 8, which are three-way interaction models*

which shows that the linear effect of X1 on Y3 depends only on X3, specifically on [0.999 + 0.878X3] with a negative sign in a statistical sense, even though it is theoretically defined that the linear effect of X1 on Y3 depends on X2 and X3. Compare this with the second output.

2. Furthermore, compare the regression with the STEPLS regression of PBV presented in Figure 8.10(a), and its reduced model in Figure 8.10(b), which are obtained using a combination of trial-and-error and stepwise estimation methods. Note that these results show two three-way interaction models, compared to the two-way interaction in Figure 8.9(a).

3. So, at the first stage, there is a choice between a two-way interaction model and a three-way interaction model. For sure, it would be very subjective. If the three-way interaction model should be selected, then the reduced model in Figure 8.10(b) would be the good fit model for the data set used. Take note of the statement "The model must follow the data, and not the other way around" (Benzecri, 1973, cited in Gifi, 1991, p. 25).

4. This example has demonstrated that several, or perhaps many, good fit linear-effect models can be obtained based on a set of two endogenous and three exogenous numerical variables, by using a multi-stage stepwise selection method with different sets of search regressors. Refer to the previous chapter and Chapter 11 in Agung (2011).

8.4 Piece-Wise Autoregressive Linear Models by Time Points

As the extension of the continuous linear-effect GLMs at a time point presented in previous subsection, this subsection presents the piece-wise linear-effect GLMs of Y on $\mathbf{X} = (X1,\ldots,Xk,\ldots,XK)$ and the discrete time variable, with the general equation specification as follows:

$$H(Y) = \sum_{t=1}^{T} F_t(X1,\ldots,Xk,\ldots,XK) \times Dt \tag{8.13}$$

where $H(Y)$ is a fixed function of Y, $F_t(\mathbf{X})$ is a function of $\mathbf{X} = (X1,\ldots,Xk,\ldots,XK)$ with a finite number of parameters, and Dt is time dummy variables, for $t = 1,2,\ldots,T$.

Note that the function $H(Y)$ could be equal to Y, $\log(Y)$, $\log(Y - L)$, $\log(U - Y)$ or $\log[(Y - L)/(U - Y)]$ where L and U, respectively, are the lower and upper bounds of Y, which are subjectively selected, in a theoretical sense.

Since, in empirical results, $F_s(\mathbf{X})$ and $F_t(\mathbf{X})$, for $s \neq t$, will not have the same set of variables, then I would recommend conducting the analysis for each time point, especially for large number of variables, using the same multistage stepwise selection method. Take note that the cross-section data at each time can be considered a random sample, but the pool data set certainly cannot be a random sample so the error terms of the model in (8.13) should be autocorrelated. For this reason the models based on the whole panel data set would be either AR(p), LV(q) or LVAR(p,q) models (see Agung, 2009a). See the following additional examples.

8.4.1   The Simplest Linear-Effect Models Based on (X_{it}, Y_{it}) by Time Points

Example 8.7   (A set of the simplest linear regressions)
For illustration, Figure 8.11 presents four alternative graphs of the simplest linear regressions in a two-dimensional space, of a numerical variable X and Y, based on the pooled data (X_{it}, Y_{it}), $i = 1, \ldots, N$, and $t = 1, \ldots, T$. Based on this figure, the following notes are presented.

1. Figure 8.11(a) presents the graph of a set of linear-effect regressions by time points, which has various intercepts α_t and slopes β_t, for $t = 1, 2, \ldots$ This set of heterogeneous regressions is commonly observed in practice. However, also note its limitation; that is that all individuals or objects at each time point are represented only by a simple linear regression in a two-dimensional space.

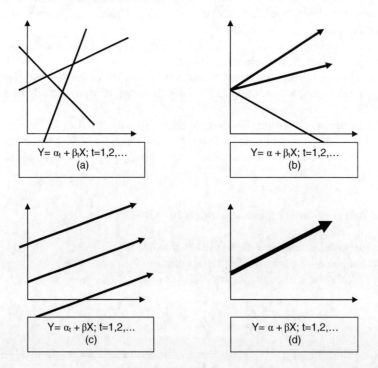

Figure 8.11   *Alternative of the simplest linear-effect regressions by time points*

2. The other three sets are its special cases, but I would not consider recommending them, with the worst possible model presented in Figure 8.11(d). This is because all research objects at all the time points are presented only by single linear regressions.

3. Even the simplest linear-effect model for a large number of individuals or research objects at each time point should be questionable, since they should belong to certain groups with distinct statuses or behaviors. However, if you have to present a simple linear regression based on a pooled data (X_{it}, Y_{it}), and Dt, $i = 1,\ldots,N$, and $t = 1,\ldots,T$, then I would recommend applying the models as follows:

$$Y_{it} = C(t1) + C(t2)^* X1_{it} + \varepsilon_{it}, \text{for each } t = 1, 2, \ldots \tag{8.14}$$

and data analysis can easily be done using the following two alternative equation specifications.

$$Y\, C\, X1^*@Expand(T)\, @Expand(T, @droptlast) \tag{8.14a}$$

$$Y\, X1^*@Expand(T)\, @Expand(T) \tag{8.14b}$$

or by using the time dummy variables: D1, D2, ..., DT, the equation specification is as follows:

$$Y = (C(10) + C(11)^* X1)^* D1 + \ldots + (C(T0) + C(T1)^* X1)^* DT \tag{8.15}$$

8.4.2 General Linear-Effect Model Based on (X_{it}, Y_{it}) by Time Points

As the extension of the set of the simplest linear regressions with the equation specification (ES) in (8.14), the following ES represents the set of general simple linear regression of H(Y) on F(X).

$$H(Y)\, F(X)^*@Expand(T)\, @Expand(T) \tag{8.16}$$

where

$H(Y)$ is a function of a numerical endogenous variable Y without a parameter, such as Y, $\log(Y - L)$, $\log(U - Y)$, and $\log[(Y - L)/(U - Y)]$, where L and U are fixed lower and upper bounds of Y, in a theoretical sense; and
$F(X)$ is a function of a numerical exogenous variable X without a parameter, such as $\log(X - \delta)$, $1/(X - \delta)$, $(X - \delta)^2$, $(X - \delta)^2(X - \gamma)$, and $(X - \delta)^\theta$, where δ, γ and θ are subjectively selected fixed numbers.

Take note that the acceptability or validity of the linear relationships between selected functions $F(X)$ and $H(Y)$ can be identified using the scatter graph with a regression line or nearest neighbor fit curve. See the following example.

For a more advanced general simple linear-effect model by times, the following ES with dummy variables can be applied.

$$\breve{G}_t(Y)\, F_1(X)^* D1\, \ldots\, F_T(X)^* DT\, @Expand(T) \tag{8.17}$$

where $F_t(X)$, $t = 1,\ldots,T$; are function of a single exogenous variable X without a parameter, and $\breve{G}_t(Y)$ is a new function of Y without a parameter, which could be different functions for each time point.

Example 8.8 A set of general simple linear regressions

Figure 8.12 presents scatter graphs of Y1 on X1, for each T = 6, 7 and 8, with their nearest neighbor fit (NNF) curves. It is very clear that the simple linear regression of Y1 on X1 is the worst regression for each time point.

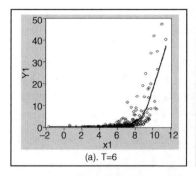 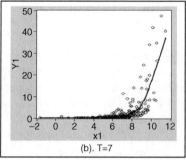 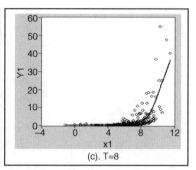

Figure 8.12   *Scatter graphs of (X1,Y1) for each T = 6, 7 and 8, with their NNF curves*

Therefore, the equation specifications in (8.14 and (8.15)), as well as the following equations, represent inappropriate or not-recommended models, for all T > 5.

$$Y1 = C(10) + C(1)^*X1^*Dt6 + C(2)^*X1)^*Dt7 + C(3)^*X1^*Dt8 \tag{8.18}$$

$$Y1 = (C(1)^*X1 + C(10)^*Dt6 + C(20)^*Dt7 + C(30)^*Dt8 \tag{8.19}$$

$$Y1 = C(1) + C(2)^*X1 \tag{8.20}$$

Take note that corresponding to the graphs in Figure 8.11, the model in (8.18) is represented by Figure 8.11b), the model in (8.19) is represented by Figure 8.11c, and the model in (8.20) is represented by Figure 8.11(d), which should be considered to be the worst in general since all individuals or the research objects at all the time points are presented using a single regression function only. However, in practice, it could be a good fit model for the data set which happens to be available to researchers.

So corresponding to the ES in (8.16), the functions G(Y1) and F(X1) should be explored. By observing each of the scatter graphs in Figure 8.13, the following alternative functions are obtained using the trial-and-error method, specific to *T* = 8. Try the other time points: *T* = 6 and *T* = 7, as an exercise.

1. *The first alternative: hyperbolic effect of X1 on Y1.*

Since each of the NNF curves shows an orthogonal hyperbolic curve with the lines X1 = 12 and Y1 = 0 as their asymptotes, then the first selected function should be G(Y1) = Y1 and F(X1) =

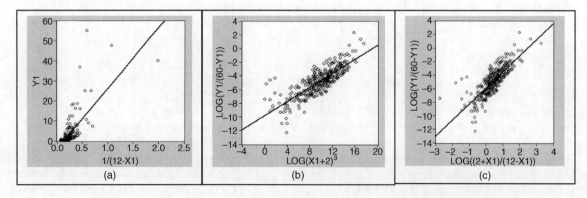

Figure 8.13   *Scatter graphs of selected (F(X1),G(Y1)), for T= 8, with their regression lines*

$1/(12 - X1)$. So, the hyperbolic fitted curve will have the general equation $Y1 = a + b/(12 - X1)$, of $(Y1 - a)(12 - X1) = b$. For this reason, the scatter graph of Y1 on $1/(12 - X1)$ with regression line is presented in Figure 8.13(a), specific for $T = 8$. Note the graph indicates that the error terms of the SLR model are heterogeneous – their variances increase with increasing score of X1. So it can be said that a simple regression line of Y1 on $1/(12 - X1)$ is an inappropriate regression.

2. *The second alternative: Bounded third degree polynomial-effect of X1 on Y1.*

It can be said that the scores of a numerical problem indicator Y1 should have lower and always upper bounds. For this reason, a bounded translog linear model (Agung, 2011, 2009a, 2008, 2006) should be applied using the dependent variable $G(Y1) = \log(Y1/(60 - Y1))$. Figure 8.13(b) presents the scatter graph of $\log(Y1/(60 - Y1))$ on $(\log(X1 + 2))^3$, with a regression line, which shows that the SLR of $\log(Y1/(60 - Y1))$ on $(\log(X1 + 2))^3$ is an acceptable model, in a statistical sense. Take note that $\log(Y1/(60 - Y1))$ is valid for all Y1, such that $0 < Y1 < 60$, and $\log(X1)$ is valid for $X1 > 0$.

3. *The third alternative: Bounded translog linear model.*

In addition to the dependent $G(Y1) = \log(Y1/(60 - Y1))$, a bounded exogenous variable is generated and $F(X1) = \log(X1 + 2)/(12 - X1)$, based on the graph in Figure 8.13(c). Note that Figure 8.13(c) shows that the simple linear regression $\log(Y1/(60 - Y1)) = a + b \times \log((X1 + 2)/(12 - X1))$ is an acceptable regression. However, based on the graph in Figure 8.12(a) and (b), the function $G(Y1) = \log(Y1/(50 - Y1))$ can be applied for $T = 6$ and $T = 7$. So to conduct the analysis based on the whole data set $T > 5$ (or $T = 6$, 7 and 8), a new dependent variable, namely *Y1_New*, should be generated using the following three equations based on the sub-samples $T = 6$, $T = 7$, and $T = 8$, respectively. Refer to the ES in (8.18) and see the following example.

$$Y1\_New = \log(Y1/(50 - Y1)), \textit{for the sub-sample}(T = 6)$$
$$Y1\_New = \log(Y1/(50 - Y1)), \textit{for the sub-sample}(T = 7) \tag{8.21}$$
$$Y1\_New = \log(Y1/(60 - Y1)), \textit{for the sub-sample}(T = 8)$$

4. *Other alternatives.*

I suggest to readers to use trial-and-error methods to define the functions H(Y) and F(X) based the Data_sesi.wf1 and on their own data sets.

8.4.2.1 Bounded Simple Regression Models

Example 8.9   Application of bounded models

For illustration of the variable Y1_New in (8.18), for $T > 5$ (or $T = 6$, 7 and 8), the following ES was applied, but an error message "*Log non positive number*" obtained, which should be related to the scores of X1, at each of the time points.

$$Y1\_New \log((2 + X1)/(12 - X1))^* Expand(T) \, @Expand(T) \tag{8.22}$$

We find that the scores of X1 at $T = 6$ have a minimum score of -2.659260, leading to $(2 + X1) < 0$. Therefore, the analysis can be done using the ES (8.22), based on the sub-sample $\{X1 > -2\}$ with the statistical results presented in Figure 8.14(a). Otherwise, for all $T > 5$; the independent variable $\log((3 + X1)/(12 - X1))$ should be used. Based on this output, the following findings and notes are presented.

1. Based on pooled data, the error terms of any basic or common models, such as in the model (8.19), should be autocorrelated, indicated by a small DW value: $DW = 0.512\,403$. Therefore, the output based on an AR(1) model is presented in Figure 8.12(b).

Dependent Variable: Y1_NEW				
Method: Least Squares				
Date: 04/07/11 Time: 12:48				
Sample: 1 2284 IF T > 5 AND X1 > -2				
Included observations: 958				

Variable	Coefficient	Std. Error	t-Statistic	Prob.
LOG((2+X1)/(12-X1))*(T=6)	2.373934	0.100071	23.72248	0.0000
LOG((2+X1)/(12-X1))*(T=7)	2.501867	0.097700	25.60770	0.0000
LOG((2+X1)/(12-X1))*(T=8)	2.376150	0.096961	24.50630	0.0000
T=6	-5.755199	0.078839	-72.99894	0.0000
T=7	-5.826874	0.081629	-71.38238	0.0000
T=8	-5.927293	0.080764	-73.39061	0.0000

R-squared	0.657128	Mean dependent var	-4.823256
Adjusted R-squared	0.655327	S.D. dependent var	2.102367
S.E. of regression	1.234275	Akaike info criterion	3.265089
Sum squared resid	1450.311	Schwarz criterion	3.295557
Log likelihood	-1557.977	Hannan-Quinn criter.	3.276693
Durbin-Watson stat	0.512403		

(a)

Dependent Variable: Y1_NEW				
Method: Least Squares				
Date: 04/07/11 Time: 12:53				
Sample: 1 2284 IF T > 5 AND X1 > -2				
Included observations: 955				
Convergence achieved after 7 iterations				

Variable	Coefficient	Std. Error	t-Statistic	Prob.
LOG((2+X1)/(12-X1))*(T=6)	2.177559	0.073242	29.73110	0.0000
LOG((2+X1)/(12-X1))*(T=7)	2.224016	0.085112	26.13057	0.0000
LOG((2+X1)/(12-X1))*(T=8)	2.213444	0.086459	25.60095	0.0000
T=6	-5.868081	0.063300	-89.54310	0.0000
T=7	-5.960333	0.079704	-71.01680	0.0000
T=8	-5.844274	0.082805	-70.57871	0.0000
AR(1)	0.722005	0.029767	24.25508	0.0000

R-squared	0.795748	Mean dependent var	-4.816454
Adjusted R-squared	0.794455	S.D. dependent var	2.100831
S.E. of regression	0.952454	Akaike info criterion	2.747754
Sum squared resid	859.9967	Schwarz criterion	2.783390
Log likelihood	-1305.053	Hannan-Quinn criter.	2.761328
Durbin-Watson stat	1.726303		

| Inverted AR Roots | .72 | | |

(b)

Figure 8.14   *Statistical results based on the model in (8.22), and its AR(1) model*

2. The characteristic differences between the two models in Figure 8.12 are presented in Figures 8.15 and 8.16. Figure 8.15 presents the scatter graph of FitY1_New on X1_New $= \log((2+X1)/(12-X1))$ which is a straight line for each $T > 5$, but the scatter graph of Fit_AR on X1_New is not, as presented in Figure 8.16.

3. Referring to the three regression lines in Figure 8.15, they seem to have the same slopes. We accept the null hypothesis H_0: $C(1) = C(2) = C(3)$ based on the F-statistic of $F_0 = 0.558\,535$ with $df = (2,952)$ and p-value $= 0.5722$, using the model in Figure 8.15(a), and using the AR(1) model in Figure 8.15(b), it is accepted based on the F-statistic of $F_0 = 0.163\,144$ with $df = (2,948)$ and p-value $= 0.8495$.

4. In a statistical sense, based on the finding previously then the model can be reduced to a homogeneous regression line, using the following ES, with the statistical results based on an OLS regression and its corresponding acceptable AR(2) model presented in Figure 8.17. Based on these results the following notes and comments are presented.

$$Y1\_New \log((X1+2)/(12-X1)) \,@Expand(T) \tag{8.23}$$

4.1 Corresponding to characteristics of the pool panel data which are time-dependent, then the OLS regression in Figure 8.17(a) has a small DW-statistic of 0.509 971. For this reason, the statistical result based on an acceptable AR(2) model, in a statistical sense, is presented in Figure 8.17(b).

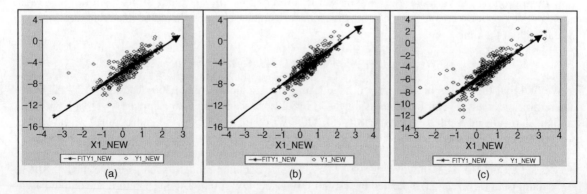

(a)   (b)   (c)

Figure 8.15   *Scatter graphs of Y1_New and FitY1_New on X1_new, based on the output in Figure 8.14(a), for T = 6, 7 and 8*

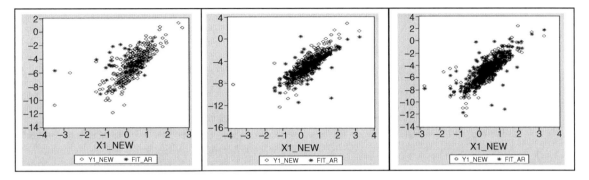

Figure 8.16 *Scatter graphs of Y1_New and Fit_AR on X1_new, based on the output of the AR(1) model in Figure 8.14(b), for T = 6, 7 and 8*

4.2 Based on the output in Figure 8.17(b), the following findings are presented.

- A set of three AR(2) homogeneous regressions as follows:

$$\text{Y1\_NEW} = 2.497\text{X1\_NEW} - 5.500 + [\text{AR}(1) = 0.490, \text{AR}(2) = -0.124], \text{for } T = 6$$
$$\text{Y1\_NEW} = 2.497\text{X1\_NEW} - 5.577 + [\text{AR}(1) = 0.490, \text{AR}(2) = -0.124], \text{for } T = 7$$
$$\text{Y1\_NEW} = 2.497\text{X1\_NEW} - 5.599 + [\text{AR}(1) = 0.490, \text{AR}(2) = -0.124], \text{for } T = 8$$

- By inserting the original variables Y1 = liability/1000 and X1 = lnSale = log(Sale), a set of complex regressions is obtained. For instance, a double logarithmic regression for *T* = 6, as follows:

$$\log\left(\frac{liability/1000}{50 - liability/1000}\right) = 2.497\log\left(\frac{2 + \log(Sale)}{12 - \log(Sale)}\right) - 5.500$$
$$+ [AR(1) = 0.490, AR(2) = -0.124]$$
(8.24)

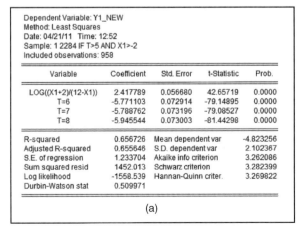

Dependent Variable: Y1_NEW
Method: Least Squares
Date: 04/21/11 Time: 12:52
Sample: 1 2284 IF T>5 AND X1>-2
Included observations: 958

Variable	Coefficient	Std. Error	t-Statistic	Prob.
LOG((X1+2)/(12-X1))	2.417789	0.056680	42.65719	0.0000
T=6	-5.771103	0.072914	-79.14895	0.0000
T=7	-5.788762	0.073196	-79.08527	0.0000
T=8	-5.945544	0.073003	-81.44298	0.0000
R-squared	0.656726	Mean dependent var		-4.823256
Adjusted R-squared	0.655646	S.D. dependent var		2.102367
S.E. of regression	1.233704	Akaike info criterion		3.262086
Sum squared resid	1452.013	Schwarz criterion		3.282399
Log likelihood	-1558.539	Hannan-Quinn criter.		3.269822
Durbin-Watson stat	0.509971			

(a)

Dependent Variable: Y1_NEW
Method: Least Squares
Date: 04/21/11 Time: 12:49
Sample: 1 2284 IF T>5 AND X1>-2
Included observations: 950
Convergence achieved after 8 iterations

Variable	Coefficient	Std. Error	t-Statistic	Prob.
LOG((X1+2)/(12-X1))	2.496757	0.078372	31.85787	0.0000
T=6	-5.499767	0.065064	-84.52849	0.0000
T=7	-5.773047	0.074084	-77.92552	0.0000
T=8	-5.993349	0.073562	-81.47311	0.0000
AR(1)	0.490182	0.031082	15.77083	0.0000
AR(2)	-0.123953	0.031909	-3.884602	0.0001
R-squared	0.748239	Mean dependent var		-4.809395
Adjusted R-squared	0.746905	S.D. dependent var		2.101660
S.E. of regression	1.057314	Akaike info criterion		2.955637
Sum squared resid	1055.311	Schwarz criterion		2.986309
Log likelihood	-1397.928	Hannan-Quinn criter.		2.967324
Durbin-Watson stat	1.372532			
Inverted AR Roots	.25-.25i	.25+.25i		

(b)

Figure 8.17 *Statistical results based on the model in (8.23a), and its AR(2) model*

- However, for $T = 8$, the following regression is obtained.

$$\log\left(\frac{liability/1000}{60 - liability/1000}\right) = 2.497\log\left(\frac{2 + \log(Sale)}{12 - \log(Sale)}\right) - 5.599$$
$$+ [AR(1) = 0.490, AR(2) = -0.124]$$

(8.25)

- Corresponding to a set of homogenous regressions, then it should be we should investigate further as to whether they could be presented as a single line, in a statistical sense. For this reason, the null hypothesis H_0: $C(2) = C(3) = C(4)$ should be considered. Finally, we find it is rejected based on the F-statistic of $F_0 = 17.46\,442$ with $df = (2,944)$ and p-value $= 0.0000$. So the data does not support a single regression line of Y1_new on X1_New, for $T > 5$. In other words, a single regression line, without the time T, is not an appropriate model for a pool data set. I would say this is an even worse statistical model for a panel data set.

5. Referring to the regression function in Figure 8.17(a), the following notes are presented.

    5.1 The graph of the regression function in the two-dimensional space with axes Y1_New and log $[(X1 + 2)/(12 - X1)]$ is a set of three parallel (homogeneous) regression lines, for $T = 6$, 7 and 8.

    5.2 The model is in fact an one-way ANCOVA model of Y1_New with a factor T, and a covariate log $[(X1 + 2)/(12 - X1)]$. The basic objective of an ANCOVA model is to study the adjusted means differences of Y1_New between time points, which in fact are the intercept differences of the three homogeneous regressions.

8.4.2.2 Problems with Outliers

Referring to the graphs in Figures 8.15 and 8.16, we recognize the regressions have outliers. So, further analysis could be done based on a sub-sample without the outliers. The problem is how many observations should be deleted, in order to obtain a sub-sample without an outlier; this cannot be predicted since we find that a sub-sample obtained by deleting some of outliers still has new outliers. For this reason, I suggest using a sub-sample indicated by selected values of the error terms, such as within the interval $(-2,2)$ of the standardized residuals as demonstrated in Agung (2011, Chapter 7), or smaller intervals. On the other hand, a sub-sample could be obtained by deleting 1(one) to 10% of the lowest and the highest scores of the numerical variables. For comparison, Rakow (2010) has been doing data analysis based on the models, where all continuous variables are winsorized at the top and bottom 1% levels, without considering whether or not the sub-sample used still has outliers. Haas and Peeters (2006) say "We carefully checked for outliers and the observations also dropped if:...", but they did not mention if the remaining sample did not have outliers. Similarly, Hameed, et al. (2010) filtered the data using only the stock's price within \$3 and \$999 each year. Alternatively, (Agung, 2011, in Chapter 12) proposes conducting an alternative analysis, that is, censored data analysis, where the outliers are considered as censored observations.

    Another alternative method recommended for a variable with a large number of outliers, is generating a group variable with at least three levels based on that variable, such as (1): the subset of the lower outliers, (3) the sub-set of the upper outliers, and (2) otherwise, which should be done using the trial-and-error method. So the differential up- and-down-stream or causal relationships between selected variables can be studied between the defined groups.

    The following illustrations present the outliers based on selected sub-samples. Figure 8.18 presents box-plots of X1 and Y1, specific for $T = 8$, based on a sub-sample by deleting 2.5% of the top and bottom levels of the scores of X1 and Y1 in the whole Data_Sesi.wf1, namely SSX1 $=$ SSY1 $= 1$. Based on these graphs the following notes are presented.

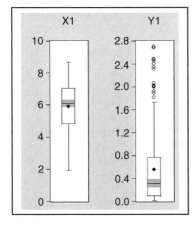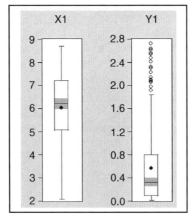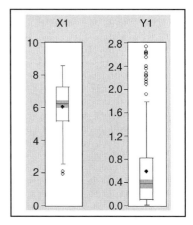

Figure 8.18   *Box-plots of X1 and Y1, based on a sub-sample SSX1 = SSY1 = 1, for T = 6, 7 and 8*

1. By deleting 2.5% of the top and bottom scores of X1 and Y1 in the whole data set, the two variables still have near outliers, specifically for T = 8. If all outliers should be deleted, then the top and bottom scores of X1 and Y1 should be deleted using different levels. This illustration shows that by deleting some outliers, the sub-sample obtained has new outliers. Examples 8.14 and 8.15 next also present problems with outliers.
2. The steps of the analysis to construct the previous graphs are as follows:
    *   The dummy variables SSX1 and SSY1 are generated using the following equations, for V = X1, and V = Y1.

$$SSV = 1^*(V > @Quantile(V, 0.025) \; and \; V =< Quantile(V, 0.925)$$

    *   Then a sub-sample is selected using the *IF condition* "SSX1 = SSY1 = 1 and T = t", for each $t = 6$, 7 and 8.
    *   Finally, by showing the variables X1 and Y1 on the screen, and then select *View/Graph . . . /Boxplot/Multiple graphs . . . OK*, the graphs in Figure 8.18 are obtained.

8.4.3   Linear-Effect Model Based on ($X1_{it}$, $X2_{it}$,Y_{it}) by Time Points

Most researchers present models where each of the numerical independent variables is the original defined variable, namely the first power of the numerical exogenous variables. In this case the model only has X1, X2 or X1*X2 as independent variables. See the following illustrative examples with special notes and comments.

8.4.3.1   Additive Models

Example 8.10   Additive linear-effect models
As an extension of the model in Figure 8.12(a), the simplest linear effect model in a three-dimensional space, of a numerical variable X1 and X2, on Y, based on the pooled data ($X1_{it}$, $X2_{it}$,Y_{it}), $i = 1,\ldots,N$, and $t = 1,\ldots,T$, for a small time points observations, can be presented as follows:

$$Y1_{it} = C(1t) + C(2t)^*X1_{it} + C(3t)^*X2_{it} + \varepsilon_{it}, \; for \; each \; t = 1, 2, \ldots \tag{8.26}$$

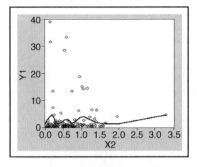 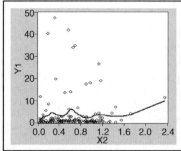 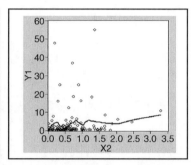

Figure 8.19   *Scatter graphs of (X2,Y1), for X2 > 0, with their NNF curves, for T = 6,7 and 8*

And the data analysis can easily be done using the following equation specification.

$$Y1\ X1^*@Expand(T)\ X2^*@Expand(T)\ @Expand(T) \tag{8.27}$$

Note that for each time point, the model is an additive model of Y1 on X1 and X2, which is acceptable, in a statistical sense. Furthermore, note that the graph of the regression functions in three-dimensional space with axes X1, X2 and Y1 is a set of (heterogeneous) planes, for all time points.

However, referring to the scatter graphs of Y1 on X1 in Figure 8.12, then the linear-effect of X1 on Y1 cannot be accepted. Corresponding to models (8.26) and (8.27), then the linear effect of X2 on Y1 should also be evaluated. For illustration, Figure 8.19 presents the scatter graph of (X2,Y1) with their NNF curves, for a sub-sample "T > 5 and SSX1 = SSX2 = SSY1 = 1 and X2 > 0", and Figure 8.20 presents the scatter graphs of G(y1) = log[y1/(60 − y1)] on F(x2) = log(1/x2) = −log(x2) with regression lines.

For further illustration, Figure 8.21(a) presents the statistical results based on an OLS regression representing the regression lines of log[[Y1/(60 − Y1)] on log(1/X2), X2 > 0, with a very small DW-statistic. So the statistical results based on an AR(1) model are presented in Figure 8.21(b).

Take note that for X2 < 0, then log(−1/X2) should be used as the independent variable of the model. If the whole sample is used, then the dummy variables generated as D1X2 = 1*(X2 < 0) and D2X2 = 1*(X2 > 0) or a dichotomous group variable, should be used as additional independent variable, which will be presented later.

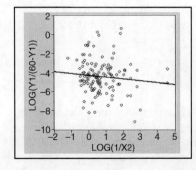 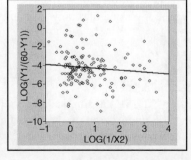 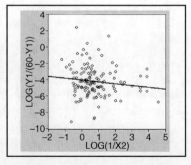

Figure 8.20   *Scatter graphs of log[Y1/(60 − Y1)] on log(1/X2), for X2 > 0, with regression lines, for T = 6,7 and 8*

| Dependent Variable: Y1_NEW |
| Method: Least Squares |
| Date: 04/15/11 Time: 11:06 |
| Sample: 1 2284 IF T>5 AND SSX1=SSX2=SSY1=1 AND X2>0 |
| Included observations: 396 |

Variable	Coefficient	Std. Error	t-Statistic	Prob.
LOG((X1+3)/(12-X1))*(T=6)	2.707531	0.136125	19.89011	0.0000
LOG((X1+3)/(12-X1))*(T=7)	2.786328	0.123603	22.54264	0.0000
LOG((X1+3)/(12-X1))*(T=8)	2.840360	0.137610	20.64067	0.0000
LOG(1/X2)*(T=6)	-0.088324	0.081276	-1.086721	0.2778
LOG(1/X2)*(T=7)	-0.177492	0.084606	-2.097865	0.0366
LOG(1/X2)*(T=8)	-0.262539	0.073303	-3.581548	0.0004
T=6	-5.959738	0.134152	-44.42514	0.0000
T=7	-6.019727	0.135899	-44.29569	0.0000
T=8	-6.087785	0.138322	-44.01170	0.0000

R-squared	0.777824	Mean dependent var	-4.174831
Adjusted R-squared	0.773231	S.D. dependent var	1.871191
S.E. of regression	0.891067	Akaike info criterion	2.629671
Sum squared resid	307.2781	Schwarz criterion	2.720157
Log likelihood	-511.6748	Hannan-Quinn criter.	2.665519
Durbin-Watson stat	0.217518		

(a)

| Dependent Variable: Y1_NEW |
| Method: Least Squares |
| Date: 04/15/11 Time: 11:10 |
| Sample: 1 2284 IF T>5 AND SSX1=SSX2=SSY1=1 AND X2>0 |
| Included observations: 341 |
| Convergence achieved after 6 iterations |

Variable	Coefficient	Std. Error	t-Statistic	Prob.
LOG((X1+3)/(12-X1))*(T=6)	2.772496	0.089992	30.80809	0.0000
LOG((X1+3)/(12-X1))*(T=7)	2.765014	0.102272	27.03590	0.0000
LOG((X1+3)/(12-X1))*(T=8)	2.719455	0.121985	22.29337	0.0000
LOG(1/X2)*(T=6)	-0.174632	0.048676	-3.587598	0.0004
LOG(1/X2)*(T=7)	-0.249341	0.060203	-4.141647	0.0000
LOG(1/X2)*(T=8)	-0.227188	0.055389	-4.101683	0.0001
T=6	-5.972312	0.091710	-65.12204	0.0000
T=7	-5.927686	0.115469	-51.33578	0.0000
T=8	-5.985332	0.132717	-45.09852	0.0000
AR(1)	0.910000	0.040827	22.28922	0.0000

R-squared	0.914568	Mean dependent var	-4.103926
Adjusted R-squared	0.912245	S.D. dependent var	1.919415
S.E. of regression	0.568597	Akaike info criterion	1.737598
Sum squared resid	107.0132	Schwarz criterion	1.849970
Log likelihood	-286.2605	Hannan-Quinn criter.	1.782369
Durbin-Watson stat	1.724332		

| Inverted AR Roots | .91 | | |

(b)

Figure 8.21 *Statistical results based on two models (a) The regression lines of log[[Y1/(60 − Y1)] on log(1/X2) in Figure 8.19, and (b) Its AR(1) model*

8.4.3.2 Interaction Models

Example 8.11 Simple interaction models

The linear effect of X1 (X2) on Y1 depends on X2 (X1), in general. For this reason, as an extension of the model in Figure 8.21, the interaction linear-effect model should be applied, using the equation specification as follows:

$$Y1\ X1^*X2^*@Expand(T)\ X1^*@Expand(T)\ X2^*@Expand(T)\ @Expand(T) \tag{8.28}$$

However, in practice one of the following alternative reduced models might be proven? to be a good fit; this is highly dependent on the data set used.

$$Y1\ X1^*X2^*@Expand(T)\ X1^*@Expand(T)\ @Expand(T) \tag{8.29}$$

$$Y1\ X1^*X2^*@Expand(T)\ X2^*@Expand(T)\ @Expand(T) \tag{8.30}$$

$$Y1\ X1^*X2^*@Expand(T)\ @Expand(T) \tag{8.31}$$

$$Y1\ X1^*@Expand(T)\ X2^*@Expand(T)\ @Expand(T) \tag{8.32}$$

Example 8.12 An advanced interaction model

As an extension of the AR(1) model presented in Figure 8.21(b), Figure 8.22 presents the statistical results based on an AR(1) interaction model and one of its possible reduced models. Based on these outputs, the following findings and notes are presented.

1. At each time point, the interaction factor $log[(X1 + 3)/(12 − X1)]^*log(1/X2)$ has an insignificant effect on Y1_New, based on the output in Figure 8.21(a). So, in general, a reduced model should be obtained

Figure (a)

Dependent Variable: Y1_NEW				
Method: Least Squares				
Date: 04/15/11 Time: 11:36				
Sample: 1 2284 IF T>5 AND SSX1=SSX2=SSY1=1 AND X2>0				
Included observations: 341				
Convergence achieved after 6 iterations				

Variable	Coefficient	Std. Error	t-Statistic	Prob.
LOG((X1+3)/(12-X1))*LOG(1/X2)*(T=6)	-0.065444	0.083728	-0.781617	0.4350
LOG((X1+3)/(12-X1))*LOG(1/X2)*(T=7)	-0.011177	0.098106	-0.113926	0.9094
LOG((X1+3)/(12-X1))*LOG(1/X2)*(T=8)	-0.134112	0.094142	-1.424577	0.1552
LOG((X1+3)/(12-X1))*(T=6)	2.832443	0.117645	24.07624	0.0000
LOG((X1+3)/(12-X1))*(T=7)	2.771200	0.125212	22.13214	0.0000
LOG((X1+3)/(12-X1))*(T=8)	2.817076	0.144874	19.44506	0.0000
LOG(1/X2)*(T=6)	-0.131658	0.072512	-1.815674	0.0703
LOG(1/X2)*(T=7)	-0.243014	0.093729	-2.592722	0.0099
LOG(1/X2)*(T=8)	-0.124710	0.090597	-1.376533	0.1696
T=6	-6.013150	0.104867	-57.34054	0.0000
T=7	-5.930262	0.126511	-46.87554	0.0000
T=8	-6.054655	0.143400	-42.22208	0.0000
AR(1)	0.908609	0.041013	22.15438	0.0000

R-squared	0.915401	Mean dependent var	-4.103926
Adjusted R-squared	0.912305	S.D. dependent var	1.919415
S.E. of regression	0.568401	Akaike info criterion	1.745399
Sum squared resid	105.9703	Schwarz criterion	1.891483
Log likelihood	-284.5906	Hannan-Quinn criter.	1.803601
Durbin-Watson stat	1.657271		

Inverted AR Roots	.91

(a)

Figure (b)

Dependent Variable: Y1_NEW				
Method: Least Squares				
Date: 04/15/11 Time: 11:45				
Sample: 1 2284 IF T>5 AND SSX1=SSX2=SSY1=1 AND X2>0				
Included observations: 341				
Convergence achieved after 6 iterations				

Variable	Coefficient	Std. Error	t-Statistic	Prob.
LOG((X1+3)/(12-X1))*LOG(1/X2)*(T=6)	-0.120498	0.081934	-1.470673	0.1423
LOG((X1+3)/(12-X1))*LOG(1/X2)*(T=7)	-0.205893	0.063750	-3.229712	0.0014
LOG((X1+3)/(12-X1))*LOG(1/X2)*(T=8)	-0.233867	0.086342	-2.708630	0.0071
LOG((X1+3)/(12-X1))*(T=6)	2.892157	0.116609	24.80216	0.0000
LOG((X1+3)/(12-X1))*(T=7)	2.918492	0.112650	25.90772	0.0000
LOG((X1+3)/(12-X1))*(T=8)	2.919633	0.139988	20.85627	0.0000
LOG(1/X2)*(T=6)	-0.073630	0.069794	-1.054958	0.2922
LOG(1/X2)*(T=8)	-0.006503	0.078854	-0.082463	0.9343
T=6	-6.066644	0.103857	-58.41370	0.0000
T=7	-6.085114	0.112460	-54.10906	0.0000
T=8	-6.154296	0.138540	-44.42253	0.0000
AR(1)	0.900753	0.041333	21.79236	0.0000

R-squared	0.913674	Mean dependent var	-4.103926
Adjusted R-squared	0.910788	S.D. dependent var	1.919415
S.E. of regression	0.573298	Akaike info criterion	1.759735
Sum squared resid	108.1327	Schwarz criterion	1.894581
Log likelihood	-288.0348	Hannan-Quinn criter.	1.813460
Durbin-Watson stat	1.662223		

Inverted AR Roots	.90

(b)

Figure 8.22 *Statistical results based on to AR(1) models (a) The AR(1) model of log[[Y1/(60 − Y1)] on log(1/X2) in Figure 8.19, and (b) Its AR(1) reduced model*

by deleting the interaction model. However, since it is defined that the effect of X1 (X2) on Y1 depends on X2 (X1), then the interaction factor should be kept in the reduced model. Therefore, by using the trial-and-error method, either one or both main factors log[(X1 + 3)/(12 − X1)] and log(1/X2) can be deleted.

2. It is surprising and unexpected: Figure 8.21(b) presents the output based on one of the possible reduced models, by deleting the independent variable log(1/X2)*(T = 7) from the full model, using the following ES.

$$Y1\_New \log((x1+3)/(12-x1))^*\log(1/x2)^*@Expand(T)$$
$$\log((x1+3)/(12-x1))^*@Expand(T) \ \log(1/x2)^*@Expand(T,@Dropt(7)) \qquad (8.33)$$
$$@Expand(T)$$

Take note of the use of @Expand(T,@Dropt(7)) to delete the log(1/X2)*(T = 7) from the full model. Even though this variable has a very small *p*-value, it should be deleted, because it correlates with log $((x1+3)/(12-x1))^*$log(1/x2)*(T = 7), which should be kept in the reduced model, since it is defined that the effect of X1 on Y1 depends on X2.

3. Various hypotheses can easily be tested using the Wald test. However, if the joint effects of all independent variables on Y1_New should be tested, the following ES with the intercept parameter can be applied, which shows that the joint effects is significant based on the *F*-statistic of $F_0 = 136.475$ with a *p*-value = 0.0000.

$$Y1\_New \ C \ \log((x1+3)/(12-x1))^*\log(1/x2)^*@Expand(T)$$
$$\log((x1+3)/(12-x1))^*@Expand(T) \ \log(1/x2)^*@Expand(T,@Dropt(7)) \qquad (8.34)$$
$$@Expand(T,@Dropt(7))$$

8.4.4   Linear Models Based on $(X1_{it}, X2_{it}, X3_{it}, Y_{it})$ by Time Points

A lot of studies present the applications of GLMs using the original defined variables, including their lagged variables, without considering the scatter graphs of the dependent variable on each or selected independent variables. This section presents the GLMs based on numerical variables X1, X2, X3 and Y by time points without considering the scatter graphs of Y on each X1, X2 and X3. However, based on my own findings, I recommend using transformed variables or a combination of original and transformed variables.

Under the assumption that X1, X2 and X3 are the three important cause (up-stream, source or exogenous) factors of an endogenous (down-stream, or impact) variable Y, then without loss of generality, X1 can be considered to be the most important cause factor. Therefore, in general, it can be defined that the linear effect of X1 on Y depends on X2 and X3 in a theoretical sense. So the proposed model at each time point should have at least one of the following sets of interaction factors as independent variables:

i. The sets of variables {X1*X2*X3, X1*X2, X1*X3}, {X1*X2*X3, X1*X2}, {X1*X2*X3, X1*X3} or {X1*X2*X3}, for a three-way interaction models.
ii. The set of variables {X1*X2, X1*X3}, for a two-way interaction model.

As the extension of the two-way interaction model presented in Example 8.11, the following equation specification presents the hierarchical three-way interaction model by times, for $T > 5$.

$$Y = F_1(X1, X2, X3)^* Dt6 + F_2(X1, X2, X3)^* Dt7 + F_3(X1, X2, X3)^* Dt8$$

where

$$F_j(X1, X2, X3) = C(j0) + C(j1)^* X1 + C(j2)^* X2 + C(j3)^* X3 + C(j4)^* X1^* X2$$
$$+ C(j5)^* X1^* X3 + C(j6)^* X2^* X3 + C(j7)^* X1^* X2^* X3, \quad (8.35)$$
$$for\ j = 1, 2, and\ 3$$

However, in practice various alternative reduced models, either non-hierarchical three-way interaction models or hierarchical and non-hierarchical two-way interaction models, could be the best fit, in a statistical sense, depending on the data set used as well as the correlations or multicollinearity between the variables X1, X2, X3, X1*X2, X1*X3, X2*X3 and X1*X2*X3. Therefore, I would say that the best fit model is an unexpected regression function.

Take note that the ES (8.34) uses time dummy variables, instead of the function @Expand(T) as in the model (8.23), since it is easier to develop alternative reduced models at each time point. Furthermore, it is much easier to make a summary of the set of the regressions, based on the output: even more so for the models with more numerical exogenous variables. See the following example.

Example 8.13   Three-way interaction models

Figure 8.23 presents the statistical results based on the model (8.35), and Table 8.6 presents the summary of the regressions by times. Based on these results the following notes and comments are presented.

1. In fact the model is a four-way interaction model indicated by the interaction X1*X2*X3*Dt, The model has a large number of variables, namely 24 independent variables, without an intercept.
2. Take note that three of the independent variables of the model at $T = 6$ only, have large p-values. Why? This finding has demonstrated that the parameters' estimates are highly dependent on the data, as well as the correlations between the variables used, which should be accepted. Therefore, in a statistical sense, a reduced model should be explored.

Table 8.6   *The summary of the regressions in Figure 8.23 by times*

Variable	T = 6 Coeff.	T = 6 Prob.	T = 7 Coeff.	T = 7 Prob.	T = 8 Coeff.	T = 8 Prob.
Intercept	−3.185	0.000	−5.902	0.000	−4.087	0.000
X1	0.675	0.000	1.164	0.000	0.814	0.000
X2	−1.443	0.002	−1.865	0.000	−1.950	0.000
X3	2.178	**0.497**	−14.398	0.000	−5.278	0.016
X1∗X2	0.296	0.000	0.320	0.000	0.403	0.000
X1∗X3	0.033	**0.913**	1.550	0.000	0.752	0.000
X2∗X3	2.431	**0.487**	−11.035	0.003	−23.061	0.000
X1∗X2∗X3	−0.004	**0.992**	1.293	0.001	2.648	0.000
R-squared	0.665		Mean dependent var			2.156
Adjusted R-squared	0.656		S.D. dependent var			5.617
S.E. of regression	3.295		Akaike info criterion			5.249
Sum squared resid	9456.820		Schwarz criterion			5.378
Log likelihood	−2325.006		Hannan–Quinn criter.			5.298
Durbin–Watson stat	0.979					

3. Corresponding to the statements in (8.33), the interaction factor X2*X3 should be deleted at the first stage, even though it has the smallest *p*-value, and then either X2 or X3, or both. Table 8.7 presents the summary of the three-way interaction regressions by times, where the regression at T = 6 is a nonhierarchical model.

4. For sure, the models might not be valid for all other panel data sets, since the estimates are highly dependent on the data sets. However, these results show that the time *T* should be taken into account in the model based on panel data.

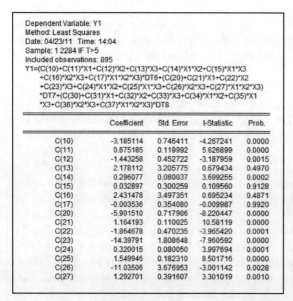
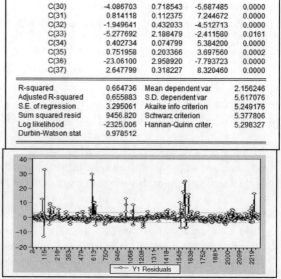

Figure 8.23   *Statistical results based on the model (8.33), with its residual graph*

Table 8.7 The summary of the regressions based a reduced model of that in Figure 8.23

Variable	T = 6 Coeff.	T = 6 Prob.	T = 7 Coeff.	T = 7 Prob.	T = 8 Coeff.	T = 8 Prob.
Intercept	−3.328	0.000	−5.902	0.000	−4.087	0.000
X1	0.703	0.000	1.164	0.000	0.814	0.000
X2	−1.479	0.001	−1.865	0.000	−1.950	0.000
X3			−14.398	0.000	−5.278	0.016
X1*X2	0.305	0.000	0.320	0.000	0.403	0.000
X1*X3	0.235	0.000	1.550	0.000	0.752	0.000
X2*X3			−11.035	0.003	−23.061	0.000
X1*X2*X3	0.228	0.000	1.293	0.001	2.648	0.000
R-squared	0.664		Mean dependent var			2.156
Adjusted R-squared	0.656		S.D. dependent var			5.617
S.E. of regression	3.293		Akaike info criterion			5.246
Sum squared resid	9466.848		Schwarz criterion			5.364
Log likelihood	2325.481		Hannan–Quinnn criter.			5.291
Durbin–Watson stat	0.961					

5. Take note that the residual graph in Figure 8.23 indicates there are outliers, and the error terms are heterogeneous as well as autocorrelated, indicated by a small DW = 0.56. So what additional analysis or modification should be done? Previous examples have presented AR models in order to increase DW-statistical values.

8.5 ANCOVA Models

Various ANCOVA models could be presented. The following sub-sections present selected alternative models with special notes and comments.

Corresponding to the GLMs in (8.53), an ANCOVA model has the general equation as follows:

$$Y_{it} = \sum_{k=1}^{K} \beta_k X_{k,it} + \sum_{gt} \delta_{gt} Dgt + \varepsilon_{it} \tag{8.36}$$

where Dgt, $g = 1,\ldots,G$, and $t = 1,\ldots,T$ are the dummy variables of the cells (g,t), which can be generated based on one or more exogenous (source, up-stream or cause) factors or variables, and the time variable. The subscript $_{it}$ is not presented in the model.

Take note that X_k is the k-th numerical exogenous (up-stream, source or cause) factor, for $k = 1,\ldots K$, and the model in (8.55) has no intercept or the constant parameter. I would say that this ANCOVA model is not recommended, since all numerical variables are defined or assumed to have the same effects within all levels or cells generated by the group and time variables. This assumption cannot be accepted in general, because the relationships between numerical variables should be dependent on the time, as well as the group of research objects (firms or individuals). So, applying the heterogeneous multiple regressions is recommended, as presented previously and in Agung (2011, Section 7.6).

In fact, the following illustrative ANCOVA model would be considered as the worst ANCOVA model, where Da1 and Db1, are the dummy variables of dichotomous factors A and B, respectively, and Dt1, Dt2

Table 8.8   Intercept parameters of the model in (8.37) by the factor CF and the time T

A	B	CF	$T=1$	$T=2$	$T=3$	$T=4$	T(1–4)	T(2–4)	T(3–4)
1	1	1	$\delta_0+\delta_1+\delta_2+\delta_3$	$\delta_0+\delta_1+\delta_2+\delta_4$	$\delta_0+\delta_1+\delta_2+\delta_5$	$\delta_0+\delta_1+\delta_2$	δ_3	δ_4	δ_5
1	2	2	$\delta_0+\delta_1+\delta_3$	$\delta_0+\delta_1+\delta_4$	$\delta_0+\delta_1+\delta_5$	$\delta_0+\delta_1$	δ_3	δ_4	δ_5
2	1	3	$\delta_0+\delta_2+\delta_3$	$\delta_0+\delta_2+\delta_4$	$\delta_0+\delta_2+\delta_5$	$\delta_0+\delta_2$	δ_3	δ_4	δ_5
2	2	4	$\delta_0+\delta_3$	$\delta_0+\delta_4$	$\delta_0+\delta_5$	δ_0	δ_3	δ_4	δ_5
CF(1-2)			δ_2	δ_2	δ_2	δ_2	0	0	0
CF(1-3)			δ_1	δ_1	δ_1	δ_1	0	0	0
CF(1-4)			$\delta_1+\delta_2$	$\delta_1+\delta_2$	$\delta_1+\delta_2$	$\delta_1+\delta_2$	0	0	0
CF(2-3)			$\delta_1-\delta_2$	$\delta_1-\delta_2$	$\delta_1-\delta_2$	$\delta_1-\delta_2$	0	0	0
CF(2-4)			δ_1	δ_1	δ_1	δ_1	0	0	0
CF(3-4)			δ_2	δ_2	δ_2	δ_2	0	0	0

and Dt3 are the first three dummy variables of the four time points.

$$Y_{it} = \sum_{k=1}^{K} \beta_k X_{k,it} + \delta_0 + \delta_1 Da1 + \delta_2 Db1 + \delta_3 Dt1 + \delta_4 Dt2 + \delta_5 Dt3 + \varepsilon_{it} \tag{8.37}$$

Note that this model should present a set of $16 = 2 \times 2 \times 4$ regressions, generated by the dichotomous factors A and B, and the time $T = 1,\ldots,4$, with the same slopes $\beta_k's$, but the model has only 6 (six) intercept parameters, indicated by the parameters $\delta_0, \delta_1, \ldots, \delta_5$. So the 16 homogenous regressions have a very special pattern of intercepts by groups and times, which are not realistic at all even though we can say that all statistical models should have some limitations and the limitations would be counted in the error terms. Table 8.8 presents the patterns of the intercepts' parameters, which shows many pairs of intercepts have equal differences.

Compared to the fixed-effects and random effects models, the model in (8.37) could be considered a *group and time fixed-effects model*.

In fact a lot more models with dummy variables are not-recommended ANCOVAs, and some of them would be the worst models to use. Refer to various ANCOVA models illustrated in Agung (2011).

Example 8.14   Application of a not-recommended ANCOVA model
For illustration, Figure 8.24 presents the statistical results based on two ANCOVA models in (8.36) using the LS estimation method, and STEPLS estimation method. Based on these results the following findings and notes are presented.

1. The output of the AR(1) model in Figure 8.24(a) could be considered an acceptable result in a statistical sense, since the regression has a large R-squared of 0.882 661, and Durbin–Watson statistic of 2.083 387. The output can be obtained using exactly the same list of variables as in the output for the equation specification (ES).
2. Compare this to the output in Figure 8.24(b), without using the AR(1) term, which shows a large R-squared of 0.808 025 but a small DW-statistical value.
3. Figure 8.24(b) shows how to apply the STEPLS estimation method using a large number of numerical independent variables and dummy variables. Refer to Agung (2011) for a lot more

Dependent Variable: Y1
Method: Least Squares
Date: 06/07/11 Time: 16:48
Sample (adjusted): 2 2283
Included observations: 1977 after adjustments
Convergence achieved after 8 iterations

Variable	Coefficient	Std. Error	t-Statistic	Prob.
C	-2.071261	0.429912	-4.817872	0.0000
X1	0.007919	0.046197	0.171412	0.8639
X2	0.479985	0.057750	8.311498	0.0000
X3	-0.416877	0.059745	-6.977643	0.0000
X4	-1.130694	0.225666	-5.010476	0.0000
X5	0.014496	0.036685	0.395139	0.6928
X6	0.023784	0.014466	1.644160	0.1003
X7	-0.268067	0.276174	-0.970648	0.3318
X8	2.226713	0.739324	3.011823	0.0026
X9	0.612220	0.096973	6.313319	0.0000
X10	0.001155	2.82E-05	41.01639	0.0000
X12	-0.001086	0.001408	-0.771678	0.4404
X13	0.000106	0.000114	0.921931	0.3567
DA1	0.917691	0.168636	5.441859	0.0000
DB1	-0.444339	0.124665	-3.564261	0.0004
DT1	0.443682	0.130574	3.397941	0.0007
DT2	0.255503	0.134421	1.900760	0.0575
DT3	0.222174	0.117271	1.894531	0.0583
AR(1)	0.630589	0.017390	36.26086	0.0000

R-squared	0.882661	Mean dependent var	1.678894
Adjusted R-squared	0.881583	S.D. dependent var	4.769420
S.E. of regression	1.641244	Akaike info criterion	3.838350
Sum squared resid	5274.230	Schwarz criterion	3.892067
Log likelihood	-3775.209	Hannan-Quinn criter.	3.858085
F-statistic	818.2635	Durbin-Watson stat	2.083387
Prob(F-statistic)	0.000000		

Inverted AR Roots	.63

(a)

Dependent Variable: Y1
Method: Stepwise Regression
Date: 06/07/11 Time: 16:58
Sample (adjusted): 1 2283
Included observations: 2058 after adjustments
Number of always included regressors: 6
Number of search regressors: 12
Selection method: Combinatorial
Number of search regressors: 9

Variable	Coefficient	Std. Error	t-Statistic	Prob.*
C	-1.649520	0.381788	-4.320517	0.0000
DA1	1.072426	0.160389	6.686411	0.0000
DB1	-0.841444	0.137995	-6.097657	0.0000
DT1	0.325208	0.159810	2.034962	0.0420
DT2	0.107688	0.152105	0.707983	0.4790
DT3	0.231383	0.149686	1.545786	0.1223
X10	0.001272	2.70E-05	47.10255	0.0000
X3	-0.737857	0.071760	-10.28235	0.0000
X8	6.800257	0.893651	7.609520	0.0000
X9	0.781402	0.095840	8.153173	0.0000
X2	0.315672	0.054615	5.779912	0.0000
X4	-1.430944	0.235341	-6.080291	0.0000
X1	-0.139848	0.038902	-3.594898	0.0003
X6	0.033976	0.016120	2.107721	0.0352
X7	-0.529904	0.343872	-1.540993	0.1235

R-squared	0.808025	Mean dependent var	1.662561
Adjusted R-squared	0.806709	S.D. dependent var	4.712499
S.E. of regression	2.071842	Akaike info criterion	4.302015
Sum squared resid	8769.638	Schwarz criterion	4.343047
Log likelihood	-4411.774	Hannan-Quinn criter.	4.317059
F-statistic	614.2146	Durbin-Watson stat	0.845680
Prob(F-statistic)	0.000000		

Selection Summary

Number of combinations compared:	220

(b)

Figure 8.24 *Statistical results based on ANOVA models in (8.36), using (a) the LS estimation method, and (b) the STEPLS estimation method*

illustrative examples, using the multi-stage STEPLS estimation method. In this example, the processes are as follows:

- The list of variables "C Da1 Db1 Dt1 Dt2 Dt3" is used as the list of always-entered regressors.
- The list of 12 variables "X1 . . . X10 X12 X13" (without X11) is used as the list of search regressors.
- The option STEPLS uses the combinatorial selection method by entering an alternative number of variables step-by-step, then nine out of 12 variables are sufficient for use.
- After getting results based on the STEPLS estimation method, the LS estimation method can be applied using the AR(*) terms.

9

Fixed-Effects Models and Alternatives

9.1 Introduction

Fixed-effects models are widely applied by most, perhaps all, researchers in finance and economics. For this reason, fixed-effects models should be acceptable. Referring to the Baltagi's books on *Econometric Analysis of Panel Data* and *A Companion to Econometric Analysis of Panel Data* (2009a, and 2009b), I would say he presents comprehensive mathematical concepts of alternative effects models, but the empirical examples presented are limited based on the additive models only, and he does not present any with the numerical time t variable. Similarly, this is the case for the fixed-effects models presented in Gujarati (2003), and Wooldridge (2002), as well as most of the models presented in the international journals, specifically the *Journal of Finance* (Dec. 2010, and Feb., April and June, 2011).

However, all students who have consulted me for their theses and dissertations do not know the detailed characteristics of fixed-effects models, let alone their limitations. In fact, a fixed model is a special type of the *least squares dummy variables* (LSDV) model, it can be a one-way fixed-effects model with the simplest one being a one-way ANOVA model as presented in (9.4), and Baltagi (2009a, Problem 2.17, p. 33). More general one-way fixed-effects models are ANCOVA models with the general equations in (9.1) as a one-way cross-section fixed-effects model, and with the general equation in (9.10) as a one-way time fixed-effects model. ANCOVA models have a special assumption or limitation: the covariates have the same effects on the dependent variable; in other words, the covariates have the same slope parameters for all levels or categories generated by the fixed-effect variables. Agung (2011) has illustrated examples of various two- and three-way ANCOVA models with the same set of variables, which can be considered the best and the worst ANCOVA models within that group.

Baltagi has mentioned a simple dynamic panel data model with heterogeneous coefficients on the lagged dependent variable and time trend presented by Winsbek and Knaap (1999, in Baltagi, 2009a, p. 168), and a random walk model with heterogeneous trend presented by Hadri (2000, in Baltagi 2009a), and Wooldridge (2002, 315–317) presented a random trend and random growth model that had the numerical time variable. In addition, Wooldridge also presented a time-related effect model: one that had an interaction independent variable with time t.

Heterogeneous regressions were introduced by Johnson and Neyman in 1936 (Huitema, 1980, p. 270). So it can be said that heterogeneous regressions models have been used for more than 75 years by a lot of

Panel Data Analysis Using EViews, First Edition. I Gusti Ngurah Agung.
© 2014 John Wiley & Sons, Ltd. Published 2014 by John Wiley & Sons, Ltd.
Companion website: www.wiley.com/go/panel_data

researchers such as Agung (2009a), Gujarati (2003) and Wooldridge (2002) based on time series data; Agung (2011), Gujarati (2003), and Neter and Wasserman (1974) based on cross-section data, and Gujarati (2003) and Wooldridge (2002) based on panel data. Therefore, it is recommended that students and researchers consider the application of heterogeneous regression interaction models in general for their studies based on panel data.

For this reason, the fixed-effect models or least squares dummy variable (LSDV) models presented in this chapter can be considered extensions of these, since they present alternative fixed-effect models with numerical time-independent variables, and interaction independent variables, in addition to additive LSDVs. For a more advanced model, various GLS regressions will be presented in Chapter 14.

9.2 Cross-Section Fixed-Effects Models

9.2.1 Individual Fixed-Effects Models

The *cross-section fixed-effects models* (CSFEMs) or *individual-fixed-effects models* (IFEMs) are in fact one-way ANCOVA models, with the following general equation.

$$H(Y_{it}) = \delta_i + F(X_{it}, t) + \mu_{it} \tag{9.1}$$

where *H(Y)* is a transformed variable of *Y*, or a defined function of *Y* without a parameter, *F(X,t)* is a function of a vector variable $X = (X1, \ldots, Xk, \ldots)$ and the time variable *t*, with a finite number of slope parameters only, and δ_i is the *i*-th individual fixed effect, for $i = 1, \ldots, N \ldots$ Note that the components of *X*, say *Xk*, can be numerical, categorical or interaction factors. Refer to all alternative functions of *H(Y)*, and $F(X_{it}, t)$, presented in previous chapters. Brooks (2008) presents the IFEM using individual dummies, which can be generalized as follows:

$$H(Y_{it}) = F(X_{it}, t) + \sum_{i=1}^{N} \delta_i Di + \mu_{it} \tag{9.2}$$

where $Di = D_i$ is the dummy variable of the *i*-th individual or research object (firm or states), for $i = 1, \ldots, N$. Compare this to the fixed-effect models, presented in Section 9.1. Note that the random effect model, as presented Figure 1.23, also presents this type of model with a single numerical independent variable, namely the time variable – see also the random effect model in Figure 1.24.

Corresponding to the IFEMs in (9.2), the following notes and comments are presented.

1. All CSFEMs presented in other books, such as Baltagi (2009a,b), and Gujarati (2003), do not have the numerical time independent variable. For this reason, the CSFEMs in (9.1) or the least squares dummy variable (LSDV) model in (9.2) should be considered extension of CSFEMs, which have been commonly applied. Note that the classical growth models with the numerical time independent variable, namely exponential and geometric growth models, have been widely applied. So the panel data models with the time trend should be valid.

2. In fact, Wooldridge (2002) presented a random trend or random growth model, and a model with a time-related effect. Wang, et al. (2010) presented two binary choice firm fixed-effects models, with time trend, which is the only paper on this topic found in the *Journal of Finance* (Dec. 2010, Feb., April and June, 2011). This chapter presents various cross-section fixed-effects models with time trend and time-related effects.

3. Furthermore, most of the fixed-effects models presented, either cross-section or period fixed-effect, are additive. Only Giroud and Mueller (2011) presented several Year-Industry fixed-effects interaction models. This chapter presents selected fixed-effects interaction models or heterogeneous regressions models as alternatives. Heterogeneous regressions were introduced for the first time in 1936 (Huitema, 1980).

4. For each i or all regressions in the model (9.1), the vector variable X_{it} and the endogenous variable Y_{it} are considered time-series variables. So all of the time series models presented in Agung (2009a, 2011) and other references should be valid for each i.

5. If the pool data is balanced or complete with $N \times T$ observations, then the models in (9.1) and (9.2) represent a set of N homogeneous time series models, where all independent variables in $F(X, t)$ have the same adjusted effects on $H(Y)$ for all $i = 1,\ldots, N$.

6. I would say that the set of N homogeneous time-series models is inappropriate for a large N in reality. In fact, even for $N = 2$ or 3, a set of heterogeneous time series models should be applied. See the following illustrative example and refer to models in Chapters 1 to 4. However, in a statistical sense various fixed-effects models have been declared as acceptable. See selected fixed-effects models in international journals. In addition, some specific general CSFEMs that could be considered are as follows.

9.2.1.1   *CSFEMs with Interaction Independent Variables*

In addition to the individual dummies or the cross-section dummies, if the models have two or more main independent (exogenous, source, cause, or up-stream) variables, in general, the effect of an independent variable should depend on at least one of the other independent variables. For this reason, then the function $F(X_{it}, t)$ in the model (9.2) should have interaction terms. Refer to the random fixed-effect model with $prog_{it} \times t$ as an independent variable presented in Wooldridge (2002, 317), and various time series models with TRE in Agung (2009a) and Bansal (2005). Researchers may not use all possible interactions, but they should subjectively select specific/very important interaction factors, corresponding to the relationships between the set of variables considered, even for $X = (X1,X2,X3)$ because it is not wise to use all possible interactions.

Therefore, a type of CSFEM with interaction independent variables considered would be that with *time-related effects* (TRE), and the general equation as follows:

$$H(Y_{it}) = F_1(X_{it}) + t \times F_2(\tilde{X}_{it}) + \sum_{i=1}^{N} \delta_i Di + \mu_{it} \tag{9.2a}$$

where $\tilde{X}$ represents selected components of the vector variable $X = (X1,\ldots, Xk,\ldots)$, and the two terms $F_1(X)$ and $[t \times F_2(\tilde{X})]$ can be a lot of possible functions with finite numbers of parameters but no constant parameter.

9.2.1.2   *CSFEMs with an Additive Function*

These models have the following general equation.

$$H(Y_{it}) = F(X_{it}) + G(t) + \sum_{i=1}^{N} \delta_i Di + \mu_{it} \tag{9.2b}$$

where $G(t)$ is a function of the time t with a finite number of parameters, without the constant parameter. For instance:

$$G(t) = \gamma \times t, G(t) = \gamma \times \log(t), \quad \text{and} \quad G(t) = \gamma_1 \times t + \gamma_2 \times t^2 + \ldots + \gamma_n \times t^n$$

9.2.1.3   CSFEMs without the Time-t Independent Variable

These models are in fact the reduced models in (9.1) and (9.2) with the following general equation.

$$H(Y_{it}) = F(X_{it}) + \sum_{i=1}^{N} \delta_i Di + \mu_{it} \tag{9.3}$$

where $H(Y_{it})$ is a function of Y_{it} without a parameter, $F(X_{it})$ can be a lot of possible functions of the vector variable $X_{it} = (X1, \ldots, Xk, \ldots, XK)_{it}$, with a finite number of parameters without a constant parameter. Note that each component Xk can be numerical or categorical variables, and can be interactions or additive. For instance, the simplest interaction and additive functions $F(X_{it})$ are the functions of $F(X1,X2) = \beta_1 X1 + \beta_2 X2 + \beta_3(X1^*X2)$ and $F(X1,X2) = \gamma_1 X1 + \gamma_2 X2$.

Most of the models presented in Baltagi (2009a,b), and Gujarati (2003), as well as the *Journal of Finance* (Dec. 2010, and Feb., April and June 2011), are additive models without the numerical time t independent variable. They present models with time dummies as independent variables and period-fixed effects.

Example 9.1   (A set of simple CSFEMs)

In order to give a better understanding to the reader on the limitations of the CSFEM, which in fact is a one-way ANCOVA model, Table 9.1 presents a summary of the statistical results based on five simple IFEMs or LSDVs, using the following equation specifications based on the Data_Sesi.wf1.

Based on the results in this table the following notes and comments are presented.

1. *Acceptability of the five models.*
   The five models are acceptable, in a statistical sense, because each of the independent variables in all of them has a significant effect on *LNSALE*, and the regressions have adjusted R-squared values greater than 0.80. So it can be said that the independent variables are good predictors for *LNSALE*. By comparing these five models, it can then be concluded that Model-5 is the best and Model-1 the worst.
   The coefficients of 348 dummies are not presented following the fixed-effects models presented in international journals.

Table 9.1   Summary of the statistical results based on five IFEMs

Dependent Variable: LNSALE										
Method: Least Squares										
Date: 08/15/11 Time: 09:50										
Sample: 1 2284										
Included observations: 2270										
	Model-1	Model-2		Model-3		Model-4		Model-5		
Variable	Coeff.	Coeff.	t-Stat.	Coeff.	t-Stat.	Coeff.	t-Stat.	Coeff.	t-Stat.	
T (Year)		0.180	27.145			0.145	21.154	-0.060	-2.347	
LNASSET				0.294	20.705	0.182	13.153	0.078	4.248	
T*LNASSET								0.031	8.354	
# of firms	348	348		348		348		348		
R-squared	0.866	0.903		0.890		0.911		0.914		
Adjusted R-squared	0.842	0.886		0.871		0.895		0.899		
Durbin-Watson stat	1.339	1.529		1.384		1.528		1.521		

2. *General Equation of Model-1*:
   This CSFEM or IFEM has the following general equation:

$$Y_{it} = C(i) + \mu_{it} = \delta_i + \mu_{it}, \quad i = 1, \ldots, N; \quad t = 1, \ldots, T_i \qquad (9.4)$$

which is the same as the model presented in Baltagi (2009a, Problem 2.17, p. 33). The statistical results in Table 9.1 obtained by using the equation specification as follows:

$$LNSALE \ @Expand(FIRM\_CODE) \qquad (9.4a)$$

This CSFEM in fact is a one-way ANOVA model of *LNSALE* on a single factor *FIRM_CODE* with 348 levels. Then the coefficients of the firm dummies are the means of *LNSALE* over the time *t* for the 348 firms. So this model is a single factor cell-means model (Agung, 2011). In order to test the means difference of *LNSALE* between all firms using the *F*-statistic, applying the following equation specification (ES) is recommended instead of using the Wald test.

$$LNSALE \ C \ @Expand(FIRM\_CODE, @Dropfirst) \qquad (9.4b)$$

or

$$LNSALE \ C \ @Expand(FIRM\_CODE, @Droplast) \qquad (9.4c)$$

We find that they have a significant difference based on the *F*-statistic of $F_0 = 35.76421$ with $df = (347, 1922)$, and *Prob(F-Stat)* $= 0.000000$.

3. *General Equation of Model-2*:
   This CSFEM has the following general equation:

$$Y_{it} = \delta_i + \beta \times t + \mu_{it}, \quad i = 1, \ldots, N; \quad t = 1, \ldots, T_i \qquad (9.5a)$$

The statistical results in Table 9.1 are obtained by using the equation specification as follows:

$$LNSALE \ T \ @Expand(FIRM\_CODE) \qquad (9.5b)$$

This CSFEM represents the classical growth models of *SALE* for the 348 firms, under the assumption that all firms have the same exponential growth rate over the observed time period, indicated by the coefficient of *T*. In this case, the exponential growth is *SALE_Growth* $= 0.180$ for the 348 firms. Refer to the classical growth models presented in Chapter 1, and in Agung (2009a), as well as the growth models by individuals, presented in Wooldridge (2002).

This model also could be viewed as a one-way ANCOVA model of *LNSALE* with a covariate being the time *t*, or a set of 348 homogeneous regression lines of *LNSALE* on the time *T*. Compare this to Model-3. Note that the time variable *t* could represent the ranks of all numerical variables which are highly correlated with *t*.

Would you agree that the 348 firms should have the same *SALE_Growth* of 1.80%? I would say that this would never be observed in real life. Figure 9.1 presents the scatter graphs of LNSALE on the time *T* with their nearest neighbor fit (NNF) curves, for three selected firms, namely _AALI, _AKKU and _INPP, which show they clearly do not have the same growth rates. Note these graphs show that _AALI and _INPP have positive growth rates, but _AKKU has a negative growth rate. Furthermore, the graph

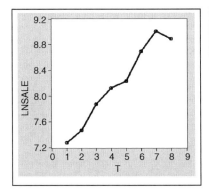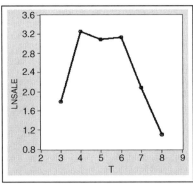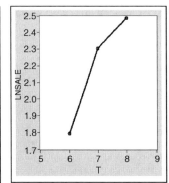

Figure 9.1   *Scatter graphs with their NNF Curves of LNSALE on T for three selected firms, namely _AALI, _AKKU and _INPP*

also indicates that the pool data is unbalanced or incomplete where the three firms have 8, 6 and 3 time-point observations.

4.   *General Equation of Model-3*

This CSFEM has the following general equation, which is the simplest model in (9.3).

$$Y_{it} = \delta_i + \beta \times X1 + \mu_{it}, \quad i = 1, \ldots, N; \quad t = 1, \ldots, T_i \tag{9.6a}$$

The statistical results in Table 9.1 were obtained by using the equation specification as follows:

$$LNSALE \; LNASSET \; @Expand(FIRM\_CODE) \tag{9.6b}$$

This model represents homogeneous regression lines of *LNSALE* on *LNASSET*, which is a set of homogeneous translog linear or Cobb–Douglas models with 348 intercepts in a two-dimensional coordinate system with the axes *LNASSET* and *LNSALE*. In this case, the time *t* is not taken into account. Note that this CSFEM is an additive model, without the time or time-period independent variables.

Refer to Figure 9.2, which presents scatter graphs with regression lines of *LNSALE* on *LNASSET* for the firms _AALI, _AKKU, and _FPNI. This figure clearly shows that the regression lines have different

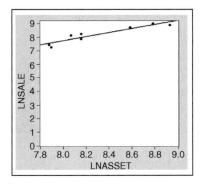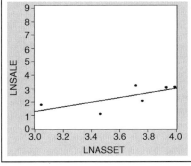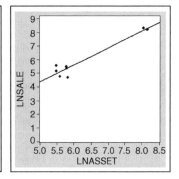

Figure 9.2   *Scatter graphs with their regression lines of LNSALE on LNASSET for three selected firms, namely _AALI, _AKKU and _FPNI*

slopes. So it can be said that heterogeneous regressions would be better to apply in general, instead of the model in (9.6).

Furthermore, the scatter graph for the firm _AKKU shows that nonlinear-effect models of *LNSALE* on *LNASSET* might be better, such as polynomial models instead of the simplest linear regressions. Also, for the firm _FPNI the eight points are divided into two groups.

On the other hand, the effect of LNASSET on LNSALE should depend on the time *t* in many cases. See the following alternative models.

5. *General Equation of Model-4*

This CSFEM has the following general equation, which is the simplest model in (9.2b).

$$Y_{it} = \delta_i + \beta \times X1 + \gamma \times t + \mu_{it}, \quad i = 1, \ldots, N; \quad t = 1, \ldots, T_i \tag{9.7a}$$

The statistical results in Table 9.1 were obtained by using the equation specification as follows:

$$LNSALE\ T\ LNASSET\ @Expand(FIRM\_CODE) \tag{9.7b}$$

This CSFEM is an additive model with *T*, *LNASSET* and 348 firm dummies as the independent variables, which represents a set of 348 homogeneous (parallel) planes in a three-dimensional space with the axes *T*, *LNASSET* and *LNSALE*. Note that this model assumes the SALE of all firms has the same exponential growth rate, indicated by the coefficient adjusted for *LNASSET*. For comparison, Parke (1994. in Wooldridge, 2002, p. 316) presents a random growth model by the individuals (or the growths of Y_{it} dependent on the individuals).

6. *General Equation of Model-5*:

This CSFEM has the following general equation, which is the simplest model in (9.2a).

$$Y_{it} = \delta_i + \alpha \times X1 + \beta \times t + \gamma(t \times X1) + \mu_{it}, \quad i = 1, \ldots, N; \quad t = 1, \ldots, T_i \tag{9.8a}$$

The statistical results in Table 9.1 were obtained by using the equation specification as follows:

$$LNSALE\ T\ LNASSET\ T^*LNASSET\ @Expand(FIRM\_CODE) \tag{9.8b}$$

Compared this to the model (9.7), this model is a two-way interaction model with *T*, *LNASSET*, *T*LNASSET* and 348 firm dummies as the independent variables, which represents a set of 348 homogeneous (parallel) curves, such as the hyperbolic curves, in a three-dimensional space with the axes *t*, *LNASSET* and *LNSALE*.

9.2.2 Group Fixed-Effects Models

If an individual fixed-effect model should be applied for a very large *N*, such as hundreds or even thousands of individuals, applying a *group-fixed-effects model(GFEM)* is suggested, with the following equation.

$$H(Y_{it}) = F(X_{it}, t) + \sum_{g=1}^{G} \delta_g Dg_{it} + \mu_{it} \tag{9.9}$$

where *Dg* is the *g*-th group dummy variable, that has *G* levels. Note that groups should be generated so that they are invariant over times. Note the differences of this model to the one in (8.56), and the empirical results in Figure 8.24. Refer to all alternative general models presented in (9.2). Furthermore, the *Y*-variable and the vector variable X could be replaced using the sample statistics within the groups, such as the mean, median and standard deviation or variance.

Example 9.2   A group fixed-effect model

For illustration, a group fixed-effects model with INDUSTRY as a group variable (see Blanconiere, et al., 2008) is selected. This is a good example of a one-way ANCOVA model with 12 numerical covariates (10 main covariates, with only one two-way and one three-way interactions of the main covariates). The model would be presented in short as follows, where $F(\mathbf{x})$ is the function of the 10 main covariates with 13 parameters (an intercept and 12 slope parameters).

$$BONUS_{i,t} = F(\mathbf{x}) + \sum_{g=1}^{11} \delta_g INDUSTRY_{i,g} + \varepsilon_{i,t} \tag{9.9a}$$

The model is in fact a three-way interaction ANCOVA model, with the assumption that the 12 covariates have the same effect on the endogenous variable: $BONUS_{i,t}$, between a set of 11 industries. Even though the assumption of the model is not realistic in general, this model by Blanconiere, Johnson and Lewis (BJL) is a good example of how researchers could subjectively define or propose the application of interaction models. Note that they (BJL) present a group fixed-effect interaction model with only one out of 45 possible two-way interactions, and only one out of 120 possible three-way interactions, which could be generated by the 10 covariates. Why? They have a specific objective(s) – refer to their paper.

9.3   Time-Fixed-Effects Models

A *time-fixed-effects model(TFEM)* or *period-fixed-effects model(PFEM)* would have the following general equation.

$$H(Y_{it}) = \tau_t + F(\mathbf{X}_{it}) + \mu_{it} \tag{9.10}$$

where τ_t is a time-fixed effect. By using time dummies, the model can be presented as follows:

$$H(Y_{it}) = F(\mathbf{X}_{it}) + \sum_{t=1}^{T} \tau_t Dt + \mu_{it} \tag{9.11}$$

where $Dt = D_t$ is the time dummy variable, for $t = 1,\ldots, T$. Note that this TFEM has similar form with the CSFEM in (9.3)

   Corresponding to TFEM, the following notes and comments are presented.

1.   For each t, the vector variable X_{it} and the endogenous variable Y_{it} are cross-section variables. So all cross-section interaction and additive models presented in Agung (2011) and other references should be valid for each time point t. However, in general, the function of the vector X_{it} should have some interaction terms of its components, which should be subjectively selected by researchers.
2.   If the pool data is balanced or complete with $N \times T$ observations, then the models in (9.10) and (9.11) represent a set of T homogeneous cross-section models, where all independent variables in $F(X)$ have the same effects on $H(Y)$ for all $t = 1,\ldots, T$.

3. I would say that the set of *T* homogeneous cross-section models is inappropriate, even for a small *T* in reality. In fact even, for *T* = 2 or 3, a set of heterogeneous cross-section models should be applied. However, in a statistical sense various time-fixed-effects models have been declared acceptable. See selected fixed-effects models from international journals presented in Section 8.8.5 and Section 9.2.
4. As *a rule of thumb*, the number of individual observations should be 10 times the number of variables in the model.
5. If a fixed-effect model should be applied for a very large *T*, such as days or 15-minute time intervals, applying a *time-period-fixed-effects model(TPFEM)* is recommended. On the other hand, as an alternative, using the time *t* as one of the numerical independent variable is suggested. Refer to the models presented in Chapter 1 to 4, and Agung (2009a).

Example 9.3   A set of TFEMs with additive or interaction exogenous variables

In order to give a basic better understanding to the reader of the limitations of TFEMs, Table 9.2 presents a summary of statistical results based on four TFEMs of NI (net-income) on LNSALE, LNLOAN and the time dummies.

Based on the results in Table 9.2, the following notes are presented.

1. Compares to the models in Table 9.1, the models in this table have intercept parameters, indicated by "C". Note that the symbol "C" presents the coefficient of the dummy variable of (*T* = 1), corresponding to the use of @*Expand(T,@Dropfirst)*, in the following equation specifications.
2. Model-1 has the following general equation, which is the simplest additive model in (9.10).

$$Y_{it} = \tau_t + \beta \times X1_{it} + \gamma \times X2_{it} + \mu_{it} \tag{9.12a}$$

The statistical result in Table 9.2 was obtained by using the ES as follows:

$$NI\ LNSALE\ LNLOAN\ C\ @Expand(T, @Dropfirst) \tag{9.12b}$$

Note that this PFEM can be considered the simplest time-fixed-effects additive model of *NI* on *LNSALE*, *LNLOAN* and time dummies, or a one-way ANCOVA model of *NI* by a single factor "Time" with the covariates *LNSALE* and *LNLOAN*, which represents 8 (eight) parallel (homogenous) planes in a three-dimensional coordinate system (*LNSALE, LNLOAN,NI*).

For comparison, Grunfeld (1958, quoted by Baltagi, 2009a, p. 24) presents an additive model of I_{it} (real gross investment) on two exogenous variables: F_{it} (real value of the firm), and C_{it} (real value of capital stock). In addition, Baltagi (2009a, Table 2.1) presents seven statistical results of the models, namely the OLS, *Within*, *Between* and GLS. Similarly so for the model of I_{it} on F_{it} an K_{it}, in Table 2.5.
3. Model-2 has the following general equation, which is the simplest two-way interaction model in (9.10).

$$Y_{it} = \tau_t + \alpha \times X1_{it} + \beta \times X2_{it} + \gamma \times X1_{it} \times X2_{it} + \mu_{it} \tag{9.12c}$$

The statistical result in Table 9.2 obtained by using the ES as follows:

$$NI\ LNSALE\ LNLOAN\ LNSALE^*LNLOAN\ C\ @Expand(T, @Dropfirst) \tag{9.12d}$$

4. Since *LNSALE*LNLOAN* in Model-2 is insignificant with a large *p*-value, but the variables *LNSALE* and *LNLOAN* are significant, then in general a reduced model would be obtained by deleting *LNSALE*LNLOAN*. However, an uncommon method is demonstrated, where the reduced Model-3 and Model-4 are obtained by deleting *LNLOAN* and *LNSALE*, respectively.

Table 9.2 Summary of statistical results based on four time fixed-effects models

Dependent Variable: NI
Method: Least Squares
Date: 08/17/11 Time: 20:45
Sample: 1 2284
Included observations: 2058

Variable	Model-1			Model-2			Model-3			Model-4		
	Coeff.	t-Stat	Prob.	Coeff.	t-Stat	Prob.	Coeff.	t-Stat	Prob.	Coeff.	t-Stat	Prob.
LNSALE	186.003	17.154	0.000	184.939	16.745	0.000	177.897	16.845	0.000	115.720	2.740	0.006
LNLOAN	-108.870	-6.677	0.000	-89.463	-2.158	0.031	-18.289	-6.332	0.000			
LNSALE*LNLOAN				-3.735	-0.509	0.611				-26.977	-3.513	0.001
C	-922.610	-10.500	0.000	-913.212	-10.170	0.000	-858.280	-9.958	0.000	132.744	1.930	0.054
T=2	-9.777	-0.111	0.912	-10.080	-0.114	0.909	-13.484	-0.153	0.879	-19.042	-0.202	0.840
T=3	-31.256	-0.357	0.721	-31.394	-0.358	0.720	-33.870	-0.386	0.699	-10.340	-0.111	0.912
T=4	-30.777	-0.356	0.722	-30.429	-0.351	0.725	-30.323	-0.350	0.727	82.641	0.898	0.369
T=5	-39.161	-0.456	0.648	-39.113	-0.455	0.649	-40.046	-0.466	0.641	69.622	0.763	0.446
T=6	43.314	0.516	0.606	43.400	0.517	0.605	43.462	0.517	0.605	174.406	1.957	0.050
T=7	-83.465	-1.001	0.317	-82.750	-0.992	0.321	-80.831	-0.968	0.333	91.432	1.036	0.300
T=8	6.201	0.074	0.941	6.489	0.078	0.938	8.041	0.096	0.923	177.337	2.012	0.044
R-squared	0.1313			0.1314			0.1295			0.0125		
Adjusted R-squared	0.1275			0.1272			0.1256			0.0081		
F-statistic	34.399			30.974			33.838			2.8697		
Prob(F-statistic)	0.0000			0.0000			0.0000			0.0023		
Durbin-Watson Stat	0.8501			0.8499			0.8467			0.7752		

Why? Because the effect of *LNSALE* on *NI* depends on *LNLOAN*. Therefore, this interaction should not be deleted in order to obtain an acceptable reduced model.

5. Since *LNSALE*LNLOAN* has a significant effect on *NI*, in Model-3 and Model-4, then both models are acceptable for presenting the effect of *LNSALE* on *NI* as significantly dependent on *LNLOAN*. In other words, both regressions or statistical results support the hypothesis stated "the effect of LNSALE on NI depends on LNLOAN". However, in a theoretical sense, which one is a better model? See their characteristics, specific for T = 1, as follows:

5.1   Model-3 can be presented as

$$NI = -858.280 + [177.897 - 18.289LNLOAN] \times LNSALE$$

which shows that the effect of *LNSALE* on *NI* depends on [177.897 − 18.289*LNLOAN*]. Note that, for various values of *LNLOAN*, this regression presents a set of heterogeneous regression lines of *NI* on *LNSALE*, in two-dimensional coordinates (*LNSALE,NI*) with a fixed intercept. See the set of regression lines in Figure 8.10(b).

The negative coefficient of *LNLOAN* × *LNSALE* should indicate that two patterns of the effects of *LNSALE* on *NI*, namely *LNSALE* have positive effects on *NI* if *LNLOAN* < 9.726 994, and non-positive effects otherwise. However, it is found that the *Max(LNSALE)* = 4.814 539. Therefore it can be concluded that the effect of *LNSALE* on *NI* increases with decreasing values of *LNLOAN*.

5.2   Model-4 can be presented as

$$NI = [132.744 + 115.720LNLOAN] - 3.513LNLOAN \times LNSALE$$

which shows that the effect of *LNSALE* on *NI* depends on [−3.513*LNLOAN*]. Note that, for various values of *LNLOAN*, this regression represents a set of heterogeneous regression lines of *NI* on *LNSALE* in two-dimensional coordinates (*LNSALE,NI*) with various intercepts and slopes indicated by [132.744 + 115.720*LNLOAN*] and [−3.513*LNLOAN*], respectively. See the set of regression lines in Figure 8.10(a).

5.3   By comparing the graphs of both regressions, which one would you chose as the better model?

6. Each of the models in Table 9.2 has a small DW-statistical value because the sets of scores at all the time points are correlated. For this reason, applying autoregressive or lagged variables models is suggested. Do it as an exercise. For comparison, various empirical models presented in Baltagi (2009a,b) have small DW-statistics, for instance, Table 2.1 in Baltagi (2009b) has a DW statistic of 0.137 461.

9.4   Two-Way Fixed-Effects Models

As the extension of the TFEMs in (9.4), *two-way fixed-effects models* (TWFEMs) could be considered with general equations as follows:

9.4.1   Two-Way Fixed-Effects Models

The TWFEMs have the general equation as follows:

$$H(Y_{it}) = \delta_i + \tau_t + F(X_{it}) + \varepsilon_{it} \tag{9.13a}$$

where δ_i and τ_t, respectively, are individual and time-fixed effects. Note that the function $F(X_{it})$ has a finite number of slope parameters only. By using individual and time dummies the general equation of the model can be presented as follows:

$$H(Y_{it}) = F(X_{it}) + \sum_{i=1}^{N} \delta_i Di + \sum_{t}^{T-1} \tau_t Dt + \varepsilon_{it}$$

or

$$H(Y_{it}) = F(X_{it}) + \sum_{i=1}^{N-1} \delta_i Di + \sum_{t}^{T} \tau_t Dt + \varepsilon_{it} \qquad (9.13\text{b})$$

or

$$H(Y_{it}) = F(X_{it}) + C + \sum_{i=1}^{N-1} \delta_i Di + \sum_{t}^{T-1} \tau_t Dt + \varepsilon_{it}$$

Note that if the following equation is applied, then the *"Near singular matrix"* error message is obtained. In fact, the design matrix of the model is a singular matrix; it is not "near singular".

$$H(Y_{it}) = F(X_{it}) + \sum_{i=1}^{N} \delta_i Di + \sum_{t}^{T} \tau_t Dt + \varepsilon_{it} \qquad (9.14)$$

Referring to the models in (9.13a), the following characteristics and limitations of the models are presented.

1. For each individual (firm) i, the model in (9.13a) would give the following regression functions.

$$\hat{H}(y) = \hat{\delta}_i + \hat{\tau}_t + F(\mathbf{x}), \quad \textit{for all } t = 1, \dots, T \qquad (9.15)$$

which represents a set of T homogeneous regressions. In other words, the T multiple regressions are assumed to have the same slope parameters, with various intercepts indicated by $\hat{\tau}_t$, since $\hat{\delta}_i$ is a fixed number.

2. For each time point t, the model in (9.13a) would give the following regression functions.

$$\hat{H}(y) = \hat{\delta}_i + \hat{\tau}_t + F(\mathbf{x}), \quad \textit{for all } i = 1, \dots, N \qquad (9.16)$$

which represents a set of N homogeneous regressions. In other words, the N multiple regressions are assumed to have the same slope parameters, with various intercepts indicated by $\hat{\delta}_i$, since $\hat{\tau}_t$ is a fixed number.

3. Finally for each cell or group (i,t), the model in (9.13a) would give a single number, not a regression function, as follows:

$$\hat{H}(y_{it}) = \hat{\delta}_i + \hat{\tau}_t + F(\mathbf{x}) \qquad (9.17)$$

where $[\hat{\delta}_i + \hat{\tau}_t]$ is a fixed intercept, and $F(\mathbf{x})$ is a value of the function $F(X)$.

4. For the reasons here, I would consider the individual time-fixed-effects models in (9.13) as acceptable in a statistical sense only, because the homogeneous regressions by individuals and time points cannot be valid in a theoretical sense. The models should only be important for scientific exercises, but they are useless for evaluation and policy analysis. Refer to the evaluation analysis presented in Chapter 5.

Table 9.3   Statistical results based on two time individual fixed-effects models

Dependent Variable: NI								
Method: Least Squares								
Date: 08/17/11 Time: 21:28								
Sample: 1 2284								
Included observations: 2058								
	Model-1				Model-2			
Variable	Coeff.	s.e.	t-Stat	Prob.				
LNSALE	41.301	25.051	1.649	0.099	41.301	25.051	1.649	0.099
LNLOAN	162.432	52.898	3.071	0.002	162.432	52.898	3.071	0.002
LNSALE*LNLOAN	-44.418	9.105	-4.878	0.000	-44.418	9.105	-4.878	0.000
Fixed effects								
C	486.397	293.102	1.659	0.097				
Time dummies	(8-1)				8			
Firm dummies	(333-1)				(333-1)			
R-squared	0.665				0.665			
Adjusted R-squared	0.598				0.598			
F-statistic	9.910							
Prob(F-statistic)	0.000							
Durbin-Watson stat	1.773				1.775			

Example 9.4   Extension of the TFEMs in Table 9.2

Table 9.3 presents statistical results based on two time-individual or (time + individual) fixed-effects models (TIFEMs). Based on this table, the following findings and notes are presented.

1. Both models in fact are the same, where Model-1 has an intercept parameter and F-statistic, but Model-2 does not.
2. Each of the numerical independent variables, namely *LNSALE*, *LNLOAN* and their interactions, have a significant adjusted effect on *NI*.
3. Furthermore, based on the F-statistic of $F_0 = 9.910$ with $df = (339,1714)$ and a p-value = 0.000 in Model-1, it can be concluded that 342 independent variables (three numerical variables and 339 dummies) have significant joint effects on *NI*.
4. These models as well as the models in Table 9.2 can be modified to FEMs as follows:
    4.1   Bounded FEMs, by using an independent variable $log[(NI-L)/(U-NI)]$, where L and U are fixed lower and upper bounds of *NI*.
    4.2   Binary choice FEMs, by using a zero-one indicator generated based on NI, as an independent variable.
    4.3   Ordered choice FEMs, by using an ordinal variable generated based on NI, as an independent variable.
5. Finally, a lot of extended FEMs with large number of numerical and categorical independent variables could easily be subjectively defined or proposed by all researchers.
6. For additional illustrations, Baltagi (2009b, Table 3.2, p. 59) presents an additive two-way fixed- (cross-section and period) effects model of *LNY* on *lnK1*, *LNK2*, *LNL*, and *U*, Gujarati (2003, p. 644) presents an additive two-way fixed-effects model of $Y = I =$ gross of investment, on $X2 = F =$ value of the firm, and $X3 = C =$ stock of plant and equipment, Dharmapala, et al. (2011) present several firm-year fixed-effects models with lagged independent variables.

Example 9.5   The simplest additive ITFEM

Figure 9.3 presents statistical results based on the simplest additive firm-time fixed effects model, which is in fact an additive cell-means model of *LNSale* on the two factors FIRM_CODE and T, based on the subsample {Firm_Code < "_AKKU", T < 3}. Compare this model to the cell-means models presented in the previous chapter and in Agung (2011). Based on these results, the following notes and comments are presented.

1. Referring to the general ITFEM in (9.13a), the simplest additive firm-time fixed-effects model can be presented as follows.

$$LNSale_{it} = \delta_i + \tau_t + \varepsilon_{it} \tag{9.18}$$

But the analysis can be done using the equation specifications

$$LNSale \ @Expand(FIRM\_CODE) \ @Expand(T, @Droplast) \tag{9.19a}$$

or

$$LNSale \ @Expand(FIRM\_CODE, @Droplast) \ @Expand(T) \tag{9.19b}$$

or

$$LNSale \ C \ @Expand(FIRM\_CODE, @Drop(^*)) \ @Expand(T, @Drop(^*)) \tag{9.19c}$$

2. Table 9.4 presents the parameter and estimated means of LNSale by the FIRM_CODE and T, based on the model in (9.19a). Based on this table, the following conditional means differences (MDs) between pairs of the cells can be derived.

$$MD_i(LNSale_{i,1} - LNSale_{i,2}) = \hat{\tau}_1 - \hat{\tau}_2 = 0.019, \forall (for \ all)i.$$
$$MD_t(LNSale_{i_1,t} - LNSale_{i_2,t}) = \hat{\delta}_{i_1} - \hat{\delta}_{i_2}, \forall t.$$

For instance, $MD_t(LNSale_{1,t} - LNSale_{2,t}) = 7.357 - 2.293 = 5.064$, for all *t*.

Figure 9.3 *Statistical results based on an additive cell-means model*

Table 9.4   The parameter and estimated means of LNSale by firm and time, based on the results in Figure 9.5

	AALI	ABBA	ACAP	ADES	ADMG	AIMS	AISA
T=1	C(1)	C(2)	C(3)	C(4)	C(5)	C(6)	C(7)
	+C(8)	+C(8)	+C(8)	+C(8)	+C(8)	+C(8)	+C(8)
T=2	C(1)	C(2)	C(3)	C(4)	C(5)	C(6)	C(7)
Means Diffrence	**C(8)**	**C(8)**	**C(8)**	**C(8)**	**C(8)**	**C(8)**	**C(8)**

	_AALI	_ABBA	_ACAP	_ADES	ADMG	_AIMS	_AISA
T=1	7.357	2.293	4.648	4.756	7.282	3.657	3.288
	+0.019	+0.019	+0.019	+0.019	+0.019	+0.019	+0.019
T=2	7.357	2.293	4.648	4.756	7.282	3.657	3.288
Means Difference	0.019	0.019	0.019	0.019	0.019	0.019	0.019

3.   The hypotheses on the means differences can easily be tested using the Wald test. However, in order to test the hypothesis on the difference between all cells, applying the equation specification (9.19c) is suggested.

9.4.2   Interaction FEMs

Interaction individual-time fixed-effects models have the general equation as follows:

$$H(Y_{it}) = \gamma_{it} + F(X_{it}) + \varepsilon_{it} \tag{9.20}$$

where γ_{it} is a fixed effect. By using the *Individual × Time* dummies, the model can be presented as follows:

$$H(Y_{it}) = F(X_{it}) + \sum_{i,t}^{NT} \gamma_{it} Dit + \varepsilon_{it}$$

or $\tag{9.21}$

$$H(Y_{it}) = F(X_{it}) + \sum_{i,t}^{NT} \gamma_{it} Di \times Dt + \varepsilon_{it}$$

However, by using this model, a problem is found: that is, two types of error messages could be obtained as presented in Figure 9.4. See the following subsections.

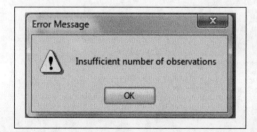
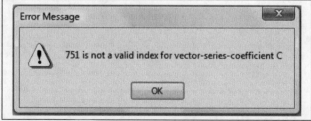

Figure 9.4   Two possible error messages might be obtained using Individual × Time *FEMs*

```
Tabulation of FIRM_CODE
Date: 08/26/11  Time: 08:58
Sample: 1 2284 IF FIRM_CODE < "_AKKU"
Included observations: 63
Number of categories: 10
```

Value	Count	Percent	Cumulative Count	Cumulative Percent
_AALI	8	12.70	8	12.70
_ABBA	8	12.70	16	25.40
_ACAP	5	7.94	21	33.33
_ACES	2	3.17	23	36.51
_ADES	8	12.70	31	49.21
_ADHI	6	9.52	37	58.73
_ADMG	8	12.70	45	71.43
_ADRO	2	3.17	47	74.60
_AIMS	8	12.70	55	87.30
_AISA	8	12.70	63	100.00
Total	63	100.00	63	100.00

(a)

```
Tabulation of FIRM_CODE and T
Date: 08/26/11  Time: 09:02
Sample: 1 2284 IF FIRM_CODE < "_AKKU" AND T < 3
Included observations: 14
Tabulation Summary
```

Variable	Categories
FIRM_CODE	7
T	2
Product of Categories	14

Count		T 1	T 2	Total
	_AALI	1	1	2
	_ABBA	1	1	2
	_ACAP	1	1	2
FIRM_CODE	_ADES	1	1	2
	_ADMG	1	1	2
	_AIMS	1	1	2
	_AISA	1	1	2
	Total	7	7	14

(b)

Figure 9.5   *Tabulation of FIRM_CODE, and two-way tabulation of the FIRM_CODE and T*

9.4.2.1   *Inappropriate Interaction FEM*

In order to understand the problem, specifically the limitations of the interaction FEMs in (9.21), see the following illustrative data analysis using a very small data set with 10 firms and two time-point observations.

Example 9.6   Frequency tables based on small incomplete pool data
Figure 9.5 presents the frequency table of the first 10 firms in the incomplete pool Data_Sesi.wf1, and the cross tabulation *Firms* × *Times*. Based on these tables, the following notes and comments are presented.

1. At the first stage, note how to select the first 10 firms. In this case, the steps are as follows:
   - Sort the data by the FIRM_CODE, then construct the tabulation of FIRM_CODE based on the whole data set, so we can identify the firm _AKKU as the eleventh observation.
   - Select the sample by entering (FIRM_CODE < "_AKKU" and T < 3) for the IF condition.
2. The cross tabulation *Firms* × *Times* presents only 14 out of 20 possible cells, and each cell has a single observation only, so that there are six empty cells. Take note that if the data is a complete or balanced pool data, then all cells will have a single observation. Now let us see the following data analysis using alternative interaction FEMs, which should be valid also for balanced pool data. In fact, the 14 cells represent a balanced pool data of the seven firms presented in Figure 9.5(b).

Example 9.7   The simplest interaction FEM
Referring to the general interaction FEMs in (9.20) and (9.21), Figure 9.6(a) presents the statistical results based on the simplest interaction FEM using the following equation specification.

$$LNSale\ @Expand(FIRM\_CODE, T) \tag{9.22}$$

and Figure 9.6(b) presents the means of *LNSale* by the *FIRM_CODE* and *T*. Based on these statistical results, the following findings and notes are presented.

1. Note that each cell has a single observation only. Refer to the two-way tabulation in Figure 9.5(b). For this reason, each cell does not have a standard error, indicated by NA in Figure 9.6(a).

Dependent Variable: LNSALE
Method: Least Squares
Date: 08/26/11   Time: 10:08
Sample: 1 2284 IF FIRM_CODE < "_AKKU" AND T < 3
Included observations: 14

Variable	Coefficient	Std. Error	t-Statistic	Prob.
FIRM_CODE=_AALI,T=1	7.273786	NA	NA	NA
FIRM_CODE=_AALI,T=2	7.459915	NA	NA	NA
FIRM_CODE=_ABBA,T=1	2.302585	NA	NA	NA
FIRM_CODE=_ABBA,T=2	2.302585	NA	NA	NA
FIRM_CODE=_ACAP,T=1	4.605170	NA	NA	NA
FIRM_CODE=_ACAP,T=2	4.709530	NA	NA	NA
FIRM_CODE=_ADES,T=1	4.663439	NA	NA	NA
FIRM_CODE=_ADES,T=2	4.867534	NA	NA	NA
FIRM_CODE=_ADMG,T=1	7.280697	NA	NA	NA
FIRM_CODE=_ADMG,T=2	7.301822	NA	NA	NA
FIRM_CODE=_AIMS,T=1	3.526361	NA	NA	NA
FIRM_CODE=_AIMS,T=2	3.806662	NA	NA	NA
FIRM_CODE=_AISA,T=1	3.761200	NA	NA	NA
FIRM_CODE=_AISA,T=2	2.833213	NA	NA	NA

R-squared	1.000000	Mean dependent var	4.763893
S.D. dependent var	1.874715	Sum squared resid	0.000000

(a)

Descriptive Statistics for LNSALE
Categorized by values of FIRM_CODE and T
Date: 08/26/11   Time: 10:15
Sample: 1 2284 IF FIRM_CODE < "_AKKU" AND T < 3
Included observations: 14

Mean		T 1	2	All
	_AALI	7.273786	7.459915	7.366851
	_ABBA	2.302585	2.302585	2.302585
	_ACAP	4.605170	4.709530	4.657350
FIRM_CODE	_ADES	4.663439	4.867534	4.765487
	_ADMG	7.280697	7.301822	7.291260
	_AIMS	3.526361	3.806662	3.666512
	_AISA	3.761200	2.833213	3.297207
	All	4.773320	4.754466	4.763893

(b)

Figure 9.6 *Statistical results based on the model of LNSale in (9.18), and the means of LNSale by the FIRM_CODE and T*

2. Therefore, the regression coefficients in Figure 9.6(a) and the means of LNSale by the FIRM_CODE and T in Figure 9.6(b) are in fact the observed scores of LNSale within the 14 cells.
3. Based on these findings, I would say the *Individual × Time* fixed effects model cannot be applied, even more so one with exogenous variables. See the following example.

9.4.2.2   *Acceptable Interaction FEMs*

The following interaction FEMs would be acceptable models, if and only if each cell generated by the interaction fixed effects has at least three observations for the simplest heterogeneous regression lines in a two-dimensional space. The general equation specification of the interaction FEMs are as follows:

1. The *individual_group × time* or *Group × T(time point)* fixed-effects models, with the general equation specification as follows:

$$H(Y)\,X1 \ldots Xk \ldots XK\ @Expand(Group, T) \tag{9.23}$$

where *H(Y)* is a constant function of an endogenous variable *Y*, such as $H(Y) = Y$, $log(Y - L)$, $log(U - Y)$, or $log[(Y - L)/(U - Y)]$, where *L* and *U* are the constant or fixed lower and upper bounds of *Y*; *Xk* can be a numerical variable, a dummy variable, a lag variable of the endogenous or exogenous variable, an environmental variable, say $Xk_{it} = Xk_t$, for all *i*, or an interaction factor, such as the interaction between numerical variables, interaction between numerical and dummy variables, and interaction between dummy variables.

2. The *individual* × *time_period* or *firm* × *TP(time − period)* fixed-effects models, with the general equation specification as follows:

$$H(Y)\,X1\ldots Xk\ldots XK\,@Expand(Firm, TP) \tag{9.24}$$

In this case, the models can be those with time trend or time-related effects.

3. The *Group* × *TP* fixed-effects models with the general equation specification as follows:

$$H(Y)\,X1\ldots Xk\ldots XK\,@Expand(Group, TP) \tag{9.25a}$$

Similar to the models in (9.24), these models can be those with time trend or time-related effects.

Example 9.8   Application of the models in (9.23) and (9.24)

Figure 9.7 presents the statistical results based on a sample (Firm_code < "_AKKU" using two interaction FEMs with the general equation specifications (9.23) and (9.24).

Based on the results in Figure 9.7, the following notes and comments are presented.

1. The output in Figure 9.7(a) is obtained by using the following equation specification, corresponding to the ES (9.23)

$$NI\ LNSALE\ LNLOAN\ LNSALE^*LANLOAN\ @Expand(Group, T) \tag{9.25b}$$

Dependent Variable: NI
Method: Least Squares
Date: 08/27/11   Time: 21:08
Sample: 1 2284 IF FIRM_CODE<"_AKKU"
Included observations: 56

Variable	Coefficient	Std. Error	t-Statistic	Prob.
LNSALE	215.8557	54.37901	3.969467	0.0003
LNLOAN	159.7301	339.3395	0.470709	0.6406
LNSALE*LNLOAN	-67.83895	48.22350	-1.406761	0.1678
GROUP=1,T=1	-1029.048	435.6436	-2.362132	0.0235
GROUP=1,T=2	-1068.419	431.7135	-2.474834	0.0180
GROUP=1,T=3	-1097.499	441.0167	-2.488566	0.0175
GROUP=1,T=4	-1205.301	489.7626	-2.460991	0.0186
GROUP=1,T=5	-1301.227	496.4086	-2.621282	0.0126
GROUP=1,T=6	-1126.391	465.0931	-2.421860	0.0205
GROUP=1,T=7	-1023.202	475.6178	-2.151312	0.0380
GROUP=1,T=8	-1066.846	500.8795	-2.129946	0.0399
GROUP=2,T=1	-841.4006	698.6100	-1.204392	0.2361
GROUP=2,T=2	-874.3508	681.8651	-1.282293	0.2077
GROUP=2,T=3	-934.9180	497.9730	-1.877447	0.0684
GROUP=2,T=4	-1168.912	489.4166	-2.388379	0.0221
GROUP=2,T=5	-1163.986	528.1790	-2.203773	0.0338
GROUP=2,T=6	-1069.275	529.0780	-2.021016	0.0506
GROUP=2,T=7	-1140.601	542.1673	-2.103781	0.0423
GROUP=2,T=8	-352.0984	553.7592	-0.635833	0.5288

R-squared	0.518871	Mean dependent var	258.5041
Adjusted R-squared	0.284808	S.D. dependent var	759.2311
S.E. of regression	642.0742	Akaike info criterion	16.03142
Sum squared resid	15253592	Schwarz criterion	16.71860
Log likelihood	-429.8798	Hannan-Quinn criter.	16.29784
Durbin-Watson stat	1.540020		

(a)

Dependent Variable: NI
Method: Least Squares
Date: 08/27/11   Time: 21:13
Sample: 1 2284 IF FIRM_CODE<"_AKKU"
Included observations: 56

Variable	Coefficient	Std. Error	t-Statistic	Prob.
LNSALE	206.5578	180.1092	1.146847	0.2590
LNLOAN	-146.3462	443.6354	-0.329879	0.7434
LNSALE*LNLOAN	5.449253	72.31573	0.075354	0.9404
FIRM_CODE=_AALI,TP=1	-1164.197	1395.778	-0.834084	0.4097
FIRM_CODE=_AALI,TP=2	112.2452	1554.804	0.072193	0.9428
FIRM_CODE=_ABBA,TP=1	-821.4049	788.8535	-1.041264	0.3047
FIRM_CODE=_ABBA,TP=2	-1082.296	967.4084	-1.118759	0.2707
FIRM_CODE=_ACAP,TP=1	-1121.899	1003.383	-1.118116	0.2709
FIRM_CODE=_ACES,TP=2	-1527.582	1321.502	-1.155944	0.2553
FIRM_CODE=_ADES,TP=1	-994.7136	898.5205	-1.107057	0.2756
FIRM_CODE=_ADES,TP=2	-983.1305	864.4532	-1.137286	0.2629
FIRM_CODE=_ADHI,TP=1	-1367.156	1428.755	-0.956887	0.3450
FIRM_CODE=_ADHI,TP=2	-1494.239	1778.912	-0.839974	0.4065
FIRM_CODE=_ADMG,TP=1	-1729.834	1542.010	-1.121805	0.2694
FIRM_CODE=_ADMG,TP=2	-1653.759	1546.097	-1.069634	0.2919
FIRM_CODE=_ADRO,TP=2	593.1445	1868.997	0.317360	0.7528
FIRM_CODE=_AIMS,TP=1	-878.8881	807.1977	-1.088814	0.2835
FIRM_CODE=_AIMS,TP=2	-912.5270	876.3424	-1.041291	0.3047
FIRM_CODE=_AISA,TP=1	-1003.606	952.5374	-1.053613	0.2991
FIRM_CODE=_AISA,TP=2	-1164.256	1126.020	-1.033957	0.3081

R-squared	0.787869	Mean dependent var	258.5041
Adjusted R-squared	0.675911	S.D. dependent var	759.2311
S.E. of regression	432.2212	Akaike info criterion	15.24821
Sum squared resid	6725346.	Schwarz criterion	15.97154
Log likelihood	-406.9497	Hannan-Quinn criter.	15.52864
Durbin-Watson stat	3.182651		

(b)

Figure 9.7   *Statistical results based on the models in (9.23) and (9.24)*

where the GROUP is generated as Group $= 1 + 1^*$(Firm_code $>$ "_ADES"), and the time $T = 1$ to 8, in the whole Data_sesi.wf1. This output represents:

1.1 A set of 16 homogenous regressions (regressions having the same slope parameters, and their intercepts are defined or considered as the fixed effects of the 16 cells of individuals indicated by (GROUP,T).

1.2 For instance, the first two regressions, the regressions in GROUP $= 1$ at the time points $T = 1$ and $T = 2$, have equations as follows:

$$NI = 215.856LNSALE + 159.730LNLOAN - 67.839LNSALE\ LNLOAN - 1029.048$$

and

$$NI = 215.856LNSALE + 159.730LNLOAN - 67.839LNSALE\ LNLOAN - 1068.419$$

where the intercepts (-1029.048) and (-106.419), respectively, are the Group $= 1$ fixed effects at the first two time points.

2. The output in Figure 9.7(b) is obtained by using the following equation specification, corresponding to the ES (9.24)

$$NI\ LNSALE\ LNLOAN\ LNSALE^*LANLOAN\ @Expand(Firm\_Code, TP) \tag{9.26}$$

where the time period, TP, is generated as TP $= 1 + 1^*(T > 5)$, and TP $= 1$ and 2, indicate the time periods before and after a new tax policy. This output represents:

2.1 A set of 17 homogenous regressions (regressions with the same slope parameters, their intercepts are defined or considered as fixed effects of the 17 cells of individuals indicated by *(Firm_Code, TP)*. Why is it an odd number? In this case, the data used is incomplete pool data.

2.2 Note that the firm _ADRO does not have an observation at TP $= 1$, with the equation of the regression at TP $= 2$ as follows:

$$NI = 206.558LNSALE - 146.346LNLOAN + 5.449LNSALE\ LNLOAN + 593.144$$

So that _ARDO only has a single fixed effect of $(+593.144)$ at TP $= 2$ (after the new tax policy).

Example 9.9   Other interaction FEMs

Table 9.5 presents a summary of the statistical results based on a *Firm_Code $\times$ Time_period* fixed-effects model and three other fixed-effects models, based on the Data_Sesi.wf1 for a comparative study. Based on the results in this table, the following notes and comments are presented.

1. Model-1 is an interaction *Firm $\times$ Time_period* fixed effects model of *NI* on *LNSale, LNLoan* and interaction between *LNSale*LNLoan*. This model represents 334×2 homogeneous regressions, for instance the first two regressions, namely for the firm _AALI at the two time periods (TP $= 1$ and 2), have the following equations.

$$NI = 74.315LNSALE + 83.414LNLOAN - 24.615LNSALE^*LNLOAN - 143.847$$

and

$$NI = 74.315LNSALE + 83.414LNLOAN - 24.615LNSALE^*LNLOAN - 251.418$$

Take note that the intercepts (-143.847) and (-251.418), respectively, are defined as the fixed effects of the firm _AALI at the two time periods. In fact they are the intercepts of the two regression functions

Table 9.5 *Statistical results based on interaction FEMs in (9.24) and alternatives FEMs*

Dependent Variable: NI (Net income)												
Method: Least Squares												
Date: 08/27/11 Time: 17:58												
Sample: 1 2284												
Included observations: 2058												

Variable	Model-1			Model-2			Model-3			Model-4		
	Coeff.	t-Stat	Prob.	Coeff.	t-Stat	Prob.	Coeff.	t-Stat	Prob.	Coeff.	t-Stat	Prob.
LNSALE	74.315	3.217	0.001	41.301	1.649	0.099	89.879	4.483	0.000	184.939	16.745	0.000
LNLOAN	83.414	1.397	0.163	162.432	3.071	0.002	161.260	3.048	0.002	-89.463	-2.158	0.031
LNSALE*LNLOAN	-24.615	-2.387	0.017	-44.418	-4.878	0.000	-44.609	-4.896	0.000	-3.735	-0.509	0.611
Fixed effects?												
FirmxTime Period (334 x 2)	Yes			No			No			No		
Firm+Time (333 + 8)	No			Yes			No			No		
Firm (334 dummies)	No			No			Yes			No		
Time (8 dummies)	No			No			No			Ye		
R-squared	0.796			0.665			0.661			0.131		
Adjusted R-squared	0.714			0.598			0.595			0.127		
Durbin-Watson stat	2.716			1.773			1.775			0.850		

Table 9.6   *Statistical results based on the models in (9.28a) and (9.29a)*

Dependent Variable: Y1						
Method: Least Squares						
Date: 08/18/11 Time: 15:19						
Sample (adjusted): 1 2283						
Included observations: 2058 after adjustments						
	Model-1			Model-2		
Variable	Coeff.	t-Stat.	Prob.	Coeff.	t-Stat.	Prob.
X1	-0.128	-2.300	0.022	-0.115	-2.078	0.038
X2	0.114	1.475	0.140	0.113	1.460	0.145
X3	1.308	15.711	0.000	1.301	15.663	0.000
X4	-5.645	-18.041	0.000	-5.591	-18.158	0.000
X5	-0.154	-2.065	0.039	-0.149	-2.006	0.045
X6	-0.002	*-0.082*	*0.934*	-0.005	*-0.195*	*0.846*
X7	-0.726	-1.484	0.138	-0.718	-1.470	0.142
X8	13.121	10.268	0.000	12.824	10.138	0.000
X9	2.956	24.969	0.000	2.945	24.989	0.000
Fixed effects?						
ABT	Yes			No		
AB	No			Yes		
T	No			Yes		
# of ABT dummies (2,2,8)	32					
# of AB dummies				3		
# of T dummies				8		
R-squared	0.623			0.619		
Adjusted R-squared	0.616			0.615		
Durbin-Watson stat	0.780			0.785		

in a three-dimensional space with the axes *LNSALE*, *LNLOAN* and *NI*. This model in fact is a two-way ANCOVA model of NI, with covariates LNSALE, LNLOAN and LNSALE*LNLOAN.

2. Model-2 is the *Firm + Time* fixed-effects model is a LSDV model with 333 dummies of the 334 firms and eight time point dummies. Another LSDV model can be presented using 334 firm dummies, and seven dummies of the eight time points. Compared to Model-1, this model is a special two-way ANCOVA model of *NI*, with the same covariates and a specific hidden assumption. Refer to the statistical results in the additive model in Figure 9.5, and the means presented in Table 9.6.

3. Model-3 is a firm fixed-effects model, and Model-4 is a time/period fixed-effects model. Note that these models are in fact one-way ANCOVA models of NI, with the same covariates.

9.5 Extended Fixed-Effects Models

9.5.1 Least Square FEMs

A lot of FEMs, such as three-way or more fixed-effects models, can easily be derived based on all models presented previously and in Chapters 6–8; as well as those models in Agung (2011, 2009a) with three or more exogenous/environmental variables. The *Journal of Finance* presents various FEMs with a large number of exogenous variables, either numeric or categorical exogenous, some selected models of which are presented in Section 9.6.

However, it should be noted that there is the possibility of obtaining a "*Near singular matrix*" error message, because the model has a large number of dummy independent variables.

Example 9.10 Extension of the model in Figure 8.24(a)

Referring to the model in Figure 8.24(a), we can say that the model is an "*additive groups and time-fixed-effects model*" with two dichotomous factors A and B, and the time T with four time points. The statistical results based on a more general fixed-effects model of Y on X_1 to X_K, with two multi-level factors A and B, and the time T, could be obtained using an ES, as follows:

$$Y \, C \, X1 \ldots XK \, @Expand(A, @Droplast)$$
$$@Expand(B, @Droplast) \, @Expand(T, @droplast) \tag{9.27}$$

where Xk can be various variables, as indicated for the model in (9.23).

For an additional illustration, Engelberg and Parsons (2011) present several FEMs with three fixed effects, namely Industry, City and Date; and four fixed-effects: Industry, Paper, City and Date.

Example 9.11 Other illustrative interaction FEMs

Referring to the model in Figure 8.24(a), it can be said that the model is an "*additive groups and time fixed-effects model*" with two dichotomous factors A and B, and the time T with four time points. Take note that the model in fact is a special ANCOVA model with additive categorical factors A, B and T, represented by their dummies. The statistical results based on a more advanced fixed-effects model of Y on X_1 to X_K, with two multi-level factors A and B, and the time T, can be considered as follows:

The models with three-way interactions $A \times B \times T$ fixed effects have the following two alternative general equation specifications.

$$Y1 \, X1 \ldots XK \, @Expand(A, B, T) \tag{9.28a}$$

or

$$Y1 \, C \, X1 \ldots XK \, @Expand(A, B, T, @Drop(^*,^*,^*)) \tag{9.28b}$$

The models with two-way interactions $(A \times B) + T$ fixed effects have the following two alternative general equation specifications.

$$Y1 \, X1 \ldots XK \, @Expand(A, B, @Droplast) \, @Expand(T) \tag{9.29a}$$

or

$$Y1 \, C \, X1 \ldots XK \, @Expand(A, B, @Drop((^*,^*)) \, @Expand(T, @Drop(^*)) \tag{9.29b}$$

with illustrative statistical results presented in Table 9.6, for $K = 9$. Based on the results in this table, the following notes and comments are presented.

Both models are additive FEMs of $Y1$ on $X1$ to $X9$, and the corresponding dummies, namely 32 ($= 2 \times 2 \times 8$) dummies of the interaction $ABT = A \times B \times T$ in Model-1, and Model-2 has $(2 \times 2 - 1) = 3$ dummies of the interaction $AB = A \times B$ and 8 dummies of the time t.

1. Instead of Model-2, the third model could be derived using 4 ($= 2 \times 2$) dummies of $AB = A \times B$ and 7 dummies of T, by using the following equation specification.

$$Y1\ X1\ldots XK\ @Expand(A,B,)\ @Expand(T,@Dropfirst) \tag{9.30}$$

2. A lot of FEMs could easily be subjectively defined, specifically in relation to the numerical variables *Y1, X1* to *XK*, such as follows:

   2.1   Various FEMs using the transformed variables of *Y1, X1* to *XK*, such as the translog linear FEMs, and FEMs with mixed independent variables.

   2.2   Binary or ordered choice FEMs, if *Y1* is a zero-one or ordinal variables.

   2.3   Bounded FEMs, with a dependent variable: *log(Y1 − L), log(U − Y1) or log[(Y1 − L)/(U − Y1)]*, where L and U are fixed lower and upper bounds of *Y1*.

   2.4   Various polynomial effects models.

   2.5   Various interaction FEMs, by using some selected two- or three-way interactions of the independent variables of all FEMs defined previously, are considered relevant in a theoretical sense. Refer to the various alternative variables *Xk* indicated for the model (9.23). Furthermore, note that based on *X1* to *X9*, there would be 36 possible two-way interactions, and 84 possible three-way interactions. So researchers should select only some of the interactions for their models. See the following example.

Example 9.12   Interaction FEMs derived from the model in Figure 8.22(a)

Table 9.7 presents a summary of the statistical results based on four models, with characteristics as follows:

1. For Model-1, the pool data is treated as a cross-section data set. Note that the model does not pass through the origin, nor it does not have an intercept. The main objective of this model is to show that the effect of *X1* on *Y1* depends on the other exogenous variables.

2. Model-2 is an IFEM with 343 firm fixed-effects, where *X1*X6* has an insignificant adjusted effect on *Y1* with a large *p*-value, so then it can be reduced.

3. Model-3 is a TFEM with eight time-fixed effects, where *X1*X6* has an insignificant adjusted effect on *Y1* with a large *p*-value, so then it can be reduced.

4. Model-4 is an individual time -fixed-effects model (ITFEM), where *X1*X6* also has an insignificant adjusted effect.

5. Note that the models have interaction independent variables with *X1* only. In fact many more advanced models could be defined by inserting additional independent variables, such as main factors, other interaction factors and other variables outside of the models.

6. Note that the small DW-statistical values indicate that a lagged variables or autoregressive model should be explored. For comparison, Baltagi (2009a,b) explores various fixed- and random effects models with small DW-statistical values. For instance, Table 2.4 (Baltagi, 2009b, p. 20) presents a random fixed-effects model with a DW-statistic of 0.315 210 for the weighted statistic and 0.026 675 for the unweighted statistic.

9.5.2   Alternative FEMs

With regards to the endogenous variable of all models presented here, various alternative FEMs can easily be derived by using other types of dependent variables, such as follows:

9.5.2.1   Quantile Regression FEMs

Data analysis based on the quantile regression FEMs can be done directly using all equation specifications presented here but only by selecting the QREG as the estimation setting method. Do it as an exercise.

Example 9.13   Unexpected quantile fixed-effects models

Table 9.8 presents the statistical results based on two unexpected quantile fixed-effects models, which are derived from the models in Chapter 8, namely the multi-factorial cell-means model of LNSALE on the

Table 9.7 Summary of the statistical results of four models, derived from the model in Figure 8.22(a)

Dependent Variable: Y1
Method: Least Squares
Date: 08/16/11 Time: 13:35
Sample (adjusted): 1 2283
Included observations: 2058 after adjustments

Variable	Model-1			Model-2			Model-3			Model-4		
	Coeff.	t-Stat	Prob.	Coeff.	t-Stat	Prob.	Coeff.	t-Stat	Prob.	Coeff.	t-Stat	Prob.
X1*X2*X3	0.059	5.353	0.000	0.011	1.308	0.191	0.069	6.148	0.000	0.011	1.376	0.169
X1*X2*X4	0.458	22.228	0.000	0.237	13.406	0.000	0.409	18.749	0.000	0.230	12.965	0.000
X1*X3*X4	1.086	17.122	0.000	0.764	14.899	0.000	1.083	17.077	0.000	0.785	15.165	0.000
X1*X2	-0.799	-20.455	0.000	-0.381	-11.153	0.000	-0.714	-17.447	0.000	-0.367	-10.704	0.000
X1*X3	-2.426	-16.132	0.000	-1.743	-14.446	0.000	-2.428	-16.153	0.000	-1.792	-14.716	0.000
X1*X4	-1.183	-35.603	0.000	-0.678	-20.540	0.000	-1.071	-28.824	0.000	-0.654	-19.644	0.000
X1*X5	0.084	5.019	0.000	0.057	5.445	0.000	0.088	5.298	0.000	0.060	5.698	0.000
X1*X6	-0.006	-1.422	0.155	0.000	-0.131	0.896	-0.006	-1.499	0.134	0.000	0.098	0.922
X1*X9	0.359	40.368	0.000	0.234	24.151	0.000	0.348	38.642	0.000	0.225	22.090	0.000
Firm fixed effects	No			Yes			No			Yes		
Time fixed effects	No			No			Yes			Yes		
# of firms	-			343			-			343		
# of time-points	-			-			8			7		
R-squared	0.728			0.927			0.734			0.928		
Adjusted R-squared	0.727			0.912			0.732			0.913		
Durbin-Watson stat	0.827			1.378			0.803			1.372		

Table 9.8 Unexpected statistical results based on two quantile fixed-effects models

Dependent Variable: LNSALE
Method: Quantile Regression (Median)
Date: 08/30/11 Time: 08:02
Sample: 1 2284
Included observations: 2270
Huber Sandwich Standard Errors & Covariance
Sparsity method: Kernel (Epanechnikov) using residuals
Bandwidth method: Hall-Sheather, bw=0.073925
Estimation successful but solution may not be unique

Variable	Coeff.	s.e	t-Stat	Prob.
C	7.533	0.116	65.120	0.000
(8-1) time dummies	Yes			
(8x2x2-1) TxAxB dummies	Yes			
Pseudo R-squared	0.395	Mean dependent var	6.00	
Adjusted R-squared	0.387	S.D. dependent var	2.00	
S.E. of regression	1.279	Objective	1062.60	
Quantile dependent var	6.096	Objective (const. only)	1757.51	
Sparsity	2.882	Quasi-LR statistic	1928.72	
Prob(Quasi-LR stat)	0.000			

Dependent Variable: Y1_NEW
Method: Quantile Regression (Median)
Date: 08/30/11 Time: 08:44
Sample: 1 2284 IF X1 > -2 AND T>5
Included observations: 958
Huber Sandwich Standard Errors & Covariance
Sparsity method: Kernel (Epanechnikov) using residuals
Bandwidth method: Hall-Sheather, bw=0.098555
WARNING: Maximum iterations exceeded (results may not be valid)

Variable	Coeff.	s.e	t-Stat	Prob.
C	-7.103	0.176	-40.370	0.000
LOG((2+X1)/(12-X1))*(T=6)	0.748	0.263	2.846	0.005
LOG((2+X1)/(12-X1))*(T=7)	0.776	0.241	3.214	0.001
LOG((2+X1)/(12-X1))*(T=8)	0.754	0.245	3.077	0.002
(3-1) time dummies	Yes			
(336-1) firm dummies	Yes			
Pseudo R-squared	0.886	Mean dependent va	-4.823	
Adjusted R-squared	0.823	S.D. dependent var	2.102	
S.E. of regression	0.545	Objective	87.931	
Quantile dependent var	-4.729	Objective (const. or	773.904	
Sparsity	0.254	Quasi-LR statistic	21646.090	
Prob(Quasi-LR stat)	0.000			

factors T, A and B, with 8, 2 and 2 levels, respectively, and the model in Figure 8.16, for the sample $X2 > -2$ and $T > 5$.

Based on the results in this table, it can be said that both outputs are not valid statistical results, since the output of the first model presents the note "*Estimation successful but solution may not be unique*", and the second model presents a warning "*results may not valid*". Why? I would say the scatter of observed scores within so many groups of individuals is the main cause of the problems. In other words, the problems are highly dependent on the data to be selected. Refer to special notes and comments presented in Agung (2009a, Section 2.14). See the unique optimal solutions based on simpler models presented in the following example.

Example 9.14   Unique optimal solutions based on quantile FEMs
Figure 9.8 presents unique optimal solutions based on two quantile FEMs, where the factor A has two levels, and the time T has three time points, namely $T = 6$, 7 and 8, corresponding to the years 2007, 2008 and 2009, in Data_Sesi.wf1. Note that these models are much simpler compared to those in Table 9.8 with the same numerical variables.

Example 9.15   Quantile and AR(2) LS polynomial group × time fixed-effects models
For a comparative study, Table 9.9 presents a summary of statistical results based on two group × time fixed-effects models, with an intercept parameter, C, using the equation specifications as follows:

$$lnSale \quad C \quad lnAsset \quad lnAsset\hat{} 2 \quad lnAsset\hat{} 3 \quad PBV \; EQUITY$$
$$LIABILITIES \quad MARKETSHARE \quad @Expand(Group, T, @Drop(1,1)) \tag{9.31}$$

Dependent Variable: LNSALE
Method: Quantile Regression (Median)
Date: 08/30/11 Time: 09:48
Sample: 1 2284
Included observations: 2270
Bootstrap Standard Errors & Covariance
Bootstrap method: XY-pair, reps=100, rng=kn, seed=241313196
Sparsity method: Kernel (Epanechnikov) using residuals
Bandwidth method: Hall-Sheather, bw=0.073925
Estimation successfully identifies unique optimal solution

Variable	Coeff.	s.e	t-Stat	Prob.
C	5.024	0.049	102.473	0.000
A=2	2.182	0.069	31.745	0.000
Pseudo R-squared	0.211	Mean dependent var		5.997
Adjusted R-squared	0.211	S.D. dependent var		1.999
S.E. of regression	1.601	Objective		1385.87
Quantile dependent var	6.096	Objective (const. only)		1757.51
Sparsity	3.666	Quasi-LR statistic		811.00
Prob(Quasi-LR stat)	0.000			

Dependent Variable: Y1_NEW
Method: Quantile Regression (Median)
Date: 08/30/11 Time: 09:52
Sample: 1 2284 IF X1 > -2 AND T >5
Included observations: 958
Huber Sandwich Standard Errors & Covariance
Sparsity method: Kernel (Epanechnikov) using residuals
Bandwidth method: Hall-Sheather, bw=0.098555
Estimation successfully identifies unique optimal solution

Variable	Coeff.	s.e	t-Stat	Prob.
C	-5.154	0.101	-50.821	0.000
LOG((2+X1)/(12-X1))*(T=6)	1.848	0.140	13.158	0.000
LOG((2+X1)/(12-X1))*(T=7)	1.870	0.112	16.767	0.000
LOG((2+X1)/(12-X1))*(T=8)	1.776	0.098	18.136	0.000
T=6	0.150	0.124	1.204	0.229
T=7	0.140	0.111	1.263	0.207
A=1	-1.366	0.099	-13.745	0.000
Pseudo R-squared	0.507	Mean dependent var		-4.823
Adjusted R-squared	0.504	S.D. dependent var		2.102
S.E. of regression	1.084	Objective		381.46
Quantile dependent var	-4.729	Objective (const. only)		773.90
Sparsity	2.537	Quasi-LR statistic		1237.57
Prob(Quasi-LR stat)	0.000			

Figure 9.8 *Unique optimal solutions based on two quantile FEMs*

Table 9.9   *Statistical results based on the Group × Time FEMs in (9.31) and (9.32)*

Dependent Variable: LNSALE					Dependent Variable: LNSALE			
Method: Quantile Regression (Median)					Method: Least Squares			
Date: 08/29/11 Time: 19:59					Date: 08/29/11 Time: 20:03			
Sample: 1 2284					Sample (adjusted): 3 2284			
Included observations: 2270					Included obs: 2246 after adjustments			
Huber Sandwich Standard Errors & Covariance					Conv. achieved after 8 iterations			
Sparsity method: Kernel (Epanechnikov) using resid.								
Bandwidth method: Hall-Sheather, bw=0.073925								
Estimation successful but solution may not be unique								
	Quantile Regression				AR(2) LS Regression			
Variable	Coeff.	s.e	t-Stat	Prob.	Coeff.	s.e	t-Stat	Prob.
C	7.566	0.222	34.017	0.000	7.172	0.241	29.819	0.000
LNASSET	-3.005	0.121	-24.863	0.000	-2.483	0.148	-16.736	0.000
LNASSET^2	0.638	0.030	21.364	0.000	0.502	0.035	14.177	0.000
LNASSET^3	-0.034	0.002	-16.079	0.000	-0.025	0.002	-10.099	0.000
PBV	0.001	0.000	3.438	0.001	0.001	0.001	*1.007*	*0.314*
EQUITY	0.000	0.000	4.282	0.000	0.000	0.000	2.162	0.031
LIABILITIES	0.000	0.000	5.060	0.000	0.000	0.000	4.044	0.000
MARKETSHARE	0.653	0.759	*0.860*	*0.390*	0.886	0.362	2.446	0.015
Fixed effects								
GroupxTimes	Yes				Yes			
(5x8 -1) dummies								
AR(1)					0.675	0.022	31.379	0.000
AR(2)					0.071	0.021	3.334	0.001
Pseudo R-squared	0.472				R-squared		0.832	
Adj. R-squared	0.461				Adjusted R-squared		0.828	
S.E. of regression	1.215				S.E. of regression		0.824	
Quantile dep. var	6.096				Sum squared resid		1490.5	
Sparsity	1.842				Log likelihood		-2726.4	
Quasi-LR statistic	3604.2				F-statistic		226.7	
Prob(Quasi-LR stat)	0.000				Prob(F-statistic)		0.000	
					Durbin-Watson stat		1.995	
					Inverted AR Roots		0.770	-0.090

and

$$lnSale\ C\ lnAsset\ lnAsset^2\ lnAsset^3\ PBV\ EQUITY\ LIABILITIES$$
$$MARKETSHARE\ @Expand(Group, T, @Drop(1, 1))\ AR(1)\ AR(2)$$

(9.32)

1. In this case, the factor GROUP has five levels, which is generated based on ROA using the equation as follows:

$$\text{GROUP} = 1 + 1^*(\text{roa} >= @\text{quantile}(\text{roa}, 0.2) + 1^*(\text{roa} >= @\text{quantile}(\text{roa}, 0.4)$$
$$+ 1^*(\text{roa} >= @\text{quantile}(\text{roa}, 0.6) + 1^*(\text{roa} >= @\text{quantile}(\text{roa}, 0.8) \tag{9.33}$$

Based on the quantile regression, only the MARKETSHARE has an insignificant adjusted effect on LNSALE, but based on the LS regression, only the PBV has an insignificant adjusted effect.

2. The intercept parameter "C" should be used for testing the hypothesis on the joint effects of all independent variables (7 numerical variables and 39 dummies) of the models on *LNSALE*. Note that the Quasi-LR statistic is used for testing based on quantile regression, and the *F*-statistic is used for testing based on the LS regression.

3. The other hypotheses on the joint effects of a sub-set of independent variables can easily be tested using the Wald test.

4. Furthermore, the parameter C is defined as the fixed effect of GROUP $= 1$, at the time $T = 1$.

5. The coefficient of each dummy variable indicates the adjusted means difference between a pair of cells using (Group $= 1$, $T = 1$) as the reference group, corresponding to the function @Drop(1,1) applied in (9.31) and (9.32).

9.5.2.2 Binary and Ordered Choice FEMs

By using a zero-one dependent variable, for all the equation specifications previously, data analysis can be done easily. Similarly, based on all binary choice models presented in Chapter 7, various FEMs can be derived easily and this is also the case for ordered choice models. See the following examples. For additional illustrations, refer to the one-way and two-way fixed-effects binary choice models presented in Baltagi (2009b, pp. 252–253), and Wang, et al. (2010) who present two binary choice firm-fixed-effects models with time trend based on 2683 firms within 18 565 observations.

Example 9.16 Binary choice fixed-effects models

Table 9.10 presents statistical results based on two binary logit FEMs, which are derived from the models in (7.16) and (7.22), respectively. Based on this table the following findings and notes are presented.

1. The first model is a binary logit *Group × Time* fixed-effects model. The characteristics of this model are as follows:

   1.1 It represents a set of 40 homogenous binary logit polynomial models of *D_Dividend on LNSALE^3* and *LNSALE^4*, and each of the independent variables has a significant effect on the logit *log [p/(1−p)]*, where $p = PR(D\_Dividend = 1)$.

   1.2 This model in fact is an ANCOVA binary logit model with a special assumption, which is not valid in many or most cases.

   1.3 This model is a reduced model of the one with a complete set of independent variables *LNSALE*, *LNSALE^2*, *LNSALE^3* and *LNSALE^4*.

2. In order to be general, the symbols *X* and *Y* are used as the variables of the second model, which is a binary probit *O5X1 × Time* fixed-effects model. This model has similar characteristics to the first; however, the following should also be noted.

   2.1 The variable *X2* has an insignificant adjusted effect with a large *p*-value. So a reduced model should be explored in a statistical sense. In general, *X2* would be deleted. Do it as an exercise.

   2.2 Various binary choice *Group × Time* fixed-effects models with interaction(s) of the exogenous variables should be applicable in most studies.

   2.3 The GROUP variable is generated based on one of the independent variable, namely X1, using its quintiles. To modify the model the GROUP variable can be generated based on one or more variables, either in or out of the model.

Table 9.10 *Statistical results based on two binary choice Group × Time FEMs*

Dependent Variable: D_DIVIDEND
Method: ML - Binary Logit (Quadratic hill climbing)
Sample: 1 2284
Included observations: 2270
Convergence achieved after 6 iterations
Covariance matrix computed using second derivatives

Variable	Coff.	s.e.	z-Stat	Prob.
LNSALE^3	0.00830	0.00172	4.83751	0.00000
LNSALE^4	-0.00061	0.00017	-3.67714	0.00020
Fixed effects				
GroupxTime	Yes			
5x8 dummies				
Mean dependent var	0.254	S.D. dependent var	0.43550	
S.E. of regression	0.387	Akaike info criterion	0.93585	
Sum squared resid	333.952	Schwarz criterion	1.04183	
Log likelihood	-1020.195	Hannan-Quinn criter.	0.97452	
Avg. log likelihood	-0.449			
Obs with Dep=0	1693	Total obs	2270	
Obs with Dep=1	577			

Dependent Variable: Y
Method: ML - Binary Probit (Quadratic hill climbing)
Sample (adjusted): 1 2283
Included observations: 2058 after adjustments
Convergence achieved after 5 iterations
Covariance matrix computed using second derivatives

Variable	Coff.	s.e.	z-Stat	Prob.
C	-2.02411	0.30291	-6.68220	0.00000
X1	0.28355	0.05655	5.01401	0.00000
X2	0.00261	0.09558	0.02733	0.97820
X1*X2	-0.05101	0.01558	-3.27503	0.00110
Fixed effects				
O5X1xTime	Yes			
5x8 dummies				
McFadden R-squared	0.15432	Mean dependent v	0.280	
S.D. dependent var	0.44905	S.E. of regression	0.412	
Akaike info criterion	1.04451	Sum squared resid	341.81	
Schwarz criterion	1.16213	Log likelihood	-1031.80	
Hannan-Quinn criter.	1.08763	Restr. log likelihood	-1220.07	
LR statistic	376.552	Avg. log likelihood	-0.501	
Prob(LR statistic)	0.00000			
Obs with Dep=0	1482	Total obs	2058	
Obs with Dep=1	576			

3. Note the first model does not have the intercept C, but the second has. So the coefficients of the dummies variables in both models have different meanings. The coefficient of the dummies in the first model indicates the fixed effects of each group of individuals at each time point. For instance, C(3) is the coefficient of the dummy *(Group = 1 and T = 1)* and indicates that C(3) is defined as the fixed effect of Group = 1 at the time point T = 1, with the assumption that LNSALE3 and LNSALE4 have the same slopes of coefficient within the 40 groups considered.

4. On the other hand, we have an intercept C(1) and the term C(5)*(Group = 1 and T = 1)*, which indicate that [C(1) + C(5)] is the fixed effect of Group = 1, at the time point $T = 1$.

Example 9.17 Advanced binary and ordered choice interaction FEMs

As the extension of the polynomial binary choice model in (7.44), Table 9.11 presents the statistical results based on advanced binary and ordered choice interaction FEMs. Both outputs are obtained using the same equation specification (ES) as follows:

$$Y \ C \ X2 \ X2\hat{\ }2 \ X3 \ X3\hat{\ }2 \ X3\hat{\ }3 \ X3\hat{\ }4 \ X1 \ X1^*X2 \ X1^*X2\hat{\ }2$$

$$X1^*X3 \ X1^*X3\hat{\ }2 \ X1^*X3\hat{\ }3 \ X1^*X3\hat{\ }4 \ @EXPAND(O5X1, T, @DROPFIRST) \tag{9.34}$$

Table 9.11 *Statistical results based on advanced binary and ordered choice interaction FEMs*

Dependent Variable: Y					Dependent Variable: Y				
Method: ML - Binary Extreme Value (Quadratic hill climbing)					*Method: ML - Ordered Extreme Value (Quadratic hill climbing)*				
Date: 08/31/11 Time: 07:37					Date: 08/31/11 Time: 07:39				
Sample (adjusted): 1 2283					Sample (adjusted): 1 2283				
Included observations: 2058 after adjustments					Included observations: 2058 after adjustments				
Convergence achieved after 9 iterations					Number of ordered indicator values: 2				
Covariance matrix computed using second derivatives					Convergence achieved after 7 iterations				
					Covariance matrix computed using second derivatives				
Variable	Coeff.	s.e.	z-Stat	Prob.	Variable	Coeff.	s.e.	z-Stat	Prob.
C	-1.324	0.302	-4.387	0.000					
X2	-0.230	0.131	-1.752	0.080	X2	-0.230	0.131	-1.752	0.080
X2^2	-0.087	0.049	-1.770	0.077	X2^2	-0.087	0.049	-1.770	0.077
X3	1.518	1.087	1.396	0.163	X3	1.518	1.087	1.396	0.163
X3^2	-0.568	0.479	-1.185	0.236	X3^2	-0.568	0.479	-1.185	0.236
X3^3	0.075	0.088	0.858	0.391	X3^3	0.075	0.088	0.858	0.391
X3^4	-0.003	0.005	-0.616	0.538	X3^4	-0.003	0.005	-0.616	0.538
X1	0.206	0.063	3.258	0.001	X1	0.206	0.063	3.258	0.001
X1*X2	-0.011	0.021	-0.514	0.607	X1*X2	-0.011	0.021	-0.514	0.607
X1*X2^2	0.005	0.009	0.629	0.529	X1*X2^2	0.005	0.009	0.629	0.529
X1*X3	-0.135	0.122	-1.112	0.266	X1*X3	-0.135	0.122	-1.112	0.266
X1*X3^2	0.058	0.054	1.072	0.284	X1*X3^2	0.058	0.054	1.072	0.284
X1*X3^3	-0.008	0.010	-0.859	0.390	X1*X3^3	-0.008	0.010	-0.859	0.390
X1*X3^4	0.000	0.001	0.672	0.502	X1*X3^4	0.000	0.001	0.672	0.502
(5x8-1) dummies:					(5x8-1) dummies:				
O5X1=1,T=2	-0.078	0.258	-0.303	0.762	O5X1=1,T=2	-0.078	0.258	-0.303	0.762
O5X1=1,T=3	-0.079	0.263	-0.300	0.764	O5X1=1,T=3	-0.079	0.263	-0.300	0.764
...					...				
McFadden R-squared	0.164	Mean dependent var		0.280	Limit Points				
S.D. dependent var	0.449	S.E. of regression		0.411	LIMIT_1:C(53)	1.324	0.302	4.387	0.000
Akaike info criterion	1.043	Sum squared resid		338.6					
Schwarz criterion	1.188	Log likelihood		-1020.1	Pseudo R-squared	0.164	Akaike info criterion		1.043
Hannan-Quinn criter.	1.096	Restr. log likelihood		-1220.1	Schwarz criterion	1.188	Log likelihood		-1020.1
LR statistic	400.0	Avg. log likelihood		-0.496	Hannan-Quinn criter.	1.096	Restr. log likelihood		-1220.1
Prob(LR statistic)	0.000				LR statistic	400.0	Avg. log likelihood		-0.496
Obs with Dep=0	1482	Total obs		2058	Prob(LR statistic)	0.000			
Obs with Dep=1	576								

Based on these models the following findings and notes are presented.

1. The basic objective of both models is to present that the linear effect of *X1* on the *logit(Y)*, or *log[p/(1 − p)]* depends on *X2* and *X3*, in the forms of polynomial functions of *X2* and *X3*, which can be presented, in the most general form, as the interaction: *X1*Pol(X2,X3)*. Refer to the notes on the model in (7.44).

    Both outputs represent a set of 40 binary or ordered choice models, but the binary choice model has an intercept parameter, and the ordered choice model does not. Refer to the notes on the ordered choice models presented in Chapter 7.

2. Other modified fixed-effects models can easily derived, by using alternative fixed-effects indicators or variables, such as *Individual/Group* or *Time* fixed effects only, and "*Individual + Time*" fixed effects.

3. Based on this example and all previous examples, it can be concluded that a lot of fixed-effects models, perhaps infinite numbers of them, could be subjectively defined based on a pool data set. Which one would be the true population model? Refer to Section 2.14 in Agung (2009a).

Example 9.18   A binary choice fixed-effects model from international journals

For an additional illustration, only Benmelech and Bergman (the *Journal of Finance*, 2011) present a binary choice additive fixed-effects model. Out of 268 articles in the *Journal of Accounting* (IJA), *Journal of Accounting and Economics* (JAE), *British Accounting Review* (BAR) and *Advances in Accounting, incorporating Advances in International Accounting* (AA), from the years 2008, 2009 and 2010 observed by Siswantoro and Agung (2010), no one else presents a binary choice fixed-effect model. See alternative equation specifications of the binary choice FEM in (9.45a) to (9.45c) for the model presented by Benmelech and Bergman (2011).

9.6   Selected Fixed-Effects Models from the *Journal of Finance*, 2011

Even though fixed effects models (FEMs) have specific or unrealistic assumptions, that is all independent variables except the fixed effects have the same slope parameters, FEMs are widely applied based on panel data sets. So the FEMs are acceptable in a statistical sense. However, I would say that they are not good models for policy analysis or evaluation studies.

Characteristics, limitations and hidden assumptions of the FEM are not mentioned in all cases. Referring to the limitations of FEMs, which have been mentioned previously, this section presents selected FEMs from the *Journal of Finance*, 2011, in order to note their characteristics, specifically their hidden assumptions, so that students and less experienced researchers can be informed of the limitations of a particular FEM, which could present a set of hundreds or thousands of homogenous regressions compared the other models.

9.6.1   Additive FEM Applied by Hendershott, et al. (2011)

Two special individual time fixed-effects models (ITFEMs) are selected here for detailed discussions, which will be presented in the following two examples.

Example 9.19   The simplest ITFEM

Hendershott et al. present a FEM that I think is the simplest additive individual-time FEM with the equation as follows:

$$M_{it} = \alpha_i + \gamma_t + \beta Q_{it} + \varepsilon_{it} \tag{9.35}$$

where M_{it} is the relevant dependent variable, for example, the number of electronic transactions per minute, Q_{it}, is the autoquote dummy set to zero before the autoquote introduction and one afterward; α_i is a stock fixed effect; and γ_t is a day dummy. The data used has 1082_*167 (stock$_*$day) observations. Based on this model the following notes and comments are presented.

1. The statistical results can be obtained using the following alternative ESs.

$$M \ @Expand(Stock, @Dropfirst) \ @Expand(Day, @Dropfirst) \ @Expand(Q) \tag{9.35a}$$

or

$$\begin{aligned} &M \ C \ @Expand(Stock, @Dropfirst) \\ &@Expand(Day, @Dropfirst) \ @Expand(Q, @Dropfirst) \end{aligned} \tag{9.35b}$$

2. The model looks the simplest with a single independent variable Q_{it}, but in fact it has thousands of parameters or independent variables: those are indicated by 1082 stock dummies, 167 day dummies and a zero-one indicator Q_{it}. This model has specific characteristics or hidden assumptions as follows:

    2.1 For each stock i, the following intercept difference between a pair of regressions is obtained, and the adjusted means difference is known, namely AMD_i of M_{it}, which is constant or invariant for all stocks. I would say that this condition would never happen in reality, but it is acceptable in a statistical sense, with an inappropriate assumption.

$$AMD_i(\hat{M}_{i,t_1} - \hat{M}_{i,t_2}) = \hat{\gamma}_{t_1} - \hat{\gamma}_{t_2} \tag{9.36}$$

    2.2 Similarly, for each time point t, there is the following AMD_t between two stocks, which is constant or invariant for all time points.

$$ADM_t(\hat{M}_{i_1,t} - \hat{M}_{i_2,t}) = \hat{\alpha}_{i_1} - \hat{\alpha}_{i_2} \tag{9.37}$$

3. Note that this cell-means model also can be presented as a pair of models as follows:

$$\begin{aligned} M_{it} &= \alpha_i + \gamma_t + \varepsilon_{it}, \quad for \ Q_{it} = 0 \\ M_{it} &= \alpha_i + \gamma_t + \beta + \varepsilon_{it}, \quad for \ Q_{it} = 1 \end{aligned} \tag{9.38}$$

So we obtain the following equation.

$$M_{it}(Q = 1) - M_{it}(Q = 0) = \beta, \quad for \ all(i, t) \tag{9.39}$$

4. This model in fact is an additive cell-means model of M_{it} on the *STOCK, DAY* and *Q* dummies, which is one out of several or many possible reduced models of a full-factorial cell-means model of M_{it} on three categorical factors. The full-factorial model can be applied using the following equation specification.

$$M \ @Expand(STOCK, DAY, Q) \tag{9.40}$$

Refer to other alternative equations for the cell-means models presented in Agung (2011, 2006). Furthermore, we find that parameter estimates of this full-factorial cell-means model are equal to sampled means of M_{it} by the three factors; *SOCK, DAY* and *Q*.

Example 9.20   An advanced ITFEM

Hendershott et al. present an advanced ITFEM, with a general equation as follows:

$$L_{it} = \alpha_i + \gamma_t + \beta A_{it} + \delta' X_{it} + \varepsilon_{it} \tag{9.41}$$

where L_{it} is a spread measure for stock i on day t, A_{it} is the AT measure $algo\_trad_{it}$ relevant independent variable, and X_{it} is a vector of control variables, including share turnover, volatility, the inverse of share price and log market cap. Compare this model with the additive model (9.16) and the empirical results in Figure 8.24. This model has similar characteristics or hidden assumptions to the model in (9.37), such as follows:

1. Similar to the equation in (9.37a), the following ADM_i obtained, which is constant or invariant for all stocks i.

$$AMD_i(\hat{L}_{i,t_1} - \hat{L}_{i,t_2}) = \hat{\gamma}_{t_1} - \hat{\gamma}_{t_2} \tag{9.41a}$$

2. Similar to the equation in (9.37b), the following ADM_t obtained, which is constant or invariant for all time points t.

$$ADM_t(\hat{L}_{i_1,t} - \hat{L}_{i_2,t}) = \hat{\alpha}_{i_1} - \hat{\alpha}_{i_2} \tag{9.41b}$$

3. This model could easily be extended to many alternative ITFEMs, by using the equation specification as follows:

$$H(L)\, @Expand(STOCK, @Drop(^*))\, @Expand(DAY)\, X1\, X2 \ldots Xk \ldots \tag{9.42a}$$

or

$$H(L)\, @Expand(STOCK)\, @Expand(DAY, @Drop(^*))\, X1\, X2 \ldots Xk \ldots \tag{9.42b}$$

or

$$H(L)\, C\, @Expand(STOCK, @Drop(^*))\, @Expand(DAY, @Drop(^*))\, X1\, X2 \ldots Xk \ldots \tag{9.42c}$$

where $H(L)$ is a certain or fixed function of L_{it}, and $Xk = X_k$ can be various alternative variables as indicated for the model in (9.23).

9.6.2   FEM Applied by Engelberg and Parsons (2011)

Compared to the interaction FEM in (9.7) with a single interaction fixed-effect indicator, Engelberg and Parsons (2011) present several FEMs. Two of the models are as follows:

9.6.2.1   *Additive FEMs*

As an extension of the models in (9.13a) or (9.41), Engelberg and Parsons present 14 FEMs with selected fixed-effects indicators, out of four; *Industry, Paper, City* and *Date fixed effects*, based on data sets containing at least 265 928 observations, which can be presented in a general form as follows:

$$Y_{it} = \alpha_i + \beta_p + \lambda_c + \gamma_t + \delta' X_{it} + \varepsilon_{it} \tag{9.43}$$

Referring to the AMD_i in (9.41a), for this model we obtain the following AMD_i, which is invariant for all industries i.

$$AMD_i[(\hat{Y}|p_1, c_1, t_1) - (\hat{Y}|p_2, c_2, t_2)] = (\beta_{p_1} - \beta_{p_2}) + (\lambda_{c_1} - \lambda_{c_2}) + (\gamma_{t_1} - \gamma_{t_2}) \tag{9.43a}$$

Similar adjusted means differences AMD_p, AMD_c and AMD_t can easily be derived.

9.6.2.2   Interaction FEMs

In addition, they also present FEMs with three interaction fixed-effects indicators, namely *Firm-date*, *Firm-city* and *City-date* fixed effects, with the following general equation.

$$Y_{it} = \alpha_{ft} + \beta_{fc} + \lambda_{ct} + \delta'X_{it} + \varepsilon_{it} \tag{9.44}$$

where α_{ft} is a firm-time (firm-date) fixed effect, β_{fc} firm-city fixed effect, and λ_{ct} is a city-time (firm-date) fixed effect.

9.6.3   Special FEMs Applied by Benmelech and Bergman (2011)

Benmelech and Bergman (2011) present many fixed-effects models using selected fixed-effects indicators. Three of them are as follows:

9.6.3.1   Forty-Two FEMs

Forty-two FEMs are presented using selected fixed-effects indicators, out of the fixed effects: *Year, Airline, Airline × Bankruptcy, Tranche, Tranche × Bankruptcy, Tranche + Year*, with 27 Tranches, 12 Airlines and 18 327 observations, from January 1990 to December 2005.

Therefore, as to which one would be the most appropriate or the best fit model depends highly upon the personal subjective judgment of researchers. However, Agung (2009a, Section 2.14, p. 204) mentions the following on the true population model: "the researchers never know the true values of the population parameters, such as the means, standard deviations, and other parameters, or the true population model".

9.6.3.2   A Binary Choice FEM

One of the models which is important to mention is an additive individual-time linear probability (binary choice) model of *Bankruptcy*, which is an *airline-year* fixed-effects model, using the first lagged independent variables: $Size_{a,t-1}$, $Leverage_{a,t-1}$, $MtoB_{a,t-1}$, $Profitabilty_{a,t-1}$, and $STDebt_{a,t-1}$. The data analysis can be done using the following alternative equation specifications.

$$\begin{aligned} & Bankruptcy\ Size(-1)\ Leverage(-1)\ MtoB(-1)\ Profitability(-1)\ STDebt(-1) \\ & \qquad @Expand(Airline, @Drop(^*))\ @Expand(Year) \end{aligned} \tag{9.45a}$$

or

$$\begin{aligned} & Bankruptcy\ Size(-1)\ Leverage(-1)\ MtoB(-1)\ Profitability(-1)\ STDebt(-1) \\ & \qquad @Expand(Airline)\ @Expand(Year, @Drop(^*)) \end{aligned} \tag{9.45b}$$

or

$$\begin{aligned} & Bankruptcy\ Size(-1)\ Leverage(-1)\ MtoB(-1)\ Profitability(-1)\ STDebt(-1) \\ & \qquad C\ @Expand(Airline, @Drop(^*))\ @Expand(Year, @Drop(^*)) \end{aligned} \tag{9.45c}$$

Based on this model, the following notes and comments are presented.

1. The model in (9.45) is a linear effects model; that is, linear in its exogenous numerical variables and dummy independent variables. A binary choice model with a numerical independent variable could have a polynomial effect or other types of models, such as presented in Chapter 7, and Agung (2011), with special notes and comments.
2. It is important to study the Bankruptcy scatter graphs on each of the independent variables, with their kernel fit or nearest neighbor fit curves, to evaluate the possibility of using a polynomial effect binary choice model.
3. In addition, alternative interaction binary choice FEMs also can be considered. See various interaction BCMs presented in Chapter 7.
4. Note that in order to be able to use the symbols $X(-1)$ for various variable X, the data should be a balanced or complete pool data set with a specific or special format, which will be presented in Part III. See also to the notes in Section 8.8.1.

9.6.3.3   A Special Interaction FEM

Finally, the third type of selected models are FEMs with interaction fixed-effects indicators *Airline* × *Bankruptcy* and *Tranche* × *Bankruptcy* each, where the variable *Bankruptcy* is used also as an independent or control variable. See alternative models presented in the following sections.

9.7   Heterogeneous Regression Models

As an extension of various fixed-effects models with the assumptions that all other independent variables have exactly the same slope parameters presented earlier, this section presents heterogeneous regressions models by individuals, groups of individuals, times or time period. We know that the effect of an exogenous variable on an endogenous variable should change over individuals or groups of individuals as well as over times or time periods in the author's opinion. In fact, a lot of heterogeneous regressions have been presented in previous chapters and Agung (2011, 2009a, 2008, 2006, 2004), as well as other references such as Gujarati (2003), Wooldridge (2002), and Neter and Wasserman (1980), by using the interactions between dummies and numerical variables as independent variables of the models.

Furthermore, the effects of interaction exogenous variables, including the dummies, on endogenous variables in accounting models have been studied by Siswantoro and Agung (2010), based on 268 papers or articles observed in four international journals; *Advanced in Accounting* (AIA), *British Accounting Review* (BAR), *TheInternational Journal of Accounting* (IJA), and *Journal of Accounting and Economics* (JAE) from the years 2008, 2009 and 2010. The findings show that very few papers present models with interaction independent dummy variables. The paper was presented at The 3rd International Accounting Conference held by the University of Indonesia, 27–28 October 2010, Bali, Indonesia.

9.7.1   Heterogeneous Regressions by Individuals

A heterogeneous regression by individuals based on panel data, best known as the CAPM (capital asset pricing model) introduced by Sharp (1964), is a set of the simplest linear regressions (SLRs) in two-dimensional space with the equation as follows:

$$R_{it} = \alpha_i + \beta_i R_{mt} + \varepsilon_{it} \tag{9.46}$$

where R_{it} is the return on the asset of the firm i at the time t, R_{mt} is the return on the market at the time t, and β_i, $i = 1, \ldots, N$ are known as the *Betas*. By using EViews the data analysis based on the model (9.46) can easily be done using the following equation specification.

$$R \ @Expand(Firm\_Code) \ Rm^* @Expand(Firm\_Code) \tag{9.47}$$

Note that this CAPM could easily be extended to more advanced models with additional independent variables, which have been presented in many studies such as the model in (10.3) presented by Frischmann, et al. (2008) and Khoon, et al. (1999), with the general equations presented in Chapter 8 such as the models in (8.12) and (8.16), up to (8.43) and (8.44).

9.7.2 Heterogeneous Classical Growth Models by Individuals or Groups

Similar to the model in (9.47), as an extension of the IFE Model-2 in Table 9.1, the heterogeneous classical growth models by individuals can be presented using the following equation specification.

$$lnSALE \ T^* @Expand(FIRM\_CODE) \ @Expand(FIRM\_CODE) \tag{9.48}$$

However, the analysis based on the whole Data_Sesi.wf1, obtains the error message: "*Near singular matrix*" because the data is an unbalanced pool of data with an unknown specific pattern. So the design matrix of the model in fact is an exact singular matrix corresponding to the dummies – it is not a near-singular matrix. I am very confident that statistical results should be obtained using its sub-samples, as well as other data sets, by using this equation specification. See the following illustrative examples.

For comparison, we find that Wooldridge (2002) presents heterogeneous regressions, namely a random trend model and a growth model by individuals.

Example 9.21 A sample of three firms

For illustration, Figure 9.9 presents the statistical results based on a sample of the three firms presented in Figure 9.2, namely _AALI, _AKKU and _FNPI using the equation specification (9.48). This sample is unbalanced pool data, since the firms _AALI and _FNPI have eight time observation, but _AKKU has only

Dependent Variable: LNSALE
Method: Least Squares
Date: 08/24/11 Time: 14:36
Sample: 1 2284 IF FIRM_CODE="_FPNI" OR FIRM_CODE="_AKKU" OR
    FIRM_CODE="_AALI"
Included observations: 22

Variable	Coefficient	Std. Error	t-Statistic	Prob.
FIRM_CODE=_AALI	7.037190	0.544804	12.91691	0.0000
FIRM_CODE=_AKKU	3.502356	0.962558	3.638592	0.0022
FIRM_CODE=_FPNI	3.689609	0.544804	6.772357	0.0000
T*(FIRM_CODE=_AALI)	0.256935	0.107887	2.381513	0.0300
T*(FIRM_CODE=_AKKU)	-0.198779	0.167138	-1.189305	0.2517
T*(FIRM_CODE=_FPNI)	0.501586	0.107887	4.649161	0.0003

R-squared	0.942840	Mean dependent var	5.798890
Adjusted R-squared	0.924978	S.D. dependent var	2.552702
S.E. of regression	0.699190	Akaike info criterion	2.349212
Sum squared resid	7.821863	Schwarz criterion	2.646769
Log likelihood	-19.84133	Hannan-Quinn criter.	2.419307
Durbin-Watson stat	1.504168		

Dependent Variable: LNSALE
Method: Least Squares
Date: 08/24/11 Time: 14:45
Sample: 1 2284 IF FIRM_CODE="_FPNI" OR FIRM_CODE="_AKKU" OR
    FIRM_CODE="_AALI"
Included observations: 22
Newey-West HAC Standard Errors & Covariance (lag truncation=2)

Variable	Coefficient	Std. Error	t-Statistic	Prob.
FIRM_CODE=_AALI	7.037190	0.063248	111.2629	0.0000
FIRM_CODE=_AKKU	3.502356	1.178959	2.970719	0.0090
FIRM_CODE=_FPNI	3.689609	0.654018	5.641448	0.0000
T*(FIRM_CODE=_AALI)	0.256935	0.017088	15.03581	0.0000
T*(FIRM_CODE=_AKKU)	-0.198779	0.197764	-1.005132	0.3298
T*(FIRM_CODE=_FPNI)	0.501586	0.136850	3.665217	0.0021

R-squared	0.942840	Mean dependent var	5.798890
Adjusted R-squared	0.924978	S.D. dependent var	2.552702
S.E. of regression	0.699190	Akaike info criterion	2.349212
Sum squared resid	7.821863	Schwarz criterion	2.646769
Log likelihood	-19.84133	Hannan-Quinn criter.	2.419307
Durbin-Watson stat	1.504168		

Figure 9.9 *Statistical results of the heterogeneous growth models of SALE using two estimation methods*

six time point observations. Two kinds of statistical results are presented; these are the LS estimation method, without and with the option: *Newey–West HAC standard errors and covariance*.

Based on the results in this figure, the following findings and notes are presented.

1. For pool data, I recommend applying the Newey–West option instead of the White, because the covariance estimator is consistent in the presence of both heteroskedasticity and autocorrelation of an unknown form.
2. Both results clearly show that the three firms have different SALE growth rates. For Newey–West statistical results, the null hypothesis H_0: $C(4) = C(5) = C(6)$ is rejected based on the F-statistic of $F_0 = 4.252\,676$ with $df = (2,16)$ and a p-value $= 0.0330$. So it can be concluded that the firms have significantly different SALE growth rates.

Example 9.22   A sample of 224 firms in balanced pool panel data
The balanced pool data in fact is a subset of the Data_Sesi.wf1, called Balanced.wf1. Figure 9.10 presents only some parts of the statistical results using the equation specification (9.44), namely the growth rates of the first 10 firms with the intercepts' estimates, the R-squared and other general statistics.

Based on the results in this figure, the following findings and notes are presented.

1. By using the Wald test, the result is obtained that the first 10 firms have significantly different SALE growth rates, based on the F-statistic of $F_0 = 4.071\,090$ with $df = (9,1335)$ and a p-value $= 0.0000$.
2. The adjusted R-squared of 0.92591, indicates that all independent variables, namely 224 interactions and 224 firm dummies, are excellent predictors for *LNSALE*.
3. In order to test the joint effects of all independent variables applying the following equation specification is recommended instead of using the Wald test.

$$\textit{lnSALE} \quad C \quad T^* @Expand(FIRM\_CODE) \quad @Expand(FIRM\_CODE, @Droplast) \tag{9.49}$$

This obtains an F-statistic of $F_0 = 50.82\,507$ and a p-value $= 0.000\,000$.

4. The DW-statistic of 1.941624 indicates that there is no autoregressive problem. If the results in some other cases show very small DW-statistical values, then using the Newey–West heteroskedasticity, AR term or lagged dependent variables option is suggested.

Dependent Variable: LNSALE
Method: Least Squares
Date: 08/24/11   Time: 15:35
Sample: 1 1792
Included observations: 1783

Variable	Coefficient	Std. Error	t-Statistic	Prob.
T*(FIRM_CODE=_AALI)	0.256935	0.081883	3.137824	0.0017
T*(FIRM_CODE=_ABBA)	0.429242	0.081883	5.242128	0.0000
T*(FIRM_CODE=_ADES)	0.031658	0.081883	0.386629	0.6991
T*(FIRM_CODE=_ADMG)	0.129177	0.081883	1.577581	0.1149
T*(FIRM_CODE=_AIMS)	0.252915	0.081883	3.088734	0.0021
T*(FIRM_CODE=_AISA)	0.456423	0.081883	5.574078	0.0000
T*(FIRM_CODE=_AKPI)	0.135959	0.081883	1.660398	0.0971
T*(FIRM_CODE=_AKRA)	0.362556	0.081883	4.427717	0.0000
T*(FIRM_CODE=_ALFA)	-0.047839	0.081883	-0.584238	0.5592
T*(FIRM_CODE=_ALKA)	0.252362	0.081883	3.081976	0.0021

FIRM_CODE=_AALI	7.037190	0.413490	17.01901	0.0000
FIRM_CODE=_ABBA	2.038992	0.413490	4.931176	0.0000
FIRM_CODE=_ADES	4.679540	0.413490	11.31718	0.0000
FIRM_CODE=_ADMG	7.367931	0.413490	17.81889	0.0000
FIRM_CODE=_AIMS	3.445989	0.413490	8.333912	0.0000
FIRM_CODE=_AISA	3.140665	0.413490	7.595506	0.0000
FIRM_CODE=_AKPI	6.309423	0.413490	15.25895	0.0000
FIRM_CODE=_AKRA	6.424660	0.413490	15.53765	0.0000
FIRM_CODE=_ALFA	8.032607	0.413490	19.42637	0.0000
FIRM_CODE=_ALKA	5.375158	0.413490	12.99949	0.0000

R-squared	0.944499	Mean dependent var	6.096442
Adjusted R-squared	0.925916	S.D. dependent var	1.949654
S.E. of regression	0.530664	Akaike info criterion	1.783782
Sum squared resid	375.9415	Schwarz criterion	3.162218
Log likelihood	-1142.242	Hannan-Quinn criter.	2.292859
Durbin-Watson stat	1.941624		

Figure 9.10   *A part of the statistical results for the heterogeneous growth models of SALE by all firms*

5. Compare this to the following three growth models, which one would you select as the best growth model? Refer to the alternative illustrative graphs presented in Figure 8.10.

$$lnSALE \ C \ T^* @Expand(FIRM\_CODE) \tag{9.50}$$

$$lnSALE \ C \ T \ @Expand(FIRM\_CODE, @Droplast) \tag{9.51}$$

$$lnSALE \ C \ T \tag{9.52}$$

6. In general, by comparing three or more similar models, all researchers can certainly subjectively select which one is the best and worst model among those in a theoretical sense.

9.7.3 Piece-Wise Heterogeneous Regressions

As an extension of the growth model in (9.49), piece-wise heterogeneous regressions can be considered, with the following alternative ESs.

$$ln \ SALE \ C \ T^* @Expand(FIRM\_CODE, TP)$$
$$@Expand(FIRM\_CODE, TP, @Drofirst) \tag{9.53}$$

$$lnSALE \ C \ T^* @Expand(GROUP, TP) \ @Expand(GROUP, TP, @Dropfirst) \tag{9.54}$$

where *TP* is a time period generated based on the time T, such as the time before and after the new tax policy, using the equation $TP = 1 + 1^*(t > 5)$, based on the Data_Seis.wf1.

However, researchers should note the possibility of getting the error message in Figure 9.9, because of an insufficient number of observations or other problems.

9.7.4 Heterogeneous Regressions with Trend by Individuals or Groups

By using the heterogeneous regressions by individuals with trend, error messages would be obtained in many cases, particularly with a large number of individuals. For this reason, I recommend applying heterogeneous regressions with trend by groups. The data analysis can be done using the general equation specification as follows:

$$H(Y) \ C \ @Expand(Group, @Dropfirst) \ t^* @Expand(Group)$$
$$X1^* @Expand(Group) \dots Xk^* Expand(Group) \ Z1 \dots Zm \tag{9.55}$$

where *H(Y)* is a transformed variable of an endogenous variable *Y*, *Xk* can be any variable as shown in the models in (9.23), and *Zm* is an exogenous, environmental or control variable which is assumed to have homogenous slopes over the groups.

Note that this model also could be extended to piece-wise regressions with trend, using the functions *@Expand(Group,TP)* as presented in the ESs in (9.53) to (9.54).

Example 9.23   A set of additive models by group

For illustration, Table 9.12 presents a summary of statistical results based on two AR(1) models in (9.55). The AR(1) models should be applied in order to reduce or overcome autoregressive problems. See the DW-statistic in the outputs. Based on this table the following findings and notes are presented.

Table 9.12   *Summary of statistical results based on AR(1) models in (9.55)*

Dependent Variable: LNSALE
Method: Least Squares
Date: 08/29/11   Time: 14:17
Sample: 1 2284
Included observations: 2270

Variable	Model-1			Model-2		
	Coeff.	t-Stat	Prob.	Coeff.	t-Stat	Prob.
C	1.464	6.731	0.000	1.910	8.935	0.000
O5ROA=2	0.926	3.819	0.000	0.851	3.641	0.000
O5ROA=3	1.268	5.160	0.000	1.349	5.702	0.000
O5ROA=4	2.094	8.463	0.000	2.025	8.461	0.000
O5ROA=5	1.249	4.737	0.000	1.560	6.013	0.000
T*(O5ROA=1)	-0.012	-0.580	0.562	-0.008	-0.369	0.712
T*(O5ROA=2)	0.071	3.480	0.001	0.072	3.671	0.000
T*(O5ROA=3)	0.087	4.295	0.000	0.092	4.705	0.000
T*(O5ROA=4)	0.101	5.461	0.000	0.103	5.833	0.000
T*(O5ROA=5)	0.102	5.739	0.000	0.095	5.405	0.000
LNASSET*(O5ROA=1)	0.663	22.208	0.000	0.560	18.587	0.000
LNASSET*(O5ROA=2)	0.470	16.058	0.000	0.385	13.338	0.000
LNASSET*(O5ROA=3)	0.449	16.952	0.000	0.333	12.324	0.000
LNASSET*(O5ROA=4)	0.339	13.435	0.000	0.249	9.885	0.000
LNASSET*(O5ROA=5)	0.503	19.297	0.000	0.357	13.030	0.000
PBV				0.001	1.249	0.212
EQUITY				0.000	3.678	0.000
LIABILITIES				0.000	6.237	0.000
MARKETSHARE				1.213	3.007	0.003
AR(1)	0.723	48.641	0.000	0.730	49.624	0.000
R-squared	0.758			0.557		
Adjusted R-squared	0.756			0.554		
F-statistic	468.268			157.443		
Prob(F-statistic)	0.000			0.000		
Durbin-Watson stat	2.059			2.056		
Inverted AR Roots	0.72			0.730		

1. In this case the group is generated based on ROA, namely O5Roa, using the following equation. Note that various other GROUP variables with various levels can be considered based on one or more variables in or out of a model. Refer to the piece-wise regressions previously.

$$O5Roa = 1 + 1^*(roa >= @quantile(roa, 0.2) + 1^*(roa >= @quantile(roa, 0.4)$$
$$+ 1^*(roa >= @quantile(roa, 0.6) + 1^*(roa >= @quantile(roa, 0.8) \qquad (9.56)$$

2. Model-1 presents a set of five additive regressions, with the first regression, for O5Roa = 1, which has the following equation with the *t*-statistic in [*].

$$LNSALE = 1.464 - 0.012T + 0.663LNASSET$$
$$[6.73] \quad [-0.58] \quad [22.208]$$

3. Based on this regression, the following findings can be presented.
   - This regression is a translog linear model of *LNSALE* on *LNASSET* with trend, and *LNASSET* has a significant positive effect on *LNSALE*, adjusted for the time variable *T*.
   - The coefficient of *T*, that is (−0.012), can be considered as the exponential growth rate of *SALE* during the observed time period, adjusted for *LNASSET*.
   - Since the null hypothesis H_0: C(6) = C(11) = 0 is rejected based on the *F*-statistic of $F_0 = 247.749$ with $df = (2,2242)$ and $p = 0.0000$, then it can be concluded that T and LNASSET have significant joint effects on LNSALE.
4. Model-2 presents a set of five additive regressions with four additional exogenous variables, namely *PBV, EQUITY, LIABILITIES* and *MARKETSHARE*, as control variables with the assumption that they have constant or invariant effects over the levels of O5Roa.

9.7.5 Heterogeneous Regressions with Time-Related Effects by Individuals or Groups

As an extension of the models with trend in (9.56), a lot of heterogeneous regressions with time-related effects can be defined, which are acceptable in a theoretical sense. Their general equation specification is as follows:

$$H(Y) C @Expand(Group, @Dropfirst) t^* @Expand(Group)$$
$$\ldots Xk^* @Expand(Group) \ldots t^* Xk^* @Expand(Group) \qquad (9.57)$$
$$\ldots Zm \ldots t^* Zm \ldots$$

where the interaction terms $t^* Xk^* @Expand(Group)$ and $t^* Zm$, for $k = 1,2\ldots, K$ and $m = 1,2\ldots, M$, indicate the time-related effects *of Xk* and *Zm* on *H(Y)*. Note that *Xk* can be any variable, as presented for the models in (9.23).

This model also could be extended to piece-wise regressions or models with trend-related effects, using the functions *@Expand(Group,TP)* as presented in the ESs in (9.53) to (9.54).

Example 9.24 Simple heterogeneous regressions with time-related effects by groups
Figure 9.11 presents statistical results based on a simple model in (9.57), and its reduced model, using the following ESs.

$$LNSALE C @Expand(O5Roa, @Dropfirst) t^* @Expand(O5Roa)$$
$$LNASSET^* @Expand(O5Roa) t^* LNASSET^* @Expand(O5Roa) AR(1) \qquad (9.58)$$

Dependent Variable: LNSALE
Method: Least Squares
Date: 08/31/11   Time: 21:40
Sample (adjusted): 2 2284
Included observations: 2258 after adjustments
Convergence achieved after 7 iterations

Variable	Coefficient	Std. Error	t-Statistic	Prob.
C	3.852400	0.393656	9.786216	0.0000
O5ROA=2	0.625508	0.480143	1.302752	0.1928
O5ROA=3	1.798239	0.461264	3.898500	0.0001
O5ROA=4	2.064304	0.457516	4.511987	0.0000
O5ROA=5	1.634470	0.457201	3.574950	0.0004
T*(O5ROA=1)	-0.547344	0.069975	-7.821976	0.0000
T*(O5ROA=2)	-0.457057	0.073268	-6.238190	0.0000
T*(O5ROA=3)	-0.638639	0.066317	-9.630039	0.0000
T*(O5ROA=4)	-0.611683	0.063597	-9.618168	0.0000
T*(O5ROA=5)	-0.623747	0.055376	-11.26381	0.0000
LNASSET*(O5ROA=1)	0.249306	0.063894	3.901870	0.0001
LNASSET*(O5ROA=2)	0.115671	0.057902	1.997712	0.0459
LNASSET*(O5ROA=3)	-0.027573	0.048202	-0.572029	0.5674
LNASSET*(O5ROA=4)	-0.037699	0.040871	-0.922394	0.3564
LNASSET*(O5ROA=5)	0.073712	0.040567	1.817044	0.0693
T*LNASSET*(O5ROA=1)	0.090032	0.011287	7.976666	0.0000
T*LNASSET*(O5ROA=2)	0.085602	0.011280	7.588633	0.0000
T*LNASSET*(O5ROA=3)	0.113764	0.009843	11.55802	0.0000
T*LNASSET*(O5ROA=4)	0.108460	0.009329	11.62650	0.0000
T*LNASSET*(O5ROA=5)	0.108315	0.007970	13.59115	0.0000
AR(1)	0.736166	0.014799	49.74438	0.0000

R-squared	0.794450	Mean dependent var		6.004704
Adjusted R-squared	0.792613	S.D. dependent var		1.993619
S.E. of regression	0.907890	Akaike info criterion		2.653869
Sum squared resid	1843.878	Schwarz criterion		2.707087
Log likelihood	-2975.218	Hannan-Quinn criter.		2.673290
F-statistic	432.3011	Durbin-Watson stat		2.034220
Prob(F-statistic)	0.000000			

Inverted AR Roots   .74

Dependent Variable: LNSALE
Method: Least Squares
Date: 08/31/11   Time: 21:38
Sample (adjusted): 2 2284
Included observations: 2258 after adjustments
Convergence achieved after 7 iterations

Variable	Coefficient	Std. Error	t-Statistic	Prob.
C	3.817583	0.390298	9.781203	0.0000
O5ROA=2	0.619347	0.480077	1.290098	0.1971
O5ROA=3	1.670644	0.394295	4.237041	0.0000
O5ROA=4	1.872582	0.392537	4.770460	0.0000
O5ROA=5	1.630149	0.456430	3.571521	0.0004
T*(O5ROA=1)	-0.541800	0.069591	-7.785467	0.0000
T*(O5ROA=2)	-0.450474	0.072620	-6.203138	0.0000
T*(O5ROA=3)	-0.610751	0.040786	-14.97465	0.0000
T*(O5ROA=4)	-0.568212	0.042221	-13.45806	0.0000
T*(O5ROA=5)	-0.617336	0.054950	-11.23450	0.0000
LNASSET*(O5ROA=1)	0.255279	0.063293	4.033322	0.0001
LNASSET*(O5ROA=2)	0.122784	0.057029	2.153025	0.0314
LNASSET*(O5ROA=5)	0.079879	0.040065	1.993746	0.0463
T*LNASSET*(O5ROA=1)	0.089081	0.011216	7.942503	0.0000
T*LNASSET*(O5ROA=2)	0.084444	0.011163	7.564626	0.0000
T*LNASSET*(O5ROA=3)	0.109117	0.004996	21.84007	0.0000
T*LNASSET*(O5ROA=4)	0.101425	0.005312	19.09189	0.0000
T*LNASSET*(O5ROA=5)	0.107338	0.007904	13.58079	0.0000
AR(1)	0.735589	0.014773	49.79171	0.0000

R-squared	0.794358	Mean dependent var		6.004704
Adjusted R-squared	0.792705	S.D. dependent var		1.993619
S.E. of regression	0.907688	Akaike info criterion		2.652547
Sum squared resid	1844.706	Schwarz criterion		2.700696
Log likelihood	-2975.725	Hannan-Quinn criter.		2.670118
F-statistic	480.4924	Durbin-Watson stat		2.033916
Prob(F-statistic)	0.000000			

Inverted AR Roots   .74

Figure 9.11   *Statistical results based on the models in (9.7) and (9.8)*

and its reduced model

$$LNSALE \; C \; @Expand(O5Roa, @Dropfirst) \; t*@Expand(O5Roa)$$
$$LNASSET*@Expand(O5Roa, @Drop(3), @Drop(4)) \tag{9.59}$$
$$t*LNASSET*@Expand(O5Roa) \; AR(1)$$

Note that the application of the function *@Expand(O5Roa, @Drop(3), @Drop(4))* deletes two of the independent variables of the model in (9.59). The indicator AR(1) should be applied to reduce or overcome their autoregressive problems. In fact, both regressions are acceptable estimates, in a statistical sense, even though two of the independent variables have large p-values. For additional examples from international journals, see Delong and Deyoung (2007) and Bansal (2005).

Example 9.25   A special illustrative model with time-related effects
Figure 9.12 presents the estimates of an AR(1) model with time-related effects by GROUP and TP (time period) having the following equation specification.

$$y1 \; c \; @expand(group, tp, @dropfirst) \; t*@expand(group, tp) \; x1*@expand(group, tp)$$
$$x2*@expand(group, tp) \; x1*x2*@expand(group, tp) \; t*x1*@expand(group, tp) \tag{9.60}$$
$$t*x2*@expand(group, tp) \; t*x1*x2*@expand(group, tp) \; x3 \; x4 \; x5 \; ar(1)$$

Dependent Variable: Y1
Method: Least Squares
Date: 09/01/11   Time: 17:16
Sample (adjusted): 2 2283
Included observations: 1977 after adjustments
Convergence achieved after 10 iterations

Variable	Coefficient	Std. Error	t-Statistic	Prob.
C	-2.483242	1.112612	-2.231903	0.0257
X3	0.608043	0.067243	9.042450	0.0000
X4	0.537478	0.120516	4.459792	0.0000
X5	-0.072504	0.035097	-2.065824	0.0390
GROUP=1,TP=2	3.497976	3.370483	1.037826	0.2995
GROUP=2,TP=1	-6.975806	2.472700	-2.821129	0.0048
GROUP=2,TP=2	16.45800	5.219278	3.153311	0.0016
T*(T*(GROUP=1),TP=1)	0.386749	0.269807	1.433429	0.1519
T*(T*(GROUP=1),TP=2)	-0.181824	0.442622	-0.410789	0.6813
T*(T*(GROUP=2),TP=1)	-0.527752	0.542467	-0.972874	0.3307
T*(T*(GROUP=2),TP=2)	-4.532623	0.689592	-6.572910	0.0000
X1*(X1*(GROUP=1),TP=1)	0.421542	0.218690	1.927580	0.0541
X1*(X1*(GROUP=1),TP=2)	-0.358963	0.651088	-0.551328	0.5815
X1*(X1*(GROUP=2),TP=1)	1.480885	0.307547	4.815153	0.0000
X1*(X1*(GROUP=2),TP=2)	-1.704095	0.669188	-2.546511	0.0110
X2*(X2*(GROUP=1),TP=1)	-0.272082	0.471936	-0.576523	0.5643
X2*(X2*(GROUP=1),TP=2)	-0.081936	1.801069	-0.045493	0.9637
X2*(X2*(GROUP=2),TP=1)	-6.022658	2.354713	-2.557704	0.0106
X2*(X2*(GROUP=2),TP=2)	-3.372743	5.407851	-0.623675	0.5329

	Coefficient	Std. Error	t-Statistic	Prob.
X1*X2*(X1*X2*(GROUP=1),TP=1)	0.148363	0.101957	1.455142	0.1458
X1*X2*(X1*X2*(GROUP=1),TP=2)	-0.054004	0.404339	-0.133562	0.8938
X1*X2*(X1*X2*(GROUP=2),TP=1)	1.050668	0.340335	3.087158	0.0020
X1*X2*(X1*X2*(GROUP=2),TP=2)	0.620893	0.716504	0.866559	0.3863
T*X1*(T*X1*(GROUP=1),TP=1)	-0.103494	0.054937	-1.883847	0.0597
T*X1*(T*X1*(GROUP=1),TP=2)	0.027869	0.091330	0.305144	0.7603
T*X1*(T*X1*(GROUP=2),TP=1)	0.058547	0.074657	0.784219	0.4330
T*X1*(T*X1*(GROUP=2),TP=2)	0.604324	0.089597	6.744912	0.0000
T*X2*(T*X2*(GROUP=1),TP=1)	0.038503	0.127334	0.302375	0.7624
T*X2*(T*X2*(GROUP=1),TP=2)	0.047127	0.256201	0.183947	0.8541
T*X2*(T*X2*(GROUP=2),TP=1)	0.081579	0.600340	0.135889	0.8919
T*X2*(T*X2*(GROUP=2),TP=2)	-0.882727	0.741885	-1.189844	0.2343
T*X1*X2*(T*X1*X2*(GROUP=1),TP=1)	-0.023657	0.027158	-0.871091	0.3838
T*X1*X2*(T*X1*X2*(GROUP=1),TP=2)	0.001827	0.057697	0.031661	0.9747
T*X1*X2*(T*X1*X2*(GROUP=2),TP=1)	-0.032065	0.084795	-0.378144	0.7054
T*X1*X2*(T*X1*X2*(GROUP=2),TP=2)	0.115868	0.097526	1.188069	0.2350
AR(1)	0.766912	0.014658	52.32003	0.0000

R-squared	0.814211	Mean dependent var	1.678894
Adjusted R-squared	0.810861	S.D. dependent var	4.769420
S.E. of regression	2.074227	Akaike info criterion	4.315096
Sum squared resid	8350.996	Schwarz criterion	4.416875
Log likelihood	-4229.473	Hannan-Quinn criter.	4.352489
F-statistic	243.0380	Durbin-Watson stat	2.060970
Prob(F-statistic)	0.000000		

Inverted AR Roots	.77

Figure 9.12   *Statistical results based on the equation specification in (9.60)*

Based this output, the following findings and notes are presented.

1. The theoretical assumptions of this model are as follows:
   - The linear effect of *X1* on *Y1* depends on *t* and *X2*. For this reason, selected interactions between *X1* and the two variables *t* and *X2* should be used as independent variables of the model. However, the defined model has all possible two- and three-way interactions. Some of the results should have large *p*-values because they are highly multi-correlated. Therefore, a reduced model should be explored. Do it as an exercise, with the reduced model in (9.59) as the reference.
   - The variables *X3, X4* and *X5* are covariates or control variables and they are assumed to have the same slopes or effects on *Y1* within the four cells generated by GROUP and TP.
   - The AR(1) term should be used to reduce or overcome the autoregressive problem. Then the DW-statistic of 2.06 is obtained.

2. The GROUP variable is a dichotomous one generated based on the exogenous variable *X1*, in order to present discontinuous regression functions of *Y1* on *X1* and in addition to piece-wise regression by the two time periods. In this case, the median is used as the cutting point. However, a dichotomous group generated by the 20–30% of the lower and upper scores of *X1* might be better to be considered.

3. All independent variables have significant joint effects on *Y1*, based on the *F*-statistic of $F_0 = 243.0380$ with a *p*-value $= 0.000\,000$, and they are good predictors for *Y1*, since the regression has the Adjusted R-squared $= 0.810\,861$.

4. Several modified or extended models can easily be derived by using the assumptions that one, two or all variables *X3, X4* and *X5* have differential effects on *Y1* by *GROUP*TP*, or by each *GROUP* and *TP* only.

10

Special Notes on Selected Problems

10.1 Introduction

Compared to time series and cross-section data analysis, I would say that panel data analysis is the most complex. Each model presented has specific limitations and hidden assumptions. In addition to my own experiments with modeling, observing various graphical pattern relationships between pairs of variables and more than 300 papers in five international journals; namely the *Journal of Finance*, 2010 and 2011, the *International Journal of Accounting* (IJA), *Journal of Accounting and Economics* (JAE), *British Accounting Review* (BAR) and *Advances in Accounting, incorporating Advances in International Accounting* (AA) from the years 2008, 2009 and 2010 (Siswantoro and Agung, 2010); I have identified some selected problems in panel data analyses which have not been mentioned to students or to statistics users in general.

So the main objectives of this chapter are to inform students and less experienced researchers about the characteristics and limitations of indicators and variables, including dummies, which have been widely used in modeling, starting with the simplest possible models.

10.2 Problems with Dummy Variables

A dummy variable (or a dummy in short) can easily be defined or generated for two groups of research objects (firms or individuals). However, based on panel data, I have found dummies can be generated based on the whole observed scores of variables in pool panel data, which would be misleading. See the following sections.

10.2.1 A Dummy of the Return Rate R_{it}

In many cases, a dummy of a return rate R_{it}, defined as $DR_{it} = 1$ if R_{it} equals and is less than zero and $DR_{it} = 0$ otherwise, has been widely used in panel data modeling.

Note that this dummy variable DR_{it} does not represent two disjointed sets of the observed firms or individuals over time, but it represents two disjointed sets of the R_{it}s scores, namely the sets: $1.\{R_{it} \leq 0\}$ and $2.\{R_{it} > 0\}$. So some or many firms should have both negative and positive scores of R_{it} over time or based

Panel Data Analysis Using EViews, First Edition. I Gusti Ngurah Agung.
© 2014 John Wiley & Sons, Ltd. Published 2014 by John Wiley & Sons, Ltd.
Companion website: www.wiley.com/go/panel_data

on all firm-time observations. In other words, the two sets of scores are not the firms' classification. In very extreme cases, all firms may have both negative and positive observed scores for a long time period of observations. Therefore, the objective of the models with the dummy variable DR_{it} would be questionable or misleading for studying the difference(s) between two sets of the observed firms.

For illustration, Figure 10.1 presents a part of the two-way tabulation of DROA and FIRM_CODE, based on an unbalanced panel data of Data_Sesi.wf1, where the firms: _ABBA, _ADES and _ADMG have both non-positive (negative or zero) and positive scores of ROA. So it can be said that the dummy variable $DROA_{it}$ would be misleading to differentiate the groups of firms.

10.2.2 Other Types of Dummy Variables

As an extension of the DR_{it}, any defined dummy $DX_{it} = 1$ if $X_{it} \leq a$, for a fixed number of a, and 0 otherwise, for any numerical variable X_{it}, could also be misleading to differentiate the groups of observed firms. For instance, He, et al. (2008) define a dummy variable $DR_{i,t} = 1$ if $R_{i,t} \leq 0$, and 0 otherwise, where $R_{i,t}$ is the cumulative return of firm i during the fiscal year t.

A student of the author, Ariastiadi (2010), presents another type of classification based on a return variable, namely *ROE*, with five levels: 1. $ROE < 0.00$; 2. $0 \leq ROE < 0.50$; 3. $0.50 \leq ROE < 1.25$; 3. $1.25 \leq ROE < 1.50$; and 5. $ROE \geq 1.50$. This could be misleading for classifying banks by their worst to best performances if the classification is generated based on a long time period of observations. However, Ariastiadi conducts evaluation analysis based on only a data set of 119 banks in two years, 2007 and 2008, so the five levels of *ROA* in the 2008 represent the five levels of the banks' performances, or the disjointed five sets of banks.

Many models have used dummy variables or ordinal categorical variables, using the cutting points: the means/averages, medians, quintiles or percentiles of selected numerical variables, which also could be misleading if they are generated based on the whole individual-time observations. For dummy variables using medians as the cutting points, refer to the models in previous chapter and Duh, et al. (2009).

```
Tabulation of DROA and FIRM_CODE
Date: 07/25/11  Time: 14:29
Sample: 1 2284
Included observations: 2284
Tabulation Summary
```

Variable	Categories
DROA	2
FIRM_CODE	348
Product of Categories	696

Count % Col		FIRM_CODE AALI	ABBA	ACAP	ACES	ADES	ADHI	ADMG
	0	8	4	5	2	3	6	6
		100.00	50.00	100.00	100.00	37.50	100.00	75.00
DROA	1	0	4	0	0	5	0	2
		0.00	50.00	0.00	0.00	62.50	0.00	25.00
	Total	8	8	5	2	8	6	8
		100.00	100.00	100.00	100.00	100.00	100.00	100.00

Figure 10.1 *An illustrative two-way tabulation of DROA and Firm_Code*

With regards to dummies or classifications mentioned previously, some researchers mention that the acceptability of the models with such dummy independent variables is made under the assumption that all individual-time observations are a random cross-section data set. In other words, each firm-year is defined as a distinct unit of analysis of a random sample. However, I would say this assumption is very unrealistic or even bad, since the same set of individuals or firms are measured or observed at several time points. For this reason, this book presents three groups of panel data analyses, presented in Part I, Part II and Part III.

On the other hand, Henderson and Hudges (2010) present another type of dummy; GOOD, NEUTRAL and BAD. These are based on a variable Regulatory climate, defined as GOOD = 1 if the rating is above average and 0 otherwise; NEUTRAL = 1 if the rating is average and 0 otherwise; and BAD = 1 if the rating is below average and 0 otherwise, in addition to a year dummy for their models. In this case, all of the firms will be in the GOOD, NEUTRAL and BAD climates.

10.3   Problems with the Numerical Variable R_{it}

10.3.1   Problem with Outliers

Since some or many of the firms have both positive and negative values in the whole pool data set, then in most cases R_{it} would have close to zero growth rates, even by deleting a sufficiently large number of outliers, as presented in Figure 10.2. In a statistical sense, the null hypotheses $H_0 : \rho(ROA, T) = 0$ are accepted based on the whole panel data, as well as the sub-data sets in Figure 10.2.

The three correlations $r(ROA,T)$s are negative, namely $-0.028\ 161$, $-0.029\ 517$ and $-0.010\ 335$, respectively, with p-values of 0.1785, 0.1589 and 0.6224, based on the three data sets.

Take note that by deleting some or many outliers does not guarantee that the sub-sample obtained does not have outliers anymore. Figure 10.3 presents the box-plots of ROA for the sub-sample $\{-1 < ROA < +1\}$ by the time $T = 1$ to 8, where each box-plot shows several or many far outliers, where the box-plot for $T = 7$ has the most outliers.

For further illustration, by deleting more outliers Figure 10.6 presents the scatter graphs of ROA on the time T based on two sub-samples $\{-0.5 < ROA < 0\}$ and $\{0 < ROA + 0.5\}$ with their nearest neighbor fit curves. Take note that these graphs also show possible outliers. So the existence of outliers should be accepted in many cases of data analysis.

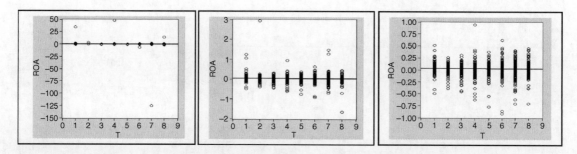

Figure 10.2   *Scatter graphs of ROA on T, with regression lines based on the whole data, and two sub-samples {−2 < ROA < 3} and {−1 < ROA < 1}, by deleting outliers*

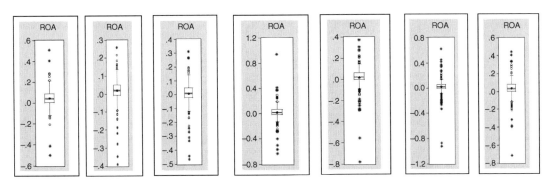

Figure 10.3   *Box-plots of ROA based on sub-sample {−1 < ROA < 1} for T = 1 to 8*

10.3.2   Problem with the Models of R_{it} with its Dummy DR_{it}

Since *DRit* does not represent two groups of the observed firms, then it can be said that the models of R_{it} and its dummy DR_{it} would be misleading. See the following example.

Example 10.1   A regression presented in Kousenidis, et al. (2009)
They quote a model from Basu (1997), as follows:

$$\frac{EPS_{i,t}}{P_{i,t-1}} = \beta_0 + \beta_1 DT_{i,t} + \beta_2 R_{i,t} + \beta_3 R_{i,t} \times DT_{i,t} + \varepsilon_{i,t} \tag{10.1}$$

where $EPS_{i,t}$ is the earning per share of the firm i at year t; $R_{i,t}$ is the annually compound stock returns of the firm i at year t; $P_{i,t-1}$ is the stock prices of the firm i and at year $(t-1)$; and $DT_{i,t}$ is a dummy variable, defined as $DT_{i,t} = 1$ if $R_{i,t} \leq 0$, and $DT_{i,t} = 0$ otherwise.

This model presents a pair of simplest linear regressions or models in a two-dimensional space, with the following equations.

$$\frac{EPS_{i,t}}{P_{i,t-1}} = \beta_0 + \beta_2 R_{i,t} + \varepsilon_{i,t}, \quad for \quad D_{i,t} = 0$$

$$\frac{EPS_{i,t}}{P_{i,t-1}} = (\beta_0 + \beta_1) + (\beta_2 + \beta_4)R_{i,t} + \varepsilon_{i,t}, \quad for \quad D_{i,t} = 1 \tag{10.2}$$

These models in fact are the simplest sets of two linear regressions, with the assumption that $R_{i,t}$ has linear effects on the dependent variable within the two groups of the scores $R_{i,t}$, namely the set of non-positive scores and the set of positive scores. So the two regressions in (10.2) do not represent two distinct groups of the firms, but the two groups or the $R_{i,t}$s scores, where some or many of the firms are in both groups.

Furthermore, take note that each of the models in (10.2) leads to a single regression function of $EPS_{i,t}/P_{i,t-1}$ on $R_{i,t}$ for all firms at all the time points, within $D_{i,t} = 0$ and $D_{i,t} = 1$, respectively. I would say that both regressions are too simplified. Furthermore, the two variables $EPS_{i,t}/P_{i,t-1}$ on $R_{i,t}$ might have nonlinear, polynomial relationships and other types of relationship, in addition to the problem with outlier(s). See the following example.

Dependent Variable: Y				
Method: Least Squares				
Date: 07/29/11 Time: 06:56				
Sample: 1 2284				
Included observations: 2284				
Variable	Coefficient	Std. Error	t-Statistic	Prob.
C	5.867515	0.049163	119.3488	0.0000
DROA	-0.501028	0.098122	-5.106181	0.0000
ROA	-0.070337	0.033443	-2.103208	0.0356
ROA*DROA	0.074884	0.037141	2.016228	0.0439
R-squared	0.012967	Mean dependent var		5.735188
Adjusted R-squared	0.011668	S.D. dependent var		2.038979
S.E. of regression	2.027049	Akaike info criterion		4.252789
Sum squared resid	9368.358	Schwarz criterion		4.262831
Log likelihood	-4852.685	Hannan-Quinn criter.		4.256452
F-statistic	9.984019	Durbin-Watson stat		0.600208
Prob(F-statistic)	0.000002			

(a)

Dependent Variable: Y				
Method: Least Squares				
Date: 07/29/11 Time: 06:58				
Sample (adjusted): 2 2284				
Included observations: 2283 after adjustments				
Convergence achieved after 6 iterations				
Variable	Coefficient	Std. Error	t-Statistic	Prob.
C	5.749074	0.105259	54.61835	0.0000
DROA	-0.055995	0.081788	-0.684638	0.4936
ROA	-0.034951	0.019364	-1.804967	0.0712
ROA*DROA	0.048452	0.021494	2.254179	0.0243
AR(1)	0.708309	0.014800	47.85945	0.0000
R-squared	0.503388	Mean dependent var		5.734564
Adjusted R-squared	0.502516	S.D. dependent var		2.039208
S.E. of regression	1.438305	Akaike info criterion		3.566996
Sum squared resid	4712.548	Schwarz criterion		3.579552
Log likelihood	-4066.726	Hannan-Quinn criter.		3.571575
F-statistic	577.2707	Durbin-Watson stat		2.081875
Prob(F-statistic)	0.000000			
Inverted AR Roots	.71			

(b)

Figure 10.4   *Statistical results based on the simplest heterogeneous model, and its AR(1) model*

Example 10.2   An application of the simplest heterogeneous regressions

To demonstrate that the model (10.1) may have problems, Figure 10.4(a) presents the statistical results based on a model similar to the model (10.1), based on Data_Sesi.wf1. A variable $Y = log$ *(liabilities)* is taken as a predicted variable, and Figure 10.4(b) presents the results based on its AR (1) model.

Based on these outputs, the following findings and notes are presented.

1. The output in Figure 10.4(a) shows an autocorrelation problem. For this reason, an AR(1) model is considered, with the output in Figure 10.4(b), which should be acceptable for showing the linear effect of *ROA* on *Y*. But the problem is whether the model is a good fit or not.
2. Both outputs in Figure 10.4 show that both regressions are acceptable models in a statistical sense, but they may have hidden problems. In addition to the problem with the dummy $DROA_{it}$, Figure 10.5 presents the problem with the outliers in the data set used. Therefore, what do you think about the regressions in Figure 10.4? Are they good fit models? Now, we could consider or evaluate the model (10.1) presented by Kousenidis, et al. However, the data is not available to do additional analysis.

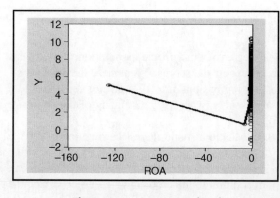 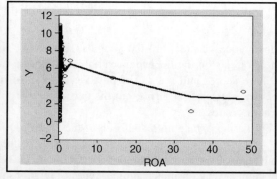

Figure 10.5   *Scatter graphs of Y on ROA with their NNF curves by the dummy DROA_{it}*

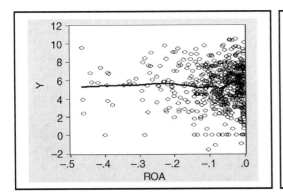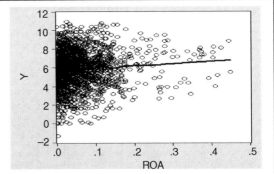

Figure 10.6 *Scatter graphs of* Y *on* ROA *with their NNF curves based on two sub-samples* {−0.5 < ROA < 0} *and* {0 < ROA < 0.5}

3. In order to have a better picture on the relationships between *Y* and *ROA* within the two groups of the *ROAs* scores, Figure 10.6 presents the scatter graphs of *Y* on *ROA* with their *Nearest Neighbor Fit* (NNF) curves based on two sub-samples {−0.5 < *ROA* < 0} and {0 < *ROA* < 0.5}. Take note that the same subset of the firms are in both sub-samples, so that the two regressions would be misleading. Furthermore, *ROA* still have many far outliers, which can easily be evaluated using the box-plots.

4. Furthermore, refer to various scatter graphs presented in the last two chapters which have demonstrated possible problems with a model based on pool panel data.

Example 10.3   Inappropriate growth model of ROA$_{it}$

For illustration, Figure 10.7 presents the scatter graphs of *ROA* on the time variable *T* with their NNF curves, based on the sub-samples {−0.5 < *ROA* < 0} and {0 < *ROA* < +0.5}, respectively. Note again that these graphs show the growths of two sets of *ROAs* scores, but they are not representing a growth model of *ROA$_{it}$*.

On the other hand, if a single growth model is fitted to all individuals or firms in the data set, the model would be inappropriate, since it is recognized that *ROA$_{it}$* should have different growth rates for all firms. Refer to the three curves in Figure 9.1.

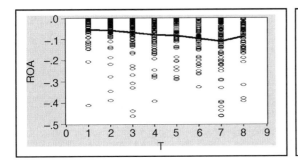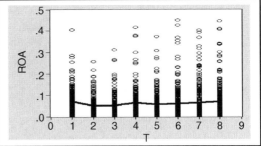

Figure 10.7 *Scatter graphs of* ROA *on* T *with their nearest neighbor fit curves for two sub-samples of* {−0.5 < ROA < 0} *and* {0 < ROA < +0.5}

Dependent Variable: LOG((ROA+0.5)/(0.5-ROA))
Method: Least Squares
Date: 09/04/11   Time: 18:55
Sample: 1 2284 IF ROA < 0.5 AND ROA > -0.5 AND FIRM_CODE <
        "_ADRO"
Included observations: 42

Variable	Coefficient	Std. Error	t-Statistic	Prob.
T*(FIRM_CODE=_AALI)	0.209472	0.054051	3.875427	0.0006
T*(FIRM_CODE=_ABBA)	0.075766	0.054051	1.401747	0.1720
T*(FIRM_CODE=_ACAP)	0.010170	0.110772	0.091807	0.9275
T*(FIRM_CODE=_ACES)	0.345083	0.495389	0.696589	0.4918
T*(FIRM_CODE=_ADES)	0.012776	0.056236	0.227193	0.8219
T*(FIRM_CODE=_ADHI)	-0.008078	0.083736	-0.096470	0.9238
T*(FIRM_CODE=_ADMG)	-0.130518	0.054051	-2.414709	0.0225
FIRM_CODE=_AALI	0.117382	0.272946	0.430056	0.6704
FIRM_CODE=_ABBA	-0.533693	0.272946	-1.955304	0.0606
FIRM_CODE=_ACAP	0.327608	0.367391	0.891716	0.3801
FIRM_CODE=_ACES	-1.726226	3.229544	-0.534511	0.5972
FIRM_CODE=_ADES	-0.120042	0.283421	-0.423544	0.6751
FIRM_CODE=_ADHI	0.162440	0.482240	0.336844	0.7387
FIRM_CODE=_ADMG	0.741613	0.272946	2.717065	0.0112

R-squared	0.754262	Mean dependent var	0.270779
Adjusted R-squared	0.640169	S.D. dependent var	0.583959
S.E. of regression	0.350293	Akaike info criterion	1.001109
Sum squared resid	3.435749	Schwarz criterion	1.580332
Log likelihood	-7.023288	Hannan-Quinn criter.	1.213417
Durbin-Watson stat	2.354810		

Wald Test:
Equation: EQ33

Test Statistic	Value	df	Probability
F-statistic	3.537046	(6, 28)	0.0099
Chi-square	21.22228	6	0.0017

Null Hypothesis Summary:

Normalized Restriction (= 0)	Value	Std. Err.
C(1) - C(7)	0.339991	0.076440
C(2) - C(7)	0.206285	0.076440
C(3) - C(7)	0.140688	0.123256
C(4) - C(7)	0.475601	0.498329
C(5) - C(7)	0.143295	0.078000
C(6) - C(7)	0.122440	0.099666

Restrictions are linear in coefficients.

Figure 10.8 *Statistical results of a heterogeneous bounded growth models of ROA based on a sub-samples of* $\{-0.5 < ROA < 0.50$ *and* $FIRM\_COD < "\_ADRO"\}$

For an additional illustration, Figure 10.8 presents the statistical results of the bounded growth models of the seven firms in DATA_Sesi.wf1 and the Wald test for testing the growths difference between the seven firms. This illustration has demonstrated that a set of seven firms has significantly different growth rates. So it can be said that all firms in a pool data set would never have a common growth rate.

Example 10.4   Application of another type of dummy
Compared to the simplest heterogeneous regressions in (10.1), Frischmann, et al. (2008) present a set of the simplest ANCOVA model with the equation as follows:

$$R_{pt} = \alpha_p + \beta_p R_{mt} + \sum_{k=1}^{K} g_{pk} D_{kt} + \varepsilon_{pt} \tag{10.3}$$

where R_{pt} is the return on portfolio p on day t $(t = 1,2,\ldots, T)$, T is the total number of daily return observations from the beginning of 2004 through March of 2007; R_{mt} is the return on the CRSP value-weighted portfolio on day t; α_p is the intercept coefficient for the portfolio p; β_p is the risk coefficient for the portfolio p; g_{pk} is the effect of event k $(k = 1,\ldots, K)$ on portfolio p's return, that is, g is an estimate of the abnormal return on the portfolio on event k. K is the total number of events examined: 11 in this study; D_{kt} is a dummy variable for the k-th event set equal to 1 during the 3-day period $(t = -1, +1$ relative to the announcement data), 0 otherwise; and ε_{pt} is a random disturbance assumed to be both normal and independent of the explanatory variables.

Take note that the model represents a set of heterogeneous regressions for all portfolios p, by the dummies D_{kt}, for $k = 1,\ldots, 11$; in two-dimensional coordinates (R_m, R_p). For instance, if $D_{kt} = 0$ for all $k > 1$, the following model will be obtained.

$$R_{pt} = \alpha_p + \beta_p R_{mt} + g_{p1} D_{1t} + \varepsilon_{pt} \tag{10.4}$$

which leads to a regression function for each portfolio p, as follows:

$$\hat{R}_p = \hat{\alpha}_p + \hat{\beta}_p R_m + \hat{g}_{p1} D_1 \tag{10.5}$$

Or

$$\hat{R}_p = \hat{\alpha}_p + \hat{\beta}_p R_m + \hat{g}_{p1}, \quad for\, D_1 = 1, and$$

$$\hat{R}_p = \hat{\alpha}_p + \hat{\beta}_p R_m, \qquad for\, D_1 = 0 \tag{10.6}$$

Take note that these two regressions are homogeneous regression lines of $\hat{R}_p$ on R_m in two-dimensional space, but there is a problem, since $\hat{R}_p$ has both negative and positive scores for the same sub-set of the firms.

For the data analysis based on the model in (10.3), using EViews, the following equation specification can be applied for each portfolio p.

$$Rp\ C\ Rm\ D1 \ldots Dk \ldots DK \tag{10.7}$$

or

$$Rp = C(p0) + c(p1)^* Rm + C(p2)^* D1 + \ldots + C(pk)^* Dk + \ldots C(p, K+1)^* DK \tag{10.8}$$

However, the set of regressions for all firms can easily be obtained by using a system equation as follows:

$$R1 = C(10) + C(11)^* Rm + C(12)^* D1 + \ldots + C(1k)^* Dk$$

$$R2 = C(20) + C(21)^* Rm + C(22)^* D1 + \ldots + C(2k)^* Dk$$

$$\ldots\ldots\ldots\ldots\ldots \tag{10.9}$$

$$Rp = C(p0) + C(p1)^* Rm + C(p2)^* D1 + \ldots + C(pk)^* Dk$$

Example 10.5   An advanced model using $DR_{i,t-1}$

Francis and Martin (2010) present several interaction models, one of the models is an advanced three-way interaction model with 23 independent variables with the dummy variable $D_{i,t-1} = DR_{i,t-1}$ as presented in (7.63), with its modification in (7.64) using EViews. I would say that many researchers have presented various models with return dummies, but they never mention the limitations or characteristics of the two groups generated by the dummies.

Example 10.6   Problems of a model with many dummies

Refer to the two models presented by Li, et al. (2010) in (7.61), and Suriawinata (2004) in (7.62), which have dummies of various dichotomous or categorical variables. The problem with the first model in (7.61) is it has additive independent variables, including the dummies of eight distinct dichotomous variables, which is a very special ANCOVA model. Take note that the eight dichotomous variables in fact generate $2^8 = 1054$ groups of research objects. So this model has some hidden assumptions, such as the homogenous slopes on the numerical independent variables, the pattern of the intercepts of the homogenous regressions and the basic assumptions of the error terms.

The second model in (7.62) has similar problems since it has nine dummies, where only one of the dummies is used in interactions with two of the numerical variables, namely $D\_RESTRU_{it}{}^*DEBT\_MBF_{it}^T$ and $D\_RESTRU_{it}{}^*SIZE\_TA_{it}$. So the variables $DEBT\_MBF_{it}^T$ and $SIZE\_TA_{it}$ are assumed to have different linear effects between the groups generated by D_RESTRU_{it}. Compare to the model in (7.64) having *11* two- and three-way interactions with a dummy $D_{i,t-1}$.

10.4 Problems with the First Difference Variable

The first difference of a variable Y_{it} is defined as $DY_{i,t} = Y_{i,t} - Y_{i,t-1}$ for the firm i, at the time point t, which can have both non-positive (negative or zero) and positive for the firm i over times. Therefore variable DY_{it} would have the same problems as the return variable R_{it}. For this reason, I will not discuss this again in detail.

However, I have found that some students misinterpret a firm, i, with a negative score of $DSALE_{i,t} = SALE_{i,t} - SALE_{i,t-1}$ as being a worse firm compared to one, j, with a positive score of $DY_{j,t}$. Take note that the $SALE_{i,t}$ and $SALE_{i,t-1}$ of the firm i can be much better than both $SALE_{j,t}$ and $SALE_{j,t-1}$, but $SALE_{i,t} = SALE_{i,t-1}-a$, for a very small positive number of a, and $SALE_{j,t} > SALE_{j,t-1}$.

Another important first difference to consider is $Dlog(Y_{i,t})$ for a positive variable $Y_{i,t}$, which is in fact an exponential growth rate variable, namely $GR_{i,t}(Y) = log(Y_{i,t}/Y_{i,t-1})$ for the firm i at the time point t.

In order to present an illustrative example, balanced or complete pool data should be used. For this reason, balanced pool data, namely BPD_1July.wf1, with 8×224 year-firm observations is developed, which is a sub-sample of the Data_Sesi.wf1, as presented in Figure 10.9.

Example 10.7   Illustration based on balanced pool data
For illustration, Figure 10.10 presents scatter graphs of $DSale = Sale - Sale(-1)$ on $SALE$ and NI (net-income), respectively with the zero-lines. Based on these graphs the following findings and notes are presented.

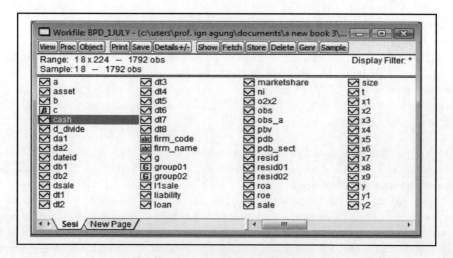

Figure 10.9   *The work file: BPD_1July.wf1*

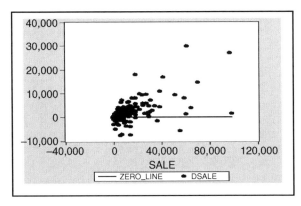 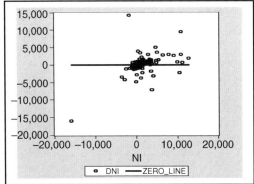

Figure 10.10 *Scatter graphs on* DSale *on* SALE, *and* DNI *on* NI, *with the* Zero_lines

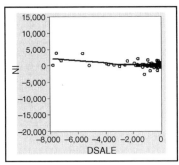 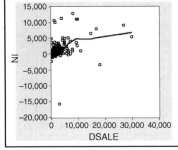 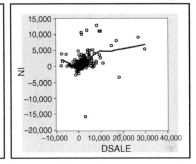

Figure 10.11 *Three alternative scatter graphs of* NI *on* DSale, *with their NNF curves*

1. Scatter graph of *DSale* on *SALE* shows that the firm "_ASII-8" with a maximum SALE of 98 526 has a relatively small positive score of *DSale* = 1462, and several of the firms with SALE > 40 000 even have negative scores of *DSale*.

2. Similarly, the scatter graph of *DNI* on *NI* shows that the firm "_TLKM-7" with a positive *NI* = 10 619 has a negative *DNI* = −2238.

3. So it can be concluded that the model with a first difference, say $DX = X − X(−1)$ for any variable *X*, would be misleading. In other words, the model would give unexpected or unpredictable estimates. For illustration, Figure 10.11 presents three scatter graphs of *NI* on *DSale* with their NNF curves, for *DSale* < 0, *DSale* > 0 and the whole sample.

10.5 Problems with Ratio Variables

Ratio variables could have problems or be misleading in many cases, such as the ratio of *SALE/Total ASSET*, namely *RSale_TA*; and *NI/TA*, namely *RNI_TA*. Take note that $SALE_{it}$ should be greater than zero in general, however, the data has a single observation of SALE = −0.30, which should be a mistake but NI_{it}

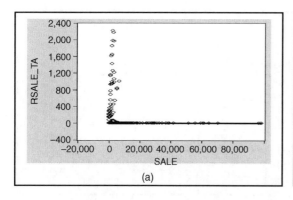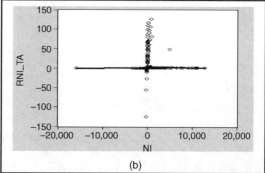

Figure 10.12 *Scatter graphs of RSale_TA on SALE, and RNI_TA on NI, with NNF curves*

can have negative scores. Figure 10.12(a) and (b) show that the scores of *RSale_TA* and *RNI_TA* are very small, respectively, for a very large *SALE* and *NI* and some or many small scores of *SALE* and *NI* correspond to very large scores of *RSale_TA* and *RNI_TA*. So the ordering or ranking of *RSale_TA* and *RNI_TA* can be unpredictable compared to *SALE* and *NI* respectively.

Further analysis based on *SALE* and *RSale_TA*, finds that the unit of observation "_ASII-8", the firm _ASII at the time $t = 8$ has a maximum score of *SALE* = 98 526, but the firm has a very small *RSale_TA* = 1.10 781. For very small scores of SALE, the firms have large scores *RSale_TA*. On the other hand, the scatter graph of *RNI_TA* on *NI* shows that the firm with a minimum negative score of $NI = -15\ 855$ is considered to be in better condition based on *RNI_TA* compared to several other firms with better *NIs* (net-incomes).

Take note that most of the ratio variables have small or very small scores, since ASSET has very large scores. For this reason, scatter graphs in Figure 10.12 show most of the scores of *RSale_TA* and *RNI_TA* are very close to the zero-line because the ratio variables have far outliers.

See the illustrative graphs of the RNI_TAs scores based on the whole data set, and two sub-samples presented in Figure 10.13. The third graph still shows some far outliers, even based on a very small interval of $\{-3 < RNI\_TA < 4\}$ compared to the first graph based on the whole sample of a very large interval of $\{-150 < RNI\_TA < 150\}$.

With ratio variables then, various or many possible models can easily be defined using these ratio variables or transformed variables in general, compared to models using original variables. But their patterns of relationships are unpredictable. See the following illustration.

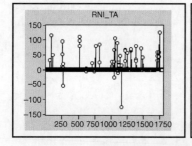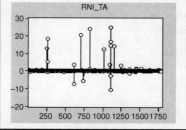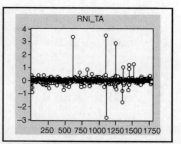

Figure 10.13 *Illustrative graphs based on three intervals of RNI_TAs scores*

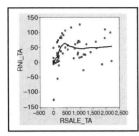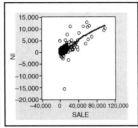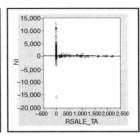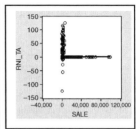

Figure 10.14   *Illustrative scatter graphs based on* NI, Sale, RNI_TA *and* RSale_TA, *with their NNF curves*

Example 10.8   Illustrative models using ratio variables

With the variables *RSale_TA* and *RNI_TA*, in addition to the original variables *SALE* and *NI*, then the effect of *SALE* on *NI* (net-income) can be studied using many of the following alternative models.

1. Linear or nonlinear-effect models of *RSale_TA* on *RNI_TA*,
2. Linear or nonlinear-effect models of *SALE* on *RNI_TA*,
3. Linear or nonlinear-effect models of *RSale_TA* on *NI*,
4. Heterogeneous regressions by individuals (or groups of individuals) and/or times (or time periods) of all possible models in points (1) to (3) here.
5. Fixed-effect models (homogeneous regressions) by individuals (or groups of individuals) and/or times (or time-periods) of all possible models in points (1) to (3) here.

Therefore, all possible models here could easily be extended to more advanced models with additional ratio variables, the transformed variables in general, as well as the original variables. However, researchers never know the true population model – refer to Agung (2009, Section 2.14).

For illustration, Figure 10.14 presents four scatter graphs based on the variables *NI*, *Sale*, *RNI_TA* and *RSale_TA*, and Figure 10.15 presents the outputs of their covariance analysis. Figure 10.14 clearly shows that the scatter graph of *RSale_TA* on *RNI_TA* has a different pattern compared to the scatter graph of *Sale* on *NI*, more so when compared to the other two scatter graphs.

Covariance Analysis: Ordinary
Date: 09/21/11 Time: 13:36
Sample: 1 8
Included observations: 1792

Correlation t-Statistic Probability	NI	SALE	RNI_TA	RSALE_TA
NI	1.000000			

SALE	0.744354	1.000000		
	47.15955	-----		
	0.0000	-----		
RNI_TA	0.030773	0.003661	1.000000	
	1.302561	0.154895	-----	
	0.1929	0.8769	-----	
RSALE_TA	-0.006908	0.003617	0.610157	1.000000
	-0.292288	0.153035	32.58284	-----
	0.7701	0.8784	0.0000	-----

(a)

Covariance Analysis: Kendall's tau
Date: 09/21/11 Time: 13:39
Sample: 1 8
Included observations: 1792

tau-a Score (S) Probability	NI	SALE	RNI_TA	RSALE_TA
NI	0.991945			
	1591810			
	—			
SALE	0.423295	0.999230		
	679277	1603501		
	0.0000	-----		
RNI_TA	0.671969	0.248019	0.999923	
	1078333	398005	1604612	
	0.0000	0.0000	-----	
RSALE_TA	0.120392	0.330084	0.217694	0.999969
	193197	529698	349341	1604686
	0.0000	0.0000	0.0000	-----

(b)

Figure 10.15   *Outputs of covariance analysis based on* NI, Sale, RNI_TA *and* RSale_TA

1. Based on the parametric statistics presented in Figure 10.15(a) the following findings and notes are presented.
   - *Sale* has a significant positive linear effect on *NI* based on a very large *t*-statistic of $t_0 = 47.1955$, and *RSale_TA* also has a significant positive linear effect on *RNI_TA* based on a smaller *t*-statistic of $t_0 = 32.58\,284$. However, their graph shows that a nonlinear model of *RNI_TA* on *RSale_TA* would be a better fit. Try it as an exercise.
   - *Sale* has an insignificant positive linear effect on *RNI_TA* based on the *t*-statistic of $t_0 = 0.154\,895$ with $df = 1790$ and a *p*-value $= 0.8759/2 = 0.43\,795$
   - In contradiction, *RSale_TA* has an insignificant negative linear effect on *NI*, based on the *t*-statistic of $t_0 = -0.292\,288$ with $df = 1790$ and a *p*-value $= 0.7701/2 = 0.38\,505$.
   - Based on these findings, it can be said that if a ratio variable is used in a model, then parameters' estimates can be misleading. So I have recommended to my students not to use ratio variables, which Panutan had for his dissertation.
2. Based on nonparametric statistics, namely the Kendall's tau that is the concordant-discordant measure of association presented in Figure 10.15(b), the following findings and notes can be presented.
   - *Sale* has a significant positive effect on *NI* based on the Kendall's tau of $\tau_0 = 0.423\,295$ with a *p*-value $= 0.0000$, and *RSale_TA* also has a significant positive effect on *RNI_TA* based on a smaller Kendall's tau of $\tau_0 = 0.217\,694$ with a *p*-value $= 0.0000$.
   - *Sale* also has a significant positive effect on *RNI_TA* based on the Kendall's tau of $\tau_0 = 0.248\,019$ with a *p*-value $= 0.0000$.
   - *RSale_TA* also has a significant positive linear effect on *NI*, based the Kendall's tau of $\tau_0 = 0.120\,392$ with a *p*-value $= 0.0000$.
   - In addition, we find that the same conclusions would be obtained using Spearman's Rank Correlation.

10.6   The CAPM and its Extensions or Modifications

Referring to the CAPM (Sharp, 1964), that is

$$R_{it} = \alpha_i + \beta_i R_{mt} + \varepsilon_{it} \qquad (10.10)$$

various models of R_{it} on R_{mt} could easily be developed by using either their transformed variables, or additional independent (exogenous, cause or up-stream) variables, to estimates Betas or β_i's.

For each firm or asset i, the estimates of the model in (10.10) is a simple linear regression having the slope $\hat{\beta}_i$, which can be presented as a scatter graph of R_{it} on R_{mt}, with a regression line, for $t = 1,2,\ldots, T$. In this case, for each firm i, the time series R_{it} and R_{mt} are considered variables of cross-section data, because the graph presents the ordered scores of R_{it} and R_{mt}, namely $R_{i,j} \le R_{i,j+1}$, for $j = 1,\ldots, T$; and $R_{m,k} \le R_{m,k+1}$, for $k = 1,\ldots, T$. This transformation might lead to an unexpected scatter graph, their relationship could be piece-wise, polynomial or nonlinear, which can easily be identified using their scatter graph with their kernel fit or nearest neighbor fit curves – refer to scatter graphs based on time-series variables presented in Chapter 1 as well as in Agung (2009a). For this reason, alternative or modified CAPMs and illustrative examples are presented in the following sections.

10.6.1   Generalized CAPM

For each asset or firm i, the CAPM in (10.10) is a time-series model. For this reason, similar models to all those presented in Chapters 1–4 and Agung (2009a) as well as other references, can easily be defined or created with the dependent variable R_{it} or its transformed variable, such as $log[(R_{it} - L)/(U - R_{it})]$, $log(R_{it} - L)$ or $log(U - R_{it})$ where L and U are the fixed lower and upper bounds of R_{it} for all i and t, which should be subjectively selected. So I would consider a generalized CAPM, namely the GCAPM, with the general equation as follows:

$$G(R_{it}) = \alpha_i + \beta_i H(R_{mt}) + \varepsilon_{it} \qquad (10.11)$$

where $G(R_{it})$ and $H(Rm)$ are fixed functions of R_{it} and R_{mt}, respectively, without a parameter, including the basic or original CAPM model in (10.10).

10.6.2 GCAPM with Exogenous Variables

In the CAPM, R_{mt} in fact is an environmental variable, and it is assumed to have a linear effect on R_{it} for each i, without taking into account the influences or effects of other exogenous variables, as well as the time variable. Therefore, as an extension of the GCAPM model in (10.11), the following model would be considered.

$$G(R_{it}) = \alpha_i + \beta_i H(R_{mt}) + F(X_{it}, t) + \varepsilon_{it} \qquad (10.12)$$

where $F(X_{it}, t)$ can be any functions of a multivariate X_{it} and the time t with a finite number of slope parameters which depend on i, but without the constant parameter. Refer to all possible functions presented in previous chapters, the specific models (9.3a) and (9.3b) in particular.

10.6.3 Advanced Models of R_{it} by Firms or Asset i

First, it is necessary to note that based on my observations in more than 300 papers presented in *The Journal of Finance*, 2010 and 2011, and *International Journal of Accounting* (IJA), *Journal of Accounting and Economics* (JAE), *British Accounting Review* (BAR) and *Advances in Accounting, incorporating Advances in International Accounting* (AA) from the years 2008, 2009, and 2010 (Siswantoro and Agung, 2010), there is only one model with the time t as a numerical independent variable presented by Ferris and Wallace (2009). However, they do present time-fixed-effects models, or models with time dummies. On the other hand, I have found models with trend or time-related effects in other journals presented by Delong and Deyoung (2007) and Bansal (2005). For this reason, the following two alternative models of R_{it} need to be considered.

10.6.3.1 A Modified CAPM with the Time-Period Dummies

Take note that the β_i in the models (10.10) and (10.11) is constant over time. In other words, it depends only on the asset i. In practice, however, time periods should be taken into account in modeling: refer to Chapters 1–4. For this reason, I would propose a modified CAPM with time-period dummies for each asset i and the following general equation.

$$G(Ri) = \sum_{tp=1}^{n} \alpha_{i,tp} D_{tp} + \sum_{tp=1}^{n} \beta_{i,tp} \times D_{tp} \times H(Rm) + \varepsilon_{i,tp} \qquad (10.13)$$

where D_{tp} is a time-period dummy, which should be identified case by case and $\beta_{i,tp}$ is a new beta, which depends on the asset i and the time-period tp.

Furthermore, corresponding to the model in (10.12), the following general model can be considered.

$$G(Ri) = \sum_{tp=1}^{n} \alpha_{i,tp}D_{tp} + \sum_{tp=1}^{n} \beta_{i,tp} \times D_{tp} \times H(Rm) + F(X_{it}, tp) + \varepsilon_{i,tp} \quad (10.14)$$

10.6.3.2   Models of R_{it} with the Time Numerical Independent Variable

Referring to any endogenous variable, including the return R_{it}, heterogeneous regressions by firms or assets considered can have the most general equation as follows:

$$G(R_{it}) = F(R_{mt}, X_{it}, t) + \varepsilon_{it} \quad (10.15)$$

where $F(R_{mt}, X_{it}, t)$ can be any functions having a finite number of parameters, which depend on the asset or firm i. Referring to the general models in (9.3a) and (9.3b), the following specific models might be considered, namely the *model-with-trend*, and the *time-related effects model* (TREM).

$$G(R_{it}) = F(R_{mt}, X_{it}) + \rho_i t + \varepsilon_{it} \quad (10.15a)$$

and

$$G(R_{it}) = F_1(R_{mt}, X_{it}) + t \times F_2(R_{mt}, X_{it}) + \varepsilon_{it} \quad (10.15b)$$

10.6.4   Illustrative Data Analysis to Generate Betas

Figure 10.16 presents the work file of my student, William Tjong, which will be used to demonstrate the data analysis based on the alternative models of R_{it}, presented above. The data contains only 4×44 firm-time observations.

The variable BAS_{it} (bid-ask-spread) is defined as the ask price of stock i minus bid price of stock i at time t, and OIB_{it} (order imbalance) is defined as the number of buyer initiated stock i minus seller initiated stock i, at time t (Cordia and Subrahmanyam, 2004). Since the variables OIB and BAS have minimum negative scores of $(-57\,276\,000)$ and (-154.7619), respectively, then the variables LNOIB and LNBAS are generated as $\log(\text{OIB-L}_{\text{OIB}})$ and $\log(\text{BAS-L}_{\text{BAS}})$ where L_{OIB} and L_{BAS} are the lower bounds of OIB and BAS. In this case, it is subjectively selected $L_{\text{OIB}} = -57\,500\,000$ and $L_{\text{BAS}} = -160$.

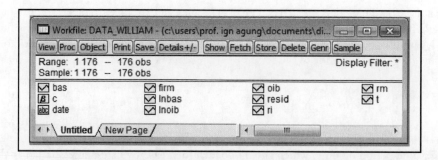

Figure 10.16   *A work file by an author's student, for his dissertation*

Take note that this data set can be considered as time series data by states/firms as presented in Part I (Chapters 1–4). So all types of models presented in Part I should be applicable for GLMs with a dependent variable R_{it} for each firm or asset i, and the Multivariate GLMs, System Equation Models or SCM (Seemingly Causal Models) with dependent variables R_{it} for some selected assets or all assets.

Furthermore, the data also can be viewed as a natural experimental data set. So all type of models based on experimental data presented in Agung (2011, Chapter 8) for each asset i, as well as various types of model, which will be presented in Part III, should also be valid for the return R_{it}. However, the following examples only present those models based on returns variables R_{it} and R_{mt} specifically to generate alternative sets of β_i's.

10.6.4.1   Simple Linear-Effect Models

Example 10.9   Simple linear-effect model of *Rm* on *Ri* or *G(Ri)*
Figure 10.17 presents statistical results based on two simple linear-effect models, namely the basic CAPM and a *semi-logarithmic (semilog)* CAPM, using the following equation specifications, respectively.

$$
\begin{aligned}
&Ri \ @Expand(Firm) \ Rm^*@Expand(Firm) \\
&and \\
&\log(Ri + 0.1) \ @Expand(Firm) \ Rm^*@Expand(Firm)
\end{aligned}
\tag{10.16}
$$

where (-0.1) is subjectively selected as a lower bound of R_{it}, for all 4×44 observed scores.

Note that these two models present two set of Betas, indicated by the parameters C(5), C(6), C(7) and C(8), as presented in Figure 10.17.

10.6.4.2   Translog Linear-Effect Models

Example 10.10   Bounded translog linear-effect models
Figure 10.18 presents the statistical results based on two *bounded trans-logarithmic (translog)* linear CAPMs, using the following equation specifications, respectively.

Dependent Variable: RI
Method: Least Squares
Date: 09/22/11   Time: 12:17
Sample: 1 176
Included observations: 175

Variable	Coefficient	Std. Error	t-Statistic	Prob.
FIRM=1	-0.000930	0.003592	-0.258780	0.7961
FIRM=2	0.001867	0.003634	0.513709	0.6081
FIRM=3	0.001095	0.003601	0.304173	0.7614
FIRM=4	0.001449	0.003593	0.403250	0.6873
RM*(FIRM=1)	0.960072	0.153418	6.257873	0.0000
RM*(FIRM=2)	0.826367	0.153419	5.386332	0.0000
RM*(FIRM=3)	0.418444	0.144277	2.900295	0.0042
RM*(FIRM=4)	0.692918	0.141782	4.887226	0.0000

R-squared	0.376427	Mean dependent var	0.001149
Adjusted R-squared	0.350289	S.D. dependent var	0.029552
S.E. of regression	0.023820	Akaike info criterion	-4.591907
Sum squared resid	0.094758	Schwarz criterion	-4.447231
Log likelihood	409.7918	Hannan-Quinn criter.	-4.533222
Durbin-Watson stat	2.236217		

Dependent Variable: LOG(RI+0.1)
Method: Least Squares
Date: 09/22/11   Time: 12:53
Sample: 1 176
Included observations: 175

Variable	Coefficient	Std. Error	t-Statistic	Prob.
FIRM=1	-2.364696	0.040391	-58.54460	0.0000
FIRM=2	-2.320405	0.040859	-56.79112	0.0000
FIRM=3	-2.345476	0.040492	-57.92420	0.0000
FIRM=4	-2.340797	0.040406	-57.93155	0.0000
RM*(FIRM=1)	11.67085	1.725126	6.765216	0.0000
RM*(FIRM=2)	9.530619	1.725137	5.524556	0.0000
RM*(FIRM=3)	3.459611	1.622332	2.132494	0.0344
RM*(FIRM=4)	7.977578	1.594276	5.003888	0.0000

R-squared	0.389383	Mean dependent var	-2.340244
Adjusted R-squared	0.363788	S.D. dependent var	0.335810
S.E. of regression	0.267852	Akaike info criterion	0.247869
Sum squared resid	11.98133	Schwarz criterion	0.392545
Log likelihood	-13.68853	Hannan-Quinn criter.	0.306554
Durbin-Watson stat	2.185198		

Figure 10.17 *Statistical results based on the models in (10.16)*

Dependent Variable: LOG((RI+0.1)/(0.2-RI))
Method: Least Squares
Date: 09/22/11   Time: 12:26
Sample: 1 176
Included observations: 175

Variable	Coefficient	Std. Error	t-Statistic	Prob.
FIRM=1	-0.743877	0.058197	-12.78205	0.0000
FIRM=2	-0.689874	0.058870	-11.71863	0.0000
FIRM=3	-0.715074	0.058305	-12.26445	0.0000
FIRM=4	-0.705332	0.058261	-12.10638	0.0000
LOG(RM+01)*(FIRM=1)	16.56173	2.476272	6.688171	0.0000
LOG(RM+01)*(FIRM=2)	13.63907	2.476276	5.507897	0.0000
LOG(RM+01)*(FIRM=3)	6.037251	2.335865	2.584589	0.0106
LOG(RM+01)*(FIRM=4)	11.79558	2.280056	5.173372	0.0000

R-squared	0.394832	Mean dependent var	-0.712897
Adjusted R-squared	0.369466	S.D. dependent var	0.486119
S.E. of regression	0.386009	Akaike info criterion	0.978722
Sum squared resid	24.88343	Schwarz criterion	1.123398
Log likelihood	-77.63817	Hannan-Quinn criter.	1.037407
Durbin-Watson stat	2.239653		

Dependent Variable: LOG((RI+0.1)/(0.2-RI))
Method: Least Squares
Date: 09/22/11   Time: 12:30
Sample: 1 176
Included observations: 175

Variable	Coefficient	Std. Error	t-Statistic	Prob.
FIRM=1	-3.678305	0.490948	-7.492245	0.0000
FIRM=2	-3.009657	0.491307	-6.125820	0.0000
FIRM=3	-1.764348	0.452410	-3.899883	0.0001
FIRM=4	-3.126090	0.466710	-6.698141	0.0000
LOG((RM+01)/(0.1-RM))*(FIRM=1)	1.256065	0.208245	6.031679	0.0000
LOG((RM+01)/(0.1-RM))*(FIRM=2)	0.992704	0.208285	4.766081	0.0000
LOG((RM+01)/(0.1-RM))*(FIRM=3)	0.448464	0.189964	2.360791	0.0194
LOG((RM+01)/(0.1-RM))*(FIRM=4)	1.034008	0.198892	5.198846	0.0000

R-squared	0.355498	Mean dependent var	-0.712897
Adjusted R-squared	0.328483	S.D. dependent var	0.486119
S.E. of regression	0.398356	Akaike info criterion	1.041693
Sum squared resid	26.50076	Schwarz criterion	1.186369
Log likelihood	-83.14816	Hannan-Quinn criter.	1.100378
Durbin-Watson stat	2.148558		

Figure 10.18   *Statistical results based on the models in (10.17)*

$$\log((Ri + 0.1)/(0.2 - Ri))\quad @Expand(Firm)\quad \log(Rm + 0.1)^* @Expand(Firm)$$
$$and$$
$$\log((Ri + 0.1)/(0.2 - Ri))\quad @Expand(Firm)$$
$$\log(Rm + 0.1)/(0.1 - Rm)^* @Expand(Firm) \tag{10.17}$$

10.6.4.3   CAPM with Additive Covariates and Trend

Based on the numerical variables available in the data set, the following additive model can be considered to generate the Betas, with the results presented in Figure 10.19.

$$ri\ c\ @expand(firm, @droplast)\quad rm^* @expand(firm)$$
$$lnbas^* @expand(firm)\quad lnoib^* @expand(firm)$$
$$t^* @expand(firm) \tag{10.18}$$

Dependent Variable: RI
Method: Least Squares
Date: 09/24/11   Time: 07:23
Sample: 1 176
Included observations: 175

Variable	Coefficient	Std. Error	t-Statistic	Prob.
C	4.739828	1.595070	2.971548	0.0034
FIRM=1	-3.713827	1.891747	-1.963174	0.0514
FIRM=2	-4.628827	1.636503	-2.828487	0.0053
FIRM=3	-4.286020	1.597423	-2.683085	0.0081
RM*(FIRM=1)	0.897535	0.152293	5.893469	0.0000
RM*(FIRM=2)	0.821799	0.142105	5.783046	0.0000
RM*(FIRM=3)	0.292637	0.148166	1.975056	0.0500
RM*(FIRM=4)	0.603876	0.135292	4.463500	0.0000

Variable	Coefficient	Std. Error	t-Statistic	Prob.
LNBAS*(FIRM=1)	-0.001364	0.003460	-0.394080	0.6941
LNBAS*(FIRM=2)	0.004520	0.014484	0.312062	0.7554
LNBAS*(FIRM=3)	-0.036542	0.012896	-2.833571	0.0052
LNBAS*(FIRM=4)	-0.001166	0.005959	-0.195584	0.8452
LNOIB*(FIRM=1)	-0.057337	0.057101	-1.004121	0.3169
LNOIB*(FIRM=2)	-0.007405	0.020132	-0.367844	0.7135
LNOIB*(FIRM=3)	-0.014723	0.003897	-3.778105	0.0002
LNOIB*(FIRM=4)	-0.264915	0.089848	-2.948490	0.0037
T*(FIRM=1)	0.000184	0.000273	0.676128	0.5000
T*(FIRM=2)	-8.97E-05	0.000271	-0.331415	0.7408
T*(FIRM=3)	0.000329	0.000273	1.208053	0.2289
T*(FIRM=4)	9.55E-05	0.000265	0.360768	0.7188

R-squared	0.509851	Mean dependent var	0.001149
Adjusted R-squared	0.449769	S.D. dependent var	0.029552
S.E. of regression	0.021921	Akaike info criterion	-4.695521
Sum squared resid	0.074483	Schwarz criterion	-4.333831
Log likelihood	430.8581	Hannan-Quinn criter.	-4.548809
F-statistic	8.485825	Durbin-Watson stat	2.335045
Prob(F-statistic)	0.000000		

Figure 10.19   *Statistical results based on the models in (10.18)*

```
Tabulation of DRI and FIRM
Date: 12/12/11  Time: 14:08
Sample: 1 176
Included observations: 176
Tabulation Summary
```

Variable	Categories
DRI	2
FIRM	4
Product of Categories	8

Count % Col		FIRM 1	2	3	4	Total
	0	18	21	24	19	82
		40.91	47.73	54.55	43.18	46.59
DRI	1	26	23	20	25	94
		59.09	52.27	45.45	56.82	53.41
	Total	44	44	44	44	176
		100.00	100.00	100.00	100.00	100.00

Figure 10.20   *Distribution of positive and negative scores of R_{it} by FIRM*

Note that even though some of the independent variables have large p-values (greater than 0.30), the model will not be reduced, because each firm should have the same independent variables to estimate the Betas.

However, the time T may be deleted from all firms, because R_{it}, for each firm i, has positive and non-positive scores over time, which leads to insignificant effects of T on R_{it} for all firms. For illustration, Figure 10.20 presents the distributions of positive and non-positive scores of R_{it} for the four firms considered.

10.6.4.4   *CAPM with Two-Way Interaction Covariates and Time-Related-Effects*

As an extension of the CAPM in (10.18), the following CAPM is a two-way interaction model with time-related effects. Its statistical results are not presented, but the data analysis based on its set of Betas will be presented in Section 10.6.4.6.

$$
\begin{aligned}
ri\ c\ @expand(firm, @droplast)\ rm^*@expand(firm) \\
lnbas^*@expand(firm)\ lnoib^*@expand(firm)\ t^*@expand(firm) \\
t^*lnbas^*@expand(firm)\ t^*lnoib^*@expand(firm) \\
lnbas^*lnoib^*@expand(firm)
\end{aligned}
\tag{10.19}
$$

10.6.4.5   *CAPM with Three-Way Interaction Covariates and Time-Related Effects*

As an extension of the CAPM in (10.19), the following CAPM is a three-way interaction model with time-related effects. Its statistical results are not presented, but the data analysis based on its set of Betas will be presented in the following section.

$$
\begin{aligned}
ri\ c\ @expand(firm, @droplast)\ rm^*@expand(firm)\ lnbas^*@expand(firm) \\
lnoib^*@expand(firm)\ t^*@expand(firm)\ t^*lnbas^*@expand(firm) \\
t^*lnoib^*@expand(firm)\ lnbas^*lnoib^*@expand(firm) \\
t^*lnbas^*lnoib^*@expand(firm)
\end{aligned}
\tag{10.20}
$$

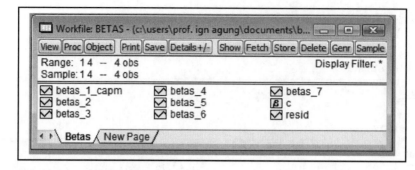

Figure 10.21   *The work file of the seven Betas*

10.6.4.6   *Analysis Based on Seven Sets of Betas*

Based on the seven CAPMs in (10.16) to (10.20), a data set of seven *Betas* is obtained, namely *Betas_1_ CAPM*, *Betas_2 to Betas_7* as presented in Figure 10.21, with their correlation matrix presented in Figure 10.22. Based on this correlation matrix, the following findings and notes are presented.

1. The original Betas, *Beta_1_CAPM*, have very large correlations with each of the others. So it can be concluded that the other Betas can be used to replace *Beta_1_CAPM* in all models having *Beta_1_ CAPM* as one of the independent variables.

```
Covariance Analysis: Ordinary
Date: 09/22/11  Time: 17:19
Sample: 1 4
Included observations: 4
```

Correlation t-Statistic Probability	BETAS_1_C...	BETAS_2	BETAS_3	BETAS_4	BETAS_5	BETAS_6	BETAS_7
BETAS_1_CAPM	1.000000 ----- -----						
BETAS_2	0.998318 24.35452 0.0017	1.000000 ------ -----					
BETAS_3	0.997713 20.87398 0.0023	0.999827 76.00062 0.0002	1.000000 ------ -----				
BETAS_4	0.954905 4.548308 0.0451	0.970503 5.692896 0.0295	0.972293 5.882062 0.0277	1.000000 ------ -----			
BETAS_5	0.991878 11.02831 0.0081	0.986544 8.533305 0.0135	0.983592 7.710317 0.0164	0.927346 3.504677 0.0727	1.000000 ------ -----		
BETAS_6	0.989716 9.784929 0.0103	0.984869 8.037055 0.0151	0.981610 7.272083 0.0184	0.927554 3.510293 0.0724	0.999727 60.48014 0.0003	1.000000 ------ -----	
BETAS_7	0.989829 9.839868 0.0102	0.983760 7.751093 0.0162	0.980545 7.064460 0.0195	0.921516 3.355879 0.0785	0.999868 86.90278 0.0001	0.999708 58.50275 0.0003	1.000000 ------ -----

Figure 10.22   *A correlation matrix of the seven Betas*

2. However, since this illustrative example is only based on four sets or firms, then further or additional data analyses should be done based on a large number of assets in order to find out the consistency of the Betas.

10.7 Selected Heterogeneous Regressions from International Journals

This section will present mainly simple heterogeneous regressions in order to show students that they can apply similar simple models for their theses and dissertations with some or small modifications.

10.7.1 Heterogeneous Regressions of Naes, Skjeltrop and Odegaard (2011)

Naes et al. present heterogeneous regressions, the basic CAPM, as follows:

$$R_{it} = \alpha_i + \beta_i R_{mt} + \varepsilon_{it} \tag{10.21}$$

where R_{it} is the return on security i at time t, R_{mt} is the market return on security at time t, $\alpha_i \& \beta_i$ are the intercept (constant) and slope (or linear effect) parameters for each i and ε_{it} is the error term.

In addition to the assumptions for the error term, it is also assumed that R_{mt} has a linear effect on R_{it} for all i. Refer to the several CAPMs presented previously and the graphs in Figure 10.22. With regards to the error term, see the notes on the residual analysis, presented in Agung (2009a, Section 2.14.3).

10.7.2 LV(1) Heterogeneous Regressions Models

10.7.2.1 LV(1) Model Presented by Laksmana and Yang (2009)

Laksmana and Yang present a set of the simplest LV(1) heterogeneous regressions with the following equation, for $i = 1, 2, \ldots, N$. Take note that this set of regressions presents a set of dynamic models. Refer to the notes presented in Section 1.3.2.

$$EARN_{i,t} = \beta_{0,i} + \beta_{1,i} EARN_{i,t-1} + \varepsilon_{i,t} \tag{10.22}$$

10.7.2.2 LV(1) Model Presented by Dedman and Lennox (2009)

They present a first-order lagged variable model, LV(1), based on a single numerical variable, namely X_{ijt}, which is generated as a different company, i's return on asset (ROA) and the mean of ROA in the i's three-digits industry in year t, within j's special group of ROA. The model can be considered to be the simplest lagged variable model, with two dummies of a dichotomous variable, with the equation as follows:

$$X_{ijt} = \beta_{0j} + \beta_{1j} D_n \times X_{ijt-1} + \beta_{2j} D_p \times X_{ijt-1} + \varepsilon_{ijt} \tag{10.23}$$

where $D_n = 1$ if $X_{ijt-1} \leq 1$ and $D_n = 0$ otherwise, and $D_p = 1$ if $X_{ijt-1} > 1$; and $D_p = 0$ otherwise. Corresponding to this model, the following notes, comments and possible alternative models are presented.

1. Take note that this type of model also can be applied to any variable X, where X_{ijt} indicates the score of a variable of i's firm in j's group/sector or region/state, at the time t, and the two dummy variables are in fact the dummies of a numerical variable. Refer to the notes and problems with the dummy variables defined based on the whole panel data, as presented in Section 10.2. Furthermore the linear causal or up- and down-stream relationship between X_{ijt} and X_{ijt-1} should be assumed. Refer to the following example.

2. This model can be presented using a pair of LV(1)_Models, as follows:

$$
\begin{aligned}
X_{ijt} &= \beta_{0j} + \beta_{1j} X_{ijt-1} + \varepsilon_{ijt}, \quad for \ (D_n = 1, D_p = 0) \\
X_{ijt} &= \beta_{0j} + \beta_{2j} X_{ijt-1} + \varepsilon_{ijt}, \quad for \ (D_n = 0, D_p = 1)
\end{aligned}
\tag{10.24a}
$$

which represent two sets of the simplest linear regressions, say for $j = 1, \ldots, J$, or *heterogeneous regressions*, but they have the same intercept parameters, namely β_{0j}, which can be considered a hidden assumption of the model (10.23a).

10.7.3 Alternative Models

Instead of the model in (10.23), a more general model with different intercept parameters is recommended, with the equation as follows:

$$
X_{ijt} = \beta_{1j} D_n \times X_{ijt-1} + \beta_{2j} D_p \times X_{ijt-1} + \delta_{1j} D_n + \delta_{2j} D_p + \varepsilon_{ijt}
\tag{10.24}
$$

which represents a pair of the simplest LV(1) models as follows:

$$
\begin{aligned}
X_{ijt} &= \delta_{1j} + \beta_{1j} X_{ijt-1} + \varepsilon_{ijt}, \quad for \ (D_n = 1, D_p = 0) \\
X_{ijt} &= \delta_{2j} + \beta_{2j} X_{ijt-1} + \varepsilon_{ijt}, \quad for \ (D_n = 0, D_p = 1)
\end{aligned}
\tag{10.24b}
$$

where the intercept parameters are $\delta_{1j} \neq \delta_{2j}$ in general, at least for one j. However, if the null hypothesis $H_0 : \delta_{1j} = \delta_{2j}, \forall j$ is accepted, then this model could be reduced to the model in (10.23) in a statistical sense.

For each j, then data analysis also can be easily done using the following equation specification in EViews.

$$
X \ Dn^* L1X \ Dp^* L1X \ Dn \ Dp
\tag{10.25}
$$

where *L1X* indicates a variable of the first-order lag of X, which should be generated based on pool data, especially for the unbalanced panel data set. Refer to Section 8.7.1. The lagged variable models based on balanced or complete pool data will be presented in Part III.

Since *Dn* and *Dp* are the dummies of a dichotomous variable, then the analysis also can be done using either one of the dummies, with the following ESs for each firm j.

$$
X \ C \ Dn \ L1X \ Dn^* L1X \ or \ X \ C \ Dp \ L1X \ Dp^* L1X
\tag{10.26}
$$

However, by generating the group of j's firm, namely Gj, the set of all regressions can easily be obtained using the following equation specification (ES), or other alternative ESs, as presented in Chapters 7 and 8.

$$X \ L1X^* @Expand(Dn, Gj) \ @Expand(Dn, Gj) \tag{10.27}$$

10.7.4 Extended Models

The effect of $L1X = X_{ijt-1}$ on X_{ijt} can change with the time, so the model (10.27) can be extended to the model using time or time-period dummy variables. Refer to the continuous regressions by groups and times presented in previous chapters. The data analysis can be done using the following ES, or its alternatives as presented in Chapters 7 and 8, where T is the time-discrete variable or a defined time-period variable.

$$X \ L1X^* @Expand(Dn, Gj, T) \ @Expand(Dn, Gj, T) \tag{10.28}$$

Furthermore, if the time observations are sufficiently large, then the model (10.28) could be modified to the growth models or models with trend, as well as the time-related effect models (TRE_Models), as presented in Agung (2009a, Chapter 2). The data analysis can be done using the equation specifications, as follows:

- An ES for the model with trend:

$$X \ L1X^* @Expand(Dn, Gj) \ T^* @Expand(Dn, Gj) \ @Expand(Dn, Gj) \tag{10.29}$$

- An ES for the TRE model:

$$\begin{aligned} X \ L1X^* @Expand(Dn, Gj) \ T^* @Expand(Dn, Gj) \\ T^* L1X^* @Expand(Dn, Gj) \ @Expand(Dn, Gj) \end{aligned} \tag{10.30}$$

Example 10.11 Inappropriate linear relationship between X_{it} and X_{it-1}

The linear relationship between X_{it} and X_{it-1} in these models can be debatable let alone for the linear relationship between Y_{it} and X_{it-1}, in general. For illustration, Figure 10.23 presents scatter graphs with regression lines between three pairs of variables: *ROA* on *ROA(−1)*, *SALE* on *SALE(−1)* and *PBV* on *PBV(−1)*, based on the Data_Sesi.wf1. These graphs clearly show that the linear relationships between the three pairs of variables are inappropriate. In this case, the main cause is the far outliers. For the notes on outliers, refer to Section 8.4.2.2.

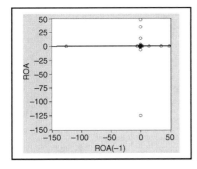 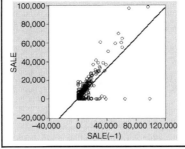 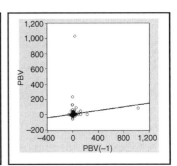

Figure 10.23 *Scatter graphs of* X_{it} *on* X_{it-1} *for three selected variables in Data Sesi-wf1*

obs	YEAR	T	FIRM_CODE	Whole sample		Sample {T > 1}	
				ROA	ROA(-1)	ROA	ROA(-1)
1	2002	1	_AALI	0.066989			
2	2003	2	_AALI	0.071564	0.066989	0.071564	0.066989
3	2004	3	_AALI	0.190054	0.071564	0.190054	0.071564
4	2005	4	_AALI	0.247494	0.190054	0.247494	0.190054
5	2006	5	_AALI	0.225050	0.247494	0.225050	0.247494
6	2007	6	_AALI	0.368578	0.225050	0.368578	0.225050
7	2008	7	_AALI	0.403528	0.368578	0.403528	0.368578
8	2009	8	_AALI	0.219390	0.403528	0.219390	0.403528
9	2002	1	_ABBA	-0.042857	0.219390		
10	2003	2	_ABBA	-0.071429	-0.042857	-0.071429	-0.042857
11	2004	3	_ABBA	-0.224490	-0.071429	-0.224490	-0.071429
12	2005	4	_ABBA	-0.067227	-0.224490	-0.067227	-0.224490
13	2006	5	_ABBA	0.006667	-0.067227	0.006667	-0.067227
14	2007	6	_ABBA	0.018868	0.006667	0.018868	0.006667
15	2008	7	_ABBA	0.008850	0.018868	0.008850	0.018868

	ROA	ROA(-1)
Mean	0.010898	0.010952
Median	0.024390	0.024390
Maximum	47.94118	47.94118
Minimum	-125.5000	-125.5000
Std. Dev.	2.922677	2.923316
Skewness	-32.01379	-32.00687
Kurtosis	1530.561	1529.896
Jarque-Bera	2.22E+08	2.22E+08
Probability	0.000000	0.000000
Sum	24.89178	25.00446
Sum Sq. Dev.	19501.47	19501.46
Observations	2284	2283

Figure 10.24   *Descriptive statistics of ROA and ROA(−1), based on the whole sample and {T > 1}*

Furthermore, Figure 10.24 shows that the regression model of *ROA* and *ROA(−1)* based on a sample of size 2283 (the whole sample minus one) is wrong, since many pairs of the observed scores belong to different firms. For instance, Figure 10.24 shows that the ABBAs score of −0.042 857 in 2002 is mismatched compared to the AALIs score of 0.219 390 in 2009. Based on this finding, it can be said that the application of lagged variable models based on pool panel data should be done with care, because all models with a dependent variable Y_{it} and independent variables X_{it-1} for all *X* will be misleading or wrong.

However, if the data analysis is done based on a sub-sample, say for the years 2003–2008 or the sample {T > 1}, and the data is a balanced pool data set, then all LV(1) models would be valid models in a theoretical sense. See the data of ROA and ROA(−1) in the sample {T > 1} presented in Figure 10.22. The applications of LV(*p*) models, for $p \geq 1$, will be presented in more detail in Part III.

10.7.5   Sets of the Simplest LV(*p*) Heterogeneous Regressions

Skinner (2008) presents three sets of the simplest LV(*p*) heterogeneous regressions where each set of the simplest regression results can be presented in a general form as follows:

$$Earnings_{t+p} = \alpha_p + \beta_p Earnings_t, \quad for \quad p = 1, 2, 3, 4, and\ 5 \tag{10.31}$$

The three sets of regressions considered are; Panel A: Regression results for the whole sample; Panel B: Regression results for the highest earnings volatility quintile; and Panel C: Regression results for the lowest earnings volatility quintile.

10.8   Models without the Time-Independent Variable

10.8.1   Interaction Models

If a model has at least two independent (cause, up-stream or exogenous) variables, I would say the effect of a certain variable on the dependent (impact, down-stream or endogenous) variable should depend on at least one of the other independent variables, even more so if the model has a dummy independent variable. Refer to the simplest interaction models with a dummy independent variable presented by Kousenidis, Ladas and Negakis (2009) in (10.1), and in (10.5) presented by Dedman and Lennox (2009).

Siswantoro and Agung (2010) discuss this further in their paper, "The Importance of the Effect of Exogenous Interaction Factors on Endogenous Variables in Accounting Models" mentioned in the introduction. Many papers also present interaction models, such as follows.

Francis and Martin (2010) present an interaction model with 23 parameters, as in (7.63), Hameed, et al. (2010) present an interaction model with a lot more parameters as in (7.62), and Duh, et al. (2009), also present an interaction model without the time-independent variable. For a detailed discussion of an interaction model without the time-independent variable, see the following example.

Example 10.12   Jermias' model for detailed discussion
Jemias (2008) presents an interaction model as follows:

$$PERFOM_{it} = \gamma_0 + \gamma_1 STRA_{it} + \gamma_2 INT_{it} + \gamma_3 LEV_{it} + \gamma_4 STRA_{it} \times LEV_{it} + \gamma_5 INT_{it} \times LEV_{it}$$
$$+ \gamma_6 SIZE_{it} + \gamma_7 INSTWON_{it} + \gamma_8 DIV_{it} + \varepsilon_{it} \qquad (10.32)$$

where $STRA_{it}$ is an indicator equal to 1 for cost leadership firms and 0 for product differentiation firms. The definitions of the other variables are not presented, since they are not important in the discussion and could be replaced by other sets of numerical variables to present models with a single dummy. Then the following notes, comments and possible modified models are presented.

1. The dummy variable *STRA* should divide the firm-time observations into two groups of firms having different characteristics as well as behavior. I assume the two groups of the firms are invariant or constant over time.
2. In a theoretical sense, I would say that each numerical independent variable should have different effects on the dependent variable between the two groups of firms. However, this model assumes that the only numerical variable *LEV* has different effects between the two groups.
3. The interaction $INT \times LEV$ indicates that the effect of the numerical variable *INT* (or *LEV*) on *PERFORM* depends on *LEV* (or *INT*).
4. Furthermore, the two interactions $ASTR \times LEV \& INT \times LEV$ indicate that the effect of *LEV* on *PERFORM* depends on both *STRA* and *INT*, adjusted for the other variables in the model.
5. The up- and down-stream relationships of the four variables *STRA*, *INT*, *LEV* and *PERFORM* would be presented by the path diagram in Figure 10.25. Take note that the broken lines from *STRA* to the other

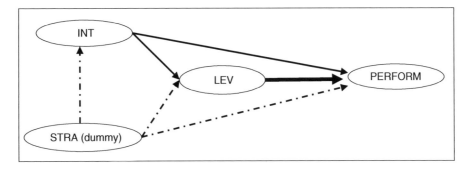

Figure 10.25 *Theoretical up- and down-stream relationships between the four variables*

three variables indicate that STRA is not a cause factor, but a classification factor, and *INT* is defined as an up-stream or causal factor of *LEV*.

6. Referring to point (3), the interaction *STRA* × *INT* × *LEV* should be considered to be an additional independent variable. Therefore, a three-way interaction model would be obtained having *STRA* × *INT* × *LEV* as an additional independent variable of Jemias' model.

7. A more advanced recommended model has all interactions between *STRA* and all numerical independent variables. Refer to illustrative three-way interaction models presented in previous chapters.

Example 10.13   Shoute's interaction model

Compared to Jemias' model, Shoute (2008) presents four interaction models with 18 numerical independent variables, but with two interactions only. The characteristics of the models are as follows:

1. The main objectives of the models are to study the effects of cost system complexity (*CS_COMP*) on cost system intensity (*CS_INTENS*), if this depends on a cost system for product planning purposes (*PRPLAN_USE*) and a cost system for cost management purposes (*CSTMAN_USE*).

2. The models have only two interaction-independent variables, namely *CS_COMP* × *PRPLAN_USE* and *CS_COMP* × *CSTMAN_USE*, in addition to their main factors and 15 other numerical variables. These interactions indicate that the effect of *CS_COMP* on *CS_INTENS* depends on *PRPLAN_USE* and *CSTMAN_USE*, adjusted for the other variables in the model, similar to Jemias' model.

3. I would present the up- and down-stream or causal relationships between the four variables, *CS_COMP*, *PRPLAN_USE*, *CSTMAN_USE* and *CS_INTENS*, as the path diagram in Figure 10.26, which is not the same as the one by Schoute. This is because I define *PRPLAN_USE* and *CSTMAN_USE* as up-stream or causal factors of *CS_COMP*, *PRPLAN_USE* and *CSTMAN_USE*, and these should be correlated in a theoretical sense.

10.8.2   Additive Models

On the other hand, Li, et al. (2010) present an additive model without the time-independent variable as presented in (7.61) with its special notes and comments, and Shah, et al. (2009) also present the model without the time-independent variable.

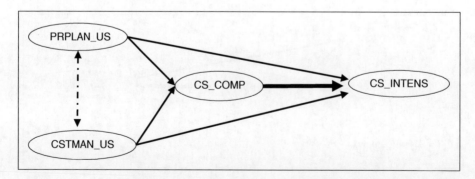

Figure 10.26   *Theoretical up- and down-stream relationships between the four variables*

10.9 Models with Time Dummy Variables

10.9.1 Interaction Models

Based on my point of view, if the time dummy variables should be used in models, then interactions between the time dummy variables with selected or specific independent variables should be used as independent variables, since the effects of these variables and their interactions on the dependent variable(s) should change over time. Refer to all interaction models with time variables presented in previous chapters. In addition see Skinner (2008) who presents two sets of regressions for each of the fiscal years 1999–2003.

10.9.2 Models with Additive Time Dummy Variables

Regarding the model in (7.62), Wysocki (2010), Feng, et al. (2009) and Zhang (2008) present models which have additive time dummy variables. Take note that such models can be considered as time-fixed-effect models with the general equations presented in (9.10) and (9.11), or one-way ANCOVA models with a time factor. So they use the assumption that all other independent variables have the same slopes at all the time points considered, or they have the same effect on the independent variable over time. See the special notes on alternative ANCOVA models in Chapters 7 and 8.

10.9.3 Models with Additive Time and Group Dummy Variables

Aboody, et al. (2010) present models with additive time and group (industry) dummy variables, which can be considered group-time fixed-effects models with the general equation presented in (9.13). This type of model has been previously mentioned as a not-recommended ANCOVA model. See the special notes on the limitation of this type of ANCOVA model and the outputs in Figure 8.24.

Similar to the model in Figure 8.24, Al-Shammari, et al. (2008) present an additive binary logit model with 15 independent variables as follows:

$$\log\left(\frac{p}{1-p}\right) = \beta_0 + \sum \beta_i X_i + \varepsilon \qquad (10.33)$$

where X_1 to X_5 are region dummy variables, X_6 to X_9 are year dummy variables, X_{15} is an industry dummy variable and five numerical independent variables. On the other hand, Brau and Johnson (2009) present an additive model with 18 year dummy variables from 1987–2005 and a VC zero-one indicator.

Markarian, et al. (2004) present the uncommon equation of an additive model as follows:

$$\begin{aligned}
Capitalization_i = {} & b_0 + b_1 ChROA_i + b_2 Leverage_i + b_3 R\mathcal{G}DTotal_i + b_4 ROA_i \\
& + b_5 Logassets_i + b_6 MB_i + b_7 Beta_i + b_8 HighCapitalizer_i \\
& + b_9 LagCapitalization_i + Industry\ Controls \\
& + Year\ Dummies + u_{it}
\end{aligned} \qquad (10.34)$$

Take note that the independent variables *Industry Controls* and *Year Dummies* do not have parameters. I would say that both variables in fact present two sets of dummies with their corresponding parameters. So this model is a group-time fixed-effects model.

10.9.4   Models with Time as a Numerical Independent Variable

Out of the 268 models observed in four journals, namely the *International Journal of Accounting* (IJA), *Journal of Accounting and Economics* (JAE), *British Accounting Review* (BAR) and *Advances in Accounting*, incorporating *Advances in International Accounting* (AA) in 2008, 2009 and 2010 (Siswantoro and Agung, 2010), only one model has a numerical time independent variable presented by Ferris and Wallace (2009). They present a pair of additive models with the equations as follows:

$$\begin{aligned}
Volatility_{it} = {} & \beta_0 + \beta_1 Sales_{it} + \beta_2 ROA_{it} + \beta_3 Return + \beta_4 MkBk_{it} \\
& + \beta_5 Industry_{it} + \beta_6 Year_{it} + \beta_7 Optionsheld_{it} + \varepsilon_{it} \\
Yield_{it} = {} & \beta_0 + \beta_1 Sales_{it} + \beta_2 ROA_{it} + \beta_3 Return + \beta_4 MkBk_{it} \\
& + \beta_5 Industry_{it} + \beta_6 Year_{it} + \beta_7 Optionsheld_{it} + \varepsilon_{it}
\end{aligned} \tag{10.35}$$

For these models, the following notes and comments are presented.

1. The variable *Year* in each model has a single parameter β_6. So the variable *Year* should be used as a numerical independent variable.
2. However, the variable *Industry* also has a single parameter β_5. For the *Industry*, I would say that it should be a dummy industry variable. Otherwise, the models are uncommon or inappropriate models with a single parameter for the independent variable *Industry*.
3. If the data analysis would be done using a system equation or a bivariate GLM in EViews, the following pair of equation specifications can be applied.

$$\begin{aligned}
Volatility = {} & C(10) + C(11)^* Sales + C(12)^* ROA + C(13)^* Return \\
& + C(14)^* MkBk + C(15)^* Industry + C(16)^* Year + C(17)^* Optionsheld \\
Yield = {} & C(20) + C(21)^* Sales + C(22)^* ROA + C(23)^* Return \\
& + C(24)^* MkBk + C(25)^* Industry + C(26)^* Year + C(27)^* Optionsheld
\end{aligned} \tag{10.36}$$

4. As a modification, a group of industries may be considered. In this case then the term $\beta_5 Industry$ will be replaced by $\sum_g \beta_{5g} DIn_{g.it}$ where $DIn_{g.it}$ is a dummy variable of the group g. Then the models would be the group-fixed-effects models with trend (Year).

10.10   Final Remarks

10.10.1   Remarks on the Interaction or Additive Models

If a model has two independent (cause, up-stream, or source) variables as well as predictors, then applying interaction models is recommended, since the effect of one variable on a dependent variable should depend on the other independent variable, in a theoretical sense. Refer to the two-way ANOVA model, the simplest model of Kousenidis, et al. (2009) in (10.1), and the model of Dedman and Lennox (2009) in (10.23).

If a model has a large number of exogenous variables, then researchers should use their best judgment to select a limited number of interactions, which are important to consider in a theoretical sense. Refer to the summary of statistical results in Table 8.8, and the outputs in Figure 8.23, which show that the effect of *X1* on the dependent variable depends on other defined variables as well as the interaction models mentioned previously, which also have selective interaction independent variables.

10.10.2 Remarks on Linear or Nonlinear Effects Models

Various linear and nonlinear effects models have been presented in Chapters 7 and 8. However, out of the 268 papers observed in the four journals (Siswantoro and Agung, 2010), there is no paper that presents nonlinear effect models, specifically the polynomial, hyperbolic and exponential effects models. And no one considers scatter graphs of a dependent variable on each of the numerical exogenous variables with regression lines, let alone with kernel fit or nearest neighbor fit curves, to try and identify possible nonlinear relationships between pairs of the numerical variables. Refer to the scatter graphs presented in Chapters 7 and 8.

10.10.3 Remarks on the True Population Model

It has been said that we will never know the true population model (Agung, 2009a, Section 2.14). So it can also be said that all estimated models or statistical results would not represent true population models but they are subjectively defined or reduced models, which happen to fit the data set available or selected by researchers. On the other hand, even though a model has been well-defined in a theoretical sense, an error message might still be obtained.

For this reason, presenting the characteristics and limitations of each of the presented models and their possible modified or reduced versions is recommended. Most students, perhaps all, never present the limitations or hidden assumptions of a model in their paper. On the other hand, they tend to derive conclusions on various testing hypotheses, including testing the basic assumptions of error terms, which should not be tested – refer to Agung (2009a, Section 2.14.3; and 2011, Chapter 1). Why? Because students are never exposed to the detailed characteristics of a model or its limitations compared to others.

11

Seemingly Causal Models

11.1 Introduction

As the extension of the univariate general linear models presented in previous chapter, this chapter presents the application of the object "*System*". The models applied are called *system equation models* (SEMs) in EViews, or Seemingly Causal Models (SCMs). Since "SEM" has been previously used for the structural equation model, I prefer to use the abbreviation "SCM". The term SCM has been used in Agung (2009a, and 2011) for time series and cross-section data sets. Take note that a SCM in fact is a set of univariate general linear models (GLMs) or multiple regressions, such as a set of ANOVA models, a set of heterogeneous models and a set of ANCOVA models. So all GLMs presented in Chapter 8 can be used as one of the models in a SCM. However, for data analysis, the equation specification of the SCM should be written using dummy variables, instead of using the function @Expand(*).

For this reason, only selected empirical statistical results using SCMs are presented in this chapter, the other models can easily be developed using sets of the same models presented in Chapter 8, such as the linear, logarithmic, hyperbolic, polynomial and nonlinear effects models. Furthermore, readers should also consider the limitations of each GLM, specifically the illustrative examples with notes and comments presented in Section 8.7.

11.2 MANOVA Models

The MANOVA model of a multivariate response variable $\mathbf{Y} = (Y_1, \ldots, Y_G)$ by a cell-factor CF having the levels $s = 1, \ldots, S$, and the time points $t = 1, \ldots, T$, can be presented using the system equation as follows:

$$Y_g = \sum_{t=1}^{T}\sum_{s=1}^{S} C(gst) \times Dst,$$

$$g = 1, 2, \ldots, G$$

(11.1a)

where Dst is the dummy variable or the zero-one indicator of the cell (CF $= s$, Time $= t$), and the parameter $C(gst)$ is the mean parameter of Y_g in the sub-population indicated by (CF $= s$, Time $= t$).

Panel Data Analysis Using EViews, First Edition. I Gusti Ngurah Agung.
© 2014 John Wiley & Sons, Ltd. Published 2014 by John Wiley & Sons, Ltd.
Companion website: www.wiley.com/go/panel_data

Table 11.1 The mean parameters of a cell-mean model in (11.1)

		CF	T = 1	T = 2	T = 3	T = 4
A = 1	B = 1	1	C(g11)	C(g12)	C(g13)	C(g14)
A = 1	B = 2	2	C(g21)	C(g22)	C(g23)	C(g24)
A = 2	B = 1	3	C(g31)	C(g32)	C(g33)	C(g34)
A = 2	B = 2	4	C(g41)	C(g42)	C(g43)	C(g44)

For illustration, Table 11.1 presents the mean parameters of a $2 \times 2 \times 4$ factorial cell-mean model (CMM). Compare this to the mean parameters of an ANOVA model presented in Table 8.1. All hypotheses on means differences between the cells (CF = s, Time = t) = (A = i, B = j, Time = t) or the conditional effects on the factors A, B, and CF, as well as the time T on Yg, can easily be tested using the Wald test. The hypotheses are as follows:

1. The effect of CF or both factors A and B on Yg, conditional for $T = 1$, with the null hypothesis H_0: $C(g11) = C(g21) = C(g31) = C(g41)$, for each g, or univariate hypothesis. For a multivariate hypothesis, namely for $g = 1$ and 2, the null hypothesis is H_0: $C(111) = C(121) = C(131) = C(141)$, $C(211) = C(221) = C(231) = C(241)$. Take note that the cell-factor CF can be generated based on a numerical cause factor, for instance, using percentiles, so the hypothesis represents the effect of the numerical cause factor on Yg, conditional for $T = 1$. So it can be said that the cell-mean model also can be used to study the effects of numerical exogenous variables on the endogenous variables.
2. The effect of the factor A on Yg, conditional for $B = 1$ and $T = 1$, with the null hypothesis H_0: $C(g11) = C(g21)$. Take note that the factor A could be generated based on a numerical causal factor with several levels, depending on the sample size.
3. The effect of the time T on Yg, conditional for $CF = 1$ ($A = 1$ and $B = 1$), with the null hypothesis H_0: $C(g11) = C(g12) = C(g21) = C(g22)$.
4. Finally, the effect of CF and T on Yg, with the null hypothesis H_0: $C(g11) = C(g12) = C(13) = C(14) = \ldots = C(g44)$. However, for this type of testing hypothesis, an alternative model with an intercept parameter, namely $C(g1)$ or using a reference cell, can be applied such as in the following system equation for $S = T = 4$.

$$Y_g = C(g1) + \sum_{t=1}^{T-1}\sum_{s=1}^{S-1} C(gst) \times Dst,$$

$$g = 1, 2, \ldots, G$$

(11.1b)

11.3 Multivariate Heterogeneous Regressions by Group and Time

Referring to the MANOVA model in (11.1), the equation specification of a multivariate heterogeneous regressions of Yg, $g = 1, \ldots, G$ on a set of numerical variables Xk, $k = 1, \ldots, K$, by CF and the time variable can be presented as follows:

$$H_g(Yg) = \sum_{t=1}^{T}\sum_{s=1}^{S} F_{gst}(X1, \ldots, Xk, \ldots, XK) \times Dst,$$

$$g = 1, 2, \ldots, G$$

(11.2)

where *Dst* is a dummy variable of the cell *(CF = s, Time = t)*, $F_{gst}(X1, \ldots, Xk, \ldots XK) = F_{gst}(X)$ is a function of *X* with a finite number of parameters for each *g*, *s* and *t*, and $H_g(Yg)$ is a fixed function (without a parameter) of *Yg*, such as $H_g(Yg) = Yg$, *log(Yg − Lgh)*, *log(Ugh-Yg)* and *log[(Yg − Lgh)/(Ugh − Yg)]*, where *Lgh* and *Ugh* are fixed lower and upper bounds of *Yg*, which should be selected for each group *h*. Refer to the bounded GLM in Agung (2011, Chapter 9, and 2006).

Take note that by taking into account the up- and down-stream causal relationships between some of the endogenous variables *Yg*, then the function $F_{gst}(X)$ should be the function of selected endogenous variables as well, namely $F_{gst}(X, Yj)$, for some $j \neq g$.

11.3.1 The Simplest MHR with an Exogenous Numerical Variable *X*

Referring to model in (11.2), the simplest MHR of *Yg* on a single numerical exogenous variable *X*, for $S = T = 2$, by CF and the time variable, will have the following general equation specification.

$$H_g(Yg) = F_{g11}(X) \times D11 + F_{g12}(X) \times D12 + F_{g21}(X) \times D21 + F_{g22}(X) \times D22$$

$$g = 1, \ldots, G$$

(11.3a)

Or, it can be presented as

$$H_g(Yg) = \sum_{h=1}^{4} F_{gh}(X) \times Dh$$

$$g = 1, \ldots, G$$

(11.3b)

where $F_{gh}(X)$ can be any function of a numerical variable *X* with a finite number of parameters, for each *g* and *h = 1,2,3* and *4*. Therefore the function $F_{gh}(X)$ can be presented as $F_{gh}(X, C(gh*))$. Refer to possible functions presented in (6.20a to (6.20f)) and in Agung (2011, Chapter 7). Then some of the alternative functions are as follows:

$$\text{The simplest function :} \quad F_{gh}(X) = C(gh1) + C(gh2)^* X \tag{11.4a}$$

$$\text{Logarithmic function :} \quad F_{gh}(X) = C(gh1) + C(gh2)^* log(X) \tag{11.4b}$$

$$\text{Hyperbolic function :} \quad F_{gh}(X) = C(gh1) + C(gh2)^* (1/X) \tag{11.4c}$$

$$\text{Exponential function :} \quad F_{gh}(X) = C(gh1) + C(gh2)^* X^{\alpha(gh)}, \alpha(gh) \neq 0$$
$$= C(gh1) + C(gh2)^* Exp(\alpha(gk)^* log(X)) \tag{11.4d}$$

$$\text{Polynomial function :} \quad F_{gh}(X) = C(gh1) + C(gh2)^* X + \ldots + C(ghn)^* X^n \tag{11.4e}$$

$$\text{Nonlinear function :} \quad F_{gh}(X) = C(gh1) + C(gh2)^* X^{C(gh3)} \tag{11.4f}$$

Note that the four $F_{g1}(X)$, $F_{g2}(X)$, $F_{g3}(X)$ and $F_{g4}(X)$ can be different types of functions. So there are a lot of possible choices, but we never know the true population model – refer to Agung (2009a, Sction 2.14). For this reason, in order to define a model, a researcher should use his/her best judgment supported by knowledge and experience in various related fields. The estimates obtained are highly dependent on the data. Refer to unexpected estimates or statistical results based on various models, which have been presented as illustrative examples in previous chapters and Agung (2011, and 2009a).

11.3.2   MHR with Two Exogenous Numerical Variables

As an extension of the MHR model in (8.16b), the MHR of Yg on two numerical exogenous variables X and Z, for $S = T = 2$, by CF and the time t, will have the following general equation specification.

$$H_g(Yg) = F_{g1}(X,Z) \times D11 + F_{g1}(X,Z) \times D12 + F_{g1}(X,Z) \times D21) + F_{g1}(X,Z) \times D22$$
$$g = 1, \ldots, G$$

(11.5)

where $F_{gh}(X,Z)$ can be any function of the numerical variable (X,Z) with a finite number of parameters, for each g and h.

In many or most cases, if we have two or more exogenous variables then the effect of an exogenous variable on the endogenous depends on at least one of the other exogenous variables. For this reason, the proposed or defined model should have at least a two-way interaction independent variable. Note that the dummy variable Dst in fact is a two-way interaction of the dummy variables $(CF = s)$ and $(time = t)$. Therefore, without loss of generality, it is assumed that the linear effect of Z on Y depends on a function of X, then the function $F_{gh}(X,Z)$ in (11.5) can be presented in general as follows:

$$F_{gh}(X,Z) = F_{gh1}(X) + Z \times F_{gh2}(X)$$

(11.6)

which indicates that the model in (11.5) will show the linear effect of Z on $H_g(Yg)$ depends on $F_{gh2}(X)$, where $F_{gh1}(X)$ and $F_{gh2}(X)$ can be any functions of the exogenous variable X with a finite number of parameters for each g and h. On the other hand, it also can be said that the effect of $F_{gh2}(X)$ on $H_g(Yg)$ depends on the variable Z.

11.3.2.1   Simple Alternative Models

The simplest possible model of all models represented by the function (11.6) uses the function as follows:

$$F_{gh}(X,Z) = [C(gh1) + C(gh2) \times X] + [C(gh3) + C(gh4) \times X] \times Z$$

(11.7a)

which indicates that the linear effect of Z on Y depends on $[C(gh3) + C(gh4) \times X]$. Note that since Dst is an interaction dummy variables of $(CF = s)$ and $(Time = t)$, then the UGLMs have four-way interaction independent variables, namely $X \times Z \times Dst$.

The parameters of the model in (11.7a) can be presented as in Table 11.2.

In practice, however, the variable Z might have other types of effects on Y, where some of those are as follows:

1.   A logarithmic effect of $Z > 0$, by using the function:

$$F_{gh}(X,Z) = [C(gh1) + C(gh2) \times X] + [C(gh3) + C(gh4) \times X] \times log(Z)$$

(11.7b)

Table 11.2   *The slope parameters of the model in (11.7a)*

	CF	Time	Intercept	X	Z	X*Z
. . .	. . .	. . .	. . .	. . .	. . .	. . .	. . .
$A = i$	$B = j$	s	t	$C(gst1) = C(gh1)$	$C(gst2) = C(gh2)$	$C(gst3) = C(gh3)$	$C(gst4) = C(gh4)$
. . .	. . .	. . .	. . .	. . .	. . .	. . .	. . .

2. A hyperbolic effect of Z, by using the function:

$$F_{gh}(X,Z) = [C(gh1) + C(gh2) \times X] + [C(gh3) + C(gh4) \times X] \times (1/Z) \tag{11.7c}$$

3. An exponential effect of Z, by using the function:

$$F_{gh}(X,Z) = [C(gh1) + C(gh2) \times X] + [C(gh3) + C(gh4) \times X] \times Z^{a(gh)} \tag{11.7d}$$

4. Another form of exponential effect:

$$F_{gh}(X,Z) = [C(gh1) + C(gh2) \times X] + [C(gh3) + C(gh4) \times X] \times Exp(Z) \tag{11.7e}$$

5. A nonlinear effect of Z, by using the function:

$$F_{gh}(X,Z) = [C(gh1) + C(gh2) \times X] + [C(gh3) + C(gh4) \times X] \times Z^{C(gh)} \tag{11.7f}$$

11.3.2.2 More Advanced Models

Referring to the functions in (11.4a to (11.4f)), then similar functions of $F_{gh1}(X)$ and $F_{gh2}(X)$ can easily be used to present more advanced models, such as the extension of the UGLM using the functions in (11.7a) to (11.7f).

Some of the advanced models can easily be obtained by using the transformed variable of X, such as $log(X)$, $1/X$, $X^{\delta(gh)}$, $X^{C(ghi)}$, for $i = 1, 2$; and $Exp(X)$ in (11.7a) to (11.7f). For instance, we may have a nonlinear MGLM by using the following function – refer to Agung (2009a, Chapter 11).

$$F_{gh}(X,Z) = [C(gh1) + C(gh2) \times X^{C(gh1)}] + [C(gh3) + C(gh4) \times X^{C(gh1)}] \times Z^{C(gh)} \tag{11.8}$$

Other advanced models could be obtained for three or more numerical exogenous variables as the extension of these illustrative UGLMs.

11.4 MANCOVA Models

Under the assumption that all numerical exogenous variables of the model in (11.3) have the same effects on each of the endogenous variable within each cell generated by *CF* and the time *t*, then a MANCOVA model is obtained with the following general equation specification.

$$H_g(Yg) = \sum_{t=1}^{T}\sum_{s=1}^{S} C(gst) \times Dst + F_{gst}(X1,\ldots,Xk,\ldots), \tag{11.9}$$

$$g = 1, 2, \ldots, G$$

For illustration, Table 11.3 presents the intercept parameters of a $2 \times 2 \times 4$ factorial MANCOVA model. Compare this to the mean parameters of the cell-mean model presented Table 11.1. All hypotheses on the adjusted means differences between the cells (CF = s, Time = t) = (A = i, B = j, Time = t), can easily be tested using the Wald test.

However, a MANCOVA model should be used with care, because the numerical independent variable(s) should have different effects on each $H_g(Yg)$ between groups of individuals as well as over times. Refer to

Table 11.3 *The intercept parameters of a MANCOVA model in (11.9)*

		CF	T=1	T=2	T=3	T=4
A=1	B=1	1	C(g11)	C(g12)	C(g13)	C(g14)
A=1	B=2	2	C(g21)	C(g22)	C(g23)	C(g24)
A=2	B=1	3	C(g31)	C(g32)	C(g33)	C(g34)
A=2	B=2	4	C(g41)	C(g42)	C(g43)	C(g44)

the special notes and comments presented in Section 8.7.3. I would say that the MANCOVA model is not recommended.

Take note that the MANCOVA model in fact is a multivariate fixed-effects model. Refer to the fixed-effects models and alternatives presented in Chapter 9.

11.5 Discontinuous and Continuous MGLM by Time

Note that for all models presented previously if the cell-factor *CF* is generated or defined based on one or two numerical exogenous variables *Xk*s, a set of piece-wise regression functions over times would be obtained, that is, polygon or step regression functions.

On the other hand, if only the dummy variables of time-points or time-periods are applied then a set of continuous regression functions would be obtained based on a MGLM by time, with the following general equation specification.

$$H_g(Yg) = \sum_{t=1}^{T} F_{gt}(X1, \ldots, Xk, \ldots, XK) \times Dt$$

$$g = 1, 2, \ldots, G$$

(11.10)

which leads to a set of $G \times T$ continuous regression functions with the equations as follows:

$$\hat{H}_g(Yg) = F_{gt}(X1, \ldots, Xk, \ldots, XK)$$

$$g = 1, 2, \ldots, G; t = 1, \ldots, T$$

(11.11)

11.6 Illustrative Linear-Effect Models by Times

This section presents illustrative models with at least three numerical exogenous (up-stream, cause or independent) variables, beside the categorical exogenous variables. Figure 11.1 presents up- and down-stream relationships between a set of numerical exogenous variables and any number of numerical endogenous

Figure 11.1 *An illustrative up- and down-stream relationship*

variable(s) as a modification of the path diagram in Figure 7.6. Take note that any variable could be used as a predicted variable, say the dependent variable *Yg*, and the other observed variables can be used as the predictors, say the X-variables. So there would be a lot of possible up- and down-stream models or seemingly causal models, even for G = 2 based on the diagram in Figure 11.1. See the illustrative examples that follow.

11.6.1   Multivariate Linear-Effect Models (LEMs) by Time

11.6.1.1   *Application of a Bivariate LEMs*

As an extension of the univariate LEM by times presented in Figure 8.22, for illustration the following examples present data analyses based on the models of a bivariate exogenous variable (Y1,Y2) and a trivariate exogenous variable (X1,X2,X3).

Example 11.1   A bivariate LEM

For illustration, Figure 11.2 presents the statistical results of a bivariate LEM by times of (Y1,Y2) on one of many possible functions F_{gst}(X1,X2,X3). Based on these results the following findings and notes are presented.

System: SYS05_5JULI
Estimation Method: Iterative Least Squares
Date: 07/04/11 Time: 10:30
Sample: 1 2284 IF T>5
Included observations: 1139
Total system (balanced) observations 1762
Convergence achieved after 8 iterations

	Coefficient	Std. Error	t-Statistic	Prob.
C(10)	-2.860711	0.519046	-5.511482	0.0000
C(11)	0.555464	0.081552	6.811150	0.0000
C(12)	-1.000309	0.306180	-3.267060	0.0011
C(13)	-3.819918	1.963235	-1.945726	0.0519
C(14)	0.212100	0.053859	3.938048	0.0001
C(15)	0.384885	0.185197	2.078250	0.0378
C(16)	-0.074512	2.089696	-0.035657	0.9716
C(17)	0.024349	0.211576	0.115083	0.9084
C(20)	-5.669315	0.582383	-9.734682	0.0000
C(21)	1.010702	0.086838	11.63895	0.0000
C(22)	-1.987564	0.347026	-5.727415	0.0000
C(23)	-6.075283	1.331416	-4.563023	0.0000
C(24)	0.393150	0.058556	6.714056	0.0000
C(25)	0.653796	0.137068	4.769871	0.0000
C(26)	-6.026952	2.360704	-2.553031	0.0108
C(27)	0.715729	0.252345	2.836315	0.0046
C(30)	-5.627931	0.619037	-9.091430	0.0000
C(31)	0.997922	0.093329	10.69254	0.0000
C(32)	-1.714058	0.349245	-4.907892	0.0000
C(33)	-2.558124	1.707811	-1.497896	0.1343
C(34)	0.397178	0.058964	6.735997	0.0000
C(35)	0.392664	0.163063	2.408059	0.0161
C(36)	-15.60280	2.191145	-7.120847	0.0000
C(37)	1.855709	0.236513	7.846135	0.0000
C(18)	0.725574	0.022335	32.48666	0.0000

C(40)	-0.075101	0.021415	-3.506898	0.0005
C(41)	0.014457	0.003437	4.206645	0.0000
C(42)	-0.010695	0.012898	-0.829188	0.4071
C(43)	0.351436	0.089704	3.917713	0.0001
C(44)	-0.001178	0.002268	-0.519496	0.6035
C(45)	-0.032056	0.008404	-3.814581	0.0001
C(46)	-0.235866	0.097239	-2.425633	0.0154
C(47)	0.022193	0.009835	2.256680	0.0242
C(50)	-0.082128	0.020875	-3.934343	0.0001
C(51)	0.017601	0.003189	5.518623	0.0000
C(52)	-0.002948	0.013483	-0.218645	0.8270
C(53)	0.402305	0.051197	7.858056	0.0000
C(54)	-0.004655	0.002291	-2.031760	0.0423
C(55)	-0.037642	0.005171	-7.279808	0.0000
C(56)	-0.250025	0.102863	-2.430667	0.0152
C(57)	0.025626	0.010963	2.337551	0.0195
C(60)	-0.062063	0.021482	-2.889016	0.0039
C(61)	0.013885	0.003337	4.161196	0.0000
C(62)	-0.023893	0.013737	-1.739357	0.0822
C(63)	0.352124	0.062675	5.618273	0.0000
C(64)	0.000396	0.002297	0.172581	0.8630
C(65)	-0.031490	0.005835	-5.396609	0.0000
C(66)	0.038404	0.084650	0.453680	0.6501
C(67)	-0.005036	0.009108	-0.552906	0.5804
C(28)	0.257012	0.033442	7.685314	0.0000

Equation: Y1 = (C(10)+C(11)*X1+C(12)*X2+C(13)*X3+C(14)*X1*X2+C(15)
*X1*X3+C(16)*X2*X3+C(17)*X1*X2*X3)*DT6+(C(20)+C(21)*X1+C(22)
*X2+C(23)*X3+C(24)*X1*X2+C(25)*X1*X3+C(26)*X2*X3+C(27)*X1*X2
*X3)*DT7+(C(30)+C(31)*X1+C(32)*X2+C(33)*X3+C(34)*X1*X2+C(35)
*X1*X3+C(36)*X2*X3+C(37)*X1*X2*X3)*DT8+[AR(1)=C(18)]

Observations: 881

R-squared	0.826931	Mean dependent var	2.180124
Adjusted R-squared	0.822078	S.D. dependent var	5.657296
S.E. of regression	2.386291	Sum squared resid	4874.395
Durbin-Watson stat	2.473893		

Equation: Y2 = (C(40)+C(41)*X1+C(42)*X2+C(43)*X3+C(44)*X1*X2+C(45)
*X1*X3+C(46)*X2*X3+C(47)*X1*X2*X3)*DT6+(C(50)+C(51)*X1+C(52)
*X2+C(53)*X3+C(54)*X1*X2+C(55)*X1*X3+C(56)*X2*X3+C(57)*X1*X2
*X3)*DT7+(C(60)+C(61)*X1+C(62)*X2+C(63)*X3+C(64)*X1*X2+C(65)
*X1*X3+C(66)*X2*X3+C(67)*X1*X2*X3)*DT8+[AR(1)=C(28)]

Observations: 881

R-squared	0.420086	Mean dependent var	0.039035
Adjusted R-squared	0.403826	S.D. dependent var	0.121113
S.E. of regression	0.093514	Sum squared resid	7.485585
Durbin-Watson stat	2.192576		

Figure 11.2   *Statistical results of a bivariate linear-effects model*

1. Take note on the symbols of the parameters used in both equation specifications, specifically the parameters of the terms $[AR(1) = C(18)]$ and $[AR(1) = C(28)]$, as presented in Figure 11.2. The equations can be written in a Word file, then use the block-copy-paste method after opening the object "*System*".

2. In the model, the up- and down-stream or causal relationship between Y1 and Y2 is not taken into account. Similarly for the variables X1, X2 and X3, but they are correlated, in a theoretical sense. Furthermore, it is defined that the effect of X1 on both Y1 and Y2 depends on X2 and X3.

3. The model is a bivariate $AR(1)$ model of (Y1,Y2) on a specific function out of a lot of possible functions F(X1,X2,X3), where each regression of Y1 and Y2 presents three continuous hierarchical three-way interaction regressions for the time points: $T = 6$, 7 and 8, as presented in Table 11.4. Refer to the alternative functions presented in Chapter 8.

4. Take note that some of the numerical variables in Table 11.4 have very large *p*-values. So the reduced models at the corresponding time points should be explored. Do it as an exercise using the trial-and-error method demonstrated in previous chapters, where a deleted variable is a less important up-stream or cause factor, even though it has significant adjusted effect. For an additional illustration, see the following example.

5. However, if the hypotheses on the effect differences of the independent variables are tested, then all regressions should have the same set of independent variables.

6. The six regressions have demonstrated that many possible acceptable reduced models can be obtained based on a three-way interaction model, either non-hierarchical or hierarchical models, at each time point, such as follows:

    6.1 By deleting at least one of the main factor or their two-way interactions, nonhierarchical three-way interaction models are obtained.

    6.2 By deleting the three-way interaction, alternative hierarchical two-way interaction models are obtained with at least one of the two-way interactions. The nonhierarchical two-way interaction model would be obtained if and only if the model does not contain at least one of the main factors of the interaction(s) in the model.

    6.3 By deleting all the interactions, an additive model is obtained.

    6.4 Therefore, there could be so many possible good fit models, either additive, two- or three-way interaction models. Finally, which one is best is highly dependent on the data set used, including the multicollinearity between independent variables and the process in developing the reduced models.

11.6.1.2 *Seemingly Causal Models by Times*

Referring to the model in Figure 11.2, as well as its possible reduced models, Figure 11.3(a) presents the path diagram of the model in Figure 11.2, where the dashed lines indicate that the up- and down-stream or causal effects between the corresponding variables are not taken into account. These path diagrams present seemingly causal models, since in general some of their relationships are not pure causal relationships, see Agung (2009a, and 2011). However, they always are correlated in a statistical sense.

For more advanced models, Figure 11.3(b–d) presents three alternative path diagrams. Take note that many other causal relationships could be defined depending on the sets of the variables, since there are 10 possible causal relationships based on five variables – compare this to the path diagram in Figure 8.1, and in Agung (2011). The general equation specification for the models based on these path diagrams are as follows:

Table 11.4 Summary of the statistical results in Figure 11.2

	Variable Dep. Y1						Variable Dep. Y2					
	T=6		T=7		T=8		T=6		T=7		T=8	
	Coeff.	Prob.	Coeff.	Prob.	Coeff.	Prob.	Coeff.	Prob.	Coeff.	Prob.	Coeff.	Prob.
Intercept	-2.861	0.000	-5.669	0.000	-5.628	0.000	-0.075	0.001	-0.082	0.000	-0.062	0.004
X1	0.555	0.000	1.011	0.000	0.998	0.000	0.014	0.000	0.018	0.000	0.014	0.000
X2	-1.000	0.001	-1.988	0.000	-1.714	0.000	-0.011	0.407	-0.003	0.827	-0.024	0.082
X3	-3.820	0.052	-6.075	0.000	-2.558	0.134	0.351	0.000	0.402	0.000	0.352	0.000
X1*X2	0.212	0.000	0.393	0.000	0.397	0.000	-0.001	0.604	-0.005	0.042	0.000	0.863
X1*X3	0.385	0.038	0.654	0.000	0.393	0.016	-0.032	0.000	-0.038	0.000	-0.031	0.000
X2*X3	-0.075	0.972	-6.027	0.011	-15.603	0.000	-0.236	0.015	-0.250	0.015	0.038	0.650
x1*X2*X3	0.024	0.908	0.716	0.005	1.856	0.000	0.022	0.024	0.026	0.020	-0.005	0.580
AR(1)	0.726	0.000	0.726	0.000	0.726	0.000	0.257	0.000	0.257	0.000	0.257	0.000

Determinant residual covariance   0.046747

Observations: 881

R-squared	0.827	R-squared	0.420
Adjusted R-squared	0.822	Adjusted R-squared	0.404
S.E. of regression	2.386	S.E. of regression	0.094
Durbin-Watson stat	2.474	Durbin-Watson stat	2.193
Mean dependent var	2.180	Mean dependent var	0.039
S.D. dependent var	5.657	S.D. dependent var	0.121
Sum squared resid	4874.395	Sum squared resid	7.486

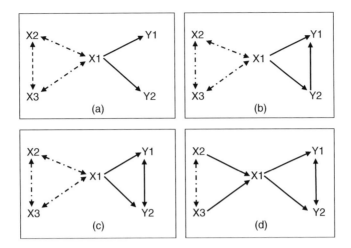

Figure 11.3   *Illustrative path diagrams for the five variables X1, X2, X3, Y1 and Y2*

1. Corresponding to the path diagram in Figure 11.3(b), the continuous models by times have the equation specification as follows:

$$Y1 = \sum_{t=1}^{T}[F_{1t}(X1, X2, X3) + Y2 \times F_{2t}(X1, X2, X3)] \times Dt$$

$$Y2 = \sum_{t=1}^{T}F_{3t}(X1, X2, X3) \times Dt$$

(11.12)

Take note that Y1 is a special case of the function F(X1, X2, X3,Y2) to present that the linear effect of Y2 on Y1 depends on X1, X2 and X3. Refer to all possible functions of $F_{jt}(X1,X2,X3)$ presented in the previous chapter and all possible reduced models mentioned earlier.

2. Referring to the first equation in (11.12), then corresponding to the path diagram in Figure 11.3c, the continuous models by times would have the equation specification as follows:

$$Y1 = \sum_{t=1}^{T}[F_{1t}(X1, X2, X3) + Y2 \times F_{2t}(X1, X2, X3)] \times Dt$$

$$Y2 = \sum_{t=1}^{T}[F_{3t}(X1, X2, X3) + Y1 \times F_{4t}(X1, X2, X3)] \times Dt$$

(11.13)

3. Corresponding to the path diagram in Figure 11.3(d), the continuous models by times would have the following general equation specification.

$$Y1 = \sum_{t=1}^{T}[F_{1t}(X1, X2, X3) + Y2 \times F_{2t}(X1, X2, X3)] \times Dt$$

$$Y2 = \sum_{t=1}^{T}[F_{3t}(X1, X2, X3) + Y1 \times F_{4t}(X1, X2, X3)] \times Dt$$

(11.14)

$$X1 = \sum F_{5t}(X2, X3) \times Dt$$

Example 11.2   A LEM in (11.12)

For illustration, Table 11.5 presents a summary of the statistical results based on a specific model in (11.13) using the following equation specification. Based on this summary, the following findings and notes are presented.

$$
\begin{aligned}
y1 = {} & (c(10) + c(11)^*x1 + c(12)^*x2 + c(13)^*x3 + c(14)^*x1^*x2 + c(15)^*x1^*x3 \\
& + c(16)^*x2^*x3 + c(17)^*x1^*x2^*x3 + \boldsymbol{c(18)^*y2})^*dt6 \\
& + (c(20) + c(21)^*x1 + c(22)^*x2 + c(23)^*x3 + c(24)^*x1^*x2 + c(25)^*x1^*x3 \\
& + c(26)^*x2^*x3 + c(27)^*x1^*x2^*x3 + \boldsymbol{c(28)^*y2})^*dt7 \\
& + (c(30) + c(31)^*x1 + c(32)^*x2 + c(33)^*x3 + c(34)^*x1^*x2 + c(35)^*x1^*x3 \\
& + c(36)^*x2^*x3 + c(37)^*x1^*x2^*x3 + \boldsymbol{c(38)^*y2})^*dt8 + [ar(1) = c(19)] \\
y2 = {} & (c(40) + c(41)^*x1 + c(42)^*x2 + c(43)^*x3 + c(44)^*x1^*x2 + c(45)^*x1^*x3 \\
& + c(46)^*x2^*x3 + c(47)^*x1^*x2^*x3 + \boldsymbol{c(48)^*y1})^*dt6 \\
& + (c(50) + c(51)^*x1 + c(52)^*x2 + c(53)^*x3 + c(54)^*x1^*x2 + c(55)^*x1^*x3 \\
& + c(56)^*x2^*x3 + c(57)^*x1^*x2^*x3 + \boldsymbol{c(58)^*y1})^*dt7 \\
& + (c(60) + c(61)^*x1 + c(62)^*x2 + c(63)^*x3 + c(64)^*x1^*x2 + c(65)^*x1^*x3 \\
& + c(66)^*x2^*x3 + c(67)^*x1^*x2^*x3 + c(68)^*y1)^*dt8 + [ar(1) = c(29)] \qquad (11.15)
\end{aligned}
$$

1. Take note that the main independent variable Y2 of the first model in (11.15) presents the linear effects of Y2 on Y1, adjusted for X1 to X1*X2*X3, at each of the three time-points, and the main independent variable Y1 of the second model represents the linear effects of Y1 on Y2, adjusted for X1 to X1*X2*X3.
2. This model also can be considered to be an extension of the model in Figure 8.22, where it is defined that the effects of X1 on Y1 and Y2 depend on a function of (X2,X3).
3. Table 11.5 presents the summary of the statistical results, which shows six AR(1) regressions, where each regression has at least one independent variable with very large p-values, say 0.346 to 0.9910. This table also shows each of the numerical independent variables has different linear effects on the corresponding dependent variables Y1 and Y2 between the three time points. Furthermore, they can easily be tested using the Wald Test.
4. However, since some of the independent variables have very large p-values, then a reduced model should be explored by deleting selected independent variables. The process of deleting selected independent variables one-by-one is as follows:
    4.1 Even though Y2 has large p-values, at T = 6 and T = 7, it should not be deleted from the model, since Y1 and Y2 have simultaneous causal effects. Since it is also defined that the effects of X1 on Y1 and Y2 depend on X2 and X3, then the six regressions should have at least one of the interaction factors X1*X2, X1*X3 or X1*X2*X3 as independent variables.
    4.2 Even though X1*X2*X3 has large p-values of 0.827 and 0.835 in two of the six regressions, however, for this illustration I have tried to keep the interaction X1*X2*X3 as an independent variable of the six reduced regressions in order to present a three-way interaction regression at each time point. The summary of the results are presented in Table 11.6. I am very sure that other reduced models, which are acceptable in a statistical sense, can be obtained.

Table 11.5 Summary of the statistical results based on the AR(1) model in (11.15)

	Dependent Variable : Y1						Dependent Variable : Y2					
	T=6		T=7		T=8		T=6		T=7		T=8	
	Coeff.	Prob.	Coeff.	Prob.	Coeff.	Prob.	Coeff.	Prob.	Coeff.	Prob.	Coeff.	Prob.
Intercept	-2.999	0.000	-5.813	0.000	-5.898	0.000	-0.090	0.000	-0.100	0.000	-0.078	0.001
X1	0.581	0.000	1.037	0.000	1.058	0.000	0.018	0.000	0.021	0.000	0.017	0.000
x2	-1.032	0.001	-2.019	0.000	-1.799	0.000	-0.017	0.207	-0.009	0.510	-0.030	0.032
X3	-3.300	0.110	-5.646	0.000	-1.138	0.531	0.349	0.000	0.362	0.000	0.329	0.000
X1*x2	0.211	0.000	0.390	0.000	0.395	0.000	0.000	0.999	-0.004	0.127	0.002	0.478
x1*x3	0.337	0.082	0.614	0.000	0.266	0.122	-0.031	0.000	-0.033	0.000	-0.029	0.000
X2*x3	-0.317	0.881	-6.342	0.008	-15.477	0.000	-0.232	0.017	-0.280	0.007	-0.023	0.805
X1*x2*x3	0.047	0.827	0.746	0.003	1.837	0.000	0.023	0.021	0.029	0.009	0.002	0.835
Y1							-0.004	0.035	-0.003	0.048	-0.003	0.044
Y2	-1.138	0.467	-0.867	0.346	-4.002	0.024						
AR(1)	0.724	0.000	0.724	0.000	0.724	0.000	0.243	0.000	0.243	0.000	0.243	0.000

Determinant residual covariance 0.0459

Observations 881

R-squared	0.828	R-squared	0.427
Adjusted R-squared	0.823	Adjusted R-squared	0.409
S.E. of regression	2.382	S.E. of regression	0.093
Durbin-Watson stat	2.481	Durbin-Watson stat	2.193
Mean dependent var	2.180	Mean dependent var	0.039
S.D. dependent var	5.657	S.D. dependent var	0.121
Sum squared resid	4841.4	Sum squared resid	7.398

Table 11.6 A set of three-way interaction regressions: a reduced model of the AR(1) model in Table 11.5

| | Dependent Variable : Y1 | | | | | | Dependent Variable : Y2 | | | | | |
| | T=6 | | T=7 | | T=8 | | T=6 | | T=7 | | T=8 | |
	Coeff.	Prob.	Coeff.	Prob.	Coeff.	Prob.	Coeff.	Prob.	Coeff.	Prob.	Coeff.	Prob.
Intercept	0.454	0.008	-4.243	0.000	-4.937	0.000	-0.083	0.000	-0.069	0.002	0.036	0.000
X1			0.786	0.000	0.913	0.000	0.016	0.000	0.016	0.000		
x2	-0.516	0.089	-1.969	0.000	-1.728	0.000	-0.016	0.003			-0.017	0.001
X3							0.341	0.000	0.315	0.000		
X1*x2	0.178	0.001	0.398	0.000	0.397	0.000			-0.005	0.000		
x1*x3	0.032	0.008	0.044	0.003	0.135	0.000	-0.030	0.001	-0.028	0.000	0.003	0.000
X2*x3			-13.018	0.000	-16.966	0.000	-0.235	0.019	-0.263	0.014		
X1*x2*x3	0.035	0.065	1.442	0.000	1.994	0.000	0.023	0.023	0.028	0.016	-0.001	0.164
y1							-0.004	0.040	-0.003	0.073	-0.002	0.059
y2	2.372	0.085	-1.777	0.048	-3.185	0.063						
AR(1)	0.746	0.000	0.746	0.000	0.746	0.000	0.249	0.000	0.249	0.000	0.249	0.000

Determinant residual covariance 0.05422

Observations: 881

R-squared	0.815	R-squared	0.374
Adjusted R-squared	0.810	Adjusted R-squared	0.359
S.E. of regression	2.466	S.E. of regression	0.097
Durbin-Watson stat	2.569	Durbin-Watson stat	2.157
Mean dependent var	2.180	Mean dependent var	0.039
S.D. dependent var	5.657	S.D. dependent var	0.121
Sum squared resid	5215.8	Sum squared resid	8.081

5. Based on the summary in Table 11.6, the following findings and notes are presented.

    5.1 We can say that the set of reduced regression functions obtained are unpredictable regressions, because of the unpredictable impacts of the multicollinearity between the independent variables – refer to notes and comments presented in Agung (2009a).

    5.2 This process has demonstrated that an acceptable reduced model, in a statistical sense, can be obtained by deleting an independent variable with a significant adjusted effect on the corresponding dependent variable.

    5.3 Two of the regressions of Y1 at $T=7$ and 8 have the same sets of independent variables, so the adjusted effects of each numerical independent variable are comparable and they can be tested using the Wald Test but they should not be compared to the regression at $T=6$ because this regression has different sets of independent variables.

    5.4 On the other hand, the three reduced regressions of Y2 have different sets of independent variables. So the adjusted effects of particular independent variables, such as X1*X3, X1*X2*X3 and Y1 on Y2, have different meanings. For instance, at $T=8$ the effect of Y1 on Y2 is *adjusted* for the variables X2, X1*X2 and X1*X2*X3, but at $T=7$ the effect of Y1 and Y2 is *adjusted* for X1, X3, X1*X2, X1*X3, X2*X3 and X1*X2*X3.

    5.5 Take note that the regression of Y1 has a very large SSR (sum of squared residuals: SSR) of 5215.8, compared to the regression of Y2 which has a very small SSR = 8.081. Why? The variable of Y1 has many very large scores compared to Y2. If the variable log(Y1) is used then a much smaller SSR would be obtained.

6. For a more detailed discussion, let us see one of the regressions, namely the reduced regression of Y1 at $T=6$. The regression function can be presented as follows:

$$Y1 = [0.454 - 0.516X2] + [0.178X2 + 0.032X3 + 0.035X2 \times X3] \times X1$$

$$+ 2.372Y2 + [AR(1) = 0.476]$$

which has the following characteristics.

    6.1 Referring to the general model in (11.3), $F_{16}(X1,X2,X3) = [0.454 - 0.516X2] + [0.178X2 + 0.032X3 + 0.035X2 \times X3] \times X1$, which shows that the linear effect of X1 on Y1 depends on the function $[0.178X2 + 0.032X3 + 0.035X2 \times X3]$.

    6.2 At the $\alpha = 0.05$ level of significance, Y2 has a positive significant linear effect on Y1 *adjusted* for the variables X2, X1*X2, X1*X3 and X1*X2*X3, with a *p*-value $0.085/2 = 0.0425 < \alpha = 0.05$.

7. Based on additional experiments, I have found other types of acceptable reduced models by deleting the interaction X1*X2*X3 with large *p*-values at the first step, then the others. In fact, all interactions of X1*X2*X3 could be deleted in order to have six two-way interaction regressions. See the summary of the statistical results of a set of two-way interaction regressions in Table 11.7. Finally, by deleting all interactions from each of the regressions, a set of additive regressions is obtained, as presented in Table 11.8, which shows that only the regression of Y1 at $T=6$ has insignificant adjusted effects. Do this as an exercise to obtain a reduced model.

11.6.1.3 Special Notes and Comments

Referring to the sets of six regressions in Tables 10.5–10.8, the following special notes and comments are presented.

1. It can be concluded many reduced models of the three-way interaction bivariate in Table 11.6 can be derived, depending on the subjective judgment of the researchers in deleting the independent variable(s) of each regression.

Table 11.7 *A set of two-way interaction regressions: a reduced model of the AR(1) model in Table 11.5*

	Dependent Variable : Y1						Dependent Variable : Y2					
	T=6		T=7		T=8		T=6		T=7		T=8	
	Coeff.	Prob.	Coeff.	Prob.	Coeff.	Prob.	Coeff.	Prob.	Coeff.	Prob.	Coeff.	Prob.
Intercep	0.493	0.006	-3.643	0.000	-2.185	0.020	-0.113	0.000	-0.101	0.000	-0.076	0.001
x1			0.731	0.000	-4.548	0.000	0.023	0.000	0.022	0.000	0.017	0.000
x2					0.864	0.000	-0.023	0.000			-0.021	0.000
x3					-1.280	0.000			0.259	0.000	0.311	0.000
x1*x2	0.116	0.000	0.127	0.000	1.369	0.000			-0.005	0.000		
x1*x3			0.058	0.000	0.381	0.000	0.002	0.008	-0.023	0.000	-0.027	0.000
X2*x3	-0.226	0.192					-0.031	0.000				
Y1							-0.004	0.043	-0.003	0.046	-0.003	0.020
y2	1.924	0.172	-2.185	0.020	-5.288	0.004						

	Determinant residual covariance	0.060060		
	Observations: 881			
R-squared	0.782		R-squared	0.4087
Adjusted R-squared	0.778		Adjusted R-squared	0.3963
S.E. of regression	2.666		S.E. of regression	0.0941
Durbin-Watson stat	2.599		Durbin-Watson stat	2.2136
Mean dependent var	2.180		Mean dependent var	0.039
S.D. dependent var	5.657		S.D. dependent var	0.1211
Sum squared resid	6149.1		Sum squared resid	7.6326

Table 11.8 A set of additive regressions: a reduced model of the AR(1) model in Table 11.5

| | Dependent variable: Y1 | | | | | | Dependent variable: Y2 | | | | | |
| | T=6 | | T=7 | | T=8 | | T=6 | | T=7 | | T=8 | |
	Coeff.	Prob.	Coeff.	Prob.	Coeff.	Prob.	Coeff.	Prob.	Coeff.	Prob.	Coeff.	Prob.
Intercept	-2.827	0.000	-5.742	0.000	-5.477	0.000	-0.119	0.000	-0.124	0.000	-0.098	0.000
x1	0.558	0.000	1.071	0.000	1.051	0.000	0.024	0.000	0.026	0.000	0.022	0.000
x2	0.146	0.280	0.264	0.064	0.508	0.003	-0.023	0.000	-0.030	0.000	-0.025	0.000
x3	-0.108	0.388	0.490	0.000	1.278	0.000	0.037	0.000	0.041	0.000	0.032	0.000
y1							-0.008	0.000	-0.008	0.000	-0.005	0.000
y2	-1.237	0.420	-2.718	0.005	-7.443	0.000						
AR(1)	0.785	0.000	0.785	0.000	0.785	0.000	0.244	0.000	0.244	0.000	0.244	0.000

Determinant residual covariance 0.06381, Obseravtions: 881

R-squared	0.776	R-squared	0.375
Adjusted R-squared	0.772	Adjusted R-squared	0.365
S.E. of regression	2.700	S.E. of regression	0.097
Durbin–Watson stat	2.661	Durbin–Watson stat	2.201
Mean dependent var	2.180	Mean dependent var	0.039
S.D. dependent var	5.657	S.D. dependent var	0.121
Sum squared resid	6306.9	Sum squared resid	8.062

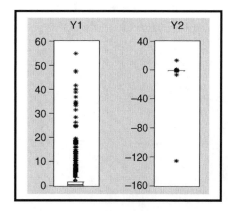 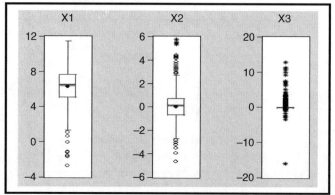

Figure 11.4   *The box-plots of the five variables of the model in Figure 11.3, for* t > 5

2. The three-way interaction model in (11.15) is a special case presented by the general SCMs in (11.13) and (11.14). So based on the set of five variables X1, X2, X3, Y1 and Y2, a lot of possible models, including LVAR(p,q) models, could be acceptable, which are highly dependent on data sets as well as multicollinearity between the variables.

3. The limitations of the models previously have not been considered or analyzed, such as follows:

   3.1 The outliers should be taken into consideration. Refer to the alternative treatment that can be done for outliers, as presented in Section 8.4.2.2. Figure 11.4 presents the box-plots of the five variables Y1, Y2, X1, X2 and X3, for t > 5. Take note that the outliers do have effects on the estimates the parameters, so they should be treated to conduct further analyses. In this case the following process is recommended.

   - At the first stage, some of the top outliers of Y1, Y2, X2 and X3, and some of the bottom outliers of Y2 and X3, should be deleted.

   - Take note that not all outliers should be deleted for further analysis, such as for the variables Y1 and X3, because they have sufficiently large numbers of outlier observations, which could provide important findings for the study.

   - For illustration, Figure 11.5 presents the box-plots of the five variables based on a sub-sample: SS = {t > 5 and y1 < 2 and y2 > −1 and y2 < 1 and x2 < 4 and x3 < 1} of size 697, compare to the box-plots in Figure 11.4 based on the sub-sample {t > 5} of size 895.

   - Take note that Y2 and X3 still have far outliers, based on a sub-sample of 72.5% of the sample {t > 5}. If all far outliers should be deleted, then a sub-sample of size 451 (41.06% of the sample {t > 5}, would be obtained, by using the trial-and-error method.

   - However, they still have near outliers. For comparison, I found a student's dissertation which presents the data analysis using a VAR model based on a sub-sample of size 122 (33.52%) out of a data set with 361 observations.

   3.2 Regard to the findings earlier, I would recommend applying SCMs by groups of individuals and times, since based on a variable with many outliers, a group variable that has three levels can easily be generated, such as; 1. The set of bottom outliers, 3. The set of top outliers, and 2. The other observations.

   3.3 The possible nonlinear relationships between each independent variable and each dependent variable should be identified using their scatter graphs. Refer to scatter graphs illustrated in the previous chapter.

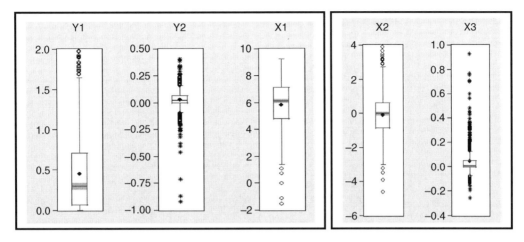

Figure 11.5 *The box-plots of the five variables of the model in Figure 11.3 based on SS = {t > 5 and y1 < 2 and y2 > −1 and y2 < 1 and x2 < 4 and x3 < 1} of size 697*

4. In addition to the multivariate linear-effect models, a lot of multivariate models of the transformed variables also could easily be defined or derived, corresponding to various models presented in Chapter 8. In addition, see the following section.

11.7   Illustrative SCMs by Group and Time

11.7.1   Linear-Effect Models by Group and Time

For illustration, based on the sub-sample SS in Figure 11.5, an ordinal variable having three levels, namely O3X3, is generated based on the variable X3 = ROA, generated as O3X3 = 1, for $X3 = NI/1000 \leq 0$, O3X3 = 2, for $0 < X3 = NI/1000 \leq 0.1$, and O3X3 = 3, for $X3 = NI/1000 > 0.1$, using the equation:

$$O3X3 = 1 + 1^*(X3 > 0) + 1^*(X3 > 0.1)$$

Then the same set of SCMs mentioned in Example 11.5 can easily be applied based on each of the sub-samples O3X3 = 1, 2 and 3. On the other hand, relevant descriptive statistical summaries and scatter graphs are very important inputs for strategic decision making, see Chapter 5, and simple statistical data analysis as illustrated in Agung (2011, Chapter 2).

Example 11.3   Summaries of descriptive statistical results
For special discussion, Table 11.9 presents the statistical summary of the variables Y1, Y2, X1, X2 and X3 by the categorical variables O3X3 and the last three time points ($T = 6$, 7 and 8) in the data set. Based on this summary, the following notes and comments are presented.

1. Take note that within each level of O3X3, the samples at $T = 6$, 7 and 8 have different sizes. This indicates that the three samples within each level of O3X3 represent different sets of firms or the research objects, in general. However, the three sets of firms should have a common sub-set, since the data is an unbalanced pool of data.

Table 11.9   *Statistical summary of the variables Y1, Y2, X1, X2 and X3, by O3X3 and T*

O3X3	T	Count	y1 Mean	y1 Std.Err.	y2 Mean	y2 Std.Err.	x1 Mean	x1 Std.Err.	x2 Mean	x2 Std.Err.	x3 Mean	x3 Std.Err.
1	6	40	0.311	0.082	-0.033	0.007	4.060	0.341	0.225	0.235	-0.033	0.007
1	7	58	0.395	0.062	-0.034	0.006	4.946	0.246	0.253	0.191	-0.034	0.006
1	8	42	0.360	0.077	-0.034	0.007	4.380	0.342	-0.110	0.300	-0.034	0.007
2	6	157	0.360	0.028	0.023	0.002	5.827	0.116	-0.163	0.092	0.023	0.002
2	7	145	0.398	0.036	0.023	0.002	5.876	0.124	-0.103	0.102	0.023	0.002
2	8	151	0.367	0.030	0.024	0.002	5.761	0.123	-0.106	0.089	0.024	0.002
3	6	30	0.956	0.095	0.231	0.032	7.276	0.186	-0.667	0.155	0.231	0.032
3	7	29	0.819	0.089	0.265	0.036	7.568	0.148	-0.530	0.153	0.265	0.036
3	8	45	0.937	0.080	0.258	0.027	7.540	0.159	-0.433	0.123	0.258	0.027
	All	697	0.452	0.018	0.046	0.004	5.812	0.067	-0.133	0.047	0.046	0.004

2. For instance, within O3X3 = 1, the sample sizes are 40, 58 and 42, at the times T = 6, 7 and 8, respectively. The sets of the firms observed at the three time points should be different sets of firms. Therefore, the three sets of firms should have different characters and behaviors in a theoretical sense. In a statistical sense, the exogenous and endogenous variables should have different patterns of relationships between the three sets of the firms. For this reason, I would say that the time dummy variables should be used as independent variables for the models.

3. Similarly, at each time point, there are different numbers of observations, within the three levels of O3X3. For instance, at T = 6, the three levels have 40, 157 and 30 observations, respectively.

4. The observed standard error, as well as other values of statistics, has been misinterpreted by users of statistics. Refer to special notes on "Misinterpretation of Selected Theoretical Concept of Statistics" presented in Agung (2011, Chapter 2, and 2004). For instance, the std.err. = 0.018 for the variable Y1 based on the sample size of 697, in Table 11.9, were interpreted as the observed mean statistic has an error of 1.8%. In fact, the std.err. (s.e.) is a sampled value of a statistic, which is computed as $(SD/\sqrt{n})$ based on a sample computed based on a sample of size n.

Example 11.4   Scatter graphs with regression lines

For a special discussion, Figure 11.6 presents the linear effects of X3 on Y1 and Y2, by O3X3, conditional for $T = 8$, in addition to their scatter graphs. Take note that whatever the scatter graph is, the linear effect of X3 on both Y1 and Y2, the linear-effects models in general, has always been considered by researchers. We (Siswantoro and Agung, 2010) observed more than 250 articles in four journals, namely the *International Journal of Counting*, *Advances in Accounting*, *British Accounting Review*, and *Journal of Accounting and Economics*, from the years 2008, 2009 and 2010, which present only linear-effects models and none mentions a scatter graph between pairs of variables. However, they do use transformed variables, such as the natural logarithms of selected variables, as well as dummies.

11.7.2   Nonlinear-Effect Models

Corresponding to the scatter graphs of Y1 and Y2 on X3 by O3X3 with regression lines conditional for $T = 8$ presented in Figure 11.6, Figure 11.7 presents the scatter graphs with their NNF (nearest neighbor fit) curves to show that X3 has nonlinear effects on Y1 and Y2, such as polynomial, logarithmic, hyperbolic or exponential effects. Refer to the alternative graphs presented in previous chapter.

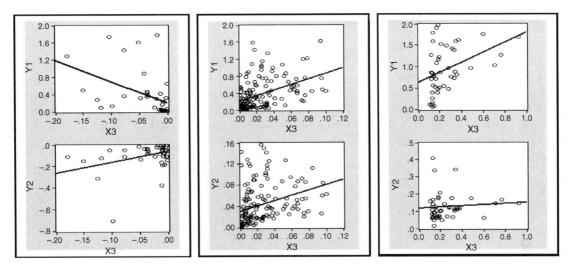

Figure 11.6 *Scatter graphs of Y1 and Y2 on X3 by O3X3, with regression lines, for* T = 8

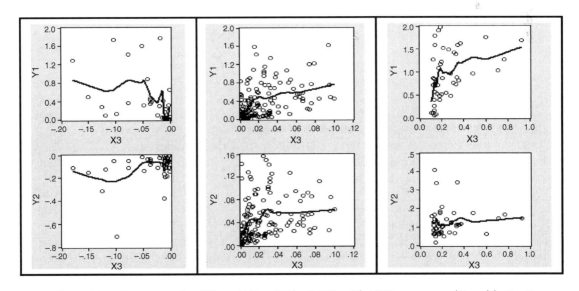

Figure 11.7 *Scatter graphs of Y1 and Y2 on X3 by O3X3, with NNF curves, conditional for* T = 8

11.7.3   Bounded SCM by Group and Time

Each endogenous numerical variable should have lower and upper bounds, in a theoretical sense, which has been illustrated in previous chapter. In this section a bounded SCM by group and time will be considered.

Example 11.5   Bounded SCM by group and time

As an extension of the bounded AR(1) model in Figure 8.22, Table 11.10 presents the summary of the statistical results based on a bounded AR(1) model of a bivariate (LBT_New, EQU_New) on X1, X2, X3 and X4, by a dichotomous factor A at the time $T = 7$ and 8, where the independent variables are generated as follows:

$$\text{LBT\_New} = \log(\text{Liability}/(60000 - \text{Liability})), \text{and}$$

$$\text{EQU\_New} = \log((\text{Equity} - 11000)/(40000 - \text{Equity}))$$

Based on this summary, the following findings and notes are presented.

1. The model applied is a simple model with the following general equation specification.

$$H_g(Yg) = \sum_{s=1}^{S} \sum_{t=1}^{T} [F_{g1}(X2, X3, X4) + F_{g2}(X2, X3, X4) \times F_{g3}(X1)] \times Dst \tag{11.16}$$

where $H_1(Y1) = \text{LBT\_New}$, $H_2(Y2) = \text{EQU\_New}$, Dst, for $S = T = 2$, represent the four dummy variables in the model previously,

$$F_{g1}(^*) = C(g0) + C(g1)^*X2 + C(g2)^*X3 + C(g3)^*X4,$$

$$F_{g2}(^*) = C(g4)^*X2 + C(g5)^*X3 + C(g6)^*X4, \text{and } F_{g3}(X1) = X1$$

Take note that the function $F_{g2}(^*)$ does not have a constant parameter, since $F_{g1}(^*)$ already has a constant parameter, namely $C(g0)$. Furthermore, within each cell (s,t), the model in (11.16) shows the effect of $F_{g3}(X1)$ on $H_g(Yg)$ depends on $F_{g2}(^*)$.

2. For a more advanced model, the functions $F_{g1}(^*)$ and $F_{g2}(^*)$ can be any functions of X2, X3 and X4 with finite number of parameters, but $F_{g3}(X1)$ is any function of X1 without a parameter – refer to alternative functions presented in the previous chapter, the model (8.44) in particular.

3. Whenever the function $F_{g3}(X1)$ also has a finite number of parameters, then the model in (11.16) would be nonlinear in parameter. For instance, the simplest one is $F_{g3}(X1) = C(g7) + C(g8)^*X1$, then $F_{g2}(^*) \times F_{g3}(X1)$ has multiplicative parameters. Refer to the nonlinear model in Figure 8.20(a).

4. The most general model of $H_g(Yg)$ on X1, X2, X3 and X4, can be presented using the following equation specification, which indicates that the effect of X1 (or a transformation of X1) on $H_g(Yg)$ depends on X2, X3 and X4.

$$H_g(Yg) = \sum_{s=1}^{S} \sum_{t=1}^{T} [F_{g1}(X2, X3, X4) + F_{g2}(X1, X2, X3, X4)] \times Dst \tag{11.17}$$

Table 11.10 Summary of the statistical results based on a bivariate AR(1) bounded model

| | Dependent variable LBT_New | | | | | | | | Dependent variable: EQU_New | | | | | | | |
| | (A=1,T=7) | | (A=1,T=8) | | (A=2,T=7) | | (A=2,T=8) | | (A=1,T=7) | | (A=1,T=8) | | (A=2,T=7) | | (A=2,T=8) | |
	Coeff.	Prob.	Coeff.	Prob.	Coeff.	Prob.	Coeff.	Prob.	Coeff.	Prob.	Coeff.	Prob.	Coeff.	Prob.	Coeff.	Prob.
Intercept	-12.27	0.000	-13.71	0.000	-19.95	0.000	-20.94	0.000	-1.032	0.000	-1.263	0.000	-3.305	0.000	-3.503	0.000
x2	0.704	0.000	0.802	0.000	0.583	0.000	0.469	0.000	-0.016	0.616	0.008	0.810	-0.023	0.619	0.005	0.906
x3	-25.002	0.000	-5.814	0.434	-0.643	0.000	-0.511	0.004	-1.829	0.530	-1.591	0.666	-0.775	0.000	-1.918	0.000
x4	3.184	0.000	4.259	0.000	7.748	0.000	8.080	0.000	-0.041	0.723	0.009	0.938	1.056	0.000	1.171	0.000
x1*x2	-0.070	0.000	-0.056	0.000	-0.024	0.056	-0.003	0.830	0.002	0.783	-0.003	0.684	-0.006	0.302	-0.012	0.040
x1*x3	3.893	0.000	1.067	0.387	0.070	0.000	0.057	0.001	0.281	0.528	0.259	0.671	0.086	0.000	0.213	0.000
x1*x4	0.095	0.000	0.018	0.282	0.024	0.022	0.028	0.007	0.004	0.584	0.001	0.857	0.017	0.001	0.003	0.507
AR(1)	0.062	0.000	0.062	0.000	0.062	0.000	0.062	0.000	0.185	0.000	0.185	0.000	0.185	0.000	0.185	0.000

Determinant residual covariance: 0.001, Observations : 599

	LBT_New		EQU_New	
R-squared	0.981		0.918	
Adjusted R-squared	0.980		0.914	
S.E. of regression	0.287		0.142	
Durbin-Watson stat	1.231		1.826	
Mean dependent var	-4.820		-1.107	
S.D. dependent var	2.030		0.4844	
Sum squared resid	46.966		11.466	

5. The output is obtained by using the following equation specification.

$$
\begin{aligned}
LBT\_NEW = (&C(10) + C(11)^*X2 + C(12)^*X3 + C(13)^*X4 \\
+ &C(14)^*X1^*X2 + C(15)^*X1^*X3 + C(16)^*X1^*X4)^*DA1^*DT7 \\
+ &C(20) + C(21)^*X2 + C(22)^*X3 + C(23)^*X4 \\
+ &C(24)^*X1^*X2 + C(25)^*X1^*X3 + C(26)^*X1^*X4)^*DA1^*DT8 \\
+ &C(30) + C(31)^*X2 + C(32)^*X3 + C(33)^*X4 \\
+ &C(34)^*X1^*X2 + C(35)^*X1^*X3 + C(36)^*X1^*X4)^*DA2^*DT7 \\
+ &C(40) + C(41)^*X2 + C(42)^*X3 + C(43)^*X4 \\
+ &C(44)^*X1^*X2 + C(45)^*X1^*X3 + C(46)^*X1^*X4)^*DA2^*DT8 \\
+ &[AR(1) = C(1)]
\end{aligned}
$$

$$
\begin{aligned}
EQU\_NEW = (&C(50) + C(51)^*X2 + C(52)^*X3 + C(53)^*X4 \\
+ &C(54)^*X1^*X2 + C(55)^*X1^*X3 + C(56)^*X1^*X4)^*DA1^*DT7 \\
+ &C(60) + C(61)^*X2 + C(62)^*X3 + C(63)^*X4 \\
+ &C(64)^*X1^*X2 + C(65)^*X1^*X3 + C(66)^*X1^*X4)^*DA1^*DT8 \\
+ &C(70) + C(71)^*X2 + C(72)^*X3 + C(73)^*X4 \\
+ &C(74)^*X1^*X2 + C(75)^*X1^*X3 + C(76)^*X1^*X4)^*DA2^*DT7 \\
+ &C(80) + C(81)^*X2 + C(82)^*X3 + C(83)^*X4 \\
+ &C(84)^*X1^*X2 + C(85)^*X1^*X3 + C(86)^*X1^*X4)^*DA2^*DT8 \\
+ &[AR(1) = C(2)]
\end{aligned}
$$

6. *Final notes.* A lot of seemingly causal models (SCMs) or system equation models have been presented in Agung (2011 and 2009a), as well as in the first four chapters of this book. Many models without the numerical time T as an independent variable have been demonstrated in Chapter 5 (Agung, 2009a), and the models by time and states have been considered in Section 3.10. For this reason, for illustration, this section only presents selected general SCMs based on the path diagram in Figure 4.32 (Agung, 2009a), as presented again in Figure 8.1. Based on this path diagram, the system equations or SCMs should have four multiple regressions having dependent variables $Y1$, $Y2$, $X1$ and $X3$, respectively, with dummy variables of time points.

Part Three

Balanced Panel Data as Natural Experimental Data

Abstract

In this chapter, panel data considered is stacked by cross-section with $N \times T$ observations, where the units of the analysis are the individual-time or firm-time observations. So the sets or multidimensional of exogenous, endogenous and environmental variables, respectively, for the firm i at the time t can be presented using the symbols $\mathbf{X}_{it} = (X1, \ldots, Xk, \ldots)_{it}$, $\mathbf{Y}_{it} = (Y1, \ldots, Yg, \ldots)_{it}$, and $\mathbf{Z}_t = (Z1, \ldots, Zj, \ldots)_t$, for $i = 1, \ldots, N$; and $t = 1, \ldots, T$. In natural experimental data analysis, the environmental variables could be represented by the time or time-period variable, say TP, such as before and after a critical time or event; before, during and after an economic crisis; and before, between and after two consecutive critical events. In addition, classification or cell-factor, namely CF, can be defined as the treatment factors, with the response variable $\mathbf{Y}_{it}$ and the covariates (causal or up-stream variables, or predictors) are the lags $\mathbf{Y}_{it}(-p)$ and $\mathbf{X}_{it}(-q)$, at least for $p = q = 1$, if the data is annual, semi-annual or a quarterly data set. Take note that by using the lags of the dependent variable as the independent variables in the models, then all models presented in this part are dynamic. Refer to the notes in Section 1.3.2.

Since, in general, a covariate, say $Xk_{it}(-1)$ or $Yg(-1)$, should have different effects on a response variable, say Yg_{it}, between the groups generated by a cell-factor CF and a time period TP, then the simplest recommended model considered is the heterogeneous regression lines of Yg on $Xk(-1)$ by CF and TP.

In addition, all models based on true experimental data presented in Agung (2011, Chapter 8) should be valid based on balanced pool data by using CF and TP as environmental treatment factors.

Panel Data Analysis Using EViews, First Edition. I Gusti Ngurah Agung.
© 2014 John Wiley & Sons, Ltd. Published 2014 by John Wiley & Sons, Ltd.
Companion website: www.wiley.com/go/panel_data

12

Univariate Lagged Variables Autoregressive Models

12.1 Introduction

The model of Y_{it} on $Y_{it}(-1)$, and $X_{it}(-1)$ for any variable Xs based on pool panel data presented in previous chapters is misleading, because many of the bivariate scores $\{X_{it}(-1),Y_{it}\}$ belong to two different individuals or research objects, more so if the pool panel data is unbalanced or incomplete.

For this reason a special format of the work file based on the balanced pool data with complete $N \times T = $ NT firm-time observations should be considered. The balanced pool panel data used for illustration is developed based on the Data_Sesi.wf1, called Balanced.wf1, with 1792 (224×8) firm-year observations as presented in Figure 12.1. Take note that the X-variables and the Y-variables are the same as in the Data_Sesi.wf1 with unbalanced 2284 firm-year observations.

For this reason, all models presented in Part II would be good or valid for this balanced pool data as well. Furthermore, EViews provides a special work file for the balanced pool data set, so that all types of models using lagged variables, such as LVAR(p,q) models which are dynamic, can be easily be applied. The following section presents how to develop a special work file.

12.2 Developing Special Balanced Pool Data

Based on the Balanced.wf1, special balanced pool data can be developed as follows:

1. Sort the data in Balanced.wf1, using the steps as follows:
   - With the Balanced.wf1 on the screen, select *View/Select all* (*except C-RESID*), then by selecting *Proc/Sort Current Page...* the window for entering the Sort key(s) shown on the screen.
   - The sort keys should be entered are "*Firm_Code t*".
2. Click the option "*Range*" at the right mouse, then the "*Workfile structure type*" shown on the screen, as presented in Figure 12.2(a).

Panel Data Analysis Using EViews, First Edition. I Gusti Ngurah Agung.
© 2014 John Wiley & Sons, Ltd. Published 2014 by John Wiley & Sons, Ltd.
Companion website: www.wiley.com/go/panel_data

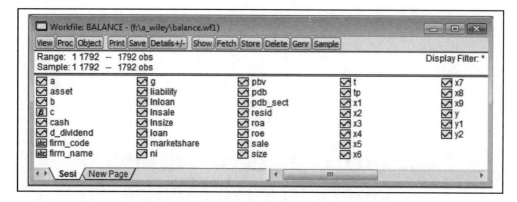

Figure 12.1 *Balanced panel data developed based on Data_Sesi.wf1*

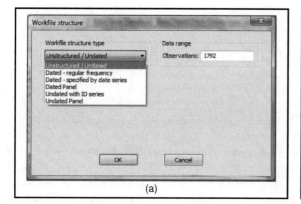
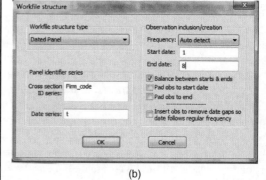

    (a)   (b)

Figure 12.2 *The windows to develop a special balanced pool data, Special_BPD.wf1*

3. Then by selecting the structure type: *Data Panel ... OK*; the options in Figure 12.2(b) are shown on the screen.
4. Finally by inserting the "*Firm_Code*" and "*T*" as the date series, the work file in Figure 12.3, called Special_BPD.wf1 is obtained.

Example 12.1 Preliminary Analysis

Figure 12.4 presents important characteristics of the data in Special_BPD.wf1, specifically its lagged variables. Various models Y_{it} or $G(Y_{it})$ on lagged independent variables, say multivariate $X(-p)$, can easily be derived using all models presented in Chapters 6 and 7, based on the sample $\{t > p\}$.

Take note that multivariate $X(-p)$ can be considered the covariate (cause, source or up-stream variable) of the natural experimental data, with categorical time t and selected group(s) of research objects (firms or individuals), which is invariant (constant) over time, as environmental treatment factors.

For this reason, all types of models for the experimental data presented in Agung (2011, Chapter 8) would be valid for this Special_BPD.wf1. The types of models are ANOVA, Heterogeneous Regressions and ANCOVA (fixed effects) models of an endogenous or response variable Y_{it} with covariates $Xk_{it}(-1)$, for $k = 1, \ldots, K$, and group and time as environmental treatment factors.

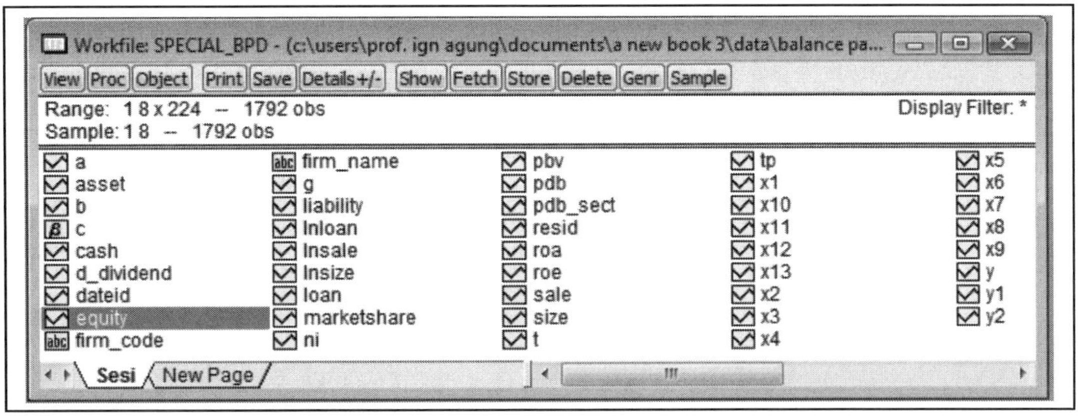

Figure 12.3 *Special_BPD.wf1 developed based on the Balanced.wf1 in Figure 10.1*

EViews - [Group: UNTITLED Workfile: SPECIAL_BPD::Sesi\]

obs	T	SALE	SALE(-1)	SALE(-2)	NI	NI(-1)	NI(-2)
_AALI - 1	1	1442.000	NA	NA	180.000	NA	NA
_AALI - 2	2	1737.000	1442.000	NA	189.000	180.000	NA
_AALI - 3	3	2627.000	1737.000	1442.000	665.000	189.000	180.000
_AALI - 4	4	3371.000	2627.000	1737.000	790.000	665.000	189.000
_AALI - 5	5	3758.000	3371.000	2627.000	787.000	790.000	665.000
_AALI - 6	6	5961.000	3758.000	3371.000	1973.000	787.000	790.000
_AALI - 7	7	8161.000	5961.000	3758.000	2631.000	1973.000	787.000
_AALI - 8	8	7224.000	8161.000	5961.000	1661.000	2631.000	1973.000
_ABBA - 1	1	10.00000	NA	NA	-3.000	NA	NA
_ABBA - 2	2	10.00000	10.00000	NA	-4.000	-3.000	NA
_ABBA - 3	3	32.00000	10.00000	10.00000	-11.000	-4.000	-3.000
_ABBA - 4	4	86.00000	32.00000	10.00000	-8.000	-11.000	-4.000
_ABBA - 5	5	95.00000	86.00000	32.00000	1.000	-8.000	-11.000
_ABBA - 6	6	112.0000	95.00000	86.00000	3.000	1.000	-8.000
_ABBA - 7	7	144.0000	112.0000	95.00000	2.000	3.000	1.000
_ABBA - 8	8	148.0000	144.0000	112.0000	1.000	2.000	3.000
_ADES - 1	1	106.0000	NA	NA	4.000	NA	NA
_ADES - 2	2	130.0000	106.0000	NA	13.000	4.000	NA
_ADES - 3	3	92.00000	130.0000	106.0000	-31.000	13.000	4.000
_ADES - 4	4	144.0000	92.00000	130.0000	-119.000	-31.000	13.000

Figure 12.4 *Some characteristics of the data in Special_BPD.wf1*

12.3 Natural Experimental Data Analysis

In a natural experiment, treatment is an environmental factor, which can be represented by time or the time-period variable, such as before and after a new tax policy; before and after a terrorist event; before, during and after a regime or economic crisis; and others. In addition various background variables, which are group or categorical variables invariant or constant over time or time periods, can be considered environmental treatment factors. Some of those variables are sectors of industries, status of the businesses (family and non-family; or public and nonpublic), and other group variables generated based on one or more numerical variables at a time point as the base or reference group of firms.

12.3.1   Developing Group-Based Variables

For illustration, two group variables based on the variable SIZE in Special_BPD.wf1 are generated, namely *M_Size*, using the median as the cutting point, and *Q_Size* with five levels with cutting points $q = 0.20, 0.40,$ 0.60 and 0.80. Other group variables based other numerical variables, such as Total ASSET and LOAN, can be generated easily. The steps of the construction are as follows:

1. Sort the whole data in Special_BPD.wf1 by the time *t* and the *Firm_code*.
2. Select the sample $\{t = 1\}$, then generate the two series, using.

$$M\_Size = 1^*(Quantile(Size, 0.5))$$

and

$$Q\_size = 1 + 1^*(Size >= @Qunatile(Size, 0.2) + 1^*(Size >= @Qunatile(Size, 0.4)$$

$$+1^*(Size >= @Qunatile(Size, 0.6) + 1^*(Size >= @Qunatile(Size, 0.8)$$

3. Select the whole sample, then present the data of the variables *t, Firm_code, M_Size*, and *Q_Size*.
4. Copy the whole data of the four variables to a spreadsheet (Excel), which shows the data of *M_Size* and *Q_size* only available for $t = 1$. Then copy the data of the two variables into time $t = 2$ up to $t = 8$.
5. Finally, the data of each of the variables *M_Size* and *Q_Size* can be copied or inserted to Special_BPD. wf1. Remember to delete the previous generated variables *M_Size* and *Q_Size* before you insert new data for the same variables.
6. Therefore, *M_Size* or *Q_Size* could be used as a treatment factor for the balanced pool data as natural experimental data. See the following examples.

Take note that the base or reference group can be generated based on data at other time points, such as one (year, semester, quarter, month or week) before a critical time point, or one year before a new tax policy or $T = 5$ based on this data set.

12.3.2   Alternative Models Applied

Various types of general linear models (GLMs) for experimental data analysis have been demonstrated using various models presented in Agung (2011, Chapter 8). I am very confident these models are acceptable for balanced pool data. The models that have been applied are as follows:

- ANOVA and MANOVA models,
- Univariate and Multivariate Heterogeneous Regressions,
- ANCOVA and MANCOVA models,
- Binary Choice (Logit, Probit, Extreme Value) models with categorical or numerical independent variables, and
- Ordered Choice Models, with categorical or numerical independent variables.

In addition, all GLMs presented in Chapters 6–8 can easily be transformed to the models with first or higher lagged independent variables to present various models based on natural experimental data. For this reason, this chapter presents only some selected illustrative empirical results, based on some heterogeneous regressions, binary or ordered choice models, fixed effects models and ANCOVA models.

12.4   The Simplest Heterogeneous Regressions

Referring to the previous notes, it is clear that the effect of a single lagged variable $X(-1)$ on Y depends on the time, for $t > 1$, as well as selected group of research objects (individuals or firms) for any variable X, including the lag(s) of endogenous variable Y. For this reason, this section presents alternative sets of the simplest models as the base models for natural experimental data analysis.

12.4.1   Sets of the Simplest Models by Time

The equation specification of the model, for $t > 1$, is as follows:

$$Y \quad C \quad @Expand(T, @dropfirst) \, X(-1)^* Expand(T) \tag{12.1a}$$

or

$$Y \quad @Expand(T) \, X(-1)^* @Expand(T) \tag{12.1b}$$

Note that this model in fact is a set of heterogeneous regressions of Y on $X(-1)$ by the time points $t > 1$. In some cases, a time-period TP should be applied, instead of the time t. If $X(-1) = Y(-1)$, then a set of special models would be obtained, namely a set of LV(1). Refer to all heterogeneous regressions presented in Chapter 9, specifically Section 9.7. Furthermore, note that the objective of this model is to study the differential linear effect of $X(-1)$ on Y over time, even though they might have a nonlinear relationship, which can easily be identified using their scatter graph. Compare this with the not-recommended model in (12.6).

The model in (12.1b) represents a set of (T-1) simple linear regression models with the following equations.

$$Y_{i,t} = \beta_{o,t-1} + \beta_{1,t-1} \times X_{i,t-1} + \varepsilon_{i,t}, \quad \text{for } t = 2, 3, \ldots, T \tag{12.1c}$$

12.4.2   The Simplest Heterogeneous Regressions by Group and Time

If the data has a large number of individual or firm observations, then the effect of $X(-1)$ on Y would depend on the time for $t > 1$, as well as the defined group(s) of research objects. For this reason, the set of simplest models considered has the following general equation specification.

$$Y \quad C \quad @Expand(Group, T, @dropfirst) \, X(-1)^* @Expand(Group, T) \tag{12.2a}$$

or

$$Y \quad @Expand(Group, T) \, X(-1)^* @Expand(Group, T) \tag{12.2b}$$

With regards to the experimental data, take note that the *group* variable should be defined or generated such that it is invariant or constant over *time*, as mentioned previously.

12.4.3   The Simplest Heterogeneous Regressions by Group with Trends

If the data has a large time observation, then the model in (12.2a) could be modified to a model with trend as follows:

$$Y \quad C \quad @Expand(Group, @dropfirst) \, X(-1)^* @Expand(Group)$$
$$T^* @Expand(Group) \tag{12.3}$$

A more advanced model would be obtained using a time period *TP* in addition to the *Group* variable, with the following equation specification. However, both factors *Group* and *TP* could be presented as a single factor, namely *CF* (*Cell-factor*) – refer to Agung (2011, 2006),

$$Y \quad C \quad @Expand(Group, TP, @dropfirst) \; X(-1)^* @Expand(Group, TP)$$
$$T^* @Expand(Group, TP) \tag{12.4a}$$

or

$$Y \quad C \quad @Expand(CF, @Dropfirst) \; X(-1)^* @Expand(CF)$$
$$T^* @Expand(CF) \tag{12.4b}$$

12.4.4   The Simplest Heterogeneous Regressions by Group with Time-Related Effects

As an extension of the model in (12.4), the set of heterogeneous regressions with time-related effects might be considered to have the equation specification as follows:

$$Y \quad C \quad @Expand(Group, TP, @dropfirst) \; X(-1)^* @Expand(Group, TP)$$
$$T^* @Expand(Group, TP) \, T^* X(-1)^* @Expand(Group, TP) \tag{12.5}$$

12.4.5   Not-Recommended or the Worst Model

Corresponding to the models in (12.1) to (12.5), for comparison, it is important to present a single model of Y_{it} on $X_{it}(-1) = X_{i,t-1}$ for the whole individual-time observations, with the equation specification as follows:

$$Y \quad C \quad X(-1) \tag{12.6}$$

This model should be considered the worst, because the behavior of a large set of research objects (firms or individuals) should change over time, as well as by their group.

12.5   LVAR(1,1) Heterogeneous Regressions

12.5.1   LVAR(1,1) Heterogeneous Regressions by Group and Time Period

As an extension of the model in (12.2), the following equation specification represents LVAR(1,1) heterogeneous regressions of $G(Y)$ on $Xk(-1)$, for $k = 1, \ldots, K$, by *Group* and *TP* (time period).

$$G(Y) \quad C \quad @Expand(Group, TP, @Dropfirst) \; G(Y(-1))^* @Expand(Group, TP)$$
$$X1(-1)^* @Expand(Group, TP) \ldots Xk(-1)^* Expand(Group, TP) \ldots AR(1) \tag{12.7}$$

where $G(Y)$ is a transformed variable of an endogenous variable Y and Xk can be any variable, either a main factor, a *lagged variable*, an interaction or various transformed variables, which are defined as the up-stream or causal factors of $G(Y)$ in a theoretical sense, and they have different effects on $G(Y)$ between group and the time periods. Take note that the variable *Group* should be invariant over *Time*, since the data is natural experimental data. Refer to the problem with dummy variables presented in the previous chapter.

By using dummies, the LV(1,1) model in (12.7) can be presented as follows:

$$G(Y) = \sum_{g,tp} \alpha_{g,tp} D_{g,tp} + \sum_{g,tp} \beta_{g,tp} G(Y(-1)) \times D_{g,tp}$$
$$+ \sum_{k,g,tp} \delta_{k,g,tp} X_k(-1) \times D_{g,tp} + AR(1) + \varepsilon$$

(12.8)

where $D_{g,tp} = 1$ for the group g within the time period *TP*, and $D_{g,tp} = 0$ otherwise.

12.5.2 LVAR(1,1) Heterogeneous Regressions by Group with Trend

As an extension of the model in (12.2), the following equation specification represents a LVAR(1,1) heterogeneous regressions of $G(Y)$ on $Xk(-1)$, for $k = 1, \ldots, K$, by group with trend.

$$G(Y) \; C \; @Expand(Group, @Dropfirst) \; G(Y(-1))^* @Expand(Group)$$
$$X1(-1)^* @Expand(Group) \ldots Xk(-1)^* @Expand(Group) \ldots$$
$$t^* @Expand(Group) \; AR(1)$$

(12.9)

By using the dummies, the LV(1,1) model in (12.9) can be written as follows:

$$G(Y) = \sum_{g} \alpha_g D_g + \sum_{g} \beta_g G(Y(-1)) \times D_g + \sum_{k,g} \delta_{k,g} X_k(-1) \times D_g$$
$$+ t \times \sum_{g} \tau_g D_g + AR(1) + \varepsilon$$

(12.10)

where $D_g = 1$ for the *Group* g, and $D_g = 0$ otherwise.

12.5.3 LVAR(1,1) Heterogeneous Regressions by Group with Trend-Related Effects

As an extension of the model in (12.9), the following equation specification represents a LVAR(1,1) heterogeneous regression of $G(Y)$ on $Xk(-1)$, for $k = 1, \ldots, K$, by group with time-related effects.

$$G(Y) \; C \; @Expand(Group, @Dropfirst) \; G(Y(-1))^* @Expand(Group)$$
$$X1(-1)^* @Expand(Group) \ldots Xk(-1)^* @Expand(Group) \ldots$$
$$t^* @Expand(Group) \ldots t^* Xk(-1)^* @Expand(Group) \ldots AR(1)$$

(12.11)

By using dummies, the LV(1,1) model in (12.11) can be written as follows:

$$G(Y) = \sum_{g} \alpha_g D_g + \sum_{g} \beta_g G(Y(-1)) \times D_g + \sum_{k,g} \delta_{k,g} X_k(-1) \times D_g$$
$$+ t \times \left[\sum_{g} \left(\tau_g + \sum_{k} \theta_{g,k} X_k(-1) \right) \times D_g \right] + AR(1) + \varepsilon$$

(12.12)

where $D_g = 1$ for the *Group* g, and $D_g = 0$ otherwise.

12.5.4   Applications of LV(p,q) Models

The LVAR(p,q) models could easily derived based on those presented in Chapter 8. However, this section presents only a limited number of LV(1), AR(1) and LVAR(1,1) models, which could easily be extended to LV(p), AR(q) and LVAR(p,q) models.

Example 12.2   The simplest LV(1) models by M_Size and time period
Table 12.1 presents the summary of statistical results of three simple LV(1) models, using the equation specifications as follows:

$$lnsale \ c \ @expand(m\_size, tp, @dropfirst) \ lnsale(-1)^* @expand(m\_size, tp) \tag{12.13}$$

$$lnsale \ c \ @expand(m\_size, tp, @dropfirst) \ lnsale(-1)^* @expand(m\_size, tp) \\ t^* @expand(m\_size, tp) \tag{12.14}$$

$$lnsale \ c \ @expand(m\_size, tp, @dropfirst) \ lnsale(-1)^* @expand(m\_size, tp) \\ t^* @expand(m\_size, tp) \ t^* lnsale(-1)^* @expand(m\_size, tp) \tag{12.15}$$

Based on the summary in Table 12.1, the following findings and notes are presented.

1. Model (12.13) is the LV(1) heterogeneous regressions of *lnSale* on *lnSale*(−1) by *M_Size* and *TP* (time period) which indicates the time before and after a new tax policy. Based on this regression, the following notes and comments are presented.
   1.1 *lnSale*(−1) has a significant *linear* effect on *lnSale* within the four cells/groups generated by *M_Size* and *TP*. This model has the objective to test the linear effect of *lnSale*(−1) on *lnSale*, even though they might have a polynomial or nonlinear relationship.
   1.2 For illustration, Figure 12.5 presents the scatter graphs of *lnSale* on *lnSale*(−1), with their kernel fit curves for the four cells. Note that the graphs in the cells (1,1), (1,2) and (2,1) show nonlinear curves. Specific to the cell (2,1) we find there are far lower outliers of *lnSale*(−1) and this can be easily checked using its box-plot.
   1.3 Based on the whole data set, the first three curves suggest use of polynomial models, such as quadratic, third degree and fourth degree respectively, since a horizontal line can have at the most two, three and four intercept points with the corresponding curves. See the following example.
   1.4 The null hypothesis H_0: $C(5) = C(6) = C(7) = C(8)$ is rejected based on the F-statistic of $F_0 = 11.15\ 761$ with $df = (3,1548)$ and a p-value $= 0.0000$. So *lnSale*(−1) has different linear effects on *lnSale* between the four cells.
2. Model (12.14) is the LV(1) heterogeneous regressions of *lnSale* on *lnSale*(−1), with *Trend*, by *M_Size* and *TP* (time period). Take note that each regression is an additive model, which can be presented using the following general equations.

$$\ln Sale_{g,tp} = \hat{\alpha}_{g,tp} + \hat{\beta}_{g,tp} \ln Sale(-1)_{g,tp} + \hat{\delta}_{g,tp} t \tag{12.16}$$

   Based on this regression, specific within the cell (*M_size* $= g = 1$, $TP = 1$), the following notes and comments are presented.
   2.1 This regression shows the time t has a very large p-value $= 0.838$, adjusted for *lnSale*(−1), or by taking into account the effect of *lnSale*(−1), or the sampled correlation between *lnSale*(−1) and t.
   2.2 However, the variables *lnSale*(−1) and time t have significant joint effects on *lnSale* based on $F_0 = 1781.027$ with $df = (2,1544)$ and a p-value $= 0.0000$.

Table 12.1 Summary of the outputs using the equation specifications (12.13)–(12.15)

Dependent Variable: LNSALE
Method: Panel Least Squares
Date: 06/24/12 Time: 19:21
Sample (adjusted): 2 8
Periods included: 7
Cross-sections included: 224
Total panel (unbalanced) observations: 1556

Variable	Model 12.13			Model 12.14			Model 12.15		
	Coef.	t-Stat.	Prob.	Coef.	t-Stat.	Prob.	Coef.	t-Stat.	Prob.
C	0.3481	3.5898	0.0003	0.3399	2.6422	0.0083	0.9602	3.0708	0.0022
M_SIZE=1 AND TP=2	0.6805	4.4289	0.0000	1.8758	5.2608	0.0000	3.3631	3.1969	0.0014
M_SIZE=2 AND TP=1	0.0528	0.2938	0.7690	-0.0672	-0.3187	0.7500	-1.3979	-2.3994	0.0165
M_SIZE=2 AND TP=2	0.4929	2.5777	0.0100	1.2430	3.6696	0.0003	-2.6915	-1.8143	0.0698
LNSALE(-1)*(M_SIZE=1 AND TP=1)	0.9525	49.1834	0.0000	0.9524	49.4682	0.0000	0.8251	13.3888	0.0000
LNSALE(-1)*(M_SIZE=1 AND TP=2)	0.8097	35.7949	0.0000	0.8197	36.2410	0.0000	0.3973	2.0760	0.0381
LNSALE(-1)*(M_SIZE=2 AND TP=1)	0.9739	45.0361	0.0000	0.9665	44.1434	0.0000	1.0706	15.0325	0.0000
LNSALE(-1)*(M_SIZE=2 AND TP=2)	0.9072	41.8507	0.0000	0.9113	42.2300	0.0000	1.3563	7.0870	0.0000
T*(M_SIZE=1 AND TP=1)				0.0025	0.0968	0.9229	-0.1860	-2.0539	0.0402
T*(M_SIZE=1 AND TP=2)				-0.1778	-3.8196	0.0001	-0.4859	-3.3244	0.0009
T*(M_SIZE=2 AND TP=1)				0.0497	1.7538	0.0797	0.2518	1.8709	0.0615
T*(M_SIZE=2 AND TP=2)				-0.1099	-2.7763	0.0056	0.3560	1.7540	0.0796
T*LNSALE(-1)*(M_SIZE=1 AND TP=1)							0.0385	2.1746	0.0298
T*LNSALE(-1)*(M_SIZE=1 AND TP=2)							0.0615	2.2225	0.0264
T*LNSALE(-1)*(M_SIZE=2 AND TP=1)							-0.0292	-1.5357	0.1248
T*LNSALE(-1)*(M_SIZE=2 AND TP=2)							-0.0625	-2.3402	0.0194
R-squared	0.8935			0.8952			0.8964		
Adjusted R-squared	0.8930			0.8945			0.8954		
F-statistic	1855.63			1199.48			888.47		
Prob(F-statistic)	0.0000			0.0000			0.0000		
Durbin-Watson stat	2.1733			2.1815			2.1673		

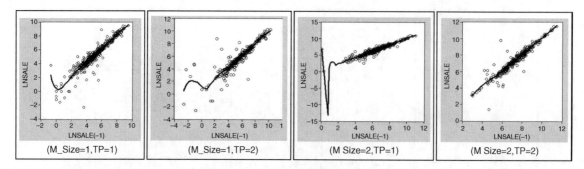

| (M_Size=1,TP=1) | (M_Size=1,TP=2) | (M Size=2,TP=1) | (M Size=2,TP=2) |

Figure 12.5   Scatter graphs with their kernel fit curves of lnsale on lnSale(−1) within three cells

2.3   If time t is deleted, then the model in this (1,1)-cell, does not have a trend.

2.4   On the other hand, for illustration, if *lnSale(−1)* is deleted from the model, the exponential growth model is obtained with the following regression function, the t-statistic in [*]. This function shows that SALE has a significant average growth rate of 18.565% during the time of observation for the group/cell ($M\_Size = 1$, $TP = 1$).

$$LNSALE = 4.48968 + 0.18565^*T$$
$$[23.747] \qquad [3.261]$$

3.   Model (12.15) is the LV(1) heterogeneous regressions of *lnSale* on *lnSale(−1)*, with the *time-related effects* (*TRE*), by *M_Size* and *TP* (time period). Take note that each regression is a two-way interaction model, which can be presented using the general equation as follows:

$$\ln Sale_{g,tp} = \hat{\alpha}_{g,tp} + \hat{\beta}_{g,tp}\ln Sale(-1)_{g,tp} + \hat{\delta}_{g,tp}t + \hat{\tau}_{g,tp}t \times \ln Sale(-1) \qquad (12.17)$$

Based on the results in Table 12.1, the following notes and comments are presented.

3.1   This interaction model is applied because the effect of *lnSale(−1)* on *lnSale* depends on the time t, in a theoretical sense. For this reason, even though the interaction $t \times \ln Sale(-1)$ has a very large p-value, it should not be deleted to obtain a reduced model.

3.2   For illustration, Table 12.2 presents the summary of the statistical results based on an acceptable reduced model, which shows the interaction $t \times \ln Sale(-1)$ has either positive or negative

Table 12.2   Summary of the statistical results based on a reduced model of the model in (12.15)

	M_Size=1 & TP=1			M_Size=1 & TP=2			M_Size=2 & TP=1			M_Size=2 & TP=2		
Variable	Coeff.	t-Stat	Prob.	Coeff.	t-Stat	Prob.	Coeff.	t-Stat	Prob.	Coeff.	t-Stat	Prob.
C	0.549	1.849	0.065	0.820	8.217	0.000	0.611	4.372	0.000	0.047	0.262	0.794
LNSALE(-1)	0.916	16.162	0.000	1.007	18.642	0.000	0.909	33.259	0.000	1.156	22.910	0.000
T	-0.128	-1.556	0.120									
T*LNSALE(-1)	0.026	1.725	0.085	-0.018	-2.492	0.013	0.007	1.831	0.067	-0.022	-3.670	0.000
R-squared				0.893								
Adjusted R-squared				0.892								
F-statistic				1069.0								
Prob(F-statistic)				0.000								
Durbin-Watson Stat				2.260								

significant adjusted effects on *lnSale* at the 5% significance level in the four cells. The output is obtained by replacing the term *@Expand(M_Size,TP)* in (12.15) with *@Expand(M_Size,TP, @drop(1,2), @drop(2,1), @Drop(2,2))*.

3.3   Other reduced models might be obtained by deleting a combination of the variables *lnSale*(−1) and the time *t*, from the model (12.15). Do it as an exercise and see the regression function presented in point (2.4) previously.

Example 12.3   Polynomial regressions

For a more advanced model, Table 12.3 presents good fit polynomial regressions in a statistical sense within each of the cells (*M_Size* = *1*,*TP* = *1*), (*M_Size* = *1*,*TP* = *2*), and (*M_Size* = *2*,*TP* = *1*).

1.   The regression in (M_Size = 1,TP = 1) has a very small DW-statistical value, so then alternative model(s) should be explored. Figure 12.6 presents the results based on two alternatives model using the trial-and-error method. Based on these results, the following notes and comments are presented.

1.1   Figure 12.6(a) presents an LVAR(1,2) model, but without the autoregressive term AR(1). If the model LVAR(1,1) was applied, then the AR(1) has a *p*-value > 0.30.

1.2   Figure 12.6(b) presents an uncommon or unexpected LV(3) quadratic model, which is a good fit model with adjusted R-squared = 0.845 662 but DW = 1.633 012. If the term AR(1) was to be inserted, then an unexpected DW-stat of 0.000 000 is obtained.

Based on this table the findings and notes are presented.

For further illustration, Figure 12.7 presents the results of a polynomial model:

2.   The regression in the cell (*M_Size* = 1,*TP* = 2) is a third-degree polynomial LV(1) model, which is a good fit. Similarly for the fourth-degree polynomial LV(1) model in the cell (*M_Size* = 2, *TP* = 1).

Table 12.3   *Summary of the statistical results based polynomial models of* lnSale *on* lnSale(−1)

Variable	M_Size=1 & TP=1 Coeff.	t-Stat	Prob.	M_Size=1 & TP=2 Coeff.	t-Stat	Prob.	M_Size=2 & TP=1 Coeff.	t-Stat	Prob.
C	2.420	34.364	0.000	1.492	8.360	0.000	3.885	10.811	0.000
LNSALE(-1)				0.272	3.190	0.002	-1.344	-4.256	0.000
LNSALE(-1)^2	0.097	45.957	0.000	0.119	5.109	0.000	0.541	5.273	0.000
LNSALE(-1)^3				-0.006	-3.502	0.001	-0.054	-4.088	0.000
LNSALE(-1)^4							0.002	3.376	0.001
R-squared	0.828			0.839			0.904		
Adjusted R-squared	0.828			0.837			0.903		
F-statistic	2112.0			576.232			1042.9		
Prob(F-statistic)	0.000			0.000			0.000		
Durbin-Watson Stat	0.988			2.380			2.555		

Dependent Variable: LNSALE
Method: Panel Least Squares
Date: 10/06/11 Time: 09:55
Sample: 1 8 IF M_SIZE=1 AND TP=1
Periods included: 2
Cross-sections included: 109
Total panel (balanced) observations: 218
Convergence achieved after 9 iterations

Variable	Coefficient	Std. Error	t-Statistic	Prob.
C	3.232151	0.283015	11.42043	0.0000
LNSALE(-1)^2	0.080736	0.005618	14.37154	0.0000
AR(2)	0.538576	0.088844	6.062043	0.0000

R-squared	0.845212	Mean dependent var	5.469507
Adjusted R-squared	0.843772	S.D. dependent var	1.971463
S.E. of regression	0.779235	Akaike info criterion	2.352657
Sum squared resid	130.5494	Schwarz criterion	2.399232
Log likelihood	-253.4396	Hannan-Quinn criter.	2.371469
F-statistic	586.9974	Durbin-Watson stat	1.628151
Prob(F-statistic)	0.000000		

Inverted AR Roots	.73	-.73

(a)

Dependent Variable: LNSALE
Method: Panel Least Squares
Date: 10/06/11 Time: 09:30
Sample: 1 8 IF M_SIZE=1 AND TP=1
Periods included: 2
Cross-sections included: 109
Total panel (balanced) observations: 218

Variable	Coefficient	Std. Error	t-Statistic	Prob.
C	1.497649	0.215645	6.944987	0.0000
LNSALE(-1)^2	0.081720	0.011046	7.398470	0.0000
LNSALE(-2)^2	-0.034770	0.015130	-2.298056	0.0225
LNSALE(-3)	0.503533	0.085498	5.889411	0.0000

R-squared	0.847796	Mean dependent var	5.469507
Adjusted R-squared	0.845662	S.D. dependent var	1.971463
S.E. of regression	0.774507	Akaike info criterion	2.344997
Sum squared resid	128.3702	Schwarz criterion	2.407098
Log likelihood	-251.6047	Hannan-Quinn criter.	2.370081
F-statistic	397.3349	Durbin-Watson stat	1.633012
Prob(F-statistic)	0.000000		

(b)

Figure 12.6   *Statistical results based on two modified models in cell (M_Size = 1, TP = 1)*

Dependent Variable: LNSALE
Method: Panel Least Squares
Date: 10/05/11 Time: 13:43
Sample: 1 8 IF M_SIZE=1 AND TP=2
Periods included: 3
Cross-sections included: 112
Total panel (balanced) observations: 336

Variable	Coefficient	Std. Error	t-Statistic	Prob.
C	2.333016	0.415931	5.609146	0.0000
LNSALE(-1)	0.279062	0.084787	3.291346	0.0011
LNSALE(-1)^2	0.118212	0.023168	5.102331	0.0000
LNSALE(-1)^3	-0.006151	0.001764	-3.487197	0.0006
T	-0.123730	0.055366	-2.234765	0.0261

R-squared	0.841284	Mean dependent var	5.762264
Adjusted R-squared	0.839366	S.D. dependent var	2.057590
S.E. of regression	0.824665	Akaike info criterion	2.467090
Sum squared resid	225.1039	Schwarz criterion	2.523892
Log likelihood	-409.4711	Hannan-Quinn criter.	2.489733
F-statistic	438.6217	Durbin-Watson stat	2.404268
Prob(F-statistic)	0.000000		

(a)

Dependent Variable: LNSALE
Method: Stepwise Regression
Date: 10/06/11 Time: 12:24
Sample: 1 8 IF M_SIZE=1 AND TP=1
Included observations: 441
Number of always included regressors: 3
Number of search regressors: 4
Selection method: Combinatorial
Number of search regressors: 3

Variable	Coefficient	Std. Error	t-Statistic	Prob.*
C	0.980436	0.341820	2.868281	0.0043
T*LNSALE(-1)^2	0.015563	0.006170	2.522214	0.0120
T*LNSALE(-1)^3	-0.001477	0.000672	-2.197657	0.0285
LNSALE(-1)	0.774732	0.093102	8.321292	0.0000
T	-0.181840	0.075057	-2.422711	0.0158
LNSALE(-1)^3	0.001606	0.001151	1.395115	0.1637

R-squared	0.887700	Mean dependent var	5.136917
Adjusted R-squared	0.886410	S.D. dependent var	1.936887
S.E. of regression	0.652793	Akaike info criterion	1.998398
Sum squared resid	185.3701	Schwarz criterion	2.054031
Log likelihood	-434.6467	Hannan-Quinn criter.	2.020343
F-statistic	687.7131	Durbin-Watson stat	1.707454
Prob(F-statistic)	0.000000		

Selection Summary	
Number of combinations compared:	4

*Note: p-values and subsequent tests do not account for stepwise selection.

(b)

Figure 12.7   *Statistical results based on two polynomial regressions, with trend and time-related effects, in cell (M_Size = 1, TP = 2)*

Example 12.4   Polynomial models with trend or trend-related effects

For illustration, Figure 12.7 presents the statistical results based on two polynomial models, namely a model with trend and a model with time-related effects, respectively, in cells $(M\_Size = 1, TP = 2)$ and $(M\_Size = 1, TP = 1)$ as the extension of the model in Table 12.3. Both models are good fits. However, they might not be the best and it is impossible to try all possible models.

Specific for the model with time-related effects in Figure 12.7(b), the final result is obtained using a multistage combinatorial selection method, such as follows:

1. At the first stage, C is used as the always included regressor, and "$t^*lnSale(-1)$ $t^*lnSale(-1)^2$ $t^*lnSale(-1)^3$" is the list of search regressors. By selecting the alternative options: 1, 2 and 3, as the alternative number of regressors to select, then the regression is obtained with two independent variables as the best fit.
2. At the second stage, the list of always-included regressors is "C $t^*lnSale(-1)^2$ $t^*lnSale(-1^3)$", and "t $lnSale(-1)$ $lnSale(-1)^2$ $lnSale(-1)^3$" used as the list of search regressors, obtained in the previous output. Note that $lnSale(-1)^3$ has a p-value $= 0.1637$, but at the 10% significance level it has a positive significant adjusted effect on $lnSale$ with a p-value $= 0.1637/2 = 0.08\,185$.

Example 12.5 An extension of the model in (12.15)

Table 12.4 presents the summary of the statistical results using the following equation specification.

$$lnsale\ c\ @expand(m\_size, tp, @dropfirst)\ lnsale(-1)^*@expand(m\_size, tp)$$
$$lnsize^*@expand(m\_size, tp)\ t^*@expand(m\_size, tp) \quad (12.18)$$
$$t^*lnsale(-1)^*@expand(m\_size, tp)\ t^*lnsize^*@expand(m\_size, tp)$$

where TP is a dichotomous variable, representing the time period before and after an Indonesian new tax policy.

Based on the output, the following findings and notes are presented.

1. Table 12.4 clearly shows a set of four heterogeneous regressions with trend and time-related effects, but each of the four regressions has at least one very large p-value. So reduced models should be explored, in a statistical sense. For how to develop possible reduced model(s), refer to the results from using combinatorial selection methods presented in Figure 12.7(b), or use the trial-and-error method. Do this as an exercise.
2. Take note this LV(1) model is one out of many possible which could be developed or subjectively defined based on the variables $SIZE_{it}$ and $SALE_{it}$, such as the LVAR(p,q) models for various integers p and q, and polynomial or nonlinear interaction models of $lnSale$ on $lnSale(-1)$, $lnSize$ or $lnSize(-1)$ and the time t. Note that the results are highly dependent on the data used.

Table 12.4 *Summary of the statistical results of the LV(1) model in (12.18)*

Variable	M_Size=1 & TP=1			M_Size=1 & TP=2			M_Size=2 & TP=1			M_Size=2 & TP=2		
	Coeff.	t-Stat	Prob.	Coeff.	t-Stat	Prob.	Coeff.	t-Stat	Prob.	Coeff.	t-Stat	Prob.
C	0.875	2.161	0.031	1.854	0.871	0.384	0.870	1.317	0.188	-0.824	-0.208	0.836
LNSALE(-1)	0.977	16.807	0.000	0.593	2.981	0.003	0.985	14.333	0.000	0.799	2.879	0.004
LNSIZE	-0.366	-1.720	0.086	0.827	0.530	*0.596*	-0.394	-1.243	0.214	1.721	0.684	*0.494*
T	-0.312	-2.543	0.011	-0.454	-1.516	0.130	-0.157	-0.707	*0.479*	-0.051	-0.091	*0.928*
T*LNSALE(-1)	0.002	0.124	*0.901*	0.016	0.563	*0.574*	-0.018	-0.928	*0.354*	0.020	0.511	*0.609*
T*LNSIZE	0.180	2.301	0.022	0.136	0.619	*0.536*	0.168	1.401	0.161	-0.126	-0.350	*0.726*
R-squared	0.9008											
Adjusted R-squared	0.8993											
F-statistic	604.34											
Prob(F-statistic)	0.000											
Durbin-Watson Stat	2.151											

Example 12.6   LVAR(1,2) model based on the model in (9.60)
Using all variables of the model in (9.60), Table 12.5 presents the summary of the statistical results of the LVAR(1,2) model with time-related effects using the following equation specification.

$$y1 \ @expand(m\_size, tp) \ y1(-1)^* @expand(m\_size, tp) \ t^*x1^*x2^* @expand(m\_size, tp)$$
$$t^*x1^* @expand(m\_size, tp) \ x1^*x2^* @expand(m\_size, tp) \ t^*x2^* @expand(m\_size, tp)$$
$$t^* @expand(m\_size, tp) \ x1^* @expand(m\_size, tp) \ x2^* @expand(m\_size, tp) \quad (12.19)$$
$$x3 \ x4 \ x5 \ ar(1) \ ar(2)$$

Based on this summary, the following findings and notes are presented.

1. In this LVAR(1,2) model, it is assumed that the covariates $X3$, $X4$ and $X5$, have invariant linear effects on $Y1$ over the Group generated by M_Size and TP. Similarly so for the terms AR(1) and AR(2).
2. The main problem of this model is that most of the independent variables (8 out of 11, including the covariates) of the regression in $(M\_Size = 1, TP = 2)$ have very large p-values. The second problem is the regression in $(M\_Size = 1, TP = 1)$, since three out of four interactions have large p-values.
3. Since it is defined that the effects of $X1$ and $X2$ on $Y1$ depend on time t in a theoretical sense, then the reduced model should be developed so that each of the four regressions in the model has at least one interaction independent variable. Refer to the output of the regression in Figure 12.7(b) using the combinatorial selection method. See the application of a *manual stepwise selection method* as an alternative, presented in the following section.
4. On the other hand, if the differential effects of one or more independent variables on $Y1$ between the four cells are tested, then each regression should have the same set of independent variables. In this case, the full model in Table 12.4 would be valid. For instance, the interaction t^*X1^*X2 has significant different adjusted effects on $Y1$, since the null hypothesis H_0: $C(9) = c(10) = C(11) = C(12)$ is rejected based on the F-statistic of $F_0 = 4.604\ 229$ with $df = (3,940)$ and a p-value $= 0.0033$. The other hypotheses can easily be tested using the Wald test.
5. On the other hand, if the four regressions have different sets of independent variables beside the interaction t^*X1^*X2, then their effects of t^*X1^*X2 on Y1 would be adjusted for different sets of variables. So the effects have different meanings.

Example 12.7   Application of a multistage combinatorial selection method
For illustration Figure 12.8 presents the statistical results based on the sample {M_Size = 1 and TP = 2} to develop a reduced model of that in Table 12.4, using a multistage combinatorial selection method. The stages of the analysis are as follows:

1. At Stage-1, click *Quick/Estimation Equation…/STEPLS*, then "*Y1 C Y1(−1)*" is inserted as the list of the dependent variables *Y1* followed by the always-included regressors, with "$t^*X1^*X2 \ t^*X1 \ X1^*X2$ t^*X2" as the list of four search regressors. By selecting the *Options/Combinatorial…*, then the output will be shown on the screen for each of number of search regressors to select: 1, 2, 3 and 4. At this stage, Figure 12.8(a) presents the best LV(1) model out of four combinations compared.
2. At Stage-2, "*Y1 C Y1(−1) $t^*X1^*X2 \ t^*X1 \ t^*X2$*" is inserted as the list of the dependent variables *Y1* followed by always-included regressors, with "*X3 X4 X5*" as the list of search regressors. At this stage, Figure 12.8(b) presents the best LV(1) model out of three combinations compared.
3. At Stage-3, "*Y1 C Y1(−1) $t^*X1^*X2 \ t^*X1 \ t^*X2 \ X3$*" is inserted as the list of the dependent variables *Y1* followed by always-included regressors, with "*t X1 X2*" as the list of search regressors. At this stage,

Table 12.5 Summary of the statistical results of the LVAR(1,1) model in (12.19)

Variable	M_Size=1 & TP=1 Coff.	t-Stat	Prob.	M_Size=1 & TP=2 Coff.	t-Stat	Prob.	M_Size=2 & TP=1 Coff.	t-Stat	Prob.	M_Size=2 & TP=2 Coff.	t-Stat	Prob.
C	-6.135	-2.299	0.022	-2.867	-1.699	0.090	-3.977	-0.824	0.410	-11.073	-4.762	0.000
Y1(-1)	0.824	13.186	0.000	0.970	26.750	0.000	1.096	46.303	0.000	1.086	76.154	0.000
T*X1*X2	-0.036	-0.682	0.496	0.009	0.319	0.750	-0.195	-1.948	0.052	-0.181	-3.945	0.000
T*X1	-0.166	-1.867	0.062	-0.032	-0.853	0.394	-0.037	-0.282	0.778	-0.211	-4.871	0.000
X1*X2	0.254	1.006	0.315	0.031	0.150	0.881	0.898	1.957	0.051	1.345	4.222	0.000
T*X2	0.122	0.433	0.665	-0.017	-0.104	0.917	0.970	1.420	0.156	1.104	3.349	0.001
T	0.801	1.469	0.142	0.089	0.373	0.710	0.374	0.388	0.698	1.214	3.741	0.000
X1	0.836	1.925	0.055	0.251	0.925	0.355	0.131	0.203	0.839	1.494	4.841	0.000
X2	-0.950	-0.708	0.479	-0.261	-0.219	0.827	-4.368	-1.401	0.162	-8.226	-3.583	0.000

Covariate	Coff.	t-Stat	Prob.
X3	-0.086	-2.297	0.022
X4	1.254	5.328	0.000
X5	0.012	0.585	0.559
AR(1)	-0.046	-1.349	0.178
AR(2)	0.179	6.098	0.000

R-squared	0.962
Adjusted R-squared	0.961
F-statistic	601.2
Prob(F-statistic)	0.000
Dubin-Watson Stat	2.143
Inverted AR Roots	0.400	-0.450

Dependent Variable: Y1
Method: Stepwise Regression
Date: 10/13/11 Time: 11:16
Sample: 1 8 IF M_SIZE=1 AND TP=2
Included observations: 313
Number of always included regressors: 2
Number of search regressors: 4
Selection method: Combinatorial
Number of search regressors: 3

Variable	Coefficient	Std. Error	t-Statistic	Prob.*
C	-0.343745	0.191639	-1.793711	0.0738
Y1(-1)	1.036824	0.030335	34.17947	0.0000
T*X1*X2	0.013007	0.002953	4.404993	0.0000
T*X2	-0.055033	0.017049	-3.227896	0.0014
T*X1	0.010074	0.004688	2.148937	0.0324

R-squared	0.868178	Mean dependent var	1.037042
Adjusted R-squared	0.866466	S.D. dependent var	2.526029
S.E. of regression	0.923068	Akaike info criterion	2.693618
Sum squared resid	262.4329	Schwarz criterion	2.753462
Log likelihood	-416.5513	Hannan-Quinn criter.	2.717533
F-statistic	507.1223	Durbin-Watson stat	2.316459
Prob(F-statistic)	0.000000		

Selection Summary
Number of combinations compared: 4

*Note: p-values and subsequent tests do not account for stepwise
   selection.

(a)

Dependent Variable: Y1
Method: Stepwise Regression
Date: 10/13/11 Time: 11:19
Sample: 1 8 IF M_SIZE=1 AND TP=2
Included observations: 313
Number of always included regressors: 5
Number of search regressors: 3
Selection method: Combinatorial
Number of search regressors: 1

Variable	Coefficient	Std. Error	t-Statistic	Prob.*
C	-0.397399	0.184946	-2.148731	0.0324
Y1(-1)	1.047092	0.029298	35.73918	0.0000
T*X1*X2	0.009599	0.002926	3.280834	0.0012
T*X2	-0.040821	0.016672	-2.448499	0.0149
T*X1	0.012204	0.004537	2.690190	0.0075
X3	-0.210364	0.042224	-4.982072	0.0000

R-squared	0.878039	Mean dependent var	1.037042
Adjusted R-squared	0.876053	S.D. dependent var	2.526029
S.E. of regression	0.889318	Akaike info criterion	2.622260
Sum squared resid	242.8023	Schwarz criterion	2.694072
Log likelihood	-404.3837	Hannan-Quinn criter.	2.650958
F-statistic	442.0391	Durbin-Watson stat	2.491507
Prob(F-statistic)	0.000000		

Selection Summary
Number of combinations compared: 3

*Note: p-values and subsequent tests do not account for stepwise
   selection.

(b)

Dependent Variable: Y1
Method: Stepwise Regression
Date: 10/13/11 Time: 11:20
Sample: 1 8 IF M_SIZE=1 AND TP=2
Included observations: 313
Number of always included regressors: 6
Number of search regressors: 3
Selection method: Combinatorial
Number of search regressors: 2

Variable	Coefficient	Std. Error	t-Statistic	Prob.*
C	-2.296913	1.433445	-1.602373	0.1101
Y1(-1)	1.028825	0.029221	35.20873	0.0000
T*X1*X2	0.010766	0.002944	3.657467	0.0003
T*X2	-0.049247	0.016914	-2.911517	0.0039
T*X1	-0.055930	0.032871	-1.701533	0.0899
X3	-0.231853	0.041689	-5.561512	0.0000
T	0.221232	0.203333	1.088028	0.2774
X1	0.539052	0.232464	2.318864	0.0211

R-squared	0.884198	Mean dependent var	1.037042
Adjusted R-squared	0.881541	S.D. dependent var	2.526029
S.E. of regression	0.869407	Akaike info criterion	2.583215
Sum squared resid	230.5398	Schwarz criterion	2.678965
Log likelihood	-396.2732	Hannan-Quinn criter.	2.621479
F-statistic	332.6879	Durbin-Watson stat	2.461861
Prob(F-statistic)	0.000000		

Selection Summary
Number of combinations compared: 3

*Note: p-values and subsequent tests do not account for stepwise
   selection.

(c)

Dependent Variable: Y1
Method: Panel Least Squares
Date: 10/13/11 Time: 11:25
Sample: 1 8 IF M_SIZE=1 AND TP=2
Periods included: 3
Cross-sections included: 106
Total panel (unbalanced) observations: 307
Convergence achieved after 56 iterations

Variable	Coefficient	Std. Error	t-Statistic	Prob.
C	-4.730313	2.153147	-2.196931	0.0288
Y1(-1)	0.925778	0.044419	20.84185	0.0000
T*X1*X2	0.014204	0.003406	4.169930	0.0000
T*X2	-0.061591	0.019670	-3.131284	0.0019
T*X1	-0.104053	0.045395	-2.292184	0.0226
X3	-0.178301	0.043106	-4.136316	0.0000
T	0.484093	0.287262	1.685197	0.0930
X1	0.992009	0.341843	2.901947	0.0040
AR(1)	0.266546	0.071175	3.744950	0.0002

R-squared	0.885677	Mean dependent var	1.049121
Adjusted R-squared	0.882608	S.D. dependent var	2.546993
S.E. of regression	0.872664	Akaike info criterion	2.594346
Sum squared resid	226.9397	Schwarz criterion	2.703602
Log likelihood	-389.2321	Hannan-Quinn criter.	2.638036
F-statistic	288.5813	Durbin-Watson stat	2.709903
Prob(F-statistic)	0.000000		

Inverted AR Roots	.27		

(d)

Figure 12.8   *Statistical results of a reduced model of the model in Table 12.4, based on the sample {M_Size= 1 and TP= 2}*

Figure 12.8(c) presents the best LV(1) model out of three combinations compared. Based on this model, the following notes and comment are made.

3.1 The time variable t has the largest p-value of 0.277, but still less than the 0.30 based on the LV (1) model. However, based on the LVAR(1,1) model in Figure 12.8(d), at the $\alpha = 0.05$ significance level, the time t has a positive significant adjusted effect on Y1 with a p-value $= 0.0930/2 = 0.0465 < \alpha = 0.05$.

3.2 Referring to the LVAR(1,2) model in Table 12.4, if an LVAR(1,2) model were applied, then the note "*Estimated AR process in nonstationary*" presented in the output. So the LVAR(1,2) model is not acceptable. On the other hand, the variable X1 and the time t have very large p-values of 0.9321 and 0.6872, respectively.

12.5.5 Special Selected LV(p) or AR(q) Models

12.5.5.1 LV(p) or AR(q) Models Without a Numerical Exogenous Variable

The two models: LV(p) model and AR(p) model have the same parameter estimates except their intercept parameters and the *Inverted AR Roots* of the AR(q) model. See the following examples.

Example 12.8 Simple linear regressions of Y1 and Y1(−1), and their correlation

Figure 12.9 presents the outputs of three statistical data analyses, using the sample {M_Size $= 1$,TP $= 2$}, such as follows:

Dependent Variable: Y1
Method: Panel Least Squares
Date: 10/13/11 Time: 15:31
Sample: 1 8 IF M_SIZE=1 AND TP=2
Periods included: 3
Cross-sections included: 112
Total panel (balanced) observations: 336

Variable	Coefficient	Std. Error	t-Statistic	Prob.
C	0.076383	0.055568	1.374573	0.1702
Y1(-1)	1.079437	0.023619	45.70282	0.0000

R-squared	0.862140	Mean dependent var	1.048467
Adjusted R-squared	0.861727	S.D. dependent var	2.530621
S.E. of regression	0.941013	Akaike info criterion	2.722214
Sum squared resid	295.7586	Schwarz criterion	2.744935
Log likelihood	-455.3320	Hannan-Quinn criter.	2.731272
F-statistic	2088.747	Durbin-Watson stat	2.293670
Prob(F-statistic)	0.000000		

Dependent Variable: Y1(-1)
Method: Panel Least Squares
Date: 10/13/11 Time: 15:33
Sample: 1 8 IF M_SIZE=1 AND TP=2
Periods included: 3
Cross-sections included: 112
Total panel (balanced) observations: 336

Variable	Coefficient	Std. Error	t-Statistic	Prob.
C	0.063143	0.047809	1.320731	0.1875
Y1	0.798694	0.017476	45.70282	0.0000

R-squared	0.862140	Mean dependent var	0.900548
Adjusted R-squared	0.861727	S.D. dependent var	2.176800
S.E. of regression	0.809444	Akaike info criterion	2.420997
Sum squared resid	218.8368	Schwarz criterion	2.443718
Log likelihood	-404.7275	Hannan-Quinn criter.	2.430054
F-statistic	2088.747	Durbin-Watson stat	2.330540
Prob(F-statistic)	0.000000		

Covariance Analysis: Ordinary
Date: 10/13/11 Time: 15:35
Sample (adjusted): 6 8
Included observations: 336 after adjustments

Correlation t-Statistic Probability	Y1	Y1(-1)
Y1	1.000000	

Y1(-1)	0.928515	1.000000
	45.70282	-----
	0.0000	-----

Figure 12.9 *Simple linear regressions and bivariate correlation of Y1′ and Y1(−1)*

1. The analysis based on the simplest LV(1) model of *Y1* on *Y1*(−1),
2. The analysis based on the simplest linear regression of Y1(−1) on Y1, and
3. The ordinary covariance analysis of Y1 and Y1(−1).

Take note that the three outputs present exactly the same *t*-test statistic value of $t_0 = 45.70\,282$ with a *p*-value = 0.0000. So the causal effect or up- and down-stream relationship of *Y1*(−1) on *Y1* and the predicted values/scores of *Y1*(−1) using *Y1* as a predictor, can be tested using ordinary covariance analysis only.

To generalize, based on balanced panel data, we can conclude that the causal effect or up- and down-stream relationship of *X*(−1) on *Y* as well as of *Y* on *X*(−1), can be tested using ordinary covariance analysis only. The main advantage in using the ordinary covariance analysis is it can be done using any set of variables. See the analyses based on the correlation matrixes presented in Figures 8.3 and 8.4.

Example 12.9   Comparison between LV(2) and AR(2) models
Figure 12.10 presents the statistical results of the LV(2) and AR(2) models of *lnSale* based on the sample {M_Size = 1 and TP = 2}. Except for their intercepts and the *Inverted AR Roots*, both outputs present exactly the same sampled statistics or estimates.

For more advanced models, Figure 12.11 presents the statistical results of the LV(3) and AR(3) models of Y1 which are good fits. However, by using AR(4) model, the output presents the note "*Estimated AR process is nonstationary*". So the AR(4) model is not acceptable. See the following example.

To generalize, an LV(p) model has the general equation, as follows:

$$Y_{i,t} = C(1) + C(2)Y_{i,t-1} + \ldots + C(p+1)Y_{i,t-p} + \varepsilon_{it} \tag{12.20}$$

and an AR(p) model has the following general equation:

$$\begin{aligned} Y_{it} &= C(1) + \mu_{it} \\ \mu_{i,t} &= \rho_1\mu_{i,t-1} + \rho_2\mu_{i,t-2}\ldots + \rho_p\mu_{i,t-p} + \varepsilon_{it} \end{aligned} \tag{12.21}$$

where the statistical results present the estimates $\hat{C}(k+1) = \hat{\rho}_k$, for $k = 1, 2, \ldots, p$.

Dependent Variable: LNSALE				
Method: Panel Least Squares				
Date: 10/13/11 Time: 13:41				
Sample: 1 8 IF M_SIZE=1 AND TP=2				
Periods included: 3				
Cross-sections included: 112				
Total panel (unbalanced) observations: 334				

Variable	Coefficient	Std. Error	t-Statistic	Prob.
C	0.681435	0.138583	4.917166	0.0000
LNSALE(-1)	0.605867	0.055448	10.92674	0.0000
LNSALE(-2)	0.308118	0.057063	5.399623	0.0000
R-squared	0.823149	Mean dependent var		5.777423
Adjusted R-squared	0.822080	S.D. dependent var		2.054050
S.E. of regression	0.866409	Akaike info criterion		2.560022
Sum squared resid	248.4699	Schwarz criterion		2.594254
Log likelihood	-424.5237	Hannan-Quinn criter.		2.573671
F-statistic	770.3161	Durbin-Watson stat		1.911060
Prob(F-statistic)	0.000000			

Dependent Variable: LNSALE				
Method: Panel Least Squares				
Date: 10/13/11 Time: 13:41				
Sample: 1 8 IF M_SIZE=1 AND TP=2				
Periods included: 3				
Cross-sections included: 112				
Total panel (unbalanced) observations: 334				
Convergence achieved after 3 iterations				

Variable	Coefficient	Std. Error	t-Statistic	Prob.
C	7.922271	0.860829	9.203076	0.0000
AR(1)	0.605867	0.055448	10.92674	0.0000
AR(2)	0.308118	0.057063	5.399623	0.0000
R-squared	0.823149	Mean dependent var		5.777423
Adjusted R-squared	0.822080	S.D. dependent var		2.054050
S.E. of regression	0.866409	Akaike info criterion		2.560022
Sum squared resid	248.4699	Schwarz criterion		2.594254
Log likelihood	-424.5237	Hannan-Quinn criter.		2.573671
F-statistic	770.3161	Durbin-Watson stat		1.911060
Prob(F-statistic)	0.000000			
Inverted AR Roots	.94	-.33		

Figure 12.10   *Outputs of LV(2) and AR(2) models of* lnSale, *using the sample {M_Size = 1, TP = 2}*

```
Dependent Variable: Y1
Method: Panel Least Squares
Date: 10/13/11   Time: 14:40
Sample: 1 8 IF M_SIZE=1 AND TP=2
Periods included: 3
Cross-sections included: 112
Total panel (balanced) observations: 336
```

Variable	Coefficient	Std. Error	t-Statistic	Prob.
C	0.094194	0.054824	1.718116	0.0867
Y1(-1)	1.155523	0.062725	18.42214	0.0000
Y1(-2)	0.142202	0.116975	1.215658	0.2250
Y1(-3)	-0.317193	0.095771	-3.311986	0.0010
R-squared	0.868079	Mean dependent var		1.048467
Adjusted R-squared	0.866887	S.D. dependent var		2.530621
S.E. of regression	0.923287	Akaike info criterion		2.690081
Sum squared resid	283.0166	Schwarz criterion		2.735523
Log likelihood	-447.9336	Hannan-Quinn criter.		2.708195
F-statistic	728.2214	Durbin-Watson stat		2.668037
Prob(F-statistic)	0.000000			

```
Dependent Variable: Y1
Method: Panel Least Squares
Date: 10/13/11   Time: 14:38
Sample: 1 8 IF M_SIZE=1 AND TP=2
Periods included: 3
Cross-sections included: 112
Total panel (balanced) observations: 336
Convergence achieved after 3 iterations
```

Variable	Coefficient	Std. Error	t-Statistic	Prob.
C	4.838422	8.270800	0.585000	0.5589
AR(1)	1.155523	0.062725	18.42214	0.0000
AR(2)	0.142202	0.116975	1.215658	0.2250
AR(3)	-0.317193	0.095771	-3.311986	0.0010
R-squared	0.868079	Mean dependent var		1.048467
Adjusted R-squared	0.866887	S.D. dependent var		2.530621
S.E. of regression	0.923287	Akaike info criterion		2.690081
Sum squared resid	283.0166	Schwarz criterion		2.735523
Log likelihood	-447.9336	Hannan-Quinn criter.		2.708195
F-statistic	728.2214	Durbin-Watson stat		2.668037
Prob(F-statistic)	0.000000			
Inverted AR Roots	.96	.68	-.49	

Figure 12.11   *Outputs of LV(3) and AR(3) models of Y1, using the sample {M_Size = 1, TP = 2}*

Example 12.10   Application of LVAR(p,q) models

For illustration Figure 12.12 presents the statistical results of an LVAR(3,3) model of *Y1* and its reduced model, namely LVAR(2,2), based on the sample {M_Size = 1, TP = 2}.

Note that the output of the LVAR(3,3) model presents a note "*Estimated AR process is nonstationary*", so it is not a good fit. On the other hand, its reduced model, an LVAR(2,2), is a good fit.

To generalize, the equation of an LVAR(p,q) model is as follows:

$$Y_{i,t} = C(1) + C(2)Y_{i,t-1} + \ldots + C(p+1)Y_{i,t-p} + \mu_{it}$$
$$\mu_{i,t} = \rho_1 \mu_{i,t-1} + \rho_2 \mu_{i,t-2} \ldots + \rho_q \mu_{i,t-q} + \varepsilon_{it}$$

(12.22)

```
Dependent Variable: Y1
Method: Panel Least Squares
Date: 10/13/11   Time: 15:05
Sample: 1 8 IF M_SIZE=1 AND TP=2
Periods included: 2
Cross-sections included: 112
Total panel (balanced) observations: 224
Convergence not achieved after 500 iterations
```

Variable	Coefficient	Std. Error	t-Statistic	Prob.
C	0.065815	0.039578	1.662891	0.0978
Y1(-1)	1.852427	0.346282	5.349468	0.0000
Y1(-2)	-0.732885	0.784808	-0.933841	0.3514
Y1(-3)	-0.222242	0.495741	-0.448303	0.6544
AR(1)	-0.740617	0.338805	-2.185965	0.0299
AR(2)	0.205963	0.564823	0.364650	0.7157
AR(3)	-0.227023	0.127058	-1.786761	0.0754
R-squared	0.821233	Mean dependent var		1.133741
Adjusted R-squared	0.816290	S.D. dependent var		2.729377
S.E. of regression	1.169849	Akaike info criterion		3.182378
Sum squared resid	296.9746	Schwarz criterion		3.288992
Log likelihood	-349.4263	Hannan-Quinn criter.		3.225412
F-statistic	166.1451	Durbin-Watson stat		2.620671
Prob(F-statistic)	0.000000			
Inverted AR Roots	.18+.41i	.18-.41i	-1.11	
Estimated AR process is nonstationary				

```
Dependent Variable: Y1
Method: Panel Least Squares
Date: 10/13/11   Time: 15:04
Sample: 1 8 IF M_SIZE=1 AND TP=2
Periods included: 3
Cross-sections included: 112
Total panel (balanced) observations: 336
Convergence achieved after 9 iterations
```

Variable	Coefficient	Std. Error	t-Statistic	Prob.
C	0.046743	0.022825	2.047929	0.0414
Y1(-1)	1.955025	0.052533	37.21550	0.0000
Y1(-2)	-1.003365	0.061550	-16.30160	0.0000
AR(1)	-0.847232	0.071671	-11.82115	0.0000
AR(2)	-0.469267	0.072539	-6.469129	0.0000
R-squared	0.878411	Mean dependent var		1.048467
Adjusted R-squared	0.876942	S.D. dependent var		2.530621
S.E. of regression	0.887732	Akaike info criterion		2.614476
Sum squared resid	260.8506	Schwarz criterion		2.671278
Log likelihood	-434.2319	Hannan-Quinn criter.		2.637119
F-statistic	597.8238	Durbin-Watson stat		2.441791
Prob(F-statistic)	0.000000			
Inverted AR Roots	-.42+.54i	-.42-.54i		

Figure 12.12   *Outputs of LVAR(3,3) and LVAR(2,2) models of Y1: sample {M_Size = 1, TP = 2}*

12.5.5.2 LV(p) or AR(q) Models with a Numerical Exogenous Variable

Example 12.11   Application of LV(2) and AR(2) models

As an extension of each of the LV(2) and AR(2) models in Figure 12.9, *lnSize(−1)* is inserted as an additional independent variable, with the statistical results presented in Figure 12.13.

Based on these results, the following findings and notes are presented.

1. Both models are acceptable, in a statistical sense, since each of their independent variables and the indicators AR(1) and AR(2) have significant effects on *lnSale*.
2. The models are LV(2) and AR(2) linear-effect models of *lnSale on lnSize(−1)*.
3. Compare these to the LV(2) and AR(2) models in Figure 12.9, note that in these models, the slope parameters of *lnSale(−1)* and *lnSale(−2)* are not equal to the autoregressive parameters of *AR(1)* and *AR(2)*.
4. For additional illustration, Figure 12.14 presents nonlinear kernel fit curves of *lnSale* on *lnSize(−1)* based on two samples {M_Size = 1,TP = 2} and {M_Size = 1,TP = 2,lnSize(−1) > 1.0}. The second sample is selected to delete some of the outliers of the variable *lnSize(−1)*. So there is possibility to apply a polynomial model of *lnSale* on *lnSize(−1)*, similar to the polynomial models of *lnSale* on *lnSale(−1)*, presented in Examples 11.2 and 11.3. See the following example.

Example 12.12   Applications of LV(2) and AR(2) polynomial-effect models

As an extension of each of the linear-effect models in Figure 12.13, Figure 12.15 presents the statistical results based on LV(2) and AR(2) polynomial models of *lnSale* on *lnSize(−1)*, based on the sample {M_Size = 1,TP = 2,lnSize(−1) > 1.0}.

Based on these results, the following findings and notes are presented.

1. Corresponding to the kernel fit curve in Figure 12.14, the variable *lnSale(−1)^3* should be used as an independent variable, because the curve has a turning point. See point (5) later.

Dependent Variable: LNSALE
Method: Panel Least Squares
Date: 10/13/11   Time: 14:30
Sample: 1 8 IF M_SIZE=1 AND TP=2
Periods included: 3
Cross-sections included: 112
Total panel (unbalanced) observations: 333

Variable	Coefficient	Std. Error	t-Statistic	Prob.
C	-0.445603	0.348746	-1.277728	0.2022
LNSIZE(-1)	0.986443	0.276032	3.573653	0.0004
LNSALE(-1)	0.529379	0.058723	9.014825	0.0000
LNSALE(-2)	0.278511	0.056939	4.891380	0.0000

R-squared	0.822909	Mean dependent var		5.799760
Adjusted R-squared	0.821294	S.D. dependent var		2.016106
S.E. of regression	0.852282	Akaike info criterion		2.530141
Sum squared resid	238.9805	Schwarz criterion		2.575884
Log likelihood	-417.2684	Hannan-Quinn criter.		2.548381
F-statistic	509.5992	Durbin-Watson stat		1.753541
Prob(F-statistic)	0.000000			

Dependent Variable: LNSALE
Method: Panel Least Squares
Date: 10/13/11   Time: 14:27
Sample: 1 8 IF M_SIZE=1 AND TP=2
Periods included: 3
Cross-sections included: 112
Total panel (unbalanced) observations: 332
Convergence achieved after 12 iterations

Variable	Coefficient	Std. Error	t-Statistic	Prob.
C	6.400111	1.170119	5.469626	0.0000
LNSIZE(-1)	0.601688	0.288198	2.087763	0.0376
AR(1)	0.593737	0.052783	11.24864	0.0000
AR(2)	0.320985	0.053352	6.016403	0.0000

R-squared	0.828972	Mean dependent var		5.808878
Adjusted R-squared	0.827408	S.D. dependent var		2.012262
S.E. of regression	0.835979	Akaike info criterion		2.491548
Sum squared resid	229.2261	Schwarz criterion		2.537392
Log likelihood	-409.5969	Hannan-Quinn criter.		2.509830
F-statistic	529.9381	Durbin-Watson stat		1.925246
Prob(F-statistic)	0.000000			

Inverted AR Roots	.94		-.34

Figure 12.13 *Statistical results of the LV(2) and AR(2) linear effect models of* lnSale *on* lnSize(−1)*, based on sample {M_Size = 1,TP = 2}*

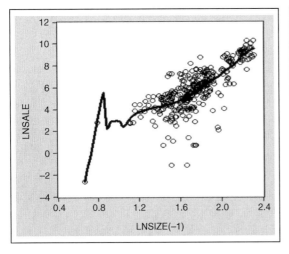 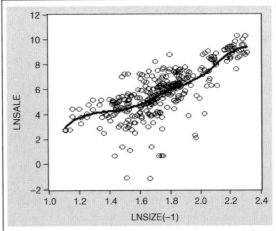

Figure 12.14 *Scatter graphs of* lnSale *on* lnSize(−1) *based on two samples: {M_Size = 1, TP = 2} and {M_Size = 1, TP = 2, lnSize(−1) > 1.0}*

Dependent Variable: LNSALE
Method: Panel Least Squares
Date: 10/14/11 Time: 22:10
Sample: 1 8 IF M_SIZE=1 AND TP=2 AND LNSIZE(-1) > 1
Periods included: 3
Cross-sections included: 111
Total panel (unbalanced) observations: 331

Variable	Coefficient	Std. Error	t-Statistic	Prob.
C	-0.281377	1.145495	-0.245638	0.8061
LNSIZE(-1)	1.796231	1.061419	1.692292	0.0916
LNSIZE(-1)^3	-0.178648	0.122049	-1.463741	0.1442
LNSALE(-1)	0.117152	0.094483	1.239934	0.2159
LNSALE(-1)^2	0.060508	0.014198	4.261649	0.0000
LNSALE(-2)	0.298650	0.091475	3.264836	0.0012
LNSALE(-2)^2	-0.015257	0.013775	-1.107567	0.2689

R-squared	0.845498	Mean dependent var	5.834461
Adjusted R-squared	0.842637	S.D. dependent var	1.960487
S.E. of regression	0.777705	Akaike info criterion	2.355983
Sum squared resid	195.9634	Schwarz criterion	2.436390
Log likelihood	-382.9152	Hannan-Quinn criter.	2.388053
F-statistic	295.5108	Durbin-Watson stat	2.002761
Prob(F-statistic)	0.000000		

(a)

Dependent Variable: LNSALE
Method: Panel Least Squares
Date: 10/14/11 Time: 22:03
Sample: 1 8 IF M_SIZE=1 AND TP=2 AND LNSIZE(-1) > 1
Periods included: 3
Cross-sections included: 111
Total panel (unbalanced) observations: 331
Convergence achieved after 11 iterations

Variable	Coefficient	Std. Error	t-Statistic	Prob.
C	7.149957	0.945762	7.559997	0.0000
LNSIZE(-1)	-0.778010	0.701850	-1.108514	0.2685
LNSIZE(-1)^3	0.216295	0.107933	2.003985	0.0459
AR(1)	0.536440	0.053317	10.06141	0.0000
AR(2)	0.349530	0.052852	6.613340	0.0000

R-squared	0.828326	Mean dependent var	5.834461
Adjusted R-squared	0.826220	S.D. dependent var	1.960487
S.E. of regression	0.817268	Akaike info criterion	2.449291
Sum squared resid	217.7442	Schwarz criterion	2.506725
Log likelihood	-400.3577	Hannan-Quinn criter.	2.472198
F-statistic	393.2368	Durbin-Watson stat	1.949291
Prob(F-statistic)	0.000000		

Inverted AR Roots	.92	-.38

(b)

Figure 12.15 *Statistical results of the LV(2) and AR(2) polynomial models of* lnSale *on* lnSize(−1), *based on the sample {M_Size = 1, TP = 2, lnsize(−1) > 1.0}*

2. The outputs are obtained using the trial-and-error method. A stepwise selection method could be another alternative method. See the examples previously.
3. Both models are acceptable models, in a statistical sense, because each of the independent variables has a significant adjusted effect, or its *p*-value < 0.30.
4. A better fit model would be obtained if *lnSale*$(-2)^2$ is deleted from the LV(2) model, and *lnSale*(-1) is deleted from the AR(2) model.

5. Based on the output in Figure 12.15(a), the regression function is obtained as follows:

$$LNSALE = -0.28138 + 1.79623^*LNSIZE(-1) - 0.17865^*LNSIZE(-1)^3$$
$$+ 0.11715LNSALE(-1) + 0.06051LNSALE(-1)^2$$
$$+ 0.29865LNSALE(-2) - 0.015257LNSALE(-2)^2$$

This regression has the following specific characteristics.

- For the partial derivative: $\frac{\partial LNSale}{\partial LNSize(-1)} = 1.79623 - 3 \times 0.17865\, LNSize(-1)^2 = 0$ the turning points are obtained for $lnSale(-1) = \pm3.351\,488$.
- For a set of discrete values of $lnSale(-1)$ and $lnSale(-2)$, a set of *homogenous third-degree polynomial regressions* is obtained as follows:

$$LNSALE = C_k + 1.79623^*LNSIZE(-1) - 0.17865^*LNSIZE(-1)^3$$

where $\{C_k, k = 1,2,\ldots\}$ is the set of their intercepts.

6. This additive polynomial LV(2) model, can be extended to a more general polynomial LV(p) model with vector exogenous variables $X(-1)$, which can be presented in general as follows:

$$Y = Pol_1(Y(-1),\ldots,Y(-p)) + Pol_2(X(-1)) + \varepsilon \qquad (12.23)$$

where $Pol_k(^*)$ is a polynomial function having a finite number of parameters.

7. Furthermore, the general additive polynomial model in (12.23) can be extended to an interaction polynomial model having the following general equation.

$$Y = Pol_1(Y(-1),\ldots,Y(-p)) + Pol_2(\boldsymbol{X}(-1)) + Pol_3(Y(-1),\ldots,Y(-p),\boldsymbol{X}(-1)) + \varepsilon \qquad (12.24)$$

where $P_3(^*) = Pol_1(Y(-1),\ldots,Y(-p)) \times Pol_2(\boldsymbol{X}(-1))$ has a finite number of different parameters. Refer to the additive and interaction polynomial models presented in Chapter 8, and see the following example.

Example 12.13   An interaction polynomial model

Corresponding to the additive polynomial model in Figure 12.15, Figure 12.16 presents the statistical results of an interaction polynomial model using a multistage combinatorial selection method based on the sample $\{M\_Size = 1, TP = 2, lnSize(-1) > 1.0\}$.

Refer to the multistage combinatorial selection method presented in Example 12.3. In this example the stages of the analysis are presented in short as follows:

1. At Stage-1, the list of variables "*lnSale C*" is inserted with "*lnsize(-1)***lnsale(-1) lnsize(-1)**lnsale (*-1)^2 lnsize(-1)***lnsale(-2) lnsize(-1)^3***lnsale(-1) lnsize(-1)^3***lnsale(-1)^2 lnsize(-1)^3**lnsale (*-2)*" as the list of six search regressors. An acceptable model obtained has three interaction independent variables.

2. At Stage-2, the list of variables "lnsale c lnsize(-1)*lnsale(-1)^2 lnsize(-1)^3*lnsale(-1)^2 lnsize (-1)^3*lnsale(-2)" is inserted, which is variables of the regression obtained at Stage-1, with "lnsize (-1) lnsize(-1)^3 lnsale(-1) lnsale(-2)" as the list of four search regressors. Note that the final regression presented has an independent variable with a p-value $= 0.3123$. A reduced model can easily be derived by deleting the variable.

Dependent Variable: LNSALE
Method: Stepwise Regression
Date: 10/15/11 Time: 10:22
Sample: 1 8 IF M_SIZE=1 AND TP=2 AND LNSIZE(-1) > 1
Included observations: 331
Number of always included regressors: 1
Number of search regressors: 6
Selection method: Combinatorial
Number of search regressors: 3

Variable	Coefficient	Std. Error	t-Statistic	Prob.*
C	2.311344	0.128573	17.97683	0.0000
LNSIZE(-1)*LNSALE(-1)^2	0.076315	0.004559	16.73817	0.0000
LNSIZE(-1)^3*LNSALE(-1)^2	-0.012466	0.001088	-11.46184	0.0000
LNSIZE(-1)^3*LNSALE(-2)	0.042357	0.008746	4.843034	0.0000

R-squared	0.848525	Mean dependent var	5.834461
Adjusted R-squared	0.847135	S.D. dependent var	1.960487
S.E. of regression	0.766510	Akaike info criterion	2.318072
Sum squared resid	192.1247	Schwarz criterion	2.364020
Log likelihood	-379.6410	Hannan-Quinn criter.	2.336398
F-statistic	610.5901	Durbin-Watson stat	1.888107
Prob(F-statistic)	0.000000		

Selection Summary	
Number of combinations compared:	20

*Note: p-values and subsequent tests do not account for stepwise selection.

Dependent Variable: LNSALE
Method: Stepwise Regression
Date: 10/15/11 Time: 10:28
Sample: 1 8 IF M_SIZE=1 AND TP=2 AND LNSIZE(-1) > 1
Included observations: 331
Number of always included regressors: 4
Number of search regressors: 4
Selection method: Combinatorial
Number of search regressors: 3

Variable	Coefficient	Std. Error	t-Statistic	Prob.*
C	5.774685	1.796904	3.213686	0.0014
LNSIZE(-1)*LNSALE(-1)^2	0.075725	0.010973	6.900807	0.0000
LNSIZE(-1)^3*LNSALE(-1)^2	-0.014034	0.002207	-6.358621	0.0000
LNSIZE(-1)^3*LNSALE(-2)	0.048639	0.011388	4.271279	0.0000
LNSIZE(-1)	-3.384969	1.726800	-1.960255	0.0508
LNSIZE(-1)^3	0.382959	0.236825	1.617059	0.1068
LNSALE(-1)	0.086290	0.085267	1.012005	0.3123

R-squared	0.851741	Mean dependent var	5.834461
Adjusted R-squared	0.848995	S.D. dependent var	1.960487
S.E. of regression	0.761833	Akaike info criterion	2.314742
Sum squared resid	188.0460	Schwarz criterion	2.395149
Log likelihood	-376.0897	Hannan-Quinn criter.	2.346811
F-statistic	310.2264	Durbin-Watson stat	1.964270
Prob(F-statistic)	0.000000		

Selection Summary	
Number of combinations compared:	4

*Note: p-values and subsequent tests do not account for stepwise selection.

Figure 12.16 *Statistical results of an interaction polynomial model of* lnSale *on* lnSize(−1), *based on the sample* {M_Size = 1, TP = 2, lnSize(−1) > 1.0}

12.5.5.3 LV(p) or AR(q) Models with a Set of Numerical Exogenous Variables

The last two examples present polynomial-effect models, which could be extended to models with any set of numerical exogenous variables.

However, based on my observations of more than 300 papers presented in various international journals mentioned earlier, all present the application of linear-effect models where all independent variables of the models are of the first power with or without their interactions, such as presented in Examples 11.5 and 11.6. In other words, they show the linear effect of a numerical independent variable either depends on the other variables or not.

Some studies even present additive models with dummies and numerical independent variables, such as the one in (7.61) presented by Li, et al. (2010) and Shim and Okamuro (2011), and see also the fixed-effect models and alternatives presented in Chapter 9. On the other hand, Uddin and Boating (2011) present six OLS regressions and Chen (2008) presents 11 binary probit models with sets of additive numerical independent variables only. Note that the main limitation of an additive model is it shows each independent variable has an invariant or constant linear effect on the dependent variable for all possible values of the other independent variables.

I found a quadratic time series model of SALE, namely the *Bass model* of diffusion for discrete time variables, which has the following equation (Bass, 1969, in Malhotra, 2007, p. 48).

$$Sale_t = a + bY_{t-1} + cY_{t-1}^2, \quad t = 2, 3, \ldots \tag{12.25}$$

Y_{t-1} is cumulative sales up to the time $(t − 1)$. Note that this model uses a single variable based on time-series data which does not take into account the effects of other cause factors. For panel data, the model can have the following general equation, for each $i = 1, 2, \ldots, N$.

$$Sale_{i,t} = \beta_{i0} + \beta_{i1} Y_{t,i-1} + \beta_{i2} Y_{i,t-1}^2 + \varepsilon_{i,t} \tag{12.26}$$

With regards to a model with a large set of numerical independent variables, applying the multistage combinatorial selection method is suggested in order to develop an acceptable model in a statistical sense, under the condition that each of independent variables should have a p-value < 0.30, because of the unpredictable impact of multicollinearity of independent variables. If the model should have dummy independent variables, stepwise regression analysis can be done based on each sub-sample generated by dummies, as demonstrated in the previous example. The following section presents an alternative manual selection method.

12.6   Manual Stepwise Selection for General Linear LV(1) Model

Referring to the models with a large number of variables using the function or independent variable @*Expand*(...), such as the one in (12.19), the data analysis to develop a reduced model cannot be done using the STEPLS estimation method. This section presents an alternative, called a *manual stepwise selection method*, to develop an acceptable model as a modification of the model in (12.19). To meet the objective, the following sets of independent variables will be considered.

1. The lagged variable $Y1(-1)$,
2. A three-way interaction factors: t^*X1^*X2,
3. A set of two-way interactions: t^*X2, t^*X1, and t^*X3,
4. A set of other main factors or covariates: $\{X3, X4, X5\}$, and
5. The main factors in the interactions: $\{X1, X2, t\}$.

For more advanced models, say the LVAR(p,q) models, these can be applied using the lagged exogenous variables or the X-variables. In other words, the variable $X_k = Xk$ is replaced by $X_k(-1)$ for some or all ks.

Example 12.14   An application of the multistage manual stepwise method
Table 12.6 presents a summary of the statistical results based on an LV(1) model which is a modification of the model in (12.19) using the eight-stage manual stepwise method. The stages of the analysis are as follows:

1. At the first stage, say Stage-1, the objective of the analysis is to develop a set of four heterogeneous regressions of $Y1$ on $Y1(-1)$ and t^*X1^*X2 by M_Size and TP, using the equation specification (ES) as follows:

$$Y1 \ @Expand(M\_Size, TP) \ Y1(-1)^* @Expand(M\_Size, TP)$$
$$t^*X1^*X2^* @Expand(M\_Size, TP) \tag{12.27}$$

Based on the output in Table 12.5, the following findings and notes are presented.
1.1   The term @*Expand*(M_Size,TP) represents the intercepts of the set of heterogeneous regressions generated by M_Size and TP. Even though they have large or very large p-values, it is recommended not to delete any to develop a reduced model.
1.2   In this case, the lagged variable $Y1(-1)$ has a significant positive linear effect on $Y1$ in the four cells. Even though they have large or very large p-values, they should not be deleted from the model since they are intended to develop a LV(1) model.
1.3   On the other hand, since $t^*X1^*X2^*(M\_Size = 2, TP = 1)$ has an insignificant adjusted effect on $Y1$ with a very large p-value $= 0.603$, so deletion is recommended in order to reduce the model. See the following stage.

Table 12.6 Summary of the results based on a LV(1) model (12.19), using a manual stepwise selection method

Variable	Stage-1 Coeff.	t-Stat.	Prob.	Stage-2 Coeff.	t-Stat.	Prob.	Stage-3 Coeff.	t-Stat.	Prob.	Stage-4 Coeff.	t-Stat.	Prob.
M_SIZE=1 AND TP=1	0.120	2.281	0.023	0.120	2.281	0.023	-0.075	-0.705	0.481	-0.075	-0.705	0.481
M_SIZE=1 AND TP=2	0.074	1.208	0.227	0.074	1.209	0.227	-0.110	-0.569	0.570	0.074	1.214	0.225
M_SIZE=2 AND TP=1	-0.083	-1.451	0.147	-0.089	-1.583	0.114	-0.436	-3.220	0.001	-0.436	-3.220	0.001
M_SIZE=2 AND TP=2	0.060	0.945	0.345	0.060	0.945	0.345	0.578	2.239	0.025	0.578	2.239	0.025
Y1(-1)*(M_SIZE=1 AND TP=1)	1.018	22.142	0.000	1.018	22.147	0.000	0.969	18.843	0.000	0.969	18.843	0.000
Y1(-1)*(M_SIZE=1 AND TP=2)	1.077	39.963	0.000	1.077	39.973	0.000	1.059	32.995	0.000	1.077	40.163	0.000
Y1(-1)*(M_SIZE=2 AND TP=1)	1.069	87.915	0.000	1.068	88.814	0.000	1.057	83.900	0.000	1.057	83.899	0.000
Y1(-1)*(M_SIZE=2 AND TP=2)	1.083	120.18	0.000	1.083	120.21	0.000	1.094	105.36	0.000	1.094	105.36	0.000
T*X1*X2*(M_SIZE=1 AND TP=1)	0.003	1.353	0.176	0.003	1.354	0.176	0.002	1.086	0.278	0.002	1.086	0.278
T*X1*X2*(M_SIZE=1 AND TP=2)	0.004	3.898	0.000	0.004	3.899	0.000	0.004	3.881	0.000	0.004	3.917	0.000
T*X1*X2*(M_SIZE=2 AND TP=1)	-0.001	-0.520	*0.603*									
T*X1*X2*(M_SIZE=2 AND TP=2)	0.002	1.800	0.072	0.002	1.800	0.072	0.002	1.939	0.053	0.002	1.939	0.053
T*X1*(M_SIZE=1 AND TP=1)							0.012	2.086	0.037	0.012	2.086	0.037
T*X1*(M_SIZE=1 AND TP=2)							0.005	1.007	*0.314*			
T*X1*(M_SIZE=2 AND TP=1)							0.015	2.815	0.005	0.015	2.815	0.005
T*X1*(M_SIZE=2 AND TP=2)							-0.011	-2.070	0.039	-0.011	-2.070	0.039
R-squared	0.949			0.949			0.949			0.949		
Adjusted R-squared	0.948			0.948			0.949			0.949		
Durbin-Watson stat	2.067			2.064			2.071			2.076		

Variable	Stage-5 Coeff.	t-Stat.	Prob.	Stage-6 Coeff.	t-Stat.	Prob.	Stage-7 Coeff.	t-Stat.	Prob.	Stage-8 Coeff.	t-Stat.	Prob.
M_SIZE=1 AND TP=1	-0.075	-0.706	0.480	-0.075	-0.710	0.478	-0.114	-1.017	0.309	-0.100	-0.898	0.369
M_SIZE=1 AND TP=2	0.076	1.238	0.216	0.074	1.224	0.221	0.049	0.802	0.423	0.077	1.260	0.208
M_SIZE=2 AND TP=1	-0.412	-3.036	0.002	-0.412	-3.038	0.002	-0.364	-2.642	0.008	-0.364	-2.664	0.008
M_SIZE=2 AND TP=2	0.485	1.885	0.060	0.485	1.886	0.060	0.434	1.689	0.092	0.434	1.703	0.089
Y1(-1)*(M_SIZE=1 AND TP=1)	0.969	18.970	0.000	0.969	18.983	0.000	0.967	18.985	0.000	0.922	17.298	0.000
Y1(-1)*(M_SIZE=1 AND TP=2)	1.076	40.299	0.000	1.077	40.462	0.000	1.075	40.511	0.000	1.092	41.022	0.000
Y1(-1)*(M_SIZE=2 AND TP=1)	1.059	83.556	0.000	1.059	83.612	0.000	1.063	83.119	0.000	1.063	83.802	0.000
Y1(-1)*(M_SIZE=2 AND TP=2)	1.094	106.02	0.000	1.094	106.10	0.000	1.090	104.65	0.000	1.090	105.51	0.000
T*X1*X2*(M_SIZE=1 AND TP=1)	0.002	0.280	*0.780*	0.002	1.094	0.274	0.007	1.425	0.154	0.008	1.820	0.069
T*X1*X2*(M_SIZE=1 AND TP=2)	0.006	0.718	*0.473*	0.004	3.947	0.000	0.011	3.651	0.000	0.007	2.367	0.018

Table 12.6 (Continued)

Variable	Coeff.	t-Stat.	Prob.	Coeff.	t-Stat.	Prob.	Coeff.	t-Stat.	Prob.	Coeff.	t-Stat.	Prob.
T*X1*X2*(M_SIZE=2 AND TP=1)	-0.042	-4.388	0.000	-0.042	-4.391	0.000	-0.030	-2.717	0.007	-0.030	-2.740	0.006
T*X1*X2*(M_SIZE=2 AND TP=2)	0.012	2.097	0.036	0.012	2.102	0.036	0.013	2.302	0.022	0.011	1.990	0.047
T*X1*(M_SIZE=1 AND TP=1)	0.015	2.765	0.006	0.015	2.767	0.006	0.017	2.384	0.017	0.013	2.403	0.016
T*X1*(M_SIZE=1 AND TP=2)	-0.009	-1.787	0.074	-0.009	-1.788	0.074	-0.008	-1.631	0.103	-0.008	-1.644	0.100
X1*X2*(M_SIZE=1 AND TP=1)	0.001	0.035	*0.972*									
X1*X2*(M_SIZE=1 AND TP=2)	-0.017	-0.273	*0.785*									
X1*X2*(M_SIZE=2 AND TP=1)	-0.009	-1.281	0.200	-0.009	-1.282	0.200	-0.040	-2.209	0.027	-0.040	-2.228	0.026
X1*X2*(M_SIZE=2 AND TP=2)	0.315	4.644	0.000	0.315	4.647	0.000	0.312	4.619	0.000	0.312	4.657	0.000
X2*(M_SIZE=1 AND TP=1)							-0.024	-1.054	0.292	-0.029	-1.260	0.208
X2*(M_SIZE=1 AND TP=2)							-0.042	-2.462	0.014	-0.026	-1.505	0.133
X2*(M_SIZE=2 AND TP=1)							0.058	1.866	0.062	0.058	1.882	0.060
X2*(M_SIZE=2 AND TP=2)							-0.086	-2.161	0.031	-0.086	-2.178	0.030
X3*(M_SIZE=1 AND TP=1)										0.414	2.638	0.008
X3*(M_SIZE=1 AND TP=2)										-0.200	-4.292	0.000
R-squared	0.950			0.950			0.951			0.951		
Adjusted R-squared	0.949			0.950			0.950			0.951		
Durbin-Watson stat	2.082			2.081			2.087			2.124		

Variable	Stage-9			Stage-10			Stage-11		
	Coeff.	t-Stat.	Prob.	Coeff.	t-Stat.	Prob.	Coeff.	t-Stat.	Prob.
M_SIZE=1 AND TP=1	-0.394	-2.521	0.012	-0.446	-2.858	0.004	-0.057	-0.313	0.754
M_SIZE=1 AND TP=2	-2.227	-4.619	0.000	-2.248	-4.689	0.000	-1.471	-2.253	0.024
M_SIZE=2 AND TP=1	-0.363	-2.721	0.007	-0.364	-2.721	0.007	1.248	1.563	0.118
M_SIZE=2 AND TP=2	-4.564	-4.775	0.000	-4.564	-4.761	0.000	-2.558	-2.160	0.031
Y1(-1)*(M_SIZE=1 AND TP=1)	0.841	14.983	0.000	0.911	17.416	0.000	0.864	16.293	0.000
Y1(-1)*(M_SIZE=1 AND TP=2)	0.991	29.749	0.000	0.990	29.686	0.000	0.993	30.107	0.000
Y1(-1)*(M_SIZE=2 AND TP=1)	1.063	85.642	0.000	1.063	85.595	0.000	1.056	74.219	0.000
Y1(-1)*(M_SIZE=2 AND TP=2)	1.059	91.069	0.000	1.059	90.794	0.000	1.058	91.484	0.000
T*X1*X2*(M_SIZE=1 AND TP=1)	0.012	2.606	0.009	0.009	2.070	0.039	0.014	3.079	0.002
T*X1*X2*(M_SIZE=1 AND TP=1)	0.010	3.151	0.002	0.010	3.125	0.002	0.010	3.273	0.001

Variable	Coefficient (1)	t-Statistic (1)	Prob. (1)	Coefficient (2)	t-Statistic (2)	Prob. (2)	Coefficient (3)	t-Statistic (3)	Prob. (3)
T*X1*X2*(M_SIZE=2 AND TP=1)	-0.026	-2.414	0.016	-0.026	-2.407	0.016	-0.157	-3.430	0.001
T*X1*X2*(M_SIZE=2 AND TP=2)	0.005	0.847	0.397	0.008	1.320	0.187	0.029	3.634	0.000
T*X1*(M_SIZE=1 AND TP=1)									
T*X1*(M_SIZE=1 AND TP=2)	0.013	2.492	0.013	0.013	2.455	0.014	0.083	2.689	0.007
T*X1*(M_SIZE=2 AND TP=1)	-0.025	-4.243	0.000	-0.025	-4.230	0.000	-0.012	-1.622	0.105
T*X1*(M_SIZE=2 AND TP=2)									
X1*X2*(M_SIZE=1 AND TP=1)									
X1*X2*(M_SIZE=1 AND TP=2)									
X1*X2*(M_SIZE=2 AND TP=1)	-0.040	-2.305	0.021	-0.040	-2.275	0.023	-0.041	-2.365	0.018
X1*X2*(M_SIZE=2 AND TP=2)	0.289	4.394	0.000	0.289	4.380	0.000	1.210	3.775	0.000
T*X2*(M_SIZE=1 AND TP=1)	-0.038	-1.705	0.088	-0.031	-1.381	0.168	-0.094	-2.238	0.025
T*X2*(M_SIZE=1 AND TP=2)	-0.037	-2.175	0.030	-0.037	-2.157	0.031	-0.040	-2.321	0.020
T*X2*(M_SIZE=2 AND TP=1)	0.058	1.921	0.055	0.058	1.922	0.055	0.058	1.960	0.050
T*X2*(M_SIZE=2 AND TP=2)	-0.086	-2.215	0.027	-0.086	-2.208	0.027	0.915	2.779	0.006
X3*(M_SIZE=1 AND TP=1)	1.035	4.246	0.000	0.398	2.591	0.010	0.349	2.293	0.022
X3*(M_SIZE=1 AND TP=2)	-0.229	-4.419	0.000	-0.221	-4.824	0.000	-0.223	-4.928	0.000
X4*(M_SIZE=1 AND TP=1)	0.258	2.975	0.003	0.269	3.095	0.002	0.320	3.665	0.000
X4*(M_SIZE=1 AND TP=2)	1.334	4.787	0.000	1.349	4.889	0.000	1.363	4.998	0.000
X4*(M_SIZE=2 AND TP=2)	2.892	5.497	0.000	2.892	5.481	0.000	2.334	4.262	0.000
X5*(M_SIZE=1 AND TP=1)	-0.107	-3.360	0.001						
X5*(M_SIZE=1 AND TP=2)	0.185	0.337	0.736						
X5*(M_SIZE=2 AND TP=1)	-0.257	-0.396	0.692						
X5*(M_SIZE=2 AND TP=2)	0.926	1.726	0.085	0.926	1.721	0.086	0.574	1.052	0.293
T*(M_SIZE=1 AND TP=1)							-0.241	-3.936	0.000
T*(M_SIZE=1 AND TP=2)							-0.115	-1.727	0.085
T*(M_SIZE=2 AND TP=1)							-0.516	-2.435	0.015
T*(M_SIZE=2 AND TP=2)							-0.217	-2.549	0.011
X2*(M_SIZE=1 AND TP=1)							0.131	1.129	0.259
X2*(M_SIZE=2 AND TP=2)							-7.055	-3.063	0.002
X1*(M_SIZE=2 AND TP=1)							-0.216	-1.813	0.070
R-squared	0.954			0.954			0.955		
Adjusted R-squared	0.953			0.953			0.954		
Durbin-Watson stat	2.064			2.068			2.099		

2. At Stage-2, the reduced model is obtained using the following ES.

$$Y1 \; @Expand(M\_Size, TP) \; Y1(-1)^* @Expand(M\_Size, TP)$$
$$t^* X1^* X2^* @Expand(M\_Size, TP, @Drop(2, 1)) \tag{12.28}$$

3. At Stage-3, the function $t^* X1^* @Expand(M\_SIze, TP)$ is inserted into the model obtained at Stage-2, which is in fact a set of four independent variables. Its output shows that there are two independent variables, which should be evaluated.
    3.1 Take note that the variable $t^* X1^* X2^* (M\_Size = 1, TP = 1)$ has a large p-value. However, following Lapin's study (1973) using the 0.15 significance level, it can be said that the variable $t^* X1^* X2^* (M\_Size = 1, TP = 1)$ has a significant positive effect on $Y1$ with a p-value $= 0.278/2 = 0.139 < 0.15$.
    3.2 So there is a reason to keep it in the model. Of course an alternative reduced model could be developed by deleting it. Do this as an exercise.
    3.3 On the other hand, since $t^* X1^* (M\_Size = 1, TP = 2)$ has a p-value $= 0.314 > 0.30$, I decided to delete it from the model.
4. At Stage-4, the reduced model is obtained using the following ES.

$$Y1 \; @Expand(M\_Size, TP) \; Y1(-1)^* @Expand(M\_Size, TP)$$
$$t^* X1^* X2^* @Expand(M\_Size, TP, @Drop(2, 1))$$
$$t^* X1^* @Expand(M\_Size, TP, @Drop(1, 2)) \tag{12.29}$$

5. At Stage-5, the function $X1^* X2^* @Expand(M\_Size, TP)$ is inserted to the previous reduced model instead of $t^* X2^* @Expand(M\_Size, TP)$, because it is defined that $X1$ is the most important cause, source or up-stream factor of $Y1$, and its effect depends on $X2$ as well as the time t. In this case, the output shows four independent variables have very large p-values, namely a pair of interactions: $X1^* X2$ and $t^* X1^* X2$ within two cells $(M\_Size = 1, TP = 1)$ and $(M\_Size = 1, TP = 2)$, respectively. So alternative reduced models can be obtained by deleting either $X1^* X2$ or $t^* X1^* X2$. See the following stage.
6. At Stage-6, in order to have a three-way interaction model the reduced model is obtained by deleting both $X1^* X2^* (M\_Size = 1, TP = 1)$ and $X1^* X2^* (M\_Size = 1, TP = 2)$ using the following ES. Note that the output shows that the interaction $t^* X1^* X2^* (M\_Size = 1, TP\_1)$ has a p-value $= 0.274 < 0.30$.

$$Y1 \; @Expand(M\_Size, TP) \; Y1(-1)^* @Expand(M\_Size, TP)$$
$$t^* X1^* X2^* @Expand(M\_Size, TP, @Drop(2, 1))$$
$$t^* X1^* @Expand(M\_Size, TP, @Drop(1, 2))$$
$$X1^* X2^* @Expand(M\_Size, TP, @Drop(1, 1), @Drop(1, 2)) \tag{12.30}$$

7. At Stage-7, by inserting the function $t^* X2^* @Expand(M\_Size, TP)$, the output shows only one independent variable, namely $t^* X2^* (M\_Size = 1, TP = 1)$ has a large p-value $= 0.292$. Since it is less than 0.30, then the full model can be considered acceptable.
8. At Stage-8, the output is obtained by inserting the function $X3^* @Expand(M\_Size, TP)$ for the selection process, then a reduced model is obtained with $X3^* @Expand(M\_Size = 1, TP = 1)$ and $X3^* @Expand(M\_Size = 1, TP = 2)$ as two additional independent variables.
9. At Stage-9, the output is obtained by inserting the function $X4^* @Expand(M\_Size, TP)$ for the selection process, and then the function $X5^* @Expand(M\_Size, TP)$ is inserted. Note that there is a problem with the cell $(M\_Size = 1, TP = 1)$, since the interaction $t^* X1^* (M\_Size = 1, TP = 1)$ has a p-value $= 0.397$,

but the main factor $X5^*(M\_Size = 1, TP = 1)$ has a p-value $= 0.001$. In general, one would delete the independent variable with a p-value. However, since t^*X1 is considered a more important variable, then $X5^*(M\_Size = 1, TP = 1)$ should be deleted.

10. The reduced model at Stage-10 is obtained, by deleting three independent variables, namely $X5^*(M\_Size = 1, TP = 2)$ and $X5^*(M\_Size = 2,\ TP = 1)$ which have large p-values, as well as $X5^*(M\_Size = 1, TP = 1)$.

11. At Stage-11, the functions $t^*@Expand(M\_Size, TP)$, $X2^*@Expand(M\_Size, TP)$ and $X1^*@Expand(M\_Size, TP)$ are inserted consecutively for the selection process. Then the final LV(1) heterogeneous regression is obtained. Based on these results, the following notes and comments are presented.

    11.1 By inserting the functions $t^*@Expand(M\_Size, TP)$, $X2^*@Expand(M\_Size, TP)$ and $X1^*@Expand(M\_Size, TP)$ with differential ordering, other forms of reduced model may be obtained. Do this for an exercise.

    11.2 The independent variables $X5^*(M\_Size = 2, TP = 2)$ and $X2^*(M\_Size = 1, TP = 1)$ have large p-values of 0.293 and 0.259, respectively. So one could reduce the model if this is desired.

    11.3 Since the regression has a Durbin–Watson statistic of 2.099, it can be said there is no auto-correlation problem and the term AR(1) should not be inserted into the model. However, if the term AR(1) is used then it has a large p-value.

    11.4 The limitations of the regression are the impact of the outliers, the unpredictable impact of mul-ticollinearity and the basic assumptions of the residual, specifically its normal distribution, which can easily be checked by selecting *View/Residual Tests/Histogram –Normality Test*. Refer to special notes presented in Section 2.14 (Agung, 2009a).

Example 12.15   A first lagged variables model, as a modification of the model in Table 12.5
It is recognized that in any natural experiment, the first lagged variables $Y1(-1)$ and all $Xk(-1)$, $k = 1, 2, \ldots$, are the cause, source or up-stream variables of $Y1$. For this reason, the following ES presents a valid model for natural experimental data.

$$
\begin{aligned}
y1\ @expand(m\_size, tp, @dropfirst)\ y1(-1)^*@expand(m\_size, tp) \\
t^*x1(-1)^*x2(-1)^*@expand(m\_size, tp)\ t^*x1(-1)^*@expand(m\_size, tp) \\
x1(-1)^*x2(-1)^*@expand(m\_size, tp)\ t^*x2(-1)^*@expand(m\_size, tp) \\
x3(-1)^*@expand(m\_size, tp)\ x4(-1)^*@expand(m\_size, tp) \\
x5(-1)^*@expand(m\_size, tp)\ t^*@expand(m\_size, tp) \\
x2(-1)^*@expand(m\_size, tp)\ x1(-1)^*@expand(m\_size, tp)
\end{aligned}
\tag{12.31}
$$

Table 12.7 presents the summary of the statistical results based on an acceptable reduced model of the model in (12.31). The output of this reduced model is obtained by inserting the terms or variables in (12.31) one-by-one for the selection process at each stage. For instance, at Stage-2, the variable $T^*X1(-1)^*X2(-1)^*@Expand(M\_Size, TP)$ is inserted for the selection process. We find that $T^*X1(-1)^*X2(-1)$ in the first two cells have large p-values, so they are not used in the model.

At Stage-3, the variable $T^*X1(-1)^*@Expand(M\_Size, TP)$ can be kept in the model, even though $T^*X1(-1)$ has a p-value $= 0.240$ in cell (2,2), since it is less than 0.30, based on the criteria. On the other hand, at Stage-4, the variable $X1(-1)^*X2(-1)$ should be deleted from the first three cells even though they have very small p-values, in order to keep all independent variables of the model obtained at Stage-3 with p-values < 0.30. Finally, the output in Table 12.6 is obtained using the ES as follows:

Table 12.7   *Summary of the statistical results of a reduced model of the model in (12.31)*

Variable	(M_Size=1,TP=1)			(M_Size=1,TP=2)			(M_Size=2,TP=1)			(M_Size=2,TP=2)		
	Coeff.	z-Stat.	Prob.	Coeff.	z-Stat.	Prob.	Coeff.	z-Stat.	Prob.	Coeff.	z-Stat.	Prob.
Intercept	0.349	2.066	0.039	-0.862	-1.770	0.077	-0.185	-1.382	0.167	-0.464	-0.381	0.703
Y1(-1)	0.851	14.803	0.000	1.075	32.095	0.000	1.091	87.225	0.000	1.072	90.724	0.000
T*X1(-1)*X2(-1)							-0.018	-2.139	0.033	-0.017	-1.842	0.066
T*X1(-1)	0.020	2.323	0.020	-0.025	-2.330	0.020	0.009	1.586	0.113	-0.009	-1.176	0.240
X1(-1)*X2(-1)										0.168	2.068	0.039
T*X2(-1)							0.083	1.576	0.115			
X3(-1)	1.058	3.915	0.000									
X4(-1)	-0.073	-1.063	0.288	0.791	2.216	0.027				1.523	2.674	0.008
X5(-1)	-0.107	-3.121	0.002	-0.693	-1.461	0.144				0.733	1.281	0.201
T	-0.132	-2.032	0.042							-0.293	-3.164	0.002
X2(-1)										-0.382	-1.589	0.112
X1(-1)				0.105	1.188	0.235						
R-squared	0.955	Mean dependent var		1.611								
Adjusted R-squared	0.955	S.D. dependent var		4.464								
S.E. of regression	0.952	Akaike info criterion		2.758								
Sum squared resid	1253.9	Schwarz criterion		2.859								
Log likelihood	-1918.8	Hannan-Quinn criter.		2.796								
F-statistic	1139.4	Durbin-Watson stat		2.216								
Prob(F-statistic)	0.000											

$$
\begin{aligned}
&y1 \ @expand(m\_size, tp)\ y1(-1)^* @expand(m\_size, tp) \\
&t^* x1(-1)^* x2(-1)^* @expand(m\_size, tp, @drop(1,1), @drop(1,2)) \\
&t^* x1(-1)^* @expand(m\_size, tp) \\
&x1(-1)^* x2(-1)^* @expand(m\_size, tp, @drop(1,1), @drop(1,2), @drop(2,1)) \\
&t^* x2(-1)^* @expand(m\_size, tp, @drop(1,1), @drop(1,2), @droonp(2,2)) \\
&x3(-1)^* @expand(m\_size, tp, @drop(1,2), @drop(2,1), @drop(2,2)) \\
&x4(-1)^* @expand(m\_size, tp, @drop(2,1))\ x5(-1)^* @expand(m\_size, tp, @drop(2,1)) \\
&t^* @expand(m\_size, tp, @drop(1,2), @drop(2,1)) \\
&x2(-1)^* @expand(m\_size, tp, @drop(1,1), @drop(1,2), @drop(2,1)) \\
&x1(-1)^* @expand(m\_size, tp, @drop(1,1), @drop(2,1), @drop(2,2))
\end{aligned}
\tag{12.32}
$$

Note that the variables $X3(-1)$, $X4(-1)$ and $X5(-1)$ are inserted before the variables t, $X2(-1)$ and $X1(-1)$ because they are not used in the interactions, and t, $X2(-1)$ and $X1(-1)$ should be highly correlated with the interactions.

Note that the model in (12.31) is in fact only one out of a lot of possible models which can easily be subjectively defined based the set of numerical variables $Y1, Y1(-1)$, $X1(-1)$ to $X5(-1)$, and the time t as a numerical variable. However, it should be understood that a good fit model, in a statistical sense, will never represent the true population model – refer to Agung (2009a, Section 2.14.2), because the model fits the data which happens to be selected or available to researchers.

12.7 Manual Stepwise Selection for Binary Choice LV(1) Models

The data analysis based on binary choice models with lagged numerical independent variables could easily be done using the same method as in Chapter 7; that is, using the nonparametric estimation method at the first stage. For this reason, this section only presents selected models.

Example 12.16 Cell proportion model of a zero-one variable D_Dividend
Figure 12.17 presents the proportion P(D_Dividend = 1) by *M_Size* and the time *t*, and Figure 12.18 presents the statistical results based on the binary logit model of *D_Dividend* on *M_Size* and the time *T*, specific for *T* < 4. All data can easily be shown. Based on these results the following notes and comments are presented.

Descriptive Statistics for D_DIVIDEND
Categorized by values of M_SIZE and T
Date: 10/02/11 Time: 08:52
Sample: 1 1792
Included observations: 1792

Mean Obs.						T				
		1	2	3	4	5	6	7	8	All
	1	0.241071	0.241071	0.250000	0.258929	0.223214	0.196429	0.196429	0.232143	0.229911
		112	112	112	112	112	112	112	112	896
M_SIZE	2	0.303571	0.285714	0.357143	0.348214	0.330357	0.258929	0.267857	0.285714	0.304688
		112	112	112	112	112	112	112	112	896
	All	0.272321	0.263393	0.303571	0.303571	0.276786	0.227679	0.232143	0.258929	0.267299
		224	224	224	224	224	224	224	224	1792

Figure 12.17 *Proportion P(D_Dividend = 1) by* M_Size *and the time* t

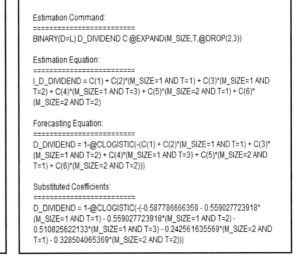

Dependent Variable: D_DIVIDEND
Method: ML - Binary Logit (Quadratic hill climbing)
Date: 10/02/11 Time: 08:56
Sample: 1 1792 IF T < 4
Included observations: 672
Convergence achieved after 3 iterations
Covariance matrix computed using second derivatives

Variable	Coefficient	Std. Error	z-Statistic	Prob.
C	-0.587787	0.197203	-2.980622	0.0029
M_SIZE=1,T=1	-0.559028	0.296126	-1.887803	0.0591
M_SIZE=1,T=2	-0.559028	0.296126	-1.887803	0.0591
M_SIZE=1,T=3	-0.510826	0.294122	-1.736779	0.0824
M_SIZE=2,T=1	-0.242562	0.284818	-0.851638	0.3944
M_SIZE=2,T=2	-0.328504	0.287470	-1.142743	0.2531

McFadden R-squared	0.007177	Mean dependent var	0.279762
S.D. dependent var	0.449217	S.E. of regression	0.448943
Akaike info criterion	1.194805	Sum squared resid	134.2321
Schwarz criterion	1.235076	Log likelihood	-395.4546
Hannan-Quinn criter.	1.210402	Restr. log likelihood	-398.3134
LR statistic	5.717572	Avg. log likelihood	-0.588474
Prob(LR statistic)	0.334677		
Obs with Dep=0	484	Total obs	672
Obs with Dep=1	188		

Estimation Command:
=========================
BINARY(D=L) D_DIVIDEND C @EXPAND(M_SIZE,T,@DROP(2,3))

Estimation Equation:
=========================
I_D_DIVIDEND = C(1) + C(2)*(M_SIZE=1 AND T=1) + C(3)*(M_SIZE=1 AND T=2) + C(4)*(M_SIZE=1 AND T=3) + C(5)*(M_SIZE=2 AND T=1) + C(6)* (M_SIZE=2 AND T=2)

Forecasting Equation:
=========================
D_DIVIDEND = 1-@CLOGISTIC(-(C(1) + C(2)*(M_SIZE=1 AND T=1) + C(3)* (M_SIZE=1 AND T=2) + C(4)*(M_SIZE=1 AND T=3) + C(5)*(M_SIZE=2 AND T=1) + C(6)*(M_SIZE=2 AND T=2)))

Substituted Coefficients:
=========================
D_DIVIDEND = 1-@CLOGISTIC(-(-0.587786666359 - 0.559027723918* (M_SIZE=1 AND T=1) - 0.559027723918*(M_SIZE=1 AND T=2) - 0.510825622133*(M_SIZE=1 AND T=3) - 0.242561635569*(M_SIZE=2 AND T=1) - 0.328504065369*(M_SIZE=2 AND T=2)))

Figure 12.18 *Statistical results of a binary logit model of D_Dividend on* M_Size *and the time* t, *for* t < 4

1. The proportion $P(D\_Dividend = 1|M\_size = 1,t) < P(D\_Dividend = 1|M\_size = F1,t)$ at all the time points. Among the 16 group of firms, the group in (M_Size = 2,T = 3) has the largest proportions of 0.357 143 in giving a dividend. So it can be said that this group is the best group for giving dividends.
2. In order to test the corresponding hypotheses, for an illustration, Figure 12.18 presents the statistical results of the binary logit model of *D_Dividend* on *M_Size* and the time *t*, only for *t* < 4, using the group of the firms (M_Size = 2, t = 3) as a referent group. Based on this output, the following findings and notes are presented.

    2.1   The negative coefficient of (M_Size = 1,T = 3), indicates the predicted probability of the group (M_Size = 1,T = 3) in giving a dividend is significantly less than the group (M_Size = 2,T = 3) based on the Z-statistic of $Z_0 = -1.736779$ with a *p*-value $= 0.0824/2 = 0.0412$.

    2.2   The other hypotheses can be tested using the Wald test. For instance, for the group (M_Size = 1, T = 1) and (M_Size = 2,T = 1), the null hypothesis H_0: C(2) = C(5) is accepted based on the F-statistic of $F_0 = 1.085\ 340$ with $df = (1,666)$ and a *p*-value $= 0.2979$. Then it can be concluded that the two groups have insignificant differences in giving dividends.

Example 12.17   Kernel fit curves of D_Dividend on lnSale(−1)

Figure 12.19 presents nonparametric regressions of *D_Dividend* on *lnSale(−1)*, in the form of the Kernel Fit curves, by *M_Size* and the time *T*, specific for *T* > 5, in order to study the differences between the curves (or the effects of lnSale(−1) on D-Dividend) between one year before (*T* = 6) and two years after (*T* = 7 and 8) the new tax policy.

Based on these graphs, the following notes and comments are presented.

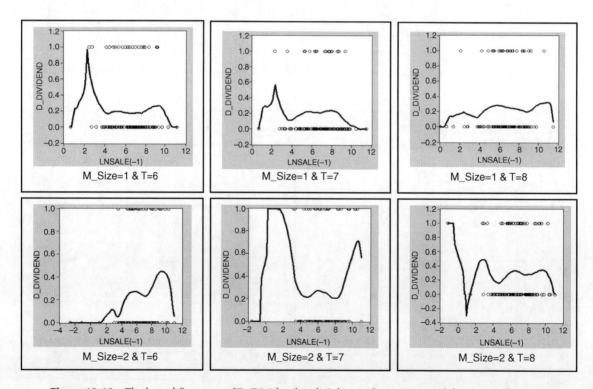

Figure 12.19   *The kernel fit curves of* D_Dividend *on* lnSale(−1) *by* M_Size *and the time* T, *for* T > 5.

1. It can be said that *lnSale*(−*1*) has different effects on *D_Dividend* between the group generated by *M_Size* and the time *T* in a nonparametric sense.
2. The graphs show that a polynomial binary choice model should be applied for each group. By observing the curve of the group (*IM_Size* = 2,*T* = 8), then the highest possible power of the polynomial model that should be applied is seven, because a horizontal line would have a maximum of seven intercepts with the curve.

   The steps of data analysis to obtain an acceptable binary logit model are the same as presented in Chapter 7, and Agung (2011), as follows:

   2.1 The first step is doing the stepwise regression analysis based on the sample or group (*M_Size* = 2, *T* = 8) with "C" as the always-included regressor, and the list of seven search regressors: "*lnSale* (−*1*) *lnSale*(−*1*)^2 *lnSale*(−*1*)^3 *lnSale*(−*1*)^4 *lnSale*(−*1*)^5 *lnSale*(−*1*)^6 *lnSale*(−*1*)^7".

   2.2 Then with respect to the combinatorial selection method, the trial-and-error method should be applied to get an acceptable stepwise regression as presented in Figure 12.20(a).

   2.3 With the output on the screen, select *Estimate/Estimation settings Method: LS-*(*Least Squares*)*...OK*, then the output of a LS regression is shown on the screen.

   2.4 Then select Estimate *Estimate/Estimation settings Method: BINARY – Binary Choice* (*logit, Probit, Extreme Value*)*/Logit...OK*, then the output of a binary logit in Figure 12.20(b) is obtained.

   2.5 Take note that each of the independent variables of the binary logit has either positive or negative significant adjusted effects at the 0.05 significance level, since its *p*-value = Prob./2 < 0.05, but the joint effects of all independent variables have insignificant effects based on the LR statistic of $G_o^2 = 0.445306$ with a *p*-value = 0.265 266. Then there is a question, *is this model a good fit model?* The answer would be very subjective. There is a residual normality test, but it is not available for the residual logistic test.

   2.6 The same outputs would be obtained using the binary probit and the binary extreme value models. Therefore, which one of the three binary choice models would be the best? The residual of the binary probit has the normal distribution assumption. Agung (2009a, Section 2.14.3) presents notes and comments that the residual assumption should not be

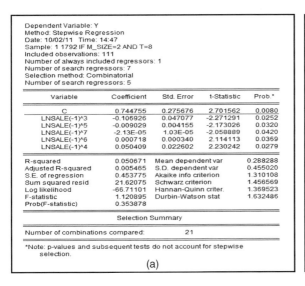

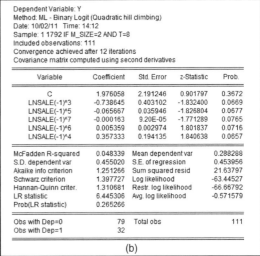

Figure 12.20 *Statistical results of a stepwise regression and a binary logit*

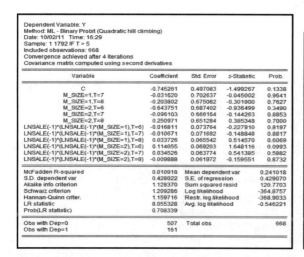
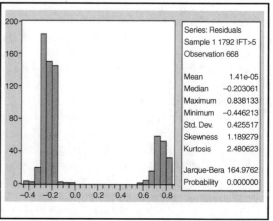

Figure 12.21   *Statistical results of a binary probit model of* D_Dividend *on* lnSale(−1) *by* M_Size *and the time* T, *for* T > 5, *and its residual histogram and normality test*

tested, because the residual distribution in fact is related to its theoretical distribution. Refer also to the normality assumption for the LR statistic, not the observed value of the LR statistic computed based on sampled data which happens to be available to researchers. See an illustration presented in point (5) later.

3. The same method could easily be done for the other group. Do it as an exercise. The other five binary choice models should have different sets of independent variables.

4. However, if the adjusted effects differences of an independent variable are tested, then the six binary logits should have the same sets of variables. Otherwise the adjusted effects of the independent variable have different meanings. But very large *p*-values for many of the variables would be obtained, except for the group (*M_Size-2,T = 8*). For this reason, I would not recommend applying the binary logit models, as well as other statistical models, by *Group* and the time *T*, with the same set of independent variables.

5. For comparison, and additional illustration, Figure 12.21 presents the output of a binary probit model of *D_Dividend* on a single independent variable *lnSale(−1)* by *M_Size* and *T*, with its residual histogram and normality test. Based on these outputs, the following notes and comments are presented.

    5.1 The independent variable *lnSale(−1)* has very large *p*-values in five out of the six models, all independent variables have insignificant joint effects based on the LR statistic of $G_o^2 = 0.055\,328$ with a *p*-value $= 0.708\,339$, and a very small R-squared $= 0.010\,918$. So it can be said that the model is not a good fit.

    5.2 On the other hand, a binary probit model has the assumption that the residual has independent identical normal distribution. Take note that the residual histogram clearly shows that the residual is not normally distributed.

Example 12.18   A LV(1) binary probit model

As a modification of the LVAR(1,1) model in (12.19), the following equation specification presents a model of a zero-one problem indicator *Q25_Y1*, which is generated as *Q25_Y1* = 1 if the scores of *Y1* = *liability/1000* is the lowest 25%, and *Q25_Y1* = 0 otherwise.

$Q25\_y1 \ @expand(m\_size, tp) \ y1(-1)^* @expand(m\_size, tp) \ t^* x1^* x2^* @expand(m\_size, tp)$

$t^* x1^* @expand(m\_size, tp) \ x1^* x2^* @expand(m\_size, tp) \ t^* x2^* @expand(m\_size, tp)$

$x3^* @expand(m\_size, tp) \ x4^* @expand(m\_size, tp) \ x5^* @expand(m\_size, tp)$ \hfill (12.33)

$t^* @expand(m\_size, tp) \ x2^* @expand(m\_size, tp) \ x1^* @expand(m\_size, tp)$

Take note that this model represents a set of four models of Q25_Y1 on 11 numerical variables:

$$y1(-1) \ t^* x1^* x2 \ t^* x1 \ x1^* x2 \ t^* x2 \ x3 \ x4 \ x5 \ t \ x1 \ x2$$

Then the model has 48 (12×4 *dummies*) independent variables. The output presents many variables with large *p*-values, and it is not an easy task to obtain an acceptable reduced model(s) by deleting the independent variables one-by-one. For this reason, applying alternative *multistage combinatorial selection method* is recommended using the means of STEPLS, as presented in Chapter 7 and Agung (2011). Do this as an exercise.

The following example presents a *manual stepwise selection method*, since EViews 6 does not provide the stepwise selection method for the binary choice model.

Example 12.19   Application of a multistage manual stepwise selection method based on a binary choice model

Similar to the multistage manual selection method presented in Example 12.6, this example presents the data analysis based on a binary choice model with a set of numerical independent variables only. The model considered is a part of the model in (12.26) specific for the group or sample ($M\_Size = 1, TP = 1$), using the following stages.

1. In order to develop an interaction binary probit model (BLM), at the first stage, the equation specification: "$Q25\_Y1 \ C \ t^*X1^*X2 \ X1^*X2 \ t^*X1 \ t^*X2$" should be applied, with the output presented in Figure 12.22(a). Take note that the interactions should be inserted at the first stage, since the effect of *X1* depends on *X2* and the time *t*. In this case, it happens that each of the independent variables has a significant adjusted effect. If one or more independent variables have large *p*-values, say greater than 0.20, 0.25 or 0.30, then a reduced model should be explored. Take note that if the variable has a *p*-value < 0.30, then it has either a positive or negative significant effect at the 0.15 significance level (Lapin, 1973).
2. At the second stage, the variables *X3*, *X4* and *X5* should be inserted as additional independent variables, with the output presented in Figure 12.22(b) and its acceptable reduced model presented in Figure 12.22(c).
3. At third stage, the variables *X1*, *X2* and *t*, should be inserted as additional independent variables of the model in Figure 12.22(c), with the output presented in Figure 12.22(d), with many independent variables having large *p*-values. By using the trial-and-error method, we find that the three variables *X1*, *X2* and *t* should be deleted.
4. As an alternative selection method, at the second stage, the variables *X1*, *X2* and *t* are inserted as additional independent variables, and the reduced model in Figure 12.22(e) is obtained. Furthermore, by inserting the variables *X3*, *X4* and *X5*, we get the reduced model, presented in Figure 12.22(f), which can be considered the best fit.
5. Finally, the LV(1) binary probit model is obtained by inserting Y1(-1) as an additional independent variable of the model in Figure 12.22(f), with the output presented in Figure 12.23(a), and one of its possible reduced models in Figure 12.23(b).

(a)

Dependent Variable: Q25_Y1
Method: ML - Binary Probit (Quadratic hill climbing)
Date: 10/04/11   Time: 10:18
Sample: 1 1792 IF M_SIZE=1 AND TP=1
Included observations: 516
Convergence achieved after 5 iterations
Covariance matrix computed using second derivatives

Variable	Coefficient	Std. Error	z-Statistic	Prob.
C	1.136149	0.130447	8.709655	0.0000
T*X1*X2	0.062319	0.012171	5.120287	0.0000
X1*X2	-0.125068	0.027962	-4.472785	0.0000
T*X1	-0.060542	0.007011	-8.635085	0.0000
T*X2	-0.230156	0.056013	-4.108991	0.0000

McFadden R-squared	0.247419	Mean dependent var		0.575581
S.D. dependent var	0.494734	S.E. of regression		0.418694
Akaike info criterion	1.045415	Sum squared resid		89.58054
Schwarz criterion	1.086560	Log likelihood		-264.7172
Hannan-Quinn criter.	1.061539	Restr. log likelihood		-351.7459
LR statistic	174.0575	Avg. log likelihood		-0.513018
Prob(LR statistic)	0.000000			

Obs with Dep=0	219	Total obs		516
Obs with Dep=1	297			

(a)

(b)

Dependent Variable: Q25_Y1
Method: ML - Binary Probit (Quadratic hill climbing)
Date: 10/04/11   Time: 10:20
Sample: 1 1792 IF M_SIZE=1 AND TP=1
Included observations: 516
Convergence achieved after 5 iterations
Covariance matrix computed using second derivatives

Variable	Coefficient	Std. Error	z-Statistic	Prob.
C	1.194528	0.166971	7.154106	0.0000
T*X1*X2	0.058130	0.012362	4.702221	0.0000
X1*X2	-0.124128	0.028097	-4.417903	0.0000
T*X1	-0.055318	0.007559	-7.318138	0.0000
T*X2	-0.213006	0.057047	-3.733867	0.0002
X3	-1.155160	0.494527	-2.335888	0.0195
X4	-0.063085	0.089513	-0.704755	0.4810
X5	0.153342	0.100132	1.531394	0.1257

McFadden R-squared	0.258060	Mean dependent var		0.575581
S.D. dependent var	0.494734	S.E. of regression		0.416657
Akaike info criterion	1.042537	Sum squared resid		88.19033
Schwarz criterion	1.108368	Log likelihood		-260.9744
Hannan-Quinn criter.	1.068334	Restr. log likelihood		-351.7459
LR statistic	181.5430	Avg. log likelihood		-0.505764
Prob(LR statistic)	0.000000			

Obs with Dep=0	219	Total obs		516
Obs with Dep=1	297			

(b)

(c)

Dependent Variable: Q25_Y1
Method: ML - Binary Probit (Quadratic hill climbing)
Date: 10/04/11   Time: 10:23
Sample: 1 1792 IF M_SIZE=1 AND TP=1
Included observations: 516
Convergence achieved after 5 iterations
Covariance matrix computed using second derivatives

Variable	Coefficient	Std. Error	z-Statistic	Prob.
C	1.122456	0.132330	8.482266	0.0000
T*X1*X2	0.058405	0.012382	4.716980	0.0000
X1*X2	-0.121585	0.027880	-4.360966	0.0000
T*X1	-0.056635	0.007322	-7.735285	0.0000
T*X2	-0.217047	0.056933	-3.812295	0.0001
X3	-1.113733	0.483165	-2.305076	0.0212
X5	0.149332	0.100963	1.479073	0.1391

McFadden R-squared	0.257361	Mean dependent var		0.575581
S.D. dependent var	0.494734	S.E. of regression		0.416843
Akaike info criterion	1.039613	Sum squared resid		88.44274
Schwarz criterion	1.097215	Log likelihood		-261.2202
Hannan-Quinn criter.	1.062186	Restr. log likelihood		-351.7459
LR statistic	181.0516	Avg. log likelihood		-0.506241
Prob(LR statistic)	0.000000			

Obs with Dep=0	219	Total obs		516
Obs with Dep=1	297			

(c)

(d)

Dependent Variable: Q25_Y1
Method: Least Squares
Date: 10/04/11   Time: 10:24
Sample: 1 1792 IF M_SIZE=1 AND TP=1
Included observations: 516

Variable	Coefficient	Std. Error	t-Statistic	Prob.
C	0.826950	0.154736	5.344253	0.0000
T*X1*X2	0.005188	0.004934	1.051441	0.2936
X1*X2	-0.026344	0.016799	-1.568196	0.1175
T*X1	-0.021825	0.007645	-2.854873	0.0045
T*X2	0.001665	0.024409	0.068202	0.9457
X3	-0.057251	0.092498	-0.618937	0.5362
X5	0.010935	0.012801	0.854242	0.3934
X1	-0.028040	0.028306	-0.990597	0.3224
X2	-0.055176	0.079098	-0.697561	0.4858
T	0.073621	0.043071	1.709304	0.0880

R-squared	0.320503	Mean dependent var		0.575581
Adjusted R-squared	0.308417	S.D. dependent var		0.494734
S.E. of regression	0.411428	Akaike info criterion		1.080825
Sum squared resid	85.65219	Schwarz criterion		1.163114
Log likelihood	-268.8528	Hannan-Quinn criter.		1.113071
F-statistic	26.51871	Durbin-Watson stat		1.820517
Prob(F-statistic)	0.000000			

(d)

Figure 12.22   *Outputs based on the model in (12.33), specific for the group (M_Size = 1, TP = 1) using a manual multistage stepwise selection method*

Based on these outputs, the following notes and comments are presented.

5.1   Take note that this reduced model is an acceptable LV(1) model, since Y1(−1) has a *p*-value = 0.2896 < 0.30.

5.2   By inserting the lagged variable *Q25_Y1*(−1), instead of *Y1*(−1), we see that *Q25_Y1*(−1) has a significant effect with *p*-value = 0.0000, but three of the other independent variables: *t*X1*, *t* and X3 have very large *p*-values of 0.9429, 0.7713 and 0.9894, respectively.

5.3   For this reason, an alternative order of variables will be inserted at the stages of analysis. By using the forward and backward selection at each stage of the analysis for each independent variable, under the condition that previously inserted variables should have *p*-values < 0.30 we

Dependent Variable: Q25_Y1
Method: Least Squares
Date: 10/04/11 Time: 10:31
Sample: 1 1792 IF M_SIZE=1 AND TP=1
Included observations: 516

Variable	Coefficient	Std. Error	t-Statistic	Prob.
C	0.892804	0.125812	7.096361	0.0000
T*X1*X2	0.008194	0.002680	3.058010	0.0023
X1*X2	-0.036822	0.007932	-4.642205	0.0000
T*X1	-0.019938	0.006726	-2.964509	0.0032
T*X2	-0.013838	0.009813	-1.410110	0.1591
X1	-0.039539	0.023975	-1.649157	0.0997
T	0.061266	0.036768	1.666279	0.0963

R-squared	0.319003	Mean dependent var	0.575581
Adjusted R-squared	0.310975	S.D. dependent var	0.494734
S.E. of regression	0.410666	Akaike info criterion	1.071402
Sum squared resid	85.84127	Schwarz criterion	1.129004
Log likelihood	-269.4217	Hannan-Quinn criter.	1.093974
F-statistic	39.73890	Durbin-Watson stat	1.816819
Prob(F-statistic)	0.000000		

(e)

Dependent Variable: Q25_Y1
Method: ML - Binary Probit (Quadratic hill climbing)
Date: 10/04/11 Time: 10:35
Sample: 1 1792 IF M_SIZE=1 AND TP=1
Included observations: 516
Convergence achieved after 6 iterations
Covariance matrix computed using second derivatives

Variable	Coefficient	Std. Error	z-Statistic	Prob.
C	1.801085	0.738262	2.439630	0.0147
T*X1*X2	0.085126	0.016589	5.131473	0.0000
X1*X2	-0.113669	0.028411	-4.000848	0.0001
T*X1	-0.100446	0.038215	-2.628484	0.0086
T*X2	-0.366748	0.083372	-4.398922	0.0000
X1	-0.192577	0.126594	-1.521213	0.1282
T	0.437124	0.206248	2.119410	0.0341
X4	-0.230435	0.102524	-2.247625	0.0246
X3	0.521134	0.458870	1.135688	0.2561

McFadden R-squared	0.323236	Mean dependent var	0.575581
S.D. dependent var	0.494734	S.E. of regression	0.397279
Akaike info criterion	0.957554	Sum squared resid	80.02009
Schwarz criterion	1.031614	Log likelihood	-238.0489
Hannan-Quinn criter.	0.986576	Restr. log likelihood	-351.7459
LR statistic	227.3940	Avg. log likelihood	-0.461335
Prob(LR statistic)	0.000000		

Obs with Dep=0	219	Total obs	516
Obs with Dep=1	297		

(f)

Figure 12.22   (Continued)

Dependent Variable: Q25_Y1
Method: ML - Binary Probit (Quadratic hill climbing)
Date: 10/04/11 Time: 13:13
Sample: 1 1792 IF M_SIZE=1 AND TP=1
Included observations: 516
Convergence achieved after 6 iterations
Covariance matrix computed using second derivatives

Variable	Coefficient	Std. Error	z-Statistic	Prob.
C	1.843217	0.745175	2.473536	0.0134
T*X1*X2	0.084240	0.016691	5.046908	0.0000
X1*X2	-0.112229	0.028636	-3.919223	0.0001
T*X1	-0.098835	0.038301	-2.580517	0.0099
T*X2	-0.364084	0.083423	-4.364332	0.0000
X1	-0.198100	0.127217	-1.557183	0.1194
T	0.428826	0.206617	2.075466	0.0379
X4	-0.230583	0.102530	-2.248931	0.0245
X3	0.516487	0.457543	1.128828	0.2590
Y1(-1)	-0.010166	0.025316	-0.401559	0.6880

McFadden R-squared	0.323465	Mean dependent var	0.575581
S.D. dependent var	0.494734	S.E. of regression	0.397621
Akaike info criterion	0.961118	Sum squared resid	79.99985
Schwarz criterion	1.043407	Log likelihood	-237.9686
Hannan-Quinn criter.	0.993365	Restr. log likelihood	-351.7459
LR statistic	227.5548	Avg. log likelihood	-0.461179
Prob(LR statistic)	0.000000		

Obs with Dep=0	219	Total obs	516
Obs with Dep=1	297		

(a)

Dependent Variable: Q25_Y1
Method: Least Squares
Date: 10/04/11 Time: 13:16
Sample: 1 1792 IF M_SIZE=1 AND TP=1
Included observations: 516

Variable	Coefficient	Std. Error	t-Statistic	Prob.
C	1.088516	0.146525	7.428891	0.0000
T*X1*X2	0.007913	0.002670	2.963556	0.0032
X1*X2	-0.038441	0.007950	-4.835513	0.0000
T*X1	-0.014250	0.007122	-2.000898	0.0459
T*X2	-0.010922	0.009830	-1.111099	0.2671
X1	-0.061452	0.025364	-2.422754	0.0158
T	0.044078	0.037430	1.177621	0.2395
X4	-0.075278	0.031005	-2.427921	0.0155
Y1(-1)	-0.005796	0.005468	-1.060008	0.2896

R-squared	0.328441	Mean dependent var	0.575581
Adjusted R-squared	0.317844	S.D. dependent var	0.494734
S.E. of regression	0.408614	Akaike info criterion	1.065198
Sum squared resid	84.65158	Schwarz criterion	1.139258
Log likelihood	-265.8210	Hannan-Quinn criter.	1.094220
F-statistic	30.99496	Durbin-Watson stat	1.883066
Prob(F-statistic)	0.000000		

(b)

Figure 12.23   *Statistical results of LV(1) BLMs based on the model in Figure 12.13(f)*

then get the two alternative acceptable models in Figure 12.24. Note that the model in Figure 12.24(a) does not have the main factor *X1*, *X2* and the time *t* as independent variables, and the other only has the time *t*.

6. The analysis for the other three groups of (*MSize,TP*) can easily be done using the same method. Do it as an exercise.

Dependent Variable: Q25_Y1
Method: ML - Binary Probit (Quadratic hill climbing)
Date: 10/04/11   Time: 21:01
Sample: 1 8 IF M_SIZE=1 AND TP=1
Included observations: 417
Convergence achieved after 6 iterations
Covariance matrix computed using second derivatives

Variable	Coefficient	Std. Error	z-Statistic	Prob.
C	4.003643	0.406084	9.859149	0.0000
Y1(-1)	-6.190445	0.756243	-8.185783	0.0000
T*X1*X2	0.037900	0.017432	2.174168	0.0297
X1*X2	-0.109956	0.047797	-2.300497	0.0214
T*X1	-0.047400	0.011982	-3.955910	0.0001
T*X2	-0.104258	0.075252	-1.385465	0.1659
X3	-2.547133	1.050493	-2.424702	0.0153
X4	-1.394876	0.192963	-7.228716	0.0000
X5	1.490530	0.798651	1.866309	0.0620

McFadden R-squared	0.488856	Mean dependent var	0.546763
S.D. dependent var	0.498406	S.E. of regression	0.324998
Akaike info criterion	0.747284	Sum squared resid	43.09456
Schwarz criterion	0.834329	Log likelihood	-146.8087
Hannan-Quinn criter.	0.781698	Restr. log likelihood	-287.2160
LR statistic	280.8145	Avg. log likelihood	-0.352059
Prob(LR statistic)	0.000000		

Obs with Dep=0	189	Total obs	417
Obs with Dep=1	228		

(a)

Dependent Variable: Q25_Y1
Method: ML - Binary Probit (Quadratic hill climbing)
Date: 10/16/11   Time: 09:35
Sample: 1 8 IF M_SIZE=1 AND TP=1
Included observations: 417
Convergence achieved after 6 iterations
Covariance matrix computed using second derivatives

Variable	Coefficient	Std. Error	z-Statistic	Prob.
C	0.083394	0.314716	0.264982	0.7910
Q25_Y1(-1)	1.826003	0.184259	9.909968	0.0000
T*X1*X2	0.036320	0.018231	1.992239	0.0463
T*X1	-0.119916	0.023417	-5.120921	0.0000
T*X2	-0.230890	0.098186	-2.351550	0.0187
X3	-1.432467	0.953200	-1.502798	0.1329
X4	-0.486867	0.124773	-3.902026	0.0001
X5	1.585359	0.855808	1.852471	0.0640
T	0.545006	0.152022	3.585050	0.0003

McFadden R-squared	0.532175	Mean dependent var	0.546763
S.D. dependent var	0.498406	S.E. of regression	0.323976
Akaike info criterion	0.687610	Sum squared resid	42.82397
Schwarz criterion	0.774655	Log likelihood	-134.3667
Hannan-Quinn criter.	0.722024	Restr. log likelihood	-287.2160
LR statistic	305.6985	Avg. log likelihood	-0.322222
Prob(LR statistic)	0.000000		

Obs with Dep=0	189	Total obs	417
Obs with Dep=1	228		

(b)

Figure 12.24   Statistical results of two alternative BLMs based on the sample {M_Size = 1, TP = 1}

Example 12.20   An alternative manual stepwise selection data analysis based on the model (12.20)
The data analysis based on the equation specification (ES) in (12.20) can be done directly using the manual stepwise selection method, as follows:

1. At Stage-1, the ES: $Q25\_Y1\ Y1(-1)^* @Expand(M\_Size, TP)$ is applied, the output presents an acceptable full model; that is, the four independent variables have p-values $= 0.0000$.
2. At Stage-2, the function $t^*X1^*X2^*@Expand(M\_Size, TP)$ is inserted, where one of the independent variables, namely $t^*X1^*X2^*(M\_size = 2, TP = 2)$, should be deleted, since it has a p-value $= 0.7836 > 0.30$.
   So the function $t^*X1^*X2^*@Expand(M\_Size, TP, @Drop(2,2))$ is inserted as additional independent variables.
3. At Stage-3, the function $t^*X1^*@Expand(M\_Size, TP, @Drop(2,1))$ is inserted as additional independent variables.
4. At Stage-4, the function $X1^*X2^*@Expand(M\_Size, TP, @Drop(2,1))$ is inserted as additional independent variables. However, $t^*X1^*(M\_size = 2, TP = 2)$ has a p-value $= 0.2453$, and the p-value of the $t^*X1^*(M\_size = 2, TP = 2)$ decreases to 0.1334.
5. At Stage-5, the function $t^*X2^*@Expand(M\_Size, TP, @Drop(1,2), @Drop(2,2))$ is inserted as additional independent variables.
6. Finally, for the main factors $X3$, $X4$, $X5$, $X1$, $X2$ and the time t, by using the trial-and-error method, we obtain: $X4^*@Expand(M\_Size, TP, @Drop(2,1), @Drop(2,2))$, $X5^*@Expand(M\_Size, TP, @Drop(2,1), @Drop(2,2))$, and $X2^*@Expand(M\_Size, TP, @Drop(1,1), @Drop(1,2), @Drop(2,2))$, which are used as additional independent variables with the summary of statistical results presented in Table 12.8. Note that other regressions can be obtained by using different ordered in inserting the functions.
7. Similar to lagged variables model in (12.31), this binary choice model can be modified or extended to various models having all lagged independent variables.

Table 12.8 Summary of a manual stepwise selection method of the binary probit model in (12.20)

Variable	M_Size=1,TP=1 Coeff.	z-Stat.	Prob.	M_Size=1,TP=2 Coeff.	z-Stat.	Prob.	M_Size=2,TP=1 Coeff.	z-Stat.	Prob.	M_Size=2,TP=2 Coeff.	z-Stat.	Prob.
Intercept	3.974	9.782	0.000	28.889	6.205	0.000	-4.411	-3.689	0.000	8.759	1.675	0.094
Y1(-1)	-6.080	-8.084	0.000	-2.554	-1.810	0.070	-1.257	-1.411	0.158	-45.470	-1.672	0.095
T*X1*X2	0.039	2.225	0.026	-0.107	-3.428	0.001	0.171	2.501	0.012			
T*X1	-0.051	-4.329	0.000	-0.052	-2.923	0.004				-0.173	-1.501	0.133
X1*X2	-0.112	-2.366	0.018	0.522	2.576	0.010				-0.235	-1.162	0.245
T*X2	-0.105	-1.399	0.162				-0.837	-1.886	0.059			
X4	-1.374	-7.086	0.000	-15.853	-5.986	0.000						
X5	0.990	1.285	0.199				10.700	2.460	0.014			
X2							-3.051	-2.496	0.013			

McFadden R-squared	0.741	Mean dependent var	0.249	
S.D. dependent var	0.433	S.E. of regression	0.208	
Akaike info criterion	0.325	Sum squared resid	61.234	
Schwarz criterion	0.413	Log likelihood	-209.1	
Hannan-Quinn criter.	0.358	Restr. log likelihood	-806.4	
LR statistic	1194.6	Avg. log likelihood	-0.146	
Prob(LR statistic)	0.000			
Obs with Dep=0	1078	Total obs	1436	
Obs with Dep=1	358			

12.8   Manual Stepwise Selection for Ordered Choice Models

For the following illustration, an ordinal variable is generated with three levels (O3NI) and a bounded logarithmic variable (BND_NI) based on the variable NI (net income) using the following equations.

$$O3NI = 1 + 1^*(NI >= Quantile(NI, 0.30) + 1^*(NI >= @Quantile(NI, 0.70)$$
$$BND\_NI = \log((NI + 16000)/(1300\text{-}NI))$$

where $L = -16\,000$ and $U = 1300$ are the lower and upper bounds of *NI*, which are subjectively selected.

Example 12.21   The simplest ordered choice LV(1) model by time
Figure 12.25 presents the statistical results of an ordered probit LV(1) model for the time $t > 1$, using the equation specification (ES) as follows:

$$O3NI\ @Expand(T, @Droplast)\ BND\_NI(-1)^*Expand(T) \tag{12.34}$$

Based on the ES in (12.34) and its results in Figure 12.25, the following notes and comments are presented.

1. Note that the function $@Expand(T, @Droplast)$ or $@Expand(T, @Drop(^*))$, in general, should be used. If the function $@Expand(T)$ is used, unacceptable output is obtained.
2. An error message would be obtained by using the default options, as presented in Figure 12.26(a), which shows "*EViews Supplied*" in the window of the *Starting coefficient values*.
3. Figure 12.26(b) presents six alternative starting coefficient values, but the output in Figure 12.25 is obtained by using only the "*Zero*" as a starting value. The others give the error message. I would say frankly that I cannot explain why.

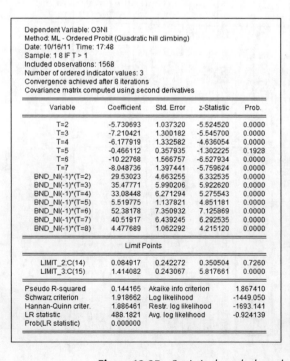

Figure 12.25   *Statistical results based on the ordered choice model in (12.34)*

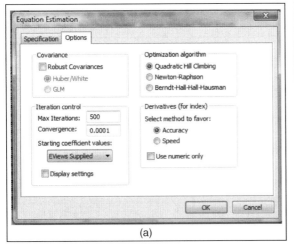 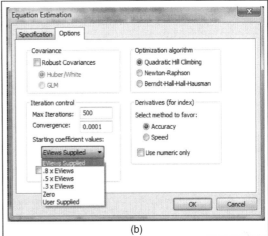

(a)   (b)

Figure 12.26 *The options for the analysis based on the ordered choice models*

4. Furthermore, there are three alternative optimization algorithms, which can easily be studied their differences. Do this as an exercise.

5. The ordered probit model has the assumption that its residuals have identical independent normal distributions, which in fact are theoretical distributions. Figure 12.27 shows the residuals' histogram, which clearly shows it is not a normal distribution, and the Jarque–Bera test also does not support the normal distribution of the residuals. The figure presents the theoretical distribution of the residuals.

6. The other ordered choice models are ordered logit and extreme value models, with the residuals assumed to have logistic and extreme value distributions, respectively. Note that there is no test available for these distributions. So it can be said that the basic assumption of the residuals of a model should be taken for granted. In other words, the basic assumption does not need to be tested. Refer to the notes and comments presented in Agung (2009a, Section 2.14.3, and 2011).

7. Table 12.9 presents a set of forecasting equations by time, with different slopes as well as intercepts. So they are heterogeneous regressions, and we find that they have significant differences based on the F-statistic of $F_0 = 22.32\,584$ with $df = (6{,}1553)$ and a p-value $= 0.0000$.

8. Therefore, the time-fixed-effect model is not appropriate for this natural experimental data.

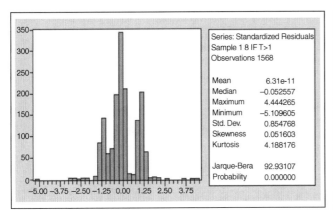 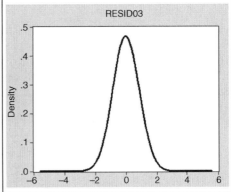

Figure 12.27 *Histogram, normality test and theoretical error distribution*

Table 12.9   *The forecasting equations of I_O3NI on BND_NI(−1) by time points*

Time Point	Intercepts				Slopes of BND_NI(-1)			
	Coeff	s.e	z-Stat,	Prob.	Coeff	s.e	z-Stat,	Prob.
T=2	-5.731	1.037	-5.525	0.000	29.530	4.663	6.333	0.000
T=3	-7.210	1.300	-5.546	0.000	35.478	5.990	5.923	0.000
T=4	-6.178	1.333	-4.636	0.000	33.084	6.271	5.276	0.000
T=5	-0.466	0.358	-1.302	0.193	5.520	1.138	4.851	0.000
T=6	-10.228	1.567	-6.528	0.000	52.382	7.351	7.126	0.000
T=7	-8.049	1.397	-5.760	0.000	40.519	6.439	6.293	0.000
T=8					4.478	1.062	4.215	0.000
Limit Points								
LIMIT_2:C(14)	0.085	0.242	0.351	0.726				
LIMIT_3:C(15)	1.414	0.243	5.818	0.000				
Pseudo R-squared	0.1442	Akaike info criterion		1.8674				
Schwarz criterion	1.9187	Log likelihood		-1449.1				
Hannan-Quinn criter.	1.8865	Restr. log likelihood		-1693.1				
LR statistic	488.18	Avg. log likelihood		-0.9241				
Prob(LR statistic)	0.0000							

Example 12.22   A polynomial ordered choice model by time

A numerical cause, source or up-stream variable might have a nonlinear effect on any endogenous variable, based on cross-sectional data, since the research objects have various differential characteristics – refer to Agung (2011). Therefore, at each time point of the panel data, $BND\_NI(-1)$ might have a nonlinear effect on $O3NI$.

By using the trial-and-error method, a reduced model of a full polynomial ordered probit is obtained for $t > 1$ with the following ES, and the summary of results is presented in Table 12.10.

$$o3ni \ @expand(t, @droplast) \ bnd\_ni(-1)^* @expand(t)$$
$$bnd\_ni(-1)^{\wedge}2^* @expand(t) \ bnd\_ni(-1)^{\wedge}3^* @expand(t) \qquad (12.35)$$

Based on this summary the following notes and comments are presented.

1. At the point $T = 3$, $BND\_NI(-1)^2$ and $BND\_NI(-1)^3$ have *p*-values of 0.301 and 0.432, respectively. In order to have a reduced model, in general one would delete $BND\_NI(-1)^3$, since it has a greater *p*-value.
2. In this case, however, $BND\_NI(-1)^2$ is deleted in order to have third-degree polynomials over time. The summary of the output presented in Table 12.11, which shows that $BND\_NI(-1)^3$ has a *p*-value $= 0.001$, at $T = 3$. So it can be said that the model in Table 12.10 is the best fit.
3. *Predicted probabilities*. To compute the predicted probabilities based on the model in Table 12.10, the steps of the analysis are as follows:

Table 12.10   *The forecasting equations of I_O3NI: a reduced model of the model (12.35)*

Time	Intercept			BND_NI(-1)			BND_NI(-1)^2			BND_NI(-1)^3		
	Coeff.	z-stat.	Prob.	Coeff.	z-stat.	Prob.	Coeff.	z-stat.	Prob.	Coeff.	z-stat.	Prob.
T=2	-59.788	-3.333	0.001	586.320	2.939	0.003	-1836.97	-2.543	0.011	1895.43	2.235	0.025
T=3	-18.393	-2.067	0.039	134.272	1.554	0.120	-264.457	-1.035	*0.301*	166.112	0.786	*0.432*
T=4	-0.443	-0.356	*0.722*	-19.064	-2.014	0.044				502.314	4.401	0.000
T=5	-11.614	-5.555	0.000	79.877	4.752	0.000	-127.905	-3.083	0.002	60.811	2.216	0.027
T=6	36.389	5.641	0.000	-333.222	-6.219	0.000				3641.7	6.842	0.000
T=7	-356.194	-3.739	0.000	4140.979	3.536	0.000	-15861.8	-3.341	0.001	20085.7	3.169	0.002
T=8				-3.868	-1.213	0.225	24.782	3.421	0.001	4.848	3.299	0.001

Limit Points

	Coeff.	z-stat.	Prob.		
LIMIT_2:C(26)	-0.518	-1.144	0.253		
LIMIT_3:C(27)	0.938	2.069	0.039		
Pseudo R-squared	0.212	Akaike info criterion		1.7	
Schwarz criterion	1.828	Log likelihood		-1334.2	
Hannan-Quinn criter.	1.770	Restr. log likelihood		-1693.1	
LR statistic	717.923	Avg. log likelihood		-0.9	
Prob(LR statistic)	0.000				

(*) BND_NI(-1) = BND_NI(T-1)

Table 12.11   *The forecasting equation of* I_O3NI: *a reduced model of the one in Table 12.10*

Time	Intercept			BND_NI(-1)			BND_NI(-1)^2			BND_NI(-1)^3		
	Coeff.	z-stat.	Prob.	Coeff.	z-stat.	Prob.	Coeff.	z-stat.	Prob.	Coeff.	z-stat.	Prob.
T=2	-59.781	-3.333	0.001	586.253	2.939	0.003	-1836.75	-2.543	0.011	1895.21	2.235	0.025
T=3	-9.199	-5.868	0.000	44.012	5.903	0.000				-39.718	-3.362	0.001
T=4	-0.443	-0.355	0.722	-19.063	-2.014	0.044				502.278	4.401	0.000
T=5	-11.613	-5.555	0.000	79.869	4.752	0.000	-127.891	-3.082	0.002	60.804	2.216	0.027
T=6	36.386	5.641	0.000	-333.191	-6.219	0.000				3641.35	6.842	0.000
T=7	-356.161	-3.739	0.000	4140.60	3.536	0.000	-15860.4	-3.341	0.001	20083.9	3.169	0.002
T=8				-3.868	-1.213	0.225	24.780	3.421	0.001	4.848	3.298	0.001
Limit Points												
LIMIT_2:C(25)	-0.51764	-1.1432	0.253									
LIMIT_3:C(26)	0.93736	2.0689	0.039									
Pseudo R-squared	0.212	Akaike info criterion			1.736							
Schwarz criterion	1.824	Log likelihood			-1334.7							
Hannan-Quinn criter.	1.769	Restr. log likelihood			-1693.1							
LR statistic	716.85	Avg. log likelihood			-0.851							
Prob(LR statistic)	0.000											

(*) BND_NI(-1) = BND_NI(T-1)

3.1   With the output on the screen, click *View/Representatives*, then the forecasting equation is as presented in Figure 12.28 on the screen.

3.2   Generate the series *I_O3NI*, by using the block-copy-paste method, namely (i) Block the estimation equation, (ii) Click "*Copy*", and (iii) Click *Quick/Generate Series…*, and then click "*Paste*"… *OK*, the variable *I_O3NI* is inserted to the file, directly.

3.3   Generate each of the series *O3NI_1*, *O3NI_2*, and *O3NI_3* using the same method, which are the predicted probabilities of $P(O3NI = 1)$, $P(O3NI = 2)$, and $P(O3NI = 3)$ with a total of one for each firm and each time point.

```
Estimation Equation:
=========================
I_O3NI = C(1)*(T=2) + C(2)*(T=3) + C(3)*(T=4) + C(4)*(T=5) + C(5)*(T=6) + C(6)*(T=7) + C(7)*BND_NI(-1)*(T=2) + C(8)*BND_NI(-1)*(T=3) +
C(9)*BND_NI(-1)*(T=4) + C(10)*BND_NI(-1)*(T=5) + C(11)*BND_NI(-1)*(T=6) + C(12)*BND_NI(-1)*(T=7) + C(13)*BND_NI(-1)*(T=8) + C(14)
*BND_NI(-1)^2*(T=2) + C(15)*BND_NI(-1)^2*(T=5) + C(16)*BND_NI(-1)^2*(T=7) + C(17)*BND_NI(-1)^2*(T=8) + C(18)*BND_NI(-1)^3*(T=2) +
C(19)*BND_NI(-1)^3*(T=3) + C(20)*BND_NI(-1)^3*(T=4) + C(21)*BND_NI(-1)^3*(T=5) + C(22)*BND_NI(-1)^3*(T=6) + C(23)*BND_NI(-1)^3*
(T=7) + C(24)*BND_NI(-1)^3*(T=8)

Forecasting Equation:
=========================
I_O3NI = C(1)*(T=2) + C(2)*(T=3) + C(3)*(T=4) + C(4)*(T=5) + C(5)*(T=6) + C(6)*(T=7) + C(7)*BND_NI(-1)*(T=2) + C(8)*BND_NI(-1)*(T=3) +
C(9)*BND_NI(-1)*(T=4) + C(10)*BND_NI(-1)*(T=5) + C(11)*BND_NI(-1)*(T=6) + C(12)*BND_NI(-1)*(T=7) + C(13)*BND_NI(-1)*(T=8) + C(14)
*BND_NI(-1)^2*(T=2) + C(15)*BND_NI(-1)^2*(T=5) + C(16)*BND_NI(-1)^2*(T=7) + C(17)*BND_NI(-1)^2*(T=8) + C(18)*BND_NI(-1)^3*(T=2) +
C(19)*BND_NI(-1)^3*(T=3) + C(20)*BND_NI(-1)^3*(T=4) + C(21)*BND_NI(-1)^3*(T=5) + C(22)*BND_NI(-1)^3*(T=6) + C(23)*BND_NI(-1)^3*
(T=7) + C(24)*BND_NI(-1)^3*(T=8)

O3NI_1 = @CNORM(C(25)-I_O3NI)
O3NI_2 = @CNORM(C(26)-I_O3NI) - @CNORM(C(25)-I_O3NI)
O3NI_3 = 1 - @CNORM(C(26)-I_O3NI)
```

Figure 12.28   *Forecasting equation of the ordered choice model in Table 12.10*

Dependent Variable: O3NI
Method: ML - Ordered Probit (Quadratic hill climbing)
Date: 10/18/11   Time: 16:10
Sample: 1 8 IF T>5
Included observations: 627
Number of ordered indicator values: 3
Convergence achieved after 8 iterations
Covariance matrix computed using second derivatives

Variable	Coefficient	Std. Error	z-Statistic	Prob.
T=6	-7.106460	1.651331	-4.303473	0.0000
T=7	-2.852384	1.439117	-1.982038	0.0475
BND_NI(-1)*(T=6)	52.40684	7.459314	7.025692	0.0000
BND_NI(-1)*(T=7)	20.95491	7.053302	2.970936	0.0030
BND_NI(-1)*(T=8)	1.972489	1.853176	1.064383	0.2872
X1(-1)*X2(-1)*(T=6)	-0.008515	0.011207	-0.759834	0.4474
X1(-1)*X2(-1)*(T=7)	-0.123410	0.050670	-2.435564	0.0149
X1(-1)*X2(-1)*(T=8)	-0.099626	0.043785	-2.275336	0.0229
X1(-1)*(T=7)	0.352495	0.071719	4.914918	0.0000
X1(-1)*(T=8)	0.586891	0.063616	9.225583	0.0000
X2(-1)*(T=7)	0.499164	0.298692	1.671163	0.0947
X2(-1)*(T=8)	0.297198	0.273884	1.085125	0.2779

Limit Points

	Coefficient	Std. Error	z-Statistic	Prob.
LIMIT_2:C(13)	2.997187	0.515575	5.813292	0.0000
LIMIT_3:C(14)	4.610427	0.527612	8.738284	0.0000

Pseudo R-squared	0.263117	Akaike info criterion	1.603425
Schwarz criterion	1.702585	Log likelihood	-488.6737
Hannan-Quinn criter.	1.641949	Restr. log likelihood	-663.1634
LR statistic	348.9795	Avg. log likelihood	-0.779384
Prob(LR statistic)	0.000000		

(a)

Dependent Variable: O3NI
Method: ML - Ordered Probit (Quadratic hill climbing)
Date: 10/18/11   Time: 16:14
Sample: 1 8 IF T>5
Included observations: 627
Number of ordered indicator values: 3
Convergence achieved after 8 iterations
Covariance matrix computed using second derivatives

Variable	Coefficient	Std. Error	z-Statistic	Prob.
T=6	4.010421	0.523702	7.657834	0.0000
T=7	-2.794386	1.441089	-1.939080	0.0525
BND_NI(-1)*(T=7)	20.63268	7.066448	2.919809	0.0035
BND_NI(-1)*(T=8)	1.979901	1.844851	1.073204	0.2832
X1(-1)*X2(-1)*(T=6)	-0.021100	0.010511	-2.007530	0.0447
X1(-1)*X2(-1)*(T=7)	-0.117767	0.050348	-2.339068	0.0193
X1(-1)*X2(-1)*(T=8)	-0.096352	0.043643	-2.207766	0.0273
X1(-1)*(T=7)	0.335438	0.071173	4.712971	0.0000
X1(-1)*(T=8)	0.565893	0.063112	8.966536	0.0000
X2(-1)*(T=7)	0.476589	0.296644	1.606600	0.1081
X2(-1)*(T=8)	0.289181	0.273032	1.059147	0.2895

Limit Points

	Coefficient	Std. Error	z-Statistic	Prob.
LIMIT_2:C(12)	2.941396	0.512706	5.737001	0.0000
LIMIT_3:C(13)	4.424581	0.522790	8.463397	0.0000

Pseudo R-squared	0.189433	Akaike info criterion	1.756103
Schwarz criterion	1.848180	Log likelihood	-537.5382
Hannan-Quinn criter.	1.791876	Restr. log likelihood	-663.1634
LR statistic	251.2504	Avg. log likelihood	-0.857318
Prob(LR statistic)	0.000000		

(b)

Figure 12.29   *Statistical results based on two reduced models of the full model in (11/30)*

3.4   Note that the predicted probabilities are not valid for the firms in the sample, they are for unobserved firms with the characteristics indicated by the same scores of the independent variables used in the model, but they have unknown levels of the *O3NI*. So it can be said that the predicted probabilities are abstract.

4. Any regression analysis using the predicted probabilities of the binary or ordered choice models, as one of the numerical independent variables called the *switching regression analysis*. For instance, the switching regression of *D_Dividend* on *O3NI_1*, *O3NI_2* or *O3NI_3*, and a set of exogenous variables – refer to Agung (2011).

Example 12.23   A more advanced ordered choice model by time

For illustration, a full model with the following specification equation (SE) is considered. However, Figure 12.29 only presents the statistical results based on two of its possible reduced models, for $T > 5$, which are obtained using the manual stepwise selection method.

$$O3NI \ @Expand(T, @Droplast) \ BND\_NI(-1)^* @Expand(T)$$
$$X1(-1)^* X2(-1)^* @Expand(T) \ X1(-1)^* @Expand(T) \ X2(-1)^* @Expand(T) \quad (12.36)$$

Based on these outputs, the following findings and notes are presented.

1. Note that the reduced model has a variable: $X1(-1)^* X2(-1)^*(T=6)$ with a large p-value $= 0.4474$ > 0.30. Should this variable be deleted? The answer is no, since we want to show that effect of $X1 = lnSale = log(SALE)$ on *O3NI* is significantly dependent on $X2 = log(LOAN)$ at each time point.

Table (a)

Dependent Variable: O3NI
Method: ML - Ordered Probit (Quadratic hill climbing)
Date: 10/18/11   Time: 16:26
Sample: 1 8 IF T>5
Included observations: 627
Number of ordered indicator values: 3
Convergence achieved after 8 iterations
Covariance matrix computed using second derivatives

Variable	Coefficient	Std. Error	z-Statistic	Prob.
T=7	-1.858395	1.439140	-1.291323	0.1966
BND_NI(-1)*(T=6)	23.50347	2.473541	9.501955	0.0000
BND_NI(-1)*(T=7)	20.89320	7.055789	2.961143	0.0031
BND_NI(-1)*(T=8)	4.292968	1.853581	2.316040	0.0206
X1(-1)*X2(-1)*(T=6)	-0.013069	0.010931	-1.195562	0.2319
X1(-1)*X2(-1)*(T=7)	-0.122295	0.050607	-2.416578	0.0157
X1(-1)*X2(-1)*(T=8)	-0.098150	0.045751	-2.145276	0.0319
X1(-1)*(T=7)	0.349119	0.071612	4.875158	0.0000
X1(-1)*(T=8)	0.656575	0.063249	10.38084	0.0000
X2(-1)*(T=7)	0.494705	0.298290	1.658470	0.0972
X2(-1)*(T=8)	0.267880	0.286539	0.934882	0.3498

Limit Points				
LIMIT_2:C(12)	3.968809	0.516759	7.680197	0.0000
LIMIT_3:C(13)	5.556842	0.531384	10.45729	0.0000

Pseudo R-squared	0.247222	Akaike info criterion		1.633859
Schwarz criterion	1.725936	Log likelihood		-499.2149
Hannan-Quinn criter.	1.669632	Restr. log likelihood		-663.1634
LR statistic	327.8972	Avg. log likelihood		-0.796196
Prob(LR statistic)	0.000000			

(a)

Table (b)

Dependent Variable: O3NI
Method: ML - Ordered Probit (Quadratic hill climbing)
Date: 10/18/11   Time: 16:35
Sample: 1 8 IF T>5
Included observations: 627
Number of ordered indicator values: 3
Convergence achieved after 8 iterations
Covariance matrix computed using second derivatives

Variable	Coefficient	Std. Error	z-Statistic	Prob.
T=7	-1.827540	1.438394	-1.270542	0.2039
BND_NI(-1)*(T=6)	23.64026	2.464637	9.591779	0.0000
BND_NI(-1)*(T=7)	20.89065	7.055892	2.960738	0.0031
BND_NI(-1)*(T=8)	4.317232	1.850442	2.333081	0.0196
X1(-1)*X2(-1)*(T=6)	-0.013031	0.010932	-1.192082	0.2332
X1(-1)*X2(-1)*(T=7)	-0.122249	0.050604	-2.415797	0.0157
X1(-1)*X2(-1)*(T=8)	-0.057031	0.012451	-4.580513	0.0000
X1(-1)*(T=7)	0.348981	0.071607	4.873522	0.0000
X1(-1)*(T=8)	0.656078	0.063072	10.40209	0.0000
X2(-1)*(T=7)	0.494522	0.298273	1.657948	0.0973

Limit Points				
LIMIT_2:C(11)	3.998745	0.514686	7.769292	0.0000
LIMIT_3:C(12)	5.585736	0.529438	10.55032	0.0000

Pseudo R-squared	0.246559	Akaike info criterion		1.632073
Schwarz criterion	1.717067	Log likelihood		-499.6548
Hannan-Quinn criter.	1.665094	Restr. log likelihood		-663.1634
LR statistic	327.0174	Avg. log likelihood		-0.796898
Prob(LR statistic)	0.000000			

(b)

Figure 12.30   *Statistical results of alternative reduced models of the model in (12.36)*

2. Then we should delete either the intercept ($T = 6$), the variable $BND\_NI(-1)^*(T = 6)$, or both. A good fit model in Figure 12.29(b) is obtained by deleting $BND\_NI(-1)^*(T = 6)$. On the other hand, Figure 12.30 shows two alternative reduced models. It could be said these are unexpected models, however, this analysis has shown or demonstrated the unpredictable impact(s) of multicollinearity between the independent variables, since a variable with a p-value $= 0.0000$ is deleted in order to have acceptable reduced models.

3. Note that a model would be considered as an acceptable or good fit in statistical sense, even though it has one or more independent variables with p-values > 0.30. Corresponding to this statement, which one among the four models in Figures 12.29 and 12.30 would you consider the best and why? I would choose the model in Figure 12.29(a), because the others have more restrictions or hidden assumptions.

4. The model in (12.36) could be extended to the model with a large number of exogenous variables with the general ES as follows:

$$O3NI \ @Expand(T, @Droplast) \ BND\_NI(-1)^* @Expand(T)$$
$$V1(-1)^* @Expand(T) \dots Vk(-1)^* @Expand(T) \dots VK(-1)^* @Expand(T)$$

(12.37)

where Vk can be a main numerical variable or an interaction between two or three main variables. The data analysis can easily be done using the manual stepwise selection method as presented earlier, because the STEPLS estimation method for the ordered choice model is not available in EViews 6.

Table 12.12 *Summary of the output of a reduced model of that in (12.38)*

Group & Time	Intercept Coeff.	z-Stat.	Prob.	BND_NI(-1) Coeff.	z-Stat.	Prob.	X1(-1)*X2(-1) Coeff.	z-Stat.	Prob.	X1(-1) Coeff.	z-Stat.	Prob.
M_SIZE=1,T=6	-29.901	-2.969	0.003	136.024	2.778	0.006	-0.028	-1.553	0.120	0.251	3.130	0.002
M_SIZE=1,T=7	-15.847	-4.406	0.000	73.230	4.316	0.000						
M_SIZE=1,T=8	-4.474	-2.923	0.004	20.224	2.846	0.004						
M_SIZE=2,T=6	-10.717	-5.192	0.000	36.366	4.173	0.000	-0.027	-1.634	0.102	0.515	4.461	0.000
M_SIZE=2,T=7	-6.760	-4.534	0.000	14.229	1.957	0.050	-0.075	-3.619	0.000	0.569	4.348	0.000
M_SIZE=2,T=8				3.375	2.344	0.019	-0.022	-1.460	0.144			
Limit Points												
LIMIT_2:C(19)	-0.872	-2.481	0.013									
LIMIT_3:C(20)	0.743	2.135	0.033									
Pseudo R-squared	0.257	Akaike info criterion					1.635					
Schwarz criterion	1.777	Log likelihood					-492.6					
Hannan-Quinn criter.	1.690	Restr. log likelihood					-663.2					
LR statistic	341.2	Avg. log likelihood					-0.786					
Prob(LR statistic)	0.000											

Example 12.24   Ordered choice model by group and time

Table 12.12 presents a summary of the statistical results of an acceptable reduced ordered probit model of the following LV(1) model.

$$O3NI \ @Expand(M\_Size, T, @Droplast)$$
$$BND\_NI(-1)^* @Expand(M\_Size, T) \ X1(-1)^* X2(-1)^* @Expand(M\_Size, T) \quad (12.38)$$
$$X1(-1)^* @Expand(M\_Size, T) \ X2(-1)^* @Expand(M\_Size, T)$$

The reduced model is obtained by using the manual stepwise selection method. The main objective of this model is to show or test that the linear effect of $X1(-1)$ on O3NI depends on $X2(-1)$, indicated by the interaction $X1(-1)^*X2(-1)$ used as an independent variable of the model within each cell.

Example 12.25   An advanced ordered choice model

As an extension the ordered choice model in (12.38), Table 12.13 presents a summary of the statistical results based on a reduced ordered probit model of the following full LV(1) model. The main objective of this model is to study or test that the linear effect of $X1(-1)$ on $O3NI$ depends on $X2(-1)$ to $X5(-1)$, M_Size and TP. Note that this model can be considered the simplest in showing the effect of $X1(-1)$ on $O3NI$ depends on the other variables, since it only has two-way interaction independent variables. The model may have three-way or higher interactions of $X1(-1)$ to $X5(-1)$.

Table 12.13   Summary of the statistical results based on a reduced model of that in (12.39)

Variable	(M_Size=1,TP=1) Coeff.	z-Stat.	Prob.	(M_Size=1,TP=2) Coeff.	z-Stat.	Prob.	(M_Size=2,TP=1) Coeff.	z-Stat.	Prob.	(M_Size=2,TP=2) Coeff.	z-Stat.	Prob.
Intercept				-2191.8	-2.571	0.010	-336.9	-5.294	0.000	-326.3	-5.173	0.000
BND_NI(-1)	-1556.8	-5.133	0.000	9000.8	2.198	0.028	66.750	1.598	0.110	12.268	1.070	0.284
X1(-1)*X2(-1)	-0.022	-2.041	0.041	-0.023	-2.121	0.034	-0.048	-5.368	0.000	-0.029	-2.623	0.009
X1(-1)*X3(-1)	1.191	1.310	0.190	-1.384	-1.285	0.199	-0.785	-1.076	0.282	-0.224	-1.329	0.184
X1(-1)*X4(-1)	-0.017	-1.331	0.183	0.141	5.884	0.000	0.014	0.732	0.464	0.201	7.833	0.000
X1(-1)*X5(-1)	0.507	2.687	0.007	0.305	1.689	0.091	2.580	3.293	0.001	1.901	7.941	0.000
X1(-1)	0.320	6.917	0.000				0.272	3.819	0.000			
X3(-1)	213.445	5.455	0.000	-1241.58	-2.187	0.029						
X5(-1)							-13.316	-2.605	0.009			
Limit Points												
LIMIT_2:C(29)	-322.06	-5.111	0.000									
LIMIT_3:C(30)	-320.28	-5.083	0.000									
Pseudo R-squared	0.310	Akaike info criterion		1.519								
Schwarz criterion	1.631	Log likelihood		-1041.6								
Hannan-Quinn criter.	1.561	Restr. log likelihood		-1509.6								
LR statistic	935.82	Avg. log likelihood		-0.738								
Prob(LR statistic)	0.000											

$$o3ni \ @expand(m\_size, tp, @drop(1,1)) \ bnd\_ni(-1)^* @expand(m\_size, tp) \ x1(-1)^* x2(-1)$$
$$^* @expand(m\_size, tp) \ x1(-1)^* x3(-1)^* @expand(m\_size, tp) \ x1(-1)^* x4(-1)$$
$$^* @expand(M\_size, TP) \ x1(-1)^* x5(-1)^* @expand(m\_size, tp) \ x1(-1)^* @expand(m\_size, tp) \quad (12.39)$$
$$x2(-1)^* @expand(m\_size, tp) \ x3(-1)^* @expand(M\_Size, TP)$$
$$x4(-1)^* @expand(m\_size, tp) \ x5(-1)^* @expand(m\_size, tp)$$

Based on this summary, the following findings and notes are presented.

1. It happens that only the regression in (M_Size = 2,T = 1) shows one of the interactions, namely $X1(-1)^*X4(-1)$ has a large p-value = 0.464. It should be kept in the model because we want to show that the linear effect of $X1(-1)$ on $O3NI$ depends on $X2(-1)$ to $X5(-1)$, within each group. And we find that the linear effect of $X1(-1)$ on $O3NI$ is significantly dependent on $X2(-1)$ to $X5(-1)$, because the null hypothesis H_0: C(10) = C(14) = C(18) = C(22) = 0 is rejected based on the F-statistic of $F_0 = 11.09$ 769 with $df = (4,1381)$ and a p-value = 0.0000.
2. On the other hand, if the joint effects of the four interactions are insignificant, then there is a reason to delete at least one of them.
3. Each of the four regressions has different sets of independent variables, therefore the joint effects of the four interactions on $O3NI$ have different special meanings, because these are adjusted for different variables. For instance, based on the regression in (M_Size = 1,TP = 1), the joint effects of the interactions are adjusted for the variables $BND\_NI(-1)$, $X1(-1)$ and $X3(-1)$.
4. As an extension, by inserting additional variables: $T^* @Expand(M\_Size, TP)$, a set of four regressions with trend would be obtained. Do it as an exercise.

Table 12.14   *Summary of the statistical results based on a reduced model of that in (12.40)*

Variable	(M_Size=1,TP=1) Coeff.	z-Stat.	Prob.	(M_Size=1,TP=2) Coeff.	z-Stat.	Prob.	(M_Size=2,TP=1) Coeff.	z-Stat.	Prob.	(M_Size=2,TP=2) Coeff.	z-Stat.	Prob.
Intercept				-2497.4	-3.016	0.003	-348.4	-6.806	0.000	-344.3	-6.727	0.000
BND_NI(-1)	-1646.0	-6.676	0.000	10373.7	2.606	0.009	34.675	6.753	0.000	5.942	3.796	0.000
T*X1(-1)*X2(-1)							0.012	1.547	0.122	-0.022	-2.000	0.046
T*X1(-1)	0.033	4.285	0.000	0.024	2.905	0.004	0.058	3.585	0.000	0.054	7.623	0.000
X1(-1)*X2(-1)	-0.052	-2.186	0.029	-0.017	-1.572	0.116	-0.092	-3.144	0.002	0.107	1.396	0.163
X3(-1)	237.2	6.742	0.000	-1442.7	-2.602	0.009						
X4(-1)	-0.243	-2.785	0.005	0.705	1.609	0.108						
X5(-1)	1.177	5.377	0.000	2.063	2.430	0.015						
X2(-1)	0.179	1.695	0.090									
T				-0.286	-2.264	0.024						
Limit Points												
LIMIT_2:C(26)	-341.790	-6.679	0.000									
LIMIT_3:C(27)	-340.143	-6.648	0.000									
Pseudo R-squared	0.261	Akaike info criterion		1.620								
Schwarz criterion	1.721	Log likelihood		-1116								
Hannan-Quinn criter.	1.658	Restr. log likelihood		-1509.6								
LR statistic	787.148	Avg. log likelihood		-0.791								
Prob(LR statistic)	0.000											

Example 12.26   An ordered probit model with time-related effects by group

As a modification of the model (12.38), Table 12.14 presents the statistical results based on a reduced ordered probit model with time-related-effects by *M_Size* and *TP* of the full LV(1) model with the following equation specification (ES).

$$o3ni \ @expand(m\_size, tp, @dropfirst) \ bnd\_ni(-1)^* @expand(m\_size, tp)$$
$$t^*x1(-1)^*x2(-1)^* @expand(m\_size, tp) \ t^*x1(-1)^* @expand(m\_size, tp)$$
$$x1(-1)^*x2(-1)^* @expand(m\_size, tp) \ t^*x2(-1)^* @expand(m\_size, tp)$$
$$x3(-1)^* @expand(m\_size, tp) \ x4(-1)^* @expand(m\_size, tp) \quad (12.40)$$
$$x5(-1)^* @expand(m\_size, tp) \ t^* @expand(m\_size, tp)$$
$$x2(-1)^* @expand(m\_size, tp) \ x1(-1)^* @expand(m\_size, tp)$$

12.9   Bounded Models by Group and Time

This section presents only two examples of simple bounded regression models, which can easily be extended to various models with a lot more exogenous variables. Refer to all models presented in previous examples, as well as those in previous chapters.

12.9.1   Bounded Polynomial LV(1) Model by Group and Time

Let $BND\_Y = LOG[(Y-L)/(U-Y)]$, where L and U are fixed lower and upper bounds of a numerical endogenous variable Y, then a simple bounded polynomial LV(1) model by *GROUP* and the time *T* considered is presented using the following general equation specification (ES), for T > 1.

$$BND\_Y \quad C \quad @Expand(Group, T, @drop(1,2))$$
$$BND\_Y(-1)^* @Expand(Group, T) \ldots BND\_Y(-1)^{\wedge}k^* @Expand(Group, T) \tag{12.41}$$

where the variable "*GROUP*" should be invariant over time, and the power k should be empirically selected, based on the data used. See the following example. For $k = 1$, the model represents a set of simple linear regressions of *BND_Y* on *BND_Y(−1)*.

This model could easily be extended to a lot of models using additional numerical exogenous variables. Then the following general ES would be considered.

$$BND\_Y \quad C \quad @Expand(Group, T, @drop(1,2))$$
$$BND\_Y(-1)^* @Expand(Group, t) \ldots BND\_Y(-1)^{\wedge}k^* @Expand(Group, t) \tag{12.42}$$
$$V1(-1)^* @Expand(Group, t) \ldots Vm(-1)^* @Expand(Group, t)$$

where $Vm = V_m$ could be any numerical exogenous variable, either a main factor, an interaction factors or a power of a variable, which is known as a cause, source or upstream variable of the variable *Y*, in a theoretical sense. So we may have the following special alternative models as follows:

1. For $k = 1$, the models can be considered are:
   1.1 A LV(1) additive model by *GROUP* and the time *T*, if and only if all variables $V_m s$ are the main factors or variables.
   1.2 A LV(1) interaction model by *GROUP* and the time *T*, if at least one of the $V_m s$ is an interaction factor between the exogenous variables, or the interaction between *BND_Y(-1)* and the exogenous variable(s).
2. For each $k > 1$, the models can be considered as multiple polynomial LV(1) models by *GROUP* and the time *T*, if the variable $V_m(-1) = V(-1)^n$ for some integers m and n. Refer to the double polynomial models presented in Chapter 7.

Furthermore, take note that the variable *GROUP* can be generated based on one or more variables and is invariant over time. If the *GROUP* is generated based on one or more exogenous variable *Vm*, then a *piecewise* or *discontinuous* LV(1) model by *Group* and *T* would be obtained.

Finally, one should always be aware that unexpected parameter estimates could be obtained based on all statistical models with at least two numerical independent variables, because of the unpredictable impact(s) of their bivariate correlation(s) or multicollinearity; even some of the independent variables are not correlated in a theoretical sense. Refer to applications of the multistage manual stepwise selection method, instead of using the STEPLS directly using all independent variables in a single step of analysis.

Example 12.27 Bounded LV(1) polynomial models
Table 12.14 presents a summary of the statistical results of a reduced model of the third degree polynomial LV(1) model of *BND_NI = log((NI + 16 000)/(1300-NI))* with the following equation specification, based on the sub-sample {t > 1}

$$bnd\_ni \ c \ @expand(m\_size, t, @drop(1, 2)) \ bnd\_ni(-1)^* @expand(m\_size, t)$$
$$bnd\_ni(-1)^{\wedge}2^* @expand(m\_size, t) \ bnd\_ni(-1)^{\wedge}3^* @expand(m\_size, t) \tag{12.43}$$

Based on the output of the reduced model, the following findings and notes are presented.

1. The reduced model is obtained using the multistage manual stepwise selection method as follows:
   1.1 At the first stage of the analysis, the following ES is inserted for the selection process. Since we find $BND\_NI(-1)^*(M\_Size=1,T=2)$ has a p-value > 0.30, it is deleted from the model.

$$bnd\_ni\ C\ @expand(m\_size, t, @drop(1,2))\ bnd\_ni(-1)^*@expand(m\_size, t)$$

   1.2 At the second stage, the term or function $bnd\_ni(-1)^{\wedge}2^*@expand(m\_size,t)$ is inserted for the selection process. We get six out of 14 variables as shown in Table 12.15, which should be used in the model.
   1.3 Finally, the term $bnd\_ni(-1)^{\wedge}3^*@expand(m\_size,t)$ is inserted for the selection process. However, only six out of the 14 variables should be kept in the model. So the stepwise selection gives only six out of 14 regressions, which are the third-degree polynomial models in a statistical sense but one of the independent variables has a large p-value $= 0.246$.
2. Take note that the final results show that the regression in $(M\_Size=1,T=2)$ only has an intercept, so the final model is not acceptable in both the theoretical and statistical sense. Based on these findings, the following notes and comments are presented.

Table 12.15   *Statistical results of a reduced model of the polynomial in (12.43)*

M_Size	T	Intercept		BND_NI(T-1)		BND_NI(T-1)^2		BND_NI(T-1)^3	
		Coeff.	Prob.	Coeff.	Prob.	Coeff.	Prob.	Coeff.	Prob.
1	2	0.213	0.000						
1	3	-0.053	0.787	1.250	0.174				
1	4	-0.047	0.636	1.282	0.006				
1	5	-0.095	0.470	1.776	0.019	-1.530	0.036	290.774	0.000
1	6	-4.151	0.001	53.979	0.000	-219.032	0.000		
1	7	0.114	0.000	0.272	0.000				
1	8	-0.082	0.514	1.655	0.030	-0.965	0.210	-0.242	0.162
2	2	0.022	0.740	0.881	0.002			-0.143	*0.264*
2	3	-0.028	0.495	1.100	0.000				
2	4	0.035	0.645	0.843	0.023			0.763	0.046
2	5	0.139	0.009	0.357	0.123				
2	6	-0.031	0.580	1.086	0.000	0.370	0.000	0.476	0.000
2	7	0.025	0.763	0.652	0.153	0.860	0.056	-0.169	0.015
2	8	0.135	0.000	0.400	0.016	0.315	0.000		
R-squared				0.646		Mean dependent var			0.238
Adjusted R-squared				0.637		S.D. dependent var			0.252
S.E. of regression				0.152		Akaike info criterion			-0.911
Sum squared resid				35.128		Schwarz criterion			-0.778
Log likelihood				753.180		Hannan-Quinn criter.			-0.861
F-statistic				73.390		Durbin-Watson stat			2.320
Prob(F-statistic)				0.000					

2.1   The impacts of multicollinearity can give unexpected parameter estimates – refer to Agung (2009a, Section 2.14). See the following illustration.

2.2   Because we find that *BND_NI* has a significant correlation with each of the variables *BND_NI* (-1), $BND\_NI(-1)^2$ and $BND\_NI(-1)^3$, then at least one of these variables should be used as an independent variable, even though this has a large *p*-value. For instance, by using only $BND\_NI(-1)^2(M\_Size=1,T=2)$ as an additional independent variable, it has a *p*-value = 0.3455, and a *p*-value = 0.3488 is obtained by using only $BND\_NI(-1)^{\wedge}3^*(M\_Size=1,T=2)$. If both variables are used as independent variables they have *p*-values of 0.9646 and 0.9755.

3.   By using the trial-and-error method, we get an alternative reduced model with its summary presented in Table 12.16. Corresponding to this summary, the following notes and comments are made.

3.1   At the first stage, the same ES is applied. Then by deleting the intercept "C", a set of 14 simple linear regressions of *BND_NI* on *BND_NI*(-1) is obtained, where each *BND_NI*(-1) has a linear significant effect on *BND_NI*, but the regression in $(M\_Size=1,T=2)$ without an intercept (through the origin).

3.2   At the second stage, the term or function $bnd\_ni(-1)^{\wedge}3^*@expand(m\_size,t)$ is inserted for the selection process. We find nine out of the 14 variables, as shown in Table 12.16, should be used in the model.

3.3   At the third stage, the term $bnd\_ni(-1)^{\wedge}2^*@expand(m\_size,t)$ is inserted for the selection process. However, only six out of the 14 variables should be kept in the model. At this stage, the *p*-value of $BND\_NI(-1)^*(M\_Size=2,T=8)$ increases to a very large value = 0.894, as shown in the table.

3.4   Since both the variables $BND\_NI(-1)^3$, and $BND\_NI(-1)^2$ have significant effects within the group (M_Size = 2, T = 8), then the final reduced model would be obtained by deleting

Table 12.16   *Statistical results of an alternative reduced model to that in (12.41)*

M_Size	T	Intercept Coeff.	Intercept Prob.	BND_NI(-1) Coeff.	BND_NI(-1) Prob.	BND_NI(-1)^3 Coeff.	BND_NI(-1)^3 Prob.	BND_NI(-1)^2 Coeff.	BND_NI(-1)^2 Prob.
1	2			0.986	0.000				
1	3	-0.053	0.786	1.250	0.172				
1	4	-0.047	0.635	1.282	0.005				
1	5	-0.021	0.829	1.147	0.014	-1.011	0.034		
1	6	-4.151	0.001	53.979	0.000	290.774	0.000	-219.032	0.000
1	7	0.114	0.000	0.272	0.000				
1	8	-0.082	0.512	1.655	0.029	-0.242	0.160	-0.965	0.208
2	2	0.533	0.118	-3.038	0.240	-3.981	0.114	8.043	0.127
2	3	-0.028	0.493	1.100	0.000				
2	4	0.035	0.644	0.843	0.022	0.763	0.045		
2	5	-0.096	0.465	1.995	0.022	1.644	0.007	-2.758	0.051
2	6	-0.070	0.139	1.340	0.000	0.107	0.000		
2	7	0.025	0.762	0.652	0.151	-0.169	0.015	0.860	0.055
2	8	0.183	0.000	-0.029	*0.894*	-0.304	0.003	1.185	0.000

R-squared	0.649	Mean dependent var	0.238
Adjusted R-squared	0.640	S.D. dependent var	0.252
S.E. of regression	0.151	Akaike info criterion	-0.917
Sum squared resid	34.774	Schwarz criterion	-0.774
Log likelihood	761.111	Hannan-Quinn criter.	-0.864
F-statistic	68.956	Durbin-Watson stat	2.321
Prob(F-statistic)	0.000		

Table 12.17   *Statistical results of another reduced model of that in (12.41)*

M_Size	T	Intercept Coeff.	Intercept Prob.	BND_NI(-1)^3 Coeff.	BND_NI(-1)^3 Prob.	BND_NI(-1)^2 Coeff.	BND_NI(-1)^2 Prob.	BND_NI(-1) Coeff.	BND_NI(-1) Prob.
1	2			-18.201	0.029	8.545	0.000		
1	3	0.148	0.004	6.667	0.187				
1	4	0.185	0.000	3.926	0.006				
1	5			-5.628	0.000	5.479	0.000		
1	6	-4.151	0.001	290.774	0.000	-219.032	0.000	53.979	0.000
1	7	0.148	0.000	-0.166	0.083	0.528	0.035		
1	8	-0.082	0.514	-0.242	0.162	-0.965	0.210	1.655	0.030
2	2	0.533	0.120	-3.981	0.115	8.043	0.128	-3.038	0.241
2	3	0.199	0.000	1.314	0.000				
2	4	0.035	0.645	0.763	0.046			0.843	0.023
2	5	0.139	0.009	0.476	0.000			0.357	0.123
2	6	0.134	0.000	-0.466	0.000	1.975	0.000		
2	7	0.025	0.763	-0.169	0.015	0.860	0.056	0.652	0.153
2	8	0.178	0.000	-0.295	0.000	1.153	0.000		

R-squared	0.648	Mean dependent var	0.238
Adjusted R-squared	0.638	S.D. dependent var	0.252
S.E. of regression	0.151	Akaike info criterion	-0.914
Sum squared resid	34.947	Schwarz criterion	-0.773
Log likelihood	757.228	Hannan-Quinn criter.	-0.861
F-statistic	70.188	Durbin-Watson stat	2.305
Prob(F-statistic)	0.000		

Note: BND_NI(-1) = BND_NI(T-1)

$BND\_NI(-1)^*(M\_Size = 2, T = 8)$. In this case, the stages of the analysis are not following the criteria of the multistage manual stepwise selection anymore.

4. For an additional illustration, Table 12.17 presents a set of 14 third-degree polynomial regressions, which is obtained as follows:

    4.1   At the first stage, the following ES specification is applied. Then by deleting two of the intercepts, a set of 14 regressions of BND_NI on $BND\_NI(-1)^3$ is obtained, where each $BND\_NI(-1)^3$ has a positive of negative significant linear effect, at the 10% significance level.

$$bnd\_ni \ @expand(m\_size, t)) \ bnd\_ni(-1)^{\wedge}3^* @expand(m\_size, t)$$

    4.2   Then, for the other stages, the standard multistage manual stepwise selection is applied.

Example 12.28   A bounded LV(1) interaction model by group and time

Table 12.18 presents a summary of the statistical results of a reduced model of a bounded LV(1) interaction model of a full model for $\{T > 1\}$, with the ES as follows:

$$bnd\_ni \ @expand(m\_size, t) \ bnd\_ni(-1)^* lnsale(-1)^* @expand(m\_size, t)$$
$$bnd\_ni(-1)^* @expand(m\_size, t) \ lnsale(-1)^* @expand(m\_size, t) \tag{12.44}$$

Note that this model is a four-way interaction model indicated by the interactions of the four variables $BND\_NI(-1)$, $lnSale(-1)$, M_Size and the time T, which could easily be extended to a lot more advanced

Table 12.18   Statistical results of a reduced model of the LV(1) model in (12.44)

M_Size	T	Intercept Coeff.	t-Stat.	Prob.	BND_NI(-1)*lnSale(-1) Coeff.	t-Stat.	Prob.	BND_NI(-1) Coeff.	t-Stat.	Prob.	lnSale(-1) Coeff.	t-Stat.	Prob.
1	2				-0.115	-1.066	*0.287*				0.066	2.706	0.007
1	3				0.179	13.156	0.000						
1	4	0.213	4.805	0.000	0.150	2.440	0.015				-0.030	-1.690	0.091
1	5	0.293	2.470	0.014	0.130	1.606	0.109	-0.452	-1.020	*0.308*	-0.025	-1.108	0.268
1	6	0.243	5.738	0.000	0.569	11.795	0.000				-0.127	-8.458	0.000
1	7	-0.024	-0.521	0.603	-0.133	-3.436	0.001	1.662	4.060	0.000			
1	8	1.366	2.746	0.006	0.661	2.563	0.011	-5.912	-2.532	0.011	-0.122	-2.170	0.030
2	2	0.177	2.453	0.014	0.059	4.373	0.000				-0.006	-0.495	*0.621*
2	3	0.192	2.653	0.008	0.112	5.512	0.000				-0.021	-1.525	0.128
2	4	0.211	2.671	0.008	0.155	8.830	0.000				-0.032	-2.204	0.028
2	5	0.280	3.232	0.001	0.137	13.173	0.000				-0.039	-2.874	0.004
2	6	0.272	3.329	0.001	0.193	29.828	0.000				-0.050	-4.128	0.000
2	7	0.254	4.700	0.000	0.149	2.616	0.009	-1.178	-1.843	0.066			
2	8	0.751	3.750	0.000	0.393	4.760	0.000	-3.277	-3.602	0.000	-0.059	-3.001	0.003

R-squared	0.615	Mean dependent var	0.238
Adjusted R-squared	0.604	S.D. dependent var	0.252
S.E. of regression	0.159	Akaike info criterion	-0.818
Sum squared resid	38.203	Schwarz criterion	-0.674
Log likelihood	680.83	Hannan-Quinn criter.	-0.765
F-statistic	59.147	Durbin-Watson stat	2.347
Prob(F-statistic)	0.000		

LV(p) interaction models by group and time, using additional numerical exogenous (source, upstream or cause) numerical variables. However, one should note the unexpected or uncertainty parameter estimates of the models because of the multicollinearity between the independent variables and the data used.

Based on this model (12.44) and the outputs of its reduced model, the following notes and comments are made.

1. I would say that this table actually presents 14 unpredictable heterogeneous regression functions even the model only has two numerical independent variables. See the following additional example.
2. The main objective of the model is to show that the effect of *BND_NI*(-1) on *BND_NI* depends on *lnSale*(-1), *M_Size* and the time *T*. So the stages of the manual stepwise selection method should be done as follows:

   2.1   At the first stage, the following ES is inserted for the selection process

$$BND\_NI \ @expand(M\_Size, T) \ BND\_NI(-1)^* lnSale(-1)^* Expand(M\_Size, T)$$

   In order to keep all interactions in the model, that is each has a *p*-value < 0.3, we find that the intercepts of the first two regressions should be deleted.

   2.2   At the second stage, *BND_NI*(-1)*@*expand*(*M_Size,T*) is inserted for the selection process, and we see five out of the 14 variables should be kept in the model.

   2.3   At the third stage, *lnSale*(-1)*@*expand*(*M_Size,T*) is inserted for the selection process, we then find 11 out of the 14 variables can be kept in the model.

3. Note that the final results show that $BND\_NI(-1)$ has a p-value > 0.30 within the cell (M_Size $= 1$, T $= 5$), however, it does not have to be deleted from the model. The main reason is its interaction $BND\_NI(-1)^*lnSale(-1)$ has a positive significant adjusted effect on BND_NI at the 10% significance level, with a p-value $= 0.109/2 = 0.0545$.

4. Furthermore, within the cell (M_Size $= 1$,T $= 5$), the variables $BND\_NI(-1)^*lnSale(-1)$, $BND\_NI(-1)$, and $lnSale(-1)$ have insignificant joint effects on BND_NI, based on the F-statistic of $F_0 = 1.701\ 714$ with $df = (3,1519)$ and a p-value $= 0.1647$, which is a relatively small p-value. The test can be done using the Wald test with the null hypothesis H_0: $C(16) = C(27) = C(34) = 0$.

12.9.2 Bounded Models by Group with the Time Numerical Variable

If panel data has a sufficiently large number of time observations, then the models considered may have the time T as a numerical independent variable – refer to the growth models by states presented in Chapter 1.

Example 12.29   Bounded LV(1) interaction models with trend

Table 12.19 presents a summary of the statistical results based on a full LV(1) interaction model and two of its reduced models. Based on this summary, the following notes and comments are made.

1. The reduced Model-1 is an expected reduced model because the two deleted independent variables have the largest p-values. On the other hand, the reduced Model-2 is unexpected, which is obtained using the trial-and-error method. See additional unexpected models presented in the following example.

Table 12.19   *Summary of the statistical results of a full LV(1) model and its two reduced models*

Dependent Variable: BND_NI									
Method: Panel Least Squares									
Date: 10/31/11 Time: 13:19									
Periods included: 7									
Cross-sections included: 224									
Total panel (unbalanced) observations: 1561									
	Full Model			Reduced Model-1			Reduced Model-2		
Variable	Coeff.	t-Stat.	Prob.	Coeff.	t-Stat.	Prob.	Coeff.	t-Stat.	Prob.
M_SIZE=1	0.269	3.222	0.001	0.237	9.093	0.000	0.177	7.297	0.000
M_SIZE=2	0.034	0.306	0.760	0.081	2.345	0.019	0.081	2.344	0.019
BND_NI(-1)*LNSALE(-1)*(M_SIZE=1)	0.060	1.472	0.141	0.045	2.866	0.004	0.013	3.918	0.000
BND_NI(-1)*LNSALE(-1)*(M_SIZE=2)	0.056	1.175	0.240	0.078	28.475	0.000	0.078	28.465	0.000
BND_NI(-1)*(M_SIZE=1)	-0.432	-1.150	0.250	-0.292	-1.968	0.049			
BND_NI(-1)*(M_SIZE=2)	0.236	0.453	*0.651*						
LNSALE(-1)*(M_SIZE=1)	-0.004	-0.406	*0.685*				0.006	1.647	0.100
LNSALE(-1)*(M_SIZE=2)	0.010	0.956	*0.339*	0.006	1.089	0.276	0.006	1.089	0.276
T*(M_SIZE=1)	-0.001	-0.412	0.680	-0.002	-0.424	0.672	-0.001	-0.352	0.725
T*(M_SIZE=2)	-0.001	-0.301	0.764	-0.001	-0.391	0.696	-0.001	-0.391	0.696
R-squared	0.410			0.410			0.409		
Adjusted R-squared	0.406			0.407			0.407		
Durbin-Watson stat	2.424			2.424			2.428		
F-statistic	119.72			154.03			153.75		
Prob(F-statistic)	0.000			0.000			0.000		

2. Note that the parameter of the variable $lnSale^*(M\_Size=2)$ has the same estimates in both reduced models. Why? Because both reduced models present exactly the same regression function within the group $M\_Size=2$.
3. Even though the time T has a very large p-value, it is not deleted from the model in order to show the trend of BND_NI is insignificant within the two groups. In fact, the coefficient of T of each regression represents the exponential growth rate of $[(NI+16\,000)/(1300\text{-}NI)]$, adjusted for $BND\_NI(-1)$ and $lnSale(-1)$.

Example 12.30   Unexpected LV(p,q) models with the time-related effects

Table 12.20 presents a summary of the statistical results based on a full LVAR(1,2) model with time-related effects, one of its possible reduced models and a full LVAR(1,3) model.

Based on this summary the following notes are presented.

1. Note that an interaction variable, namely $T^*BND\_NI(-1)^*(M\_Size=2)$ of the LV(1,2) model has a large p-value $=0.374$. So that, in general, a reduced model is obtained by deleting this variable.
2. Since the interaction factors are considered the most important set of cause or up-stream variables within each group generated by M_Size, then the reduced model should be obtained by deleting at least

Table 12.20   *Summary of the statistical results of three unexpected LVAR(p,q) models*

Variable	LV(1,2) Model			A Reduced Model			LVAR(1,3) Model		
	Coeff.	t-Stat.	Prob.	Coeff.	t-Stat.	Prob.	Coeff.	t-Stat.	Prob.
C	2.201	3.436	0.001	2.169	3.270	0.001	2.458	3.727	0.000
M_SIZE=2	-1.206	-1.554	0.121	-2.749	-4.045	0.000	-1.389	-1.850	0.065
T*BND_NI(-1)*LNSALE(-1)*(M_SIZE=1)	-0.245	-3.718	0.000	-0.242	-3.539	0.000	-0.278	-3.982	0.000
T*BND_NI(-1)*LNSALE(-1)*(M_SIZE=2)	-0.057	-1.776	0.076	0.049	3.195	0.001	-0.073	-2.529	0.012
T*BND_NI(-1)*(M_SIZE=1)	1.818	3.296	0.001	1.769	3.111	0.002	2.139	3.648	0.000
T*BND_NI(-1)*(M_SIZE=2)	0.301	0.890	0.374	-0.822	-4.450	0.000	0.565	1.869	0.062
T*LNSALE(-1)*(M_SIZE=1)	0.052	3.651	0.000	0.051	3.466	0.001	0.058	3.929	0.000
T*LNSALE(-1)*(M_SIZE=2)	0.021	3.068	0.002	-0.002	-1.296	0.195	0.021	3.332	0.001
BND_NI(-1)*LNSALE(-1)*(M_SIZE=1)	1.466	3.904	0.000	1.452	3.732	0.000	1.602	4.155	0.000
BND_NI(-1)*LNSALE(-1)*(M_SIZE=2)	0.516	2.549	0.011	-0.178	-2.116	0.035	0.543	3.129	0.002
BND_NI(-1)*(M_SIZE=1)	-9.419	-3.116	0.002	-9.253	-2.965	0.003	-10.699	-3.419	0.001
BND_NI(-1)*(M_SIZE=2)	-2.486	-1.172	0.241	4.839	4.816	0.000	-3.283	-1.838	0.067
LNSALE(-1)*(M_SIZE=1)	-0.311	-3.836	0.000	-0.309	-3.649	0.000	-0.339	-4.111	0.000
LNSALE(-1)*(M_SIZE=2)	-0.150	-3.507	0.001				-0.141	-3.737	0.000
T*(M_SIZE=1)	-0.382	-3.284	0.001	-0.373	-3.096	0.002	-0.447	-3.630	0.000
T*(M_SIZE=2)	-0.120	-1.714	0.087	0.123	4.812	0.000	-0.154	-2.418	0.016
AR(1)	-0.587	-9.981	0.000	-0.513	-9.483	0.000	-0.790	-11.081	0.000
AR(2)	-0.218	-2.349	0.019	-0.120	-1.575	0.116	-0.656	-4.695	0.000
AR(3)							-0.715	-5.188	0.000
R-squared	0.495			0.490			0.511		
Adjusted R-squared	0.487			0.482			0.500		
Durbin-Watson stat	2.482			2.495			2.742		
F-statistic	62.844			65.527			50.141		
Prob(F-statistic)	0.000			0.000			0.000		
Inverted AR Roots	-.29+.36i		-.29-.36i	-.26+.23i		-.26-.23i	.07-.88i	.07+.88i	-0.92

one of the main factors, $T^*(M\_Size=2)$, $BND\_NI(-1)^*(M\_Size=2)$, and $lnSale(-1)^*(M\_Size=2)$. By using the trial-and-error method, a good fit reduced model is obtained by deleting $lnSale(-1)^*(M\_Size=2)$ which has a very small p-value $=0.001$.

3. This process has demonstrated the following characteristics.
   - A good fit reduced model might be obtained by deleting an independent variable which has a significant adjusted effect even it has a very small p-value.
   - The uncertainty of the parameters' estimates of all models, where the most important variable in a theoretical sense might have a significant adjusted effect.
   - In general, the *manual stepwise selection method* should be applied, in addition to the STEPLS estimation method in EViews.

13

Multivariate Lagged Variables Autoregressive Models

13.1 Introduction

As the extension of the univariate lagged variables autoregressive (LVAR) models presented in Chapter 12 based on natural experimental data, this chapter presents the application of the multivariate lagged variables autoregressive (MLVAR) models. Since a MLVAR model represents a set of LVAR models or a set of multiple regressions, then all models presented in Chapter 12, could be used as a member of the MLAR model set, or a multiple regression in the MLAR model.

The data analysis can easily be done using the object "*System*". Instead of using the name System Equation Model (SEM) to present causal or up- and down-stream relationships between a set of numerical and categorical variables, Agung (2009a) has introduced the term SCM (*seemingly causal model*) since SEM has been used for the Structural Equation Model.

The applications of the object "*System*" have been presented in Chapter 11 based on pool panel data, specifically for unbalanced or incomplete pool panel data, and based on the time series by states in Chapter 3. Refer to various *MLVAR(p,q)* models presented in Chapter 3. This chapter presents the application of the object *System* for balanced or complete pool panel data, which in fact is natural experimental data. For this reason, this chapter presents illustrative examples on multivariate (or multiple responses) natural experimental data analysis. Referring to "Multivariate Time Series Models by States" presented in Chapter 3, this chapter presents the "Multivariate Natural Experimental or Cross-Section Models by Times".

13.2 Seemingly Causal Models

Similar to the data analysis based on the multivariate time series models by states presented in Chapter 3, the data analysis based on multivariate (multiple responses) natural experimental models, can be done using two alternative estimation methods. The same as presented in Chapter 12, the treatment factors considered are the time T or the time-period (TP) variable, and a classification-factor or cell-factor CF, which is

Panel Data Analysis Using EViews, First Edition. I Gusti Ngurah Agung.
© 2014 John Wiley & Sons, Ltd. Published 2014 by John Wiley & Sons, Ltd.
Companion website: www.wiley.com/go/panel_data

generated so that its groups of research objects are invariant or constant over time. Note that CF can be generated based on one or more categorical or numerical variables. Refer to Section 12.3.1.

In this case, $Y_{it} = (Y_1, \ldots, Y_G)$ would be considered multivariate response or problem indicators of the natural experiment, with covariates (causal or up-stream variables) $Y_{it}(-1) = (Y_1(-1), \ldots, Y_G(-1))$ and theoretically defined exogenous multivariates $X_{it} = (X_1, \ldots, X_K)$, as well as $X_{it}(-1) = (X_1(-1), \ldots, X_K(-1))$.

Then the natural experimental models would have the following general equation.

$$H_g(Y_g) = \sum_{s=1}^{S}\sum_{t=2}^{T} F_{gst}(X, X(-1), Y(-1)) \times Dst + \varepsilon_g$$

$$for\ g = 1, \ldots, G$$

(13.1)

where $H_g(Y_g)$ is a fixed function of Y_g, such as Y_g, $log(Y_g - L_g)$, $log(U_g - Y_g)$, $log\{(Y_g - L_g)/(U_g - Y_g)\}$, where L_g and U_g are fixed lower and upper bounds of Yg; Dst is a dummy variable or a zero-one indicator of the cell $(CF = s)$ at the time t, or the objects in $(CF = s, Time = t) = (s,t)$, $F_{gst}(X, X(-1), Y(-1))$ is a function of X, $X(-1)$ and $Y(-1)$ with a finite number of parameters for each (g,s,t), $g = 1 \ldots, G$, $s = 1, \ldots, S$, and $t = 2, \ldots, T$; and ε_g is the multivariate random error or residual of the model, which is assumed to have specific basic assumptions.

Note that the time points in the model (13.1) should be $t = 2, \ldots, T$; corresponding to the first lagged variables model, namely the LV(1) model. This model could easily be extended to a LV(p) model, with time points $t = (p+1), \ldots, T$. For other cases, the time points can be replaced by a time-period (TP) variable, such as before and after IPO; before and after a critical event; before, during and after economic crises.

Furthermore, note that $F_{gst}(X, X(-1), Y(-1))$ might be a function of different sub-sets of components of the multivariate X and $X(-1)$: For instance, a function of the multivariate $X1$ and $X2(-1)$, with $X1 \neq X2$.

Two specific general models that it might be important to consider are as follows:

1. The SCMs based on endogenous variables only, with the following general equation. In this case, the number of multiple regressions in a SCM equals the number of the variables considered. However, note that an exogenous variable also could be treated as an endogenous variable.

$$H_g(Y_g) = \sum_{s=1}^{S}\sum_{t=2}^{T} F_{gst}(Y(-1)) \times Dst + \varepsilon_g$$

$$for\ g = 1, \ldots, G$$

(13.1a)

In some cases an endogenous variable Y_h, for $h \neq g$, is more appropriate to use as the causal factor of Y_g, instead of $Y_h(-1)$, for instance $Y_g = Net\_income$, and $Y_h = Sale$. On the other hand, both Y_h and $Y_h(-1)$, should be used as the cause factors of an Y_g, such as for $Y_g = Return$, and $Y_h = Price$, $Sale$, or $Net-income$.

2. The SCMs present effects of exogenous variable $X(-1)$ only, or X and $X(-1)$ on the endogenous variable Y, with the following general equations. In these cases, the number of multiple regressions in the SCM also equal G, because an exogenous variable will not be used as a dependent variable.

$$H_g(Y_g) = \sum_{s=1}^{S}\sum_{t=2}^{T} F_{gst}(X(-1)) \times Dst + \varepsilon_g$$

$$for\ g = 1, \ldots, G$$

(13.1b)

and

$$H_g(Y_g) = \sum_{s=1}^{S}\sum_{t=2}^{T} F_{gst}(X, X(-1)) \times Dst + \varepsilon_g$$

(13.1c)

$$for \, g = 1,..., G$$

Note that these models, in general, would have autocorrelation problems, which can be overcome by using the AR term, or the lag(s) of the dependent variable as additional independent variables of each regression.

So there would be a lot of models, such as additive, two- and three-way interaction models, which can be defined based on the original and transformed variables. However, the following sections present only various alternative simple linear-effects multivariate models, or SCMs, starting with the SCM based on two endogenous variables: *(Y1,Y2)*.

For other effects models, such as logarithmic, hyperbolic, polynomial and nonlinear effects models, refer to various models presented in previous chapter, and Agung (2011, 2009a).

Furthermore, note that for the alternative functions presented previously, they can be represented by $F(V) = F(V_1, V_2, \ldots, V_m)$. In general, under the assumption that V_1 is the most important cause, source or up-stream factor, the functions $F(V)$ should have at least two-way interaction terms, say $V_1 * V_j$, as well as $V_i * V_j$ for some i and j, which should be selected, based on a good theoretical reason. In addition, the functions may have few selected three-way interaction terms, namely $V_i * V_j * V_k$. See the following simple illustrative models.

13.2.1   Specific Characteristics of the Model (13.1)

Some specific characteristics of the model in (13.1) that should be noted are as follows:

1. Whenever the research objects can be considered as a single group, then the multivariate models by only time or time-period with the following general equation can be applied.

$$H_g(Y_g) = \sum_{t=2}^{T} F_{gt}(X, X(-1), Y(-1)) \times Dt + \varepsilon_g$$

(13.2)

$$for \, g = 1, \ldots, G$$

2. The estimates or statistical results of the model in (13.1) represent a set of $G \times S \times (T-1)$ continuous multiple regression functions:

$$\hat{H}_g(Y_g) = \hat{F}_{gst}(X, X(-1), Y(-1))$$

$$for \, g = 1, \ldots, G; s = 1, \ldots, S; t = 2, \ldots, T$$

(13.3)

3. For each *(s,t)*, the estimates or statistical results of the model in (13.1) represent a continuous multivariate regression function:

$$\hat{H}_g(Y_g) = \hat{F}_{gst}(X, X(-1), Y(-1))$$

$$for \, g = 1, \ldots, G$$

(13.4)

4. Whenever the cell-factor CF is generated based on one or more exogenous variables, the estimates or statistical results of the model in (13.1) represent a set of piece-wise or discontinuous regression functions.

13.2.2 Alternative or Modified Models

Whenever, the data has a large time-point observation then time t may be used as numerical independent variable. Referring to the model in (13.1), in this case, the models considered would have the general equation as follows:

$$H_g(Y_g) = \sum_{s=1}^{S} F_{gs}(X, X(-1), Y(-1), t) * Ds + \varepsilon_g$$

$$for\ g = 1, \ldots, G$$

(13.5)

where the function $F_{gs}(X, X(-1), Y(-1), t)$ is a *continuous function* in the time t, with a finite number of parameters within the cell $CF = s$, represented by the dummy variable Ds for each $g = 1, \ldots, G$.

Refer to the specific models in (13.1a)–(13.1c). In addition, two alternative models with numerical time independent variables should also be considered, as presented in the following sections.

13.2.2.1 Models with Trend

For the model with trend, the function $F_{gs}(X, X(-1), Y(-1), t)$ in (13.5) is a specific additive function as follows:

$$F_{gs}(X, X(-1), Y(-1), t) = F_{gs1}(X, X(-1), Y(-1)) + C(gs) * t$$

(13.6)

where $F_{gs1}(X, X(-1), Y(-1))$ shows continuous functions with a finite number of parameters for each (g,s), and $C(gs)$ is the coefficient parameter or trend of time t. For $H_g(Yg) = log(Yg)$, the model (13.5) would be a multivariate growth model of Yg, with growth rate of $C(gs)$ adjusted for the other independent variables in the model.

For a more advanced model, the function is as follows:

$$F_{gs}(X, X(-1), Y(-1), t) = F_{gs1}(X, X(-1), Y(-1)) + F_{gs2}(t)$$

(13.7)

where $F_{gs2}(t)$ is a continuous function with a finite number of parameters for each (g,s). Refer to various possible functions presented in Chapters 1–4, and in Agung (2009a). For the function $F_{gs1}(X, X(-1), Y(-1))$ refer to all possible specific functions in (13.1a) to (13.1d).

13.2.2.2 Time-Related Effects Models

For time-related effects, the function $F_{gs}(X, X(-1), Y(-1), t)$ in (13.5) has the following general form.

$$F_{gs}(X, X(-1), Y(-1), t) = F_{gs1}(X, X(-1), Y(-1)) + t * F_{gs2}(X, X(-1), Y(-1))$$

(13.8)

where $F_{gs1}(X, X(-1), Y(-1))$ and $F_{gs2}(X, X(-1), Y(-1))$ are continuous functions with finite numbers of parameters for each (g,s). Refer to all possible specific functions in (13.1a–d). Note that the parameters in $t*F_{gs2}(X, X(-1), Y(-1))$ represent the time-related effects of the corresponding independent variables.

13.2.2.3   A Not-Recommended Model

Based on my point of view, the following model should be considered inappropriate because each multiple regression estimate presents a single continuous function only for all firm-time observations.

$$H_g(Y_g) = F_g(X, X(-1), Y(-1), t) + \varepsilon_g$$
$$for\ g = 1, \dots, G \tag{13.9}$$

where $F_g(X,X(-1),Y(-1),t)$ is a continuous function of $(X,X(-1),Y(-1),t)$ with a finite number of parameters for each g. Refer to specific functions in (13.6)–(13.8). Many papers present models without the time numerical variables, so that $F_g(X,X(-1),Y(-1),t) = F_g(X,X(-1),Y(-1))$.

For example, Filip and Raffourmier (2010) present a simple linear regression of P_{jt} (year-end market price of firm j) on BV_{jt} (year-and book value of the firm j) and some other continuous multiple regressions without using the time independent variable. However, they also present the results of regressions – changes over times (Table 7, in their paper) for the years 1998–2004.

13.3   Alternative Data Analyses

13.3.1   Regression Analysis Based on Each Yg

For large number of exogenous and endogenous variables, the equation specification used to conduct the analysis based on the set of all multivariates X, $X(-1)$, Y, and $Y(-1)$ would be a mess, because such a large set of multiple regressions with many independent variables is being considered. The independent variables of the regressions would be numerical or categorical main variables (factors), as well as their interactions. In general, such a large number of independent variables cause uncertainty of parameter estimates, which are the unpredicted impacts of multicollinearity between independent variables. Therefore, obtaining a good fit reduced model will not be an easy task. For this reason, the first alternative data analysis based on a multivariate general model is to conduct the analysis based on each multiple regression of the endogenous variable Yg.

Regression analysis, such as LS regression, binary choice, ordered choice and quantile regression, can be easily done for each endogenous variable Y_g, or $H_g(Y_g)$ in general, under the following conditions.

1. $H_g(Y_g)$, $g = 1, \dots, G$ are uncorrelated variables in a theoretical sense, or they are assumed to be uncorrelated in a statistical sense.
2. If the differential effects of an independent variable or a sub-set of independent variables on $H_g(Y_g)$, $g = 1, \dots, G$ do not need to be tested.
3. The set of G-regression functions in the model or reduced model should have different sets of independent variables

In these cases, for each Y_g, conducting data analysis using the following general equation specification is recommended for the sample $\{T > 1\}$ since it would be simpler as presented in the previous chapter. So a set of G-regressions analyses should be conducted:

$$Yg\ C@Expand(CF, T, @Droplast)$$
$$Y1(-1) * @Expand(CF, T) \dots Yg(-1) * @Expand(CF, T) \dots \tag{13.10}$$
$$V1(-1) * @Expand(CF, T) \dots Vm(-1) * Expand(CF, T) \dots$$

where *Vm* can be a main exogenous variable *Xk* or its transformation, an interaction between selected *Xk(−1)*s, and an interaction between *Xk(−1)* and *Yg(−1)*. Refer to the application of the multistage manual stepwise selection method presented in the previous chapter. To generalize, the transformed endogenous variable $H_g(Yg)$ can be applied, and the time-point variable can be replaced by a time-period variable.

13.3.2   Multivariate Data Analysis Based on Each Sub-Sample

Whenever hypotheses on the effects differences of an independent variable or a subset of the independent variables on *Yg* or $H_g(Yg)$ is tested, then a multivariate regression analysis can be done within each sub-sample *(CF = s,Time = t)* or some selected sub-samples.

In this case, the *G*-regression functions should have exactly the same sets of independent variables. Otherwise, adjusted effects of an independent variable, say *X1(−1)*, could have different meanings because effects are adjusted for different sets or sub-sets of the independent variables.

13.4   SCMs Based on (*Y1,Y2*)

For a basic illustration, Figure 13.1 presents three alternative path diagrams based on a pair of endogenous variables, namely (*Y1,Y2*), which are valid for all cells or groups generated by *CF* and the time *t*, or a defined time period. These path diagrams have characteristics as follows.

1. Note these path diagrams are similar to the path diagrams in Figure 1.11, based on two time-series variables *Gdp_US(Y2)* and *Gdp_Can(Y1)*, but these are path diagrams for cross-section variables, within each groups of the firms generated by *CF* and the time *t*.
2. Figure 13.1(a) presents the case where the endogenous variables *Y1* and *Y2* do not have a causal relationship in a theoretical sense. In this case, both *Y1(−1)* and *Y2(−1)* are up-stream variables of both *Y1* and *Y2*. The dashed line between *Y1(−1)* and *Y2(−1)* shows that their correlation has an unpredictable impact on parameter estimates, even though they are not correlated in a theoretical sense.
3. Figure 13.1(b) shows that *Y2* is defined as the cause factor of *Y1* in a theoretical sense, so then *Y2(−1)* also should be the cause factors of both *Y1(−1)* and *Y1*. The arrow with the dashed line from between *Y2(−1)* and *Y1(−1)* indicates their causal relationship is not presented explicitly in the model.
4. An extension of the path diagram in Figure 13.1(c) would be the model(s) where *Y1* and *Y2* have simultaneous causal effects, then *Y1(−1)* and *Y2(−1)* also should have simultaneous causal effects and both are the cause or up-stream factors for both *Y1* and *Y2*.

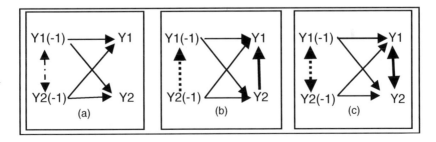

Figure 13.1   *Path diagrams presenting two possible causal relationships between* Y1 *and* Y2

13.4.1   SCMs by CF and the Time *t* Based on Figure 13.1(a)

For simple illustrations, see the model in (13.1), for $g = 1,2$; $s = 1,2$; and $t = 1,2$; corresponding to the diagram in Figure 13.1(a) which shows that the data analysis can be done using the following general equation specification. See the following example.

$$
\begin{aligned}
H_1(Y1) &= F_{113}(V(-1)) \times D11 + F_{123}(V(-1)) \times D12 \\
&\quad + F_{114}(V(-1)) \times D21 + F_{124}(V(-1)) \times D22 \\
H_2(Y2) &= F_{213}(V(-1)) \times D11 + F_{223}(V(-1)) \times D12 \\
&\quad + F_{223}(V(-1)) \times D21 + F_{224}(V(-1)) \times D22
\end{aligned}
\tag{13.11a}
$$

or, using a simpler form

$$
\begin{aligned}
H_1(Y1) &= F_{11}(V(-1)) \times D1 + F_{12}(V(-1)) \times D2 \\
&\quad + F_{13}(V(-1)) \times D3 + F_{14}(V(-1)) \times D4 \\
H_2(Y2) &= F_{21}(V(-1)) \times D1 + F_{22}(V(-1)) \times D2 \\
&\quad + F_{23}(V(-1)) \times D3 + F_{24}(V(-1)) \times D4
\end{aligned}
\tag{13.11b}
$$

where $V(-1) = (Y1(-1), Y2(-1))$, and $D1$, $D2$, $D3$ and $D4$ are the four dummy variables of the four groups of the firms generated by *CF* and the time *t*.

The two simplest possible models would be linear-effects models, namely an additive model and a two-way interaction model, respectively, such as follows:

13.4.1.1   *Additive Linear-Effects Models by* CF *and the Time*-t

The function $F_{gk}(V(-1))$ for each (g,k) would be

$$
F_{gk}(V(-1)) = C(gk0) + C(gk1) \times Y1(-1) + C(gk2) \times Y2(-1)
\tag{13.12a}
$$

Note that the LS regression function obtained within each cell generated by *CF* and the time *t* would have the following equation.

$$
\hat{H}_g(Yg) = \hat{C}(gk0) + \hat{C}(gk1) \times Y1(-1) + \hat{C}(gk2) \times Y2(-1)
\tag{13.12b}
$$

The characteristics of this regression function are as follows:

1. This regression is the simplest in a three-dimensional space with the coordinate axes: $H_g(Yg)$, $Y1(-1)$ and $Y2(-1)$, and its graph is a plane.
2. For all scores of $Y2(-1)$, this function presents *homogenous regressions* of $\hat{H}_g(Yg)$ on $Y1(-1)$. In a statistical sense, it can be said that $Y1(-1)$ has a constant linear effect on $H_g(Yg)$, adjusted for $Y2(-1)$, indicated by $C(gk1)$. In other words, the adjusted effect of $Y1(-1)$ on $H_g(Yg)$ is *constant*.
3. Similarly, this function also presents a set of *homogenous regressions* $\hat{H}_g(Yg)$ on $Y2(-1)$ for all scores of $Y1(-1)$.
4. These characteristics should be considered limitations or hidden assumptions of additive models with numerical or categorical independent variables. Refer to the notes on alternative ANCOVA and fixed-effects models presented in previous chapters.

13.4.1.2 Interaction Linear Effects Models by CF and the Time t

The function $F_{gk}(V(-1))$ for each (g,k) would be.

$$F_{gk}(V(-1)) = C(gk0) + C(gk1) \times Y1(-1) + C(gk2) \times Y2(-1) \\ + C(gk3) \times Y1(-1) \times Y2(-1)$$

(13.12c)

Note that the LS regression function obtained within each cell generated by *CF* and the time t would have the following equation.

$$\hat{H}_g(Yg) = [\hat{C}(gk0) + \hat{C}(gk1) \times Y1(-1)] \\ + [\hat{C}(gk2) + \hat{C}(gk3) \times Y1(-1)] \times Y2(-1)$$

(13.12d)

The characteristics of this regression function are as follows:

1. For all scores of $Y1(-1)$, this function presents a set of *heterogeneous regressions* of $\hat{H}_g(Yg)$ on $Y2(-1)$, with intercepts and slopes dependent on $Y1(-1)$. Similarly, this function also presents *heterogeneous regressions* of $\hat{H}_g(Yg)$ on $Y1(-1)$ for all scores of $Y2(-1)$.
2. The graph of this function in a three-dimensional space with the coordinate axes: $H_g(Yg)$, $Y1(-1)$, and $Y2(-1)$, is a *hyperboloid*.

13.4.2 SCMs by CF and the Time t Based on Figure 13.1(b)

As an extension of the model in (10.11b), the model based on the path diagram in Figure 13.1(b) would have the following general equation specification.

$$H_1(Y1) = F_{11}(V(-1), Y2) \times D1 + F_{12}(V(-1), Y2) \times D2 \\ + F_{13}(V(-1), Y2) \times D3 + F_{14}(V(-1), Y2) \times D4 \\ H_2(Y2) = F_{21}(V(-1)) \times D1 + F_{22}(V(-1)) \times D2 \\ + F_{23}(V(-1)) \times D3 + F_{24}(V(-1)) \times D4$$

(13.13)

where $V(-1) = (Y1(-1), Y2(-1))$, and Dj is a dummy variable of the j-th group of the firms generated by *CF* and the time t.

13.4.2.1 Additive Linear-Effects Models by CF and the Time t

The function $F_{gk}(*)$ for each (g,k) would be

$$F_{gk}(V(-1), Y2) = C(gk0) + C(gk1) \times Y1(-1) + C(gk2) \times Y2(-1) + C(gk3) \times Y2 \\ F_{gk}(V(-1)) = C(gk0) + C(gk1) \times Y1(-1) + C(gk2) \times Y2(-1)$$

(13.13a)

Since $Y2(-1)$ and $Y2$ are highly correlated, then unexpected parameter estimates would be obtained in many cases. In such a case, the model should be reduced by deleting one of the variables. It is recommended to delete $Y2$, because $Y2(-1)$ is a more important independent variable since it is the up-stream variable and cause factor for $Y1$. Then the reduced model would be the same as the additive model in (13.12).

13.4.2.2   Two-Way Interaction Linear Effects Models by CF and the Time t

The function $F_{gk}(*)$ for each (g,k) would be

$$
\begin{aligned}
F_{gk}(V(-1),Y2) &= C(gk0) + C(gk1) \times Y1(-1) + C(gk2) \times Y2(-1) + C(gk3) \times Y1(-1) \times Y2(-1) \\
&\quad + [C(gk4) + C(gk5) \times Y1(-1) + C(gk6) \times Y2(-1))] \times \mathbf{Y2} \\
F_{gk}(V(-1)) &= C(gk0) + C(gk1) \times Y1(-1) + C(gk2) \times Y2(-1) \\
&\quad + C(gk3) \times Y1(-1) \times Y2(-1)
\end{aligned}
$$

$$(13.13b)$$

Compare this to the model in (13.13a) where only two variables $Y2(-1)$ and $Y2$ can give unexpected parameter estimates of the first regression; in this model there are five highly correlated variables: $Y2(-1)$, $Y1(-1)*Y2(-1)$, $Y2$, $Y1(-1)*Y2$ and $Y2(=1)*Y2$ to be considered, in addition to the three variables $Y1(-1)$, $Y1(-1)*Y2(-1)$ and $Y1(-1)*Y2$.

Then there would be many possible good fit reduced models obtained, which are highly dependent on the data used. Two types of reduced models are in (13.12), and (13.13a).

13.4.2.3   Three-Way Interaction Linear-Effects Models by CF and the Time t

As an extension of the function (13.13b), the function $F_{gk}(V(-1),Y2)$ is a three-way interaction function as follows:

$$
\begin{aligned}
F_{gk}(V(-1),Y2) &= C(gk0) + C(gk1) \times Y1(-1) + C(gk2) \times Y2(-1) + C(gk3) \times Y1(-1) \times Y2(-1) \\
&\quad + [C(gk4) + C(gk5) \times Y1(-1) + C(gk6) \times Y2(-1) + C(gk7) \times Y1(-1) \times Y2(-1)] \times \mathbf{Y2} \\
Fgk(V(-1)) &= C(gk0) + C(gk1) \times Y1(-1) + C(gk2) \times Y2(-1) \\
&\quad + C(gk3) \times Y1(-1) \times Y2(-1)
\end{aligned}
$$

$$(13.13c)$$

Refer to the notes presented for the models previously, which in fact are reduced general equations of this model.

13.4.3   SCMs by CF and the Time *t* Based on Figure 13.1(c)

As a modification of the models presented in previous Section 13.4.2, the model based on the path diagram in Figure 13.1(c) would have the following general equation specification.

$$
\begin{aligned}
H_1(Y1) &= F_{11}(V(-1),Y2) \times D1 + F_{12}(V(-1),Y2) \times D2 \\
&\quad + F_{13}(V(-1),Y2) \times D3 + F_{14}(V(-1),Y2) \times D4 \\
H_2(Y2) &= F_{21}(V(-1),Y1) \times D1 + F_{22}(V(-1),Y1) \times D2 \\
&\quad + F_{23}(V(-1),Y1) \times D3 + F_{24}(V(-1),Y1) \times D4
\end{aligned}
$$

$$(13.14)$$

where $V(-1)=(Y1(-1),Y2(-1))$, and Dj is a dummy variable of the j-th group of the firms generated by CF and the time t. Refer to the notes presented for the models in (13.1a).

13.4.3.1   *Additive Linear-Effects Models by* **CF** *and the Time* t

The functions $F_{gk}(*)$ for each (g,k) would be

$$F_{gk}(V(-1), Y2) = C(gk0) + C(gk1) \times Y1(-1) + C(gk2) \times Y2(-1) + C(gk3) \times \mathbf{Y2}$$
$$F_{gk}(V(-1), Y1) = C(gk0) + C(gk1) \times Y1(-1) + C(gk2) \times Y2(-1) + C(gk3) \times \mathbf{Y1}$$

(13.14a)

13.4.3.2   *Two-Way Interaction Linear-Effects Models by* **CF** *and the Time* t

The functions $F_{gk}(*)$ for each (g,k) would be

$$F_{gk}(V(-1), Y2) = C(gk0) + C(gk1) \times Y1(-1) + C(gk2) \times Y2(-1) + C(gk3) \times Y1(-1) \times Y2(-1)$$
$$+ [C(gk4) + C(gk5) \times Y1(-1) + C(gk6) \times Y2(-1)] \times \mathbf{Y2}$$
$$F_{gk}(V(-1), Y1) = = C(gk0) + C(gk1) \times Y1(-1) + C(gk2) \times Y2(-1) + C(gk3) \times Y1(-1) \times Y2(-1)$$
$$+ [C(gk4) + C(gk5) \times Y1(-1) + C(gk6) \times Y2(-1)] \times \mathbf{Y1}$$

(13.14b)

13.4.3.3   *Three-Way Interaction Linear Effects Models by* **CF** *and the Time* t

The function $F_{gk}(*)$ for each (g,k) would be

$$F_{gk}(V(-1), Y2) = C(gk0) + C(gk1) \times Y1(-1) + C(gk2) \times Y2(-1) + C(gk3) \times Y1(-1) \times Y2(-1)$$
$$+ [C(gk4) + C(gk5) \times Y1(-1) + C(gk6) \times Y2(-1) + C(gk7) \times Y1(-1) \times Y2(-1)] \times \mathbf{Y2}$$
$$F_{gk}(V(-1), Y1) = C(gk0) + C(gk1) \times Y1(-1) + C(gk2) \times Y2(-1) + C(gk3) \times Y1(-1) \times Y2(-1)$$
$$+ [C(gk4) + C(gk5) \times Y1(-1) + C(gk6) \times Y2(-1) + C(gk7) \times Y1(-1) \times Y2(-1)] \times \mathbf{Y1}$$

(13.14c)

13.4.4   Empirical Results Based on the Models of (Y1,Y2)

13.4.4.1   *Estimation Method using the Function @Expand(CF,T)*

If a large number of groups of the firms by CF and the time t are considered, it is better to conduct the analysis based on each univariate regression using the function *@Expand(CF,T)* to present its equation specification, instead of using a large number of dummy variables. Various analyses have been presented in the previous chapters. For an additional illustration, see the following example, which presents only the additive models by a *CF* and the time *T*.

For all other models presented earlier this can easily be done, using the same method or the multistage manual stepwise selected method.

Example 13.1   Application of the simplest model in (13.14)

For an illustration, based on a small size {T = 2 or T = 3} Table 13.1 presents the summary of the statistical results of the first model in (13.14a), using the following equation specification, and two of its reduced models.

$$y1 \; y1(-1) * @expand(m\_size, t) \; y2(-1) * @expand(m\_size, t)$$
$$y2 * @expand(m\_size, t) \; @expand(m\_size, t)$$

(13.15a)

Table 13.1   *Summary of the statistical results of the full model in (13.12a) and its reduced models*

Dependent Variable: Y1									
Method: Panel Least Squares									
Date: 12/28/11 Time: 11:46									
Sample: 1 8 IF T =2 OR T=3									
Periods included: 2									
Cross-sections included: 224									
Total panel (balanced) observations: 448									

	Full Model			Reduced Model-1			Reduced Model-2		
Variable	Coeff.	t-Stat.	Prob.	Coeff.	t-Stat.	Prob.	Coeff.	t-Stat.	Prob.
Y1(-1)*(M_SIZE=1 AND T=2)	-0.059	-0.178	*0.859*	-0.059	-0.179	*0.858*			
Y1(-1)*(M_SIZE=1 AND T=3)	1.137	15.792	0.000	1.135	15.842	0.000	1.133	16.030	0.000
Y1(-1)*(M_SIZE=2 AND T=2)	0.911	49.700	0.000	0.911	49.750	0.000	0.911	49.919	0.000
Y1(-1)*(M_SIZE=2 AND T=3)	1.157	58.895	0.000	1.157	58.954	0.000	1.157	59.154	0.000
Y2(-1)*(M_SIZE=1 AND T=2)	-0.027	-0.050	*0.960*	-0.027	-0.050	*0.960*			
Y2(-1)*(M_SIZE=1 AND T=3)	-0.399	-0.311	*0.756*	-0.131	-0.125	*0.901*			
Y2(-1)*(M_SIZE=2 AND T=2)	-1.376	-1.469	0.143	-1.376	-1.470	0.142	-1.376	-1.475	0.141
Y2(-1)*(M_SIZE=2 AND T=3)	-0.292	-1.086	0.278	-0.292	-1.087	0.278	-0.292	-1.091	0.276
Y2*(M_SIZE=1 AND T=2)	2.082	1.686	0.093	2.082	1.688	0.092	2.008	1.940	0.053
Y2*(M_SIZE=1 AND T=3)	0.347	0.364	*0.716*						
Y2*(M_SIZE=2 AND T=2)	-0.386	-1.308	0.192	-0.386	-1.309	0.191	-0.386	-1.314	0.190
Y2*(M_SIZE=2 AND T=3)	2.889	2.780	0.006	2.889	2.783	0.006	2.889	2.792	0.006
M_SIZE=1 AND T=2	0.369	4.181	0.000	0.369	4.185	0.000	0.359	4.828	0.000
M_SIZE=1 AND T=3	0.031	0.390	0.696	0.031	0.387	0.699	0.030	0.378	0.705
M_SIZE=2 AND T=2	0.089	0.846	0.398	0.089	0.847	0.398	0.089	0.850	0.396
M_SIZE=2 AND T=3	-0.163	-1.880	0.061	-0.163	-1.882	0.061	-0.163	-1.889	0.060
R-squared	0.940			0.940			0.940		
Adjusted R-squared	0.938			0.938			0.938		
Durbin-Watson stat	1.482			1.482			1.480		

Based on this summary, the following findings and notes are presented.

1. Since the numerical independent variables of the full model have large p-values, say greater than 0.30, then an acceptable reduced model should be explored. Even though $Y2*(M\_Size = 1, T = 3)$ has the smallest p-value, I decided to delete it in order to have a set of four regressions of $Y1$ on $Y1(-1)$ and $Y2(-1)$, which will be compared to the VAR model presented next.

2. Finally, the reduced Model-2 is obtained by deleting three additional independent variables, which is an acceptable model in a statistical sense. Note that two of the intercept parameters have a p-value > 0.30 so could be deleted from the model. Do it as an exercise.

3. The data analysis based on the second regression in (13.14a) can easily be done using the same method. Do it as an exercise, using the following equation specification.

$$y2 \; y1(-1) * @expand(m\_size, t) \; y2(-1) * @expand(m\_size, t)$$
$$y1 * @expand(m\_size, t) \; @expand(m\_size, t) \tag{13.15b}$$

4. Note that the pair of equation specifications in (13.15a) and (13.15b) show that Y1 and Y2 have a simultaneous causal relationship.

13.4.4.2   Estimation Method Using the Object "System"

It is well-known that the system equation of a model should be written using dummy variables, as presented in previous chapters. So this section will present only one example, based on a SCM with the multiple regressions having different independent sets of independent variables.

Example 13.2   Application of the object *System*

Table 13.2 presents a summary of the statistical results of a SCM, and its reduced model, using the following system equation.

$$
\begin{aligned}
Y1 = &\ (C(110) + C(111)*Y1(-1) + C(112)*Y2(-1) + C(113)*Y2)*D1 \\
&+ (C(120) + C(121)*Y1(-1) + C(122)*Y2(-1) + C(123)*Y2)*D2 \\
&+ (C(130) + C(131)*Y1(-1) + C(132)*Y2(-1) + C(133)*Y2)*D3 \\
&+ (C(140) + C(141)*Y1(-1) + C(142)*Y2(-1) + C(143)*Y2)*D4 \\
Y2 = = &\ (C(210) + C(211)*Y1(-1) + C(212)*Y2(-1) + C(213)*Y1)*D1 \\
&+ (C(220) + C(221)*Y1(-1) + C(222)*Y2(-1) + C(223)*Y1)*D2 \\
&+ (C(230) + C(231)*Y1(-1) + C(232)*Y2(-1) + C(233)*Y1)*D3 \\
&+ (C(240) + C(241)*Y1(-1) + C(242)*Y2(-1) + C(243)*Y1)*D4
\end{aligned}
\tag{13.16}
$$

Table 13.2   *Summary of the statistical results of the full model in (13.16) and its reduced model*

System: SYS17
Estimation Method: Least Squares
Date: 12/28/11 Time: 13:23
Sample: 1 8 IF T=2 OR T=3
Included observations: 448
Total system (balanced) observations 896

| | Dpendent Variable : Y1 | | | | | | | Dpendent Variable : Y2 | | | | | |
| | Full Model | | | Reduced Model | | | | Full Model | | | Reduced Model | | |
Par.	Coeff.	t-Stat.	Prob.	Coeff.	t-Stat.	Prob.	Par.	Coeff.	t-Stat.	Prob.	Coeff.	t-Stat.	Prob.
C(110)	0.369	4.181	0.000	0.359	4.828	0.000	C(210)	-0.004	-0.232	0.817	-0.001	-0.047	0.962
C(111)	-0.059	-0.178	*0.858*				C(211)	0.005	0.088	*0.930*			
C(112)	-0.027	-0.050	*0.960*				C(212)	0.217	2.653	0.008	0.222	2.911	0.004
C(113)	2.082	1.686	0.092	2.008	1.940	0.053	C(213)	0.007	0.543	0.587			
C(120)	0.031	0.390	0.696	0.030	0.378	0.705	C(220)	-0.001	-0.085	0.932	-0.003	-0.258	0.796
C(121)	1.137	15.792	0.000	1.133	16.030	0.000	C(221)	-0.016	-0.503	*0.615*			
C(122)	-0.399	-0.311	*0.756*				C(222)	0.775	4.124	0.000	0.760	4.101	0.000
C(123)	0.347	0.364	*0.716*				C(223)	0.008	0.315	*0.753*			
C(130)	0.089	0.846	0.398	0.089	0.850	0.396	C(230)	0.140	8.106	0.000	0.140	8.140	0.000
C(131)	0.911	49.700	0.000	0.911	49.919	0.000	C(231)	0.053	2.910	0.004	0.053	2.923	0.004
C(132)	-1.376	-1.469	0.142	-1.376	-1.475	0.141	C(232)	-1.405	-9.225	0.000	-1.405	-9.264	0.000
C(133)	-0.386	-1.308	0.191	-0.386	-1.314	0.189	C(233)	-0.053	-2.714	0.007	-0.053	-2.726	0.007
C(140)	-0.163	-1.880	0.060	-0.163	-1.889	0.059	C(240)	0.015	0.937	0.349	0.017	1.111	0.267
C(141)	1.157	58.895	0.000	1.157	59.154	0.000	C(241)	-0.027	-1.401	0.161	-0.026	-1.360	0.174
C(142)	-0.292	-1.086	0.278	-0.292	-1.091	0.276	C(242)	0.038	0.789	*0.430*			
C(143)	2.889	2.780	0.006	2.889	2.792	0.005	C(243)	0.023	1.376	0.169	0.022	1.331	0.184

Based on the results in Table 13.2, the following notes and comments are made.

1. The full and reduced multiple regressions of *Y1* are exactly the same as the full and reduced Model-2 in Table 13.1.
2. The reduced model of *Y2* is obtained by deleting four numerical independent variables from the full model with *p*-values greater than 0.30. It is a very basic method.
3. However, based on my experimentation, other forms of reduced model(s) can be obtained by using the multistage manual stepwise selection method and inserting the independent variables, which are subjectively selected as the most important causal or up-stream variable(s) at the first stage of the analyses. Refer to the previous illustrative examples.

13.4.4.3   *Special Estimation Using the Object "VAR"*

The following examples present a special estimation process using the object "VAR" based on dummy variables models presented previously. I would say this is a special estimation process, because it has not been presented in other papers or books based on my observations. On the other hand, compared to the application of the object "*System*", the VAR model has a disadvantage, because all multiple regressions in the VAR model should have the same set of independent variables. So the object "VAR" cannot be applied based on the SCM in (13.17), and most of other SCMs, such as all SCMs based on Figure 13.1(b) and (c).

Example 13.3   A special estimation method using of the object "VAR"
For illustration, let's see the interaction model indicated by the function in (13.12d). The summary of its statistical results are presented in Table 13.3. The stages of the estimation process are as follows:

1. Click *Object/New Objects/OK* . . . , then
2. Insert the endogenous variables *"Y1 Y2"*; the lag interval for endogenous "0 0", and the following list of exogenous variables, in addition to the parameter *C*, then the result is obtained by clicking . . . *OK*.

$$
\begin{aligned}
&D1 * Y1(-1)\ \ D1 * Y2(-1)\ \ D1 * Y1(-1) * Y2(-1) \\
&D2\ \ D2 * Y1(-1)\ \ D2 * Y2(-1)\ \ D2 * Y1(-1) * Y2(-1) \\
&D3\ \ D3 * Y1(-1)\ \ D3 * Y2(-1)\ \ D3 * Y1(-1) * Y2(-1) \\
&D4\ \ D4 * Y1(-1)\ \ D4 * Y2(-1)\ \ D4 * Y1(-1) * Y2(-1)
\end{aligned}
\tag{13.17}
$$

Note that these variables are a modification of the independent variables of the first model in (13.16), or the regression of *Y1* without the dummy variable *D1*, because the parameter "*C*" should be in the list and the variable *Y2* is replaced by *Y1*(−1)∗*Y2*(−1). So both regressions of *Y1* and *Y2* have the same sets of independent variables.

Based on this summary, the following notes and comments are made.

1. Since *D1*∗*Y1*(−1) has an insignificant adjusted effect on both *Y1* and *Y2*; in a statistical sense they can be deleted from the model in order to obtain a reduced model.
   * However, even though the interactions *D2*∗*Y1*(−1)∗*Y2*(−1) have insignificant effects on both *Y1* and *Y2*, they should not be deleted, because the effect of *Y1*(−1) on *Y2* depends on *Y2*(−1) (or the effect of *Y2*(−1) on *Y2* depends on *Y1*(−1))) in a theoretical sense, then the reduced model should be obtained by deleting the independent variable(s): *D2*∗*Y1*(−1), or *D2*∗*Y2*(−1), or both.

Table 13.3 Summary of the statistical results of the model in (13.17), using the object "VAR"

Vector Autoregression Estimates
Date: 01/04/12 Time: 14:16
Sample: 1 8 IF T=2 OR T=3
Included observations: 448
Standard errors in ( ) & t-statistics in [ ]

Variable	Y1	Y2	Variable	Y1	Y2	Variable	Y1	Y2	Variable	Y1	Y2
C	0.335	-0.009	D2	-0.304	0.008	D3	-0.353	0.134	D4	-0.413	0.026
	-0.091	-0.016		-0.121	-0.021		-0.134	-0.024		-0.126	-0.022
	[3.69]	[-0.55]		[-2.52]	[0.36]		[-2.64]	[5.65]		[-3.28]	[1.18]
D1*Y1(-1)	0.133	0.050	D2*Y1(-1)	1.127	-0.003	D3*Y1(-1)	0.907	0.004	D4*Y1(-1)	1.111	-0.006
	-0.358	-0.063		-0.133	-0.024		-0.018	-0.003		-0.025	-0.004
	[0.37]	[0.79]		[8.46]	[-0.13]		[49.84]	[1.18]		[44.26]	[-1.33]
D1*Y2(-1)	1.077	0.380	D2*Y2(-1)	-0.162	0.791	D3*Y2(-1)	-1.546	-1.533	D4*Y2(-1)	-1.032	-0.063
	-0.680	-0.121		-1.151	-0.204		-0.877	-0.156		-0.404	-0.072
	[1.58]	[3.15]		[-0.14]	[3.87]		[-1.76]	[-9.86]		[-2.56]	[-0.88]
D1*Y1(-1)*Y2(-1)	-1.348	-0.335	D2*Y1(-1)*Y2(-1)	0.171	-0.096	D3*Y(-1)*Y2(-1)	4.832	1.203	D4*Y1(-1)*Y2(-1)	0.795	0.091
	-1.059	-0.188		-2.720	-0.483		-1.705	-0.303		-0.289	-0.051
	[-1.27]	[-1.78]		[0.06]	[-0.20]		[2.83]	[3.98]		[2.75]	[1.78]
						R-squared				0.941	0.251
						Adj. R-squared				0.939	0.225
						F-statistic				456.58	9.652

```
Estimation Proc:
================================
LS 0 0 Y1 Y2 @ C D1*Y1(-1) D1*Y2(-1) D1*Y1(-1)*Y2(-1)
D2 D2*Y1(-1) D2*Y2(-1) D2*Y1(-1)*Y2(-1)
D3 D3*Y1(-1) D3*Y2(-1) D3*Y(-1)*Y2(-1)
D4 D4*Y1(-1) D4*Y2(-1) D4*Y1(-1)*Y2(-1)

VAR Model:
================================
Y1 = C(1,1) + C(1,2)*D1*Y1(-1) + C(1,3)*D1*Y2(-1) + C(1,4)*D1*Y1(-1)*Y2(-1)
     + C(1,5)*D2 + C(1,6)*D2*Y1(-1) + C(1,7)*D2*Y2(-1) + C(1,8)*D2*Y1(-1)*Y2(-1)
     + C(1,9)*D3 + C(1,10)*D3*Y1(-1) + C(1,11)*D3*Y2(-1) + C(1,12)*D3*Y(-1)*Y2(-1)
     + C(1,13)*D4 + C(1,14)*D4*Y1(-1) + C(1,15)*D4*Y2(-1) + C(1,16)*D4*Y1(-1)*Y2(-1)

Y2 = C(2,1) + C(2,2)*D1*Y1(-1) + C(2,3)*D1*Y2(-1) + C(2,4)*D1*Y1(-1)*Y2(-1)
     + C(2,5)*D2 + C(2,6)*D2*Y1(-1) + C(2,7)*D2*Y2(-1) + C(2,8)*D2*Y1(-1)*Y2(-1)
     + C(2,9)*D3 + C(2,10)*D3*Y1(-1) + C(2,11)*D3*Y2(-1) + C(2,12)*D3*Y(-1)*Y2(-1)
     + C(2,13)*D4 + C(2,14)*D4*Y1(-1) + C(2,15)*D4*Y2(-1) + C(2,16)*D4*Y1(-1)*Y2(-1)
```

Figure 13.2   *The general equation of the VAR model in Table 13.3*

- On the other hand, even though $D2*Y1(-1)$ has an insignificant effect on $Y2$ only, it cannot be deleted using the object VAR. In order to get reduced models of $Y1$ and $Y2$ with different sets of independent variables, the object "*System*" should be applied as presented in Table 7.2.
2. Figure 13.2 presents the general equation of the VAR model in the work file, with the symbol $C(g,k)$ for its parameters. EViews does not provide a method for testing the differential effects of an independent variable between the defined sub-populations. So an alternative special estimation method is presented in the following example.

Example 13.4   Another special estimation method using the object "VAR"

For testing the differential effects of an independent variable between the defined sub-populations, Table 13.4 presents the summary of statistical results based on the model in Figure 13.2, but using different forms of the equation specification.

The stages of the estimation process are as follows:

1. Click *Object/New Objects/OK* ..., then
2. Insert the endogenous variables "*Y1 Y2*"; the lag interval for endogenous "1 1" and the following list of exogenous variables, in addition to the parameter C, then the result is obtained by clicking ... *OK*.

$$Y1(-1) * Y2(-1)$$
$$D2 \ D2 * Y1(-1) \ D2 * Y2(-1) \ D2 * Y1(-1) * Y2(-1)$$
$$D3 \ D3 * Y1(-1) \ D3 * Y2(-1) \ D3 * Y1(-1) * Y2(-1)$$
$$D4 \ D4 * Y1(-1) \ D4 * Y2(-1) \ D4 * Y1(-1) * Y2(-1)$$

$$(13.18)$$

Based on the summary in Table 13.4 and the equation of the VAR model, after an adjustment in Figure 13.3 the following findings and notes are presented.

1. Note that the first two regressions in both Tables 13.3 and 13.4 are exactly the same function.
2. Table 13.4 clearly shows that the first two out of the eight regressions do not have the dummy variable *D1*, which indicates that first regressions of *Y1* and Y2 are used as reference regressions. In addition, Figure 13.3 presents the adjusted positions of the model parameters $C(g,k)$.

Table 13.4 *Summary of statistical results from the model in (13.18)*

Vector Autoregression Estimates
Date: 01/04/12 Time: 14:16
Sample: 1 8 IF T=2 OR T=3
Included observations: 448
Standard errors in ( ) & t-statistics in [ ]

Variable	Y1	Y2	Variable	Y1	Y2	Variable	Y1	Y2
Y1(-1)	-0.036	-0.005	D2	-0.333	-0.002	D3	-0.399	0.138
	-0.330	-0.059		-0.119	-0.021		-0.138	-0.025
	[-0.12]	[-0.08]		[-2.81]	[-0.08]		[-2.89]	[5.64]
Y2(-1)	0.475	0.185	D2*Y1(-1)	1.164	0.002	D3*Y1(-1)	0.953	0.002
	-0.470	-0.083		-0.356	-0.063		-0.329	-0.058
	[1.01]	[2.22]		[3.27]	[0.03]		[2.90]	[0.03]
C	0.364	0.000	D2*Y2(-1)	-0.636	0.605	D3*Y2(-1)	-1.813	-1.881
	-0.088	-0.016		-1.245	-0.221		-1.090	-0.193
	[4.13]	[0.02]		[-0.51]	[2.74]		[-1.66]	[-9.74]
Y1(-1)*Y2(-1)	-0.092	0.072	D2*Y1(-1)*Y2(-1)	0.263	-0.167	D3*Y(-1)*Y2(-1)	4.897	1.152
	-0.257	-0.046		-2.737	-0.485		-1.718	-0.305
	[-0.36]	[1.57]		[0.10]	[-0.34]		[2.85]	[3.78]

Variable	Y1	Y2
D4	-0.442	0.017
	-0.124	-0.022
	[-3.56]	[0.78]
D4*Y1(-1)	1.147	-0.001
	-0.331	-0.059
	[3.47]	[-0.02]
D4*Y2(-1)	-1.507	-0.248
	-0.620	-0.110
	[-2.43]	[-2.26]
D4*Y1(-1)	0.887	0.020
	-0.387	-0.069
	[2.29]	[0.29]
R-squared	0.940	0.250
Adj. R-squared	0.938	0.224
F-statistic	454.91	9.589

```
Estimation Proc:
===================================
LS 1 1 Y1 Y2 @ C  Y1(-1)*Y2(-1)
                 D2 D2*Y1(-1) D2*Y2(-1) D2*Y1(-1)*Y2(-1)
                 D3 D3*Y1(-1) D3*Y2(-1) D3*Y(-1)*Y2(-1)
                 D4 D4*Y1(-1) D4*Y2(-1) D4*Y1(-1)*Y2(-1)


VAR Model:
===================================
Y1 =  C(1,3)    + C(1,1)*Y1(-1)    + C(1,2)*Y2(-1)    + C(1,4)*Y1(-1)*Y2(-1)
     + C(1,5)*D2 + C(1,6)*D2*Y1(-1) + C(1,7)*D2*Y2(-1) + C(1,8)*D2*Y1(-1)*Y2(-1)
     + C(1,9)*D3 + C(1,10)*D3*Y1(-1) +C(1,11)*D3*Y2(-1) + C(1,12)*D3*Y(-1)*Y2(-1)
     + C(1,13)*D4 + C(1,14)*D4*Y1(-1) +C(1,15)*D4*Y2(-1) + C(1,16)*D4*Y1(-1)*Y2(-1)


Y2 =  C(2,3)    + C(2,1)*Y1(-1)    + C(2,2)*Y2(-1)    + C(2,4)*Y1(-1)*Y2(-1)
     + C(2,5)*D2 + C(2,6)*D2*Y1(-1) + C(2,7)*D2*Y2(-1) + C(2,8)*D2*Y1(-1)*Y2(-1)
     + C(2,9)*D3 + C(2,10)*D3*Y1(-1) +C(2,11)*D3*Y2(-1)+C(2,12)*D3*Y(-1)*Y2(-1)
     + C(2,13)*D4 +C(2,14)*D4*Y1(-1) +C(2,15)*D4*Y2(-1)+C(2,16)*D4*Y1(-1)*Y2(-1)
```

Figure 13.3   *The equation of the VAR model in Table 13.4 after an adjustment*

3. Then the t-statistic in Table 13.4 can be applied to test the hypotheses that the differential adjusted effects of an independent variable, specifically $Y1(-1)*Y2(-1)$ on either $Y1$ or $Y2$ between pairs of sub-populations using $\{M\_Size=1,T=2\}=\{D1=1\}$ act as the reference group. For instance, the parameter $C(1,8)$ indicates the effect difference of $Y1(-1)*Y2(-1)$ on $Y1$ between the groups $\{D1=1\}$ and $\{D2=1\}$ with the t-statistic of $t_0 = 0.09\,618 < 0.10$ presented in the table. So it can be concluded they have a significant difference at the 10% level of significance.

4. This example is presented just to show how to test some hypotheses using the VAR model. However, if the hypotheses on the effect differences would be tested?, conducting analysis using the object "*System*" is recommended, so that various hypotheses can easily be tested using the Wald test. See the following example.

Example 13.5   Application of the model in Figure 13.3 using the object "*System*"

Figure 13.4 presents the statistical results of the model, using object *System*, with its parameters presented in Table 13.5.

Based on this table various hypotheses on the differential adjusted effects of independent variables on $Y1$ or $Y2$ between the four sub-populations can easily be tested using the t-statistic presented in the output or the Wald test, such as follows:

1. The hypotheses based on each parameters $C(ijk)$, for $i=1,2; j=2,3,4; k=1,2,3;$ can be tested using the t-statistic in the output, the same as presented in Table 13.4, that is the output based on the VAR model. For instance, the null hypothesis H_0: $C(123)=0$ is accepted based on the t-statistic of $t_0 = 0.516$ with a p-value $= 0.606$. So the adjusted effect of $Y1(-1)*Y2(-1)$ on $Y1$ has an insignificant difference between the subpopulation $\{D1=1\}$ and $\{D2=1\}$.

2. However, we find that parameter estimates are not the same as those in Table 13.3 using the object *VAR*. I do not really know why. I expected the objects VAR and System to give the same results, so I communicated by e-mail with IHS EViews about my finding. Corresponding to this communication, I would like to present the following special notes.

3. *Special notes on unexpected findings.* In response my e-mail, the Senior Principal Economist, IHS EViews, Dr Gareth Thomas (January, 16, 2011) gave the statement:

Table 13.5 *Parameters of the model in Figure 13.4 by the variables and sub-samples*

Variable	{D1=1}	{D2=1}	{D3=1}	{D4=1}
		Dependent variable : Y1		
Intercept	c(110)	c(110)+c(120)	c(110)+c(130)	c(110)+c(140)
Y1(-1)	c(111)	c(111)+c(121)	c(111)+c(131)	c(111)+c(141)
y2(-1)	c(112)	c(112)+c(122)	c(112)+c(132)	c(112)+c(142)
y1(-1)*y2(-1)	c(113)	c(113)+c(123)	c(113)+c(133)	c(113)+c(143)
		Dependent variable : Y2		
Intercept	c(210)	c(210)+c(220)	c(210)+c(230)	c(210)+c(240)
Y1(-1)	c(211)	c(211)+c(221)	c(211)+c(231)	c(211)+c(241)
y2(-1)	c(212)	c(212)+c(222)	c(212)+c(232)	c(212)+c(242)
y1(-1)*y2(-1)	c(213)	c(213)+c(223)	c(213)+c(233)	c(213)+c(143)

"Our best guess, much like it was when your stepwise results changed, is that the underlying data has changed. Obviously it is impossible for us to prove this though".

As a result, I tried to use a small work file containing only the endogenous variables *Y1*, *Y2*, the dummy variables *D1* to *D4* and the time *t*. It is surprising that the object VAR and the System present the same outputs. Table 13.6 presents the summary of a new output using the object VAR, which equals the output in Figure 13.5 using the object *System*.

System: SYS18
Estimation Method: Least Squares
Date: 01/05/12   Time: 16:45
Sample: 1 8 IF T=2 OR T=3
Included observations: 448
Total system (balanced) observations 896

	Dependent variable: Y1				Dependent variable: Y2		
Par.	Coeff.	t-Stat.	Prob.	Par.	Coeff.	t-Stat.	Prob.
C(110)	0.335	3.653	0.000	C(210)	-0.009	-0.546	0.585
C(111)	0.133	0.370	0.712	C(211)	0.050	0.781	0.435
C(112)	1.077	1.569	0.117	C(212)	0.380	3.116	0.002
C(113)	-1.348	-1.262	0.207	C(213)	-0.335	-1.764	0.078
C(120)	-0.304	-2.500	0.013	C(220)	0.008	0.353	0.724
C(121)	0.994	2.581	0.010	C(221)	-0.053	-0.775	0.438
C(122)	-1.239	-0.918	0.359	C(222)	0.410	1.711	0.087
C(123)	1.519	0.516	0.606	C(223)	0.239	0.457	0.648
C(130)	-0.288	-2.032	0.043	C(230)	0.168	6.674	0.000
C(131)	0.769	2.124	0.034	C(231)	-0.057	-0.890	0.374
C(132)	-2.090	-1.659	0.097	C(232)	-2.016	-9.001	0.000
C(133)	1.417	1.287	0.198	C(233)	0.451	2.306	0.021
C(140)	-0.413	-3.249	0.001	C(240)	0.026	1.167	0.243
C(141)	0.978	2.703	0.007	C(241)	-0.056	-0.871	0.384
C(142)	-2.109	-2.642	0.008	C(242)	-0.443	-3.123	0.002
C(143)	2.143	1.935	0.053	C(243)	0.427	2.166	0.031

Determinant residual covariance		0.01118			

Equation: Y1 = (C(110)+C(111)*Y1(-1)+C(112)*Y2(-1)+C(113)*Y1(-1)*Y2(-1))
    +(C(120)+C(121)*Y1(-1)+C(122)*Y2(-1)+C(123)*Y1(-1)*Y2(-1))*D2
    +(C(130)+C(131)*Y1(-1)+C(132)*Y2(-1)+C(133)*Y1(-1)*Y2(-1))*D3
    +(C(140)+C(141)*Y1(-1)+C(142)*Y2(-1)+C(143)*Y1(-1)*Y2(-1))*D4

R-squared	0.93957	Mean dependent var	1.26461
Adjusted R-squared	0.93747	S.D. dependent var	3.141
S.E. of regression	0.78542	Sum squared resid	266.493
Durbin-Watson stat	1.43276		

Equation: Y2 = (C(210)+C(211)*Y1(-1)+C(212)*Y2(-1)+C(213)*Y1(-1)*Y2(-1))
    +(C(220)+C(221)*Y1(-1)+C(222)*Y2(-1)+C(223)*Y1(-1)*Y2(-1))*D2
    +(C(230)+C(231)*Y1(-1)+C(232)*Y2(-1)+C(233)*Y1(-1)*Y2(-1))*D3
    +(C(240)+C(241)*Y1(-1)+C(242)*Y2(-1)+C(243)*Y1(-1)*Y2(-1))*D4

R-squared	0.23438	Mean dependent var	0.02228
Adjusted R-squared	0.2078	S.D. dependent var	0.15687
S.E. of regression	0.13962	Sum squared resid	8.42164
Durbin-Watson stat	1.84136		

Figure 13.4 *Statistical results of the model in Figure 13.3 using the object "System"*

Table 13.6 *Summary of a new output of the same model as in Figure 13.3*

Vector Autoregression Estimates
Date: 01/16/12 Time: 15:06
Sample: 1 1792 IF T=2 OR T=3
Included observations: 448
Standard errors in ( ) & t-statistics in [ ]

	Y1	Y2		Y1	Y2		Y1	Y2		Y1	Y2
Y1(-1)	0.133	0.050	D2*Y1(-1)	-0.304	0.008	D3*Y1(-1)	-0.288	0.168	D4*Y1(-1)	-0.413	0.026
	(0.361)	(0.064)		(0.122)	(0.022)		(0.142)	(0.025)		(0.127)	(0.023)
	[0.369]	[0.781]		[-2.50]	[0.353]		[-2.032]	[6.674]		[-3.249]	[1.167]
Y2(-1)	1.077	0.380	D2*Y2(-1)	0.994	-0.053	D3*Y2(-1)	0.769	-0.057	D4*Y2(-1)	0.978	-0.056
	(0.687)	(0.122)		(0.385)	(0.068)		(0.362)	(0.064)		(0.362)	(0.064)
	[1.569]	[3.116]		[2.581]	[-0.775]		[2.124]	[-0.890]		[2.703]	[-0.871]
C	0.335	-0.009	D2	-1.239	0.410	D3	-2.090	-2.016	D4	-2.109	-0.443
	(0.092)	(0.016)		(1.349)	(0.240)		(1.260)	(0.224)		(0.798)	(0.142)
	[3.653]	[-0.546]		[-0.918]	[1.711]		[-1.659]	[-9.001]		[-2.642]	[-3.123]
Y1(-1)*Y2(-1)	-1.348	-0.335	D2*Y1(-1)*Y2(-1)	1.519	0.239	D3*Y1(-1)*Y2(-1)	1.417	0.451	D4*Y1(-1)*Y2(-1)	2.143	0.427
	(1.069)	(0.190)		(2.945)	(0.524)		(1.101)	(0.196)		(1.108)	(0.197)
	[-1.262]	[-1.764]		[0.516]	[0.457]		[1.287]	[2.306]		[1.935]	[2.166]
R-squared	0.940	0.234									
Adj. R-squared	0.937	0.208									
F-statistic	447.797	8.817									

After getting the output in Table 13.6, I tried again to use my original (large) work file by doing a completely new step-by-step process. Surprisingly the output is exactly the same as the output in Table 13.6. I e-mailed these findings to Dr Thomas, with an additional note as follows:

"I would like to present also what I have found in my book with a special note. Because someone may be facing the same problem, in the future."

Then Dr Thomas (January 20, 2011) replied with the following:

"Having the incorrect data probably accounts for over half of the 'errors' that get reported to us. It is, of course, vitally important when doing any econometric/statistic work that the user ensures that they have the correct data. I think it would be a good idea to point this out in a text book, yes."

So there is a serious problem, at least for me. Corresponding to my findings, I have an unanswered question.

"Why can the data change to unexpected and unknown new data during the data analysis at a time point, and then change back to the original data at another time point?"

I will never know whether the data had changed during the data analysis or if I did not apply two different estimation methods based on exactly the same econometric models. So, now I am wondering whether or not an output presented in this book as well as in Agung (2011, 2009a) based on the original data set, changed whenever the output was obtained by using a single estimation method.

Following a discussion with my college Bambang Hermanto, PhD I found out he had the same experience while he was doing his dissertation using EViews 3. We have an agreement that an econometric model would present different outputs or parameter estimates because of the initial parameter values used in the recursive iteration estimation process.

4. Now return to the results using the system equation previously, Some of the hypotheses which can be tested using the Wald test are as follows:

   4.1 Testing a univariate hypothesis, at the 15% level of significance
   - The null hypothesis H_0: $C(123) = C(133) = C(143) = 0$ is rejected based on the Chi-squared statistic of $\chi_0^2 = 5.975749$ with $df = 3$ and a p-value $= 0.1128$. So the adjusted effect of $Y1(-1)*Y2(-1)$ on $Y1$ has significant differences between the four sub-populations.
   - The null hypothesis H_0: $C(123) = C(133) = C(143)$ is accepted based on the Chi-squared statistic of $\chi_0^2 = 3.397532$ with $df = 2$ and a p-value $= 0.1829$. So the adjusted effect of $Y1(-1)*Y2(-1)$ on $Y1$ has insignificant differences between the three subpopulations, $\{D1 = 2\}$, $\{D3 = 1\}$ and $\{D4 = 1\}$.

   4.2 Testing the multivariate hypothesis at the 15% level of significance
   - The null hypothesis H_0: $C(123) = C(133) = C(143) = C(223) = C(233) = C(243) = 0$ is rejected based on the Chi-squared statistic of $\chi_0^2 = 11.43897$ with $df = 6$ and a p-value $= 0.0757$. So the adjusted effect of $Y1(-1)*Y2(-1)$ on $(Y1,Y2)$ has significant differences between the four sub-populations.
   - The null hypothesis H_0: $C(123) = C(133) = C(143)$, $C(223) = C(233) = C(443)$, is accepted based on the Chi-squared statistic of $\chi_0^2 = 3.691152$ with $df = 4$ and a p-value $= 0.4494$. So

the adjusted effect of $Y1(-1)*Y2(-1)$ on $(Y1,Y2)$ has insignificant differences between the three sub-populations, *{D1 = 2}*, *{D3 = 1}* and *{D4 = 1}*.

13.4.4.4   *Application of the STEPLS Estimation Method*

Example 13.6   Application of the STEPLS estimation method

To demonstrate unexpected good fit models in a statistical sense, obtained based on any models with interaction numerical independent variables, Table 13.7 presents a summary of statistical results of the four regressions in (13.14a) based on the sub-samples generated by M_Size = {1,2} and T = {2,3}, using the STEPLS estimation method.

This summary clearly shows four unexpected STEPLS regressions, which highly depend on the data, as well as multicollinearity between the dependent variables. Note that the sub-sample *{M_Size = 1,T = 2}* gives a regression with only one out of the seven independent variables, and the sub-sample *{M_Size = 2, T = 3}* gives a regression that has all independent variables, even though one of them has a large *p*-value = 0.446, using the default stopping criteria: *p*-value forward/backward = 0.5/0.5. By using stopping criteria 0.5/0.4, based on the sub-sample *{M_Size = 2,t = 3}*, a regression with five independent variables is obtained, where *Y2* has a *p*-value = 0.1830.

Example 13.7   Application of the combinatorial selection method

For an additional illustration, Figure 13.5 presents the statistical results of the three-way interaction model in (13.14a) with endogenous variables $Y1 = Ya = lnSALE$ and $Y2 = Yb = lnSIZE$, based on only two intentionally selected sub-samples *{M_Size = 1,T = 2}* and *{M_Size = 2,T = 3}*.

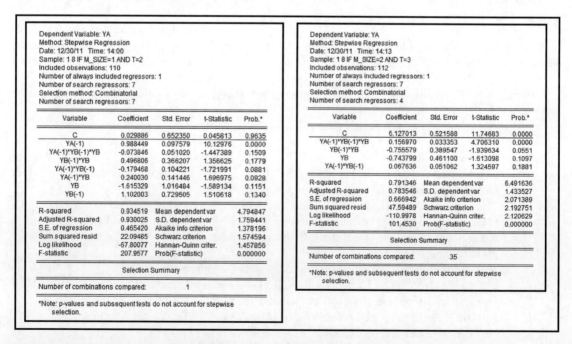

Figure 13.5   *Statistical results of two three-way interaction models based on sub-samples* {M_Size = 1,T = 2} *and* {M_Size = 2,T = 3}

Table 13.7 Summary of statistical results of the four regressions in (13.15a) based on the sub-samples generated by $M\_Size = \{1,2\}$ and $T = \{2,3\}$, using the STEPLS estimation method

Dependent Variable: Y1												
Method: Stepwise Regression												
Date: 12/30/11 Time: 18:36												
Included observations: 112												
Number of always included regressors: 1												
Number of search regressors: 7												
Selection method: Stepwise forwards												
Stopping criterion: p-value forwards/backwards = 0.5/0.5												
	M-SIZE=1 & T=2			M-SIZE=1 & T=3			M-SIZE=2 & T=2			M-SIZE=2 & T=3		
Variable	Coeff.	t-Stat	Prob.*	Coeff.	t-Stat	Prob.*	Coeff.	t-Stat	Prob.*	Coeff.	t-Stat	Prob.*
C	0.359	3.614	0.001	0.036	0.691	0.491	0.083	0.829	0.409	-0.021	-0.235	0.815
Y1(-1)				1.116	22.931	0.000	0.906	33.266	0.000	1.106	32.457	0.000
Y1(-1)*Y2							-0.830	-2.853	0.005	1.251	3.089	0.003
Y1(-1)*Y2(-1)										-1.366	-2.199	0.030
Y1(-1)*Y2(-1)*Y2				15.148	1.082	0.282	0.532	1.972	0.051	13.387	3.233	0.002
Y2	2.008	1.452	0.149				2.266	1.732	*0.862*	1.364	0.765	*0.446*
Y2(-1)							-3.179	-2.588	0.011	1.596	2.091	0.039
Y2(-1)*Y2				-2.792	-0.707	*0.481*	3.444	1.188	0.237	-36.477	-2.490	0.014
R-squared	0.019			0.849		0.460	0.971		2.024	0.974		2.195
Adjusted R-squared	0.010			0.845		1.287	0.969		3.785	0.972		4.445
F-statistic	2.109			202.665			586.884			559.313		
Prob(F-statistic)	0.149			0.000			0.000			0.000		
*Note: p-values and subsequent tests do not account for stepwisen selection.												

Based on these results, the following notes and comments are made.

1. The same as the regressions presented in Table 13.3, both three-way interaction models should also be considered unexpected regressions because all independent variables are inserted as a single set of search regressors.
2. The STEPLS regression obtained using a single set of search regressors would be an acceptable good fit model under the assumption that the search regressors are equally important cause factors for the dependent variable. However, I would say that this assumption is an inappropriate or a worse assumption to make. So I recommend applying the multistage STEPLS estimation method.

13.4.5 SCM with the Time Numerical Independent Variable

Whenever the data has a sufficiently large number of the time-point observations, then using the time t as a numerical independent variable is suggested instead of time dummy variables. The SCM can easily be derived from all models presented in previous sections. However, for illustration, models are derived from the general model in (13.14) only. Based on this general model, then those with the time numerical independent variable would have the following general specific equation.

$$H_1(Y1) = F_{11}(V(-1), Y2, t) \times DS1 + F_{12}(V(-1), Y2, t) \times DS2$$
$$H_2(Y2) = F_{21}(V(-1), Y1, t) \times DS1 + F_{22}(V(-1), Y1, t) \times DS2$$

$$(13.19)$$

where $V(-1) = (Y1(-1), Y2(-1))$, and *DS1*, and *DS2* are the dummy variables of the dichotomous variable *M_Size*.

For an illustrative example, the Special_BPD.wf1 will be used even though it has only eight time-point observations.

13.4.5.1   The Simplest SCM with Trend

The simplest SCM in (13.19) is an additive model of the variable *(Y1,Y2)* by *M_Size*, having the following equation specification.

$$
\begin{aligned}
Y1 = {}& (C(10) + C(11) * Y1(-1) + C(12) * Y2(-1) + C(13) * Y2 + C(14) * t) * DS1 \\
& + (C(20) + C(21) * Y1(-1) + C(22) * Y2(-1) + C(23) * Y2 + C(24) * t) * DS2
\end{aligned}
$$

(13.20)

$$
\begin{aligned}
Y2 = {}& (C(30) + C(31) * Y1(-1) + C(32) * Y2(-1) + C(33) * Y1 + C(34) * t) * DS1 \\
& + (C(40) + C(41) * Y1(-1) + C(42) * Y2(-1) + C(43) * Y1 + C(44) * t) * DS2
\end{aligned}
$$

where *DS1* and *DS2* are the *M_Size* dummy variables.

13.4.5.2   The Simplest Time-Related Effects SCM

As an extension of the SCM with trend in (13.20), the simplest time-related effects SCM considered is a two-way interaction model by the *M_Size*, with the following ES.

$$
\begin{aligned}
Y1 = {}& (C(10) + C(11) * Y1(-1) + C(12) * Y2(-1) + C(13) * Y2) * DS1 \\
& + (C(14) + C(15) * Y1(-1) + C(16) * Y2(-1) + C(17) * Y2) * t * DS1 \\
& + (C(20) + C(21) * Y1(-1) + C(22) * Y2(-1) + C(23) * Y2) * DS2 \\
& + (C(24) + C(25) * Y1(-1) + C(26) * Y2(-1) + C(27) * Y2) * t * DS2
\end{aligned}
$$

(13.21)

$$
\begin{aligned}
Y2 = {}& (C(30) + C(31) * Y1(-1) + C(32) * Y2(-1) + C(33) * Y1) * DS1 \\
& + (C(34) + C(35) * Y1(-1) + C(36) * Y2(-1) + C(37) * Y2) * t * DS1 \\
& + (C(40) + C(41) * Y1(-1) + C(42) * Y2(-1) + C(43) * Y1 + C(44) * t) * DS2 \\
& + (C(44) + C(45) * Y1(-1) + C(46) * Y2(-1) + C(47) * Y2) * t * DS2
\end{aligned}
$$

Example 13.8   LS regressions based on the model in (13.21)

For an illustration, this example only presents the statistical results based on a more complex model (13.21), which are presented in Table 13.8. Based on this summary, the following findings and notes are presented.

1. This example has demonstrated that the variable $Y1(-1)$ with the parameter C(11)) is deleted, even though it has the smallest *p*-value, in order to keep the interaction $t*Y1(-1)$ in the model. In this case, $t*Y1(-1)$ happens to have a *p*-value $= 0.000$ in the reduced model.
2. However, in general, a reduced model is obtained by deleting an independent variable (or parameter) from the full model with a large or largest *p*-value, as presented for the other variables.

Table 13.8   *Summary of the statistical results of the model in (13.17) and its reduced model*

System: SYS17_TRE												
Sample: 2 8												
Included observations: 1568												
Total system (balanced) observations 3136												
Dependent Variable: Y1							Dependent Variable: Y2					
	Full Model			Reduced Model				Full Model			Reduced Model		
Par.	Coeff.	t-Stat.	Prob.	Coeff.	t-Stat.	Prob.	Par.	Coeff.	t-Stat.	Prob.	Coeff.	t-Stat.	Prob.
C(10)	0.184	1.783	0.075	0.562	5.493	0.000	C(30)	0.002	0.200	0.841	0.001	0.266	0.790
c(11)	1.107	12.209	*0.000*				C(31)	-0.004	-0.365	*0.715*			
C(12)	0.112	0.312	*0.755*				C(32)	0.196	4.957	0.000	0.196	4.979	0.000
C(13)	0.362	1.373	0.170	0.435	1.595	0.111	C(33)	0.010	2.006	0.045	0.009	2.103	0.036
C(14)	-0.018	-0.911	*0.362*	-0.074	-3.818	0.000	C(34)	0.000	-0.089	*0.929*			
C(15)	-0.006	-0.439	*0.661*	0.155	45.290	0.000	C(35)	-0.002	-1.112	0.266	-0.002	-2.992	0.003
C(16)	-0.023	-0.319	*0.750*				C(36)	-0.039	-5.009	0.000	-0.039	-5.031	0.000
C(17)	-0.091	-1.397	0.162	-0.109	-1.628	0.104	C(37)	0.246	429.0	0.000	0.246	429.7	0.000
C(20)	-0.139	-1.216	0.224	-0.139	-1.216	0.224	C(40)	0.047	3.775	0.000	0.048	4.283	0.000
C(21)	1.026	47.149	0.000	1.026	45.154	0.000	C(41)	0.001	0.189	*0.851*			
C(22)	0.057	0.117	*0.907*				C(42)	-0.090	-1.655	0.098	-0.089	-1.654	0.098
C(23)	-0.094	-0.208	*0.835*				C(43)	0.000	-0.130	*0.897*			
C(24)	0.022	1.079	0.281	0.022	1.055	0.292	C(44)	-0.004	-3.087	0.002	-0.007	-3.480	0.001
C(25)	0.008	2.170	0.030	0.008	2.082	0.037	C(45)	0.000	-0.111	*0.912*			
C(26)	-0.007	-0.112	*0.911*				C(46)	0.013	1.948	0.052	0.013	1.948	0.052
C(27)	0.014	0.213	*0.831*				C(47)	0.143	1114.5	0.000	0.143	1116.5	0.000
R-squared			0.946			0.941	R-squared			0.999			0.999
Adjusted R-squared			0.946			0.941	Adjusted R-squared			0.999			0.999
Durbin-Watson stat			2.019			1.821	Durbin-Watson stat			1.744			1.744

3. I have done experiments using alternative manual stepwise selection methods with unexpected final regression functions, which are presented in Agung (2011), as well as in this book. See the following example.

4. In order to test hypotheses on the differential effects of an independent variable on *Y1* or *Y2* between the subpopulations of the firms $\{DS1 = 1, DS2 = 0\}$ and $\{DS1 = 0, DS2 = 1\}$, the full model should be used because effects should be adjusted for the same set of other independent variables. Similarly so for the hypothesis on the joint effects of a sub-set of independent variables. The testing can easily be done using the Wald test, for instance: the following hypotheses.

   4.1 The null hypothesis H_0: $C(17) = C(27)$ is considered for testing differential linear effects of *Y2∗t* on *Y1* between the two sub-populations of firms: $\{DS1 = 1, DS2 = 0\}$ and $\{DS1 = 0, DS2 = 1\}$, adjusted for *Y1(−1), Y2, Y2(−1), t, t∗Y1(−1)* and *t∗Y2(−1)*.

   4.2 The null hypothesis H_0: $C(13) = C(23)$, $C(17) = C(27)$ is considered for testing differential *joint effects of Y2 and Y2∗t* on *Y1* between the two sub-populations of firms: $\{DS1 = 1, DS2 = 0\}$ and $\{DS1 = 0, DS2 = 1\}$ adjusted for *Y1(−1), Y2(−1), t, t∗Y1(−1)* and *t∗Y2(−1)*.

   4.3 The null hypothesis H_0: $C(14) = C(24)$, $C(15) = C(25)$, $C(16) = C(26)$, $C(17) = C(27)$ is considered for testing the differential *joint effects* of *t, Y1(−1)∗t, Y2(−1)∗t* and *Y2∗t* on *Y2* on *Y1* between the two subpopulations of firms: $\{DS1 = 1, DS2 = 0\}$ and $\{DS1 = 0, DS2 = 1\}$, adjusted for $Y1(-1)$, $Y2(-1)$ and $Y2$.

Example 13.9   STEPLS estimation for the model in (13.20)

The main objective of this example is to present different estimates obtained for the model in (13.20) by using the STEPLS estimation method with several alternative options. However, in order to apply the STEPLS estimation for the MAR(1) model in (13.20), the data analysis would have to be done four times based on a set of four multiple regressions of $Y1$ and $Y2$, based on each of the two sub-samples represented by the dummy variables $DS1$ and $DS2$.

For an illustration, Figure 13.6 only presents the statistical results based on the full model of $Y1$ for the sub-sample $\{DS1 = 1\}$ and one of its possible reduced models using the combinatorial selection method. Based on these results, the following findings and notes are presented.

1. The stages of the analysis are as follows:
   1.1   After selecting the sample $\{DS1 = 1\}$, click *Quick/Estimate Equation . . . /STEPLS*.
   1.2   Enter the equation specification: namely "*Y1 C*", and the list of all search regressors: "*Y1(−1) Y2 (−1) Y2 t Y1(−1)∗t Y2(−1)∗t Y2∗t*".
   1.3   Click *Options/Combinatorial Selection Method*, and after entering "7" as the number of regressors to select, press ... *OK*, then the results in Figure 13.6(a) are obtained.
   1.4   Finally, the reduced model is obtained by entering "5" as the number of regressors to select with the results in Figure 13.6(b), which is the best fit model out of 21 possible models with five numerical independent variables.
2. These two regressions clearly show different parameter estimates compared to the parameters $C(10)$ to $C(17)$ in Table 13.7 for the regression of $Y1$ based on the sample $\{DS1 = 1\}$. Note that the reduced model in Figure 13.6 is worse than the reduced model in Table 13.7.

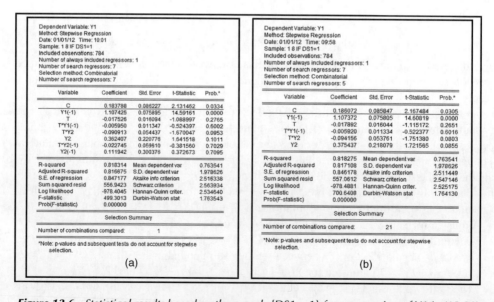

Figure 13.6   *Statistical results based on the sample $\{DS1 = 1\}$ for a regression of Y1 in (13.20)*

13.5   Advanced Autoregressive SCMs

All models presented in Section 13.4 could be extended to more advanced heterogeneous regressions by *CF* and the time *t* based on multivariate $Y = (Y_1, Y_2, \ldots, Y_G)$, for $G > 2$, with the up-stream or cause factors $Y(-p)$, for $p = 1, 2, \ldots$. Referring to the model in (13.1a), then $Y(-p)$, for $p = 1, 2 \ldots$; should have effects on all components of Y. The basic model considered would have the following general equation.

$$H_g(Y_g) = \sum_{s=1}^{S} \sum_{t=p+1}^{T} F_{gst}(Y(-1), \ldots, Y(-p)) \times Dst + \varepsilon_g \tag{13.22}$$

$$for\ g = 1, \ldots, G$$

where $H_g(Y_g)$ is a fixed function of Y_g, *Dst* is a dummy variable or a zero-one indicator of the cell $(CF = s)$ at the time *t*, or the objects in $(CF = s, Time = t) = (s, t)$, $F_{gst}(Y(-1), \ldots Y(-p))$ is a function of $(Y(-1), \ldots, Y(-p))$ with a finite number of parameters for each (g, s, t), $g = 1 \ldots, G$, $s = 1, \ldots, S$, and $t = p + 1, \ldots, T$; and ε_g is the random error or residual of the model, which is assumed to have specific basic assumptions.

There would be a lot of possible functions $F_{gst}(Y(-1), \ldots, Y(-p))$ for each (g, s, t). Note that for $G = 3$ and $p = 2$ only, this function is a function of six numerical variables, namely $Y1(-1)$, $Y2(-1)$, $Y3(-1)$, $Y1(-2)$, $Y2(-2)$ and $Y3(-2)$, and can be an additive or two and three-way interaction function, which should be subjectively selected by researchers. In addition refer to the notes presented for the alternative models in (13.1a).

13.5.1   SCMs Based on (Y1,Y2,Y3)

13.5.1.1   *Application of the Object "System"*

For illustration, Figure 13.7 presents two alternative path diagrams of the association between the set of variables for $G = 3$, having the following characteristics.

1. The path diagram in Figure 13.7a has specific characteristics as follows:
    1.1   It presents the effects of $Y(-p) = Y_{i,t-p} = (Y1_{i,t-p}, Y2_{i,t-p}, Y3_{i,t-p})$ on $Y_{it} = (Y1_{it}, Y2_{it}, Y3_{it})$. Since $Y_{i,t-p}$ is the up-stream variable of Y_{it} then the three components of $Y_{i,t-p}$, for all *p* should have effects on each of the components of Y_{it}.
    1.2   The dashed lines between pairs of the variables $Y1(-p)$, $Y2(-p)$ and $Y3(-p)$ indicates that their correlations should have unpredictable impacts on the parameters' estimates, even though they are not correlated in a theoretical sense.

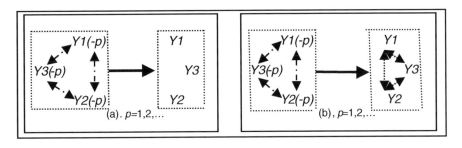

Figure 13.7   *Two path diagrams to present the relationships between Y1, Y2 and Y3*

1.3   The causal relationships between the variables *Y1*, *Y2* and *Y3* are not considered. They should be correlated because they have the same set of up-stream variables.

1.4   Therefore, the general model in (13.22) can be used to present this path diagram.

2.   In addition to the lines used in Figure 13.7a, the path diagram in Figure 13.7b has dotted lines between pairs of *Y1*, *Y2* and *Y3*, to indicate that the pairs may have unidirectional or simultaneous causal relationships, which are highly dependent on the set of selected variables. Then based on this path diagram, the basic general model in (13.22) should be modified by inserting additional independent variable(s) *Y1*, *Y2* or *Y3*.

3.   Several possible causal relationships between *Y1*, *Y2* and *Y3* can be found in practice. For instance, such as follows:

3.1   If *Y3* is the cause factor of both *Y1* and *Y2*, the general equation of the model is

$$H_g(Y_g) = \sum_{s=1}^{S} \sum_{t=p+1}^{T} F_{gst}(V(-p), Y3) \times Dst + \varepsilon_g, \quad for\ g = 1, and\ 2$$

$$\hspace{6cm}(13.23)$$

$$H_g(Y_g) = \sum_{s=1}^{S} \sum_{t=p+1}^{T} F_{gst}(V(-p)) \times Dst + \varepsilon_g, \quad for\ g = 3$$

where $V(-p) = [Y1(-1), \ldots, Y1(-p), Y2(-1), \ldots Y2(-p), Y3(-1), \ldots, Y3(-p)]$. So a lot of possible functions of $F_{gst}(V(-p), Y3)$ could be subjectively defined, either additive or interaction functions of the components of the vector variable $(V(-p), Y3)$, specifically the interaction(s) between *Y3* and the components of $V(-p)$, to indicate that the effect *Y3* on *Y1* and *Y2* depends on selected components of $V(-p)$. However, refer to the special notes presented for the models in (13.13). Similarly for the following cases:

3.2   If *Y2* and *Y3* are the cause factors of *Y1*, and *Y3* is the cause factor of *Y2*, the general equation of the model is

$$H_g(Y_g) = \sum_{s=1}^{S} \sum_{t=p+1}^{T} F_{gst}(V(-p), Y2, Y3) \times Dst + \varepsilon_g, \quad for\ g = 1,$$

$$H_g(Y_g) = \sum_{s=1}^{S} \sum_{t=p+1}^{T} F_{gst}(V(-p), Y3) \times Dst + \varepsilon_g, \quad for\ g = 2,$$

$$\hspace{6cm}(13.24)$$

$$H_g(Y_g) = \sum_{s=1}^{S} \sum_{t=p+1}^{T} F_{gst}(V(-p)) \times Dst + \varepsilon_g, \quad for\ g = 3$$

3.3   If *Y1* and *Y2* have simultaneous causal relationships, and *Y3* is the cause factor of *Y2*, the general equation of the model would be as follows:

$$H_g(Y_g) = \sum_{s=1}^{S} \sum_{t=p+1}^{T} F_{gst}(V(-p), Y2) \times Dst + \varepsilon_g, \quad for\ g = 1,$$

$$H_g(Y_g) = \sum_{s=1}^{S} \sum_{t=p+1}^{T} F_{gst}(V(-p), Y1, Y3) \times Dst + \varepsilon_g, \quad for\ g = 2,$$

$$\hspace{6cm}(13.25)$$

$$H_g(Y_g) = \sum_{s=1}^{S} \sum_{t=p+1}^{T} F_{gst}(V(-p)) \times Dst + \varepsilon_g, \quad for\ g = 3$$

Example 13.10   Application of a simple model in (13.22)
For an illustration, Figure 13.8 presents the statistical results of a simple model in (13.22), for $G = 3$ and $p = 1$, based on a sub-sample $\{T = 3\}$, using the object *System*.
  Based on this output, the following notes and comments are made.

1. This model is an additive MAR(1) of a trivariate *(Y1,Y2,Y3)* by a dichotomous variable *M_Size*, using both of its dummy variables *DS1* and *DS2*. This type of model should be applied if the reduced model is explored because the regression within each group has its own specific equation.
2. Note that based on this model all hypotheses on the differences between the levels of *M_Size* should be tested using the Wald test. For this reason, applying an alternative model is recommended using either one of the dummy variables. Table 13.9 presents a summary of the output of the model using only the dummy variable *DS2*. Based on this summary, compared to the output in Figure 13.8 the following findings and notes are presented.

    2.1   Compare to the output in Figure 13.8, the estimates of the parameters *C(gk)*, for $g = 1, 3$, and 5; $k = 0,1,2$ and 3; are exactly the same. In other words, they represent the three regressions of *Y1*, *Y2* and *Y3*, within the level $M\_Size = 1$.

    2.2   The parameters *C(gk)*, for $g = 2, 4$ and 6; $k = 1, 2$ and 3 represent the effects differences of an independent variable on each of *Y1*, *Y2* and *Y3*, between the two levels of *M_Size*. For instance, the regression of Y1 is obtained as follows:

    a. For $DS2 = 0$: $Y1 = [C(10) + C(20)] + [C(11) + C(21)]*Y1(-1)$

    b. For $DS2 = 1$:

$$Y1 = [C(10) + C(20)] + [C(11) + C(21)] * Y1(-1)$$
$$+ [C(12) + C(22)] * Y2(-1) + [C(13) + C(23)] * Y3(-1)$$

```
System: SYS17_3JAN12
Estimation Method: Least Squares
Date: 01/03/12   Time: 16:47
Sample: 1 8 IF T=3
Included observations: 224
Total system (balanced) observations 672
```

	Coefficient	Std. Error	t-Statistic	Prob.
C(10)	0.024936	0.064898	0.384236	0.7009
C(11)	1.035823	0.089621	11.55780	0.0000
C(12)	-0.503547	0.897523	-0.561041	0.5750
C(13)	1.265533	0.867713	1.458469	0.1452
C(20)	-0.071554	0.071162	-1.005509	0.3150
C(21)	1.072047	0.022858	46.90007	0.0000
C(22)	-0.477823	0.225692	-2.117149	0.0346
C(23)	0.754827	0.148725	5.075314	0.0000
C(30)	-0.001977	0.007172	-0.275724	0.7828
C(31)	-0.025077	0.009904	-2.531995	0.0116
C(32)	0.705225	0.099186	7.110157	0.0000
C(33)	0.232638	0.095891	2.426062	0.0155
C(40)	0.017864	0.007864	2.271516	0.0234
C(41)	-0.009925	0.002526	-3.929036	0.0001
C(42)	0.002613	0.024941	0.104775	0.9166
C(43)	0.082268	0.016436	5.005450	0.0000
C(50)	0.007841	0.018583	0.421967	0.6732
C(51)	-0.129471	0.025662	-5.045222	0.0000
C(52)	-0.218923	0.256997	-0.851851	0.3946
C(53)	2.195530	0.248462	8.836494	0.0000
C(60)	-0.013716	0.020377	-0.673133	0.5011
C(61)	-0.017225	0.006545	-2.631771	0.0087
C(62)	-0.302363	0.064625	-4.678743	0.0000
C(63)	1.184945	0.042586	27.82473	0.0000

```
Determinant residual covariance          4.74E-05
```

```
Equation: Y1=(C(10)+C(11)*Y1(-1)+C(12)*Y2(-1)+C(13)*Y3(-1))*DS1
         +(C(20)+C(21)*Y1(-1)+C(22)*Y2(-1)+C(23)*Y3(-1))*DS2
Observations: 224
```

R-squared	0.965079	Mean dependent var	1.327436
Adjusted R-squared	0.963947	S.D. dependent var	3.378456
S.E. of regression	0.641489	Sum squared resid	88.88583

```
Equation: Y2=(C(30)+C(31)*Y1(-1)+C(32)*Y2(-1)+C(33)*Y3(-1))*DS1
         +(C(40)+C(41)*Y1(-1)+C(42)*Y2(-1)+C(43)*Y3(-1))*DS2
Observations: 224
```

R-squared	0.316376	Mean dependent var	0.008079
Adjusted R-squared	0.294222	S.D. dependent var	0.084384
S.E. of regression	0.070891	Sum squared resid	1.085521

```
Equation: Y3=(C(50)+C(51)*Y1(-1)+C(52)*Y2(-1)+C(53)*Y3(-1))*DS1
         +(C(60)+C(61)*Y1(-1)+C(62)*Y2(-1)+C(63)*Y3(-1))*DS2
Observations: 224
```

R-squared	0.872159	Mean dependent var	0.084612
Adjusted R-squared	0.868016	S.D. dependent var	0.505605
S.E. of regression	0.183684	Sum squared resid	7.287834

Figure 13.8   *Statistical results of a simple model in (13.18), for* G = 3, p = 1 *and* t = 3

Table 13.9   *Summary of the statistical results of the model in Figure 13.8, using DS2 only*

Sample: 1 8 IF T=3											
Included observations: 224											
Total system (balanced) observations 672											
Dependent variable: Y1				Dependent variable: Y2				Dependent variable: Y3			
Par.	Coeff.	t-Stat.	Prob.	Par.	Coeff.	t-Stat.	Prob.	Par.	Coeff.	t-Stat.	Prob.
C(10)	0.025	0.384	0.701	C(30)	-0.002	-0.276	0.783	C(50)	0.008	0.422	0.673
C(11)	1.036	11.558	0.000	C(31)	-0.025	-2.532	0.012	C(51)	-0.129	-5.045	0.000
C(12)	-0.504	-0.561	0.575	C(32)	0.705	7.110	0.000	C(52)	-0.219	-0.852	0.395
C(13)	1.266	1.458	0.145	C(33)	0.233	2.426	0.016	C(53)	2.196	8.836	0.000
C(20)	-0.096	-1.002	0.317	C(40)	0.020	1.864	0.063	C(60)	-0.022	-0.782	0.435
C(21)	0.036	0.392	0.695	C(41)	0.015	1.482	0.139	C(61)	0.112	4.238	0.000
C(22)	0.026	0.028	0.978	C(42)	-0.703	-6.870	0.000	C(62)	-0.083	-0.315	0.753
C(23)	-0.511	-0.580	0.562	C(43)	-0.150	-1.546	0.123	C(63)	-1.011	-4.009	0.000
A-squared	0.965				0.316				0.872		
Adjusted R-squared	0.964				0.294				0.868		
Equation: Y1=(C(10)+C(11)*Y1(-1)+C(12)*Y2(-1)+C(13)*Y3(-1))+(C(20)+C(21)*Y1(-1)+C(22)*Y2(-1)+C(23)*Y3(-1))*DS2											
Equation: Y2=(C(30)+C(31)*Y1(-1)+C(32)*Y2(-1)+C(33)*Y3(-1))+(C(40)+C(41)*Y1(-1)+C(42)*Y2(-1)+C(43)*Y3(-1))*DS2											
Equation: Y3=(C(50)+C(51)*Y1(-1)+C(52)*Y2(-1)+C(53)*Y3(-1))+(C(60)+C(61)*Y1(-1)+C(62)*Y2(-1)+C(63)*Y3(-1))*DS2											

where $C(21)$, $C(22)$ and $C(23)$, respectively, represents the adjusted linear effect difference of $Y1(-1,Y2(-1)$ and $Y3(-1)$ on $Y1$, between of $M\_ize = 1$ and $M\_Size = 2$, which can be tested using the *t*-statistic in the output, either for two or one-sided hypotheses.

Example 13.11   Application of a three-way interaction model in (13.22)

As an extension of the model in Figure 13.8, Table 13.10 presents the statistical results of a three-way interaction SCM by the dichotomous M_Size, presented by the dummy variables $DS1$ and $DS2$, using the equation specification as presented at the bottom of the table, based on a sub-sample $\{T = 3\}$. Based on this table, the following findings and notes are presented.

1. It can be said the reduced model is unexpected, specifically for the regression of $Y1$. Note that the full model has four independent variables with large *p*-values. So a reduced model should be explored. How?
2. The model is presented to study that the effect of $Y1(-1)$ on $Y1$ depends on $Y2(-1)$ and $Y3(-1)$, represented by the interactions $Y1(-1)*Y2(-1)$, $Y1(-1)*Y3(-1)$ and $Y1(-1)*Y2(-1)*Y3(-1)$. For this reason, using the trial-and-error method, at the first stage the main variable $Y1(-1)*DS1$ should be deleted from the model, even though it has a *p*-value $= 0.000$, because it is highly correlated with each of the interactions. In general, the interactions of the reduced model would have small *p*-values. In this case, it happens the reduced model is acceptable in both the theoretical and statistical sense. Otherwise, the main variables $Y2(-1)*DS1$ or $Y3(-1)*DS1$, should be deleted using the trial-and-error method.
3. The reduced model of $Y2$ is obtained by deleting two out of the three main factors, namely $Y1(-1)*DS1$ and $Y3(-1)*DS1$.
4. With regards to the sub-sample $\{T = 3\}$, the variables $Yg(-1)$, for each $g = 1$, 2 and 3 represent the observed scores of Yg, at the time $t = 2$. So the SCM in fact represents the effects of the $Y1$, $Y2$ and $Y3$, at the time $t = 2$, on each of the variables at the time $t = 3$.

Table 13.10 Summary of the statistical results based on a three-way interaction model, as an extension of the SCM in Figure 13.8

System: SYS20_3WAY_INTERACTION
Estimation Method: Least Squares
Date: 01/09/12 Time: 14:47
Sample: 1 8 IF T=3
Included observations: 224
Total system (balanced) observations 672

Dependent variable: Y1

Par.	Full Model Coeff.	Full Model Prob.	Reduced Model Coeff.	Reduced Model Prob.
C(10)	-0.012	0.847	0.179	0.012
C(11)	1.175	0.000		
C(12)	-0.600	0.567	-2.408	0.055
C(13)	2.778	0.016	2.073	0.135
C(14)	-1.657	0.704	12.147	0.016
C(15)	-1.182	0.289	4.473	0.000
C(16)	12.295	0.402	-67.426	0.000
C(20)	-0.145	0.039	-0.145	0.088
C(21)	1.071	0.000	1.071	0.000
C(22)	3.675	0.000	3.675	0.000
C(23)	1.870	0.000	1.870	0.000
C(24)	-3.518	0.000	-3.518	0.000
C(25)	0.183	0.000	0.183	0.000
C(26)	-1.420	0.000	-1.420	0.000

Dependent variable: Y2

Par.	Full Model Coeff.	Full Model Prob.	Reduced Model Coeff.	Reduced Model Prob.
C(30)	-0.002	0.782	0.001	0.882
C(31)	-0.013	0.331		
C(32)	1.018	0.000	1.084	0.000
C(33)	0.397	0.003		
C(34)	-2.146	0.000	-1.931	0.000
C(35)	0.276	0.030	0.291	0.007
C(36)	-1.985	0.236	-1.735	0.221
C(40)	0.004	0.613	0.004	0.620
C(41)	-0.009	0.001	-0.009	0.001
C(42)	0.266	0.000	0.266	0.000
C(43)	0.279	0.000	0.279	0.000
C(44)	-0.256	0.000	-0.256	0.000
C(45)	0.006	0.099	0.006	0.105
C(46)	-0.091	0.008	-0.091	0.009

Dependent variable: Y3

Par.	Full Model Coeff.	Full Model Prob.	Reduced Model Coeff.	Reduced Model Prob.
C(50)	0.001	0.952	0.001	0.953
C(51)	-0.121	0.001	-0.120	0.000
C(52)	-0.074	0.809		
C(53)	2.824	0.000	2.815	0.000
C(54)	-1.794	0.160	-1.983	0.049
C(55)	0.610	0.062	0.629	0.046
C(56)	-9.832	0.022	-9.876	0.021
C(60)	-0.022	0.281	-0.022	0.280
C(61)	-0.018	0.007	-0.018	0.007
C(62)	0.537	0.004	0.537	0.004
C(63)	1.311	0.000	1.311	0.000
C(64)	-0.677	0.000	-0.677	0.000
C(65)	0.042	0.000	0.042	0.000
C(66)	-0.296	0.001	-0.296	0.001

Adjusted R-squared. Full model: 0.97094 & Reduced Model: 957344

Adjusted R-squared. Full model: 0.39415 & Reduced Model: 0.369594

Adjusted R-squared. Full model: 0.88904 & Reduced Model: 0.889539

Eq: Y1=(C(10)+C(11)*Y1(-1)+C(12)*Y2(-1)+C(13)*Y3(-1)+C(14)*Y1(-1)*Y2(-1)+C(15)*Y1(-1)*Y3(-1)+C(16)*Y1(-1)*Y2(-1)*Y3(-1))*DS1
+(C(20)+C(21)*Y1(-1)+C(22)*Y2(-1)+C(23)*Y3(-1)+C(24)*Y1(-1)*Y2(-1) +C(25)*Y1(-1)*Y3(-1)+C(26)*Y1(-1)*Y2(-1)*Y3(-1))*DS2

Eq: Y2=(C(30)+C(31)*Y1(-1)+C(32)*Y2(-1)+C(33)*Y3(-1) +C(34)*Y1(-1)*Y2(-1)+C(35)*Y1(-1)*Y3(-1)+C(36)*Y1(-1)*Y2(-1)*Y3(-1))*DS1
+(C(40)+C(41)*Y1(-1)+C(42)*Y2(-1)+C(43)*Y3(-1) +C(44)*Y1(-1)*Y2(-1)+C(45)*Y1(-1)*Y2(-1)+C(46)*Y1(-1)*Y2(-1)*Y3(-1))*DS2

Eq: Y3=(C(50)+C(51)*Y1(-1)+C(52)*Y2(-1)+C(53)*Y3(-1) +C(54)*Y1(-1)*Y2(-1)+C(55)*Y1(-1)*Y3(-1)+C(56)*Y1(-1)*Y2(-1)*Y3(-1))*DS1
+(C(60)+C(61)*Y1(-1)+C(62)*Y2(-1)+C(63)*Y3(-1) +C(64)*Y1(-1)*Y2(-1)+C(65)*Y1(-1)*Y3(-1)+C(66)*Y1(-1)*Y2(-1)*Y3(-1))*DS2

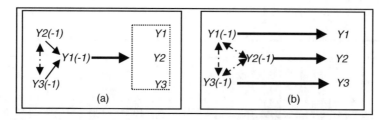

Figure 13.9   *Two path diagrams to present the relationships between* Y1, Y2 *and* Y3

5. Exactly the same data analysis can be done based on the cross-section data set at each of other time points. On the other hand, the models at all the time points can be presented as a single model by using the dummy variables *Dst*, for $s = 1,2$; and $t = 2,3,\ldots,T$.

6. Furthermore, the model could easily be extended to the multivariate AR(p) models, for p >1, by any classification factor, categorical cause factor, or cell factor, *CF* and the time t, by using the dummy variable *Dst*, for $s = 1,\ldots,S$; and $t = 2,\ldots,T$.

7. Note that this model is defined under the assumption that the linear effect of *Y1(−1)* on each Y_g, $g = 1$, 2, and 3, depends on *Y2(−1)* and *Y3(−1)*. For this reason the three full regressions in the SCM have the same set of interaction factors, namely *Y1(−1)*Y2(−1)*, *Y1(−1)*Y3(−1)*, and *Y1(−1)*Y2(−1)*Y3(−1)*, which could be represented using the path diagram in Figure 13.9(a). This path diagram shows that *Y2(−1)* and *Y3(−1)* are defined as the cause factors for *Y1(−1)*. If this is the case, then *Y2* and *Y3* also should be the cause factors for *Y1*. However, their causal relationships are not indicated or presented in the model. So this model could be extended to the models where causal effects are taken into account.

8. For a comparison, Figure 13.9(b) presents a path diagram with the following characteristics.

   8.1   The arrows between *Y1(−1)*, *Y2(−1)* and *Y3(−1)* with dashed lines indicate that their correlations as a set of independent variables have unpredictable impact(s) on each of the dependent variables *Y1*, *Y2* and *Y3*, even though they may not be correlated in theoretical sense. However, it always can be said that the effect of one of them on the corresponding dependent variable depends on the other two, more so whenever they have a type of causal relationship. All researchers can easily define the causal relationships between any set of three selected endogenous variables *Y1*, *Y2* and *Y3*.

   8.2   The three arrows with the bold lines present:
   - The effect of *Y1(−1)* on *Y1* depends on *Y2(−1)* and *Y3(−1)*, so that the full regression for *Y1* should have the interactions *Y1(−1)*Y2(−1)*, *Y1(−1)*Y3(−1)*, and *Y1(−1)*Y2(−1)*Y3(−1)* as the independent variables.
   - The effect of *Y2(−1)* on *Y2* depends on *Y1(−1)* and *Y3(−1)*, so the full regression model for *Y1* should have the interactions *Y1(−1)*Y2(−1)*,Y2(−1)*Y3(−1)*, and *Y1(−1)*Y2(−1)* *Y3(−1)* as the independent variables.
   - The effect of *Y3(−1)* on *Y3* depends on *Y1(−1)* and *Y2(−1)*, so the full regression model for *Y1* should have the interactions *Y1(−1)*Y3(−1)*,Y2(−1)*Y3(−1)*, and *Y1(−1)*Y2(−1)* *Y3(−1)* as the independent variables.

   8.3   Similar to the model in Table 13.9, this model also could be extended to models where the causal or up- and down-stream relationships between *Y1*, *Y2* and *Y3* are taken into account. See the following example, where one of their possible causal relationships is presented.

Example 13.12   Application of a model in (13.24)

Table 13.11 presents a summary of the statistical results of a model in (13.24), as an extension of the SCM in Table 13.9, with one of its possible reduced models. Based on this table the following notes are presented.

Table 13.11 Summary of the statistical results of a simple model in (13.22)

Estimation Method: Least Squares
Sample: 1 8 IF T=3

| | Dependent Variable: Y1 | | | | | Dependent Variable: Y2 | | | | | Dependent Variable: Y3 | | | |
| | Full Model | | Reduced Model | | | Full Model | | Reduced Model | | | Full Model | | Reduced Model | |
Par.	Coeff.	Prob.	Coeff.	Prob.	Par.	Coeff.	Prob.	Coeff.	Prob.	Par.	Coeff.	Prob.	Coeff.	Prob.
C(10)	-0.012	0.833	-0.001	0.985	C(30)	-0.002	0.763	-0.003	0.667	C(50)	0.001	0.952	0.001	0.953
C(11)	1.209	0.000	1.246	0.000	C(31)	-0.005	0.718			C(51)	-0.121	0.001	-0.120	0.000
C(12)	-0.543	0.661			C(32)	1.023	0.000	1.016	0.000	C(52)	-0.074	0.809		
C(13)	1.978	0.200			C(33)	0.209	0.237	0.203	0.233	C(53)	2.824	0.000	2.815	0.000
C(14)	-1.215	0.784			C(34)	-2.027	0.000	-1.991	0.000	C(54)	-1.794	0.160	-1.983	0.049
C(15)	-1.348	0.216	-1.293	0.204	C(35)	0.235	0.058	0.141	0.001	C(55)	0.610	0.062	0.629	0.046
C(16)	15.060	0.295	16.301	0.251	C(36)	-1.331	0.420			C(56)	-9.832	0.022	-9.876	0.021
C(17)*Y2	-0.035	0.962	0.621	0.020	C(37)*Y3	0.067	0.134	0.074	0.068	C(60)	-0.022	0.281	-0.022	0.280
C(18)*Y3	0.288	0.459	-0.093	0.163	C(40)	0.007	0.337	0.007	0.340	C(61)	-0.018	0.007	-0.018	0.007
C(20)	-0.093	0.165	1.070	0.000	C(41)	-0.006	0.022	-0.006	0.022	C(62)	0.537	0.004	0.537	0.004
C(21)	1.070	0.000	3.733	0.000	C(42)	0.186	0.009	0.192	0.001	C(63)	1.311	0.000	1.311	0.000
C(22)	3.733	0.000	0.657	0.222	C(43)	0.085	0.166	0.087	0.138	C(64)	-0.677	0.000	-0.677	0.000
C(23)	0.657	0.224	-3.302	0.000	C(44)	-0.156	0.024	-0.160	0.011	C(65)	0.042	0.000	0.042	0.000
C(24)	-3.302	0.000	0.132	0.000	C(45)	-0.001	0.889			C(66)	-0.296	0.001	-0.296	0.001
C(25)	0.132	0.000	-1.249	0.000	C(46)	-0.047	0.161	-0.051	0.000					
C(26)	-1.249	0.000	-3.664	0.001	C(47)*Y3	0.148	0.000	0.147	0.000					
C(27)*Y2	-3.664	0.001	1.706	0.000										
C(28)*Y3	1.706	0.000												

Equations of the Full Models

Eq: Y1=(C(10)+C(11)*Y1(-1)+C(12)*Y2(-1)+C(13)*Y3(-1))+(C(14)*Y1(-1)+C(15)*Y1(-1)*Y2(-1)+C(16)*Y1(-1)*Y2(-1)*Y3(-1))*DS1
+(C(17)*Y2+C(18)*Y3)*DS1
+(C(20)+C(21)*Y1(-1)+C(22)*Y2(-1)+C(23)*Y3(-1)+C(24)*Y1(-1)*Y2(-1)+C(25)*Y1(-1)*Y2(-1)*Y3(-1))*DS2
+(C(27)*Y2+C(28)*Y3)*DS2

Adjusted R-squared. Full model: 0.97422 & Reduced Model: 0.97448

Eq: Y2=(C(30)+C(31)*Y1(-1)+C(32)*Y2(-1)+C(33)*Y3(-1)+C(34)*Y1(-1)*Y3(-1)+C(35)*Y1(-1)*Y2(-1)*Y3(-1))*DS1
+C(37)*Y3*DS1
+(C(40)+C(41)*Y1(-1)+C(42)*Y2(-1)+C(43)*Y3(-1)+C(44)*Y1(-1)*Y2(-1)+C(45)*Y1(-1)*Y2(-1)*Y3(-1))*DS2
+C(47)*Y3*DS2

Adjusted R-squared. Full model: 0.45328 & Reduced Model: 0.45926

Eq: Y3=(C(50)+C(51)*Y1(-1)+C(52)*Y2(-1)+C(53)*Y3(-1)+C(54)*Y1(-1)*Y2(-1)+C(55)*Y1(-1)*Y3(-1)+C(56)*Y1(-1)*Y2(-1)*Y3(-1))*DS1
+(C(60)+C(61)*Y1(-1)+C(62)*Y2(-1)+C(63)*Y3(-1)+C(64)*Y1(-1)*Y2(-1)+C(65)*Y1(-1)*Y3(-1)+C(66)*Y1(-1)*Y2(-1)*Y3(-1))*DS2

Adjusted R-squared. Full model: 0.88904 & Reduced Model: 0.88954

1. Note that the regression equation of *Y1* has two additional independent variables *Y2* and *Y3*, but the regression of *Y2* has only *Y3* as an additional independent variable, presented using bold in the table.
2. Compared to the regression of the full model in Table 13.9 has the same set of independent variables and the three regressions in this model have different sets of independent variables.
3. This model should be considered as a simple model in (13.24). Various more advanced interaction models with a lot more independent variables could easily be defined. However, the good fit model is unpredictable.

13.5.1.2 Application of VAR(p) Models

If the multiple regressions in SCMs should have the same set of independent variables, conducting the data analysis using the object "VAR" is recommended, especially for $p > 2$.

Example 13.13 Application of a three-way interaction VAR model
For an illustration of the model in (13.21) for $p = 2$, Figure 13.10 presents the statistical results based on a three-way interaction VAR model, based on a sub-sample *{M_Size = 1,T = 3}*.

Based on these results, the following findings and notes are made.

1. Even though *Y1(−2)* has insignificant effects on *Y1*, *Y2* and *Y3*, it cannot be deleted from the model because *Y2(−2)* has significant effects on *Y2* and *Y3*.
2. On the other hand, *Y1(−1)∗Y2(−1)∗Y3(−1)* can be deleted because it has insignificant effects on *Y1*, *Y2* and *Y3*. However, we should consider the following two alternatives.

Vector Autoregression Estimates
Date: 01/11/12 Time: 10:43
Sample: 1 8 IF M_SIZE=1 AND T=3
Included observations: 112
Standard errors in ( ) & t-statistics in [ ]

	Y1	Y2	Y3
Y1(-1)	1.284891	-0.014107	0.068583
	(0.13306)	(0.01729)	(0.02050)
	[9.65632]	[-0.81571]	[3.34559]
Y1(-2)	-0.173467	0.049023	0.017045
	(0.24516)	(0.03186)	(0.03777)
	[-0.70757]	[1.53852]	[0.45129]
Y2(-1)	-0.920974	1.263949	-0.335957
	(0.98960)	(0.12862)	(0.15246)
	[-0.93066]	[9.82703]	[-2.20363]
Y2(-2)	0.342235	-0.243469	0.486137
	(0.40400)	(0.05251)	(0.06224)
	[0.84711]	[-4.63671]	[7.81067]
Y3(-1)	3.960753	0.255197	4.765621
	(1.33761)	(0.17385)	(0.20607)
	[2.96106]	[1.46790]	[23.1261]
Y3(-2)	-1.201848	-0.023270	-2.348570
	(1.00958)	(0.13122)	(0.15553)
	[-1.19044]	[-0.17734]	[-15.1000]
C	-0.002414	0.001362	-0.014916
	(0.06074)	(0.00789)	(0.00936)
	[-0.03974]	[0.17258]	[-1.59406]
Y1(-1)*Y2(-1)	-1.503648	-1.630206	-3.143242
	(4.19285)	(0.54495)	(0.64595)
	[-0.35862]	[-2.99147]	[-4.86611]
Y1(-1)*Y3(-1)	-1.780056	0.265653	0.115272
	(1.12511)	(0.14623)	(0.17333)
	[-1.58211]	[1.81665]	[0.66503]
Y1(-1)*Y2(-1)*Y3(-1)	19.45264	-2.110214	-2.172132
	(14.0758)	(1.82946)	(2.16851)
	[1.38199]	[-1.15346]	[-1.00167]
R-squared	0.861199	0.571450	0.887685
Adj. R-squared	0.848952	0.533637	0.877775
Sum sq. resids	25.53894	0.431421	0.606144
S.E. equation	0.500382	0.065035	0.077088
F-statistic	70.31853	15.11245	89.57311
Log likelihood	-76.13662	152.3924	133.3505
Akaike AIC	1.538154	-2.542722	-2.202688
Schwarz SC	1.780877	-2.299999	-1.959965
Mean dependent	0.460105	0.004405	0.037808
S.D. dependent	1.287491	0.095233	0.220499

Determinant resid covariance (dof adj.)	5.90E-06
Determinant resid covariance	4.45E-06
Log likelihood	213.2425
Akaike information criterion	-3.272188
Schwarz criterion	-2.544018

Figure 13.10 *Statistical results of a three-way interaction VAR model*

2.1 In general, the three-way interaction $Y1(-1)*Y2(-1)*Y3(-1)$ would be deleted in order to get a simpler model, namely a two-way interaction model. We find both $Y1(-1)*Y2(-1)$ have significant effects on $Y2$ and $Y3$, based on the t-statistics of $-2.89\,407$ and $-4.78\,489$, respectively; and $Y1(-1)*Y3(-1)$ has a significant effect on $Y2$, based on the t-statistic of $2.12\,448$.

2.2 If we are interested in having a three-way interaction model, then $Y1(-1)*Y2(-1)$ or $Y1(-1)*Y2(-1)$, or both would be deleted instead of $Y1(-1)*Y2(-1)*Y3(-1)$, since the three interactions are highly correlated, using the trial-and-error method . In this case, both should be deleted in order to obtain $Y1(-1)*Y2(-1)*Y3(-1)$ with a negative significant effect on $Y3$ based on the t-statistic of $-6.77\,058$.

3. The process of data analysis is as follows:

3.1 Click *Object/New Object* . . . */VAR/OK*,

3.2 Enter the endogenous variables: $Y1$, $Y2$ and the lag interval for endogenous: 1 2,

3.3 Finally, enter the exogenous variables: $Y1(-1)*Y2(-1)$ $Y1(-1)*Y3(-1)Y1(-1)*Y2(-1)*Y3(-1)$, in addition to the intercept "*C*", . . . *OK*, the output is obtained.

13.5.1.3 Not-Recommended Models

Corresponding to the model in Table 13.9, the following two SCMs can easily be derived or written using the block-copy-paste method. However, these models are not recommended, as well as other similar models, because of the following reasons.

1. The SCM in (13.26) is applied without using the time dummy variables. So the SCM presents a set of six continuous regression functions by *M_Size* for all time observations, two continuous regressions for each endogenous variable.

$$
\begin{aligned}
y1 = {} & (c(10)+c(11)*y1(-1)+c(12)*y2(-1)+c(13)*y3(-1)\\
& +c(14)*y1(-1)*y2(-1)+c(15)*y1(-1)*y3(-1)+c(16)*y1(-1)*y2(-1)*Y3(-1))Ds1\\
& +c(20)+c(21)*y1(-1)+c(22)*y2(-1)+c(23)*y3(-1)\\
& +c(24)*y1(-1)*y2(-1)+c(25)*y1(-1)*y3(-1)+c(26)*y1(-1)*y2(-1)*Y3(-1))*Ds2
\end{aligned}
$$

$$
\begin{aligned}
y2 = {} & (c(30)+c(31)*y1(-1)+c(32)*y2(-1)+c(33)*y3(-1)\\
& +c(34)*y1(-1)*y2(-1)+c(35)*y1(-1)*y3(-1)+c(36)*y1(-1)*y2(-1)*Y3(-1))Ds1\\
& +c(40)+c(41)*y1(-1)+c(42)*y2(-1)+c(43)*y3(-1)\\
& +c(44)*y1(-1)*y2(-1)+c(45)*y1(-1)*y3(-1)+c(46)*y1(-1)*y2(-1)*Y3(-1))*Ds2
\end{aligned}
$$

$$
\begin{aligned}
y3 = {} & (c(50)+c(51)*y1(-1)+c(52)*y2(-1)+c(53)*y3(-1)\\
& +c(54)*y1(-1)*y2(-1)+c(55)*y1(-1)*y3(-1)+c(56)*y1(-1)*y2(-1)*Y3(-1))Ds1\\
& +c(60)+c(61)*y1(-1)+c(62)*y2(-1)+c(63)*y3(-1)\\
& +c(64)*y1(-1)*y2(-1)+c(65)*y1(-1)*y3(-1)+c(66)*y1(-1)*y2(-1)*Y3(-1))*Ds2
\end{aligned}
$$

$$(13.26)$$

In other words, all firm-time observations within each *M_Size* $= 1$ and *M_Size* $= 2$ are presented or estimated using only a single multiple regression function. For instance, the multiple regression of *Y1* within the sub-sample *{M_Size* $= 1$} $=$ *{DS1* $= 1$} has the following equation.

$$\hat{y}1 = \hat{c}(10) + \hat{c}(11)y1(-1) + \hat{c}(12)y2(-1) + \hat{c}(13)y3(-1)$$
$$+ \hat{c}(14)y1(-1)y2(-1) + \hat{c}(15)y1(-1)y3(-1) + \hat{c}(16)y(-1)y2(-1)y3(-1) \quad (13.27)$$

For this reason, I would not recommend this SCM because the effect(s) of an independent variable on the corresponding dependent variable in general should change over time in a theoretical sense.

2. This is more so for the model without dummies *DS1* and *DS2* and time dummy variables, with the equation specification (13.28).

$$y1 = (c(10) + c(11) * y1(-1) + c(12) * y2(-1) + c(13) * y3(-1) + c(14) * y1(-1) * y2(-1)$$
$$c(15) * y1(-1) * y3(-1) + c(16) * y1(-1) * y2(-1) * Y3(-1))$$
$$y2 = (c(30) + c(31) * y1(-1) + c(32) * y2(-1) + c(33) * y3(-1) + c(34) * y1(-1) * y2(-1)$$
$$c(35) * y1(-1) * y3(-1) + c(36) * y1(-1) * y2(-1) * Y3(-1))$$
$$y3 = (c(50) + c(51) * y1(-1) + c(52) * y2(-1) + c(53) * y3(-1) + c(54) * y1(-1) * y2(-1)$$
$$c(55) * y1(-1) * y3(-1) + c(56) * y1(-1) * y2(-1) * Y3(-1))$$

$$(13.28)$$

Compared to the SCM in (13.26), this SCM is a worse model, because a single continuous regression function is used for all firm-time observations. For instance, the endogenous variable *Y1* is estimated using the continuous regression in (13.27) based on all firm-time observations. Refer to the special notes for fixed-effects or ANCOVA models, presented in previous Chapter 9, as well as Agung (2011).

13.6   SCMs Based on (*Y1,Y2*) with Exogenous Variables

13.6.1   SCMs Based on (X1,Y1,Y2)

The SCMs based on *(X1,Y1,Y2)* can easily be derived as the modification of all models based on *(Y1,Y2,Y3)* by using the first two regressions in the SCMs, namely the regressions *Y1* and *Y2*, but the variable *X1* and/or *X1(−1)* should be used to replace *Y3*. Then based on the model in (13.22) the following general bivariate SCM would be considered.

$$H_g(Y_g) = \sum_{s=1}^{S} \sum_{t=p+1}^{T} F_{gst}(V(-p), X1, X1(-1)) \times Dst + \varepsilon_g, \quad for \ g = 1, and \ 2 \quad (13.29)$$

where $V(-p) = [Y1(-1), \ldots, Y1(-p), Y2(-1), \ldots Y2(-p)]$. So a lot of possible functions of $F_{gst}(V(-p), X1, X1(-1))$ could be subjectively defined, either additive or interaction functions of the components of the vector variable *(V(−p),X1,X1(−1))*, specifically between *X1* or *X(−1)* and selected component(s) of V(−p), to indicate that the effect(s) *X1* or *X1(−1)* on *Y1* and *Y2* depend on selected component(s) of *V(−p)*: This is similar for the following cases.

Following the special notes presented for the model (13.13a), the two variables *X1* and *X1(−1)* are highly correlated, so then unexpected parameter estimates would be obtained in many cases. In such cases, the model should be reduced by deleting *X1* because its lag *X1(−1)* is a more important independent variable since it is the *"upstream variable and the cause factor"* for *Y1*. See the following simple alternative SCMs.

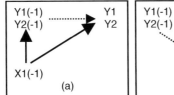 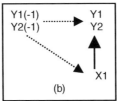 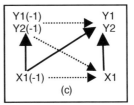

Figure 13.11 *Path diagrams of three alternative SCMs based on* X1, Y1 *and* Y2

Figure 13.11 presents three alternative path diagrams of simple theoretical causal associations of the variables *X1*, *Y1* and *Y2*, where *X1* is the cause or source variable/factor for both variables *Y1* and *Y2*. The characteristics of each path diagram and its corresponding SCM are as follows:

1. The arrows with bold lines indicate that the corresponding variables are defined as having causal relationships, and the arrows with dotted lines indicate that the corresponding variables have up- and down-stream relationships.
2. The path diagram in Figure 13.11a indicates that $X1(-1)$, $Y1(-1)$ and $Y2(-1)$ are the up-stream variables of $(Y1,Y2)$. Since X1 is a cause factor of $(Y1,Y2)$ then the linear effect of $X1(-1)$ on $(Y1,Y2)$ should depend on $(Y1(-1),Y2(-1))$, In this case, the function $F_{gst}(*)$ in (13.29) would have the most general form as follows:

$$F_{gst}(Y1(-1), Y2(-1), X1(-1)) = C(gst0) + C(gst1) * Y1(-1) + C(gst2) * Y2(-1)$$
$$+ X1(-1) * (C(gst3) + C(gst4) * Y1(-1) + C(gst5) * Y2(-1) + C(gst5) * Y1(-1) * Y2(-1))$$
$$(13.29a)$$

Note that this function shows the linear adjusted effect of $X1(-1)$ on $H_g(Y_g)$ depends on $(C(gst3) + C(gst4)*Y1(-1) + C(gst5)*Y2(-1) + C(gst5)*Y1(-1)*Y2(-1))$.

3. Similarly, the path diagram in Figure 13.11b also indicates that the effect of *X1* on $(Y1,Y2)$ depends $(Y1(-1),Y2(-1))$, In this case, the function $F_{gst}(*)$ in (13.29) would have the most general form as follows:

$$F_{gst}(Y1(-1), Y2(-1), X1(-1)) = C(gst0) + C(gst1) * Y1(-1) + C(gst2) * Y2(-1)$$
$$+ X1 * (C(gst3) + C(gst4) * Y1(-1) + C(gst5) * Y2(-1) + C(gst5) * Y1(-1) * Y2(-1))$$
$$(13.29b)$$

Note that this function shows that the linear adjusted effect of *X1* on $H_g(Y_g)$ depends on $(C(gst3) + C(gst4)*Y1(-1) + C(gst5)*Y2(-1) + C(gst5)*Y1(-1)*Y2(-1))$.

4. Finally, the diagram in Figure 13.11c indicates that the effect of both *X1* and $X1(-1)$ on $(Y1,Y2)$ depends $(Y1(-1),Y2(-1))$, In this case, the function $F_{gst}(*)$ in (13.29) would have the most general form as follows:

$$F_{gst}(Y1(-1), Y2(-1), X1, X1(-1)) = C(gst0) + C(gst1) * Y1(-1) + C(gst2) * Y2(-1)$$
$$+ X1 * (C(gst3) + C(gst4) * Y1(-1) + C(gst5) * Y2(-1) + C(gst5) * Y1(-1) * Y2(-1))$$
$$+ X1(-1) * (C(gst6) + C(gst7) * Y1(-1) + C(gst8) * Y2(-1) + C(gst9) * Y1(-1) * Y2(-1))$$
$$(13.29c)$$

Note that the models in (13.29a) and (13.29b) are two out of a lot of possible reduced models of this one. For instance, for the simplest reduced model is additive represented by the following function $F_{gst}(*)$.

$$F_{gst}(Y1(-1), Y2(-1), X1, X1(-1)) = C(gst0) + C(gst1) * Y1(-1) + C(gst2) * Y2(-1)$$
$$+ C(gst3) * X1 + C(gst4) * X1(-1) \tag{13.29d}$$

13.6.2   Illustrative Data Analyses

13.6.2.1   Application of the Object "System"

Example 13.14   Statistical results based on (X1,Y1,Y2)

With regards to the ES in (13.29), a lot of models by a cell-factor CF and the discrete time T-variable can easily be defined. A good fit model obtained would present a set of regressions with different sets of independent variables.

For this reason, it is recommended to do data analysis based on each sub-sample generated by CF and T. For an illustration, Table 13.12 presents a summary of the statistical results based on an hierarchical three-way interaction model of *(Y1,Y2)* on *(X1(−1),Y1(−1),Y2(−1))* and two of its reduced models, which are nonhierarchical based on a sub-sample {M_Size = 1,T = 2}. The analyses based on the other three sub-samples can easily be done using the same process.

1.   For comparison, however, the reduced Model-2 of *Y1* is obtained by deleting *Y1(−1)* and *X1(−1)∗ Y1(−1)*, which have the two largest *p*-values.

The hierarchical three-way interaction models should be applied. Based on this summary, the following findings and notes are presented.

1.   Note that the table presents the equations of the full models of *Y1* and *Y2* at the bottom of the table, which are hierarchical.
2.   For this model it is defined that the linear effect of *X1(−1)* on both *Y1* and *Y2* depends on *Y1(−1)* and *Y2(−1)*, which is indicated by the three interaction independent variables *X1(−1)∗Y1(−1)∗Y2(−1)*, *X1(−1)∗Y1(−1)* and *X1(−1)∗Y2(−1)*.
3.   Note that the interaction *X1(−1)∗Y1(−1)* has a large *p*-value = 0.685. In general, this interaction should be deleted in order to have a reduced model. Since it is defined that the linear effect of *X1(−1)* on both *Y1* and *Y2* depends of *Y1(−1)* and *Y2(−1)*, then the main factors *X1(−1)* or *Y1(−1)* should be deleted using the trial-and-error method. Note that the reduced Model-1 of *Y1* is obtained by deleting two of the main factors, namely *X1(−1)* and *Y1(−1)* even though *X1(−1)* has a *p*-value = 0.000.
4.   Both reduced models are acceptable or valid models to present that the linear effect of *X1(−1)* on *(Y1,Y2)* depends on both *Y1(−1)* and *Y2(−1)*. However, the reduced Model-2 is a better fit in a statistical sense, since it has a greater adjusted R-squared. In addition, the two interactions *X1(−1)∗Y1(−1)∗Y2(−1)*, and *X1(−1)∗Y2(−1)* are sufficient to show that the linear effect of *X1(−1)* on both *Y1* and *Y2* are significantly dependent on *Y1(−1)* and *Y2(−1)*.
5.   For comparison, Table 13.13 presents the summary of the statistical results of a similar model using the cause factor *X1* instead of *X1(−1)*, with the system specification presented at the bottom of the table. Based on this summary, the following findings and notes are presented.
  5.1   It is important to note that the reduced Model-1 of *Y1* is obtained by deleting the main factor *X1*, even though it has a *p*-value = 0.000. Note that the reduced model is acceptable in a statistical sense. However, it is unexpected reduced model since it has a very small adjusted R-squared value of 0.078 786 compared to 0.266 628 for the full model. Why?
  5.2   In order to study possible cause of the finding previously, the correlation analysis has been done based on the variables *Y1, X1, X1∗Y1(−1), X1∗Y2(−1)* and *X1∗Y1(−1)∗Y2(−1)*, with the output presented in Figure 13.12. Based on this output, the following notes and comments are made.

Table 13.12   *Summary of the statistical results of a full three-way interaction models, and its reduced models, based on a sub-sample {M_Size = 1, T = 2}*

	Full Model			Reduced Model-1			Reduced Model-2		
	Coeff.	t-Stat.	Prob.	Coeff.	t-Stat.	Prob.	Coeff.	t-Stat.	Prob.
C(10)	-1.072	-3.835	0.000	0.044	0.335	0.738	-1.115	-4.421	0.000
C(11)	-5.281	-1.904	0.058	-8.662	-3.249	0.001	-4.878	-3.377	0.001
C(12)	0.165	0.407	*0.685*	0.274	2.475	0.014			
C(13)	2.215	3.385	0.001	3.095	4.442	0.000	2.216	3.941	0.000
C(14)	18.628	2.016	0.045	27.142	3.192	0.002	16.860	3.381	0.001
C(15)	0.290	4.783	0.000				0.295	5.785	0.000
C(16)	-1.078	-0.438	*0.662*						
C(17)	-7.659	-3.198	0.002	-9.174	-3.491	0.001	-7.749	-3.353	0.001
C(20)	-0.009	-0.520	0.604	-0.016	-2.225	0.027	-0.016	-2.225	0.027
C(21)	-0.558	-3.419	0.001	-0.499	-3.514	0.001	-0.499	-3.514	0.001
C(22)	0.069	2.893	0.004	0.064	2.925	0.004	0.064	2.925	0.004
C(23)	0.148	3.850	0.000	0.126	7.839	0.000	0.126	7.839	0.000
C(24)	1.627	2.994	0.003	1.430	3.033	0.003	1.430	3.033	0.003
C(25)	-0.002	-0.565	*0.573*						
C(26)	-0.303	-2.097	0.037	-0.290	-2.087	0.038	-0.290	-2.087	0.038
C(27)	-0.077	-0.545	*0.586*						
Det. Resid. Covariance			0.001831		0.002296			0.001845	

Equation: Y1 = C(10)+C(11)*X1(-1)*Y1(-1)*Y2(-1)+C(12)*X1(-1)*Y1(-1)
 +C(13)*X1(-1)*Y2(-1)+C(14)*Y1(-1)*Y2(-1)+C(15)*X1(-1)+C(16)*Y1(-1)+C(17)*Y2(-1)

R-squared		0.338		0.172		0.337
Adjusted R-squared		0.293		0.132		0.305
S.E. of regression		0.887		0.982		0.879
SSR		80.267		100.383		80.42

Equation: Y2 = C(20)+C(21)*X1(-1)*Y1(-1)*Y2(-1)+C(22)*X1(-1)*Y1(-1)
 +C(23)*X1(-1)*Y2(-1)+C(24)*Y1(-1)*Y2(-1)+C(25)*X1(-1)+C(26)*Y1(-1)+C(27)*Y2(-1)

R-squared		0.505		0.502		0.502
Adjusted R-squared		0.471		0.478		0.478
S.E. of regression		0.052		0.052		0.052
SSR		0.278		0.279		0.279

System: SYS01
Estimation Method: Least Squares
Date: 11/07/11   Time: 09:48
Sample: 1 8 IF M_SIZE=1 AND T=2
Included observations: 110
Total system (balanced) observations 220

- $X1$ has a significant linear effect on $Y1$, based on the t-statistic of $t_0 = 5.562$ with $df = 108$ and a p-value $= 0.000$.
- At the 10% level of significance, each of the interactions $X1*Y1(-1)$, $X1*Y2(-1)$ and $X1*Y1$ $(-1)*Y2(-1)$ has an insignificant linear effect on $Y1$. But based on the reduced Model-1, unexpected findings are obtained too, where each of these interactions has a significant adjusted linear effect on $Y1$. For this reason, this reduced Model-1 can be considered acceptable in a statistical sense to show that the linear effect of $X1$ on $Y1$ depends on $Y1(-1)$, $Y2(-1)$ and $Y1(-1)*Y2(-1)$.
- Another unexpected finding should be noted; the correlation between $Y1$ and $X1*Y1(-1)*$ $Y2(-1)$ has a very large p-value $= 0.9102$. Why can it have a significant adjusted effect on $Y1$ in the reduced Model-1? This finding demonstrates the unknown and unpredictable impacts of multicollinearity between the dependent variables.

5.3 For a comparative study a reduced Model-2 is presented in Table 13.13. This model has a greater adjusted R-squared of 0.278 616 compared to the full model, even though two of the independent variables have been deleted.

Table 13.13 *Summary of statistical results based on an alternative full model, presented at the bottom of the table, and two of its reduced models*

Par	Full Model			Reduced Model-1			Reduced Model-2		
	Coeff.	t-Stat.	Prob.	Coeff.	t-Stat.	Prob.	Coeff.	t-Stat.	Prob.
C(10)	-1.138	-4.069	0.000	0.172	1.210	0.228	-1.115	-4.466	0.000
C(11)	-2.893	-1.196	*0.233*	-7.215	-2.830	0.005	-3.846	-2.896	0.004
C(12)	-0.052	-0.144	*0.885*	0.692	1.856	0.065			
C(13)	1.345	2.837	0.005	1.800	3.444	0.001	1.464	3.536	0.001
C(14)	9.926	1.300	0.195	22.177	2.722	0.007	12.546	2.871	0.005
C(15)	0.313	5.257	**0.000**				0.299	5.969	0.000
C(16)	-0.045	-0.020	*0.984*	-3.130	-1.294	0.197			
C(17)	-4.515	-2.714	0.007	-3.555	-1.918	0.057	-4.541	-2.760	0.006
C(20)	-0.014	-0.946	0.345	-0.014	-2.014	0.045	-0.014	-2.014	0.045
C(21)	-0.620	-4.735	0.000	-0.622	-5.074	0.000	-0.622	-5.074	0.000
C(22)	0.092	4.724	0.000	0.093	5.176	0.000	0.093	5.176	0.000
C(23)	0.102	3.960	0.000	0.102	4.055	0.000	0.102	4.055	0.000
C(24)	1.783	4.313	0.000	1.788	4.564	0.000	1.788	4.564	0.000
C(25)	0.000	0.040	*0.968*						
C(26)	-0.452	-3.727	0.000	-0.453	-3.892	0.000	-0.453	-3.892	0.000
C(27)	0.172	1.908	0.058	0.172	1.934	0.055	0.172	1.934	0.055
Equation: Y1 = C(10)+C(11)*X1*Y1(-1)*Y2(-1)+C(12)*X1*Y1(-1)+C(13)*X1*Y2(-1)									
+C(14)*Y1(-1)*Y2(-1)+C(15)*X1+C(16)*Y1(-1)+C(17)*Y2(-1)									
Adjusted R-squared; Full Model: 0.266628; Reduced Model-1: 0.078786 & Reduced Model-2: 0.278616									
Equation: Y2 = C(20)+C(21)*X1*Y1(-1)*Y2(-1)+C(22)*X1*Y1(-1)+C(23)*X1*Y2(-1)									
+C(24)*Y1(-1)*Y2(-1)+C(25)*X1+C(26)*Y1(-1)+C(27)*Y2(-1)									
Adjusted R-squared; Full Model: 0.534142; & Reduced Model-1 = Reduced Model-2: 0.538614									

```
Covariance Analysis: Ordinary
Date: 01/16/12  Time: 13:12
Sample (adjusted): 2 2
Included observations: 111 after adjustments
Balanced sample (listwise missing value deletion)
```

Correlation t-Statistic Probability	Y1	X1	X1*Y1(-1)	X1*Y2(-1)	X1*Y1(-1)...
Y1	1.000000				

X1	0.470152	1.000000			
	5.561530	-----			
	0.0000	-----			
X1*Y1(-1)	0.038760	0.406334	1.000000		
	0.404966	4.642813	-----		
	0.6863	0.0000	-----		
X1*Y2(-1)	0.143304	0.161526	0.252115	1.000000	
	1.511738	1.708820	2.720021	-----	
	0.1335	0.0903	0.0076	-----	
X1*Y1(-1)*Y2(-1)	0.010827	0.044361	0.428357	0.784672	1.000000
	0.113046	0.463601	4.949242	13.21513	-----
	0.9102	0.6439	0.0000	0.0000	-----

Figure 13.12 *Statistical results based on the correlation analysis*

5.4   These findings have shown that a reduced model can have either greater or smaller adjusted R-squared than its full version.

13.6.2.2 *Application of the Object VAR*

The object VAR can provide two basic VAR type models, namely the unrestricted VAR model, and the *Vector Error Correction* (VEC) model. Refer to more detailed illustrative examples in Agung (2009a).

Example 13.15   Statistical results of an unrestricted VAR model based on (X1,Y1,Y2)

As an alternative data analysis based on the hierarchical model presented in previous example, Figure 13.13 presents the statistical results using the object VAR, specifically the basic unrestricted VAR model of the endogenous variables "y1 y2", with Lag Interval for Endogenous: "1 1", and Exogenous Variables: "*c x1* (−1)∗y1(−1)∗y2(−1) x1(−1)∗y1(−1) x1(−1)∗y2(−1) y1(−1)∗y2(−1) x1(−1)*", based on a sub-sample {M_Size = 1, T = 2}.

Based on these results the following findings and notes are presented.

1. Compared to the SCM, the set of multiple regressions of a VAR model as well as its reduced models should have exactly the same set of independent variables. For instance, *Y1(−1)* has an insignificant effect on *Y1*, but it has a significant effect on *Y2*. A reduced model should be obtained by deleting *Y1(−1)* from the first regression. For this reason, I would recommend applying the object "*System*", instead of the object "VAR".
2. By using the VEC model, an error message "*Insufficient number of observations*" is shown on the screen. Refer to the VAR and VEC models presented in Agung (2009a) and the following example.

Example 13.16   Statistical results of a VEC model based on (X1,Y1,Y2)

For illustration, data analysis using the VEC model based on a larger sub-sample of {M_Size = 1} with statistical results is presented in Figure 13.14. Based on these results the following findings and notes are presented.

Vector Autoregression Estimates		
Date: 11/07/11 Time: 11:40		
Sample: 1 8 IF M_SIZE=1 AND T=2		
Included observations: 110		
Standard errors in ( ) & t-statistics in []		
	Y1	**Y2**
Y1(-1)	-1.077715	-0.303467
	(2.46102)	(0.14472)
	[-0.43791]	[-2.09691]
Y2(-1)	-7.658729	-0.076743
	(2.39466)	(0.14082)
	[-3.19825]	[-0.54498]
C	-1.071786	-0.008545
	(0.27951)	(0.01644)
	[-3.83454]	[-0.51989]
X1(-1)*Y1(-1)*Y2(-1)	-5.280532	-0.557635
	(2.77345)	(0.16309)
	[-1.90396]	[-3.41911]
X1(-1)*Y1(-1)	0.164565	0.068804
	(0.40450)	(0.02379)
	[0.40684]	[2.89252]

X1(-1)*Y2(-1)	2.215453	0.148174
	(0.65456)	(0.03849)
	[3.38467]	[3.84954]
Y1(-1)*Y2(-1)	18.62759	1.627236
	(9.24094)	(0.54342)
	[2.01577]	[2.99446]
X1(-1)	0.289942	-0.002014
	(0.06062)	(0.00356)
	[4.78267]	[-0.56506]
R-squared	0.338096	0.504708
Adj. R-squared	0.292671	0.470717
Sum sq. resids	80.26716	0.277569
S.E. equation	0.887092	0.052166
F-statistic	7.442987	14.84844
Log likelihood	-138.7517	172.9359
Akaike AIC	2.668212	-2.998835
Schwarz SC	2.864610	-2.802436
Mean dependent	0.382290	0.011032
S.D. dependent	1.054771	0.071704
Determinant resid covariance (dof adj.)		0.002130
Determinant resid covariance		0.001831
Log likelihood		34.48727
Akaike information criterion		-0.336132
Schwarz criterion		0.056665

Figure 13.13   *Statistical results of the hierarchical model in Table 13.11 using the object VAR*

1. The output shows a cointegrating equation,

$$CointEq1 = Y(-1) + 3.492330Y2(-1) - 0.957240$$

which has an insignificant negative effect on $D(Y1) = Y1_{i,t} - Y1_{i,t-1}$, and a significant negative effect on $D(Y2)$.

2. Figure 13.14 also presents the equations of the VEC model.
3. Since the exogenous variable or covariate $X1(-1)*Y2(-1)$ has significant effects on both $D(Y1)$ and $D(Y2)$, then a reduced model could be obtained by deleting this interaction. Do it as an exercise.
4. On other hand, even though $D(Y1(-1))$ and $D(Y2(-1))$ have insignificant effects on both $D(Y1)$ and $DY(2)$, they should not be deleted from the model, because "1 1" is used for the lag interval for the endogenous.
5. If these variables would be deleted then "0 0" should be applied for the lag interval for endogenous.

13.6.2.3   *Causal Relationships between Endogenous Variables*

As an extension of the SCM in Table 13.12, and the path diagrams in Figure 13.11(a), Figure 13.15 presents two modified SCMs. Based on these diagrams, the following notes are presented.

Figure 13.15(a) shows that $Y2$ has a direct effect on $Y1$, then referring to the simple model in (13.11b) a simple SCM considered would have the following general equation specification.

```
Vector Error Correction Estimates
Date: 11/07/11   Time: 14:34
Sample: 1 8 IF M_SIZE=1
Included observations: 669
Standard errors in ( ) & t-statistics in [ ]
```

Cointegrating Eq:	CointEq1
Y1(-1)	1.000000
Y2(-1)	3.492330
	(0.66869)
	[5.22266]
C	-0.957240

Error Correction:	D(Y1)	D(Y2)
CointEq1	-0.022334	-0.273096
	(0.02193)	(0.05275)
	[-1.01860]	[-5.17715]
D(Y1(-1))	0.064272	0.011266
	(0.04485)	(0.10790)
	[1.43309]	[0.10441]
D(Y2(-1))	0.001984	-0.001868
	(0.01620)	(0.03896)
	[0.12253]	[-0.04794]
C	-0.100924	-0.169876
	(0.10104)	(0.24308)
	[-0.99885]	[-0.69884]
X1(-1)*Y1(-1)*Y2(-1)	-0.456412	-0.041954
	(0.15403)	(0.37058)
	[-2.96307]	[-0.11321]

X1(-1)*Y1(-1)	0.007561	0.029970
	(0.00374)	(0.00899)
	[2.02367]	[3.33419]
X1(-1)*Y2(-1)	0.000653	-0.014394
	(0.02143)	(0.05156)
	[0.03048]	[-0.27917]
Y1(-1)*Y2(-1)	3.806098	0.394614
	(1.40217)	(3.37338)
	[2.71443]	[0.11698]
X1(-1)	0.034180	0.001130
	(0.01836)	(0.04416)
	[1.86201]	[0.02559]

R-squared	0.075313	0.498209
Adj. R-squared	0.064104	0.492127
Sum sq. resids	405.0665	2344.523
S.E. equation	0.783414	1.884757
F-statistic	6.719356	81.91123
Log likelihood	-781.4402	-1368.751
Akaike AIC	2.363050	4.118836
Schwarz SC	2.423666	4.179452
Mean dependent	0.124902	0.000590
S.D. dependent	0.809799	2.644708

Determinant resid covariance (dof adj.)	2.179815
Determinant resid covariance	2.121560
Log likelihood	-2150.134
Akaike information criterion	6.487696
Schwarz criterion	6.622399

```
Estimation Proc:
=============================
EC(C,1) 1 1 Y1 Y2 @ X1(-1)*Y1(-1)*Y2(-1) X1(-1)*Y1(-1) Y1(-1)*Y2(-1) X1(-1)

VAR Model:
=============================
D(Y1) = A(1,1)*(B(1,1)*Y1(-1) + B(1,2)*Y2(-1) + B(1,3)) + C(1,1)*D(Y1(-1)) + C(1,2)*D(Y2(-1)) + C(1,3) + C(1,4)*X1(-1)*Y1(-1)*Y2(-1) +
C(1,5)*X1(-1)*Y1(-1) + C(1,6)*Y1(-1)*Y2(-1) + C(1,7)*X1(-1)

D(Y2) = A(2,1)*(B(1,1)*Y1(-1) + B(1,2)*Y2(-1) + B(1,3)) + C(2,1)*D(Y1(-1)) + C(2,2)*D(Y2(-1)) + C(2,3) + C(2,4)*X1(-1)*Y1(-1)*Y2(-1) +
C(2,5)*X1(-1)*Y1(-1) + C(2,6)*Y1(-1)*Y2(-1) + C(2,7)*X1(-1)

VAR Model - Substituted Coefficients:
=============================
D(Y1) = - 0.0197943182615*( Y1(-1) + 3.8218158209*Y2(-1) - 0.981275578153 ) + 0.0640071977731*D(Y1(-1)) +
0.0019244700716*D(Y2(-1)) - 0.0986002234325 - 0.456519128303*X1(-1)*Y1(-1)*Y2(-1) + 0.00729199587633*X1(-1)*Y1(-1) +
3.80769447495*Y1(-1)*Y2(-1) + 0.0340846637679*X1(-1)

D(Y2) = - 0.26166115332*( Y1(-1) + 3.8218158209*Y2(-1) - 0.981275578153 ) + 0.0111102355221*D(Y1(-1)) - 0.00194933575546
*D(Y2(-1)) - 0.177746553292 - 0.0242288992469*X1(-1)*Y1(-1)*Y2(-1) + 0.0284522681939*X1(-1)*Y1(-1) + 0.224138707459*Y1(-1)
*Y2(-1) + 0.00310186941094*X1(-1)
```

Figure 13.14 *Statistical results of the hierarchical model in Table 13.11 using a VEC model based on a subsample {M_Size = 1}*

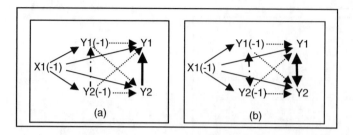

Figure 13.15   *Path diagrams of two special cases of SCMs based on* X1, Y1 *and* Y2

$$
\begin{aligned}
H_1(Y1) = {}& F_{11}(V(-1), Y2) \times D1 + F_{12}(V(-1), Y2) \times D2 \\
& + F_{13}(V(-1), Y2) \times D3 + F_{14}(V(-1), Y2) \times D4 \\
H_2(Y2) = {}& F_{21}(V(-1)) \times D1 + F_{22}(V(-1)) \times D2 \\
& + F_{23}(V(-1)) \times D3 + F_{24}(V(-1)) \times D4
\end{aligned}
\tag{13.30}
$$

where $V(-1) = (X1(-1), Y1(-1), Y2(-1))$.

Figure 13.15(b) shows that *Y1* and *Y2* have simultaneous causal effects, so a simple SCM considered should have the following general equation specification.

$$
\begin{aligned}
H_1(Y1) = {}& F_{11}(V(-1), Y2) \times D1 + F_{12}(V(-1), Y2) \times D2 \\
& + F_{13}(V(-1), Y2) \times D3 + F_{14}(V(-1), Y2) \times D4 \\
H_2(Y2) = {}& F_{21}(V(-1), Y1) \times D1 + F_{22}(V(-1), Y1) \times D2 \\
& + F_{23}(V(-1), Y1) \times D3 + F_{24}(V(-1), Y1) \times D4
\end{aligned}
\tag{13.31}
$$

Refer to the three possible functions of $F_{gk}(*)$ in (13.12), (13.13) and (13.14), then the functions $F_{gk}(*)$ in (13.30) and (13.31) also can be additive, two-way and three-way interaction functions of *(X1(−1), Y1(−1),Y2,Y2(−1))* and *(X1(−1),Y1,Y1(−1),Y2(−1))*: they could even be four-way. So based on the only three variables *X1*, *Y1* and *Y2*, a lot of acceptable models with unpredictable estimates can be obtained based on various data sets. See the following example.

Example 13.17   A SCM based on the path diagram in Figure 13.15(b)
Table 13.14 presents a summary of the statistical results of a full model using the following system equation, and one of its possible reduced models based on a sub-sample {M_Size = 1,t = 2}.

$$
\begin{aligned}
Y1 = {}& C(10) + C(11) * X1(-1) * Y1(-1) + C(12) * X1(-1) * Y2(-1) + C(13) * Y1(-1) \\
& + C(14) * Y2(-1) + C(15) * X1(-1) + \mathbf{C(16)} * \mathbf{X1(-1)} * \mathbf{Y2} + C(17) * \mathbf{Y2}
\end{aligned}
\tag{13.32a}
$$

$$
\begin{aligned}
Y2 = {}& C(20) + C(21) * X1(-1) * Y1(-1) + C(22) * X1(-1) * Y2(-1) + C(23) * Y1(-1) \\
& + C(24) * Y2(-1) + C(25) * X1(-1) + \mathbf{C(26)} * \mathbf{X1(-1)} * \mathbf{Y1} + C(27) * \mathbf{Y1}
\end{aligned}
\tag{13.32b}
$$

In addition to the simultaneous causal effects of *Y1* and *Y2*, the three two-way interaction independent variables are applied in order to test the hypotheses that the linear effect of *X1(−1)* on *Y1* depends on *Y1(−1)*, *Y2* and *Y2(−1)*; and the linear effect of *X1(−1)* on *Y2* depends on *Y1, Y1(−1)* and *Y2(−1)*.

Table 13.14 *Summary of the statistical results of the model (13.29) and its reduced model, based on the sub-sample {M_Size = 1,T = 2}*

System: SYS06											
Estimation Method: Least Squares											
Sample: 1 8 IF M_SIZE=1 AND T=2											
Included observations: 110											
Total system (balanced) observations 220											
	Dependent Variable: Y1						Dependent Variable: Y2				
	Full Model			Reduced Model			Full Model			Reduced Model		
Variable	Coeff.	t-Stat.	Prob.	Coeff.	t-Stat.	Prob.	Coeff.	t-Stat.	Prob.	Coeff.	t-Stat.	Prob.
Intercept	-1.221	-4.462	0.000	-1.126	-4.383	0.000	-0.017	-0.910	0.364	-0.006	-0.860	0.391
X1(-1)*y1(-1)	-0.480	-1.772	0.078	-0.191	-3.051	0.003	0.038	2.184	0.030	0.005	1.239	0.217
x1(-1)*y2(-1)	0.736	1.256	0.210	0.213	1.452	0.148	0.067	2.054	0.041	0.098	3.244	0.001
y1(-1)	1.963	1.116	0.266				-0.245	-2.131	0.034			
y2(-1)	-2.659	-1.047	0.297				0.001	0.004	0.997	-0.166	-1.349	0.179
x1(-1)	0.330	5.905	0.000	0.321	5.921	0.000	0.004	0.917	0.360			
x1(-1)*y2	2.573	2.288	0.023	2.963	3.234	0.001						
y2	-12.456	-2.266	0.025	-14.620	-3.139	0.002						
X1(-1)*y1							0.011	0.899	0.370	0.021	1.842	0.067
y1							-0.086	-0.913	0.362	-0.164	-1.820	0.070
Det. Resid. Cov	0.002094			0.002269								
R-squared	0.344405			0.333864			0.430859			0.393368		
Adjusted R-squared	0.299413			0.301838			0.391801			0.364203		
S.E. of regression	0.882855			0.881326			0.05592			0.057174		
Mean dependent var	0.38229			0.38229			0.011032			0.011032		
S.D. dependent var	1.054771			1.054771			0.071704			0.071704		
Sum squared resid	79.50214			80.78041			0.318955			0.339965		

Based on this summary, the following findings and notes are presented.

1. The process for obtaining the reduced model of *Y2* is an uncommon process, deleting two of the main independent variables: *X1(−1)* and *Y1(−1)* with *p*-values of 0.368 and 0.034, respectively, compared to the variable *Y2(−1)* with a *p*-value = 0.997 because *Y2(−1)* is considered the best predictor for *Y2*. Based on the output of the reduced model the following conclusions can be derived at the 5% level of significance.

   1.1 The main variable *Y1* has a negative significant adjusted effect on *Y2* with a *p*-value = 0.070/2 = 0.035.

   1.2 The interaction *X1(−1)∗Y1* has a positive significant adjusted effect on *Y2* with a *p*-value = 0.067/2 = 0.0335.

   1.3 Even though *X1(−1)∗Y1(−1)* has a large *p*-value of 0.217 > 0.05, the three interactions *X1(−1)∗ Y1(−1)*, *X1(−1)∗Y2(−1)* and *X1(−1)∗Y1* have significant joint effects on *Y2*, based on the Chi-square statistic of 16.18 358 with *df* = 3 and *p*-value = 0.0010.

   1.4 Therefore it can be concluded that the effect of *X1* on *Y2* is significantly dependent on *Y1*, *Y1(−1)* and *Y2(−1)*. In other words, the data supports the hypothesis.

2. On the other hand, based on the full model of *Y1*, the following findings and notes are presented.

   2.1 The three interactions *X1(−1)∗Y1(−1)*, *X1(−1)∗Y2(−1)* and *X1(−1)∗Y2* has a significant joint effects on *Y1* based on the Chi-square statistic of 13.80525 with *df* = 3 and *p*-value = 0.0032.

2.2   At the 5% level of significance, *Y2* has a negative significant adjusted effect on *Y1*, with a *p*-value $= 0.025/2 = 0.0125$, and *X1(−1)*∗*Y2* has a positive significant adjusted effect on *Y2* with a *p*-value $= 0.023/2 = 0.0115$.

2.3   By using a pair comprising the full model of *Y1* and the reduced model of *Y2*, it can be shown that *Y1* and *Y2* already have significant simultaneous causal effects.

3.   The data analyses based on the other groups of the firms, say *{M_Size = k, Time = t}*, can easily be done using the same method. However, we may find unpredictable estimates might be obtained.

14

Applications of GLS Regressions

14.1 Introduction

Corresponding to the applications of all OLS (Ordinary Least Squares) regressions, including the instrumental variables models presented in previous chapters, this chapter presents their extensions and modifications by using the GLS (General Least Squares) estimation method. Based on unstacked panel data analysis, as presented in Chapter 1, a random effects multivariate classical growth model is the worst, because all states are represented by a single growth model only. However, based on special structure panel data as presented in Table 12.3, various types of GLS random effects models, either *Heterogeneous Cross-Section Random Effects Models* (CSREMs) or *Heterogeneous Period Random Effects Models* (PEREMs) can easily be applied. The heterogeneous regressions model is known as the Johnson–Neyman model (1936, quoted by Huitema, 1980). In addition, all OLS fixed-effects models or least squares dummy variable (LSDV) models presented in Chapter 9 and instrumental variables models can also be modified to GLSs.

This chapter also presents various effects models with numerical time-independent variables, as the extension of those presented in previous chapters and Agung (2009a). For comparison, refer to a random effect model with trend and time-related effects (TRE) presented in Wooldridge (2002, 315–317), and the time series models with trend and TRE presented by Bansal (2005). On the other hand, even though Baltagi (2009a) does not present any empirical statistical results of a model with the numerical time variable, he discussed a simple dynamic panel data model with heterogeneous coefficients on the lagged dependent variable and the time trend as presented by Winsbek and Knaap (1999, in Baltagi, 2009a, p. 168), and a random walk model with heterogeneous trend presented by Hadri (2000, in Baltagi 2009a).

14.2 Cross-Section Random Effects Models (CSREMs)

14.2.1 General Equation of CSREM

As an extension or modification of the cross-section random effects model presented in Baltagi (2009a), Gujarati (2003, 648) came up with the unobserved effects model (UEM) presented in Wooldridge (2002, p. 251, Eq. (10.11)); the CSREMs considered are LV(p) CSREMs with the

Panel Data Analysis Using EViews, First Edition. I Gusti Ngurah Agung.
© 2014 John Wiley & Sons, Ltd. Published 2014 by John Wiley & Sons, Ltd.
Companion website: www.wiley.com/go/panel_data

following general equation.

$$Y_{it} = RE(i) + \sum_{j=1}^{p} C(1j)Y_{i,t-j} + \sum_{k=1}^{K} C(2k)Xk_{it} + \mu_{it} \qquad (14.1a)$$

where *RE(i)* is defined as a *random effect (RE)* variable, for $i = 1,\ldots,N$, with the mean or expected value of *RE*, so it can be defined as the following mean model.

$$RE(i) = RE + \varepsilon_i \qquad (14.1b)$$

Therefore, the general LV(*p*) CSREM in (14.1) can be presented as follows:

$$Y_{it} = RE + \sum_{j=1}^{p} C(1j)Y_{i,t-j} + \sum_{k=1}^{K} C(2k)Xk_{it} + \varepsilon_i + \mu_{it} \qquad (14.1c)$$

with the following assumptions

$$\varepsilon_i, i = 1,\ldots,N \text{ have } i.i.d.N(0,\sigma_\varepsilon^2)$$
$$\mu_{it}, i = 1,\ldots,N; t = 1,\ldots,T \text{ have } i.i.d.N(0,\sigma_\mu^2)$$
$$E(\varepsilon_i\mu_{it}) = 0; \quad E(\varepsilon_i\varepsilon_j) = 0, (i \neq j) \qquad (14.1d)$$
$$E(\mu_{is}\mu_{it}) = E(\mu_{it}\mu_{jt}) = E(\mu_{is}\mu_{js}) = 0, (i \neq j; s \neq t)$$

that is, individual error components have normal distributions with zero means and variances σ_ε^2 and σ_μ^2, respectively, they are not correlated with each other and are not autocorrelated across both cross-section and time series units.

Take note there are two error terms in this model, ε_i which are the cross-section, or individual specific; and the time-series error, μ_{it}, which is the combined time-series and cross-section error component. Overall, three errors here are assumed to have zero means and an individual error is not correlated with combination error. As a result of the assumptions stated previously, it follows:

$$E(w_{it}) = E(\varepsilon_i + \mu_{it}) = 0$$
$$\text{var}(w_{it}) = \sigma_\varepsilon^2 + \sigma_\mu^2 \qquad (14.1e)$$
$$\rho(rho) = corr(w_{is}, w_{it}) = \frac{\sigma_\varepsilon^2}{\sigma_\varepsilon^2 + \sigma_\mu^2}$$

Referring to the specific characteristics of the models presented here, a lot of possible random effects models can easily be defined, even based on only three to five variables. Furthermore, all LS regressions with numerical dependent variables presented in previous chapters can be transformed to the CSREMs. However, the following sections and examples present only selected CSREMs, which can easily be extended to more advanced models.

Specific Characteristics

1. The lags $Y_{i,t-j}$, $j = 1,\ldots,p$, must be used as independent variables to anticipate or overcome the auto-correlation problem, where the integer should be selected during the process of the analysis, which is highly dependent on the data used.
2. The main factor, Xk_{it} can be several types of variable, such as
   - a numerical variable (exogenous or another endogenous variable),
   - a lag of the numerical independent variables,
   - a transformed variable of the numerical variable,

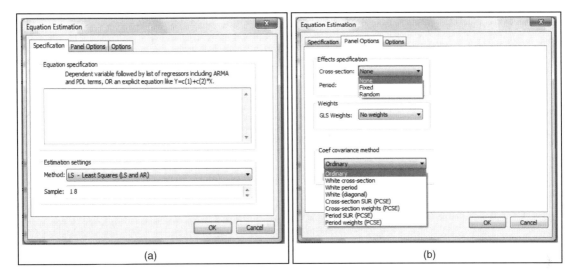

Figure 14.1   *The options for doing GLS regression analysis based on panel data*

- a dummy variable,
- an environmental variable, namely $Xk_{it} = Xk_t, \forall i$.
- the numerical time t.

3. The models that should be applied are those with selected interactions of the main factors indicated in point (2), because, in general, the effect of a main factor on the dependent variable depends on at least one of the other main factors, more so for the models with categorical independent variable(s) including the interaction between $Y_{i,t-j}$ and other exogenous variables. Refer to all two-way and higher interaction models presented in previous chapters, as well as the following examples.

4. Similar models can easily be defined using the transformed endogenous variable, such as $log(Y_{it})$, $log(Y_{it}-L_i)$, $log(U_i-Y_{it})$, or, $log[(Y_{it}-L_i)/(U_i-Y_{it})]$, where L_i and U_i are the fixed lower and upper bound of Y_{it}, at the time-point t, as well as the transformation of Xk.

14.2.2   Estimation Method

The steps of the GLS estimation method for univariate linear models are as follows:

1. With the work file on the screen, click *Quick/Estimation Equation* . . . , the block in Figure 14.1(a) is shown on the screen. Then the equation estimation can be entered.

2. By clicking the *Panel Options*, the options in Figure 14.1(b) can be selected, specifically the *Random Effects Specification*, and a *Coef Covariance Method*.

3. In addition, by clicking the *Options*, there are three additional options of the *Random Effects Methods*; namely *Swamey–Arora, Wallace–Hussain* and *Wansbeek–Kaptyen*, and these are shown on the screen. Note that which one would be the best possible good fit cannot be predicted.

14.3   LV(1) CSREMs by Group or Time

This section presents a comparative study between selected simple LV(1) CSREMs, which can be acceptable, not-recommended, or the worst CSREM. Refer to the heterogeneous regressions, homogeneous regressions or ANCOVA models and the worst models presented in previous chapters. We expect these simple CSREMs are a good guide for readers to develop more advanced CSREMs.

14.3.1   ANOVA CSREMs

Example 14.1   Application of ANOVA CSREMs of lnSALE
Figure 14.2 presents statistical results based on two simple ANOVA CSREMs. Based on these results the following notes and comments are presented.

1. The results in Figure 14.2 are obtained by using the following equation specification (ES), respectively,

$$lnSale \quad C \ @Expand(M\_Size, TP, @Dropfirst) \tag{14.2a}$$

$$lnSale \quad C \ @Expand(Q\_Size, Dropfirst) \tag{14.2b}$$

where M_Size and Q_Size are the based groups at the time $T = 1$, which are invariant over times, generated based on the variable SIZE as presented in Part II.

2. The very small DW-statistics values indicate that both models have autoregressive problems. So the lagged variables model should be applied as has been indicated in the general CSREM in (14.1).

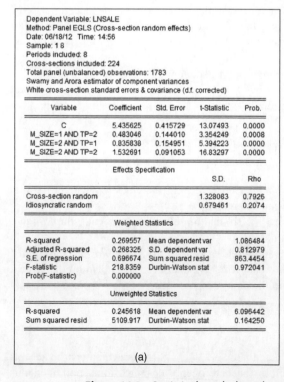

(a)   (b)

Figure 14.2   *Statistical results based on the CSREM models in (14.2a), and (14.2b)*

Cross-section Random Effects

	FIRM_CODE	Effect
1	_AALI	1.608005
2	_ABBA	-1.594031
3	_ADES	-0.769584
4	_ADMG	1.371572
5	_AIMS	-0.999942
6	_AISA	-0.662918
7	_AKPI	0.376148
8	_AKRA	1.576285
9	_ALFA	1.472070
10	_ALKA	0.865697

(a)

	RE_1
Mean	1.04E-15
Median	-0.050718
Maximum	11.79110
Minimum	-5.701472
Std. Dev.	1.048862
Skewness	4.263854
Kurtosis	41.49346
Jarque-Bera	72541.84
Probability	0.000000
Sum	1.20E-12
Sum Sq. Dev.	1231.026
Observations	1120

(b)

Figure 14.3   *List of cross-section random effects of the CSREM in Figure 14.2(a)*

3. The results show the estimates of two *Rho*-statistics, namely the *Cross section random* and *Idiosyncratic random* with a total of 1. Refer to the formula in (14.1e), which indicates the proportion of σ_ε^2 in the var $(w_{it}) = \sigma_\varepsilon^2 + \sigma_\mu^2$.

4. By selecting *View/Fixed/Random Effects* . . . *Cross-section Effects*, the list of Cross-section Random Effects of the CSREM in Figure 14.3(a), namely RE_1, and its statistics are obtained as presented in Figure 14.3(b). Based on these results the following findings and notes are presented.

4.1 Referring to the mean model of *RE(i) = REi* in (4.1b), we obtain RE = 1.04E-15 → RE = 0.000 000. Since *RE = 0.000 000*, then the equation of the regression in Figure 14.3(a), can be presented as follows:

$$LNSALE = 5.435624 + 0.483046 * D12 + 0.835838 * D21 + 1.532691 * D22$$

where *Dij* is the zero-one indicator of $(M\_Size = i, TP = j)$. However, by selecting the option *Representative*, the regression function is presented as follows:

$$LNSALE = 5.435624 + 0.483046 * D12 + 0.835838 * D21 + 1.532691 * D22 + [CX = R]$$

where $[CX = R]$ should indicate that the model is a CSREM.

4.2 For comparison, the following function is a LS regression having the Adjusted $R^2 = 0.361\,017$, compare to the Adjusted $R^2 = 0.268\,325$ for the CSREM above. Which one would you prefer? Do the analysis for a cross-section fixed effects model as an exercise.

$$LNSALE = 4.890289 + 0.1612521 * D12 + 2.070398 * D21 + 2.714367 * D22$$

5. These basic models are acceptable models in a statistical sense, even though they have small DW-statistics. Each model will have some limitations or hidden assumptions, which should be noted by all researchers. For a comparison, various CSREMs presented in Baltagi (2009) have small DW-statistics. For instance, the CSREM in Table 2.4 (Baltagi, 2009a, p. 20) has a DW-statistic of 0.026 675 for the unweighted statistic. To anticipate the small DW-statistic, the models should be modified to the lagged CSREM presented in the following example.

6. The CSREM in Figure 14.3(b) has similar characteristics. So these are not presented again.

Dependent Variable: LNSALE
Method: Panel EGLS (Cross-section random effects)
Date: 06/22/12   Time: 13:56
Sample (adjusted): 2 8
Periods included: 7
Cross-sections included: 224
Total panel (unbalanced) observations: 1556
Swamy and Arora estimator of component variances
White cross-section standard errors & covariance (d.f. corrected)
WARNING: estimated coefficient covariance matrix is of reduced rank

Variable	Coefficient	Std. Error	t-Statistic	Prob.
C	0.348134	0.275805	1.262247	0.2071
M_SIZE=1 AND TP=2	0.680546	0.356459	1.909182	0.0564
M_SIZE=2 AND TP=1	0.052777	0.267199	0.197520	0.8434
M_SIZE=2 AND TP=2	0.492919	0.320758	1.536735	0.1246
LNSALE(-1)*(M_SIZE=1 AND TP=1)	0.952514	0.040576	23.47458	0.0000
LNSALE(-1)*(M_SIZE=1 AND TP=2)	0.809716	0.028012	28.90629	0.0000
LNSALE(-1)*(M_SIZE=2 AND TP=1)	0.973908	0.016420	59.31177	0.0000
LNSALE(-1)*(M_SIZE=2 AND TP=2)	0.907200	0.029833	30.40884	0.0000

Effects Specification

	S.D.	Rho
Cross-section random	0.000000	0.0000
Idiosyncratic random	0.576572	1.0000

Weighted Statistics

R-squared	0.893516	Mean dependent var	6.190340
Adjusted R-squared	0.893035	S.D. dependent var	1.949122
S.E. of regression	0.637471	Sum squared resid	629.0604
F-statistic	1855.631	Durbin-Watson stat	2.173326
Prob(F-statistic)	0.000000		

Unweighted Statistics

R-squared	0.893516	Mean dependent var	6.190340
Sum squared resid	629.0604	Durbin-Watson stat	2.173326

Dependent Variable: LNSALE
Method: Panel EGLS (Cross-section random effects)
Date: 10/11/12   Time: 09:40
Sample (adjusted): 2 8
Periods included: 7
Cross-sections included: 224
Total panel (unbalanced) observations: 1556
Swamy and Arora estimator of component variances
White cross-section standard errors & covariance (d.f. corrected)
WARNING: estimated coefficient covariance matrix is of reduced rank

Variable	Coefficient	Std. Error	t-Statistic	Prob.
C	0.611199	0.236304	2.586489	0.0098
Q_SIZE=2	-0.258320	0.334885	-0.771369	0.4406
Q_SIZE=3	-0.264607	0.179042	-1.477905	0.1396
Q_SIZE=4	0.347964	0.516967	0.673088	0.5010
Q_SIZE=5	-0.399843	0.177796	-2.248882	0.0247
LNSALE(-1)*(Q_SIZE=1)	0.931336	0.032528	28.63192	0.0000
LNSALE(-1)*(Q_SIZE=2)	0.948034	0.054609	17.36046	0.0000
LNSALE(-1)*(Q_SIZE=3)	0.958333	0.021556	44.45787	0.0000
LNSALE(-1)*(Q_SIZE=4)	0.877465	0.074246	11.81840	0.0000
LNSALE(-1)*(Q_SIZE=5)	0.990101	0.015467	64.01475	0.0000

Effects Specification

	S.D.	Rho
Cross-section random	0.000000	0.0000
Idiosyncratic random	0.609499	1.0000

Weighted Statistics

R-squared	0.889935	Mean dependent var	6.190340
Adjusted R-squared	0.889294	S.D. dependent var	1.949122
S.E. of regression	0.648521	Sum squared resid	650.2168
F-statistic	1388.915	Durbin-Watson stat	2.264305
Prob(F-statistic)	0.000000		

Unweighted Statistics

R-squared	0.889935	Mean dependent var	6.190340
Sum squared resid	650.2168	Durbin-Watson stat	2.264305

Figure 14.4   *Statistical results based on the CSREM models in (14.3a), and (14.3b)*

14.3.2   Simple LV(1) CSREMs by Groups

Example 14.2   The simplest LV(1) CSREM of lnSALE by groups

As an extension of the CSREMs in (4.2), the two following ESs represent two LV(1) CSREMs with the statistical results presented in Figure 14.4. Based on these results the following findings and notes are presented.

$$lnSale \quad C \; @Expand(M\_Size, TP, @dropfirst) \; lnSale(-1) * @Expand(M\_Size, TP) \qquad (14.3a)$$

and

$$lnSale \quad C \; @Expand(Q\_Size, @dropfirst) \; lnSale(-1) * @Expand(Q\_Size) \qquad (14.3b)$$

1. These models present heterogeneous regressions of *lnSale* on *lnSale(−1)* by the defined groups of the firms.
2. The objectives of analysis using this model are to test the linear effects differences of lnSale(−1) on lnSale between the cells generated by M_Size and TP. We find the null hypothesis H_0: $C(5) = C(6) = C(7) = C(8)$ is rejected based on the F-statistic of $F_0 = 22.71072$ with $df = (3,1548)$ and a p-value $= 0.0000$. This conclusion indicates that the model cannot be reduced to an ANCOVA or fixed-effects model having lnSale(−1) as a numerical independent variable.
3. For additional discussions, Table 14.1 presents the parameters of the model in Figure 14.3(a).

Table 14.1 *Summary of the statistical results in Figure 14.3(a)*

		Constant				lnSale(-1)			
M_Size	TP	Parameter	Coef.	t-Stat.	Prob.	Parameter	Coef.	t-Stat.	Prob.
1	1	C(1)	0.34813	1.59208	0.11160	C(5)	0.95251	23.15489	0.00000
1	2	C(2)	0.68055	1.39470	0.16330	C(6)	0.80972	10.11516	0.00000
2	1	C(3)	0.05278	0.21424	0.83040	C(7)	0.97391	64.07554	0.00000
2	2	C(4)	0.49292	1.16820	0.24290	C(8)	0.90720	20.05087	0.00000
Weighted Statistics									
R-squared			0.89352	Durbin-Watson stat			2.17333		
Adjusted R-squared			0.89304	F-statistic			1855.631		
S.E. of regression			0.63747	Prob(F-statistic)			0.00000		

3.1 Take note that C(1) is the intercept of the simple linear regression in the cell (M_Size = 1, TP = 1), and the intercepts of the other regressions are C(1) + C(2), C(1) + C(3) and C(1) + C(4), respectively. For example, the equations of the first two regressions are as follows:

$$\ln SALE = 0.34813 + 0.95251 \times \ln SALE(-1) + [CX = R], \text{in (M_Size} = 1, \text{TP} = 1),$$

and

$$\ln SALE = (0.34813 + 0.68055) + 0.80972 \times \ln SALE(-1) + [CX = R], \text{in (M_Size} = 1, \text{TP} = 2)$$

3.2 For this model, unexpected output is obtained of $RE(i) = 0.000\,000$ for all $i = 1, \ldots, N$, as presented in Figure 14.5. Similarly this is the case for the model in Figure 14.3(b). I have also found this for many other CSREMs as presented in the following examples.

3.3 Since C(3) has a very large p-value $= 0.83\,040$, do you want to reduce the model? Do it as an exercise. If it should be deleted, then the two regressions in (M_Size = 1,TP = 1) and (M_Size = 2,TP = 1) would have the same intercept.

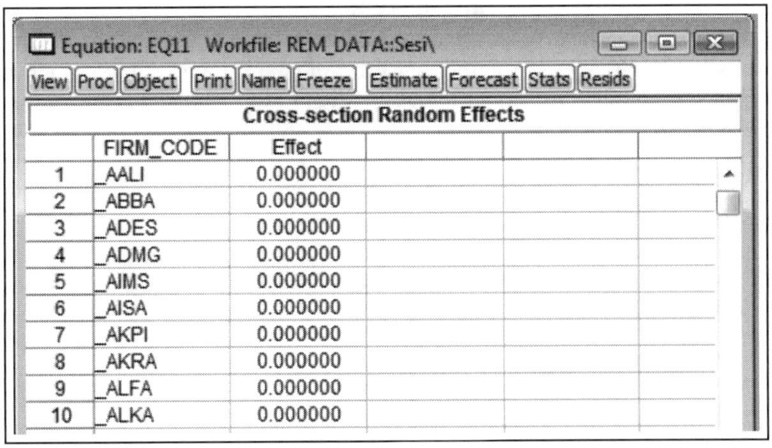

	FIRM_CODE	Effect
1	_AALI	0.000000
2	_ABBA	0.000000
3	_ADES	0.000000
4	_ADMG	0.000000
5	_AIMS	0.000000
6	_AISA	0.000000
7	_AKPI	0.000000
8	_AKRA	0.000000
9	_ALFA	0.000000
10	_ALKA	0.000000

Figure 14.5 *The first 10 of the cross-section random effects of the model in Figure 14.3(a)*

14.3.3   LV(1) CSREM by Time

If the individuals can be considered as a single or homogenous group, then the LV(1) CSREM by time would be an acceptable model, which can be represented using the following ES, for $T > 1$.

$$lnSale \quad C \ @Expand(T, @dropfirst) \ lnSale(-1) * @Expand(T) \tag{14.4}$$

Note that this model is a heterogeneous CSREM, which shows the linear effects of $lnSale(-1)$ on $lnSale$ changing over time. Do it as an exercise, for any variable Y_{it} and test the effects differences of $Y(-1)$ on Y, using the Wald test.

For heterogeneous groups of individuals, then the suggested CSREM is as follows:

$$lnSale \quad C \ @Expand(Group, T, @dropfirst) \ lnSale(-1) * @Expand(Group, T) \tag{14.5}$$

14.3.4   Not-Recommended or Worst LV(1) CSREMs

Homogeneous CSREMs or ANCOVA LV(1) CSREMs were applied in many studies, using the following ES. However, I would not recommend this model, since the linear effect of $lnSale(-1)$ on $lnSale$ should change over group and time t or the time period (TP) in general.

$$lnSale \quad C \ lnSale(-1)@Expand(Group, T, @dropfirst) \tag{14.6}$$

In addition to this CSREM or ANCOVA model by $Group \times Time$, the following ES also presents an CSREM or ANCOVA model by $Group + Time$, which is not recommended because it has additional hidden assumptions on intercepts by $Group$ and the time $t > 1$ and because it has a special pattern of intercepts.

$$lnSale \quad C \ lnSale(-1)@Expand(Group, @Dropfirst)@Expand(T, @Dropfirst) \tag{14.7}$$

Finally, the following ES presents an even worse LV(1) CSREM, because the scores of all individuals at all the time points (or all individual time observations) are presented only by a single regression function.

$$lnSale \quad C \ lnSale(-1) \tag{14.8}$$

14.4   CSREMs with the Numerical Time Variable

This section presents CSREMs with the numerical time independent variable, which are derived from selected LS regressions presented in previous chapters, especially Chapter 12. Refer to the notes on similar models presented in the introduction.

14.4.1   CSREMs with Trend

Example 14.3   An CSREM of the AR(p) multivariate growth model in Figure 1.8

Note that Figure 1.8 presents a set of four AR(p) growth models by the time period TP, then as its modification, Figure 14.6 presents the statistical results of a LV(1) CSREM with Trend by *M_Size* and *TP* using the ES as follows:

$$\begin{aligned} &lnsale \quad c \ @expand(m\_size, tp, @dropfirst) \\ &lnsale(-1) * @expand(m\_size, tp) \ t * @expand(m\_size, tp) \end{aligned} \tag{14.9}$$

Dependent Variable: LNSALE
Method: Panel EGLS (Cross-section random effects)
Date: 07/03/12   Time: 13:15
Sample (adjusted): 2 8
Periods included: 7
Cross-sections included: 224
Total panel (unbalanced) observations: 1556
Swamy and Arora estimator of component variances
White diagonal standard errors & covariance (d.f. corrected)

Variable	Coefficient	Std. Error	t-Statistic	Prob.
C	0.339881	0.218305	1.556906	0.1197
M_SIZE=1 AND TP=2	1.875818	0.618933	3.030729	0.0025
M_SIZE=2 AND TP=1	-0.067188	0.244312	-0.275010	0.7833
M_SIZE=2 AND TP=2	1.242996	0.323663	3.840402	0.0001
LNSALE(-1)*(M_SIZE=1 AND TP=1)	0.952431	0.041387	23.01275	0.0000
LNSALE(-1)*(M_SIZE=1 AND TP=2)	0.819733	0.080159	10.22630	0.0000
LNSALE(-1)*(M_SIZE=2 AND TP=1)	0.966456	0.015828	61.05938	0.0000
LNSALE(-1)*(M_SIZE=2 AND TP=2)	0.911264	0.047059	19.36414	0.0000
T*(M_SIZE=1 AND TP=1)	0.002541	0.029142	0.087204	0.9305
T*(M_SIZE=1 AND TP=2)	-0.177812	0.056543	-3.144751	0.0017
T*(M_SIZE=2 AND TP=1)	0.049705	0.016881	2.944521	0.0033
T*(M_SIZE=2 AND TP=2)	-0.109940	0.041586	-2.643671	0.0083

Effects Specification		S.D.	Rho
Cross-section random		0.000000	0.0000
Idiosyncratic random		0.562341	1.0000

Weighted Statistics			
R-squared	0.895238	Mean dependent var	6.190340
Adjusted R-squared	0.894492	S.D. dependent var	1.949122
S.E. of regression	0.633114	Sum squared resid	618.8863
F-statistic	1199.475	Durbin-Watson stat	2.181539
Prob(F-statistic)	0.000000		

Unweighted Statistics			
R-squared	0.895238	Mean dependent var	6.190340
Sum squared resid	618.8863	Durbin-Watson stat	2.181539

Figure 14.6   *Statistical results of the LV(1) CSREM with trend in (14.9)*

Based on these results the following notes and comments are presented.

1. Compared to the set of four AR(1) Growth models in Figure 1.8, the model in Figure 14.6 also presents a set of four LV(1) models with trend, which is represented by a LV(1) CSREM with trend.
2. Take note that this model also can be considered an extension of the LV(1) CSREM in Figure 14.3. A question should arise, *why do both results show the Rho* $= 0$? Several other models also present the Rho $= 0$.
3. Furthermore, take note that all AR(p) growth models also could easily be modified to the LV(p) CSREMs, based on structured panel data. Do this as an exercise.

14.4.2   CSREMs with Time-Related Effects (TRE)

Example 14.4   The CSREM of the LV(1) model (12.15)

Table 14.2 presents the statistical results of the CSREM of the LV(1) model in (12.15), and one of its reduced models. Compared to LS regression in Table 12.1, the following notes and comments are presented.

1. The parameter estimates of the model in (12.15) show that each of the independent variables has either positive or negative adjusted effects at the $\alpha = 0.10$ level of significance. Note that t*lnSale(−1)* (M_Size $= 2$, TP $= 1$) has the greatest Prob. of 0.1248, which indicates that it has a significant negative effect with a *p*-value $= 0.1248/2 = 0.0624 < \alpha = 0.10$.
2. The reduced model is obtained by deleting the variable *lnSale(−1)*(M_Size $= 1$, TP $= 1$), which has a *p*-value $= 0.0000$ to demonstrate that an acceptable reduced model cannot be obtained by deleting an independent variable with a very large or largest *p*-value, which has been presented in previous chapters for the LS regressions. However, other acceptable reduced models in a statistical sense can be obtained. Do it as an exercise.
3. Note that two of the independent variables of the full CSREM in Table 14.2 have very large Prob. of 0.3702 and 0.5293. So it can be said that the LS regression is better than the full CSREM or the GLS regression in a statistical sense.
4. Note that the output shows a *WARNING*. So we could try other method(s), as presented in Figure 14.1. However, we cannot predict which one would give the best fit. See the following example.

Table 14.2   *Statistical results based on the CSREM of the LV(1) model in 12.15 and its reduced model*

	Full REM			Reduced REM		
Dependent Variable: LNSALE						
Method: Panel EGLS (Cross-section random effects)						
Date: 06/18/12 Time: 14:59						
Sample (adjusted): 2 8						
Periods included: 7						
Cross-sections included: 224						
Total panel (unbalanced) observations: 1556						
Swamy and Arora estimator of component variances						
White cross-section standard errors & covariance (d.f. corrected)						
WARNING: estimated coefficient covariance matrix is of reduced rank						
Variable	Coef.	t-Stat.	Prob.	Coef.	t-Stat.	Prob.
C	0.96019	1.12368	0.26130	4.95426	20.37539	0.00000
M_SIZE=1 AND TP=2	3.36309	3.43358	0.00060	-0.63098	-1.17544	0.24000
M_SIZE=2 AND TP=1	-1.39789	-2.59429	0.00960	-5.39195	-27.35967	0.00000
M_SIZE=2 AND TP=2	-2.69146	-3.10953	0.00190	-6.68552	-23.92117	0.00000
LNSALE(-1)*(M_SIZE=1 AND TP=1)	0.82507	6.53470	0.00000			
LNSALE(-1)*(M_SIZE=1 AND TP=2)	0.39733	6.35042	0.00000	0.39733	6.35248	0.00000
LNSALE(-1)*(M_SIZE=2 AND TP=1)	1.07058	28.77049	0.00000	1.07058	28.77983	0.00000
LNSALE(-1)*(M_SIZE=2 AND TP=2)	1.35630	40.32446	0.00000	1.35630	40.33755	0.00000
T*(M_SIZE=1 AND TP=1)	-0.18596	-0.62914	*0.52930*	-1.28411	-9.41794	0.00000
T*(M_SIZE=1 AND TP=2)	-0.48587	-6.58476	0.00000	-0.48587	-6.58690	0.00000
T*(M_SIZE=2 AND TP=1)	0.25177	1.66416	0.09630	0.25177	1.66470	0.09620
T*(M_SIZE=2 AND TP=2)	0.35601	16.69726	0.00000	0.35601	16.70268	0.00000
T*LNSALE(-1)*(M_SIZE=1 AND TP=1)	0.03855	0.89629	*0.37020*	0.26412	8.85928	0.00000
T*LNSALE(-1)*(M_SIZE=1 AND TP=2)	0.06146	6.23948	0.00000	0.06146	6.24151	0.00000
T*LNSALE(-1)*(M_SIZE=2 AND TP=1)	-0.02925	-2.27127	0.02330	-0.02925	-2.27201	0.02320
T*LNSALE(-1)*(M_SIZE=2 AND TP=2)	-0.06247	-14.71500	0.00000	-0.06247	-14.71978	0.00000
R-squared	0.89642			0.88436		
Adjusted R-squared	0.89541			0.88331		
F-statistic	888.474			841.760		
Prob(F-statistic)	0.00000			0.00000		
Durbin-Watson stat	2.16730			2.04955		

Example 14.5   An alternative GLS regression

With regards to the warning previously, I tried applying another option, namely the *cross-section SUR (PCSE) standard errors and covariance (d.f. corrected)*, with the results presented in Figure 14.7. Based on these results the following findings and notes are presented.

1. Even though eight out of 12 numerical independent variables have large *p*-values, say greater than 0.30, it can be said the model is an acceptable model in theoretical sense for showing that the effect of *lnSale* (−1) on *lnSale* by *M_Size* and *TP* depends on the time *t*.
2. However, in a statistical sense, a reduced CSREM should be explored. How can we reduce the model because many independent variables have large *p*-values? Note that two of the intercepts also have *p*-values > 0.30. We find that the manual stepwise selection method is the best. Table 14.3 presents the

```
Dependent Variable: LNSALE
Method: Panel EGLS (Cross-section random effects)
Date: 06/25/12  Time: 07:34
Sample (adjusted): 2 8
Periods included: 7
Cross-sections included: 224
Total panel (unbalanced) observations: 1556
Swamy and Arora estimator of component variances
Cross-section SUR (PCSE) standard errors & covariance (d.f. corrected)
```

Variable	Coefficient	Std. Error	t-Statistic	Prob.
C	0.960192	0.650481	1.476125	0.1401
M_SIZE=1 AND TP=2	3.363085	3.886648	0.865292	0.3870
M_SIZE=2 AND TP=1	-1.397888	0.876340	-1.595144	0.1109
M_SIZE=2 AND TP=2	-2.691457	3.050481	-0.882306	0.3777
LNSALE(-1)*(M_SIZE=1 AND TP=1)	0.825072	0.119726	6.891321	0.0000
LNSALE(-1)*(M_SIZE=1 AND TP=2)	0.397326	0.723839	0.548914	0.5831
LNSALE(-1)*(M_SIZE=2 AND TP=1)	1.070576	0.133272	8.033037	0.0000
LNSALE(-1)*(M_SIZE=2 AND TP=2)	1.356295	0.383784	3.534006	0.0004
T*(M_SIZE=1 AND TP=1)	-0.185960	0.207939	-0.894304	0.3713
T*(M_SIZE=1 AND TP=2)	-0.485868	0.555248	-0.875047	0.3817
T*(M_SIZE=2 AND TP=1)	0.251766	0.257672	0.977080	0.3287
T*(M_SIZE=2 AND TP=2)	0.356012	0.425563	0.836568	0.4030
T*LNSALE(-1)*(M_SIZE=1 AND TP=1)	0.038549	0.039505	0.975798	0.3293
T*LNSALE(-1)*(M_SIZE=1 AND TP=2)	0.061456	0.104331	0.589053	0.5559
T*LNSALE(-1)*(M_SIZE=2 AND TP=1)	-0.029246	0.035855	-0.815688	0.4148
T*LNSALE(-1)*(M_SIZE=2 AND TP=2)	-0.062472	0.055151	-1.132753	0.2575

Effects Specification		S.D.	Rho
Cross-section random		0.000000	0.0000
Idiosyncratic random		0.560022	1.0000

Weighted Statistics			
R-squared	0.896416	Mean dependent var	6.190340
Adjusted R-squared	0.895407	S.D. dependent var	1.949122
S.E. of regression	0.630363	Sum squared resid	611.9313
F-statistic	888.4744	Durbin-Watson stat	2.167301
Prob(F-statistic)	0.000000		

Unweighted Statistics			
R-squared	0.896416	Mean dependent var	6.190340
Sum squared resid	611.9313	Durbin-Watson stat	2.167301

Figure 14.7   *Statistical results of the full CSREM in Table 14.2 using the PCSE option*

Table 14.3   *Statistical results of the first two stages of data analysis based on the model in Figure 14.7*

Dependent Variable: LNSALE						
Method: Panel EGLS (Cross-section random effects)						
Periods included: 7						
Total panel (unbalanced) observations: 1556						
Swamy and Arora estimator of component variances						
Cross-section SUR (PCSE) standard errors & covariance (d.f. corrected)						
	Stage-1			Stage-2		
Variable	Coef.	t-Stat.	Prob.	Coef.	t-Stat.	Prob.
C	2.7895	4.2268	0.0000	4.9543	14.7772	0.0000
M_SIZE=1 AND TP=2	-1.1206	-0.8756	0.3814	1.3423	1.1600	0.2462
M_SIZE=2 AND TP=1	1.9504	5.7608	0.0000	1.8237	10.6152	0.0000
M_SIZE=2 AND TP=2	0.1211	0.0727	0.9421	3.4000	3.3901	0.0007
T*LNSALE(-1)*(M_SIZE=1 AND TP=1)	0.1289	3.6745	0.0002	0.2641	9.1639	0.0000
T*LNSALE(-1)*(M_SIZE=1 AND TP=2)	0.0975	3.6068	0.0003	0.1185	8.9032	0.0000
T*LNSALE(-1)*(M_SIZE=2 AND TP=1)	0.0931	2.6937	0.0071	0.2433	7.4259	0.0000
T*LNSALE(-1)*(M_SIZE=2 AND TP=2)	0.0895	2.8941	0.0039	0.1255	7.1108	0.0000
T*(M_SIZE=1 AND TP=1)				-1.2841	-7.0544	0.0000
T*(M_SIZE=1 AND TP=2)				-0.7703	-4.3931	0.0000
T*(M_SIZE=2 AND TP=1)				-1.6060	-6.5334	0.0000
T*(M_SIZE=2 AND TP=2)				-1.0440	-5.4848	0.0000
	Weighted Statistics					
R-squared	0.6797			0.8655		
Adjusted R-squared	0.6782			0.8645		
F-statistic	469.275			903.158		
Prob(F-statistic)	0.0000			0.0000		
Durbin-Watson stat	1.1983			2.0288		

statistical results of the first two stages of the data analysis. Based on these results the following notes and comments are made.

2.1 At the first stage, the following list of independent variables is inserted:

$$C \ @Expand(M\_Size, TP, @Dropfirst) \ t * lnsale(-1) * @Expand(M\_Size.TP)$$

Unexpected results are obtained where all numerical independent variables have very small p-values. So we do not have a problem with the numerical independent variables. Even though one has a large p-value, say greater than 0.30, the interaction should not be deleted, since it is defined that the effect of $lnSale(-1)$ on $lnSale$ depends on the time t.

2.2 However, one of the three intercepts differences has a very large p-value $= 0.9421$ and another has a p-value $= 0.3814$, indicated by the parameters C(4) and C(2), respectively. So we may delete C(4) or C(4) and C(2). In this case I will not reduce the model.

2.3 At the second stage, we have two choices; to enter either $lnSale(-1)*@Expand(M\_Size, TP)$, or $t*@Expand(M\_Size, TP)$. In this case, we find a better fit model by entering $t*@Expand(M\_Size, TP)$. An unexpected result is obtained, where all numerical independent variables have very small p-values, the p-value of C(4) decreases from 0.9421 to 0.0007 and the p-value of C(2) decreases from 0.3814 to $0.2462 < 0.30$.

2.4 At the third stage, if the variable $lnSale(-1)*@Expand(M\_Size, TP)$ is entered, then the results in Figure 14.7 are obtained, where all interactions $t*lnSale(-1)$ by M_Size and TP have very large p-values.

2.5 So the last CSREM in Table 14.3 should be considered the best fit model.

Example 14.6 The CSREM of the LV(1) model in (12.18)

Table 14.4 presents the statistical results of the full CSREM of the LV(1) model in (12.18) and one of its reduced models. Based on the results in this table, the following findings and notes are presented.

1. The output presents an acceptable reduced CSREM; however, it shows the *WARNING*. For this reason I applied another method, namely the *white diagonal standard errors and covariance (d.f. corrected)*, instead of the cross-section SUR applied earlier.

2. The full CSREM obtained shows no *WARNING*, but many independent variables have large p-values. In a statistical sense it is an acceptable CSREM with the Adjusted R^2 of 0.899 357 and a DW-statistic of 2.105 874.

3. In fact a reduced model can be explored using the manual stepwise selection method (MSSM). The statistical results of the first stage of the analysis are presented in Figure 14.8. Note that all interactions have very small p-values so the model is acceptable in both theoretical and statistical senses, since it can show that the joint effects of $lnSale(-1)$ and $lnSize$ are significantly dependent on the time t.

4. For the second stage, choose any one of the variables $lnSale(-1)*@Expand(M\_Size, TP)$, $lnSize*@Expand(M\_Size, TP)$ and $t*@Expand(M\_Size, TP)$ in order to obtain the best possible model.

5. At the third stage there are two possible choices of variables, which were not selected at the second stage and then followed by the final stage. Do it as an exercise.

Table 14.4 *Statistical results of the CSREM of the model in (12.18) and one of its reduced CSREMs*

Dependent Variable: LNSALE
Method: Panel EGLS (Cross-section random effects)
Date: 06/18/12 Time: 15:03
Sample (adjusted): 2 8
Periods included: 7
Cross-sections included: 224
Total panel (unbalanced) observations: 1555
Swamy and Arora estimator of component variances
White cross-section standard errors & covariance (d.f. corrected)
WARNING: estimated coefficient covariance matrix is of reduced rank

Variable	Full REM			Reduced REM		
	Coef	t-Stat.	Prob.	Coef	t-Stat.	Prob.
C	1.56800	1.46665	0.143	4.86775	18.98406	0.000
M_SIZE=1 AND TP=2	0.91953	0.82039	0.412	-2.38022	-5.62854	0.000
M_SIZE=2 AND TP=1	-1.34350	-1.05851	0.290	-5.36504	-25.71683	0.000
M_SIZE=2 AND TP=2	1.18469	0.54500	0.586	-2.11506	-1.10846	0.268
LNSALE(-1)*(M_SIZE=1 AND TP=1)	0.83449	7.09692	0.000			
LNSALE(-1)*(M_SIZE=1 AND TP=2)	0.07304	1.82333	0.068	0.07304	1.82511	0.068
LNSALE(-1)*(M_SIZE=2 AND TP=1)	1.09973	17.00867	0.000	1.07536	28.49638	0.000
LNSALE(-1)*(M_SIZE=2 AND TP=2)	1.46510	32.51432	0.000	1.46510	32.54616	0.000
LNSIZE*(M_SIZE=1 AND TP=1)	-0.41565	-1.81645	0.070			
LNSIZE*(M_SIZE=1 AND TP=2)	2.08612	11.20942	0.000	2.08612	11.22040	0.000
LNSIZE*(M_SIZE=2 AND TP=1)	-0.43559	-0.79822	*0.425*			
LNSIZE*(M_SIZE=2 AND TP=2)	-2.58066	-2.20535	0.028	-2.58066	-2.20751	0.027
T*(M_SIZE=1 AND TP=1)	-0.40960	-1.11408	0.265	-1.14096	-10.23552	0.000
T*(M_SIZE=1 AND TP=2)	-0.36470	-8.37899	0.000	-0.36470	-8.38720	0.000
T*(M_SIZE=2 AND TP=1)	-0.27193	-1.36409	0.173	-0.08918	-1.33051	0.184
T*(M_SIZE=2 AND TP=2)	-0.68501	-2.57953	0.010	-0.68501	-2.58206	0.010
T*LNSALE(-1)*(M_SIZE=1 AND TP=1)	0.03226	0.81503	*0.415*	0.26374	10.06415	0.000
T*LNSALE(-1)*(M_SIZE=1 AND TP=2)	0.09789	14.91104	0.000	0.09789	14.92565	0.000
T*LNSALE(-1)*(M_SIZE=2 AND TP=1)	-0.05052	-2.07443	0.038	-0.04426	-2.54804	0.011
T*LNSALE(-1)*(M_SIZE=2 AND TP=2)	-0.09251	-13.54010	0.000	-0.09251	-13.55336	0.000
T*LNSIZE*(M_SIZE=1 AND TP=1)	0.15941	1.61096	0.107	-0.07542	-2.57330	0.010
T*LNSIZE*(M_SIZE=1 AND TP=2)	-0.18204	-6.35124	0.000	-0.18204	-6.35746	0.000
T*LNSIZE*(M_SIZE=2 AND TP=1)	0.33255	1.65999	0.097	0.22182	3.06055	0.002
T*LNSIZE*(M_SIZE=2 AND TP=2)	0.61466	3.70074	0.000	0.61466	3.70437	0.000
R-squared	0.90085			0.88946		
Adjusted R-squared	0.89936			0.88802		
F-statistic	604.771			617.153		
Prob(F-statistic)	0.00000			0.00000		
Durbin-Watson stat	2.10587			2.01787		

Dependent Variable: LNSALE				
Method: Panel EGLS (Cross-section random effects)				
Date: 06/25/12 Time: 18:01				
Sample (adjusted): 2 8				
Periods included: 7				
Cross-sections included: 224				
Total panel (unbalanced) observations: 1555				
Swamy and Arora estimator of component variances				
White diagonal standard errors & covariance (d.f. corrected)				

Variable	Coefficient	Std. Error	t-Statistic	Prob.
C	3.506245	0.141538	24.77247	0.0000
M_SIZE=1 AND TP=2	-0.049086	0.391174	-0.125483	0.9002
M_SIZE=2 AND TP=1	3.093126	0.193939	15.94898	0.0000
M_SIZE=2 AND TP=2	2.642016	0.472319	5.593711	0.0000
T*LNSALE(-1)*(M_SIZE=1 AND TP=1)	0.196953	0.011714	16.81400	0.0000
T*LNSALE(-1)*(M_SIZE=1 AND TP=2)	0.112805	0.012151	9.283338	0.0000
T*LNSALE(-1)*(M_SIZE=2 AND TP=1)	0.276677	0.011370	24.33420	0.0000
T*LNSALE(-1)*(M_SIZE=2 AND TP=2)	0.132849	0.013648	9.733957	0.0000
T*LNSIZE*(M_SIZE=1 AND TP=1)	-0.349162	0.037327	-9.354245	0.0000
T*LNSIZE*(M_SIZE=1 AND TP=2)	-0.201077	0.050844	-3.954812	0.0001
T*LNSIZE*(M_SIZE=2 AND TP=1)	-0.876546	0.048240	-18.17035	0.0000
T*LNSIZE*(M_SIZE=2 AND TP=2)	-0.379413	0.079675	-4.762022	0.0000

Effects Specification		
	S.D.	Rho
Cross-section random	0.000000	0.0000
Idiosyncratic random	0.580210	1.0000

Weighted Statistics		
R-squared	0.794981	Mean dependent var 6.196031
Adjusted R-squared	0.793520	S.D. dependent var 1.936774
S.E. of regression	0.880072	Sum squared resid 1195.095
F-statistic	543.9220	Durbin-Watson stat 1.599692
Prob(F-statistic)	0.000000	

Unweighted Statistics		
R-squared	0.794981	Mean dependent var 6.196031
Sum squared resid	1195.095	Durbin-Watson stat 1.599692

Figure 14.8 *Statistical results of the first stage in developing an alternative reduced REM in Table 4.4*

14.5 CSREMs by Time or Time Period

Under the assumption that the research objects (individuals) can be considered a single group of individuals or a homogeneous group, then continuous CSREMs can be applied at each time point. See the following examples.

14.5.1 LV(1) CSREM with Exogenous Variables

Example 14.7   A LV(1) CSREM derived from the two-way interaction model in (8.28)
Table 14.5 presents the summary of statistical results of a reduced model of the LV(1) CSREM with the following ES, which is derived from the model (8.28). Based on this summary, the following notes and comments are made.

$$y1 \ c \ @expand(t, @dropfirst)y1(-1) * @expand(t)$$
$$x1 * x2 * @expand(t) \ x1 * @expand(t) \ x2 * @expand(t) \tag{14.10}$$

1. The reduced model is obtained using the *manual stepwise selection method* (MSSM) as follows:
   1.1 At the first stage, the list of variables "*Y1 C @expand(t,@dropfirst) Y1(−1)*@Expand(t)*" is used as the ES (equation specification).
   1.2 At the second stage, the variables *X1*X2*@Expand(t)* are inserted as additional variables. It is found that two of the variables, namely *X1*X2*(T = 5)* and *X1*X2*(T = 8)* have large *p*-values of 0.9029 and 0.3523, respectively. I decided to keep them in the model, since the linear effect of *X1* on *Y1* depends on *X2* in a theoretical sense.
   1.3 At the third stage, the variables *X2*@Expand(t)* are inserted as additional variables. It is found only five out of the seven variables can be used in the model, namely *X2*(T = 2)* and the four variables shown in the table, in order have the previous inserted independent variables with the same conditions.
   1.4 At the fourth stage, the variables *X1*@Expand(t)* are inserted as additional variables. We find only four out of the seven variables can be used in the model, but in addition the variable *X2*(T = 2)*, inserted at the third stage, should be deleted.
   1.5 Note that the interaction *X1*X2*(T = 5)* in the final model has a very large *p*-value = 0.9564. Then it can be concluded that the adjusted effect of *X1* on *Y1* is insignificantly dependent on *X2*, conditional for *T = 5*. On the other hand, if it should be deleted, then inserting the variable *X2*(T = 5)* is suggested, so that at the time *T = 5* the model is an additive model of *Y1(−1)*, *X1* and *X2*.

Table 14.5 *Statistical results based on a reduced LV(1) CSREM of the model in (14.10)*

Dependent Variable: Y1							
Method: Panel EGLS (Cross-section random effects)							
Date: 06/26/12 Time: 11:41							
Sample: 1 8 IF T > 1							
Periods included: 7							
Cross-sections included: 218							
Total panel (unbalanced) observations: 1436							
Swamy and Arora estimator of component variances							
White diagonal standard errors & covariance (d.f. corrected)							
Variable	Coef.	t-Stat.	Prob.	Variable	Coef.	t-Stat.	Prob.
C	-0.5888	-2.5063	0.0123	X1*X2*(T=2)	-0.0143	-2.0349	0.0421
T=3	0.3489	1.2225	0.2217	X1*X2*(T=3)	-0.0098	-1.2449	0.2134
T=4	-0.4501	-1.0712	0.2843	X1*X2*(T=4)	0.0696	2.3294	0.0200
T=5	0.6757	2.6892	0.0072	X1*X2*(T=5)	-0.0005	-0.0547	*0.9564*
T=6	-0.1144	-0.3021	0.7626	X1*X2*(T=6)	0.1246	2.1411	0.0324
T=7	0.5571	2.3050	0.0213	X1*X2*(T=7)	0.0570	1.8442	0.0654
T=8	0.5311	2.2385	0.0253	X1*X2*(T=8)	0.0476	1.1211	0.2625
Y1(-1)*(T=2)	0.8627	13.1524	0.0000	X2*(T=4)	-0.3637	-2.5289	0.0116
Y1(-1)*(T=3)	1.1631	17.8646	0.0000	X2*(T=6)	-0.6488	-2.1974	0.0282
Y1(-1)*(T=4)	0.9562	19.1668	0.0000	X2*(T=7)	-0.2190	-1.5960	0.1107
Y1(-1)*(T=5)	1.1544	18.9672	0.0000	X2*(T=8)	-0.2182	-1.1946	0.2325
Y1(-1)*(T=6)	1.0711	15.0899	0.0000	X1*(T=2)	0.1350	2.4756	0.0134
Y1(-1)*(T=7)	1.1993	43.9988	0.0000	X1*(T=3)	0.0382	1.0918	0.2751
Y1(-1)*(T=8)	0.9956	54.6420	0.0000	X1*(T=4)	0.1812	2.8700	0.0042
				X1*(T=5)	-0.0300	-1.4092	0.1590
				X1*(T=6)	0.1367	2.5850	0.0098
R-squared					0.9586		
Adjusted R-squared					0.9578		
F-statistic					1122.78		
Prob(F-statistic)					0.0000		
Durbin-Watson stat					2.0155		

Example 14.8   A polynomial LV(1) CSREM by time

For an illustration, Table 14.6 presents the summary of the statistical results of a third-degree polynomial LV(1) CSREM, derived from the model in (12.44), using the following ES for T > 1.

$$bnd\_ni \; c \; @expand(t, @dropfirst) \; bnd\_ni(-1) * @expand(t)$$
$$bnd\_ni(-1)^{\wedge}2 * @expand(t) \; bnd\_ni(-1)^{\wedge}3 * @expand(t) \tag{14.11}$$

Table 14.6 Summary of the statistical results of the CSREM in (14.11)

Dependent Variable: BND_NI
Method: Panel EGLS (Cross-section random effects)
Date: 06/25/12   Time: 18:39
Sample: 1 8 IF T >1
Periods included: 7
Cross-sections included: 224
Total panel (balanced) observations: 1568
Swamy and Arora estimator of component variances
White diagonal standard errors & covariance (d.f. corrected)

Variable	Parameter	Coef.	t-Stat.	Prob.	Variable	Parameter	Coef.	t-Stat.	Prob.
C	C(1)	0.4406	2.5228	0.0117	$BND\_NI(-1)^2*(T=2)$	C(15)	6.7373	2.0490	0.0406
T=3	C(2)	-0.5367	-2.2541	0.0243	$BND\_NI(-1)^2*(T=3)$	C(16)	-2.7880	-0.9049	*0.3657*
T=4	C(3)	-0.3665	-2.0644	0.0391	$BND\_NI(-1)^2*(T=4)$	C(17)	1.2483	1.4818	0.1386
T=5	C(4)	-0.6205	-3.1544	0.0016	$BND\_NI(-1)^2*(T=5)$	C(18)	-4.3009	-3.6209	0.0003
T=6	C(5)	-0.9307	-2.1943	0.0284	$BND\_NI(-1)^2*(T=6)$	C(19)	-2.9635	-0.9178	*0.3589*
T=7	C(6)	-0.6046	-2.9756	0.0030	$BND\_NI(-1)^2*(T=7)$	C(20)	-0.7987	-3.1404	0.0017
T=8	C(7)	-0.3313	-1.8479	0.0648	$BND\_NI(-1)^2*(T=8)$	C(21)	0.2150	4.6762	0.0000
$BND\_NI(-1)*(T=2)$	C(8)	-2.3605	-1.6363	0.1020	$BND\_NI(-1)^3*(T=2)$	C(22)	-3.3744	-2.0636	0.0392
$BND\_NI(-1)*(T=3)$	C(9)	1.9279	1.4667	0.1427	$BND\_NI(-1)^3*(T=3)$	C(23)	2.3406	1.1104	0.2670
$BND\_NI(-1)*(T=4)$	C(10)	0.4410	1.5187	0.1290	$BND\_NI(-1)^3*(T=4)$	C(24)	-0.1633	-0.2716	*0.7860*
$BND\_NI(-1)*(T=5)$	C(11)	2.6746	4.1772	0.0000	$BND\_NI(-1)^3*(T=5)$	C(25)	2.3757	4.4492	0.0000
$BND\_NI(-1)*(T=6)$	C(12)	3.9190	1.5962	0.1106	$BND\_NI(-1)^3*(T=6)$	C(26)	0.8876	0.9788	*0.3279*
$BND\_NI(-1)*(T=7)$	C(13)	1.8006	3.3921	0.0007	$BND\_NI(-1)^3*(T=7)$	C(27)	0.1038	3.4254	0.0006
$BND\_NI(-1)*(T=8)$	C(14)	0.5094	2.5880	0.0097	$BND\_NI(-1)^3*(T=8)$	C(28)	0.0233	1.4749	0.1405

Weighted Statistics

R-squared	0.6034		Mean dependent var	0.2376
Adjusted R-squared	0.5965		S.D. dependent var	0.2516
F-statistic	86.791		Sum squared resid	39.339
Prob(F-statistic)	0.0000		Durbin-Watson stat	2.3198

Based on the results in Table 14.6 the following findings and notes are presented.

1. This model is a bounded polynomial LV(1) CSREM, since $BND\_NI = log((NI - L)/(U - NI))$ where L and U are lower and upper bounds of *NI* (net income).
2. Since, only four out of 21 numerical independent variables have *p*-values > 0.30, it is not so difficult to explore a reduced model using the trial-and-error method just by deleting the four variables one-by-one.

14.5.2 LV(*p*) CSREMs

Example 14.9 LV(2) three-way interaction CSREM by time
For an illustration, Table 14.7 presents the summary of a LV(2) three-way CSREM, conditional for $T = 6$ and $T = 7$, which indicate the years before and after a new tax policy. The model is derived from the model in Figure 8.22, using the ES as follows:

$$y1 \, c \, @expand(t, @dropfirst) \, y1(-1) * @expand(t) \, y1(-2) * @expand(t)$$

$$x1 * x2 * x3 * @expand(t) \, x1 * x2 * @expand(t) \, x1 * x3 * @expand(t) \, x2 * x3 * @expand(t) \quad (14.12)$$

$$x1 * @expand(t) \, x2 * @expand(t) \, x3 * @expand(t)$$

Based on the results in Table 14.7, the following notes and comments are presented.

1. Only two of the independent variables have *p*-values > 0.30, namely *X1*∗*X2*∗*(T=6)* and *X2*∗ *(T=6)*. So in general, a reduced model could be explored by deleting the two variables. However, since both variables *X1*∗*X2* and *X2* are the independent variables of a regression, then the first trial *X2* should be deleted using the following ES because the effect of *X1* on *Y1* depends on *X2*, as well as *X3*:

$$y1 \, c \, @expand(t, @dropfirst) \, y1(-1) * @expand(t) \, y1(-2) * @expand(t)$$

$$x1 * x2 * x3 * @expand(t) \, x1 * x2 * @expand(t) \, x1 * x3 * @expand(t) \, x2 * x3 * @expand(t)$$

$$x1 * @expand(t) \, \textbf{x2} * \textbf{@expand}(t, \textbf{@Drop}(6)) \, x3 * @expand(t)$$

$$(14.13)$$

2. It so happens that a good fit reduced model is obtained, but the output of the full model is not presented. However, the statistical results of an alternative ES are presented in Figure 14.7. Some of the results need to be considered with special notes are as follows:
   2.1 The *p*-value of X1∗X2∗(T=6) reduces from 0.6675 to 0.1418. So at the 10% level of significance it has a significant positive effect on Y1 with a *p*-value $0.1418/2 = 0.0709$
   2.2 The *p*-value of Y1(−2)∗(T=6) reduces from 0.2644 to 0.2525, which has the greatest *p*-value in the model
   2.3 On the other hand, the *p*-value of X1∗X2∗(T=7) increases from 0.1999 to 0.2002.
   2.4 These contradictory findings indicate the unpredicted impact of multicollinearity between independent variables.
3. In addition, Figure 14.9 presents the statistical results of the cross-section random effects of the CSREM in (14.13), namely RE_14_13, Based on the results in this figure and Table 14.7, the following findings and notes are presented.
   3.1 Compared to several CSREMs which have Rho $= 0.000\,000$, this CSREM has a positive Rho $= 0.0028$. In fact, the full model in (14.12) also has a positive Rho $= 0.0097$.

Table 14.7   *Summary of the statistical results of the CSREM in (14.12)*

Dependent Variable: Y1							
Method: Panel EGLS (Cross-section random effects)							
Date: 07/07/12 Time: 20:21							
Sample: 1 8 IF T=6 OR T=7							
Periods included: 2							
Cross-sections included: 212							
Total panel (unbalanced) observations: 418							
Swamy and Arora estimator of component variances							
White period standard errors & covariance (d.f. corrected)							
Variable	Coef.	t-Stat.	Prob.	Variable	Coef.	t-Stat.	Prob.
C	0.4472	1.8261	0.0686	X1*X3*(T=6)	-1.1884	-4.1577	0.0000
T=7	-0.9149	-2.6854	0.0075	X1*X3*(T=7)	0.2535	2.0416	0.0418
Y1(-1)*(T=6)	1.1118	10.1881	0.0000	X2*X3*(T=6)	11.3426	2.7875	0.0056
Y1(-1)*(T=7)	1.3551	10.2366	0.0000	X2*X3*(T=7)	-10.0874	-1.9669	0.0499
Y1(-2)*(T=6)	0.1553	1.1393	0.2552	X1*(T=6)	-0.0907	-1.9908	0.0472
Y1(-2)*(T=7)	-0.3780	-1.7603	0.0791	X1*(T=7)	0.0822	2.1611	0.0313
X1*X2*X3*(T=6)	-1.0739	-2.7890	0.0055	X2*(T=6)			
X1*X2*X3*(T=7)	1.0853	2.0000	0.0462	X2*(T=7)	-0.2775	-1.4472	0.1486
X1*X2*(T=6)	0.0117	1.4724	0.1417	X3*(T=6)	12.4520	3.9287	0.0001
X1*X2*(T=7)	0.0491	1.2831	0.2002	X3*(T=7)	-2.4672	-2.0933	0.0370
Effects Specification							
			S.D.	Rho			
Cross-section random			0.0301	0.0028			
Idiosyncratic random			0.5694	0.9972			
Weighted Statistics							
R-squared		0.9730		Mean dependent var		2.0180	
Adjusted R-squared		0.9717		S.D. dependent var		5.2114	
S.E. of regression		0.8760		Sum squared resid		306.194	
F-statistic		797.700		Durbin-Watson stat		2.0609	
Prob(F-statistic)		0.0000					
Unweighted Statistics							
R-squared		0.973038		Mean dependent var		2.023596	
Sum squared resid		307.0168		Durbin-Watson stat		2.055355	

3.2   Based on the results in Figure 14.9, we have var $(w_{it}) = \sigma_\varepsilon^2 + \sigma_\mu^2 = (0.002619)^2 = 6.85916\text{E-}06$, and from Table 14.7, we have $\rho(rho) = \sigma_\varepsilon^2/(\sigma_\varepsilon^2 + \sigma_\mu^2) = 0.002800$, then it can be computed:

$$\sigma_\varepsilon^2 = \rho \times \text{var}(w_{it}) = 1.92057\text{E-}08, \text{ and}$$
$$\sigma_\mu^2 = (6.85916\text{E-}06) - (1.92057\text{E-}08) = 6.83996\text{E-}06$$

Test for Equality of Variances of RE_14_13
Categorized by values of T
Date: 07/07/12   Time: 20:51
Sample: 1 8 IF T=6 OR T=7
Included observations: 52

Method	df	Value	Probability
F-test	(25, 25)	2.303943	0.0416
Siegel-Tukey		0.356873	0.7212
Bartlett	1	4.149924	0.0416
Levene	(1, 50)	1.011804	0.3193
Brown-Forsythe	(1, 50)	0.901376	0.3470

Category Statistics

T	Count	Std. Dev.	Mean Abs. Mean Diff.	Mean Abs. Median Diff.	Mean Tukey- Siegel Rank
6	26	0.003117	0.001901	0.001873	25.73077
7	26	0.002053	0.001332	0.001331	27.26923
All	52	0.002619	0.001616	0.001602	26.50000

Bartlett weighted standard deviation: 0.002639

Descriptive Statistics for RE_14_13
Categorized by values of T
Date: 07/07/12   Time: 20:57
Sample: 1 8 IF T=6 OR T=7
Included observations: 52

T	Mean	Max	Min.	Std. Dev.	Obs.
6	-0.000319	0.010962	-0.005113	0.003117	26
7	1.67E-05	0.004159	-0.005745	0.002053	26
All	-0.000151	0.010962	-0.005745	0.002619	52

Figure 14.9   *Statistical results based on the cross-section random effects of the model (14.13)*

3.3   Corresponding to the mean model in (14.1b), based on the mean statistic of $-0.000\,151$ in Figure 14.9, then RE $= -0.000\,151$, and corresponding to the CSREM in (14.1c), based on the results in Table 14.7, we have the following regression function:

$$\hat{Y}1 = -\boldsymbol{0.000151} + 0.447 - 0.915(T=7) + 1.112Y1(-1)*(T=6) + 1.355Y1(-1)*(T=7)$$
$$+ 0.155Y1(-2)*(T=6) - 0.378Y1(-2)*(T=7) - 1.074X1*X2*X3*(T=6)$$
$$+ 1.085X1*X2*X3*(T=7) + 0.012X1*X2*(T=6) + 0.049X1*X2*(T=7) - 1.188X1*X3*(T=6)$$
$$+ 0.253X1*X3*(T=7) + 11.3426X2*X3*(T=6) - 10.087X2*X3*(T=7) - 0.091X1*(T=6)$$
$$+ 0.082X1*(T=7) - 0.277X2*(T=7) + 12.452X3*(T=6) - 2.467X3*(T=7)$$

Then the regression functions for $T=6$ and $T=7$ can easily be written.

4.   Take note that this model could easily be extended to LV(p) CSREMs for $p > 2$, but the acceptable value of p depends highly on the data.

Example 14.10   An alternative ES of the ES in (14.13)
Figure 14.10 presents the statistical results based on the model (14.13), using an alternative ES as follows:

$$\begin{aligned}
&y1\ c\ y1(-1)*dt6\ y1(-2)*dt6\ x1*x2*x3*dt6\\
&\quad x1*x2*dt6\ x1*x3*dt6\ x2*x3*dt6\ \boldsymbol{x1*dt6}\ \boldsymbol{x3*dt6}\\
&\quad dt7\ y1(-1)*dt7\ y1(-2)*dt7\ x1*x2*x3*dt7\\
&\quad x1*x2*dt7\ x1*x3*dt7\ x2*x3*dt7\ \boldsymbol{x1*dt7}\ \boldsymbol{x2*dt7}\ \boldsymbol{x3*dt7}
\end{aligned} \tag{14.14}$$

Dependent Variable: Y1
Method: Panel EGLS (Cross-section random effects)
Date: 07/02/12   Time: 09:27
Sample: 1 8 IF T=6 OR T=7
Periods included: 2
Cross-sections included: 212
Total panel (unbalanced) observations: 418
Swamy and Arora estimator of component variances
White diagonal standard errors & covariance (d.f. corrected)

Variable	Coefficient	Std. Error	t-Statistic	Prob.
C	0.447188	0.245011	1.825178	0.0687
Y1(-1)*DT6	1.111759	0.109105	10.18982	0.0000
Y1(-2)*DT6	0.155283	0.136288	1.139372	0.2552
X1*X2*X3*DT6	-1.073928	0.385130	-2.788483	0.0055
X1*X2*DT6	0.011668	0.007927	1.471924	0.1418
X1*X3*DT6	-1.188432	0.285900	-4.156803	0.0000
X2*X3*DT6	11.34257	4.069971	2.786892	0.0056
X1*DT6	-0.090663	0.045559	-1.989998	0.0473
X3*DT6	12.45201	3.170259	3.927758	0.0001
DT7	-0.914886	0.324487	-2.819485	0.0050
Y1(-1)*DT7	1.355134	0.132369	10.23756	0.0000
Y1(-2)*DT7	-0.377974	0.214695	-1.760515	0.0791
X1*X2*X3*DT7	1.085306	0.542732	1.999711	0.0462
X1*X2*DT7	0.049069	0.038245	1.283006	0.2002
X1*X3*DT7	0.253479	0.124113	2.042332	0.0418
X2*X3*DT7	-10.08744	5.129177	-1.966678	0.0499
X1*DT7	0.082224	0.038063	2.160186	0.0314
X2*DT7	-0.277459	0.191739	-1.447070	0.1487
X3*DT7	-2.467210	1.178253	-2.093956	0.0369

(a)

Effects Specification		
	S.D.	Rho
Cross-section random	0.030073	0.0028
Idiosyncratic random	0.569449	0.9972

Weighted Statistics			
R-squared	0.972963	Mean dependent var	2.017987
Adjusted R-squared	0.971743	S.D. dependent var	5.211356
S.E. of regression	0.876016	Sum squared resid	306.1940
F-statistic	797.6997	Durbin-Watson stat	2.060878
Prob(F-statistic)	0.000000		

Unweighted Statistics			
R-squared	0.973038	Mean dependent var	2.023596
Sum squared resid	307.0168	Durbin-Watson stat	2.055355

(b)

Figure 14.10   *Statistical results based on the model in (14.14)*

Based on these results, the following notes and comments are made.

1. Compared to the ES in (14.13), an advantage in using this ES (14.14) is it is easier to construct a summary table using Microsoft Excel as presented in Table 14.8.

   However, the disadvantage in using the dummy variables is the ES would be a long list of variables if a large number of groups of firms will be considered.

Table 14.8   *Summary of the statistical results in Figure 14.7*

	T=6				T=7			
Variable	Parameter	Coef.	t-Stat.	Prob.	Parameter	Coef.	t-Stat.	Prob.
Constant	C(1)	0.4472	1.8252	0.0687	C(10)	-0.9149	-2.8195	0.0050
Y1(-1)	C(2)	1.1118	10.1898	0.0000	C(11)	1.3551	10.2376	0.0000
Y1(-2)	C(3)	0.1553	1.1394	0.2552	C(12)	-0.3780	-1.7605	0.0791
X1*X2*X3	C(4)	-1.0739	-2.7885	0.0055	C(13)	1.0853	1.9997	0.0462
X1*X2	C(5)	0.0117	1.4719	0.1418	C(14)	0.0491	1.2830	0.2002
X1*X3	C(6)	-1.1884	-4.1568	0.0000	C(15)	0.2535	2.0423	0.0418
X2*X3	C(7)	11.3426	2.7869	0.0056	C(16)	-10.0874	-1.9667	0.0499
X1	C(8)	-0.0907	-1.9900	0.0473	C(17)	0.0822	2.1602	0.0314
X2					C(18)	-0.2775	-1.4471	0.1487
X3	C(9)	12.4520	3.9278	0.0001	C(19)	-2.4672	-2.0940	0.0369
Weighted Statistics					Effects Specification			
R-squared		0.9730						
Adjusted R-squared		0.9717					S.D.	Rho
S.E. of regression		0.8760			Cross-section rand.		0.03007	0.0028
F-statistic		797.700			Idiosyncratic random		0.56945	0.9972
Prob(F-statistic)		0.0000			Durbin-Watson stat		2.06088	

Example 14.11 LV(3) CSREMs by the time-period TP

For a simple illustrative CSREM, Figure 14.11 presents the statistical results of a LV(3) CSREM of Y1 by TP and its reduced model. Based on these results, the following notes and comments are presented.

1. I wondered why many of the statistical results of the CSREMs have Rho $= 0.0000$ based on the Special_BPD.wf1. So I did the following additional analyses:
2. *Residual Analysis.* By selecting *Proc/Make Residual/Series . . . /Ordinary . . . OK*, the scores of RESID01 as presented in Figure 14.10(a), with the test for equality of variances by the time period TP are obtained. *View/Fixed/Random Effects/Cross-section Effects . . .* , the results in Figure 14.12 are obtained.
3. Referring to the formulas of the residual w_{it} in (14.1e), based on these statistical results, the following findings and notes are presented.

$$\bar{w} = 1.06E - 15 \approx 0, and \text{ var}(w) = 1.048862^{\wedge}2 = 1.100111$$

$$\hat{\rho}(rho) = corr(\hat{w}_{is}, \hat{w}_{it}) = \frac{\hat{\sigma}_{\varepsilon}^2}{\hat{\sigma}_{\varepsilon}^2 + \hat{\sigma}_{\mu}^2} = 0, then \widehat{\sigma_{\varepsilon}^2} = 0$$

4. In addition, the results in Figure 14.13 show that the correlations between pairs of the residuals: RESID01 and RESID01(-1); RESID01(-2) and RESID01(-3), which show they are not equal to zero. So these correlations cannot represent the $corr(\hat{w}_{is}, \hat{w}_{it}) = 0$. What is the difference? It can't be explained yet.

Dependent Variable: Y1
Method: Panel EGLS (Cross-section random effects)
Date: 07/03/12 Time: 15:56
Sample (adjusted): 4 8
Periods included: 5
Cross-sections included: 224
Total panel (balanced) observations: 1120
Swamy and Arora estimator of component variances
White cross-section standard errors & covariance (d.f. corrected)
WARNING: estimated coefficient covariance matrix is of reduced rank

Variable	Coefficient	Std. Error	t-Statistic	Prob.
C	0.018633	0.060810	0.306412	0.7593
TP=2	0.046536	0.077113	0.603476	0.5463
Y1(-1)*(TP=1)	1.113255	0.033408	33.32340	0.0000
Y1(-1)*(TP=2)	1.016059	0.096857	10.49028	0.0000
Y1(-2)*(TP=1)	0.091457	0.064001	1.429002	0.1533
Y1(-2)*(TP=2)	0.140661	0.137374	1.023932	0.3061
Y1(-3)*(TP=1)	-0.169819	0.050619	-3.354875	0.0008
Y1(-3)*(TP=2)	-0.068877	0.068343	-1.007806	0.3138

Effects Specification

		S.D.	Rho
Cross-section random		0.000000	0.0000
Idiosyncratic random		0.847046	1.0000

Weighted Statistics

R-squared	0.951446	Mean dependent var	1.820116
Adjusted R-squared	0.951140	S.D. dependent var	4.747736
S.E. of regression	1.049454	Sum squared resid	1224.705
F-statistic	3112.880	Durbin-Watson stat	2.039417
Prob(F-statistic)	0.000000		

Unweighted Statistics

R-squared	0.951446	Mean dependent var	1.820116
Sum squared resid	1224.705	Durbin-Watson stat	2.039417

Dependent Variable: Y1
Method: Panel EGLS (Cross-section random effects)
Date: 07/03/12 Time: 15:59
Sample (adjusted): 4 8
Periods included: 5
Cross-sections included: 224
Total panel (balanced) observations: 1120
Swamy and Arora estimator of component variances
White cross-section standard errors & covariance (d.f. corrected)
WARNING: estimated coefficient covariance matrix is of reduced rank

Variable	Coefficient	Std. Error	t-Statistic	Prob.
C	0.018633	0.060756	0.306688	0.7591
TP=2	0.047904	0.078844	0.607588	0.5436
Y1(-1)*(TP=1)	1.113255	0.033378	33.35335	0.0000
Y1(-1)*(TP=2)	1.082830	0.056240	19.25377	0.0000
Y1(-2)*(TP=1)	0.091457	0.063943	1.430287	0.1529
Y1(-3)*(TP=1)	-0.169819	0.050573	-3.357890	0.0008

Effects Specification

		S.D.	Rho
Cross-section random		0.000000	0.0000
Idiosyncratic random		0.849099	1.0000

Weighted Statistics

R-squared	0.951195	Mean dependent var	1.820116
Adjusted R-squared	0.950976	S.D. dependent var	4.747736
S.E. of regression	1.051214	Sum squared resid	1231.026
F-statistic	4342.309	Durbin-Watson stat	2.136236
Prob(F-statistic)	0.000000		

Unweighted Statistics

Figure 14.11 *Statistical results based on a full LV(1) CSREM of Y1 by TP and its reduced model*

Series: RESID01 Workfile: REM_DATA::Sesi\		
View Proc Object Properties Print Name Freeze Default		
RESID01		
	Last updated: 07/04/12 - 09:20	
	Modified: 1 8 // eq11.makeresid	
_AALI - 1	NA	
_AALI - 2	NA	
_AALI - 3	NA	
_AALI - 4	-0.918071	
_AALI - 5	0.172704	
_AALI - 6	0.371961	
_AALI - 7	-0.129874	
_AALI - 8	-0.202525	
_ABBA - 1	NA	
_ABBA - 2	NA	
_ABBA - 3	NA	
_ABBA - 4	0.001674	
_ABBA - 5	0.006465	
_ABBA - 6	-0.065839	
_ABBA - 7	-0.077336	
_ABBA - 8	-0.070921	
_ADES - 1	NA	
_ADES - 2	NA	
_ADES - 3	NA	
_ADES - 4	0.235841	

Test for Equality of Variances of RESID01
Categorized by values of TP
Date: 07/04/12   Time: 09:28
Sample (adjusted): 4 8
Included observations: 1120 after adjustments

Method	df	Value	Probability
F-test	(447, 671)	1.453282	0.0000
Siegel-Tukey		0.035168	0.9719
Bartlett	1	18.17094	0.0000
Levene	(1, 1118)	0.967151	0.3256
Brown-Forsythe	(1, 1118)	0.406678	0.5238

Category Statistics

TP	Count	Std. Dev.	Mean Abs. Mean Diff.	Mean Abs. Median Diff.	Mean Tukey-Siegel Rank
1	448	0.930381	0.351217	0.348527	560.9174
2	672	1.121593	0.409706	0.386747	560.2217
All	1120	1.048862	0.386310	0.371459	560.5000

Bartlett weighted standard deviation: 1.049331

Descriptive Statistics for RESID01
Categorized by values of TP
Date: 07/04/12   Time: 09:33
Sample (adjusted): 4 8
Included observations: 1120 after adjustments

TP	Mean	Max	Min.	Std. Dev.	Obs.
1	1.56E-15	7.331429	-5.701472	0.930381	448
2	7.19E-16	11.79110	-3.804135	1.121593	672
All	1.06E-15	11.79110	-5.701472	1.048862	1120

Figure 14.12   *Statistical results of the RESID01 of the reduced CSREM in Figure 14.11*

Covariance Analysis: Ordinary
Date: 07/04/12   Time: 10:49
Sample (adjusted): 7 8
Included observations: 448 after adjustments
Balanced sample (listwise missing value deletion)

Correlation t-Statistic Probability	RESID01	RESID01(-1)	RESID01(-2)	RESID01(-3)
RESID01	1.000000			

RESID01(-1)	-0.008930	1.000000		
	-0.188595	-----		
	0.8505	-----		
RESID01(-2)	0.019009	0.052360	1.000000	
	0.401518	1.107300	-----	
	0.6882	0.2688	-----	
RESID01(-3)	0.028405	0.217708	-0.046457	1.000000
	0.600124	4.710703	-0.982166	-----
	0.5487	0.0000	0.3266	-----

Figure 14.13   *Correlation matrix between selected residuals of the reduced CSREM in Figure 14.11*

14.6 Period Random Effects Models (PEREMs)

14.6.1 General Equation

Similar to the general LV(p) CSREM in (14.1), the lagged variables period random effects model, namely LV(p) PEREM, would have the following general equation.

$$Y_{it} = PE(t) + \sum_{j=1}^{p} C(1j)Y_{i,t-j} + \sum_{k=1}^{K} C(2k)Xk_{it} + \mu_{it} \qquad (14.15a)$$

where *PE(t)* is defined as a random variable, for $t = 1, \ldots, T$, with the mean or the expected value of *PE*, which could be presented as a cell-mean model:

$$PE(t) = PE + \upsilon_t \qquad (14.15b)$$

Therefore, the LV(p) PEREM model in (14.15a) can be presented as follows:

$$Y_{it} = PE + \sum_{j=1}^{p} C(1j)Y_{i,t-j} + \sum_{k=1}^{K} C(2k)Xk_{it} + \upsilon_t + \mu_{it} \qquad (14.15c)$$

with the following assumptions

$$
\begin{aligned}
&\upsilon_t, t = 1, \ldots, T \text{ have } i.i.d.N(0, \sigma_{\upsilon}^2) \\
&\mu_{it}, i = 1, \ldots, N; t = 1, \ldots, T \text{ have } i.i.d.N(0, \sigma_{\mu}^2) \\
&E(\upsilon_t \mu_{it}) = 0; \quad E(\upsilon_s \upsilon_t) = 0, (s \neq t) \\
&E(\mu_{is}\mu_{it}) = E(\mu_{it}\mu_{jt}) = E(\mu_{is}\mu_{js}) = 0, (i \neq j; s \neq t)
\end{aligned}
\qquad (14.15d)
$$

that is, the individual error components have normal distribution with zero means and variances σ_{υ}^2 and σ_{μ}^2 respectively; they are not correlated with each other and are not autocorrelated across both cross-section and time series units.

Note that there are two error terms in this model, υ_t which are the time-series, or time-specific and cross-section error, and μ_{it}, which is the combined time-series and cross-section error component. Overall, three errors here are assumed to be zero and the time-series error is not correlated with combination error. As a result of the assumptions stated previously, it follows:

$$
\begin{aligned}
E(w_{it}) &= E(\upsilon_t + \mu_{it}) = 0 \\
\text{var}(w_{it}) &= \sigma_{\upsilon}^2 + \sigma_{\mu}^2 \\
\rho(rho) &= corr(w_{it}, w_{is}) = \frac{\sigma_{\upsilon}^2}{\sigma_{\upsilon}^2 + \sigma_{\mu}^2}
\end{aligned}
\qquad (14.15e)
$$

In the process of data analysis, the error message presented in Figure 14.14 might be obtained, which indicates that the number of cross-sections or the time-point observations should be sufficiently large in order to have significant statistical results. See the following simple examples. I am very sure that the data also should have a sufficiently large number of individual observations in order to estimate PEREMs.

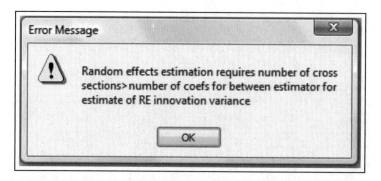

Figure 14.14   *A possible error message for random effects models*

14.6.2   Illustrative Analysis Based on Special_BPD.wf1

Example 14.12   PEREMs using the equation specifications (14.2a) and (14.2b)
Compared to the statistical results of the ANOVA REMs in Figure 14.2, using the equation specifications (ESs) in (14.2a) and (14.2b), Figure 14.15 presents their statistical results of the ANOVA PEREMs. Note the regressions have very small DW-statistics. So the LV(1) PEREMs should be applied.

But by using the following ESs in (14.16a) and (14.16b), the error message in Figure 14.14 was obtained, since the two LV(1) PEREMs have eight and 10 parameters, respectively, compared to only seven time

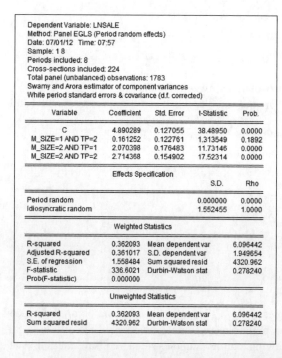

Figure 14.15   *Statistical results of PEREMs using the ESs in (14.16a) and (14.16b)*

Dependent Variable: LNSALE
Method: Panel EGLS (Period random effects)
Date: 07/01/12   Time: 08:07
Sample (adjusted): 2 8
Periods included: 7
Cross-sections included: 224
Total panel (unbalanced) observations: 1556
Swamy and Arora estimator of component variances
White period standard errors & covariance (d.f. corrected)

Variable	Coefficient	Std. Error	t-Statistic	Prob.
C	0.508410	0.110501	4.600932	0.0000
M_SIZE=1 AND TP=2	-0.032241	0.079698	-0.404542	0.6859
M_SIZE=2 AND TP=1	0.257959	0.066041	3.906025	0.0001
M_SIZE=2 AND TP=2	0.240162	0.104531	2.297534	0.0217
LNSALE(-1)	0.919815	0.022673	40.56902	0.0000

Effects Specification

	S.D.	Rho
Period random	0.048232	0.0060
Idiosyncratic random	0.622884	0.9940

Weighted Statistics

R-squared	0.892890	Mean dependent var	4.053028
Adjusted R-squared	0.892614	S.D. dependent var	1.926652
S.E. of regression	0.631325	Sum squared resid	618.1842
F-statistic	3232.376	Durbin-Watson stat	2.209653
Prob(F-statistic)	0.000000		

Unweighted Statistics

R-squared	0.891233	Mean dependent var	6.190340
Sum squared resid	642.5473	Durbin-Watson stat	2.232527

(a)

Dependent Variable: LNSALE
Method: Panel EGLS (Period random effects)
Date: 07/09/12   Time: 13:13
Sample (adjusted): 2 8
Periods included: 7
Cross-sections included: 224
Total panel (unbalanced) observations: 1556
Swamy and Arora estimator of component variances
White diagonal standard errors & covariance (d.f. corrected)

Variable	Coefficient	Std. Error	t-Statistic	Prob.
C	0.552410	0.156911	3.520523	0.0004
Q_SIZE=2	-0.168661	0.070650	-2.387259	0.0171
Q_SIZE=3	-0.112162	0.058169	-1.928218	0.0540
Q_SIZE=4	-0.005632	0.055207	-0.102017	0.9188
Q_SIZE=5	0.027676	0.053559	0.516741	0.6054
LNSALE(-1)	0.941696	0.022194	42.42972	0.0000

Effects Specification

	S.D.	Rho
Period random	0.000000	0.0000
Idiosyncratic random	0.628231	1.0000

Weighted Statistics

R-squared	0.889231	Mean dependent var	6.190340
Adjusted R-squared	0.888873	S.D. dependent var	1.949122
S.E. of regression	0.649753	Sum squared resid	654.3782
F-statistic	2488.605	Durbin-Watson stat	2.259827
Prob(F-statistic)	0.000000		

Unweighted Statistics

R-squared	0.889231	Mean dependent var	6.190340
Sum squared resid	654.3782	Durbin-Watson stat	2.259827

(b)

Figure 14.16   *Statistical results of LV(1) PEREMs of the PEREMs in Figure 14.15*

point observations, that is (8–1) because of using the LV(1) model.

$$lnSale \ C \ @Expand(M\_Size, TP, @dropfirst) \ lnSale(-1) * @Expand(M\_Size, TP) \qquad (14.16a)$$

and

$$lnSale \ C \ @Expand(Q\_Size, @dropfirst) \ lnSale(-1) * @Expand(Q\_Size) \qquad (14.16b)$$

Then, for illustration, Figure 14.16 presents the statistical results of two LV(1) PEREMs, which are the ANCOVA PEREMs with a covariate LNSALE(−1) as the extension of the ANOVA PREMs in Figure 14.15. Note that both models have five and six parameters respectively. Based on these illustrative findings, it can be concluded that the Special_BPD.wf1 cannot be used to present illustrative empirical results of the PEREMs with a total number of parameters above seven.

14.7   Illustrative Panel Data Analysis Based on CES.wf1

Since the Special_BPD.wf1 cannot be used to present empirical results of PEREMs with a large number of parameters, I decided to use one of the work files available in EViews, namely the CES_wf1. The data has 28×82 firm-time observations. So this can be used to present illustrative examples based on a LV(p) PEREMs with the number of parameters below (28-p). For illustration, a sub-set of the CES.wf1 is considered, as presented in Figure 14.17. See the following examples.

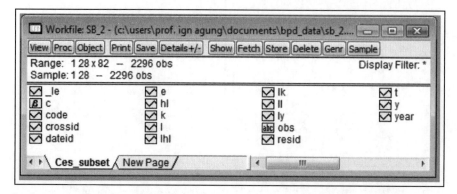

Figure 14.17   *Selected subset of the CES.wf1, provided in EViews*

14.7.1   Period Random Effects Models

Example 14.13   A LV(2) PEREM of LY by Code of the Firms

For illustration, a sample CODE < 5 is taken for data analysis. The statistical results in Figure 14.18 are obtained using the ES as follows:

$$LY \; C \; @Expand(Code, @Dropfirst) \; LY(-1) * @Expand(Code) \; LY(-2) * @Expand(Code) \quad (14.17)$$

Based on these results, the following findings and notes are presented.

1. Note that the model in (14.17) is a translog (trans-logarithmic) linear models of the first four selected firms, namely Code $= 1$ to 4.

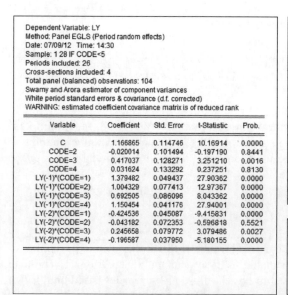

Figure 14.18   *Statistical results based on the model in (14.17)*

```
Correlated Random Effects - Hausman Test
Equation: EQ01
Test period random effects
```

Test Summary	Chi-Sq. Statistic	Chi-Sq. d.f.	Prob.
Period random	0.000000	11	1.0000

```
* Period test variance is invalid. Hausman statistic set to zero.
** WARNING: robust standard errors may not be consistent with
    assumptions of Hausman test variance calculation.
```

Figure 14.19 *A result of the Correlated Random Effects/Hausman Test*

2. The mean of the period random effects PER(t) $= -5.75$E-15 $= 0.000\,000$
3. The estimates of the error variances are as follows:

$$\text{var}(w_{it}) = \sigma_v^2 + \sigma_\mu^2 = (0.020547)^2$$

$$\rho(rho) = corr(w_{it}, w_{is}) = \frac{\sigma_v^2}{\sigma_v^2 + \sigma_\mu^2} = 0.3493$$

$$\Rightarrow \sigma_v^2 = (0.020547)^2 \times 0.3493 = 0.000147$$

$$\sigma_\mu^2 = (0.020547)^2 - 0.000147 = 0.000275$$

4. The null hypothesis H_0: $C(9) = C(10) = C(11) = C(12)$ is rejected based on the F-statistic of $F_0 = 115.7572$ with $df = (6,92)$ and a p-value $= 0.00002$. So it can be concluded that $LY(-2)$ has a significant different effects on LY between the first four firms.
5. By selecting *View_Fixed/Random Effects/Correlated Random Effects – Hausman Test*, the result in Figure 14.19 was obtained, which shows "*Period test variance is invalid*" and a *WARNING*. With regard to these findings, I had done the same Hausman test for all the CSREMs presented previously, as well as some other REMs, and the same notes were obtained. I would say frankly that I cannot modify the models in order to obtain a valid test, so I suggest these are limitations of the REMs.
6. Since $LY(-2)*(Code = 2)$ has a p-value $= 0.5521$, the model could be reduced, by using the following ES. However, its statistical results are not presented.

$$LY \ C \ @Expand(Code, @Dropfirst) \ LY(-1) * @Expand(Code)$$
$$LY(-2) * @Expand(Code, @Drop(2)) \tag{14.18}$$

Example 14.14   A LV(2) PEREM of LY by group and time period
As an extension of the model in (14.17), this example presents the analysis based on a LV(2) PEREM by the categorical variables *Group* (groups of the firms), and TP (time-period), using the ES as follows:

$$LY \ C \ @Expand(Group.TP, @Dropfirst) \ LY(-1) * @Expand(Group, TP)$$
$$LY(-2) * @Expand(Group, TP) \tag{14.19}$$

Dependent Variable: LY				
Method: Panel EGLS (Period random effects)				
Date: 07/10/12 Time: 12:46				
Sample (adjusted): 3 28				
Periods included: 26				
Cross-sections included: 82				
Total panel (balanced) observations: 2132				
Swamy and Arora estimator of component variances				
White period standard errors & covariance (d.f. corrected)				

Variable	Coefficient	Std. Error	t-Statistic	Prob.
C	0.034668	0.037754	0.918253	0.3586
GROUP=1 AND TP=2	0.045517	0.043936	1.035981	0.3003
GROUP=2 AND TP=1	-0.057070	0.062736	-0.909680	0.3631
GROUP=2 AND TP=2	-0.162667	0.065474	-2.484468	0.0131
LY(-1)*(GROUP=1 AND TP=1)	1.108899	0.095163	11.65261	0.0000
LY(-1)*(GROUP=1 AND TP=2)	1.288069	0.026932	47.82732	0.0000
LY(-1)*(GROUP=2 AND TP=1)	1.046343	0.088724	11.79326	0.0000
LY(-1)*(GROUP=2 AND TP=2)	1.170756	0.119484	9.798458	0.0000
LY(-2)*(GROUP=1 AND TP=1)	-0.108439	0.095179	-1.139320	0.2547
LY(-2)*(GROUP=1 AND TP=2)	-0.290540	0.026775	-10.85117	0.0000
LY(-2)*(GROUP=2 AND TP=1)	-0.043246	0.088143	-0.490638	0.6237
LY(-2)*(GROUP=2 AND TP=2)	-0.163849	0.120794	-1.356440	0.1751

Effects Specification		
	S.D.	Rho
Period random	0.007944	0.0236
Idiosyncratic random	0.051080	0.9764

Weighted Statistics			
R-squared	0.999283	Mean dependent var	13.63252
Adjusted R-squared	0.999279	S.D. dependent var	1.902208
S.E. of regression	0.051066	Sum squared resid	5.528295
F-statistic	268620.9	Durbin-Watson stat	2.035393
Prob(F-statistic)	0.000000		

Unweighted Statistics			
R-squared	0.999280	Mean dependent var	23.54663
Sum squared resid	5.646105	Durbin-Watson stat	2.030273

Figure 14.20   *Statistical results of the simplest LV(2) PEREM in (14.19)*

For the simplest model, dichotomous variables *Group* and TP are generated using the equations Group $= 1 + 1*(code > 40)$ and $TP = 1 + 1*(T > 14)$. Then the statistical results in Figure 14.20 are obtained. Additional findings are as follows:

1. The null hypothesis H_0: $C(5) = C(6) = C(7) = C(8)$, $C(9) = C(10) = C(11) = C(12)$ is rejected based on the *F*-statistic of $F_0 = 4\,357\,365$ with $df = (62\ 120)$ and a *p*-value $= 0.0002$. So it can be concluded that *LY(−1)* and *LY(−2)* have different joint effects on *LY* between the groups of firms generated by Group and TP.
2. This model also presents the Hausman Test with the same notes as presented in Figure 14.19.

Example 14.15   A LV(2) PEREM of LY with exogenous variables by group and TP

As an extension of the model in (14.18), this example presents analysis based on a LV(2) PEREM with exogenous variables *LK* and *LL*, by Group and TP (time-period), using the ES as follows:

$$LY\ C\ @Expand(Group.TP, @Dropfirst)\ LY(-1) * @Expand(Group, TP)$$
$$LY(-2) * @Expand(Group, TP)\ LK * @Expand(Group, TP)\ LL * @Expand(Group, TP) \tag{14.20}$$

Note that this model is a translog linear or additive model by *Group* and *TP*. Based on this ES the error message presented in Figure 14.13 is obtained. For this reason, Figure 14.21 presents the statistical results based on one of its reduced models, which represents a set of four interaction PEREMs by the dichotomous Group and TP.

14.8   Two-Way Effects Models

14.8.1   Two-Way Random Effects Models

As an extension of the one-way CSREMs and PEREMs, various *two-way random effects models* (TWREMs) also can easily be applied. See the following example.

Dependent Variable: LY
Method: Panel EGLS (Period random effects)
Date: 07/12/12 Time: 10:12
Sample (adjusted): 3 28
Periods included: 26
Cross-sections included: 82
Total panel (balanced) observations: 2132
Swamy and Arora estimator of component variances
White period standard errors & covariance (d.f. corrected)

Variable	Coefficient	Std. Error	t-Statistic	Prob.
C	0.226180	0.261630	0.864502	0.3874
GROUP=1 AND TP=2	0.276903	0.255302	1.084608	0.2782
GROUP=2 AND TP=1	-0.395951	0.518937	-0.763004	0.4455
GROUP=2 AND TP=2	0.363347	0.531694	0.683376	0.4944
LY(-1)*(GROUP=1 AND TP=1)	1.104436	0.094893	11.63873	0.0000
LY(-1)*(GROUP=1 AND TP=2)	1.287718	0.029011	44.38728	0.0000
LY(-1)*(GROUP=2 AND TP=1)	1.031618	0.091157	11.31698	0.0000
LY(-1)*(GROUP=2 AND TP=2)	1.158129	0.115758	10.00470	0.0000
LY(-2)*(GROUP=1 AND TP=1)	-0.099507	0.095721	-1.039559	0.2987
LY(-2)*(GROUP=1 AND TP=2)	-0.306269	0.024873	-12.31347	0.0000
LY(-2)*(GROUP=2 AND TP=1)	-0.042979	0.091939	-0.467470	0.6402
LY(-2)*(GROUP=2 AND TP=2)	-0.136769	0.117885	-1.160190	0.2461
LK*(GROUP=1 AND TP=1)	-0.013622	0.014031	-0.970887	0.3317
LK*(GROUP=1 AND TP=2)	-0.002936	0.010739	-0.273379	0.7846
LK*(GROUP=2 AND TP=1)	0.021030	0.019152	1.098049	0.2723
LK*(GROUP=2 AND TP=2)	-0.040236	0.019868	-2.025165	0.0430
LL*(GROUP=1 AND TP=1)	-0.008922	0.018145	-0.491695	0.6230
LL*(GROUP=1 AND TP=2)	-0.024884	0.008270	-3.009120	0.0027
LL*(GROUP=2 AND TP=1)	0.007207	0.029398	0.245160	0.8064
LL*(GROUP=2 AND TP=2)	-0.049805	0.031006	-1.606321	0.1084

LK*LL*(GROUP=1 AND TP=1)	0.000463	0.000706	0.654865	0.5126
LK*LL*(GROUP=1 AND TP=2)	0.001082	0.000337	3.211040	0.0013
LK*LL*(GROUP=2 AND TP=1)	-0.000347	0.001238	-0.280619	0.7790
LK*LL*(GROUP=2 AND TP=2)	0.001850	0.001207	1.531977	0.1257

Effects Specification

	S.D.	Rho
Period random	0.003974	0.0061
Idiosyncratic random	0.050888	0.9939

Weighted Statistics

R-squared	0.999291	Mean dependent var	19.22494
Adjusted R-squared	0.999283	S.D. dependent var	1.910117
S.E. of regression	0.051146	Sum squared resid	5.514414
F-statistic	129133.4	Durbin-Watson stat	2.035515
Prob(F-statistic)	0.000000		

Unweighted Statistics

R-squared	0.999289	Mean dependent var	23.54663
Sum squared resid	5.573062	Durbin-Watson stat	2.032870

Figure 14.21 *Statistical results based on a reduced PEREM of the model in (14.20)*

Example 14.16   A heterogeneous two-way random effects model

For an illustration, Figure 14.22 presents the statistical results of a heterogeneous TWREM of *LY* on *LY* (−1), *LV*(−2), *LK*, *LL* and *LK*LL*, by two dichotomous factors, namely *Group* and *Time Period*(TP), using the ES of model in Figure 14.21.

Dependent Variable: LY
Method: Panel EGLS (Two-way random effects)
Date: 07/12/12 Time: 10:16
Sample (adjusted): 3 28
Periods included: 26
Cross-sections included: 82
Total panel (balanced) observations: 2132
Swamy and Arora estimator of component variances
White cross-section standard errors & covariance (d.f. corrected)
WARNING: estimated coefficient covariance matrix is of reduced rank

Variable	Coefficient	Std. Error	t-Statistic	Prob.
C	0.226270	0.145401	1.556175	0.1198
GROUP=1 AND TP=2	0.276336	0.269865	1.023979	0.3060
GROUP=2 AND TP=1	-0.396016	0.356654	-1.110365	0.2670
GROUP=2 AND TP=2	0.361589	0.273860	1.320342	0.1869
LY(-1)*(GROUP=1 AND TP=1)	1.104282	0.099775	11.06773	0.0000
LY(-1)*(GROUP=1 AND TP=2)	1.287059	0.074956	17.17084	0.0000
LY(-1)*(GROUP=2 AND TP=1)	1.031486	0.053966	19.11347	0.0000
LY(-1)*(GROUP=2 AND TP=2)	1.158015	0.074023	15.64389	0.0000
LY(-2)*(GROUP=1 AND TP=1)	-0.099340	0.095039	-1.045255	0.2960
LY(-2)*(GROUP=1 AND TP=2)	-0.305734	0.079065	-3.866843	0.0001
LY(-2)*(GROUP=2 AND TP=1)	-0.042836	0.053248	-0.804479	0.4212
LY(-2)*(GROUP=2 AND TP=2)	-0.136710	0.075056	-1.821449	0.0687
LK*(GROUP=1 AND TP=1)	-0.013636	0.010231	-1.332902	0.1827
LK*(GROUP=1 AND TP=2)	-0.002800	0.017815	-0.157194	0.8751
LK*(GROUP=2 AND TP=1)	0.021021	0.017150	1.225731	0.2204
LK*(GROUP=2 AND TP=2)	-0.040107	0.009694	-4.137469	0.0000
LL*(GROUP=1 AND TP=1)	-0.008928	0.010379	-0.860207	0.3898
LL*(GROUP=1 AND TP=2)	-0.024867	0.014006	-1.775446	0.0760
LL*(GROUP=2 AND TP=1)	0.007201	0.021984	0.327550	0.7433
LL*(GROUP=2 AND TP=2)	-0.049742	0.015457	-3.218025	0.0013

LK*LL*(GROUP=1 AND TP=1)	0.000463	0.000357	1.296221	0.1950
LK*LL*(GROUP=1 AND TP=2)	0.001082	0.000539	2.005973	0.0450
LK*LL*(GROUP=2 AND TP=1)	-0.000347	0.001016	-0.341801	0.7325
LK*LL*(GROUP=2 AND TP=2)	0.001847	0.000628	2.941078	0.0033

Effects Specification

	S.D.	Rho
Cross-section random	0.000000	0.0000
Period random	0.004248	0.0074
Idiosyncratic random	0.049040	0.9926

Weighted Statistics

R-squared	0.999291	Mean dependent var	18.52701
Adjusted R-squared	0.999283	S.D. dependent var	1.908985
S.E. of regression	0.051108	Sum squared resid	5.506052
F-statistic	129176.3	Durbin-Watson stat	2.035408
Prob(F-statistic)	0.000000		

Unweighted Statistics

R-squared	0.999289	Mean dependent var	23.54663
Sum squared resid	5.573098	Durbin-Watson stat	2.032371

Figure 14.22 *Statistical results of a two-way random effects model of the model in Figure (14.20)*

The stages of the analysis are as follows:

1. Select the option "*Random*" for both cross-section and period effects.
2. Select "*Coef covariance method: White cross-section*" . . . *OK*, the statistical results are obtained. Note that we can choose any one of the other methods.
3. Based on these results, the following notes and comments are presented.
   3.1 Similar to most CSREMs, this model also presents the Rho $= 0.0000$ for the cross-section random, then estimated variance: $\hat{\sigma}_{\varepsilon}^2 = 0.0000$.
   3.2 For the period *random*, Rho $= 0.0074$, compared to Rho $= 0.0061$ of the PEREM in Figure 14.21. With the SD $= 0.004\,248$, then its estimated variance $\hat{\sigma}_{\nu}^2 = (0.004248)^2 = 1.8E - 05$.
   3.3 Since some of the independent variables have large *p*-values, then a reduced model can be explored. Do it as an exercise.
   3.4 Note that the output presents the *WARNING*. However, it is surprising that no warning was obtained by using any of the other methods.

14.8.2 Two-Way Fixed Effects Model

Several two-way fixed effects models have been illustrated in Chapter 9, which are those with time and individual (firm) dummy variables. Based on this special structure panel data, this section will present additional *two-way fixed effects models* (TWFEMs).

Example 14.17 A TWFEM and a LSDV model

For illustration, Figure 14.23 presents the statistical results based on a two-way fixed model, using the equation specification (ES) in (14.21), and a part of the statistical results of a LSDV model using the ES

Dependent Variable: LY
Method: Panel Least Squares
Date: 10/10/12 Time: 15:21
Sample (adjusted): 3 28
Periods included: 26
Cross-sections included: 82
Total panel (balanced) observations: 2132

Variable	Coefficient	Std. Error	t-Statistic	Prob.
C	1.233757	0.519152	2.376484	0.0176
LY(-1)	1.064259	0.020974	50.74212	0.0000
LY(-2)	-0.133358	0.021858	-6.101222	0.0000
LK	0.017670	0.021102	0.837387	0.4025
LL	-0.007574	0.031603	-0.239664	0.8106
LK*LL	0.000295	0.001224	0.240608	0.8099

Effects Specification

Cross-section fixed (dummy variables)
Period fixed (dummy variables)

R-squared	0.999355	Mean dependent var	23.54663
Adjusted R-squared	0.999320	S.D. dependent var	1.918040
S.E. of regression	0.050019	Akaike info criterion	-3.101717
Sum squared resid	5.053880	Schwarz criterion	-2.804128
Log likelihood	3418.430	Hannan-Quinn criter.	-2.992802
F-statistic	28211.24	Durbin-Watson stat	2.041178
Prob(F-statistic)	0.000000		

(a)

Dependent Variable: LY
Method: Panel Least Squares
Date: 10/10/12 Time: 15:24
Sample: 1 28 IF YEAR >1961
Periods included: 26
Cross-sections included: 82
Total panel (balanced) observations: 2132

Variable	Coefficient	Std. Error	t-Statistic	Prob.
C	1.351059	0.537397	2.514081	0.0120
LY(-1)	1.064259	0.020974	50.742120	0.0000
LY(-2)	-0.133358	0.021858	-6.101222	0.0000
LK	0.017670	0.021102	0.837387	0.4025
LL	-0.007574	0.031603	-0.239664	0.8106
LK*LL	0.000295	0.001224	0.240608	0.8099
Code Dumies (81)	Yes			
Year Dummies	Yes			
R-squared	0.999355	Mean dependent var	23.54663
Adjusted R-squared	0.999320	S.D. dependent var	1.91804
S.E. of regression	0.050019	Akaike info criterion	-3.101717
Sum squared resid	5.053880	Schwarz criterion	-2.804128
Log likelihood	3418.430	Hannan-Quinn criter.	-2.992802
F-statistic	28211.240	Durbin-Watson stat	2.041178
Prob(F-statistic)	0.000000		

(b)

Figure 14.23 *Statistical results of a two-way fixed-effects model (14.21) and a LSDV (14.22)*

in (14.22) based on the sample {Year > 1961}, for a comparison.

$$LY \ C \ LY(-1) \ LY(-2) \ LK \ LL \ LK * LL \tag{14.21}$$

$$LY \ C \ LY(-1) \ LY(-2) \ LK \ LL \ LK * LL$$
$$@Expand(Code, @Droplast) \ @Expand(Year, @Droplast) \tag{14.22}$$

Based on these results, the following notes and comments are presented.

1. For the TWFEM model, the statistical results are obtained by selecting the options *Fixed* for both cross-section and period effect specifications.
2. Even though each of the independent variables LK, LL and LK*LL has large *p*-values, the results are acceptable estimates. The three variables are highly correlated so the reduced model should be explored. By deleting both LK and LL, a reduced model is obtained where LK*LL is a significant adjusted effect with a *p*-value = 0.0274. Note that the interaction LK*LL should be kept in the model, since it is known, in a theoretical sense, that the effect of LK (LL) on Y should depend on LL (LK).
3. In this case, the estimation method is "*Panel Least Squares*": it is not a GLS estimation method. Additional analysis should be done for exercises.
4. Note the model in Figure 14.23 represents an additive two-way ANCOVA model generated by 82(firms) × 28(time points) = 2296 cells. However, the regressions of *LY* on the numerical variables *LY(−1)*, *LY(−2)*, *LK*, *LL* and *LK*LL*, only have (82 + 28 − 1) intercept parameters, indicated by the additive 82 firm dummies, 26 period dummies and the parameter "C" in the model. This model is acceptable in a statistical sense, but I wouldn't consider this TWFEM appropriate in a theoretical sense because the numerical independent variables have the same effects within the 2296 cells, with a specific hidden assumption. Refer to the parameters of an additive model presented in Table 7.4, as well as special notes and comments for the fixed-effects models, presented in Chapter 9.
5. For a comparison, Figure 14.23 also presents a part of the statistical results of LSDV model using the ES in (14.22), based on the sample {Year > 1961}, because years of observations are 1960–1987 and the model is a LV(2) model. Note that estimates of the first six parameters of the model are exactly the same as the estimates of the parameters of the TWFEM in (9.21).
6. The advantages in using LSDV models are that we can apply multiple fixed-effects models, which have been presented in Chapter 9, such as the models in (9.2) and (9.28), and the models presented by Engelberg and Parsons (2011).
7. Note that the three independent variables LK, LL and LK*LL are kept in the model, even though each of them has very large *Prob*, since this model will be used for several testing hypotheses in Section 14.9.
8. To generalize, the LV(p) TWFEMs can be presented using the following ES. For various alternative models refer to the special notes on the exogenous variables of the general models in (14.1).

$$Y \ C \ Y(-1) \ldots Y(-p) \ X1 \ldots Xk \ldots XK \tag{14.23}$$

where *Xk* can be various alternative variables, as indicated for the general model in (14.1).

Example 14.18   An application
Corresponding to the TWREM in Figure 14.22, a *cross-section fixed- and period random effects model* (CSF-PEREM) is considered, with the statistical results presented in Figure 14.24.
   Based on these results, the following notes and comments are presented.

1. The results obtained by selecting the option *Fixed* for the Cross-section Effect Specification, and the option *Random* for the Period Effect Specification

Dependent Variable: LY
Method: Panel EGLS (Period random effects)
Date: 07/12/12 Time: 12:00
Sample (adjusted): 3 28
Periods included: 26
Cross-sections included: 82
Total panel (balanced) observations: 2132
Swamy and Arora estimator of component variances
White cross-section standard errors & covariance (d.f. corrected)

Variable	Coefficient	Std. Error	t-Statistic	Prob.
C	1.986941	0.824706	2.409273	0.0161
TP=2	0.620861	0.277256	2.239305	0.0252
LY(-1)*(TP=1)	0.947968	0.073824	12.84091	0.0000
LY(-1)*(TP=2)	1.158402	0.062609	18.50205	0.0000
LY(-2)*(TP=1)	-0.018102	0.067560	-0.267941	0.7888
LY(-2)*(TP=2)	-0.224992	0.072155	-3.118163	0.0018
LK*(TP=1)	-0.009667	0.036347	-0.265965	0.7903
LK*(TP=2)	-0.038853	0.036031	-1.078317	0.2810
LL*(TP=1)	-0.062979	0.053623	-1.174468	0.2403
LL*(TP=2)	-0.095187	0.054344	-1.751568	0.0800
LK*LL*(TP=1)	0.002406	0.002225	1.081255	0.2797
LK*LL*(TP=2)	0.003729	0.002136	1.745726	0.0810

Effects Specification	S.D.	Rho
Cross-section fixed (dummy variables)		
Period random	0.007055	0.0199
Idiosyncratic random	0.049522	0.9801

Weighted Statistics			
R-squared	0.999348	Mean dependent var	23.54663
Adjusted R-squared	0.999319	S.D. dependent var	1.903168
S.E. of regression	0.049677	Sum squared resid	5.031897
F-statistic	33974.41	Durbin-Watson stat	2.066081
Prob(F-statistic)	0.000000		

Unweighted Statistics			
R-squared	0.999341	Mean dependent var	23.54663
Sum squared resid	5.164815	Durbin-Watson stat	2.051215

Figure 14.24   *Statistical results of a CSF_PEREM in (14.23)*

2. If the option *Fixed* is used for both Cross-section and Period Effect specifications, then its design matrix would be singular, caused by the variable *TP*, which represents the time variable or DATEID in Figure 14.23.

3. Note that both LY(−2)*(TP = 1) and LK*(TP = 1) have large *p*-values and they are in the same time period, so then in order to obtain a reduced model the trial-and-error method should be applied to delete either one at the first two steps. Otherwise, we should delete both of the variables. In this case we find that both variables should be deleted.

4. This model could easily be extended to the time period with three or more levels, which should be considered the relevant one.

5. To generalize the model in (4.20), the LV(p) CSF_PEREMs by the time period (TP), we would have the following general ES: refer to the special notes earlier.

$$LY \quad C \quad @Expand(TP, @Dropfirst) \ LY(-1) * @Expand(TP) \dots LY(-p) * @Expand(TP)$$
$$X1 * @Expand(TP) \ \dots \ Xk * @Expand(TP) \ \dots \ XK * @Expand(TP) \tag{14.24}$$

where *Xk* can be various alternative variables, as indicated for the general model in (14.1), based on the sample {Time-t > p}.

14.8.3   Two-Way CSRandom-PEFixed-Effects Model

Example 14.19   An application
Corresponding to the TWREM in Figure 14.22, a *cross-section random and period fixed-effects model* (CSR_PEFEM) by GROUP would be presented using the following ES, with the statistical results presented in Figure 14.25.

$$LY \ C \ @Expand(Group, @Dropfirst) \ LY(-1) * @Expand(Group) \ LY(-2) * @Expand(Group)$$
$$LK * @Expand(Group) \ LL * @Expand(Group) \ LK * LL * @Expand(Group) \tag{14.25}$$

Based on these results, the following notes and comments are presented.

1. The results obtained by selecting the option *Random* for the Cross-section Effect Specification, and the option *Fixed* for the Period Effect Specification.

Dependent Variable: LY
Method: Panel EGLS (Cross-section random effects)
Date: 07/12/12   Time: 12:05
Sample (adjusted): 3 28
Periods included: 26
Cross-sections included: 82
Total panel (balanced) observations: 2132
Swamy and Arora estimator of component variances

Variable	Coefficient	Std. Error	t-Statistic	Prob.
C	0.292672	0.176230	1.660740	0.0969
GROUP=2	-0.033145	0.298202	-0.111151	0.9115
LY(-1)*(GROUP=1)	1.225202	0.030541	40.11709	0.0000
LY(-1)*(GROUP=2)	1.099891	0.025270	43.52491	0.0000
LY(-2)*(GROUP=1)	-0.229469	0.031089	-7.381074	0.0000
LY(-2)*(GROUP=2)	-0.095915	0.025745	-3.725635	0.0002
LK*(GROUP=1)	-0.007865	0.008911	-0.882635	0.3775
LK*(GROUP=2)	-0.010653	0.010524	-1.012196	0.3116
LL*(GROUP=1)	-0.012433	0.011367	-1.093816	0.2742
LL*(GROUP=2)	-0.026596	0.015648	-1.699644	0.0893
LK*LL*(GROUP=1)	0.000590	0.000441	1.336970	0.1814
LK*LL*(GROUP=2)	0.000966	0.000626	1.543389	0.1229

Effects Specification		S.D.	Rho
Cross-section random		0.000000	0.0000
Period fixed (dummy variables)			
Idiosyncratic random		0.049554	1.0000

Weighted Statistics			
R-squared	0.999299	Mean dependent var	23.54663
Adjusted R-squared	0.999287	S.D. dependent var	1.918040
S.E. of regression	0.051223	Sum squared resid	5.496807
F-statistic	82940.21	Durbin-Watson stat	2.027945
Prob(F-statistic)	0.000000		

Unweighted Statistics			
R-squared	0.999299	Mean dependent var	23.54663
Sum squared resid	5.496807	Durbin-Watson stat	2.027945

Figure 14.25   *Statistical results of a CSR_PEREM in (14.25)*

2. If the option *Fixed* is used for both Cross-section and Period Effect specifications, then its design matrix would be singular, caused by the variable *Group*, which represents the firm/individual variable or CODE in Figure 14.23.
3. To generalize the model in (14.25), the LV(p) CSF_PEREMs by GROUP would have the following general ES: refer to the special notes earlier.

$$LY \ C \ @Expand(Group, @Dropfirst)$$
$$LY(-1) * @Expand(Group) \ldots LY(-p) * @Expand(Group) \tag{14.26}$$
$$X1 * @Expand(Group) \ldots Xk * @Expand(Group) \ldots XK * @Expand(Group)$$

where Xk can be various alternative variables, as indicated for the general model in (14.1) based on the sample {Time $t > $ p}.

14.9   Testing Hypotheses

EViews provides Fixed/Random Effects Testing, namely the *Redundant Fixed Effects* – Likelihood Ratio and the *Correlated Random Effects* – Hausman Tests in addition to the *Coefficient Tests* and *Residual Tests*. However, it should be noted that the conclusion of testing a hypotheses does not directly mean it represents the true conditions in the corresponding population, since we never know the true characteristic(s) of that population, and testing uses sampled data, which happens to be selected or available to researchers. Refer to special notes and comments presented in Agung (2009a, Section 2.14).

14.9.1   Redundant Fixed Effects

The objective of this test is to explore whether or not a variable or a set of variables should be inserted as additional independent variables of the model considered. This test also can be applied to develop a model using the manual stepwise selection method.

Example 14.20   Redundant fixed-effect testing for the TWFEM using the ES in (14.20)
The steps of analysis are as follows:

1. Conduct the data analysis based on the TWFEM model in (14.20). The statistical results are presented in Figure 14.23(a)
2. Get the output on the screen, by selecting the objects: *View, Fixe/Random Effects Testing*, and *Redundant Effects – Likelihood Ratio*, four sets of the statistical results are obtained, as presented in Figure 14.26(a–d).

Redundant Fixed Effects Tests
Equation: EQ10_FF
Test cross-section and period fixed effects

Effects Test	Statistic	d.f.	Prob.
Cross-section F	2.655974	(81,2020)	0.0000
Cross-section Chi-square	215.766160	81	0.0000
Period F	3.149997	(25,2020)	0.0000
Period Chi-square	81.537029	25	0.0000
Cross-Section/Period F	3.194295	(106,2020)	0.0000
Cross-Section/Period Chi-square	330.393279	106	0.0000

(a)

Cross-section fixed effects test equation:
Dependent Variable: LY
Method: Panel Least Squares
Date: 10/11/12   Time: 13:15
Sample (adjusted): 3 28
Periods included: 26
Cross-sections included: 82
Total panel (balanced) observations: 2132

Variable	Coefficient	Std. Error	t-Statistic	Prob.
C	0.196160	0.145914	1.344358	0.1790
LY(-1)	1.163432	0.020368	57.11991	0.0000
LY(-2)	-0.168159	0.020524	-8.193203	0.0000
LK	-0.003615	0.006643	-0.544158	0.5864
LL	-0.008842	0.009148	-0.966553	0.3339
LK*LL	0.000455	0.000369	1.233232	0.2176

Effects Specification

Period fixed (dummy variables)

R-squared	0.999287	Mean dependent var	23.54663
Adjusted R-squared	0.999277	S.D. dependent var	1.918040
S.E. of regression	0.051591	Akaike info criterion	-3.076498
Sum squared resid	5.592128	Schwarz criterion	-2.994130
Log likelihood	3310.547	Hannan-Quinn criter.	-3.046352
F-statistic	98110.74	Durbin-Watson stat	2.037734
Prob(F-statistic)	0.000000		

(b)

Period fixed effects test equation:
Dependent Variable: LY
Method: Panel Least Squares
Date: 10/11/12   Time: 13:15
Sample (adjusted): 3 28
Periods included: 26
Cross-sections included: 82
Total panel (balanced) observations: 2132

Variable	Coefficient	Std. Error	t-Statistic	Prob.
C	0.851587	0.473247	1.799455	0.0721
LY(-1)	1.076333	0.020784	51.78771	0.0000
LY(-2)	-0.140266	0.021705	-6.462479	0.0000
LK	0.023712	0.021228	1.117041	0.2641
LL	0.006835	0.030900	0.221189	0.8250
LK*LL	2.32E-07	0.001233	0.000188	0.9998

Effects Specification

Cross-section fixed (dummy variables)

R-squared	0.999330	Mean dependent var	23.54663
Adjusted R-squared	0.999302	S.D. dependent var	1.918040
S.E. of regression	0.050672	Akaike info criterion	-3.086925
Sum squared resid	5.250906	Schwarz criterion	-2.855762
Log likelihood	3377.662	Hannan-Quinn criter.	-3.002321
F-statistic	35478.76	Durbin-Watson stat	2.043876
Prob(F-statistic)	0.000000		

(c)

Cross-section and period fixed effects test equation:
Dependent Variable: LY
Method: Panel Least Squares
Date: 10/11/12   Time: 13:15
Sample (adjusted): 3 28
Periods included: 26
Cross-sections included: 82
Total panel (balanced) observations: 2132

Variable	Coefficient	Std. Error	t-Statistic	Prob.
C	0.278498	0.148400	1.876669	0.0607
LY(-1)	1.189940	0.020200	58.90720	0.0000
LY(-2)	-0.189741	0.020470	-9.269207	0.0000
LK	-0.011446	0.006677	-1.714229	0.0866
LL	-0.013386	0.009312	-1.437474	0.1507
LK*LL	0.000619	0.000376	1.648588	0.0994

R-squared	0.999247	Mean dependent var	23.54663
Adjusted R-squared	0.999246	S.D. dependent var	1.918040
S.E. of regression	0.052684	Akaike info criterion	-3.046185
Sum squared resid	5.901018	Schwarz criterion	-3.030243
Log likelihood	3253.234	Hannan-Quinn criter.	-3.040351
F-statistic	564466.3	Durbin-Watson stat	2.037681
Prob(F-statistic)	0.000000		

(d)

Figure 14.26   The redundant fixed-effects tests for a TWFEM in (14.20)

Based on these statistical results, the following conclusions are obtained.

1. Based on the two *F*-statistics presented in Figure 14.26(a) respectively, can be concluded that the cross-section and period dummy variables have significant adjusted effects on *LY*.
2. Note that the three Figure 14.26(b), (c) and (d), respectively, have the titles (b) *Cross-section fixed test equation*, (c) *Period fixed test equation*, and (d) *Cross-section and period fixed test equation*, with their own *F*-statistics.
   - The *F*-statistic in Figure 14.26(b) is for testing the joint effects of the five numerical variables and the period dummies on LY.
   - The *F*-statistic in Figure 14.26(c) is for testing the joint effects of the five numerical variables and the cross-section dummies on LY.
   - The *F*-statistic in Figure 14.26(d) is for testing the joint effects of the five numerical variables on LY.
3. At the $\alpha = 0.10$ level of significance, each of the numerical independent variables has either positive or negative significant effects on LY if and only if the *p*-value = Prob(t-stat)/2 $< \alpha = 0.10$. For instance, in the fourth set of statistical results the variable LL has a negative significant adjusted effect on LY with a *p*-value = $0.1507/2 = 0.075\,035 < \alpha = 0.10$.

Example 14.21   Redundant fixed-effect testing for the CSF_PEREM using the ES in (14.20)

By using the same steps as in the previous example, the set of statistical results in Figure 14.27 is obtained. Based on the statistical results in this figure the flowing notes and conclusions are presented.

1. Figure 14.27(a) presents the output of the CSF_PEREM based on the CES.wf1.
2. Based on the *F*-statistic of $F_0 = 2.916\,865$ with $df = (81.2045)$ and $p = 0.0000$, in Figure 14.27(b), it can be concluded that the set of cross-section dummies has a significant effect on LY. Note that the "81-df" for the *F*-statistic represents the 82 cross-section dummies.
3. Figure 14.27(c) presents the output with the title "*Cross-Section fixed tests equations*". Based on this output, the following findings and notes are presented.
   - The *Period* random and *Idiosyncratic* random effects specification have exactly the same values as the output in Figure 14.27(a).
   - But these have different values for the other statistics, specifically for the *t*-statistic and the *F*-statistic. Therefore, it can be said that the *t*-statistic and *F*-statistic in the two outputs should be used to test different types of effects.
   - Note the model in this output has five numerical independent variables and a period random variable, but the model in Figure 14.27(a) has five numerical independent variables, the period random variable and a set of cross-section dummies.

Example 14.22   Redundant fixed-effect testing for the CSR_PEFEM using the ES in (14.20)

By using the same steps as in the previous example, the set of statistical results in Figure 14.28 is obtained. Based on the statistical results in this figure, the following notes and conclusions are presented.

1. Figure 14.28(a) presents the output of the CSR_PEFEM, based on the CES.wf1.
2. Based on the *F*-statistic of $F_0 = 4.642\,089$ with $df = (25.2101)$ and $p = 0.0000$ in Figure 14.28(b), it can be concluded that the set of period dummies has a significant effect on LY. Note that the "25-df" for the *F*-statistic represents the $(28 - 2 \text{ lags}) = 26$ time effects.
3. Figure 14.28(c) presents the output with the title "*Cross-Section fixed tests equations*". Based on this output, the following findings and notes are presented.
   - The cross-section random and idiosyncratic random effects specification have exactly the same values as the output in Figure 14.28(a).

Dependent Variable: LY
Method: Panel EGLS (Period random effects)
Date: 10/11/12 Time: 14:25
Sample (adjusted): 3 28
Periods included: 26
Cross-sections included: 82
Total panel (balanced) observations: 2132
Swamy and Arora estimator of component variances

Variable	Coefficient	Std. Error	t-Statistic	Prob.
C	0.938890	0.474103	1.980351	0.0478
LY(-1)	1.070264	0.020746	51.58840	0.0000
LY(-2)	-0.135804	0.021605	-6.285605	0.0000
LK	0.021554	0.020997	1.026539	0.3048
LL	0.003447	0.030632	0.112546	0.9104
LK*LL	0.000147	0.001221	0.120391	0.9042

Effects Specification

		S.D.	Rho
Cross-section fixed (dummy variables)			
Period random		0.005682	0.0127
Idiosyncratic random		0.050019	0.9873

Weighted Statistics

R-squared	0.999334	Mean dependent var	23.54663
Adjusted R-squared	0.999306	S.D. dependent var	1.905809
S.E. of regression	0.050190	Sum squared resid	5.151350
F-statistic	35704.81	Durbin-Watson stat	2.042072
Prob(F-statistic)	0.000000		

Unweighted Statistics

R-squared	0.999330	Mean dependent var	23.54663
Sum squared resid	5.251517	Durbin-Watson stat	2.031379

(a)

Redundant Fixed Effects Tests
Equation: EQ10_FR
Test cross-section fixed effects

Effects Test	Statistic	d.f.	Prob.
Cross-section F	2.916865	(81,2045)	0.0000

(b)

Cross-section fixed effects test equation:
Dependent Variable: LY
Method: Panel EGLS (Period random effects)
Date: 10/13/12 Time: 14:27
Sample (adjusted): 3 28
Periods included: 26
Cross-sections included: 82
Total panel (balanced) observations: 2132
Use pre-specified random component estimates
Swamy and Arora estimator of component variances

Variable	Coefficient	Std. Error	t-Statistic	Prob.
C	0.238779	0.141178	1.691330	0.0909
LY(-1)	1.177228	0.019459	60.49786	0.0000
LY(-2)	-0.179396	0.019665	-9.122800	0.0000
LK	-0.007676	0.006388	-1.201595	0.2297
LL	-0.011196	0.008855	-1.264335	0.2062
LK*LL	0.000540	0.000357	1.512019	0.1307

Effects Specification

	S.D.	Rho
Period random	0.005682	0.0127
Idiosyncratic random	0.050019	0.9873

Weighted Statistics

R-squared	0.999258	Mean dependent var	16.41378
Adjusted R-squared	0.999256	S.D. dependent var	1.905809
S.E. of regression	0.051990	Sum squared resid	5.746504
F-statistic	572280.6	Durbin-Watson stat	2.037555
Prob(F-statistic)	0.000000		

Unweighted Statistics

R-squared	0.999247	Mean dependent var	23.54663
Sum squared resid	5.905256	Durbin-Watson stat	2.010206

(c)

Figure 14.27 *Statistical results based on the CSF_PEREM in (14.20) and its redundant fixed-effects tests*

- But these have different values for the other statistics, specifically for the t-statistic and the F-statistic. Therefore, it can be said that the t-statistic and F-statistic in the two outputs should be used to test different type of effects.
- Note the model in this output has five numerical independent variables and a cross-section random variable, but the model in Figure 14.28(a) has the five numerical independent variables, the cross-section random variable and a set of cross-section dummies.

14.9.2   Hausman Test: Correlated Random Effects

Example 14.23   Hausman Test for the TWREM using the ES in (4.20)

With the statistical results of the TWREM in Figure 14.29(a), the statistical results of its Hausman test can easily be obtained, as presented in Figure 14.29(b–e).

Dependent Variable: LY
Method: Panel EGLS (Cross-section random effects)
Date: 10/11/12   Time: 12:37
Sample (adjusted): 3 28
Periods included: 26
Cross-sections included: 82
Total panel (balanced) observations: 2132
Swamy and Arora estimator of component variances

Variable	Coefficient	Std. Error	t-Statistic	Prob.
C	0.196160	0.141468	1.386608	0.1657
LY(-1)	1.163432	0.019748	58.91505	0.0000
LY(-2)	-0.168159	0.019899	-8.450695	0.0000
LK	-0.003615	0.006441	-0.561260	0.5747
LL	-0.008842	0.008869	-0.996929	0.3189
LK*LL	0.000455	0.000358	1.271989	0.2035

Effects Specification

	S.D.	Rho
Cross-section random	0.000000	0.0000
Period fixed (dummy variables)		
Idiosyncratic random	0.050019	1.0000

Weighted Statistics

R-squared	0.999287	Mean dependent var	23.54663
Adjusted R-squared	0.999277	S.D. dependent var	1.918040
S.E. of regression	0.051591	Sum squared resid	5.592128
F-statistic	98110.74	Durbin-Watson stat	2.037734
Prob(F-statistic)	0.000000		

Unweighted Statistics

R-squared	0.999287	Mean dependent var	23.54663
Sum squared resid	5.592128	Durbin-Watson stat	2.037734

(a)

Redundant Fixed Effects Tests
Equation: EQ10_RF
Test period fixed effects

Effects Test	Statistic	d.f.	Prob.
Period F	4.642089	(25,2101)	0.0000

(b)

Period fixed effects test equation:
Dependent Variable: LY
Method: Panel EGLS (Cross-section random effects)
Date: 10/13/12   Time: 15:28
Sample (adjusted): 3 28
Periods included: 26
Cross-sections included: 82
Total panel (balanced) observations: 2132
Use pre-specified random component estimates
Swamy and Arora estimator of component variances

Variable	Coefficient	Std. Error	t-Statistic	Prob.
C	0.278498	0.140893	1.976664	0.0482
LY(-1)	1.189940	0.019178	62.04595	0.0000
LY(-2)	-0.189741	0.019434	-9.763098	0.0000
LK	-0.011446	0.006339	-1.805568	0.0711
LL	-0.013386	0.008841	-1.514067	0.1302
LK*LL	0.000619	0.000357	1.736430	0.0826

Effects Specification

	S.D.	Rho
Cross-section random	0.000000	0.0000
Idiosyncratic random	0.050019	1.0000

Weighted Statistics

R-squared	0.999247	Mean dependent var	23.54663
Adjusted R-squared	0.999246	S.D. dependent var	1.918040
S.E. of regression	0.052684	Sum squared resid	5.901018
F-statistic	564466.3	Durbin-Watson stat	2.037681
Prob(F-statistic)	0.000000		

Unweighted Statistics

R-squared	0.999247	Mean dependent var	23.54663
Sum squared resid	5.901018	Durbin-Watson stat	2.037681

(c)

Figure 14.28   *Statistical results based on the CSR_PEFEM in (14.20) and its redundant fixed-effects tests*

Based on these statistical results, the following notes and comments are presented.

1. Statistical results of the TWREM in Figure 14.29(a), show that each of the numerical independent variables has either positive or negative significant adjusted effects at the $\alpha = 0.15$ level of significance. For instance, LK has a negative significant adjusted effect on LY with a p-value $= 0.2478/2 = 0.1239 < \alpha = 0.15$.

2. Figure 14.29(b) presents the correlated random effects – Hausman test, namely the test cross-section and random effects using the Chi-square statistic, and the cross-section random effects comparison using the t-statistic.

3. It is unexpected: this Hausman test also presents the output in Figure 14.29(c), which is exactly the same as the output in Figure 14.27(a) based on the CSF_PEREM, the output in Figure 14.29(d), which is exactly the same as the output in Figure 14.28(a) based on the CSR_PEFEM and the output in Figure 14.29(e), which is exactly the same as the output in Figure 14.30(a) based on the CSF_PEFEM.

```
Dependent Variable: LY
Method: Panel EGLS (Two-way random effects)
Date: 10/11/12   Time: 13:38
Sample (adjusted): 3 28
Periods included: 26
Cross-sections included: 82
Total panel (balanced) observations: 2132
Swamy and Arora estimator of component variances
```

Variable	Coefficient	Std. Error	t-Statistic	Prob.
C	0.238779	0.146741	1.627215	0.1038
LY(-1)	1.177228	0.020226	58.20450	0.0000
LY(-2)	-0.179396	0.020439	-8.776972	0.0000
LK	-0.007676	0.006640	-1.156045	0.2478
LL	-0.011196	0.009204	-1.216406	0.2240
LK*LL	0.000540	0.000371	1.454701	0.1459

Effects Specification

		S.D.	Rho
Cross-section random		0.000000	0.0000
Period random		0.005682	0.0127
Idiosyncratic random		0.050019	0.9873

Weighted Statistics

R-squared	0.999258	Mean dependent var	16.41378
Adjusted R-squared	0.999256	S.D. dependent var	1.905809
S.E. of regression	0.051990	Sum squared resid	5.746504
F-statistic	572280.6	Durbin-Watson stat	2.037555
Prob(F-statistic)	0.000000		

Unweighted Statistics

R-squared	0.999247	Mean dependent var	23.54663
Sum squared resid	5.905256	Durbin-Watson stat	2.010206

(a)

```
Correlated Random Effects - Hausman Test
Equation: EQ10_RR
Test cross-section and period random effects
```

Test Summary	Chi-Sq. Statistic	Chi-Sq. d.f.	Prob.
Cross-section random	1222.406363	5	0.0000
Period random	0.000000	5	1.0000
Cross-section and period random	501.256959	5	0.0000

* Period test variance is invalid. Hausman statistic set to zero.
** WARNING: estimated cross-section random effects variance is zero.

Cross-section random effects test comparisons:

Variable	Fixed	Random	Var(Diff.)	Prob.
LY(-1)	1.070264	1.177228	0.000021	0.0000
LY(-2)	-0.135804	-0.179396	0.000049	0.0000
LK	0.021554	-0.007676	0.000397	0.1423
LL	0.003447	-0.011196	0.000854	0.6162
LK*LL	0.000147	0.000540	0.000001	0.7353

(b)

```
Cross-section random effects test equation:
Dependent Variable: LY
Method: Panel EGLS (Period random effects)
Date: 10/11/12   Time: 15:15
Sample (adjusted): 3 28
Periods included: 26
Cross-sections included: 82
Total panel (balanced) observations: 2132
Swamy and Arora estimator of component variances
```

Variable	Coefficient	Std. Error	t-Statistic	Prob.
C	0.938890	0.474103	1.980351	0.0478
LY(-1)	1.070264	0.020746	51.58840	0.0000
LY(-2)	-0.135804	0.021605	-6.285605	0.0000
LK	0.021554	0.020997	1.026539	0.3048
LL	0.003447	0.030632	0.112546	0.9104
LK*LL	0.000147	0.001221	0.120391	0.9042

Effects Specification

		S.D.	Rho
Cross-section fixed (dummy variables)			
Period random		0.005682	0.0127
Idiosyncratic random		0.050019	0.9873

Weighted Statistics

R-squared	0.999334	Mean dependent var	23.54663
Adjusted R-squared	0.999306	S.D. dependent var	1.905809
S.E. of regression	0.050190	Sum squared resid	5.151350
F-statistic	35704.81	Durbin-Watson stat	2.042072
Prob(F-statistic)	0.000000		

Unweighted Statistics

R-squared	0.999330	Mean dependent var	23.54663
Sum squared resid	5.251517	Durbin-Watson stat	2.031379

(c)

Figure 14.29   *Statistical results based on the TWREM in (14.20) and its Hausman test*

14.9.3   Coefficient Tests

For illustration, the summary of statistical results based on the four two-way effects models, namely the two-way random effects model (TWREM), two-way cross-section random and period fixed-effects model (CSR_PEFEM), two-way cross-section fixed- and period random effects model (CSF_PEREM) and the two-way period fixed effects model (TWFEM), is presented for comparative study.

Period random effects test comparisons:

Variable	Fixed	Random	Var(Diff.)	Prob.
LY(-1)	1.163432	1.177228	-0.000019	NA
LY(-2)	-0.168159	-0.179396	-0.000022	NA
LK	-0.003615	-0.007676	-0.000003	NA
LL	-0.008842	-0.011196	-0.000006	NA
LK*LL	0.000455	0.000540	-0.000000	NA

Period random effects test equation:
Dependent Variable: LY
Method: Panel EGLS (Cross-section random effects)
Date: 10/11/12   Time: 15:15
Sample (adjusted): 3 28
Periods included: 26
Cross-sections included: 82
Total panel (balanced) observations: 2132
Swamy and Arora estimator of component variances

Variable	Coefficient	Std. Error	t-Statistic	Prob.
C	0.196160	0.141468	1.386608	0.1657
LY(-1)	1.163432	0.019748	58.91505	0.0000
LY(-2)	-0.168159	0.019899	-8.450695	0.0000
LK	-0.003615	0.006441	-0.561260	0.5747
LL	-0.008842	0.008869	-0.996929	0.3189
LK*LL	0.000455	0.000358	1.271989	0.2035

Effects Specification

	S.D.	Rho
Cross-section random	0.000000	0.0000
Period fixed (dummy variables)		
Idiosyncratic random	0.050019	1.0000

Weighted Statistics

R-squared	0.999287	Mean dependent var	23.54663
Adjusted R-squared	0.999277	S.D. dependent var	1.918040
S.E. of regression	0.051591	Sum squared resid	5.592128
F-statistic	98110.74	Durbin-Watson stat	2.037734
Prob(F-statistic)	0.000000		

Unweighted Statistics

R-squared	0.999287	Mean dependent var	23.54663
Sum squared resid	5.592128	Durbin-Watson stat	2.037734

(d)

Cross-section and period random effects test comparisons:

Variable	Fixed	Random	Var(Diff.)	Prob.
LY(-1)	1.064259	1.177228	0.000031	0.0000
LY(-2)	-0.133358	-0.179396	0.000060	0.0000
LK	0.017670	-0.007676	0.000401	0.2057
LL	-0.007574	-0.011196	0.000914	0.9047
LK*LL	0.000295	0.000540	0.000001	0.8334

Cross-section and period random effects test equation:
Dependent Variable: LY
Method: Panel Least Squares
Date: 10/11/12   Time: 15:15
Sample (adjusted): 3 28
Periods included: 26
Cross-sections included: 82
Total panel (balanced) observations: 2132

Variable	Coefficient	Std. Error	t-Statistic	Prob.
C	1.233757	0.519152	2.376484	0.0176
LY(-1)	1.064259	0.020974	50.74212	0.0000
LY(-2)	-0.133358	0.021858	-6.101222	0.0000
LK	0.017670	0.021102	0.837387	0.4025
LL	-0.007574	0.031603	-0.239664	0.8106
LK*LL	0.000295	0.001224	0.240608	0.8099

Effects Specification

Cross-section fixed (dummy variables)
Period fixed (dummy variables)

R-squared	0.999355	Mean dependent var	23.54663
Adjusted R-squared	0.999320	S.D. dependent var	1.918040
S.E. of regression	0.050019	Akaike info criterion	-3.101717
Sum squared resid	5.053880	Schwarz criterion	-2.804128
Log likelihood	3418.430	Hannan-Quinn criter.	-2.992802
F-statistic	28211.24	Durbin-Watson stat	2.041178
Prob(F-statistic)	0.000000		

(e)

Figure 14.29 (Continued)

14.9.3.1 Omitted Variables

Having the output of a model on the screen, the omitted variables test can easily be done by selecting *View/Coefficient Tests/Omitted Variables* then inserting a variable or a set of variables, which should be considered the most important cause (source or up-stream) variable(s) to be inserted, in a theoretical sense, as additional independent variable(s). Refer to the manual stepwise selection method, which has been demonstrated in previous chapters.

Example 14.24   Application of the models using the ES in (14.20)
For a comparative study, Table 14.9 presents the summary of statistical results of the omitted variables tests, based on the four two-way effects models using the ES in (4.20).

Table 14.9   *Summary of the omitted variables tests based on TWREM, CSR_FEM, CSF_PREM and TWFEM*

Dependent Variable: LY											
Equation Specification : C LY(-1) LY(-2) LK LL LK*LL											
Omitted Variables: LHL LYF LHL*LYF											
	TWREM			CSR_FEM			CSF_PEFM			TWFEM		
	Value	df	Prob	Value	df	Prob	Value	df	Prob	Value	df	Prob
F-Stat.	12.4656	(3,2123)	0.0000	11.5103	(3,2098)	0.0000	35.1809	(3,2042)	0.0000	72.285	(3,2017)	0.0000

Based on this summary, it can be concluded that the set of variables *LHL*, *LYL* and *LHL∗LYF* should be inserted, in a statistical sense, as additional independent variables of the four models since they have significant joint effects. Therefore, by inserting these independent variables, four new models are obtained with the summary of a part of their statistical results presented in Table 14.10. Based on this summary, compared to the statistical results of their original models, the following findings and notes are presented.

1. In general, it can be said that unexpected statistical results would be obtained by inserting an additional independent variable into a model, more so for a set of variables, because of the unpredictable impacts of multicollinearity between independent variables. See the statistical results in Table 14.10.
2. Each of the independent variables of the TWREM has a Prob.(*t*-stat) less than 0.02, compared to the original model in Figure 14.29(a), where LK, LL and LK∗LL, respectively, have Prob.(*t*-stat)s of 0.2478, 0.2248 and 0.1450. Therefore, the new model should be considered better in a statistical sense.
3. Each of the independent variables of the CSR_PEFM has a Prob.(*t*-stat) less than 0.005, compared to the original model in Figure 14.29(a), where LK, LL and LK∗LL, respectively, have Prob.(*t*-stat)s of 0.5147, 0.3187 and 0.2095. Therefore, the new model should be considered better in a statistical sense.
4. Each of the independent variables LK, LL, and LK∗LL of the CSF_PEREM has a very large Prob.(*t*-stat) in both the new and original models. Note that five independent variables of the new model have a very large Prob.(*t*-stat). Therefore, a reduced model should be explored. In this case, reducing the original model is recommended at the first stage, then conducting the omitted variable test. See the following example.

Table 14.10   *Summary of a part of the statistical results from the four new models*

	Dependent Variable: LY											
	TWREM			CSR_PEFEM			CSF_PEREM			TWFEM		
Variable	Coeff.	t-Stat.	Prob.	Coeff.	t-Stat.	Prob.	Coeff.	t-Stat.	Prob.	Coeff.	t-Stat.	Prob.
C	0.2120	1.4184	0.1562	0.1372	1.0038	0.3156	0.4754	0.9906	0.3220	-3.8799	-6.1740	0.0000
LY(-1)	1.1754	58.3338	0.0000	1.1516	61.0211	0.0000	1.0263	51.1572	0.0000	0.9609	45.4009	0.0000
LY(-2)	-0.2186	-10.4662	0.0000	-0.1914	-9.7762	0.0000	-0.1734	-8.3849	0.0000	-0.1787	-8.4679	0.0000
LK	-0.0473	-2.7803	0.0055	-0.0459	-2.9594	0.0031	0.0147	0.3206	0.7485	0.0201	0.4368	0.6623
LL	-0.0695	-2.6612	0.0078	-0.0747	-3.1351	0.0017	-0.0509	-0.7463	0.4556	-0.0166	-0.2418	0.8089
LK*LL	0.0029	2.6475	0.0082	0.0031	3.0894	0.0020	-0.0006	-0.2133	0.8311	-0.0005	-0.1751	0.8610
LHL	0.0510	2.4299	0.0152	0.0612	3.1921	0.0014	-0.0124	-0.2118	0.8323	-0.0025	-0.0422	0.9664
LYF	0.0857	4.6439	0.0000	0.0827	4.9157	0.0000	0.1580	3.1611	0.0016	0.3815	6.9064	0.0000
LHL*LYF	-0.0023	-2.4506	0.0143	-0.0026	-3.0728	0.0021	0.0005	0.1745	0.8615	0.0002	0.0622	0.9504

Omitted Variables: LHL*LYF			
F-statistic	11.01875 Prob. F(1,2045)		0.0009

Test Equation:
Dependent Variable: LY
Method: Panel EGLS (Period random effects)
Date: 10/15/12   Time: 13:15
Sample: 3 28
Periods included: 26
Cross-sections included: 82
Total panel (balanced) observations: 2132
Use pre-specified random component estimates
Swamy and Arora estimator of component variances

Variable	Coefficient	Std. Error	t-Statistic	Prob.
C	1.932435	0.314976	6.135180	0.0000
LY(-1)	1.062016	0.020913	50.78221	0.0000
LY(-2)	-0.131363	0.021557	-6.093785	0.0000
LK*LL	0.000931	0.000479	1.941974	0.0523
LL	-0.066486	0.023212	-2.864316	0.0042

(a)

Effects Specification		S.D.	Rho
Cross-section fixed (dummy variables)			
Period random		0.008385	0.0273
Idiosyncratic random		0.050016	0.9727

Weighted Statistics			
R-squared	0.999339	Mean dependent var	23.54663
Adjusted R-squared	0.999311	S.D. dependent var	1.901428
S.E. of regression	0.049892	Sum squared resid	5.090464
F-statistic	35966.13	Durbin-Watson stat	2.037954
Prob(F-statistic)	0.000000		

Unweighted Statistics			
R-squared	0.999334	Mean dependent var	23.54663
Sum squared resid	5.220183	Durbin-Watson stat	2.025020

(b)

Figure 14.30   *An output of an omitted variables test of a reduced CSF_PEREM*

5. Similarly, each of the independent variables LK, LL and LK*LL, of the TWFEM has a very large Prob.(*t*-stat) in both the new and original models.

Example 14.25   Omitted variables test based on a reduced CSF_PEREM
Figure 14.30 presents the statistical results of a omitted variables test based on a reduced CSF_PEREM of LY on LY(−1), LY(−2), LK*LL and LL, with LHL*LYF as an omitted variable. Based on this output, the following notes and comments are presented.

1. The reduced model of the CSF_PEREM is obtained by inserting the interaction LK*LL after the lags LY(−1) and LY(−2), because the linear effect of LK on LY depends on LL. Then, one of the main factors LK and LL is inserted using the trial-and-error method. In this stage, we find LL has a Prob.(*t*-stat) = 0.2429.
2. The first omitted variable considered is LHL*LYF, since it is defined that the linear effect of LHL (LYF) on LY depends on LYF (LHL). Then no one of the main factors LHL and LYF can be selected as an additional omitted variable.
3. Then the good fit CSF_PEREM obtained is a model of LY on LY(−1), LY(−2), LK*LL, LL and LHL*LYF.

14.9.3.2   Redundant Variables or Wald Tests

Get the output of a model on the screen by selecting *View/Coefficient Tests/Redundant Variables* and then insert a set of independent variables of the model. Then by clicking *OK*, the statistical results of the test are obtained.

Example 14.26   A redundant variable test based on TWREM in (14.20)
Getting the output of the TWREM on the screen, the statistical result in Figure 14.31 obtained for testing the redundant variables: LK, LL, and LK*LL. Based on this output, the following findings and notes are presented.

1. At the $\alpha = 0.10$ level of significance, LK, LL and LKJ*LL have insignificant joint adjusted effects on LY based on the *F*-statistic of $F_0 = 1.967531$ with $df = (3.2126)$ and $p = 0.1136$.

Redundant Variables: LK LL LK*LL			
F-statistic	1.967531	Prob. F(3,2126)	0.1168

Test Equation:
Dependent Variable: LY
Method: Panel EGLS (Two-way random effects)
Date: 10/15/12   Time: 14:18
Sample: 3 28
Periods included: 26
Cross-sections included: 82
Total panel (balanced) observations: 2132
Use pre-specified random component estimates
Swamy and Arora estimator of component variances

Variable	Coefficient	Std. Error	t-Statistic	Prob.
C	0.034501	0.014006	2.463264	0.0138
LY(-1)	1.178704	0.020227	58.27269	0.0000
LY(-2)	-0.178782	0.020236	-8.834720	0.0000

Effects Specification		S.D.	Rho
Cross-section random		0.000000	0.0000
Period random		0.005682	0.0127
Idiosyncratic random		0.050019	0.9873

Weighted Statistics			
R-squared	0.999255	Mean dependent var	16.41378
Adjusted R-squared	0.999255	S.D. dependent var	1.905809
S.E. of regression	0.052025	Sum squared resid	5.762459
F-statistic	1428751.	Durbin-Watson stat	2.034832
Prob(F-statistic)	0.000000		

Unweighted Statistics			
R-squared	0.999245	Mean dependent var	23.54663
Sum squared resid	5.921151	Durbin-Watson stat	2.007693

Figure 14.31   *Output of a redundant variables test of the TWREM in (4.20)*

Wald Test: Equation: EQ10_RR			
Test Statistic	Value	df	Probability
F-statistic	1.967531	(3, 2126)	0.1168
Chi-square	5.902593	3	0.1164

Null Hypothesis Summary:		
Normalized Restriction (= 0)	Value	Std. Err.
C(4)	-0.007676	0.006640
C(5)	-0.011196	0.009204
C(6)	0.000540	0.000371

Restrictions are linear in coefficients.

Figure 14.32   *Output of the Wald test for the joint effects of LK, LL, and LK∗LL*

2. So a reduced model should be explored, in general. However, I use the criteria of a Prob.(t-stat) < 0.30 in order to keep an independent variable in the model, because such a variable would have either positive or negative effect on the dependent variable at the 15% level of significance (Lapin, 1973).
3. This output also presents the output of the TWREM of LY on the remaining independent variables in the model, namely LY(-1) and LY(-2).
4. For a comparison, Figure 14.32 presents the output for testing the joint effects of LK, LL and LK∗LL using the Wald test.

14.10   Generalized Method of Moments/Dynamic Panel Data

To select a set of appropriate or good instrumental variables is not an easy task, because any variable(s) could be used as instrumental variables in addition to the lags of the independent variables of the mean model: refer to the notes on the instrumental variables model presented in Chapter 1, and Agung (2009a). The very basic criteria in defining an instrumental variables model is that the number of instrumental variables should be greater than the number of variables in the mean model.

Furthermore, it has been found that a subset of the independent variables, or the lag(s) of the dependent and independent variables can be used as instruments in addition to external variables, as presented in the EViews 6 User's Guide II (pp. 39, 557 and 561), and Baltagi (2009a, pp. 128 and 129); as well as the following examples.

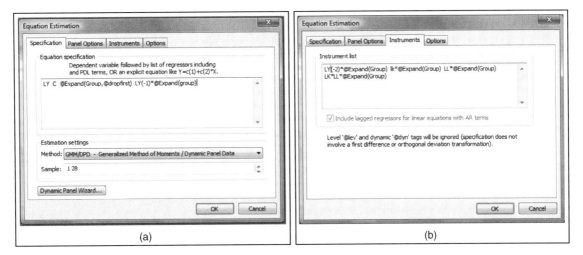

Figure 14.33   *The options for the GMM estimation method*

The steps of the analysis are as follows:

1. By selecting *Quick/Estimate Equation . . . /GMM/DPD*, then the block or box in Figure 14.33(a) is shown on the screen for entering the list of dependent and independent variables of the mean model.
2. By selecting the object *Instruments*, then the block in Figure 14.33(b) is shown on the screen for entering the list of instrumental variables.
3. By selecting the object *Panel Options*, the three sets of options for the *Effect specification*, *GMM Weights* and *Coef covariance method*, as presented in Figure 14.34, can be identified.
4. Note that each set has several options, for instance, there are four options for the cross-section effects specification, namely fixed, random, difference and orthogonal deviation. So we have to do exercises to apply the several options and compare their statistical results, then we can select the best among those. It is impossible to conduct the data analysis for all alternative options as well as all possible sets of the instrumental variables.
5. On the other hand, by doing experiments using several alternative models with the same selected set of instrumental variables, I have found the "*Near Singular Matrix*" error message, the error message as presented in Figure 14.14, or the error message in Figure 14.35.
6. So a researcher should use the trial-and-error method in order to obtain the best possible model to fit the data used.

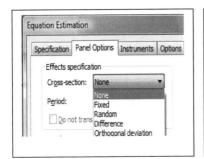
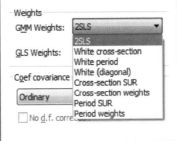
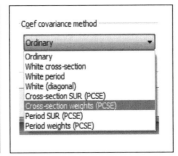

Figure 14.34   *The sets of* Panel Options *for the GMM estimation method*

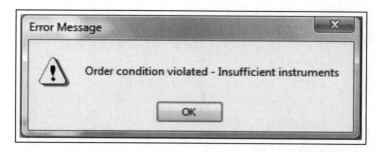

Figure 14.35   *An error message of an instrumental variables model*

14.10.1   Application of GLS Method

Similar to various effects models presented previously, a lot of possible effects models can also be easily presented using the GMM estimation method. However, this section presents only some illustrative examples as follows:

Example 14.27   A GMM estimation method for the TWREM in Figure 14.33(a)
Referring to the TWREM in Figure 14.33(a), this example presents a GMM estimation method using the following mean model.

$$LY \quad C \quad LY(-1) \quad LY(-2) \quad LK \quad LL \quad LK * LL \tag{14.27}$$

Furthermore, a lot of possible sets of instrumental variables could be applied, as long as the number of instrumental variables is greater than the number of variables in the mean model. So students and researchers can easily use any sets of instruments in order to get statistical results, but we will never know the true instrumental variables.

For illustration, Figure 14.36 presents two out of a lot of possible sets of instrumental variables. Based on these outputs, the following findings and notes are presented.

1. Note that the sub-sets of instrumental variables are the independent variables of the main model in (14.27) with several additional variables, as presented in Figure 14.36(a). But the instruments of the model in Figure 14.36(b) contain all independent variables, including the constant parameter "C".
2. The sets of instrumental variables are in fact selected using the trial-and-error method in order to obtain the regressions of the main model that are acceptable in a statistical sense. So both outputs can be considered to be unexpected statistical results.
3. Note that the instrument rank $= 7.0$ for the model in Figure 14.36(a), which is exactly the same as the number of the instruments, but the model in Figure 14.36(b) has nine instruments, it presents the instrument rank $= 8.0$. This indicates that the design matrix of the instruments is singular.
4. It is surprising that Figure 14.37 presents different GMM estimates of the model in Figure 14.36(b) using the same set of instrumental variables, but with different ordering. Why? I guess that the estimation method should be done step-by-step by entering sub-sets of instrumentals, so their ordering has unpredicted impact. On the other hand, I am very sure that no other researchers have ever found these contradictory results.
5. This example has demonstrated that it is not an easy task to obtain the best possible set of instrumental variables, let alone the true instrumentals which are never known. Refer to the special notes on the true population model presented in Agung (2009a, Section 2.14). For comparison, readers can test applications of various sets of instrumental variables, including a set of polynomial and transformed variables.

Dependent Variable: LY
Method: Panel GMM EGLS (Two-way random effects)
Date: 10/16/12   Time: 11:38
Sample (adjusted): 3 28
Periods included: 26
Cross-sections included: 82
Total panel (balanced) observations: 2132
2SLS instrument weighting matrix
Swamy and Arora estimator of component variances
Instrument list: C LY(-1) LY(-2) LK LL LHL LYF

Variable	Coefficient	Std. Error	t-Statistic	Prob.
C	3.723868	1.016208	3.664473	0.0003
LY(-1)	1.174292	0.023093	50.85140	0.0000
LY(-2)	-0.174678	0.023356	-7.478781	0.0000
LK	-0.151511	0.041436	-3.656485	0.0003
LL	-0.229146	0.063641	-3.600570	0.0003
LK*LL	0.009385	0.002585	3.630463	0.0003

Effects Specification

	S.D.	Rho
Cross-section random	0.000000	0.0000
Period random	0.000000	0.0000
Idiosyncratic random	0.120846	1.0000

Weighted Statistics

R-squared	0.999055	Mean dependent var	23.54663
Adjusted R-squared	0.999052	S.D. dependent var	1.918040
S.E. of regression	0.059045	Sum squared resid	7.411886
Durbin-Watson stat	1.596759	J-statistic	15.76077
Instrument rank	7.000000		

Unweighted Statistics

R-squared	0.999055	Mean dependent var	23.54663
Sum squared resid	7.411886	Durbin-Watson stat	1.596759

(a)

Dependent Variable: LY
Method: Panel GMM EGLS (Two-way random effects)
Date: 10/16/12   Time: 12:20
Sample (adjusted): 3 28
Periods included: 26
Cross-sections included: 82
Total panel (balanced) observations: 2132
2SLS instrument weighting matrix
Swamy and Arora estimator of component variances
Instrument list: C LY(-1) LY(-2) LHL LYF LE LK LL LK*LL

Variable	Coefficient	Std. Error	t-Statistic	Prob.
C	0.223211	0.146120	1.527585	0.1268
LY(-1)	1.172206	0.020235	57.93105	0.0000
LY(-2)	-0.175307	0.020427	-8.582290	0.0000
LK	-0.006195	0.006627	-0.934808	0.3500
LL	-0.010336	0.009163	-1.128038	0.2594
LK*LL	0.000509	0.000370	1.377187	0.1686

Effects Specification

	S.D.	Rho
Cross-section random	0.000000	0.0000
Period random	0.008414	0.0275
Idiosyncratic random	0.050019	0.9725

Weighted Statistics

R-squared	0.999262	Mean dependent var	12.92178
Adjusted R-squared	0.999260	S.D. dependent var	1.901394
S.E. of regression	0.051724	Sum squared resid	5.688850
Durbin-Watson stat	2.037580	J-statistic	28.42438
Instrument rank	8.000000		

Unweighted Statistics

R-squared	0.999246	Mean dependent var	23.54663
Sum squared resid	5.909248	Durbin-Watson stat	1.998709

(b)

Figure 14.36   *Two GMM estimates for the TWREM in Figure 14.33(a)*

6. Note that the mean model (14.28) is a continuous two-way interaction model, since it has numerical dependent and independent variables. The following example presents a simple heterogeneous mean model.

7. Similar to the testing presented earlier, various testing can also easily be done for this model. Do it as an exercise.

Dependent Variable: LY
Method: Panel GMM EGLS (Two-way random effects)
Date: 10/16/12   Time: 12:53
Sample (adjusted): 3 28
Periods included: 26
Cross-sections included: 82
Total panel (balanced) observations: 2132
2SLS instrument weighting matrix
Swamy and Arora estimator of component variances
Instrument list: C LY(-1) LY(-2) LK LL LK*LL LHL LYF LE

Variable	Coefficient	Std. Error	t-Statistic	Prob.
C	0.204023	0.145403	1.403157	0.1607
LY(-1)	1.165988	0.020244	57.59564	0.0000
LY(-2)	-0.170242	0.020410	-8.341040	0.0000
LK	-0.004365	0.006610	-0.660372	0.5091
LL	-0.009277	0.009114	-1.017838	0.3089
LK*LL	0.000471	0.000368	1.280512	0.2005

Effects Specification

	S.D.	Rho
Cross-section random	0.000000	0.0000
Period random	0.018116	0.1160
Idiosyncratic random	0.050019	0.8840

Weighted Statistics

R-squared	0.999267	Mean dependent var	6.867299
Adjusted R-squared	0.999265	S.D. dependent var	1.896217
S.E. of regression	0.051414	Sum squared resid	5.619810
Durbin-Watson stat	2.037674	J-statistic	26.85470
Instrument rank	8.000000		

Unweighted Statistics

R-squared	0.999245	Mean dependent var	23.54663
Sum squared resid	5.915995	Durbin-Watson stat	1.983994

Figure 14.37   *A GMM estimates of the model in Figure 14.36(b) using the same set of instrumental variables, but with different ordering*

Example 14.28   A CSR_PEFEM by group using the GMM estimation method

As an extension of the continuous mean model in (14.28), this example presents a heterogeneous mean model, or a mean model by group. By using the trial-and-error method, statistical results are obtained based on several effects models, where many of them have a Rho = 0.0000. One of the statistical results of a CSR_PEFEM by *Group* using the GMM estimation method is presented in Figure 14.38. Based on these results, the following notes and comments are presented.

1. The statistical results are obtained using the mean model by *Group* as follows:

$$LY \; LY(-1) * @Expand(Group) \; LY(-2) * @Expand(Group) \; LK * @Expand(Group)$$
$$LL * @Expand(Group) \; LK * LL * @Expand(Group) \tag{14.28}$$

with the following set of instrumental variables.

$$LY(-1) * @Expand(Group) \; LY(-2) * @Expand(Group) \; LK * @Expand(Group)$$
$$LL * @Expand(Group) \; LK * LL * @Expand(Group) \tag{14.29}$$
$$LHL * @Expand(Group) \; LYF * @Expand(Group)$$

2. Note that all independent variables in (14.28) are used as the instruments, in addition to two sets of external variables, namely *LHL∗@Expand(Group) LYF∗@Expand(Group)*, which represent four independent variables in Figure 14.38.

3. The model has very a large R^2 and Adjusted R^2, both greater than 0.99. It can then be said that the set of independent variables and instrumental variables are the best possible predictors for *LY*. It has been found that lagged variables or autoregressive models will have a very large Adjusted R^2.

Dependent Variable: LY
Method: Panel GMM EGLS (Cross-section random effects)
Date: 10/16/12   Time: 14:16
Sample (adjusted): 3 28
Periods included: 26
Cross-sections included: 82
Total panel (balanced) observations: 2132
2SLS instrument weighting matrix
Swamy and Arora estimator of component variances
Instrument list: C LY(-1)*@EXPAND(GROUP) LY(-2)*@EXPAND(GROUP)
    LK*@EXPAND(GROUP) LL*@EXPAND(GROUP) LK*LL
    @EXPAND(GROUP) LHL@EXPAND(GROUP) LYF
    *@EXPAND(GROUP)

Variable	Coefficient	Std. Error	t-Statistic	Prob.
C	0.281150	0.142494	1.973063	0.0486
LY(-1)*(GROUP=1)	1.225384	0.030503	40.17309	0.0000
LY(-1)*(GROUP=2)	1.099844	0.025267	43.52945	0.0000
LY(-2)*(GROUP=1)	-0.229674	0.031040	-7.399177	0.0000
LY(-2)*(GROUP=2)	-0.095862	0.025740	-3.724237	0.0002
LK*(GROUP=1)	-0.007384	0.007785	-0.948420	0.3430
LK*(GROUP=2)	-0.011540	0.006865	-1.680847	0.0929
LL*(GROUP=1)	-0.011710	0.009322	-1.256183	0.2092
LL*(GROUP=2)	-0.027970	0.009591	-2.916245	0.0036
LK*LL*(GROUP=1)	0.000561	0.000358	1.568512	0.1169
LK*LL*(GROUP=2)	0.001022	0.000374	2.730304	0.0064

Effects Specification		S.D.	Rho
Cross-section random		0.000000	0.0000
Period fixed (dummy variables)			
Idiosyncratic random		0.049554	1.0000

Weighted Statistics			
R-squared	0.999299	Mean dependent var	23.54663
Adjusted R-squared	0.999287	S.D. dependent var	1.918040
S.E. of regression	0.051211	Sum squared resid	5.496838
Durbin-Watson stat	2.028006	J-statistic	34.59961
Instrument rank	40.000000		

Unweighted Statistics			
R-squared	0.999299	Mean dependent var	23.54663
Sum squared resid	5.496838	Durbin-Watson stat	2.028006

Figure 14.38   *Statistical results of a two-way random effects model, using the GMM estimation method*

Cross-section random effects test equation:
Dependent Variable: LY
Method: Panel Generalized Method of Moments
Date: 10/16/12   Time: 14:34
Sample (adjusted): 3 28
Periods included: 26
Cross-sections included: 82
Total panel (balanced) observations: 2132
2SLS instrument weighting matrix
Instrument list: C LY(-1)*@EXPAND(GROUP) LY(-2)*@EXPAND(GROUP)
  LK*@EXPAND(GROUP) LL*@EXPAND(GROUP) LK*LL
  @EXPAND(GROUP) LHL@EXPAND(GROUP) LYF
  *@EXPAND(GROUP)

Variable	Coefficient	Std. Error	t-Statistic	Prob.
C	1.880910	0.526570	3.572008	0.0004
LY(-1)*(GROUP=1)	1.123446	0.032683	34.37402	0.0000
LY(-1)*(GROUP=2)	0.997982	0.026941	37.04314	0.0000
LY(-2)*(GROUP=1)	-0.220432	0.034486	-6.391955	0.0000
LY(-2)*(GROUP=2)	-0.056953	0.027976	-2.035820	0.0419
LK*(GROUP=1)	0.079690	0.031327	2.543824	0.0110
LK*(GROUP=2)	-0.038921	0.028699	-1.356172	0.1752
LL*(GROUP=1)	0.103796	0.045716	2.270431	0.0233
LL*(GROUP=2)	-0.167282	0.044426	-3.765405	0.0002
LK*LL*(GROUP=1)	-0.004016	0.001755	-2.288168	0.0222
LK*LL*(GROUP=2)	0.004323	0.001712	2.525928	0.0116

Effects Specification

Cross-section fixed (dummy variables)
Period fixed (dummy variables)

R-squared	0.999369	Mean dependent var	23.54663
Adjusted R-squared	0.999333	S.D. dependent var	1.918040
S.E. of regression	0.049554	Sum squared resid	4.948007
Durbin-Watson stat	2.029238	J-statistic	194.1734
Instrument rank	121.000000		

Correlated Random Effects - Hausman Test
Equation: EQ14_HET
Test cross-section random effects

Test Summary	Chi-Sq. Statistic	Chi-Sq. d.f.	Prob.
Cross-section random	214.394851	10	0.0000

** WARNING: estimated cross-section random effects variance is zero.

Cross-section random effects test comparisons:

Variable	Fixed	Random	Var(Diff.)	Prob.
LY(-1)*(GROUP=1)	1.123446	1.225384	0.000138	0.0000
LY(-1)*(GROUP=2)	0.997982	1.099844	0.000087	0.0000
LY(-2)*(GROUP=1)	-0.220432	-0.229674	0.000226	0.5385
LY(-2)*(GROUP=2)	-0.056953	-0.095862	0.000120	0.0004
LK*(GROUP=1)	0.079690	-0.007384	0.000921	0.0041
LK*(GROUP=2)	-0.038921	-0.011540	0.000777	0.3258
LL*(GROUP=1)	0.103796	-0.011710	0.002003	0.0099
LL*(GROUP=2)	-0.167282	-0.027970	0.001882	0.0013
LK*LL*(GROUP=1)	-0.004016	0.000561	0.000003	0.0077
LK*LL*(GROUP=2)	0.004323	0.001022	0.000003	0.0481

(a)   (b)

Figure 14.39 *The Correlated Random – Hausman Test for the CSR_PEFEM in Figure 14.38*

4. By selecting *View/Fixed/Random Effects Testing/Correlated Random Effect – Hausman test*, the statistical results in Figure 14.39 are obtained. However, the *Redundant Fixed Effects – Likelihood Ratio* is not available with this estimation method.

5. *Application of a Wald Test.*

Specific for the GROUP = 1, the null hypothesis H_0: $C(6) = C(8) = C(10) = 0$ is considered to test the joint effects of *LK*, *LL* and *LK∗LL*, with the statistical results presented in Figure 14.40. However, the *Omitted* and *Redundant Variables Tests* are not available with this estimation method. Therefore, in a statistical sense, a reduced model should be explored for the GROUP = 1. Do it as an exercise.

Wald Test:
Equation: EQ14_HET

Test Statistic	Value	df	Probability
F-statistic	0.963196	(3, 2096)	0.4092
Chi-square	2.889589	3	0.4090

Null Hypothesis Summary:

Normalized Restriction (= 0)	Value	Std. Err.
C(6)	-0.007384	0.007785
C(7)	-0.011540	0.006865
C(8)	-0.011710	0.009322

Restrictions are linear in coefficients.

Figure 14.40 *The Wald Test for the joint effects of LK, LL, and LK∗LL, specific for Group = 1*

Example 14.29   Application of a mean model by group and time period

As an extension of the mean model by group in (14.29a), this is an example of a mean model by group and time period (TP). For illustration, Figure 14.41 presents the statistical results of the mean model:

$$ly \ ly(-1) * @expand(group, tp) \ ly(-2) * @expand(group, tp) \ lk * @expand(group, tp)$$
$$ll * @expand(group, tp) \ lk * ll * @expand(group, tp) \tag{14.30a}$$

with the set of instrumental variables as follows:

$$y(-1) * @expand(group, tp) \ ly(-2) * @expand(group, tp) \ lk * @expand(group, tp)$$
$$ll * @expand(group, tp) \ lk * ll * @expand(group, tp) \tag{14.30b}$$
$$lhl * @expand(group, tp) \ lyf * @expand(group, tp)$$

Based on these results the following findings, notes and comments are presented.

1. Compared to other random effects models, where most of the models have the Rho = 0, this TWREM has $\rho_{CS} = 0.0176$, for the cross-section random, and $\rho_{PE} = 0.5396$, for the period random.
2. This output presents the instrument rank = 29, which is exactly the same as the total number of instruments, that is $(1 + 4 \times 7)$.

	Effects Specification		
		S.D.	Rho
Cross-section random		0.009776	0.0176
Period random		0.054123	0.5396
Idiosyncratic random		0.049030	0.4428

Dependent Variable: LY
Method: Panel Two-Stage EGLS (Two-way random effects)
Date: 10/16/12   Time: 15:34
Sample (adjusted): 3 28
Periods included: 26
Cross-sections included: 82
Total panel (balanced) observations: 2132
Swamy and Arora estimator of component variances
Instrument list: C LY(-1)*@EXPAND(GROUP,TP) LY(-2)
   @EXPAND(GROUP,TP) LK@EXPAND(GROUP,TP) LL
   *@EXPAND(GROUP,TP) LK*LL*@EXPAND(GROUP,TP) LHL
   @EXPAND(GROUP,TP) LYF@EXPAND(GROUP,TP)

Weighted Statistics			
R-squared	0.998618	Mean dependent var	2.331982
Adjusted R-squared	0.998605	S.D. dependent var	1.333367
S.E. of regression	0.049803	Sum squared resid	5.235907
F-statistic	76269.08	Durbin-Watson stat	2.027025
Prob(F-statistic)	0.000000	Second-Stage SSR	5.235907
Instrument rank	29.000000		

Unweighted Statistics			
R-squared	0.999285	Mean dependent var	23.54663
Sum squared resid	5.601795	Durbin-Watson stat	1.949243

Variable	Coefficient	Std. Error	t-Statistic	Prob.
C	0.329372	0.207171	1.589855	0.1120
LY(-1)*(GROUP=1 AND TP=1)	1.048665	0.055819	18.78688	0.0000
LY(-1)*(GROUP=1 AND TP=2)	1.259622	0.038058	33.09711	0.0000
LY(-1)*(GROUP=2 AND TP=1)	1.012982	0.035059	28.89363	0.0000
LY(-1)*(GROUP=2 AND TP=2)	1.109795	0.039150	28.34721	0.0000
LY(-2)*(GROUP=1 AND TP=1)	-0.048027	0.056176	-0.854927	0.3927
LY(-2)*(GROUP=1 AND TP=2)	-0.285456	0.039363	-7.251972	0.0000
LY(-2)*(GROUP=2 AND TP=1)	-0.026425	0.035144	-0.751896	0.4522
LY(-2)*(GROUP=2 AND TP=2)	-0.094579	0.040626	-2.328044	0.0200
LK*(GROUP=1 AND TP=1)	-0.014404	0.011397	-1.263859	0.2064
LK*(GROUP=1 AND TP=2)	0.009594	0.013162	0.728940	0.4661
LK*(GROUP=2 AND TP=1)	0.002248	0.010412	0.215925	0.8291
LK*(GROUP=2 AND TP=2)	-0.025587	0.010796	-2.370142	0.0179
LL*(GROUP=1 AND TP=1)	-0.013435	0.013985	-0.960656	0.3368
LL*(GROUP=1 AND TP=2)	-0.011510	0.013673	-0.841830	0.4000
LL*(GROUP=2 AND TP=1)	-0.024066	0.014778	-1.628542	0.1036
LL*(GROUP=2 AND TP=2)	-0.029503	0.014038	-2.101685	0.0357
LK*LL*(GROUP=1 AND TP=1)	0.000704	0.000535	1.315046	0.1886
LK*LL*(GROUP=1 AND TP=2)	0.000641	0.000513	1.249505	0.2116
LK*LL*(GROUP=2 AND TP=1)	0.000956	0.000570	1.675593	0.0940
LK*LL*(GROUP=2 AND TP=2)	0.001137	0.000535	2.124538	0.0337

A Part of the Effects			
Cross-Section		Time	
CODE	Effect	DATEID	Effect
1	-0.00208		
2	-0.00406		
3	-0.00081	3	-0.005
4	-0.00099	4	0.005678
5	0.005409	5	-0.00058
6	-0.00339	6	-0.0002
7	-4.38E-05	7	-0.00634
8	0.001612	8	-0.01774
9	-0.00379	9	0.002248
10	0.005186	10	0.009459

Figure 14.41   *Statistical results of a TWREM in (14.30), using the GMM estimation method*

3. Even though five out of 20 numerical independent variables have Prob. > 0.30, I would say that the estimates are acceptable statistical results. However, reduced models could be explored by using the manual stepwise selection or trial-and-error methods.
4. Note that the period random effects starting at DATEID $= 3$, which corresponds to the use of LY(-2) as an independent variable.

14.11   More Advanced Interaction Effects Models

14.11.1   Interaction Effects Models Based on (X1,X2,Y1)

Referring to the model in (4.20) which can be considered the simplest two-way interaction model, since it has only a single two-way interaction, namely LK∗LL, this section presents more advanced interaction models based on the trivariate (LK,LL,LY), which can be generalized to the trivariate (X1,X2,Y1). Then corresponding to the three-way interaction models of Y1 on X1(-1), X2(-1) and Y1(-1) presented in Table 12.22, the path diagrams in Figure 14.42 present four alternative causal or up- and down-stream relationships based on the trivariate (X1,X2,Y1) out of a lot of possible relationships, using the higher order lags of the variables X1, X2 and Y1. Refer to various models presented by the general equations earlier and in previous chapters. The path diagrams have characteristics as follows, under the assumptions that X1 and X2 are cause (source, or up-stream) factors/variables of Y1.

1. *Path diagram in Figure 14.42(a)*
   Since X1 and X2 are the cause factors of Y1, then X1(-1) and X2(-1) should be the cause factors of both Y1 and Y1(-1). In addition, the arrow from X2(-1) to X1(-1) shows X2 is an up-stream variable of X1. Note that this path diagram can be considered the presentation of the causal or up- and down-stream relationships between the numerical variables in Table 12.22.
   Note that the arrows in this diagram show that the effect of Y1(-1) on Y(1) depends on two variables X1(-1) and X2(-1), then the model should have the interactions Y1(-1)∗X1(-1)∗X2(-1), Y1(-1)∗X1(-1) and Y1(-1)∗X2(-1) as independent variables. In addition, the impact of X1(-1) on Y1 depends on X2(-1), so then model should have also X1(-1)∗X2(-1) as an independent variable. So the full interaction mean model considered would have the equation specification (ES) as follows:

$$Y1 \ C \ Y1(-1) * X1(-1) * X2(-1) \ Y1(-1) * X1(-1) \ Y1(-1) * X2(-1)$$
$$X1(-1) * X2(-1) \ Y1(-1) \ X1(-1) \ X2(-1) \tag{14.31}$$

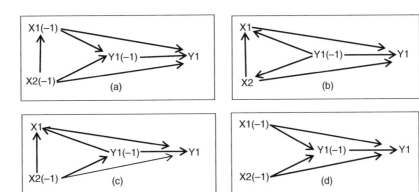

Figure 14.42   *Four alternative up- and down-stream relationships based on a trivariate (X1,X2,Y1)*

which is a hierarchical model. However, unpredictable good fit models, either hierarchical or nonhierarchical, would be obtained which are highly dependent on the data used as well as the unpredictable impacts of multicollinearity.

2. *Path diagram in Figure 14.42(b)*

This path diagram shows that $Y1(-1)$ is a cause, source or up-stream variable of $X1$, $X2$ and $Y1$. In addition, the effect of $X1$ on $Y1$ depends on both variables $X2$ and $Y1(-1)$, then the model should have the interactions $Y1(-1)*X1*X2$, $Y1(-1)*X1$ and $X1*X2$ as independent variables. The impact of $X2$ on $Y1$ depends on $Y1(-1)$, so then model should have also $Y1(-1)*X2$ as an independent variable. So the full interaction mean model considered would have the equation specification (ES) as follows:

$$Y1 \ C \ Y1(-1)*X1*X2 \ Y1(-1)*X1 \ X! *X2$$
$$Y1(-1)*X2 \ Y1(-1) \ X1 \ X2 \tag{14.32}$$

3. *Path diagram in Figure 14.42(c)*

This path diagram shows that the effect of $X1$ on $Y1$ depends on both variables $X2(-1)$ and $Y1(-1)$, so then the model should have the interactions $Y1(-1)*X1*X2(-1)$, $Y1(-1)*X1$ and $X1*X2(-1)$ as independent variables. In addition, the impact of $Y1(-1)$ on $Y1$ depends on $X2(-1)$, then model should have also $Y1(-1)*X2(-1)$ as an independent variable. So the full interaction mean model considered would have the equation specification (ES) as follows:

$$Y1 \ C \ Y1(-1)*X1*X2(-1) \ Y1(-1)*X1 \ X1*X2(-1)$$
$$Y1(-1)*X2(-1) \ Y1(-1) \ X1 \ X2(-1) \tag{14.33}$$

4. *Path diagram in Figure 14.42(d)*

This path diagram has the assumption that the variable $X1$ and $X2$ are the cause, source or up-stream variables of $Y1$, but they do not have a causal relationship, represented by $X1(-1)$ and $X2(-1)$ without an arrow. However, their correlations have unpredictable impacts on the parameter estimates. Compare this to the full model based on the path diagram in Figure 14.42(a); the model based on this path diagram would be its reduced hierarchical two-way interaction model with ES as follows:

$$Y1 \ C \ Y1(-1)*X1(-1) \ Y1(-1)*X2(-1) \ Y1(-1) \ X1(-1) \ X2(-1) \tag{14.34}$$

Example 14.30   TWREMs in (14.32), based on (LK,LL,LY)

Figure 14.43 presents the statistical results of a full three-way interaction TWREM in (14.32) based on the trivariate (LK,LL,LY) and one of its reduced models. Based on these statistical results, the following notes and comments are presented.

1. Note that the reduced model is obtained by deleting the main variables, even though LL has a smaller *p*-value compared to all interactions, in order to have an interaction reduced model. Do this as an exercise by deleting some of the interactions.
2. The reduced model is an acceptable model in a statistical sense, to show that the effect of $LY(-1)$ on LY is significantly dependent on LK, LL and LK*LL. But its limitation is this regression is a single continuous function to present the total of 2214 firm-time observations. For this reason

Dependent Variable: LY
Method: Panel EGLS (Two-way random effects)
Date: 10/18/12   Time: 14:06
Sample (adjusted): 2 28
Periods included: 27
Cross-sections included: 82
Total panel (balanced) observations: 2214
Swamy and Arora estimator of component variances

Variable	Coefficient	Std. Error	t-Statistic	Prob.
C	0.401190	2.280107	0.175952	0.8603
LY(-1)*LK*LL	-3.10E-05	0.000237	-0.130855	0.8959
LY(-1)*LK	-0.001860	0.003960	-0.469744	0.6386
LY(-1)*LL	0.002806	0.007118	0.394116	0.6935
LK*LL	0.002469	0.006114	0.403834	0.6864
LY(-1)	1.017497	0.116626	8.724444	0.0000
LK	0.012059	0.098356	0.122608	0.9024
LL	-0.108318	0.140175	-0.772734	0.4398

Effects Specification

	S.D.	Rho
Cross-section random	0.011586	0.0457
Period random	0.007223	0.0178
Idiosyncratic random	0.052449	0.9365

Weighted Statistics

R-squared	0.998217	Mean dependent var	11.95431
Adjusted R-squared	0.998211	S.D. dependent var	1.265224
S.E. of regression	0.053508	Sum squared resid	6.316107
F-statistic	176441.0	Durbin-Watson stat	1.686089
Prob(F-statistic)	0.000000		

Unweighted Statistics

R-squared	0.999164	Mean dependent var	23.52420
Sum squared resid	6.817365	Durbin-Watson stat	1.588467

Dependent Variable: LY
Method: Panel EGLS (Two-way random effects)
Date: 10/18/12   Time: 14:03
Sample (adjusted): 2 28
Periods included: 27
Cross-sections included: 82
Total panel (balanced) observations: 2214
Swamy and Arora estimator of component variances

Variable	Coefficient	Std. Error	t-Statistic	Prob.
C	-1.209788	0.354138	-3.416148	0.0006
LY(-1)*LK*LL	0.000133	3.64E-05	3.637935	0.0003
LY(-1)*LK	-0.004204	0.001119	-3.837589	0.0001
LY(-1)*LL	-0.006470	0.001752	-3.693299	0.0002
LK*LL	0.003108	0.000928	3.350943	0.0008
LY(-1)	1.158215	0.042175	27.46239	0.0000

Effects Specification

	S.D.	Rho
Cross-section random	0.011707	0.0465
Period random	0.007215	0.0176
Idiosyncratic random	0.052537	0.9359

Weighted Statistics

R-squared	0.998198	Mean dependent var	11.93148
Adjusted R-squared	0.998194	S.D. dependent var	1.259214
S.E. of regression	0.053513	Sum squared resid	6.322854
F-statistic	244631.9	Durbin-Watson stat	1.681032
Prob(F-statistic)	0.000000		

Unweighted Statistics

R-squared	0.999161	Mean dependent var	23.52420
Sum squared resid	6.840733	Durbin-Watson stat	1.579891

Figure 14.43   *Statistical results of a TWREM in (4.32) and its reduced model based on (LK,LL,LY)*

applying the models by Group, or by Group and Time Period is recommended, as presented in previous examples.

3. For the other effects models with various path diagrams, do this step (2) as an exercise.

Example 14.31   GMM estimation method for the two TWREMs in Figure 14.43

For additional illustration, Figure 14.44 presents the statistical results of the two models in Figure 14.43 using the GMM estimation method, with the instruments are as follows:

$$c \; ly(-1) * lk * ll \; ly(-1) * lk \; ly(-1) * ll \; lk * ll \; ly(-1) \; lhl \; lhl(-1) \; lyf \; lyf(-1) \qquad (14.35)$$

Based on these results, the following findings and notes are presented.

1. Compare to the full model in Figure 14.43, the full model in this figure is an acceptable instrumental variables TWREM in a statistical sense, because each of the independent variables of the mean model has a significant adjusted effect. However, it has a small DW-statistical value, and the Rho = 0 for both cross-section and period random effects. This is an unexpected result.
2. The output of the reduced model shows that it is better, with the DW-statistic = 1.668 680, and the Rhos $<> 0$, namely. $\rho_{CS} = 0.0470$, and $\rho_{PE} = 0.0747$.
3. For the other effects models, either one-way or two-way effects models, using an alternative set of instruments could easily be done: do it as an exercise.

   For more advanced models, the ESs in (14.31) to (14.35) can be extended to the models by Group, and by (Group,TP), as presented in the Examples 14.28 and 14.29.

Dependent Variable : LY				
2SLS instrument weighting matrix				
Swamy and Arora estimator of component variances				
Instrument list: C LY(-1)*LK*LL LY(-1)*LK LY(-1)*LL LK*LL LY(-1) LHL LHL(-1) LYF LYF(-1)				

Variable	Coefficient	Std. Error	t-Statistic	Prob.
C	71.08890	10.65715	6.670536	0.0000
LY(-1)*LK*LL	-0.007319	0.001101	-6.645689	0.0000
LY(-1)*LK	0.115370	0.018213	6.334571	0.0000
LY(-1)*LL	0.235356	0.029736	7.914784	0.0000
LK*LL	0.132194	0.035504	3.723331	0.0002
LY(-1)	-2.675369	0.470800	-5.682602	0.0000
LK	-2.096873	0.578376	-3.625450	0.0003
LL	-4.540058	0.651188	-6.971966	0.0000

Effects Specification			
		S.D.	Rho
Cross-section random		0.000000	0.0000
Period random		0.000000	0.0000
Idiosyncratic random		0.299015	1.0000

Weighted Statistics			
R-squared	0.998527	Mean dependent var	23.54663
Adjusted R-squared	0.998522	S.D. dependent var	1.918040
S.E. of regression	0.073732	Sum squared resid	11.54690
Durbin-Watson stat	0.882710	J-statistic	90.60609
Instrument rank	10.000000		

Unweighted Statistics			
R-squared	0.998527	Mean dependent var	23.54663
Sum squared resid	11.54690	Durbin-Watson stat	0.882710

Dependent Variable :LY				
2SLS instrument weighting matrix				
Swamy and Arora estimator of component variances				
Instrument list: C LY(-1)*LK*LL LY(-1)*LK LY(-1)*LL LK*LL LY(-1) LHL LHL(-1) LYF LYF(-1)				

Variable	Coefficient	Std. Error	t-Statistic	Prob.
C	-1.212197	0.347870	-3.484631	0.0005
LY(-1)*LK*LL	0.000122	3.57E-05	3.408528	0.0007
LY(-1)*LK	-0.004127	0.001097	-3.760405	0.0002
LY(-1)*LL	-0.006341	0.001719	-3.689820	0.0002
LK*LL	0.003293	0.000912	3.610286	0.0003
LY(-1)	1.153347	0.041388	27.86675	0.0000

Effects Specification			
		S.D.	Rho
Cross-section random		0.011652	0.0470
Period random		0.014685	0.0747
Idiosyncratic random		0.050361	0.8783

Weighted Statistics			
R-squared	0.998319	Mean dependent var	7.694800
Adjusted R-squared	0.998315	S.D. dependent var	1.235312
S.E. of regression	0.050715	Sum squared resid	5.468018
Durbin-Watson stat	1.799135	J-statistic	157.7579
Instrument rank	10.000000		

Unweighted Statistics			
R-squared	0.999224	Mean dependent var	23.54663
Sum squared resid	6.081056	Durbin-Watson stat	1.658680

Figure 14.44   *Statistical results of the TWREMs in Figure 14.43, using the instruments in (14.35)*

14.11.2   Interaction Effects Models Based on (X1, X2, X3, Y1)

Note that the models presented above are the LV(1) models with only two exogenous variables, X1 and X2. Those models can easily be extended to LV(p) models having three or more exogenous variables. For an illustration, Figure 14.45 presents the path diagrams of four alternative up-and-down stream relationships based on the variables X1, X2, X3, and Y1, under the assumption that X1, X2, and X3 are the cause (source or upstream) variables for the endogenous (impact or downstream) variable Y1. So that X1(−1), X2(−1), X3(−1), and Y1(−1) would be the upstream (cause or source) variables of *Y1*. Note that the arrows using the dotted lines are representing that the corresponding pairs of variables do have up-and-down or causal relationships, in theoretical sense, but it will not be taken account in the models considered. However, their correlations have unpredictable impacts of the parameters' estimates of the model For sure, the models can be extended to more complex models if those causal relationship would be taken into account. Based on each path diagram di following alternative general model would be considered.

1.   **Models based on Figure 14.45(a)**

This path diagram presents that the effect of Y1(−1) on Y1 depends on the three variables Xk(−1), k = 1, 2, and 3, and Xk(−1), k = 1, 2, 3, have direct effects on Y1. In addition, the effect of the three variables Xk(−1)'s on Y1 depend on Y1(−2). Then the general equation of the base mean models

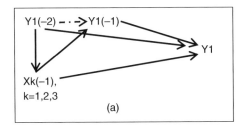 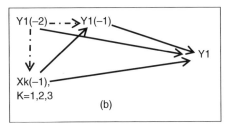

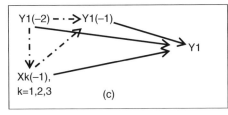 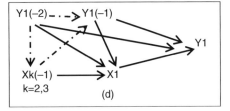

Figure 14.45 *Four alternative up-and down-stream relationships based on (X1,X2,X3,Y1)*

is as follows:

$$
\begin{aligned}
Y_{it} = {} & C(1) + C(2) * Y1(-1) + C(3) * Y1(-2) + Y1(-1) * f_1(Xk(-1), k = 1, 2, 3) \\
& + Y1(-2) * f_2(Xk(-1), k = 1, 2, 3) + f_3(Xk(-1), k = 1, 2, 3) + \mu_{it}
\end{aligned} \tag{14.36}
$$

where $F_j(*)$, $j = 1$, 2, and 3 can be various functions of $(*)$ having finite numbers of parameters, without the constant parameters, which are highly dependent on the up-and-down stream relationships defined, in a theoretical sense, between the $Xk(-1)$'s; $Y1(-1)*F_1(*)$ indicates that the linear effect of $Y1(-1)$ on Y1 depends of the function $F_1(*)$, and $Y1(-2)*F_2(*)$ indicates that the effects of $F_2(*)$ on Y1 depends on $Y1(-2)$.

If the causal relationships between $Xk(-1)$s are not taken into account, then the $F_j(*)$ is an additive function as follows:

$$
F_j(X1(-1), X2(-1), X3(-1)) = C(j1) * X1(-1) + C(j2) * X2(-1) + C(j3) * X3(-1) \tag{14.36a}
$$

On the other hand, if their causal relationships are taken into account, the function should have at least an interaction term. Under the assumption that the $Xk(-1)$s have a complete association or up-and-down stream relationships, then the following full three-way interaction functions should be considered.

$$
\begin{aligned}
F_j(X1(-1), X2(-1), X3(-1)) = {} & C(j1) * X1(-1) * X2(-1) * X3(-1) \\
& + C(j2) * X1(-1) * X2(-1) + C(j3) * X1(-1) * X3(-1) + C(j4) * X2(-1) * X3(-1) \\
& + C(j5) * X1(-1) + C(j6) * X2(-1) + C(j7) * X3(-1)
\end{aligned} \tag{14.36b}
$$

However, by having such a large number of independent variables, there would be great uncertainty of the parameters' estimates, because of the multicollinearity between the independent variables. So that expert's judgments should be used to select the subsets of independent variables, having different levels of importance as the cause factors. Refer to the manual stepwise selection method demonstrated in Section 7.6.

2. **Models based on Figure 14.45(b)**

This path diagram presents the base mean models, which are the reduced models based on the path diagram in Figure 14.30(a), with the general equation is as follows:

$$Y_{it} = C(1) + C(2) * Y1(-1) + C(3) * Y1(-2) + Y1(-1) * F_1(Xk(-1), k = 1, 2, 3) + F_3(Xk(-1), k = 1, 2, 3) + \mu_{it} \tag{14.37}$$

3. **Models based on Figure 14.45(c)**

This path diagram presents the base mean models, which are the reduced models based on the path diagram in Figure 14.30(b), with the general equation is as follows:

$$Y_{it} = C(1) + C(2) * Y1(-1) + C(3) * Y1(-2) + F_3(Xk(-1), k = 1, 2, 3) + \mu_{it} \tag{14.38}$$

4. **Models based on Figure 14.45(d)**

This path diagram presents special mean models, which shows that the effect of X1 on Y1 depends on Y1(−1), Y1(−2), and a function $F_1(X2(-1), X3(-1))$. The general equation of the mean models is as follows:

$$Y_{it} = C(1) + C(2) * Y1(-1) + C(3) * Y1(-2) + C(4) * X1 + C(5) * Y1(-1) * X1 + C(6) * Y1(-2) * X1 + X1 * F_1(X2(-1), X3(-1)) + F_2(X2(-1), X3(-1)) + \mu_{it} \tag{14.39}$$

where $F_k(X2(-1), X3(-1))$ can be any function of $X2(-1)$ and $X3(-1)$, for each $k = 1$ and 2, having finite numbers of paramters, but $F_2(X2(-1), X3(-1))$ does not have a constant parameter.

Example 14.32   Application of a model in (14.37)

Figure 14.46 presents the statistical results of a TWREM in (14.37), using the GMM estimation method, based on the variables Y1 = LY, X1 = LK, X2 = LL, and X3 = LE, in CES.wf1. Based on this output, the following findings and notes are presented.

1. At the $\alpha = 0.10$ level of significance, the four-way interaction has a negative significant effect on LY, based on the t-statistic of $t_0 = -1.606124$ with $df=$ and a p-value $= 0.1084/2 = 0.0542$.

Dependent Variable: LY
Method: Panel GMM EGLS (Two-way random effects)
Date: 10/22/12   Time: 15:22
Sample (adjusted): 3 28
Periods included: 26
Cross-sections included: 82
Total panel (balanced) observations: 2132
2SLS instrument weighting matrix
Swamy and Arora estimator of component variances
Instrument list: C LY(-1) LY(-2) LY(-1)*LK(-1)*LL(-1)*LE(-1) LY(-1)*LK(-1)
    *LE(-1) LY(-1)*LL(-1)*LE(-1) LK(-1)*LL(-1)*LE(-1) LK(-1)*LL(-1) LK(-1)
    *LE(-1) LL(-1)*LE(-1) LK(-1) LL(-1) LE(-1) LHL(-1) LYF(-1)

Variable	Coefficient	Std. Error	t-Statistic	Prob.
C	-0.415207	0.374967	-1.107314	0.2683
LY(-1)	1.136857	0.020892	54.41703	0.0000
LY(-2)	-0.142120	0.020747	-6.850117	0.0000
LY(-1)*LK(-1)*LL(-1)*LE(-1)	-9.94E-05	6.19E-05	-1.606124	0.1084
LY(-1)*LK(-1)*LE(-1)	0.000146	0.000971	0.150277	0.8806
LY(-1)*LL(-1)*LE(-1)	0.002819	0.000609	4.631922	0.0000
LK(-1)*LL(-1)*LE(-1)	0.004742	0.003234	1.466248	0.1427
LK(-1)*LL(-1)	-0.001126	0.001006	-1.119275	0.2631
LK(-1)*LE(-1)	-0.051103	0.050278	-1.016408	0.3096
LL(-1)*LE(-1)	-0.122848	0.047193	-2.603103	0.0093
LK(-1)	0.024504	0.017205	1.424239	0.1545
LL(-1)	0.025982	0.023916	1.086398	0.2774
LE(-1)	1.131569	0.687566	1.645760	0.1000

Effects Specification		
	S.D.	Rho
Cross-section random	0.000000	0.0000
Period random	0.162748	0.9141
Idiosyncratic random	0.049881	0.0859

Weighted Statistics			
R-squared	0.999280	Mean dependent var	0.796506
Adjusted R-squared	0.999276	S.D. dependent var	1.894203
S.E. of regression	0.050954	Sum squared resid	5.501642
Durbin-Watson stat	2.032499	J-statistic	29.06247
Instrument rank	14.000000		

Unweighted Statistics			
R-squared	0.999258	Mean dependent var	23.54663
Sum squared resid	5.815174	Durbin-Watson stat	1.975927

Figure 14.46   *Statistical results of a reduced TWREM in (14.37), using the GMM estimation method*

2. Since the interaction LY(−1)∗LK(−1)∗LE(−1) has a very large Prob., then a reduced model might be explored. However, I would say this TWREM is an acceptable model, in both theoretical and statistical senses.

3. If a reduced model should be explored, then the simplest method is to delete the interaction LY(−1)∗ LK(−1)∗LE(−1). It will be obtained a nonhierarchical reduced model, where only two of the independent variables have p-values > 0.20, namely Lk(−1)∗LL(−1), and ll(−1), having the *p*-values of 0.2642, and 0.2792, respectively.

4. For a more advanced models, the ESs in (14.36) to (14.39) can be extended to the models by Group, or by (Group, TP), as presented in the Examples 14.28, and 14.29. Do it for the exercises, using either the one- or two-way effects models.

5. *Application of Omitted Variables – Likelihood RatioTest*
   Figure 14.47 presents two alternative LR-Tests for the model above, which shows that the interactions: *LY(−2)∗LK(−1), LY(−2)∗LL(−1)*, and *LY(−2)∗LE(−1)* can be inserted in the model, to improve the good fit of the model.

However, the interactions *LY(−1)∗LK(−1), LY(−1)∗LL(−1)*, and *LY(−1)∗LE(−1)* do not need to be inserted as additional independent variables, because they are highly correlated with some or all independent variables in the model.

14.11.3 Models with Numerical Time Independent Variables

The interaction models above can easily be modified to the models, specifically the CS (Cross-section), Group, and (Group, TP) effects models, having the numerical time independent variable, or the models with the time-related-effects (TRE). See the following example, which presents only an extension of the model in Figure 14.46.

Example 14.33 CSFEM and CSREM with trend using the GMM estimation method)
For the illustration, Figure 14.48 presents the statistical results of two CSFEMs with trends, which are derived from the model in Figure 14.46. Based on these outputs, the following findings and notes are presented.

1. Based on the output in Figure 14.48(a), the following notes and comments are presented.
   1.1 The statistical results obtained by using Cross-section fixed (dummy variables) model, or LSDV model.
   1.2 Even though 3 out of 13 independent variables have very large p-values, this output is an acceptable statistical results. So that we do not need to explore a reduced model.
   The *LY* has a positive significant time trend, adjusted for the other independent variables in the model. In other words, *LY* has a positive growth rate over the time-period $t > 2$, adjusted for the other independent variables in the model, including a set of 81 firm dummies, where their coefficients are representing the unobservable firms' effects (Baltagi, 2009, Wooldridge, 2002).

Omitted Variables: LY(-2)*LK(-1) LY(-2)*LL(-1) LY(-2)*LE(-1)			
F-statistic	9.733668	Prob. F(3,2116)	0.0000

Omitted Variables: LY(-1)*LK(-1) LY(-1)*LL(-1) LY(-1)*LE(-1)			
F-statistic	0.985907	Prob. F(3,2116)	0.3984

Figure 14.47 *Outputs of two omitted variables – LR-tests, based on the model in Figure 14.32*

Dependent Variable: LY
Method: Panel Generalized Method of Moments
Date: 10/23/12   Time: 15:18
Sample: 1 28 IF T>2
Periods included: 26
Cross-sections included: 82
Total panel (balanced) observations: 2132
2SLS instrument weighting matrix
Instrument list: C LY(-1) LY(-2) LY(-1)*LK(-1)*LL(-1)*LE(-1) LY(-1)*LK(-1)
 *LE(-1) LY(-1)*LL(-1)*LE(-1) LK(-1)*LL(-1)*LE(-1) LK(-1)*LL(-1) LK(-1)
 *LE(-1) LL(-1)*LE(-1) LK(-1) LL(-1) LE(-1) LHL(-1) LYF(-1)

Variable	Coefficient	Std. Error	t-Statistic	Prob.
C	7.843139	1.613165	4.861956	0.0000
LY(-1)	1.028497	0.023392	43.96757	0.0000
LY(-2)	-0.113459	0.023422	-4.844173	0.0000
LY(-1)*LK(-1)*LL(-1)*LE(-1)	-4.28E-05	0.000172	-0.248856	0.8035
LY(-1)*LK(-1)*LE(-1)	-0.010551	0.002912	-3.623086	0.0003
LY(-1)*LL(-1)*LE(-1)	0.019133	0.001962	9.751760	0.0000
LK(-1)*LL(-1)*LE(-1)	-0.001506	0.008448	-0.178285	0.8585
LK(-1)*LL(-1)	0.008767	0.004315	2.031676	0.0423
LK(-1)*LE(-1)	0.258455	0.136111	1.898864	0.0577
LL(-1)*LE(-1)	-0.414602	0.116244	-3.566641	0.0004
LK(-1)	-0.135397	0.068399	-1.979520	0.0479
LL(-1)	-0.379861	0.104287	-3.642478	0.0003
LE(-1)	0.119357	1.695552	0.070394	0.9439
T	0.009612	0.000821	11.71449	0.0000

Effects Specification

Cross-section fixed (dummy variables)

R-squared	0.999274	Mean dependent var	23.54663
Adjusted R-squared	0.999240	S.D. dependent var	1.918040
S.E. of regression	0.052873	Sum squared resid	5.694552
Durbin-Watson stat	1.832866	J-statistic	0.025058
Instrument rank	95.000000		

(a)

Dependent Variable: LY
Method: Panel GMM EGLS (Cross-section random effects)
Date: 10/24/12   Time: 10:11
Sample (adjusted): 3 28
Periods included: 26
Cross-sections included: 82
Total panel (balanced) observations: 2132
2SLS instrument weighting matrix
Swamy and Arora estimator of component variances
Instrument list: C LY(-1) LY(-2) LY(-1)*LK(-1)*LL(-1)*LE(-1) LY(-1)*LK(-1)
 *LE(-1) LY(-1)*LL(-1)*LE(-1) LK(-1)*LL(-1)*LE(-1) LK(-1)*LL(-1) LK(-1)
 *LE(-1) LL(-1)*LE(-1) LK(-1) LL(-1) LE(-1) LHL(-1) LYF(-1)

Variable	Coefficient	Std. Error	t-Statistic	Prob.
C	-1.914944	0.480813	-3.982719	0.0001
LY(-1)	1.039175	0.028935	35.91445	0.0000
LY(-2)	-0.051358	0.027965	-1.836515	0.0664
LY(-1)*LK(-1)*LL(-1)*LE(-1)	-0.000321	7.69E-05	-4.176894	0.0000
LY(-1)*LK(-1)*LE(-1)	0.003967	0.001241	3.196360	0.0014
LY(-1)*LL(-1)*LE(-1)	-0.000560	0.000892	-0.627672	0.5303
LK(-1)*LL(-1)*LE(-1)	0.019903	0.004425	4.497654	0.0000
LK(-1)*LL(-1)	-0.005323	0.001309	-4.066958	0.0000
LK(-1)*LE(-1)	-0.258355	0.065358	-3.952907	0.0001
LL(-1)*LE(-1)	-0.275736	0.057029	-4.834997	0.0000
LK(-1)	0.096498	0.022395	4.308823	0.0000
LL(-1)	0.130647	0.031666	4.125837	0.0000
LE(-1)	3.861671	0.882323	4.376710	0.0000
T	-0.009902	0.001681	-5.889682	0.0000

Effects Specification

	S.D.	Rho
Cross-section random	0.000000	0.0000
Idiosyncratic random	0.052860	1.0000

Weighted Statistics

R-squared	0.998298	Mean dependent var	23.54663
Adjusted R-squared	0.998287	S.D. dependent var	1.918040
S.E. of regression	0.079376	Sum squared resid	13.34440
Durbin-Watson stat	0.769932	J-statistic	0.001091
Instrument rank	14.000000		

Unweighted Statistics

R-squared	0.998298	Mean dependent var	23.54663
Sum squared resid	13.34440	Durbin-Watson stat	0.769932

(b)

Figure 14.48   *Statistical results of two CSFEMs with trend*

2. For a comparison, the output in Figure 14.48(b) obtained by using the Cross-section random effects model, which show only one of the independent variables has very large p-value. However, the output presents a small DW-statistics of 0.769932, which shows the autocorrelation problem of the model. Another limitation of this model is it presents only a single continuous regression function for all firm-time observations. I would consider this model is the worst panel data model. It is recommended to apply the random effects model by Group, or (Goup, TP) as presented in the Examples 14.28 and 14.29. See the following example.

Example 14.34   CSREM in Figure 14.48(b) by (Group,TP)

Figure 14.49 presents the statistical results of the CSREM in Figure 14.48(b) by (Goup, TP), using the following equation specification (ES), for the mean model.

$$y\ c\ ly(-1) * @expand(group, tp)\ ly(-2) * @expand(group, tp)$$
$$ly(-1) * lk(-1) * ll(-1) * le(-1) * @expand(group, tp)\ ly(-1) * lk(-1) * le(-1) * @expand(group, tp)$$
$$ly(-1) * ll(-1) * le(-1) * @expand(group, tp)\ lk(-1) * ll(-1) * le(-1) * @expand(group, tp)$$
$$lk(-1) * ll(-1) * @expand(group, tp)\ lk(-1) * le(-1) * @expand(group, tp)$$
$$ll(-1) * le(-1) * @expand(group, tp)\ lk(-1) * @expand(group, tp)$$
$$ll(-1) * @expand(group, tp)\ le(-1) * @expand(group, tp)\ t * @expand(group, tp) \quad (14.40a)$$

Dependent Variable: LY
Method: Panel GMM EGLS (Cross-section random effects)
Date: 10/24/12   Time: 11:26
Sample (adjusted): 3 28
Periods included: 26
Cross-sections included: 82
Total panel (balanced) observations: 2132
2SLS instrument weighting matrix
Swamy and Arora estimator of component variances
Instrument list: C LY(-1)*@EXPAND(GROUP,TP) LY(-2)
   *@EXPAND(GROUP,TP) LY(-1)*LK(-1)*LL(-1)*LE(-1)
   *@EXPAND(GROUP,TP) LY(-1)*LK(-1)*LE(-1)*@EXPAND(GROUP,TP)
   LY(-1)*LL(-1)*LE(-1)*@EXPAND(GROUP,TP) LK(-1)*LL(-1)*LE(-1)
   *@EXPAND(GROUP,TP) LK(-1)*LL(-1)*@EXPAND(GROUP,TP) LK(-1)
   LE(-1)@EXPAND(GROUP,TP) LL(-1)*LE(-1)*@EXPAND(GROUP,TP)
   LK(-1)*@EXPAND(GROUP,TP) LL(-1)*@EXPAND(GROUP,TP) LE(-1)
   @EXPAND(GROUP,TP) T@EXPAND(GROUP,TP) LHL(-1)
   @EXPAND(GROUP,TP) LYF(-1)@EXPAND(GROUP,TP)

Variable	Coefficient	Std. Error	t-Statistic	Prob.
C	-0.501961	1.560661	-0.321634	0.7478
LY(-1)*(GROUP=1 AND TP=1)	0.926557	0.061723	15.01157	0.0000
LY(-1)*(GROUP=1 AND TP=2)	1.215989	0.055333	21.97589	0.0000
LY(-1)*(GROUP=2 AND TP=1)	0.945582	0.036580	25.84997	0.0000
LY(-1)*(GROUP=2 AND TP=2)	0.992064	0.046384	21.38792	0.0000
LY(-2)*(GROUP=1 AND TP=1)	0.040662	0.058812	0.691389	0.4894
LY(-2)*(GROUP=1 AND TP=2)	-0.316698	0.041456	-7.639361	0.0000
LY(-2)*(GROUP=2 AND TP=1)	-0.011219	0.035835	-0.313072	0.7543
LY(-2)*(GROUP=2 AND TP=2)	-0.056684	0.042814	-1.323951	0.1857
LY(-1)*LK(-1)*LL(-1)*LE(-1)*(GROUP=1 A...	0.000190	0.000277	0.685994	0.4928
LY(-1)*LK(-1)*LL(-1)*LE(-1)*(GROUP=1 A...	-0.000181	0.000244	-0.743309	0.4574
LY(-1)*LK(-1)*LL(-1)*LE(-1)*(GROUP=2 A...	-0.000341	0.000484	-0.703990	0.4815
LY(-1)*LK(-1)*LL(-1)*LE(-1)*(GROUP=2 A...	-0.000184	0.000371	-0.495357	0.6204
LY(-1)*LK(-1)*LE(-1)*(GROUP=1 AND TP=1)	-0.004918	0.003466	-1.418738	0.1561
LY(-1)*LK(-1)*LE(-1)*(GROUP=1 AND TP=2)	0.002008	0.003647	0.550516	0.5820
LY(-1)*LK(-1)*LE(-1)*(GROUP=2 AND TP=1)	0.005018	0.006496	0.772446	0.4399
LY(-1)*LK(-1)*LE(-1)*(GROUP=2 AND TP=2)	-2.45E-05	0.004839	-0.005066	0.9960
LY(-1)*LL(-1)*LE(-1)*(GROUP=1 AND TP=1)	0.002463	0.003844	0.640701	0.5218
LY(-1)*LL(-1)*LE(-1)*(GROUP=1 AND TP=2)	0.001396	0.004029	0.346389	0.7291
LY(-1)*LL(-1)*LE(-1)*(GROUP=2 AND TP=1)	0.001493	0.007021	0.212578	0.8317
LY(-1)*LL(-1)*LE(-1)*(GROUP=2 AND TP=2)	0.007248	0.005587	1.297316	0.1947
LK(-1)*LL(-1)*(GROUP=1 AND TP=1)	-0.003299	0.004486	-0.735363	0.4622
LK(-1)*LL(-1)*(GROUP=1 AND TP=2)	-0.005131	0.004402	-1.165623	0.2439
LK(-1)*LL(-1)*(GROUP=2 AND TP=1)	-0.003597	0.004341	-0.828530	0.4075
LK(-1)*LL(-1)*(GROUP=2 AND TP=2)	-0.004169	0.004256	-0.979634	0.3274
LK(-1)*LE(-1)*(GROUP=1 AND TP=1)	0.179972	0.187369	0.960519	0.3369
LK(-1)*LE(-1)*(GROUP=1 AND TP=2)	-0.123638	0.181453	-0.681376	0.4957
LK(-1)*LE(-1)*(GROUP=2 AND TP=1)	-0.240964	0.291138	-0.827662	0.4080
LK(-1)*LE(-1)*(GROUP=2 AND TP=2)	-0.074337	0.233206	-0.318763	0.7499
LL(-1)*LE(-1)*(GROUP=1 AND TP=1)	0.003805	0.239639	0.015879	0.9873
LL(-1)*LE(-1)*(GROUP=1 AND TP=2)	-0.277733	0.212213	-1.308745	0.1908
LL(-1)*LE(-1)*(GROUP=2 AND TP=1)	-0.227956	0.319482	-0.713518	0.4756
LL(-1)*LE(-1)*(GROUP=2 AND TP=2)	-0.254590	0.268470	-0.948300	0.3431
LK(-1)*(GROUP=1 AND TP=1)	0.021424	0.071660	0.298972	0.7650
LK(-1)*(GROUP=1 AND TP=2)	0.038305	0.078869	0.485684	0.6272
LK(-1)*(GROUP=2 AND TP=1)	0.063810	0.069588	0.916959	0.3593
LK(-1)*(GROUP=2 AND TP=2)	0.057452	0.070101	0.819565	0.4126
LL(-1)*(GROUP=1 AND TP=1)	0.132244	0.108599	1.217727	0.2235
LL(-1)*(GROUP=1 AND TP=2)	0.256843	0.110220	2.330276	0.0199
LL(-1)*(GROUP=2 AND TP=1)	0.116111	0.101818	1.140380	0.2543
LL(-1)*(GROUP=2 AND TP=2)	0.137801	0.101941	1.351774	0.1766
LE(-1)*(GROUP=1 AND TP=1)	-1.453500	2.782906	-0.522296	0.6015
LE(-1)*(GROUP=1 AND TP=2)	2.601306	2.615763	0.994473	0.3201
LE(-1)*(GROUP=2 AND TP=1)	2.967772	3.606178	0.822969	0.4106
LE(-1)*(GROUP=2 AND TP=2)	1.888587	3.083244	0.612533	0.5403
T*(GROUP=1 AND TP=1)	0.002106	0.000796	2.646597	0.0082
T*(GROUP=1 AND TP=2)	-0.000778	0.000725	-1.073068	0.2834
T*(GROUP=2 AND TP=1)	-3.32E-05	0.000863	-0.038424	0.9694
T*(GROUP=2 AND TP=2)	-0.001497	0.000817	-1.832498	0.0670

Effects Specification		S.D.	Rho
Cross-section random		0.076927	0.7108
Idiosyncratic random		0.049065	0.2892

Weighted Statistics			
R-squared	0.986364	Mean dependent var	2.922578
Adjusted R-squared	0.986023	S.D. dependent var	0.413383
S.E. of regression	0.048871	Sum squared resid	4.965508
Durbin-Watson stat	2.020791	J-statistic	178.0948
Instrument rank	57.000000		

Unweighted Statistics			
R-squared	0.998910	Mean dependent var	23.54663
Sum squared resid	8.546143	Durbin-Watson stat	1.174127

Figure 14.49   *Output of the CSREM in (14.40), using the GMM estimation method*

with the instruments as follows:

$$ly(-1) * @expand(group, tp)\ldots\ldots\ldots\ldots\ldots t * @expand(group, TP)$$
$$\textbf{\textit{hl}}(-1) * \textbf{\textit{@expand}}(\textbf{\textit{group}}, \textbf{\textit{tp}})\ \textbf{\textit{lyf}}(-1) * \textbf{\textit{@expand}}(\textbf{\textit{group}}, \textbf{\textit{tp}}) \tag{14.40b}$$

where the first line indicates the list of all independent variables of the mean model.

Based on this output, the following findings and notes are presented.

1. Compare to the CSREM in Figure 14.48(b) having a very small DW-statistic, and Rho = 0.0000, this CSREM has a sufficiently large DW-statistic of 2.020791, and Rho (CSREM) = 0.7108. So that this model is a better model, since it does not have autocorrelation problem.
2. This output is representing a set of four heterogeneous multiple regressions, having the same set of independent variables.

3. However, since many of the independent variables have large p-values, then the reduced model should be explored. It is recommended to apply the manual stepwise selection method, which has be demonstrated in Section 7.6. Do it for an exercise.

4. It is important to note how the instruments can easily be selected, that is by using all independent variables of the mean model plus one or more external variables. For a comparison, refer to the empirical statistical results of the instrumental variables models presented in EViews6 User's Guide II, and Baltagi (2009a).

Example 14.35   (Special CSFEM and CSREM with TRE using the GMM estimation method)

For the illustration, Figure 14.50 presents the statistical results of two special CSFEM and CSREM with TREs, which are derived from the model in Figure 14.46. Based on these outputs, the following findings and notes are presented.

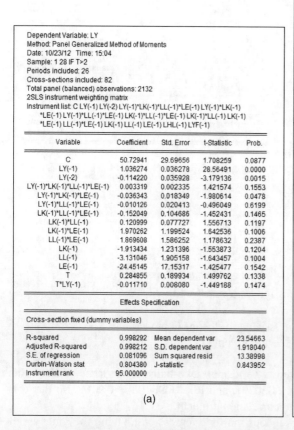

Figure 14.50   *Statistical results of CSFEM and CSREM with the time-related-effects (TRE)*

Dependent Variable: LY
Method: Panel GMM EGLS (Cross-section random effects)
Date: 10/24/12 Time: 15:29
Sample (adjusted): 3 28
Periods included: 26
Cross-sections included: 82
Total panel (balanced) observations: 2132
2SLS instrument weighting matrix
Swamy and Arora estimator of component variances
Instrument list: C LY(-1) LY(-2) LY(-1)*LK(-1)*LL(-1)*LE(-1) LY(-1)*LK(-1)
  *LE(-1) LY(-1)*LL(-1)*LE(-1) LK(-1)*LL(-1)*LE(-1) LK(-1)*LL(-1) LK(-1)
  *LE(-1) LL(-1)*LE(-1) LK(-1) LL(-1) LE(-1) LHL(-1) LYF(-1) LHL(-1)*LYF(
  -1)

Variable	Coefficient	Std. Error	t-Statistic	Prob.
C	0.831169	0.544062	1.527712	0.1267
LY(-1)	1.203657	0.030438	39.54430	0.0000
LY(-2)	-0.155768	0.024483	-6.362246	0.0000
LY(-1)*LK(-1)*LL(-1)*LE(-1)	-0.000539	0.000108	-4.979954	0.0000
LY(-1)*LK(-1)*LE(-1)	0.008245	0.001894	4.353431	0.0000
LY(-1)*LL(-1)*LE(-1)	-0.000546	0.000896	-0.608992	0.5426
LK(-1)*LL(-1)*LE(-1)	0.026557	0.005414	4.905499	0.0000
LK(-1)*LL(-1)	0.004729	0.001871	2.526723	0.0116
LK(-1)*LE(-1)	-0.397286	0.085544	-4.644225	0.0000
LL(-1)*LE(-1)	-0.317476	0.062081	-5.113902	0.0000
LK(-1)	-0.071871	0.031386	-2.289929	0.0221
LL(-1)	-0.123872	0.046706	-2.652180	0.0081
LE(-1)	4.830679	1.014058	4.763709	0.0000
T	0.073706	0.016645	4.428147	0.0000
T*LY(-1)	-0.003234	0.000698	-4.635216	0.0000

Effects Specification		S.D.	Rho
Cross-section random		0.000000	0.0000
Idiosyncratic random		0.052813	1.0000

Weighted Statistics			
R-squared	0.998754	Mean dependent var	23.54663
Adjusted R-squared	0.998746	S.D. dependent var	1.918040
S.E. of regression	0.067927	Sum squared resid	9.768076
Durbin-Watson stat	1.183680	J-statistic	0.023109
Instrument rank	15.000000		

Unweighted Statistics			
R-squared	0.998754	Mean dependent var	23.54663
Sum squared resid	9.768076	Durbin-Watson stat	1.183680

Figure 14.51 *Statistical results of a CSREM with an unintentionally set of instruments*

1. Based on the output in Figure 14.50(a) the following findings and notes are presented.

   1.1 This model is a special model with TRE, since it has only a single interaction of the time variable, namely $t*LY(-1)$. The main reason to select this interaction for the first additional independent variable is the main variable $LY(-1)$ is defined or assumed to be the most important upstream variable of *LY*. Thence the linear effect of the time-*t* on *LY* should depend on $LY(-1)$.

   1.2 This model could easily be extended using an additional interaction or set of interactions, such as the two-way interactions $t*LK(-1)$, $t*LL(-1)$, and $t*LE(-1)$, or higher interactions. Do it for an exercise.

   1.3 The need for entering additional interaction(s) can be explored using the trial-and-error method, because the *Omitted Variables-Likelihood Ratio Test* is not available with this estimation method. Do for the exercises. Note that by inserting additional independent variables, you might need to insert additional instruments. Otherwise, the error message *"Order condition violated – Insufficient Instruments"* would be obtained.

   1.4 At the 10% level of significance, each of the independent variables has either positive or negative significant adjusted effect on LY, except two interactions: $LY(-1)*LL(-1)*LE(-1)$, and $LL(-1)*LE(-1)$.

   1.5 Referring to the independent variables t and $t*LY(-1)$ only, we could consider the following relationship.

$$LY = (0.284855 - 0.011710 * LY(-1)) * t$$

which give a turning point at $LY(-1) = 24.32579 = 0.284855/0.11710$, for the time trend of *LY*. Its trend is positive if $LY(-1) < 24.32579$, and negative if otherwise. Note that the variable $LY(-1)$ has $Min(LY(-1)) = 23.48461$, and $Max(LY(-1)) = 29.08964$.

2. For an additional illustration, Figure 14.50(b) presents an unexpected output of a CSREM, where all independent variables have very large *p*-values. Why? I would say that the main causes should be the sampled data, as well as the multicollinearity between the independent variables, or the instruments. See the following example.

Example 14.36   (An unexpected impact of the instruments)
As an extension or modification of the model in Figure 14.50(b), Figure 14.51 presents an unexpected statistical results of the model by inserting an additional instrument, namely *LHL(−1)*LYF(−1)*, in the instruments (14.40b). Based on this output, the following notes and comments are presented.

1. In order to study the unknown impact of a modified instruments on the statistical results in Figure 14.50(b), I have been using the trial-and-error method to modify the instruments in (14.40b). It is unexpected the output presented in Figure 14.51 obtained by inserting an additional instrument, namely *LHL(−1)*LYF(−1)*.
2. Compare to the output in Figure 14.50(b), this output should be considered as a much better statistical results.
3. For the comparison, I have tried also to insert alternative additional instruments in (14.40b), such as (a). LK(−1)*LHL(−1); (b). LK(−1)*LHL(−1), and LK(−1)*LYF(−1); (c). LK(−1)*LHL(−1), LL(−1)*LHL(−1), and LE(−1)*LHL(−1); (d). LK(−1)^2, and Lk(−1)^3; and (e). LHL(−1)^2, and LYF(−1)^2. Do it for an exercise, and try to insert other instruments.
4. Based on these findings, it can be concluded that the researcher would never know the best possible instruments, more over the true set of instruments, for a defined mean model.

References

Aboody, D., Johnson, N.B., and Kasznik, R. (2010) Employee stock options and future firm performance: evidence from option repricings. *Journal of Accounting and Economics*, **50**, 74–92.

Agresti, A. (1990) *Categorical Data Analysis*, John Wiley & Sons, Inc., New York.

Agresti, A. (1984) *Analysis of Ordinal Categorical Data*, John Wiley & Sons, Inc., New York.

Agung, I.G.N. (2011) *Cross Section and Experimental Data Analysis Using EViews*. John Wiley & Sons, Inc., Hoboken, NJ.

Agung, I.G.N. (2011a) *Cross Section and Experimental Data Analysis Using EViews*, John Wiley & Sons, Inc., Singapore.

Agung, I.G.N. (2011b) *Manajemen Penulisan Skripsi, Tesis Dan Disertasi Statistika: Kiat-Kiat Untuk Mempersingkat Waktu Penulisan Karya Ilmiah Yang Bermutu*, 4th edn, PT RajaGrafindo Persada, Jakarta.

Agung, I.G.N. (2009a) *Time Series Data Analysis Using EViews*, John Wiley & Sons. Inc., Singapore.

Agung, I.G.N. (2009b) Simple quantitative analysis but very important for decision making in business and management. *The Ary Suta Center Series on Strategic Management*, **3**, 173–198.

Agung, I.G.N. (2009c) What should have a great leader done with statistics? *The Ary Suta Center Series on Strategic Management*, **4**, 37–47.

Agung, I.G.N. (2008) Simple quantitative analysis but very important for decision making in business and management. Presented at the Third International Conference on Business and Management Research, 27–29 August, 2008. Sanur Paradise, Bali, Indonesia.

Agung, I.G.N. (2006) *Statistika: Penerapan Model Rerata-Sel Multivariat Dan Model Ekonometri Dengan SPSS*, Yayasan Sad Satria Bhakti, Jakarta.

Agung, I.G.N. and Bustami, D. (2004) Issues related to gender equity and socio-economic aspect in Indonesia, in *Empowerment of Indonesian Women: Family, Reproductive Health, Employment and Migration* (eds S.H. Hatmadji and I.D. Utomo), Ford Foundation & Demographic Institute, FEUI, pp. 44–60.

Agung, I.G.N. (2004) *Statistika: Penerapan Metode Analisis Untuk Tabulasi Sempurna Dan Tak-sempurna*. Cetakan Kedua. PT RajaGrafindo Persada, Jakarta.

Agung, I.G.N. (2002) *Statistika: Analisis Hubungan Kausal Berdasarkan Data Katagorik*. Cetakan Kedua. PT Raja Grafindo Persada, Jakarta.

Agung, I.G.N., Coordinator (2001) Evaluating of KDP Impacts on Community Organization and Household Welfare. Kerjanasama LD-FEUI, PMD, Bappenas dan Bank Dunia.

Agung, I.G.N., Coordinator (2000a) Based-line Survey on the Kecamatan Development Program. Kerjanasama LD-FEUI, PMD, Bappenas dan Bank Dunia.

Agung, I.G.N. (2000b) Analisis statistik sederhana untuk pengambilan keputusan, in *Populasi: Buletin Penelitian dan Kebijakan Kependudukan*, vol. **11**, Nomor 2, Tahun 2000, PPK UGM., Yogyakarta.

Agung, I.G.N. (1999a) Generalized exponential growth functions. *Journal of Population*, **5**(2), 1–20.

Agung, I.G.N., Coordinator (1996b) The Impact of Economic Crisis on Familiy Planning and Health. LDFEUI, BKKBN and Policy Project (USAID).

Agung, I.G.N. (1999c) *Faktor Interaksi: Pengertian Secara Substansi dan Statistika*, Lembaga Penerbit FEUI, Jakarta.

Agung, I.G.N. (1998) Metode Penelitian Sosial, Bagian 2 (Unpublished).

Agung, I.G.N. (1996) The development of composite indices for the quality of life and human resources using factor analysis. *Journal of Population*, **2**(2), 207–217.

Agung, I.G.N. (1992a) *Metode Penelitian Sosial*, Bagian 1, PT. Gramedia Utama, Jakarta.

Agung, I.G.N. (1992b) *Analisis Regresi Ganda untuk Data Kependudukan*, 4th edn, Bagian 1. PPK UGM, Yogyakarta.

Agung, I.G.N. (1992c) *Analisis Regresi Ganda untuk Data Kependudukan*, 3rd edn., Bagian 2. PPK UGM, Yogyakarta.

Agung, I.G.N. (1988) *Garis Patah Paritas: Pengembangan Suatu Metode Untuk Memperkirakan Fertilitas*, 2nd edn, (Kedua). PPK UGM, Yogyakarta.

Agung, I.G.N. (1981) *Some Nonparametric Procedures for Right Censored Data*, Memeo Series No. 1347, Institute of Statistics, Chapel Hill, North Carolina.

Agung, I.G.N. and Pasay danSugiharso, H.A. (2008) *Teori Ekonomi Mikro: Suatu Analisis Produksi Terapan*, PT RajaGrafindo Persada, Jakarta.

Agung, I.G.N. and Pasay danSugiharso, H.A. (1994) *Teori Ekonomi Mikro: Suatu Analisis Produksi Terapan*, Lembaga Demografi dan Lembaga Penerbit FEUI, Jakarta.

Agung, I.G.N. and Siswantoro, D. (2012) Heterogeneous Regressions, Fixed Effects and Random Effects Models, presented in the 4th IACSF International Accounting Conference. 22-23 November 2012, Bidakara Hotel, Jakarta - Indonesia.

Allawadi, K.L., Pauwels, K., and Steenkamp, E.M. (2008) Private-label use and store loyalty. *Journal of Marketing*, **72**(6), 19–30.

Alamsyah, C. (2007) Factor-faktor yang Mempengaruhi Kualitas Pengetahuan Yang iDialihkanpada Perusahaan yang Melakukan Aliansi Stratejik. Dissertation, Graduate School of Management, Faculty of Economics, University of Indonesia.

Al-Shammari, B., Brown, P., and Tarca, A. (2008) An investigation of compliance with international accounting standards by listed companies in the Gulf Co-Operation Council member states. *The International Journal of Accounting*, **43**, 425–447.

Andreasen, A.R. (1988) *Cheap but Good Marketing Research*, IRWIN, Burr Ridge, Illinois.

Ary Suta, I.P.G. (2005) Market Performance of Indonesian Public Companies. Dissertation, Graduate School of Management, Faculty of Economics, University of Indonesia. Publishers: Sad Satria Bhakti Foundation, Jakarta, Indonesia Foundation, Jakarta.

Asher, J.J. and Sciarrino, J.A. (1974) Realistic work sample tests. *A Review: Personnel Psychology*, **27**, 519–533.

Ariastiadi (2010) Pengaruh Sistem Pengendalian Risiko, Persepsi risiko, dan Keunggulan Daya Saing terhadap Prilaku Pengambilan Keputusan Stratejik Berisiko dan Dampaknya terhadap Kinerja: Studi Industri Perbankan Indonesia. Dissertation, Graduate School of Management, Faculty of Economics, University of Indonesia.

Banbeko, I., Lemmon, M., and Tserlukevich, Y. (2011) Employee stock options and investment. *The Journal of Finance*, **LXVI**(3), 981–1009.

Baltagi, B.H. (2009a) *Econometric Analysis of Panel Data*, John Wiley & Sons, Ltd., Chichester.

Baltagi, B.H. (2009b) *A Companion to Econometric Analysis of Panel Data*, John Wiley & Sons, Ltd., Chichester.

Bansal, P. (2005) Evolving sustainably: a longitudinal study of corporate sustainable development. *Strategic Management Journal*, **26**, 197–218.

Bansal, H.S., Taylor, S.F., and James, Y.S. (2005) "Migrating" to new service providers: toward a unifying framework of consumer's switching behaviours. *Journal of the Academy of Marketing Science*, **33**(1), 96–115.

Barthelemy, J. (2008) Opportunism, knowledge, and the performance of franchise chains. *Strategic Management Journal*, **29**, 1451–1463.

Bass, F.M. (1969) A new product growth model for consumer durables. *Management Science*, **15**(5), 215–227.

Benmelech, E. and Berman, N.K. (2011) Bankruptcy and collateral channel. *The Journal of Finance*, **LXVI**(2), 337–378.

Benzecri, J.P. (1973) *Analyse des Donnes, 2 vols*. Dunod, Paris.

Bertrand, M., Schoar, A., and Thesmar, D. (2007) Banking deregulation and industry structure: evidence from the French banking reforms of 1985. *The Journal of Finance*, **LXII**(2), 597–628.

Billett, M.T., King, Tao.-Hsien.Dolly., and Maucer, D.C. (2007) Growth opportunities and the choice of leverage, debt maturity, and covenants. *The Journal of Finance*, **LXII**(2), 697–730.

Binsbergen, J.H., Graham, J.R., and Yang, J. (2010) The cost of debt. *The Journal of Finance*, **LXV**(6), 2089–2136.

Bishop, Y.M.M., Fienberg, S.E., and Holland, P.W. (1976) *Discrete Multivariate Analysis: Theory and Practice*, The MIT Press Cambridge, Massachusetts, and London.

Blanconiere, W.J., Johson, M.F., and Lewis, M.F. (2008) The role of tax regulation and compensation contracts in the decision of voluntarily expense employee stock options. *Journal of Accounting and Economics*, **46**, 101–111.

Brau, J.C. and Johnson, P.M. (2009) Earnings management in IPOIs: Pst-engagement third-party mitigation or issuer signalling? *Advance in Accounting, incorporating Advances in International Accounting*, **25**, 125–135.

Brehanu, A. and Fufa, B. (2008) Repayment rate of loans from semi-formal financial institution among small-scale farmers in Ethiopia: two-limit tobit analysis. *Journal of Socio-Economics. Greenwich*, **37**(6), 2221.

Brooks, C. (2008) *Introduction Econometric for Finance*, Cambridge University Press.

Buchori, N.S. (2008) Pengaruh Karakrteristik Demografi, Sosial dan Ekonomi Terhadap Pengetahuan, Sikap Interaksi dan Praktek Perbankan, di DKI Jakarta dan Sumatera Barat. Thesis, Graduate Program on Population and Manpower, University of Indonesia.

Campion, J.E. (1972) Work sampling for personnel selection. *Journal for Applied Psychology*, **56**, 40–44.

Chalmers, K., Koh, P.-S., and Stapledon, G. (2006) The determinants of CEO compensation: rent extraction or labour demand? *The British Accounting Review*, **38**, 259–275. Available online at www.sciencedirect.com.

Chandra, M.A. and Primasari, M. (2009) The Australian Tourist Perception on the Tourism Products and Services in West Sumatra. Presented in the 7th Asia-Pacific Council on Hotels Restaurants and Institutional Education Conference, 28–31 May 2009, Singapore.

Chandra, M.A. (2008) The Role of Public Relations: Perceptions of Jakarta Travel Agents. Presented in the 6th Asia Pacific CHRIE (APacCHRIE) 2008 Conference and THE-ICE Panel of Experts Forum 2008, 21–24 May 2008, Perth, Australiag.

Chatterji, A.K. (2009) Spawned with a silver spoon? Entrepreneurial performance and innovation in the medical device industry. *Strategic Management Journal*, **30**, 185–206.

Chen, S.F.S. (2008) The motives for international acquisitions: capability procurements, strategic considerations, and the role of ownership structures. *Journal of International Business Studies*, **39**, 454–471.

Coelli, T., Prasada Rao, D.s., and George, E.B. (2001) *An Introduction to Efficiency and Productivity Analysis*, Kluwer Academic Publishers, Boston.

Cochran, W.G. (1952) The χ^2 test of goodness of fit. *Annals of Mathematical Statistics*, **23**, −345 (4.2, 4.5).

Collins, T.A., Rosewnberg, L., Makambi, K., Plmer, J.R., and Campbell, L.A. (2009) Dietary pattern and breast cancer risk in women participating in the Black Women's Health Study. *The American Journal of Clinical Nutrition*, **90**(3), 621–628.

Conover, W.J. (1980) *Practical Nonparametric Statistics*, John Wiley & Sons, Inc., New York.

Coombs, J.E. and Gilley, K.M. (2005) Stakeholder management as a predictor of CEO compensation: main effects and interactions with financial performance. *Strategic Management Journal*, **26**, 827–840.

Cooper, W.W., Seiford, L.M., and Tone, K. (2000) *Data Envelopment Analysis*, Kluwer Academic Publishers, Boston.

Cordia, T. and Subrahmanyam, A. (2004) Order imbalance and individual stock returns: theory and evidence. *Journal of Financial Economics*, **72**, 485–518.

Crenn, P. *et al.* (2009) Plasma citruline is a biomarker of enterocyte mass and an indicator of parenteral nutrition in HIV-infected patients. *The American Journal of Clinical Nutrition*, **90**(3), 587–594.

Dharmapala, D., Foley, C.F., and Forres, K.J. (2011) Watch what I do, not what I say: the unintended consequences of the homeland investment act. *The Journal of Finance*, **66**(3), 753–787.

Dedman, E. and Lennox, C. (2009) Perceived competition, profitability and with holding of information about sales and cost of sales. *Journal of Accounting and Economics*, **48**, 210–230.

Delong, G. and Deyoung, R. (2007) Learning by observing; information spillovers in the execution and valuation of commercial bank M & As. *The Journal of Finance*, **1**(1), 181–216.

Do, A.D. (2006) Strategic Management at the Bottom Level: the Role of Leadership in Developing Organizational Citizenship Behaviour and its Implications for Organization Performance. Dissertation, Graduate School of Management, Faculty of Economics, University of Indonesia.

Doskeland, T.M. and Hvide, H.K. (2011) Do individual investors have asymmetric information based on work experience? *The Journal of Finance*, **LXVI**(3), 1011–1041.

Duh, R.-R., Lee, W.-W., and Lin, C.-C. (2009) Reversing an impairment loss and earnings management: the role of corporate governance. *The International Journal of Accounting*, **44**, 113–137.

Dunnette, M.D. (1972) *Validity Study Results for Jobs Relevant to Petroleum Refining Company*, American Petroleum Institute., Washington, DC.

Engelberg, J.E. and Parsons, C.A. (2011) The causal impact of media in financial markets. *The Journal of Finance*, **LXVI**(1), 67–97.

Enders, W. (1995) *Applied Econometric Time Series*, John Wiley & Sons, Inc., New York.

Faad, H.M. (2008) Pengaruh Metode Mengajar dan Umpan Balik Penilaian terhadap Hasil Belajar Matematika dengan memperhitungkan Kovariat Minat dan Pengetahuan dasar Siswa. Disertasi, Program Pascasarjana, Universitas Negeri Jakarta.

Fang, E. (2008) Customer participation and the trade-off between new product innovativeness and speed to market. *Journal of Marketing*, **72**(4), 90–104.

Feng, M., Li, C., and McVay, S. (2009) Internal control and management guidance. *Journal of Accounting and Economics*, **46**, 190–209.

Ferris, K.R. and Wallace, J.S. (2009) IRC Section 162(m) and the law of unintended consequences. *Advance in Accounting, incorporating Advances in International Accounting*, **25**, 147–155.

Filip, A. and Raffournier, B. (2010) The value relevance of earning in a transition economy: the case Romania. *The International Journal of Accounting*, **45**, 77–103.

Fitriwati, L. (2004) Faktor-faktor yang Mempengaruhi Status Kesehatan Individu. Thesis, Graduate Program on Population and Manpower, University of Indonesia.

Freund, J.E., Williams, F.J., and Peters, B.M. (1993) *Elementary Business Statistics*, Prentice Hall, Inc., New Jersey.

Frischmann, P.J., Shevlin, T., and Wilson, R. (2008) Economic consequences of increasing the conformity in accounting for uncertain tax benefit. *Journal of Accounting and Economics*, **46**, 261–278.

Gardner, H. (2006) *Multiple Intelligences*, Basic Books, New York.

Garcia, M.T.C. (2009) The impact of securities regulation on the earnings properties of European cross-listed firms. *The International Journal of Accounting*, **44**, 279–304.

Giroud, X. and Mueller, H.M. (2011) Corporate governance, product market competition, and equity prices. *The Journal of Finance*, **66**(2), 563–600.

Gifi, A. (1991) *Nonlinear Multivariate Analysis*, John Wiley & Sons, Inc., New York.

Golder, P.N. and Tellis, G.J. (2004) Going, going, gone: cascadesw, diffusion, and turning points of the product life-cycle. *Marketing Science*, **23**(2), 207–218.

Gourierroux, C. and Manfort, A. (1997) *Time Series and Dynamic Models*, Cambridge University Press, Cambridge.

Govindarajan, V. and Fisher, J. (1990) Strategy, control systems, and resource sharing: effects on business-unit performance. *Academy of Management Journal*, **33**(2), 259–285.

Grant, E.L. and Leavenworth, R.S. (1988) *Statistical Quality Control*, McGraw-Hill, New York.

Gregory, R.J. (2000) *Psychological Testing: History, Principles, and Applications*, Allyn and Bacon, Boston.

Grunfeld, Y. (1958) The Determinants of Corporate Investment. PhD thesis, Department of Economics, University of Chicago.

Garybill, F.A. (1976) *Theory and Application of the Linear Model*, Duxbury Press, Belmont, California.

Gujarati, D.N. (2003) *Basic Econometric*, McGraw-Hill, Boston.

Gulati, R., Lawrence, P.R., and Puranam, P. (2005) Adaptation in vertical relationships: beyond incentive conflict. *Strategic Management Journal*, **26**, 415–440.

Gunawan, F.A. (2004) Analisis Nilai Tukar Valuta Asing Intrahari Dengan Pendekatan Model Hybrid. Dissertation, Graduate School of Management, Faculty of Economics, University of Indonesia.

Guo, S., Hotchkiss, E.S., and Song, W. (2011) Do buyouts (still create value? *The Journal of Finance*, **LXVI**(2), 479–518.

Haas, R. de. and Peeters, M. (2006) The dynamic adjusted toward target capital structures of of firms in transition economics. *Economics of Transition*, **14**(1), 133–169.

Hadri, K. (2000) Testing for the stationary in heterogeneous panel data. *Econometrics Journal*, **3**, 148–161.

Hair, J.F. Jr, Black, W.C., Babin, B.J., *et al.* (2006) *Multivariate Analysis*, Prentice-Hall International, Inc.

Hameed, A., Kang, W., and Viswanathan, S. (2010) Stock market declines and liquidity. *The Journal of Finance*, **LXV**(1)

Hank, J.E. and Reitsch, A.G. (1989) *Business Forecasting*, Allyn and Bacon, Boston.

Hank, J.E. and Reitsch, A.G. (1992) *Business Forecasting*, 2nd edn, Allyn and Boston, Boston.

Hardle, W. (1999) Applied nonparametric regression, in *Economic Society Monographs*, Cambridge University Press, Cambridge.

Harford, J. and Li, K. (2007) Decoupling CEO wealth and firm performance: the case of acquiring CEOs. *The Journal of Finance*, **LXII**(2), 917–949.

Harris, R.J. (1975) *A Primer of Multivariate Analysis*, Academic Press, New York.

Haryanto, J.O. (2008) Analisis Intensi Mengkonsumsi Lagi pada Anak dalam Membangun Kekuatan Mempengaruhi, Pembelian Impulsif dan Autobiographical Memory. Dissertation, Graduate School of Management, Faculty of Economics, University of Indonesia.

He, H., EI-Masry, EI.-H., and Wu, Y. (2008) Accounting conservatism of cross-listing firms in the pre and post-Oxley periods. *Advances in Accounting, incorporating Advances in International Accounting*, **24**, 237–242.

Heckman, J.J. (1979) Sample selection bias as a specification error. *Econometrica*, **47**(1), 153–161.

Heitzman, S., Wasley, C., and Zimmerman, J. (2010) The joint effects of materiality thresholds and voluntary disclosure incentives on firms' disclosure decisions. *Journal of Accounting and Economics*, **49**, 109–132.

Henders, B.C. and Hughes, K.E. II (2010) Valuation implications of regulatory climate for utilities facing future environmental costs. *Advances in Accounting, incorporating Advances in International Accounting*, **26**, 13–24.

Hendershott, T., Jones, C.M., and Menkveld, A.J. (2011) Does algorithmic trading improve liquidity? *The Journal of Finance*, **LXVI**(1), 1–33.

Henning-Thurau, T., Henning, V., and Sattle, H. (2007) Consumer file sharing of motion pictured. *Journal of Marketing*, **71**, 1–18.

Herman (2010) Studi Tentang Persepsi Mahasiswa Peserta Tutorial Terhadap Pelaksanaan Tutorial Tatap Muka di Universitas Terbuka. Dissertation, Jakarta State University.

Hill, R.C., Griffiths, W.E., and Judge, G.G. (2001) *Using EViews For Undergraduate Econometrics*, 2nd edn, John Wiley & Sons, Inc., New York.

Homburg, C., Droll, M., and Totzek, D. (2008) Customer prioritization: does it pay off, and how should it be implemented? *Journal of Marketing*, **72**(5), 110–130.

Hosmer, D.W. Jr and Lemesshow, S. (2000) *Applied Logistic Regression*, John Wiley & Sons, Inc., New York.

Hugo, A. and Muslu, V. (2010) Market demand for conservative analysis. *Journal of Accounting and Economics*, **50**, 42–57.

Huitema, B.E. (1980) *The Analysis of Covariance and Alternatives*, John Wiley & Sons, Inc., New York.

Insaf, S. (2010) Faktor-faktor yang mempengaruhi Keputusan Antara Tahun 2010–2007.

Jemias, J. (2008) The relative influence of competitive intensity and business strategy on the relationship between financial leverage and performance. *The British Accounting Review*, **40**, 71–86.

Johnson, P.O. and Neyman, J. (1936) Test of certain linear hypotheses and their application to some education problems. *Statistical Research Memoirs*, **1**, 57–93.

Jotikasthira, K., Lundblad, C., and Ramadorai, T. (2012) Asset fire sales and purchases and the international transmission of funding shocks. *The Journal of Finance*, **LXVII**(6)

Kacperczyk, A. (2009) With greater power comes greater responsibility? Takeover protection and corporate attention to stakeholders. *Strategic Management Journal*, **30**, 261–285.

Kaplan, R.M. and Saccuzzo, D.P. (2005) *Psychological Testing Principles, Application and Issues*, 6th edn, Thomson Wadworth, USA.

Kementa, J. (1980) *Elements of Econometrics*, Macmillan Publishing Company, New York.

Kendal, M.G. (1938) A new measure of rank correlation. *Biomterika*, **30**, 81–93.

Kernen, K.A. (2003) Dampak Pengelolaan Aset Perusahaan Terbuka Indonesia Periode 1990-n1997 Pada Kinerja Keuangan Dikaji Dari Teori Governan Korporat. Dissertation, Faculty of Economic, University of Indonesia.

Khoon, C.H., Santa, A.U., and Gupta, G.S. (1999) *Malaysian Management Journal*, **3**(2), 49–72.

Kirsch, D., Goldfarb, B., and Gera, Z. (2009) Form or substance: the role of business plans in venture capital decision making. *Strategic Management Journal*, **30**, 487–515.

Kish, L. (1965) *Survey Sampling*, John Wiley & Sons, Inc., New York.

Korteweg, A. (2010) The net benefits to leverage. *The Journal of Finance*, **LXV**(6), 2137–2170.

Kousenidis, D.V., Lagas, A.C., and Negakis, C.I. (2009) Value relevance of conservative and non-conservative accounting information. *The International Journal of Accounting*, **44**, 219–238.

Kruskal, W.H. and Wallis, W.A. (1952) Use of ranks on one-criterion variance analysis. *Journal of the American Statistical Association*, **47**, 583–621 (correction appears in Vol. 48, pp. 907–911 (5.2).

Laksmana, I. and Yang, Y. (2009) Corporate citizenship and earnings attributes. *Advances in Accounting, incorporating Advances in International Accounting*, **25**, 40–48.

Lapin, L.L. (1973) *Statistics for Modern Business Decisions*, Harcourt Brace Jovanovich, Inc.

Leiblein, M.J. and Miller, D.J. (2003) An empirical examination of transaction and firm-level influences on the vertical boundaries of the firm. *Strategic Management Journal*, **24**, 839–859.

Li, C., Sun, L., and Ettredge, M. (2010) Financial executive qualifications, financial executive turnover, and adverse SOX 404 opinions. *Journal of Accounting & Economics*, **50**, 93–110.

Lindawati, G. (2002) A Contingency Approach to strategy Implementation at the Business Unit Level: Integrating Organizational Design and Management Accounting System with Strategy. Dissertation, Faculty of Economic, University of Indonesia.

Lindstrom, M. (2009) Social capital, political trust and daily smoking and smoking cessation: a population-based study in southern Sweden. *The Journal of Public Health, Elsevier*, **123**(7), 496–501.

Li, T. and Zheng, X. (2008) Semiparametric Bayesian inference for dynamic tobit panel data models with unobserved heterogeneity. *Journal of Applied Econometrics*, **23**(6), 699.

Maddala, G.S. (1989) *Limited Dependent and Qualitative Variables in Econometrics*, Cambridge University Press, Cambridge.

Malhotra, N.K. (ed.) (2007) *Review of Marketing Research*, vol. **3**, M. E. Sharpe, Inc., New York.

Mann, H.B. and Whitney, D.R. (1947) On a test of whether one of two random variables is stochastically larger than the other. *The Annual Statistical Association*, **59**, 935–959 (3.4).

Markarian, G., Pozza, L., and Prencipe, A. (2008) Capitalization of R&D costs and earnings management Evidence from Italian listed companies. *The International Journal of Accountings*, **43**, 246–267.

McDonald, J. (2009) Using least squares and tobit in second stage DEA efficiency analysis. *European Journal of Operational Research*, **197**(2), 792.

Meyer, K.E., Estrin, S., Bhaumik, S.K., and Peng, M.W. (2009) Institution, resources, and entry strategies in emerging economies. *Strategic Management Journal*, **30**, 61–80.

Mohammad, H. (2006) The Effect of Paradoxical Strategies on Firm Performance: An Empirical Study of Indonesian Banking Industry. Dissertation, Graduate School of Management, Faculty of Economics, University of Indonesia.

Mohammad, H. and Agung, I.G.N. (2007) Paradoxical strategies and firm performance: the case of Indonesian banking industry. *The South East Asian Journal of Management*, **1**(1), 43–61.

Naes, R., Skjeltorp, J.A., and Odegaard, B.A. (2011) Stock market liquidity and the business cycle. *The Journal of Finance*, **LXVI**(1), 139–176.

Narindra, I.M.D. (2006) Pengaruh Struktur Dan Skala Perusahaan Terhadap Profitalibitas Perushaan. Thesis, Faculty of Economic, University of Indonesia.

Neter, J. and Wasserman, W. (1974) *Applied Statistical Models*, Richard D. Irwin, Inc, Homewood, Illinois 60430.

Novarudin, J.P. (2010) Pengaruh Pendidikan Terhadap Unemployment dan Underemployment di Provinsi Nusa Tenggara Barat. Thesis, Thesis, Graduate Program in Population and Labour Force, University of Indonesia, Jakarta.

Palmatier, R.W., Dant, R.P., and Grewal, D. (2007) Consumer file sharing of motion pictured. *Journal of Marketing*, **71**, 172–194.

Park, L.E. (1994) The policy and urban development: evidence from the Indiana Enterprise Zone Program. *Journal of Public Economics*, **54**, 37–49.

Park, K. and Jang, S. (2011) Mergers and acquisitions and firm growth: investing restaurant firms. *International Journal of Hospitality Management*, **30**, 141–149.

Parzen, E. (1960) *Modern Probability Theory and Its Applications*, John Wiley & Sons, Inc., New York.

Pearson, K. (1900) On the criterion that a given system of deviations from the probable in the case of a correlated system of variables is such that it can reasonably be supposed to have arisen from random sampling. *Philosophical Magazine*, **50**(5), 157–175, (4.5).

Qin, Yu, Xia, M., Ma, J., Hao, Y.T., Liu, J., Mou, H.Y., Cao, Li., and Ling, W.H. (2009) Anthocyanin supplementation improves serum LDL- and HDL-cholesterol concentrations associated with the inhibition of cholesterol ester transfer protein in dyslipidaemic subjects. *The American Journal of Clinical Nutrition*, **90**(3), 485–492.

Rakow, K.C. (2010) The effect of management earnings forecast characteristics on cost of equity capital. *Advances in Accounting, incorporating Advances in International Accounting*, **26**, 37–46.

Rahman, M., Khan, A.R., and Islam, N. (2008) Influences of selected socio-economic variables on the age of first birth in Rajshahi district of Bangladesh. *The Journal of Population*, **14**(1), 101–117.

Reisman, D. (2009) Economics and old age: the Singapore experience, in *The Older Persons in Southeast Asia*, ISEAS, Singapore, 71–96.

Ruslan (2008) Studi Tentang Kinerja Dosen Berdasarkan Kepuasan Mahasiswa dan Pengaruhnya Terhadap Perilaku Pascakiliah di FMIPA Universitas Negeri Makassar, Dissertation, Jakarta State University.

Saunders, A. (1999) *Credit Risk Measurement*, John Wiley & Sons, Inc., Canada.

Schlesselman, J.J. (1982) *Case-control Studies: Design, Conduct, Analysis*, Oxford University Press, Oxford.

Schoute, M. (2009) The relationship between cost system complexity, purposes of use, and cost system effectiveness. *The British Accounting Review*, **41**, 208–226.

Schumaker, R.E. and Lomax, R.G. (1996) *A Beginner's Guide to Structural Equation Modeling*, Lawrence Erlbaum Associates, Publishers, New Jersey.

Shah, S.Z.A., Stark, A.W., and Akbar, S. (2009) The value relevance of major media advertising expenditure: some UK evidence. *The International of Accounting*, **44**, 187–206.

Shannon, R.E. (1975) *System Simulation: The Art and Science*, Prentice-Hall, Inc., Englewood Cliffs, NJ.

Sharp, W.J. (1964) Capital asset prices: a theory of market equilibrium under conditions of risk. *Journal of Finance*, **19**, 425–442.

Shim, J. and Okamuro, H. (2011) Does ownership matter in merger? A comparative study of the causes and consequences of mergers by family and non-family firms. *Journal of Banking & Finance*, **35**, 193–203.

Simonin, B.L. (1999) Ambiguity and the process of knowledge transfer in strategic alliances. *Strategic Management Journal*, **20**, 595–623.

Sinang, R. (2010) Studi Tentang Masalah Perceraian Wanita di Indonesia. Analisis Data Sakerti Tahun 2000 dan 2007.

Siswantoro, D. and Agung, I.G.N. (2010) The importance of the effect of exogenous interaction factors on endogenous variables in accounting modeling. *International Journal of Finance and Accounting*, **1** (6), 194–197.

Skinner, D.J. (2008) The rise of deferred tax assets in Japan: the role of deferred tax accounting in the Japanese banking crisis. *Journal of Accounting and Economics*, **46**, 218–239.

Sudarwati (2009) Studi Tentnag Putus Sekolah Anak Usia 7-15 Tahun di Indonesia (analisis Data Susenas Tahun 2006). Thesis, Graduate Program on Population and Manpower, University of Indonesia.

Suk, Kim Sung (2006) Hubungan Simultan Antara Struktur Kepemilikan, Corporate Governace, Dan Nilai Perusahaan Dari Perusahaan Di Bursa Efek Jakarta. Dissertation, Graduate School of Management, Faculty of Economics, University of Indonesia.

Sulfitera (2008) Faktor-faktor yang Mempengaruhi Pemberian Imunisasi Lengakp di Inodnsia. Thesis, Graduate Program on Population and Manpower, University of Indonesia.

Supriyono, R.A. (2003) Hubungan Partiisipasi Penganggaran Dan Kinerja Manajer, Peran Kecukupan Anggaran, Komitmen Organisasi, Asimetri Informasi, Sllak Anggaran Dan Peresponan Keinginan Sosial. Dissertation, Graduate School of Management, Faculty of Economics, University of Indonesia.

Suriawinata, I.S. (2004) Studi Tentang Perilaku Hedging Perusahaan denganInstrument Derivatif Valuta Asing. Dissertation, the Graduate School of Management, Faculty of Economic, University of Indonesia.

Thomopoulos, N.T. (1980) *Applied Forecasting Methods*, Prentice-Hall, Inc., Englewood Cliffs, New Jersey, 07632.

Timm, N.H. (1975) *Multivariate Analysis with Applications in Education and Psychology*, Brooks/Cole Publishing Company, Monterey, California.

Triyanto, P. (2009) Analisis Kualitas Layanan Perguruan Tinggi dan Harapan Mahasiswa Setelah Menyelesaikan Studi di Universitas Negeri Makassar. Dissertation, Jakarta State University.

Tsay, R.S. (2002) *Analysis of Financial Time Series*, John Wiley & Sons, Inc., New York.

Tukey, J.W. (1962) The future of data analysis. *Annals of Mathematical Statistics*, **33**, 1–67.

Uddin, M. and Boateng, A. (2011) Explaining the trends in UK cross-border mergers & acquisitions: an analysis of macro-economic factors. *International Business Review*, **20**, 547–556.

van der Waerden, B.L. (1952) A simple statistical significance test. *Rhodesia Agricultural Journal*, **49**, 96–104 (5.1).

van der Waerden, B.L. ((1952)/1953) Order test for the two-sample problem and their power. *Proceedings Koninklijke Nederlandse Akademic van Wetenschappen (A)*, **55** (Indagationes Mathematical 14), 453–458 and 56 (Indagationes Mathematical 15), 303–316 (correction appears in Vol. 56, p. 80) (5.10).

van der Waerden, B.L. (1953) Testing a distribution function. *Proceedings Koninklijke Nederlandse Akademic van Wetenschappen (A)*, **56**, 201–207 (Indagationes Mathematical 15) (6.1).

Vose, D. (2000) *Risk Analysis*, John Wiley & Sons, Ltd, London.

Wang, T., Winton, A., and Yu, X. (2010) Corporate fraud and business conditions: evidence from IPOs. *The Journal of Finance*, **65** (6), 2255–2292.

Winsbek, T.J. and Knaap, T. (1999) Estimating a dynamic panel data model with heterogeneous trend. *Annales d'Economie et de Statistique*, **55-56**, 331–349.

Watson, C.J. *et al.* (1993) *Statistics for Management and Economics*, Allyn and Bacon, Boston Singapore.

Wilhelm, M.O. (2008) Practical considerations for choosing between tobit and SCLS or CLAD estimators for censored regression models with an application to charitable giving. *Oxford Bulletin of Economics and Statistics*, **70**(4), 559.

Wilson, J.H. and Keating, B. (1994) *Business Forecasting*, 2nd edn, Richard D. Irwin, Inc., Burr Ridge, Illinois.

Wilks, S.S. (1962) *Mathematical Statistics*, John Wiley & Sons, Inc., New York.

Winer, B.J. (1971) *Statistical Principles in Experimental Design*, McGraw-Hill, Kogakusha, Ltd.

Wood, A. (2009) Capacity rationalization and exit strategies. *Strategic Management Journal*, **30**, 25–44.

Wooldridge, J.M. (2002) *Econometric Analysis of Cross Section and Panel Data*, The MIT Press Cambridge, Massachusetts.

Widyastuty, U., Blomdine, ChP., and Yuniati, R.A. (2008) The effect of pH and storage temperature on larvicidal activity of *bacillus sphaericus* 2362. *Bulletin of Health Studies*, **36**(1), 33–47.

Wright, J.F. (2002.) *Monte Carlo Risk Analysis and Due Diligence of New Business Ventures*, AMACOM, a division of American Management Association, New York.

Wysocki, P. (2010) Corporate compensation policies audit fees. *Journal of Accounting and Economics*, **49**, 155–160.

Yaffee, R. and McGee, M. (2000.) *Introduction to Time Series Analysis and Forecasting with Application of SAS and SPSS*, Academic Press, Inc., New York.

Yuping, L. (2007) The long-term impact of loyalty programs on consumer purchase behaviour and loyalty. *Journal of Marketing*, **71**, 19–35.

Zhang, Y. (2008) Analyst responsiveness and the post-earnings-announcement drift. *Journal of Accounting and Economics*, **46**, 201–215.

Index

Panel Data Analysis Using EViews, First Edition. I Gusti Ngurah Agung.
© 2014 John Wiley & Sons, Ltd. Published 2014 by John Wiley & Sons, Ltd.
Companion website: www.wiley.com/go/panel_data

Printed and bound by CPI Group (UK) Ltd, Croydon, CR0 4YY

17/04/2025

14658915-0001